Springer Series in Translational Stroke Research

Series Editor
John Zhang

For further volumes:
http://www.springer.com/series/10064

Eng H. Lo • Josephine Lok • MingMing Ning •
Michael J. Whalen
Editors

Vascular Mechanisms in CNS Trauma

Springer

Editors
Eng H. Lo
Departments of Neurology
and Radiology
Massachusetts General Hospital
Harvard Medical School
Boston, MA, USA

Josephine Lok
Department of Pediatrics
Massachusetts General Hospital
Harvard Medical School
Boston, MA, USA

MingMing Ning
Department of Neurology
Massachusetts General Hospital
Harvard Medical School
Boston, MA, USA

Michael J. Whalen
Department of Pediatrics
Massachusetts General Hospital
Harvard Medical School
Boston, MA, USA

ISBN 978-1-4614-8689-3 ISBN 978-1-4614-8690-9 (eBook)
DOI 10.1007/978-1-4614-8690-9
Springer New York Heidelberg Dordrecht London

Library of Congress Control Number: 2013953546

Printed on acid-free paper

Springer is part of Springer Science+Business Media (www.springer.com)

Foreword

Recognizing that the modern history of the study of traumatic brain and spinal cord injury has spanned less than 50 years, it is not surprising that our current understanding of the pathobiology and treatment of these devastating neurological disorders remains incomplete and in some cases, controversial. When first addressed in the early 1970s, the focus of neurotrauma research took a distinct vascular focus, with many believing that the most important mechanism underlying the morbidity associated with CNS injury was the occurrence of traumatically induced vasogenic edema. It was assumed that the mechanical forces of injury disrupted the blood–brain barrier leading to the extravasation of serum proteins and other damaging agents capable of recruiting increased water within the central nervous system thereby elevating intra-parenchymal/intracranial pressure. Further, it was assumed that this elevated pressure alone triggered all the damaging cascades of injury associated with tissue compression and subsequent CNS tissue damage. Others suggested that the same forces of injury altered vascular function leading to impaired autoregulation and/or impaired vasoreactivity to normal physiological challenges, predisposing the CNS to further injury when a secondary insult, such as hypoxia or hypotension ensued. This early vascular focus was solidified by the recognition that the same forces of injury, particularly those on the more severe end of the spectrum, invariably caused overt vascular damage reflected in local spinal cord hemorrhage followed by tissue cavitation or the occurrence of hemorrhagic contusional change within the cerebral cortex. In part, because of the failure of early clinical trials targeting vascular mechanisms to achieve therapeutic benefit and in part, due to the frustration of some over this vascular-dominated approach to neurotrauma research, the subsequent 2–3 decades of neurotrauma discovery took a decidedly neuronal centric focus. During this period, there was a dramatic shift in neurotrauma research to the considerations of altered neurotransmission/neuroexcitation, neuronal cell death framed in the context of apoptosis versus necrosis, axonal dysfunction and disconnection, and the sequelae of such disconnection in terms of CNS tissue deafferentation and neuroplastic change. These neuronal focused studies were interfaced with parallel metabolic, behavioral, and targeted therapeutic studies conducted both in animals and humans who had sustained traumatic injury to either the brain or spinal cord. Unfortunately, as appreciated by all, this period of discovery, although highly significant, also failed to generate a full understanding of the pathobiology of CNS injury and/or lead to the development of more rational therapeutic approaches to improve outcome in those who have sustained traumatic injury to either the brain or spinal cord. In light of these limitations and clinical trial failures, the last decade has witnessed a more integrative approach, coupling the interaction of the CNS and its intrinsic vasculature, to better understand the overall pathogenesis of CNS injury and its potential therapeutic targeting. Now framed in the context of the neurovascular unit, contemporary research has begun to focus on the cell-specific responses therein while also exploring the various complex interactions between these neuronal vascular and their related glial pathways. In this context, renewed emphasis has been placed on the understanding of the blood–brain/blood-spinal cord barrier, not only in terms of its alteration following traumatic insult but also in the context of its ability to modulate nutrient transport as well as the passage of various purported neuroprotective agents.

Against this backdrop of discovery, the current text edited by Lo and colleagues frames our contemporary understanding of the vascular sequelae of traumatic injury to the CNS in an effort to achieve a more comprehensive understanding of CNS/neuronal responses to injury. The 29 chapters contained therein provide important insight into the complex and diverse vascular changes associated with traumatic CNS injury, with a good mix of both basic science and clinical discovery. The authors who have participated in the generation of this book provide detailed insight into virtually all of the important components of the vascular sequelae of injury and their CNS/neuronal interactions. The chapters in this text focus on important themes considering traumatically altered cerebral blood flow, autoregulation, vasoreactivity, coagulation, and blood–brain barrier status, while also addressing the importance of the neuronal/glial vascular unit. These studies are complemented by parallel considerations of more contemporary diagnostic and therapeutic approaches ranging from the use of proteomics and advanced imaging to the modern tools of electrophysiology and microdialysis-based assessment.

Collectively, the editors and authors are to be congratulated on their success in producing this important text. It will serve as an invaluable reference for those first entering the field as well as the seasoned investigator who wishes to update his or her understanding of the complex vascular sequelae associated with traumatic injury to the central nervous system.

Richmond, VA John T. Povlishock, PhD

Preface

The vasculature of the central nervous system performs the vital tasks of perfusing the brain and spinal cord, and maintaining barrier functions that ensure a proper environment for neuronal activity. Additionally, the vasculature participates in other key functions, including the production of vasoconstricting and vasodilating substances, the provision of trophic support to the neuronal and glial parenchyma, the response to inflammatory stimuli, and the regulation of tissue remodeling and repair after injury. Hence, a rigorous understanding of vascular mechanisms is essential for the development of therapeutic strategies for brain and spinal cord trauma.

It is now an opportune time to incorporate the study of the vasculature in the field of CNS trauma science. Our hope in putting together this book is to provide a reference for clinicians and researchers who are undertaking further explorations of the CNS vasculature within the complex pathophysiology of injury and disease. We dedicate this book to investigators who are studying ways to treat patients and to improve the lives of survivors of brain and spinal cord trauma. And most importantly, we thank the many patients and families whose efforts to reclaim their lives after CNS trauma provide the inspiration for our work.

Boston, MA — Eng H. Lo
Boston, MA — Josephine Lok
Boston, MA — MingMing Ning
Boston, MA — Michael J. Whalen

Contents

Part I Molecular Mechanisms

Part III Clinical Challenges and Opportunities

Contributors

Sandra Acosta Department of Neurosurgery and Brain Repair, University of South Florida College of Medicine, Tampa, FL, USA

Bradley P. Ander Department of Neurology and the MIND Institute, University of California at Davis, Sacramento, CA, USA

Ken Arai Neuroprotection Research Laboratory, Departments of Neurology and Radiology, Massachusetts General Hospital, Harvard Medical School, Boston, MA, USA

William M. Armstead Departments of Anesthesiology and Critical Care and Pharmacology, University of Pennsylvania, Philadelphia, PA, USA

David A. Boas Department of Radiology, Massachusetts General Hospital, Athinoula A. Martinos. Center for Biomedical Imaging, Boston, MA, USA

Michael J. Bell Departments of Critical Care Medicine, Neurological Surgery, Pediatrics and the Safar Center for Resuscitation Research, University of Pittsburgh School of Medicine, Pittsburgh, PA, USA

Rebecca Bell Department of Pediatrics, Division of Pediatric Critical Care Medicine, Massachusetts General Hospital and Harvard Medical School, Boston, MA, USA

Chetan Bhupali Department of Pediatrics, Division of Pediatric Critical Care Medicine, Massachusetts General Hospital and Harvard Medical School, Boston, MA, USA

Cesar V. Borlongan Department of Neurosurgery and Brain Repair, University of South Florida College of Medicine, Tampa, FL, USA

Helen M. Bramlett Bruce W. Carter Department of Veterans Affairs Medical Center, Miami, FL, USA

Robert S.B. Clark Department of Pediatric Critical Care Medicine, Safar Center for Resuscitation Research, University of Pittsburgh School of Medicine, Pittsburgh, PA, USA

Sheng Chen Department of Physiology and Pharmacology, Loma Linda University, Loma Linda, CA, USA

Adam Chodobski Department of Emergency Medicine, Neurotrauma and Brain Barriers Research Laboratory, Alpert Medical School of Brown University, Providence, RI, USA

Michael Chopp Department of Physics, Henry Ford Hospital, Detroit, MI, USA

Oakland University, Rochester, MI, USA

Sherry H.-Y. Chou Department of Neurology, Brigham and Women's Hospital, Boston, MA, USA

Brian M. Cummings Department of Pediatrics, Division of Pediatric Critical Care Medicine, Massachusetts General Hospital and Harvard Medical School, Boston, MA, USA

Marek Czosnyka Academic Neurosurgery Division, University of Cambridge, Cambridge, UK

Travis Dailey Department of Neurosurgery and Brain Repair, University of South Florida College of Medicine, Tampa, FL, USA

Kunjan R. Dave The Cerebral Vascular Disease Research Laboratories, Department of Neurology, Leonard M. Miller School of Medicine, University of Miami, Miami, FL, USA

W. Dalton Dietrich Department of Neurological Surgery and Miami Project to Cure Paralysis, University of Miami Miller School of Medicine, Miami, FL, USA

Ann-Christine Duhaime Department of Neurosurgery, Massachusetts General Hospital, Harvard Medical School, Boston, MA, USA

Philip E. Empey Department of Pharmacy and Therapeutics, School of Pharmacy, University of Pittsburgh, Pittsburgh, PA, USA

Nikki Miller Ferguson Department of Critical Care Medicine and the Safar Center for Resuscitation Research, University of Pittsburgh School of Medicine, Pittsburgh, PA, USA

Mutsumi Fujii Department of Physiology and Pharmacology, Loma Linda University, Loma Linda, CA, USA

Volodymyr Gerzanich Department of Neurosurgery, University of Maryland School of Medicine, Baltimore, MD, USA

Thomas C. Glenn Departments of Neurosurgery and of Molecular and Medical Pharmacology, David Geffen School of Medicine at UCLA, Los Angeles, CA, USA

Olena Glushakova Banyan Biomarkers, Inc., Alachua, FL, USA

Eugene V. Golanov Feisntein Institute for Medical Research, Manhasset, NY, USA

Gerald A. Grant Division of Neurosurgery, Department of Surgery, Duke University School of Medicine, Durham, NC, USA

Rao P. Gullapalli Department of Diagnostic Radiology & Nuclear Medicine, University of Maryland School of Medicine, Baltimore, MD, USA

Shu-zhen Guo Radiology Department, Neuroprotection Research Lab, Massachusetts General Hospital, Charlestown, MA, USA

Theo Hagg Department of Neurological Surgery, Kentucky Spinal Cord Injury Research Center, University of Louisville, Louisville, KY, USA

Ronald L. Hayes Banyan Biomarkers, Inc., Alachua, FL, USA

David A. Hovda Departments of Neurosurgery and of Molecular and Medical Pharmacology, David Geffen School of Medicine at UCLA, Los Angeles, CA, USA

Philip H. Iffland II Departments of Neurosurgery and Cellular and Molecular Medicine, Cleveland Clinic Foundation, Cleveland, OH, USA and School of Biomedical Sciences, Kent State University, Kent, OH, USA

Hiroto Ishikawa Department of Neurosurgery and Brain Repair, University of South Florida College of Medicine, Tampa, FL, USA

Damir Janigro Cerebrovascular Research, Departments of Neurosurgery and Cellular and Molecular Medicine, Department of Molecular Medicine, Cleveland Clinic Lerner College of Medicine, Cleveland, OH, USA and Flocel, Inc., Cleveland, OH, USA

Glen C. Jickling Department of Neurology and the MIND Institute, University of California at Davis, Sacramento, CA, USA

Kristopher T. Kahle Department of Neurological Surgery, Massachusetts General Hospital, Boston, MA, USA

Brian Kalish Harvard Medical School, Boston, MA, USA

Yuji Kaneko Department of Neurosurgery and Brain Repair, University of South Florida College of Medicine, Tampa, FL, USA

Yoichi Katayama Department of Neurological Surgery, Nihon University School of Medicine, Tokyo, Japan

Tatsuro Kawamata Department of Neurological Surgery, Nihon University School of Medicine, Tokyo, Japan

Arjun Khanna Department of Neurological Surgery, Massachusetts General Hospital, Boston, MA, USA

Damon Klebe Department of Physiology and Pharmacology, Loma Linda University, Loma Linda, CA, USA

Patrick M. Kochanek Department of Critical Care Medicine and Safar Center for Resuscitation Research, University of Pittsburgh School of Medicine, Pittsburgh, PA, USA

Monisha A. Kumar Departments of Neurology, Neurosurgery, Anesthesiology and Critical Care, Perelman School of Medicine, University of Pennsylvania, Philadelphia, PA, USA

Sang Mi Lee Department of Neurosurgery, University of California, San Francisco, CA, USA

Wendy Leung Neuroprotection Research Laboratory, Departments of Neurology and Radiology, Massachusetts General Hospital, Harvard Medical School, Boston, MA, USA

Philip M. Lewis Department of Neurosurgery, Alfred Hospital, Prahran, Australia

Klaus van Leyen Neuroprotection Research Laboratory, Departments of Neurology and Radiology, Massachusetts General Hospital, Harvard Medical School, Boston, MA, USA

Zhengbu Liao Department of Neurosurgery, The First Affiliated Hospital of Chongqing Medical University, Chonqing, China

Neuroprotection Research Laboratory, Departments of Neurology and Radiology, Massachusetts General Hospital, Harvard Medical School, Charlestown, MA, USA

Teck Chuan Lim Massachusetts Institute of Technology, Harvard-MIT Division of Health Sciences and Technology, Cambridge, MA, USA

Hung W. Lin The Cerebral Vascular Disease Research Laboratories, Department of Neurology, Leonard M. Miller School of Medicine, University of Miami, Miami, FL, USA

Da Zhi Liu Department of Neurology and the MIND Institute, University of California at Davis, Sacramento, CA, USA

Qi Liu Neuroprotection Research Laboratory, Departments of Neurology and Radiology, Massachusetts General Hospital, Harvard Medical School, Boston, MA, USA

Eng H. Lo Departments of Neurology and Radiology, Massachusetts General Hospital, Harvard Medical School, Boston, MA, USA

Josephine Lok Neuroprotection Research Laboratory and Department of Pediatrics, Division of Pediatric Critical Care Medicine, Massachusetts General Hospital and Harvard Medical School, Boston, MA, USA

Takeshi Maeda Departments of Neurological Surgery & Anesthesiology, Nihon University School of Medicine, Tokyo, Japan

Asim Mahmood Department of Neurosurgery, Henry Ford Hospital, Detroit, MI, USA

Takakuni Maki Neuroprotection Research Laboratory, Departments of Neurology and Radiology, Massachusetts General Hospital, Harvard Medical School, Boston, MA, USA

Zenab Mansoor Department of Pediatrics, Division of Pediatric Critical Care Medicine, Massachusetts General Hospital and Harvard Medical School, Boston, MA, USA

Stefania Mondello Department of Neuroscience, University of Messina, Messina, Italy

Jennifer C. Munoz Pareja Department of Pediatric Critical Care Medicine, Massachusetts General Hospital, Harvard Medical School, Boston, MA, USA

Sarah A. Murphy Department of Pediatrics, Division of Pediatric Critical Care Medicine, Massachusetts General Hospital and Harvard Medical School, Boston, MA, USA

Deepti Navaratna Neuroprotection Research Laboratory, Departments of Neurology and Radiology, Massachusetts General Hospital, Harvard Medical School, Boston, MA, USA

Jake T. Neumann The Cerebral Vascular Disease Research Laboratories, Department of Neurology, Leonard M. Miller School of Medicine, University of Miami, Miami, FL, USA

MingMing Ning Neuroprotection Research Laboratory, Departments of Neurology and Radiology, Massachusetts General Hospital, Harvard Medical School, Boston, MA, USA

Linda J. Noble-Haeusslein Department of Neurosurgery, Department of Physical Therapy and Rehabilitation Science, University of California, San Francisco, CA, USA

Natan Noviski Department of Pediatrics, Division of Pediatric Critical Care Medicine, Massachusetts General Hospital and Harvard Medical School, Boston, MA, USA

Mibel M. Pabón Department of Neurosurgery and Brain Repair, University of South Florida College of Medicine, Tampa, FL, USA

Miguel A. Perez-Pinzon The Cerebral Vascular Disease Research Laboratories, Department of Neurology, Leonard M. Miller School of Medicine, University of Miami, Miami, FL, USA

John D. Pickard Academic Neurosurgery Division, University of Cambridge, Cambridge, UK

Claudia S. Robertson Department of Neurosurgery, Baylor College of Medicine, Houston, TX, USA

J. Marc Simard Departments of Neurosurgery, Pathology and Physiology, University of Maryland School of Medicine, Baltimore, MD, USA

Frank R. Sharp Department of Neurology and the MIND Institute, University of California at Davis, Sacramento, CA, USA

Steven L. Shein Department of Critical Care Medicine and the Safar Center for Resuscitation Research, University of Pittsburgh School of Medicine, Pittsburgh, PA, USA

Kazutaka Shinozuka Department of Neurosurgery and Brain Repair, University of South Florida College of Medicine, Tampa, FL, USA

Piotr Smielewski Academic Neurosurgery Division, University of Cambridge, Cambridge, UK

Douglas H. Smith Department of Neurosurgery, Perelman School of Medicine, University of Pennsylvania, Philadelphia, PA, USA

Sherman C. Stein Department of Neurosurgery, Perelman School of Medicine, University of Pennsylvania, Philadelphia, PA, USA

Boryana Stamova Department of Neurology and the MIND Institute, University of California at Davis, Sacramento, CA, USA

Chandler Sours Department of Diagnostic Radiology & Nuclear Medicine, University of Maryland School of Medicine, Baltimore, MD, USA

Myron Spector Department of Orthopedic Surgery, Brigham and Women's Hospital, Boston, MA, USA

Xiaochuan Sun Department of Neurosurgery, The First Affiliated Hospital of Chongqing Medical University, Chonqing, China

Joanna Szmydynger-Chodobska Department of Emergency Medicine, Neurotrauma and Brain Barriers Research Laboratory, Alpert Medical School of Brown University, Providence, RI, USA

Naoki Tajiri Department of Neurosurgery and Brain Repair, University of South Florida College of Medicine, Tampa, FL, USA

John W. Thompson The Cerebral Vascular Disease Research Laboratories, Department of Neurology, Leonard M. Miller School of Medicine, University of Miami, Miami, FL, USA

Jose Alberto Toranzo Baylor College of Medicine, Houston, TX, USA

Alpa Trivedi Department of Neurosurgery, University of California, San Francisco, CA, USA

Alexander Vakhmyanin Department of Physiology and Pharmacology, Loma Linda University, Loma Linda, CA, USA

Monica S. Vavilala Departments of Anesthesiology and Pediatrics, Harborview Medical Center, University of Washington, Seattle, WA, USA

Brian P. Walcott Department of Neurological Surgery, Massachusetts General Hospital, Boston, MA, USA

Xiaoshu Wang Department of Neurosurgery, The First Affiliated Hospital of Chongqing Medical University, Chongqing, China

Neuroprotection Research Laboratory, Departments of Neurology and Radiology, Massachusetts General Hospital, Harvard Medical School, Charlestown, MA, USA

Xiaoying Wang Neuroprotection Research Laboratory, Departments of Neurology and Radiology, Massachusetts General Hospital, Harvard Medical School, Charlestown, MA, USA

Michael J. Whalen Department of Pediatric Critical Care Medicine, Massachusetts General Hospital and Harvard Medical School, Boston, MA, USA

F. Anthony Willyerd Department of Critical Care Medicine, Children's Hospital of Pittsburgh, Pittsburgh, PA, USA

Limin Wu Department of Neurology, Jilin University, Changchun, Jilin, China

Changhong Xing Neuroprotection Research Laboratory, Departments of Neurology and Radiology, Massachusetts General Hospital, Harvard Medical School, Boston, MA, USA

Ye Xiong Department of Neurosurgery, Henry Ford Hospital, Detroit, MI, USA

Phoebe H. Yager Department of Pediatric Critical Care Medicine, Massachusetts General Hospital and Harvard Medical School, Boston, MA, USA

Atsuo Yoshino Department of Neurological Surgery, Nihon University School of Medicine, Tokyo, Japan

Zhanyang Yu Neuroprotection Research Laboratory, Departments of Neurology and Radiology, Massachusetts General Hospital, Harvard Medical School, Charlestown, MA, USA

Xinhua Zhan Department of Neurology and the MIND Institute, University of California at Davis, Sacramento, CA, USA

Haoqian Zhang Department of Neurosurgery, University of California, San Francisco, CA, USA

John H. Zhang Department of Physiology and Pharmacology, Loma Linda University, Loma Linda, CA, USA

Yanlu Zhang Department of Neurosurgery, Henry Ford Hospital, Detroit, MI, USA

Zheng Gang Zhang Department of Neurology, Henry Ford Hospital, Detroit, MI, USA

Brian J. Zink Department of Emergency Medicine, Neurotrauma and Brain Barriers Research Laboratory, Alpert Medical School of Brown University, Providence, RI, USA

Part I
Molecular Mechanisms

Chapter 1
CNS Barriers in Neurotrauma

Adam Chodobski, Brian J. Zink, and Joanna Szmydynger-Chodobska

Abstract Despite significant advances in designing neuroprotective therapies, multiple clinical trials in head trauma and spinal cord injury have produced mixed results. A better understanding of the pathophysiology of central nervous system (CNS) barriers may improve clinical translation of therapies for injury of the CNS. The blood–brain barrier and blood-spinal cord barrier are formed by specialized CNS endothelial cells. These barriers play a fundamental role in restricting the entry of blood-borne factors into the CNS. However, they also function as the gateway for the delivery of neuroprotective drugs. Neurotrauma changes the properties of CNS barriers, which may significantly affect the efficacy of neuroprotective therapies. The endothelial barriers of the CNS together with the blood-cerebrospinal fluid barrier, which is predominantly formed by the epithelial cells of the choroid plexus, also restrict the influx of circulating leukocytes into the CNS. Dysfunction of these barriers resulting from injury plays a key role in the initiation and progression of inflammation that accompanies neurotrauma.

1.1 Introduction

Traumatic brain injury (TBI) and spinal cord injury (SCI) represent a considerable public health problem. The complexity and interdependent nature of pathophysiological mechanisms underlying secondary injury—a constellation of pathophysiological processes triggered by primary injury, which include the formation of edema, excitotoxicity, ischemia, oxidative stress, and neuroinflammation—create

---

A. Chodobski (✉)
Neurotrauma and Brain Barriers Research Laboratory, Department of Emergency Medicine, Alpert Medical School of Brown University, Coro Center West, Room 112, 1 Hoppin Street, Providence, RI 02903, USA
e-mail: Adam_Chodobski@Brown.edu

E.H. Lo et al. (eds.), *Vascular Mechanisms in CNS Trauma*, Springer Series in Translational Stroke Research 5, DOI 10.1007/978-1-4614-8690-9_1,

formidable therapeutic challenges. Despite significant advances in designing neuroprotective therapies and encouraging preclinical data, multiple clinical trials in TBI and SCI have produced mixed results [1–4]. A better understanding of the pathophysiology of central nervous system (CNS) barriers may improve clinical translation of therapies for injury of the CNS. The blood–brain barrier (BBB) and blood-spinal cord barrier (BSCB) are formed by specialized CNS endothelial cells. They play a fundamental role in restricting the entry of blood-borne factors into the CNS. However, the close distance between the CNS endothelium and neuronal perikarya (on average 10–12 μm and no further than 28–36 μm apart in various regions of the rat brain; Fenstermacher, personal communication) makes the BBB and BSCB the most effective routes for the delivery of neuroprotective drugs. Neurotrauma significantly changes the properties of CNS barriers, which may have an important effect on the efficacy of neuroprotective therapies. As we will discuss in detail below, neurotrauma is associated with increased permeability of CNS barriers, augmented production of proteolytic enzymes, and altered transport activities at the CNS endothelium, all of which may affect the CNS concentrations and biological activities of neuroprotectants. The endothelial barriers of the CNS not only selectively control the movement of molecules but they also restrict the influx of circulating immune cells. Injury may trigger the invasion of leukocytes across CNS barriers. Invading inflammatory cells, such as neutrophils and monocytes, carry powerful proteases [5, 6] that may reduce the biological activity of peptide- or protein-based neuroprotective drugs.

It is important to note that the CNS endothelium is closely associated with other types of cells, such as pericytes, astrocytes, and microglia [7–9], which functionally interact with endothelial cells in a paracrine manner (Fig. 1.1). Both pericytes and astroglia play critical roles in inducing and maintaining the "tight" phenotype of CNS endothelium [10, 11]. Astrocytes and microglia are particularly responsive to blood-borne proteins that may leak through disrupted endothelial barrier and to various factors released from parenchymal cells after injury [12]. This has a significant effect on function of CNS endothelium and may play an important role in initiating the inflammatory cascade in the injured neural tissue. To emphasize the intimate anatomical and functional relationship between pericytes, glia, and the endothelial cells, and their association with nearby neurons, a more general term—the *neurovascular unit*—is now frequently used [10]. The analysis of CNS barriers in neurotrauma should also include the discussion on the blood-cerebrospinal fluid (CSF) barrier (BCSFB). Unlike the BBB and BSCB, the BCSFB is predominantly formed by the epithelial cells of the choroid plexus (Fig. 1.1). Increasing evidence indicates that the choroid plexus has the ability to produce proinflammatory mediators and that the BCSFB plays an important role in posttraumatic invasion of leukocytes. Accordingly, in this chapter, we will review the available literature on how neurotrauma changes the function of these three barriers.

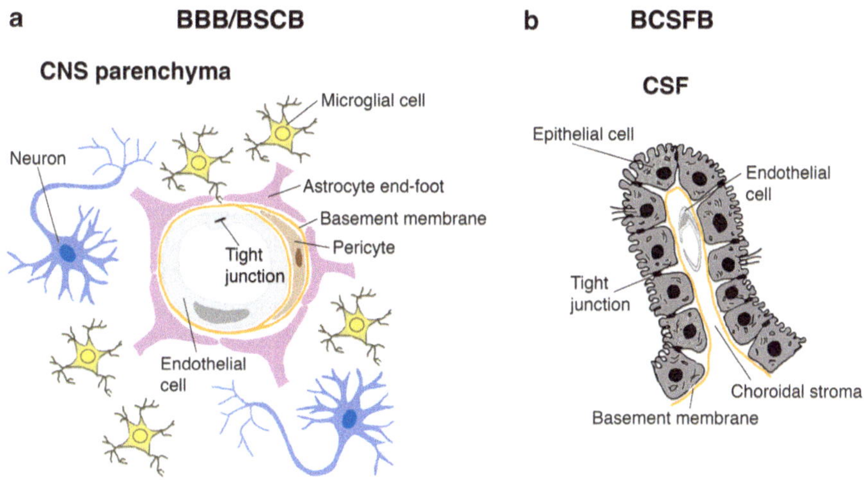

Fig. 1.1 (**a**) Schematic representation of the blood–brain barrier (BBB) and blood-spinal cord barrier (BSCB). Both the BBB and BSCB are formed by specialized CNS endothelial cells connected by tight junctions. Endothelial cells are closely associated with pericytes and astrocytes, which play critical roles in inducing and maintaining the "tight" phenotype of CNS endothelium. The BBB differs from the BSCB with regard to pericyte coverage and the level of expression of tight junction proteins, which results in the higher paracellular permeability of the BSCB compared to the BBB. Microglial processes are also closely associated with endothelial cells in 4–13 % of microvessels. A broader term—the *neurovascular unit*—is now commonly used to reflect an intimate anatomical and functional relationship between the endothelium and glia, and to emphasize a close distance between the endothelial cells and neurons. (**b**) Schematic representation of the blood-cerebrospinal fluid (CSF) barrier (BCSFB). The BCSFB primarily resides in the choroid plexus, a highly vascularized tissue located in all four cerebral ventricles. The BCSFB is formed by tight junctions connecting cuboidal epithelial cells, which enclose fenestrated blood microvessels. Although there are some similarities in the composition of the tight junction complexes between the BBB and BCSFB, there are also some distinctive differences between these two barriers. These differences may be responsible for the less "tight" phenotype of the BCSFB compared to the BBB. The BBB and BCSFB appear to complement each other functionally with regard to their ability to clear drugs and drug metabolites from the brain

1.2 Differences Between the BSCB and BBB, and Characteristic Features of the BCSFB

Although both the BBB and BSCB are similarly formed by the CNS endothelium, with adjacent endothelial cells connected by tight junctions, there are some distinct differences in the properties of these two barriers. As mentioned above, pericytes play a key role in inducing and maintaining the integrity of CNS barriers [13, 14]. Accordingly, the smaller number of pericytes covering the capillaries of the spinal cord compared to those found in the brain is probably responsible for reduced expression of tight junction protein occludin and the tight junction-associated protein zonula occludens (ZO)-1 at the BSCB versus BBB [15, 16]. The pericyte coverage varies among spinal cord regions, and an inverse correlation between the regional pericyte coverage in the spinal cord and the capillary permeability has been found. The differences in the pericyte number and in the level of expression of tight junction proteins between the BSCB and BBB appear to be responsible for the higher paracellular permeability of the BSCB to both the low and high molecular weight markers [15, 17].

The efficacy of small-molecule neuroprotective drugs may depend, sometimes considerably, on the activity of xenobiotic efflux transporters expressed at CNS barriers. A recent preliminary report [18] has demonstrated that the expression levels and the activity of multidrug resistance protein 1 (MDR1/ABCB1), also known as P-glycoprotein, multidrug resistance-associated protein (MRP)-2 (ABCC2), and breast cancer resistance protein (ABCG2), the members of the superfamily of ATP-binding cassette transporters, are similar at the BBB and BSCB, although another study [16] has shown the reduction in MDR1 expression at the BSCB compared to the BBB. The effect of neurotrauma on the activity of transporters and their therapeutic targeting will be discussed later in this review.

The BCSFB primarily resides in the choroid plexus, a highly vascularized tissue located in all four cerebral ventricles. The BCSFB is formed by tight junctions connecting cuboidal epithelial cells, which enclose the leaky choroidal blood microvessels [19]. Although there are some similarities in the composition of the tight junction complexes between the BBB and BCSFB, there are also some distinctive differences between these two barriers. For example, tight junction protein claudin (CLDN)-2 is only expressed at the BCSFB, whereas CLDN5 is predominantly expressed at the BBB [20]. Claudin 2 is not only a tight junction protein but it can also form the cation-selective channels [21], which may explain, at least in part, why the BCSFB is less "tight" than the BBB.

The BBB and BCSFB appear to complement each other functionally with regard to their ability to clear drugs and drug metabolites from the brain. With the total apical (CSF-facing) surface area of the choroid plexus being comparable to that of the cerebral capillaries [22] and the high level of blood flow [23], the choroid plexus is well positioned to effectively clear drugs from the CSF. Although MDR1 is predominantly expressed at the BBB, another important efflux transporter MRP1 (ABCC1) is only expressed at the BCSFB [24]. Similarly, a proton-coupled oligopeptide transporter PEPT2 is only expressed at the BCSFB [25]. By comparison, MRP4 (ABCC4) is expressed at both the BBB and BCSFB [26].

1.3 Neurotrauma and Disruption of CNS Barriers

The brain is a highly heterogeneous organ. Different brain structures have distinctive viscoelastic properties and a different degree of attachment to each other and to the skull. Consequently, certain brain structures move faster than others when a direct impact or acceleration-deceleration forces are applied to the head. This may generate considerable shear, tensile, and compressive forces within the brain. Common pathoanatomical consequences of head trauma, especially those found in patients sustaining moderate to severe TBI, include hematoma, subarachnoid hemorrhage (SAH), contusion, and diffuse axonal injury [27]. In the lateral fluid percussion model of moderate TBI in rats, hemorrhagic contusions at the gray-white interfaces and isolated petechial hemorrhages scattered throughout the brain have been shown [28]. Consistent with these histological findings, an electron microscopic analysis has demonstrated disrupted endothelial walls and extravasation of erythrocytes. These observations indicate that impact-induced shearing stresses in TBI may result in primary vascular damage leading to the leakage of blood-borne proteins and extravasation of red blood cells.

The most common type of SCI in humans is contusion. Using a contusive model of SCI in rats, Noble and Wrathall [29] have demonstrated that the impact produces evolving hemorrhage, which at the epicenter predominantly occupies the gray matter and, to a variable degree, the adjacent white matter (in the form of petechial or larger hemorrhages). The initial size of the hemorrhage, which depends on the amount of energy dissipated within the neural tissue, is maximal at the epicenter and extends into the dorsal columns in both the rostral and caudal directions. At sites distant from the epicenter, the hemorrhage usually occupies the central part of the dorsal column. Over time the boundaries of hemorrhage extend markedly in both radial and longitudinal directions. This progressive secondary hemorrhage leads to necrosis of neural tissue. This not only occurs after SCI, but is also observed in TBI [30, 31]. Simard, Gerzanich, and colleagues have proposed that progressive secondary hemorrhage results from oncotic swelling and subsequent death of endothelial cells, which involves the sulfonylurea receptor 1 (SUR1)-regulated Ca^{2+}-activated, ATP-sensitive, nonselective cation ($NC_{Ca\text{-}ATP}$) channel [32, 33]. The pathophysiological role of this SUR1-regulated $NC_{Ca\text{-}ATP}$ channel in neurotrauma will be analyzed later in this review.

1.4 An Increase in the Permeability of CNS Barriers Resulting from Injury

An immediate and delayed destruction of vascular walls caused by injury disrupts the integrity of CNS barriers; however, an increase in the permeability of CNS barriers observed in the pericontusional areas likely results from a combination of some structural and predominantly functional changes occurring within the neurovascular unit. In the controlled cortical impact model of TBI in rats, which produces contusive CNS injury, a biphasic increase in the BBB permeability to a high molecular weight marker albumin (MW = ~70 kDa) has been shown, with two peaks occurring at 4–6 h and 2–3 days post injury [34]. It is important to note, however, that in a rat model of diffuse TBI uncomplicated by contusion and hemorrhage, a transient increase in the permeability of the BBB to albumin has also been observed [35]. The problem with the use of high molecular weight markers, such as albumin, is that the results obtained may not only provide information about the paracellular permeability of endothelial barriers but may also reflect changes in pinocytotic activity in the CNS endothelium [29]. In fact, there is evidence for increased pinocytosis in the cerebrovascular endothelium of TBI patients [36]. When low (sucrose; MW = ~300 Da) to moderate (inulin; MW = ~5 kDa) molecular weight markers were used in a rat model of mechanical brain injury, an increase in the permeability of the BBB was observed for at least 2 weeks after the surgery [37]. It is also worth noting that the long-term increase in the permeability of the BBB to gadolinium-DTPA has been observed in TBI patients [38]. Similar to TBI, in a rodent model of contusive SCI, a biphasic opening of the BSCB to luciferase (MW = ~60 kDa) has been shown [39]. The second peak in the BSCB permeability occurred at 3–7 days after injury and was interpreted to reflect the intrinsic adaptive angiogenesis. However, when post-SCI changes in the permeability of the BSCB were assessed using a low molecular weight marker α-aminoisobutyric acid (MW = ~100 Da), an increase in the BCSB permeability for at least a month after injury was observed [40]. This prolonged increase in the paracellular permeability of CNS barriers may allow for a small but functionally important leakage of plasma proteins into the CNS parenchyma. There is evidence that thrombin (derived from circulating precursor prothrombin) is neurotoxic [41, 42], and a gradual deposition of fibrin (derived from circulating fibrinogen) accelerates neurovascular damage and promotes neuroinflammation [43]. This may result in progressive neurodegenerative changes in the injured CNS and, consequently, poor neurological recovery.

Data obtained from animal studies suggest the formation of vasogenic edema early after injury. However, the significance of vasogenic edema in posttraumatic brain swelling, especially in patients with TBI, has been questioned [44]. This may be, at least in part, related to the relatively small and heterogeneous groups of TBI patients studied, and it is likely that both vasogenic and cytotoxic/cellular edema coexist in TBI. Interestingly, magnetic resonance imaging studies of SCI patients have demonstrated that low values of apparent diffusion coefficient (ADC) for water, which suggest cytotoxic edema, correlate with poor functional outcome, whereas high ADC values, which are indicative of vasogenic edema, are associated with better recovery [45].

The pathophysiological mechanisms leading to increased permeability of CNS barriers after neurotrauma are not completely understood. An increase in endothelial permeability may result from the loss of pericytes covering CNS capillaries [46]; however, the available data suggest that changes in expression, distribution, and/or function of tight junction proteins are predominantly involved. There are a large number of putative factors that may increase the permeability of CNS barriers after injury. The possible involvement of glutamate, reactive oxygen species (ROS), vascular endothelial growth factor A, matrix metalloproteinases (MMPs), and proinflammatory cytokines has been recently discussed [12]. Here we will extend our analysis of the possible role of proteases in increasing the permeability of CNS barriers and describe how apolipoprotein E (ApoE) may affect function of endothelial cells.

1.5 Proteases and Dysfunction of CNS Barriers After Injury

Increasing evidence indicates that various proteases that are produced by parenchymal cells and/or carried by invading leukocytes, such as MMPs and neutrophil elastase, play an important role in promoting the disruption of CNS barriers observed after injury [12]. These proteases may represent promising targets for therapeutic intervention. However, it should be kept in mind that proteases have diverse biological actions and an indiscriminate use of protease inhibitors or inappropriate timing of intervention may produce adverse therapeutic effects (see below). In this section, the pathophysiological role of MMPs in increasing the permeability of both the BBB and BSCB will be covered and we will also discuss the possible effect of tissue-type plasminogen activator (tPA) on function of CNS barriers in the traumatized neural tissue.

1.5.1 The Mediatory Role of MMPs in Increasing the Permeability of CNS Barriers

MMPs are a family of zinc-dependent endopeptidases with the ability to cleave not only the components of the extracellular matrix but also a large number of other unrelated proteins. These proteinases can disrupt the integrity of CNS barriers by attacking basal lamina proteins and degrading the components of the tight junction complexes, leading to the formation of vasogenic edema [47–49]. In the injured CNS, MMPs, such as MMP2, MMP3, MMP9, and MMP12, are produced by various types of cells, including the endothelial cells, astrocytes, microglia, and neurons [50–52]. Pericytes can also produce MMP9 when exposed to the proinflammatory cytokine tumor necrosis factor-α (TNF-α) [53]. They appear to respond to TNF-α with increased synthesis of MMP9 more vigorously than other components of the neurovascular unit, such as the endothelial cells and astrocytes. In addition, MMPs are carried by neutrophils (predominantly MMP9) [5] and monocytes (multiple MMPs) [6], which can invade the injured parenchyma. These MMPs may play an important role in facilitating the migration of leukocytes across CNS barriers [54], and may also contribute to the overall MMP activity in the injured neural tissue. In a rat model of TBI, a rapid (within 4 h post-TBI) increase in MMP9 activity in the traumatized brain parenchyma was observed, whereas the activity of MMP2 increased with a 24-h delay [55]. The augmented activity of these MMPs persisted for 5 days after injury (the longest observation period in this study). In a mouse model of SCI, a maximum increase in MMP9 activity was observed at 24 h post injury (the earlier time points post-SCI were not analyzed) and lasted until 3 days after SCI [56]. By comparison, the active form of MMP2 was detected in the injured spinal cord parenchyma a week after SCI and persisted until 3 weeks post injury. A comprehensive analysis of post-SCI changes in expression of MMPs has shown an increase in mRNA levels for several members of the MMP family, but a decrease in the message for some MMPs was also observed [52]. These data, however, do not provide information about the actual activity of these MMPs.

The deletion of the *Mmp9* or *Mmp12* gene, or the pharmacological inhibition of MMP activity limited to the first 3 days after injury was found to significantly decrease the permeability of CNS barriers, reduce the extent of neural tissue damage, and improve functional recovery in murine models of neurotrauma and ischemic brain injury [51, 52, 57, 58]. However, it is important to note that MMPs may also play significant roles in repair processes in the injured CNS. For example, diminished white matter sparing, more extensive astrocytic scar, and reduced functional recovery was observed in MMP2-deficient mice compared to wild-type animals [59]. Because of their putative role in neuronal repair after CNS injury, delayed inhibition of activity of MMPs may have adverse rather than beneficial therapeutic effects [60].

Clinical observations and laboratory studies have demonstrated that advanced age is associated with a higher mortality rate and worse functional outcome after TBI [61–64]. The pathophysiological mechanisms involved are not fully understood, but may be related to the exacerbation of brain inflammatory response to injury, which is observed in aged rodents [65]. A recent study [66] suggests that MMPs may also play a part in age-dependent increase in mortality and morbidity after neurotrauma. Aged mice have higher basal levels of activity of MMP2 and MMP9 in their brains, and a much larger increase in the activity of MMP9 in aged versus young adult animals was observed after injury. These differences in MMP activity between aged and young mice may be responsible, at least in part, for a significantly larger increase in the permeability of the BBB in aged rodents sustaining TBI.

1.5.2 tPA and Dysfunction of CNS Barriers After Injury

Tissue plasminogen activator encoded by the *PLAT* gene is a serine protease best known for its ability to convert the circulating plasminogen to plasmin. Recombinant tPA has been successfully used for treatment of acute ischemic stroke, but not without side effects, such as the opening of the BBB [67]. The multiple biological effects of tPA on the CNS may or may not be related to proteolytic processing of plasminogen or other tPA protein substrates, and include its involvement in excitotoxic neuronal death, as well as its role as a modulator of neurotransmission and synaptic plasticity [68, 69]. Although tPA can cross the intact BBB, a process mediated by the low-density lipoprotein receptor-related protein (LRP) [70], its major source in the CNS appears to be the CNS endothelium [71]. Interestingly, under normal conditions, both pericytes and astrocytes through their close contact with CNS endothelium significantly inhibit endothelial production of tPA [72, 73]. Acting in a paracrine manner, astrocytes concomitantly decrease the endothelial level of mRNA for tPA and increase that for the endogenous inhibitor of tPA plasminogen activator inhibitor-1 (PAI-1), which results in diminished tPA activity. Astrocytes themselves are the source of tPA, and proinflammatory cytokines TNF-α and interleukin (IL)-1β have been shown to dramatically reduce the astrocytic tPA activity, possibly by upregulating the synthesis of PAI-1 [74]. An astrocytic uptake of tPA from the extracellular space mediated by LRP1 was also described [75]. Part of this endocytosed tPA could be released, which is modulated by glutamate through its kainate receptor. Tissue plasminogen activator is also produced by other components of the neurovascular unit, such as neurons and microglia [76].

The intracerebroventricular administration of tPA in mice has been shown to increase the permeability of the BBB [77, 78], and studies of rodent models of neurotrauma and cerebral ischemia have provided evidence for an important mediatory role of tPA in increasing the permeability of the BBB in the injured brain [77–80]. This protease may disrupt the integrity of CNS barriers through multiple pathophysiological mechanisms that may be unrelated to its proteolytic activity and may involve various components of the neurovascular unit. Tissue plasminogen activator can augment the production of MMPs by astrocytes and the cerebrovascular endothelium [81–83]. The induction of synthesis of MMPs by tPA appears to be independent of its proteolytic activity [84], even though this protease has the ability to shed the ectodomain of LRPs expressed on astrocytes [79]. Instead, tPA acts by binding to LRP1, followed by a rapid tyrosine phosphorylation of its β subunit and the activation of extracellular signal-regulated kinase (ERK) 1 and 2. It has been proposed that it is the tPA-PAI-1 complex, which through its binding to LRP and the resulting increase in production of MMP3, increases the permeability of endothelial barriers in the injured CNS [80]. This hypothesis was supported by observation that CSF levels of tPA-PAI-1 complex and MMP3 are increased in patients with TBI. Another pathophysiological mechanism by which tPA may increase the permeability of CNS barriers could involve its interaction with platelet-derived growth factor-CC (PDGF-CC) [78]. Tissue plasminogen activator can activate a latent dimeric PDGF-CC [85], which is then able to bind to its receptor PDGFR-α [86]. Both tPA and PDGF-CC bind to LRP, which facilitates activation of PDGF-CC by tPA [78]. It has been demonstrated that PDGFR-α is expressed on astrocyte processes closely associated with cerebral microvessels. By binding to its receptor, PDGF-CC increases the permeability of the BBB, but the molecular mechanisms underlying this increase in endothelial permeability remain to be established. Although this may not directly affect the permeability of CNS barriers, it is also important to note that tPA, acting synergistically with lipopolysaccharide (LPS), a ligand for toll-like receptor (TLR)-4, causes significant activation of microglia [87]. This tPA action is unrelated to its proteolytic activity and involves the interaction of its finger domain with annexin II expressed on microglial cells.

Tissue plasminogen activator may play a role in facilitating leukocyte invasion observed after injury. It has been demonstrated that a direct contact of monocytes with brain endothelium results in a ROS-dependent release of tPA from endothelial cells [88]. This tPA then acts on brain endothelium, causing the activation of the ERK signaling cascade and degradation of tight junction protein occludin, which facilitates the migration of monocytes across the endothelial barrier. In addition, thanks to its ability to generate plasmin, tPA may enhance the chemotactic activity of the major monocyte chemoattractant CCL2 [89]. This chemokine can also increase the permeability of the BBB by inducing the formation of actin stress fibers and causing the redistribution of tight junction proteins occludin and CLDN5 and the tight junction-associated proteins ZO1 and ZO2 [90]. Importantly, the plasmin-cleaved fragment of CCL2 is more potent in increasing the permeability of endothelial barrier than the full-length CCL2 [91].

A rapid (within 1 h after injury), but short-lasting (24 h post injury), increase in tPA activity has been observed in mice sustaining TBI [92]. By comparison, in a rodent model of SCI, a significant increase in tPA activity in the injured spinal cord tissue was not seen until 2 weeks after injury [93]. These results are, however, inconsistent with immunohistochemical findings in a similar model of contusive SCI [94]. Sashindranath et al. [80] have shown a larger posttraumatic increase in the permeability of the BBB in transgenic mice overexpressing tPA when compared to wild-type controls. These transgenic mice also had larger volumes of posttraumatic lesion and worse functional outcome than control animals. In contrast, *Plat*$^{-/-}$ mice were protected from posttraumatic disruption of the BBB. Somewhat different results were reported by another group [95]. These investigators have shown a similar posttraumatic increase in the permeability of the BBB in *Plat*$^{-/-}$ and wild-type mice. However, control animals had a larger magnitude of posttraumatic brain edema and increased loss of neural tissue when compared to tPA-deficient mice. No difference in functional outcome between *Plat*$^{-/-}$ and wild-type mice was found. In a contusive model of SCI, *Plat*$^{-/-}$ mice have been shown to have a better functional recovery than wild-type animals, which correlated with a smaller damage of the spinal cord tissue observed in the former experimental group [94]. However, another study [93] has suggested that tPA may also play a role in neuronal repair processes after SCI and its deficiency may adversely affect functional recovery.

1.6 Apolipoprotein E, CNS Barriers, and Neurotrauma

Apolipoprotein E, which binds to the members of the family of low-density lipoprotein receptors, plays an essential role in regulating the metabolism of lipids. It is synthesized by multiple organs, with the highest expression found in the liver and CNS [96]. Astrocytes and neurons appear to be the major source of ApoE in the brain [97]. Compared to rodents, which carry one form of ApoE, the human *APOE* gene is polymorphic and three isoforms of ApoE—ApoE2, ApoE3, and ApoE4—exist in humans. The E3 allele of the *APOE* gene is most commonly represented, with 9–98 % frequency worldwide [98]. Several epidemiological studies have demonstrated that the carriers of the E4 allele of the *APOE* gene (0–49 % frequency worldwide) have poorer outcome after mild and severe TBI than those without the

E4 allele [99–101]. In SCI, the presence of the E4 allele has also been associated with worse recovery of motor function [102, 103]. However, other studies involving both children and adults with mild TBI, and a large study conducted on a population of adult patients with severe TBI did not confirm these findings [104–106]. Further work is clearly needed to resolve this issue. The presence of the E4 allele is one of the major risk factors for the late onset or sporadic Alzheimer's disease [107]. An increased deposition of amyloid-β (Aβ)—a common feature of Alzheimer's disease—may play a pathophysiological role in TBI [108, 109]. To determine whether there is a link between the presence of ApoE4, the deposition of Aβ, and outcome after TBI, transgenic mice carrying the *APOE4* or *APOE3* gene, and ApoE-deficient mice were subjected to a controlled cortical impact injury [110]. These studies showed increased deposition of Aβ in *APOE4* transgenic mice; however no differences in the extent of neuronal loss were found among the experimental groups studied. There is evidence suggesting that other pathophysiological mechanisms may be responsible for an adverse effect of ApoE4 on outcome after neurotrauma. These mechanisms could be related to anti-inflammatory properties of ApoE and to its role in maintaining the proper function of endothelial barrier.

In a rat model of TBI, a delayed (at 4 days post-TBI) increase in ApoE mRNA in the cerebral cortex and subcortical structures was observed [97]. This increase in ApoE synthesis lasted for up to 1 month after injury. At the protein level, ApoE was found to be predominantly expressed in astroglia and neurons. Similarly, in a mouse model of SCI, an increase in ApoE mRNA in the injured spinal cord tissue was not observed until 4 days post-SCI, and lasted for at least 3 weeks after injury [111]. Early on, ApoE was localized to invading neutrophils and macrophages, and later was localized to astroglia and white matter tracks. Laboratory studies involving *Apoe*$^{-/-}$ mice have demonstrated that ApoE-deficiency exacerbates motor and cognitive deficits occurring after TBI [112], suggesting that ApoE may play a part in mitigating secondary injury and/or in promoting neuronal repair after neurotrauma. These findings inspired the idea of using mimetic peptides derived from the receptor-binding region of ApoE for treatment in neurotrauma. In rodent models of TBI, these mimetic peptides reduced the loss of neural tissue and improved functional recovery after injury [113–115]. These peptides have also been found to decrease the posttraumatic production of TNF-α [113], which was consistent with previous observation that ApoE deficiency is accompanied by increased TNF-α synthesis in the injured brain [116]. Further studies, in which mice with targeted replacement of the murine *Apoe* gene with human *APOE* isoforms (TR-*APOE*) were used, have demonstrated that, when compared with ApoE3, ApoE4 is associated with a more vigorous brain inflammatory response to the peripheral challenge with LPS [117]. This suggests that ApoE3 has an anti-inflammatory effect, whereas ApoE4 does not. Given the potentially detrimental role of neuroinflammation in CNS injury, this may explain, at least in part, why the presence of the E4 allele of the *APOE* gene increases the risk of poorer outcome after neurotrauma. The possible link between the E4 allele and inflammation in the context of function of the neurovascular unit will be discussed further at the end of this section.

An adverse affect of ApoE4 on outcome after injury may also be related to the inability of this isoform of ApoE to support the normal function of the endothelial barrier. An increase in the permeability of the BBB, especially in cerebellum, has been observed in *Apoe*$^{-/-}$ mice [118]. It has also been reported that the age-dependent leakage of the BBB is enhanced in ApoE-deficient mice [119]. The use of an in vitro model of the neurovascular unit in which the endothelial cells and pericytes were obtained from wild-type mice, and astrocytes (the major source of ApoE in the CNS) were harvested from wild-type, *Apoe*$^{-/-}$, or TR-*APOE* mice, provided a mechanistic insight into the ApoE-dependent control of endothelial function [120]. The reconstitution of the neurovascular unit with astrocytes from ApoE-deficient or TR-*APOE4* mice significantly increased the permeability of endothelial monolayers, whereas when astrocytes from TR-*APOE3* mice were used, the "tightness" of the endothelial barrier was similar to that observed with wild-type astrocytes. The lack of ApoE or the presence of ApoE4 was associated with a low level of phosphorylation of occludin at its threonine residues, suggesting the disruption of integrity of tight junctions [121]. These findings were confirmed by in vivo studies [122] in which TR-*APOE* mice were also used. Both TR-*APOE2* and TR-*APOE3* mice had normal BBB, whereas in TR-*APOE4* and *Apoe*$^{-/-}$ mice, the BBB was leaky. These observations indicate that unlike ApoE4, ApoE2 and ApoE3 can effectively maintain the integrity of CNS barriers. An increase in the permeability of the BBB observed in TR-*APOE4* mice was found to be mediated by the upregulation of expression of cyclophilin A (CyPA), the activation of nuclear factor-κB (NF-κB), and increased synthesis of MMP9 in pericytes. Cyclophilin A could be targeted by cyclosporine A, which binds intracellular CyPA and inhibits its biological activity [123]. When administered to TR-*APOE4* mice, cyclosporine A reversed the increased permeability of the BBB observed in these animals. This suggests that under normal conditions, ApoE2 or ApoE3 inhibits the synthesis of CyPA within the neurovascular unit.

Cyclophilin A is not only the intracellular protein but may also behave as a proinflammatory cytokine [124]. In addition, CyPA can play a role of chemokine acting synergistically with the neutrophil chemoattractant CXCL2 [125]. These findings provide further support for the link between ApoE4 and inflammation.

1.7 CNS Barriers and Neuroinflammation

Neuroinflammation is a significant part of secondary injury processes, and increasing evidence supports an important role of neuroinflammation in delayed neuronal death after neurotrauma. The pathophysiological mechanisms by which the inflammatory cascade in the injured CNS is initiated are not well defined, but recent investigations suggest that CNS barriers are involved in this process. The disruption of integrity of vascular walls caused by injury forces allows plasma proteins, such as albumin, fibrinogen, and thrombin (cleaved from circulating prothrombin by Factor Xa), to enter CNS parenchyma. These proteins have the ability to activate astrocytes and microglia—the integral components of the neurovascular unit—causing an increase in synthesis of proinflammatory cytokines, including TNF-α, IL-1β, and IL-6, CXC and CC chemokines, and MMPs [12]. The cellular damage caused by injury results in release of various intracellular factors, collectively called damage-associated molecular patterns (DAMPs), such as RNA, DNA, heat shock proteins, and high mobility group box 1 (HMGB1), which are the ligands for TLRs [126]. By binding to TLRs, DAMPs trigger the activation of the NF-κB and mitogen-activated protein kinase signaling cascades, leading to increased synthesis of proinflammatory mediators. Other intracellular factors that may also play a part in initiating the inflammatory cascade are nucleotides, such ATP, UTP, or their analogues [126]. There has been a considerable interest in the DNA-binding protein HMGB1 because of its putative pathophysiological role in various CNS and peripheral diseases with inflammatory component, such as neurotrauma, cerebral ischemia, and sepsis. An increase in CSF level of HMGB1 has been observed in TBI patients [127]. Although HMGB1 is known to be released after the cellular damage, it has been demonstrated that this protein could be secreted by macrophages in response to the TLR4 ligand LPS [128]. Further studies have shown that HMGB1 functionally interacts with TLR2 and TLR4, which are expressed on CNS endothelium, astrocytes, and microglia [129, 130]. Toll-like receptors may play an important role in posttraumatic increase in the permeability of endothelial barriers. Sumi et al. [131] have demonstrated that in co-cultures of brain endothelial cells and microglia from the rat, the activation of microglial LTR4 results in significant increase in the permeability of endothelial monolayers. The pathophysiological role of TLRs in brain injury is supported by studies in which TLR4-mutant or TLR4-null and TLR2-deficient mice were subjected to TBI or cerebral ischemia. These studies have demonstrated a significant reduction in the magnitude of post-ischemic neuroinflammation and a considerable neuroprotection in these two types of brain injury [132–136]. In contrast, TLR4-mutant and TRL2-deficient mice subjected to SCI showed increased demyelination, astrogliosis, and macrophage activation, and worse recovery of motor function when compared to wild-type animals [137]. Further studies will be needed to determine the therapeutic potential of targeting TLRs in neurotrauma.

Both TBI and SCI are associated with a rapid increase in synthesis of proinflammatory cytokines, such as TNF-α, IL-1β, and IL-6, [12, 138]. These cytokines exert multiple important biological effects on function of the neurovascular unit, including an increase in endothelial permeability, production of chemokines, and expression of cell adhesion molecules on the luminal surface of endothelial cells. These actions of proinflammatory cytokines, as well as the role of the BBB in the influx of leukocytes occurring after TBI, have recently been reviewed [12]. Here we will extend our discussion on the pathophysiological consequences of invasion of leukocytes to cover both TBI and SCI, and describe the role of the BCSFB in posttraumatic neuroinflammation.

1.7.1 Invasion of Leukocytes After Neurotrauma

In rodent models of TBI, neutrophil invasion is observed within hours after injury, whereas monocytes infiltrate the traumatized parenchyma within days after TBI [139, 140]. In mice sustaining SCI, there is also an early influx of neutrophils peaking at 24 h after injury, which is followed by a secondary gradual increase in the number of infiltrating neutrophils during the subacute/chronic stage of SCI [141]. The magnitude of post-SCI influx of monocytes initially peaks at 1 week after injury, but a much greater secondary peak in the influx of these inflammatory cells is observed between 2 and 3 months after SCI. Laboratory studies of TBI have shown that there is an association between the magnitude of posttraumatic influx of neutrophils and monocytes, and the formation of cerebral edema, the extent of loss of neural tissue, and functional outcome after injury [142–147]. Similarly, in rodent models of SCI, the depletion of circulating neutrophils and/or monocytes or prevention of their recruitment has been shown to result in the reduction in BSCB permeability, a better sparing of white matter tracks, and improved functional outcome after injury [148–151]. Some controversy has arisen, however, regarding the pathophysiological role of neutrophils in SCI [152]. It is also important to note that the indiscriminate targeting of circulating monocytes may not be the most optimal therapeutic approach in neurotrauma. Monocytes entering the injured CNS parenchyma are thought to differentiate into macrophages. Macrophages could be generally classified as classically activated M1 type and alternatively activated M2 type with significantly different physiological functions [153]. Kigerl et al. [154] have demonstrated that M1 macrophages dominate over M2 macrophages in the injured spinal cord tissue, especially at later time points after SCI. These investigators have also shown that M1 macrophages are neurotoxic, whereas M2 macrophages promote regeneration, suggesting that the high M1/M2 macrophage ratio observed in the injured tissue has a negative impact on neuronal repair processes.

In contrast to neutrophils and monocytes, T- and B-lymphocytes do not appear to play any significant pathophysiological roles in TBI [155]. In SCI, a delayed gradual increase in the number of T cells accumulating in the injured spinal cord tissue is observed [141]. There is evidence that SCI may be associated with the production of autoreactive T-lymphocytes directed against CNS-specific antigens, such as myelin basic protein (MBP) [156]. In transgenic mice in which the majority of $CD4^+$ T-lymphocytes were reactive against MBP, a poorer preservation of myelinated axons and worse recovery of locomotor function after SCI were observed when compared to control animals [157]. However, the opposite effects of autoreactive T cells have also been shown. Hauben et al. [158, 159] have demonstrated a better tissue preservation and an improved functional outcome after SCI in rats that were either immunized with MBP or injected systemically with MBP-reactive T cells. Further studies may provide an explanation for these discrepant results.

1.7.2 The Role of the BCSFB in Posttraumatic Neuroinflammation

Increasing evidence indicates that the BCSFB plays an important role in posttraumatic neuroinflammation. Studies of a rat model of TBI have shown that injury results in a rapid increase in production of proinflammatory cytokines TNF-α and IL-1β in the choroid plexus in a manner similar to that observed in the injured cerebral cortex and hippocampus [160]. Proinflammatory cytokines are strong inducers of epithelial synthesis of chemokines. Consequently, a gradual increase in the choroidal expression of CXC and CC chemokines is observed after TBI [160, 161]. The immunohistochemical analysis of the choroidal tissue has demonstrated that these chemokines are produced by the choroidal epithelium and do not appear to be synthesized by any other type of cells normally present in the choroid plexus, such as the endothelial and epiplexus cells or stromal macrophages. CXC and CC chemokines are secreted across both the apical (CSF-facing) and basolateral (stroma/blood-facing) membranes of the choroidal epithelium [160, 161], and studies of other epithelia have demonstrated that bidirectional secretion of chemokines by the epithelial cells is a prerequisite for leukocyte migration across the epithelial barrier [162, 163].

The time course of migration of neutrophils across the BCSFB differs slightly from that found for the BBB [160]. In the traumatized brain parenchyma, significant numbers of invading neutrophils are observed between 6 and 8 h post-TBI, whereas these inflammatory cells are only sporadically found to accumulate in the choroid plexus during this time period. The magnitude of influx of neutrophils across the BCSFB peaks at 24 h after injury, and no neutrophils could be found to accumulate in the choroid plexus at 2 days post-TBI. Monocytes cross the BBB and BCSFB in large numbers at 1 day after injury [142, 161]. However, unlike neutrophils, monocytes continue to migrate across the BCSFB for at least 2 weeks after injury (the longest observation period; Szmydynger-Chodobska and Chodobski, unpublished data). The analysis of CSF samples serially collected from TBI patients indicates that CSF levels of CCL2 decrease quite rapidly within 3 days after injury [145]. Therefore, it is rather unlikely that CCL2 would drive the monocyte influx across the BCSFB beyond 3 days post-TBI. Further studies will be needed to identify the potential chemokine(s) that would promote monocyte invasion across the BCSFB during the subacute/chronic stage of injury.

Electron microscopic studies [160, 161] have provided evidence for the migration of neutrophils and monocytes across the BCSFB. The movement of monocytes is frequently associated with the widening of space between invading inflammatory cells and the adjacent epithelial cells, which is not observed when neutrophils migrate across the choroidal epithelial barrier. Invading neutrophils sometimes move in tandem with monocytes. After crossing the BCSFB, leukocytes enter the CSF space. This raises the question as to how they then invade the brain parenchyma. One possible mechanism may involve their movement along the perivascular Virchow-Robin space. Indeed, both neutrophils and monocytes were found to move along the Virchow-Robin space entering from the subarachnoid CSF space near the injury site and from the cistern of velum interpositum located ipsilateral to the injury (Szmydynger-Chodobska and Chodobski, unpublished observations).

1.8 Neurotrauma and ATP-Binding Cassette Transporters

We have recently discussed the therapeutic targeting of ion transporters, such as the Na^+-K^+-$2Cl^-$ cotransporter and Na^+/H^+ exchanger—both expressed at CNS barriers—in neurotrauma [12]. We have also recently reviewed the role of selected ATP-binding cassette transporters expressed in CNS endothelium in the context of brain injury [12]. Here we will update this review to include the most recent information pertinent to ATP-binding cassette transporters and neurotrauma.

1.8.1 ABCB1

The presence of xenobiotic efflux transporters at CNS barriers creates a major obstacle for delivery of neuroprotective drugs to the CNS. One of the most extensively studied xenobiotic efflux transporters is ABCB1 (MDR1 or P-glycoprotein). It has recently been shown that the expression of ABCB1 at the BSCB is significantly upregulated during both acute and chronic phases of SCI [164]. Using the anti-inflammatory agent licofelone, which targets both cyclooxygenases and 5-lipoxygenase, these investigators were able to reduce the level of endothelial expression of ABCB1, consequently increasing the concentration of systemically administered neuroprotective drug riluzole in the injured spinal cord tissue. These results are in line with previously reported upregulation of ABCB1 expression at the BBB in mice subjected to ischemic brain injury [165]. In contrast to those findings, rats sustaining TBI at postnatal day 17 had reduced levels of ABCB1 expression at the BBB 2 months after injury [166]. Similarly, a significant decrease in ABCB1 mRNA in the injured cerebral cortex and hippocampus was found in adult rats as early as 6 h after TBI (Szmydynger-Chodobska and Chodobski, unpublished observations). Given the fact that both neurotrauma and cerebral ischemia are accompanied by increased release of glutamate and augmented production of proinflammatory mediators and ROS, which are known to upregulate the expression/activity of ABCB1 [167, 168], it is unclear why discrepant results in different models of CNS injury were obtained. It is also not known what would cause long-term changes in expression of ABCB1 at CNS barriers observed after injury.

1.8.2 *ABCC8*

Sulfonylurea receptor 1 or ABCC8 is an atypical ATP-binding cassette protein, which acts as a regulatory subunit of $NC_{Ca\text{-}ATP}$ channel. It has been proposed that a member of the mammalian transient receptor potential superfamily of nonselective cation channels—transient receptor potential melastatin 4 (TRPM4)—represents the SUR1-regulated $NC_{Ca\text{-}ATP}$ channel [169]. This has recently been confirmed by co-expression of SUR1 and TRPM4, combined with the fluorescence resonance energy transfer and co-immunoprecipitation assays [170]. However, other investigators did not corroborate these findings [171]. An increase in expression of both ABCC8 and TRPM4 in CNS endothelium has been demonstrated in SCI, cerebral ischemia, and SAH [32, 172–174]. It has been hypothesized that uncontrolled activation of SUR1-regulated TRPM4 channel occurring after injury results in oncotic death of CNS endothelium, leading to ischemia and disintegration of neural tissue. This suggested that this channel could be therapeutically targeted in neurotrauma. Consistent with this hypothesis, an antidiabetic drug glibenclamide, a potent blocker of SUR1-regulated TRPM4 channel, has been shown to reduce edema and the loss of neural tissue, and to improve functional outcome in diverse forms of CNS injury, such as SCI, TBI, ischemic stroke, and SAH [32, 33, 172, 173, 175, 176].

1.9 Future Directions

Significant progress in our understanding of the role and function of CNS barriers in neurotrauma has been made in the last decades. However, further research into this area is needed to develop therapeutic tools to effectively target these barriers. Although this review has provided some explanation for the observed phenomena, it also raises several unanswered questions. Among them is the question about the nature of the pathophysiological processes underlying the long-term increase in the permeability of CNS barriers observed after neurotrauma, the possible adverse effects of this increased endothelial permeability on neuronal repair, and how to therapeutically approach this problem. As we described above, various proteases can disrupt the integrity of CNS barriers, especially at the early stage of injury, but it has also been shown that some proteases may play an important role in neuronal regeneration. This suggests that for therapies involving protease inhibitors, the timing of intervention may be one of the key issues. This also emphasizes the importance of designing the inhibitors that are sufficiently selective.

Barriers of the CNS also play a key role in initiating and propagating neuroinflammation, which frequently accompanies injury. Since this secondary injury process progresses at a relatively slow pace, there is an extended window of opportunity for anti-inflammatory intervention in neurotrauma. However, more research into the pathophysiological role of leukocytes invading neural tissue after injury is needed. For example, the development of therapeutic tools to selectively target different monocyte/macrophage populations may allow us to reduce the damage of neural tissue and, at the same time, enhance regenerative processes. Additional investigations are also required to clarify the discrepant results concerning the T cell-dependent damage of spinal cord tissue after SCI. Increasing evidence supports the pathophysiological role of the choroid plexus in neurological disease [177], which warrants further studies of function of the BCSFB in neurotrauma.

CNS barriers themselves may represent important targets for therapeutic intervention. As described above, promising data were obtained in preclinical studies of neurotrauma in which ABCC8 transporter was targeted. Of concern are the discrepant results on expression/activity of xenobiotic efflux transporters in various forms of CNS injury. These transporters play a critical role in regulating the parenchymal levels of many neuroprotective drugs. Therefore, more effort is needed to enhance our understanding of how the expression/activity of these transporters is regulated and how to utilize this knowledge to increase the efficacy of neuroprotectants. It may also be possible to engage CNS barriers in neuronal repair processes by stimulating the endothelial production of neuroprotective growth factors [178, 179]. These growth factors could then act on closely located neurons. With this therapeutic approach, potential problems associated with delivering neuroprotective drugs across CNS barriers could be avoided.

References

1. Marklund N, Bakshi A, Castelbuono DJ et al (2006) Evaluation of pharmacological treatment strategies in traumatic brain injury. Curr Pharm Des 12:1645–1680
2. Schouten JW (2007) Neuroprotection in traumatic brain injury: a complex struggle against the biology of nature. Curr Opin Crit Care 13:134–142
3. Tator CH (2006) Review of treatment trials in human spinal cord injury: issues, difficulties, and recommendations. Neurosurgery 59:957–982, discussion 982–957
4. Hawryluk GW, Rowland J, Kwon BK et al (2008) Protection and repair of the injured spinal cord: a review of completed, ongoing, and planned clinical trials for acute spinal cord injury. Neurosurg Focus 25:E14
5. Borregaard N, Sørensen OE, Theilgaard-Monch K (2007) Neutrophil granules: a library of innate immunity proteins. Trends Immunol 28:340–345
6. Newby AC (2008) Metalloproteinase expression in monocytes and macrophages and its relationship to atherosclerotic plaque instability. Arterioscler Thromb Vasc Biol 28: 2108–2114
7. Balabanov R, Dore-Duffy P (1998) Role of the CNS microvascular pericyte in the blood–brain barrier. J Neurosci Res 53:637–644
8. Mathiisen TM, Lehre KP, Danbolt NC et al (2010) The perivascular astroglial sheath provides a complete covering of the brain microvessels: an electron microscopic 3D reconstruction. Glia 58:1094–1103
9. Lassmann H, Zimprich F, Vass K et al (1991) Microglial cells are a component of the perivascular glia limitans. J Neurosci Res 28:236–243

10. Abbott NJ, Rönnbäck L, Hansson E (2006) Astrocyte-endothelial interactions at the blood–brain barrier. Nat Rev Neurosci 7:41–53
11. Winkler EA, Bell RD, Zlokovic BV (2011) Central nervous system pericytes in health and disease. Nat Neurosci 14:1398–1405
12. Chodobski A, Zink BJ, Szmydynger-Chodobska J (2011) Blood–brain barrier pathophysiology in traumatic brain injury. Transl Stroke Res 2:492–516
13. Daneman R, Zhou L, Kebede AA et al (2010) Pericytes are required for blood–brain barrier integrity during embryogenesis. Nature 468:562–566
14. Armulik A, Genové G, Mäe M et al (2010) Pericytes regulate the blood–brain barrier. Nature 468:557–561
15. Winkler EA, Sengillo JD, Bell RD et al (2012) Blood-spinal cord barrier pericyte reductions contribute to increased capillary permeability. J Cereb Blood Flow Metab 32:1841–1852
16. Ge S, Pachter JS (2006) Isolation and culture of microvascular endothelial cells from murine spinal cord. J Neuroimmunol 177:209–214
17. Prockop LD, Naidu KA, Binard JE et al (1995) Selective permeability of [^{3}H]-D-mannitol and [^{14}C]-carboxyl-inulin across the blood–brain barrier and blood-spinal cord barrier in the rabbit. J Spinal Cord Med 18:221–226
18. Campos CR, Schröter C, Wang X et al (2012) ABC transporter function and regulation at the blood-spinal cord barrier. J Cereb Blood Flow Metab 32:1559–1566
19. Strazielle N, Ghersi-Egea JF (2000) Choroid plexus in the central nervous system: biology and physiopathology. J Neuropathol Exp Neurol 59:561–574
20. Kratzer I, Vasiljevic A, Rey C et al (2012) Complexity and developmental changes in the expression pattern of claudins at the blood-CSF barrier. Histochem Cell Biol 138:861–879
21. Amasheh S, Meiri N, Gitter AH et al (2002) Claudin-2 expression induces cation-selective channels in tight junctions of epithelial cells. J Cell Sci 115:4969–4976
22. Keep RF, Jones HC (1990) A morphometric study on the development of the lateral ventricle choroid plexus, choroid plexus capillaries and ventricular ependyma in the rat. Brain Res Dev Brain Res 56:47–53
23. Szmydynger-Chodobska J, Chodobski A, Johanson CE (1994) Postnatal developmental changes in blood flow to choroid plexuses and cerebral cortex of the rat. Am J Physiol 266: R1488–R1492
24. Gazzin S, Strazielle N, Schmitt C et al (2008) Differential expression of the multidrug resistance-related proteins ABCB1 and ABCC1 between blood–brain interfaces. J Comp Neurol 510:497–507
25. Shen H, Smith DE, Keep RF et al (2004) Immunolocalization of the proton-coupled oligopeptide transporter PEPT2 in developing rat brain. Mol Pharm 1:248–256
26. Leggas M, Adachi M, Scheffer GL et al (2004) Mrp4 confers resistance to topotecan and protects the brain from chemotherapy. Mol Cell Biol 24:7612–7621
27. Saatman KE, Duhaime AC, Bullock R et al (2008) Classification of traumatic brain injury for targeted therapies. J Neurotrauma 25:719–738
28. Dietrich WD, Alonso O, Halley M (1994) Early microvascular and neuronal consequences of traumatic brain injury: a light and electron microscopic study in rats. J Neurotrauma 11:289–301
29. Noble LJ, Wrathall JR (1989) Distribution and time course of protein extravasation in the rat spinal cord after contusive injury. Brain Res 482:57–66
30. Oertel M, Kelly DF, McArthur D et al (2002) Progressive hemorrhage after head trauma: predictors and consequences of the evolving injury. J Neurosurg 96:109–116
31. Kurland D, Hong C, Aarabi B et al (2012) Hemorrhagic progression of a contusion after traumatic brain injury: a review. J Neurotrauma 29:19–31
32. Simard JM, Tsymbalyuk O, Ivanov A et al (2007) Endothelial sulfonylurea receptor 1-regulated $NC_{Ca\text{-}ATP}$ channels mediate progressive hemorrhagic necrosis following spinal cord injury. J Clin Invest 117:2105–2113

33. Simard JM, Kilbourne M, Tsymbalyuk O et al (2009) Key role of sulfonylurea receptor 1 in progressive secondary hemorrhage after brain contusion. J Neurotrauma 26:2257–2267
34. Başkaya MK, Rao AM, Doğan A et al (1997) The biphasic opening of the blood–brain barrier in the cortex and hippocampus after traumatic brain injury in rats. Neurosci Lett 226:33–36
35. Kelley BJ, Lifshitz J, Povlishock JT (2007) Neuroinflammatory responses after experimental diffuse traumatic brain injury. J Neuropathol Exp Neurol 66:989–1001
36. Vaz R, Sarmento A, Borges N et al (1997) Ultrastructural study of brain microvessels in patients with traumatic cerebral contusions. Acta Neurochir (Wien) 139:215–220
37. Preston E, Webster J, Small D (2001) Characteristics of sustained blood–brain barrier opening and tissue injury in a model for focal trauma in the rat. J Neurotrauma 18:83–92
38. Tomkins O, Shelef I, Kaizerman I et al (2008) Blood–brain barrier disruption in post-traumatic epilepsy. J Neurol Neurosurg Psychiatry 79:774–777
39. Whetstone WD, Hsu JY, Eisenberg M et al (2003) Blood-spinal cord barrier after spinal cord injury: relation to revascularization and wound healing. J Neurosci Res 74:227–239
40. Popovich PG, Horner PJ, Mullin BB et al (1996) A quantitative spatial analysis of the blood-spinal cord barrier. I. Permeability changes after experimental spinal contusion injury. Exp Neurol 142:258–275
41. Xue M, Hollenberg MD, Yong VW (2006) Combination of thrombin and matrix metalloproteinase-9 exacerbates neurotoxicity in cell culture and intracerebral hemorrhage in mice. J Neurosci 26:10281–10291
42. Rao HV, Thirumangalakudi L, Desmond P et al (2007) Cyclin D1, cdk4, and Bim are involved in thrombin-induced apoptosis in cultured cortical neurons. J Neurochem 101: 498–505
43. Paul J, Strickland S, Melchor JP (2007) Fibrin deposition accelerates neurovascular damage and neuroinflammation in mouse models of Alzheimer's disease. J Exp Med 204:1999–2008
44. Marmarou A, Signoretti S, Fatouros PP et al (2006) Predominance of cellular edema in traumatic brain swelling in patients with severe head injuries. J Neurosurg 104:720–730
45. Endo T, Suzuki S, Utsunomiya A et al (2011) Prediction of neurological recovery using apparent diffusion coefficient in cases of incomplete spinal cord injury. Neurosurgery 68: 329–336
46. Dore-Duffy P, Owen C, Balabanov R et al (2000) Pericyte migration from the vascular wall in response to traumatic brain injury. Microvasc Res 60:55–69
47. Cunningham LA, Wetzel M, Rosenberg GA (2005) Multiple roles for MMPs and TIMPs in cerebral ischemia. Glia 50:329–339
48. Yang Y, Estrada EY, Thompson JF et al (2007) Matrix metalloproteinase-mediated disruption of tight junction proteins in cerebral vessels is reversed by synthetic matrix metalloproteinase inhibitor in focal ischemia in rat. J Cereb Blood Flow Metab 27:697–709
49. Rosenberg GA, Yang Y (2007) Vasogenic edema due to tight junction disruption by matrix metalloproteinases in cerebral ischemia. Neurosurg Focus 22:E4
50. Rosenberg GA, Cunningham LA, Wallace J et al (2001) Immunohistochemistry of matrix metalloproteinases in reperfusion injury to rat brain: activation of MMP-9 linked to stromelysin-1 and microglia in cell cultures. Brain Res 893:104–112
51. Noble LJ, Donovan F, Igarashi T et al (2002) Matrix metalloproteinases limit functional recovery after spinal cord injury by modulation of early vascular events. J Neurosci 22: 7526–7535
52. Wells JE, Rice TK, Nuttall RK et al (2003) An adverse role for matrix metalloproteinase 12 after spinal cord injury in mice. J Neurosci 23:10107–10115
53. Takata F, Dohgu S, Matsumoto J et al (2011) Brain pericytes among cells constituting the blood–brain barrier are highly sensitive to tumor necrosis factor-α, releasing matrix metalloproteinase-9 and migrating *in vitro*. J Neuroinflammation 8:106
54. Reijerkerk A, Kooij G, van der Pol SM et al (2006) Diapedesis of monocytes is associated with MMP-mediated occludin disappearance in brain endothelial cells. FASEB J 20: 2550–2552

55. Truettner JS, Alonso OF, Dalton Dietrich W (2005) Influence of therapeutic hypothermia on matrix metalloproteinase activity after traumatic brain injury in rats. J Cereb Blood Flow Metab 25:1505–1516
56. Goussev S, Hsu JY, Lin Y et al (2003) Differential temporal expression of matrix metalloproteinases after spinal cord injury: relationship to revascularization and wound healing. J Neurosurg 99:188–197
57. Wang X, Jung J, Asahi M et al (2000) Effects of matrix metalloproteinase-9 gene knock-out on morphological and motor outcomes after traumatic brain injury. J Neurosci 20:7037–7042
58. Asahi M, Asahi K, Jung JC et al (2000) Role for matrix metalloproteinase 9 after focal cerebral ischemia: effects of gene knockout and enzyme inhibition with BB-94. J Cereb Blood Flow Metab 20:1681–1689
59. Hsu JY, McKeon R, Goussev S et al (2006) Matrix metalloproteinase-2 facilitates wound healing events that promote functional recovery after spinal cord injury. J Neurosci 26: 9841–9850
60. Zhao BQ, Wang S, Kim HY et al (2006) Role of matrix metalloproteinases in delayed cortical responses after stroke. Nat Med 12:441–445
61. Livingston DH, Lavery RF, Mosenthal AC et al (2005) Recovery at one year following isolated traumatic brain injury: a Western Trauma Association prospective multicenter trial. J Trauma 59:1298–1304, discussion 1304
62. Rapoport MJ, Herrmann N, Shammi P et al (2006) Outcome after traumatic brain injury sustained in older adulthood: a one-year longitudinal study. Am J Geriatr Psychiatry 14: 456–465
63. Hamm RJ, Jenkins LW, Lyeth BG et al (1991) The effect of age on outcome following traumatic brain injury in rats. J Neurosurg 75:916–921
64. Onyszchuk G, He YY, Berman NE et al (2008) Detrimental effects of aging on outcome from traumatic brain injury: a behavioral, magnetic resonance imaging, and histological study in mice. J Neurotrauma 25:153–171
65. Kumar A, Stoica BA, Sabirzhanov B et al (2013) Traumatic brain injury in aged animals increases lesion size and chronically alters microglial/macrophage classical and alternative activation states. Neurobiol Aging 34:1397–1411
66. Lee P, Kim J, Williams R et al (2012) Effects of aging on blood brain barrier and matrix metalloproteases following controlled cortical impact in mice. Exp Neurol 234:50–61
67. Kidwell CS, Latour L, Saver JL et al (2008) Thrombolytic toxicity: blood brain barrier disruption in human ischemic stroke. Cerebrovasc Dis 25:338–343
68. Yepes M, Roussel BD, Ali C et al (2009) Tissue-type plasminogen activator in the ischemic brain: more than a thrombolytic. Trends Neurosci 32:48–55
69. Samson AL, Medcalf RL (2006) Tissue-type plasminogen activator: a multifaceted modulator of neurotransmission and synaptic plasticity. Neuron 50:673–678
70. Benchenane K, Berezowski V, Ali C et al (2005) Tissue-type plasminogen activator crosses the intact blood–brain barrier by low-density lipoprotein receptor-related protein-mediated transcytosis. Circulation 111:2241–2249
71. Zlokovic BV, Wang L, Sun N et al (1995) Expression of tissue plasminogen activator in cerebral capillaries: possible fibrinolytic function of the blood–brain barrier. Neurosurgery 37:955–961
72. Kim JA, Tran ND, Li Z et al (2006) Brain endothelial hemostasis regulation by pericytes. J Cereb Blood Flow Metab 26:209–217
73. Tran ND, Schreiber SS, Fisher M (1998) Astrocyte regulation of endothelial tissue plasminogen activator in a blood–brain barrier model. J Cereb Blood Flow Metab 18:1316–1324
74. Faber-Elman A, Miskin R, Schwartz M (1995) Components of the plasminogen activator system in astrocytes are modulated by tumor necrosis factor-alpha and interleukin-1 beta through similar signal transduction pathways. J Neurochem 65:1524–1535
75. Cassé F, Bardou I, Danglot L et al (2012) Glutamate controls tPA recycling by astrocytes, which in turn influences glutamatergic signals. J Neurosci 32:5186–5199

76. Tsirka SE, Rogove AD, Bugge TH et al (1997) An extracellular proteolytic cascade promotes neuronal degeneration in the mouse hippocampus. J Neurosci 17:543–552
77. Yepes M, Sandkvist M, Moore EG et al (2003) Tissue-type plasminogen activator induces opening of the blood–brain barrier via the LDL receptor-related protein. J Clin Invest 112: 1533–1540
78. Su EJ, Fredriksson L, Geyer M et al (2008) Activation of PDGF-CC by tissue plasminogen activator impairs blood–brain barrier integrity during ischemic stroke. Nat Med 14:731–737
79. Polavarapu R, Gongora MC, Yi H et al (2007) Tissue-type plasminogen activator-mediated shedding of astrocytic low-density lipoprotein receptor-related protein increases the permeability of the neurovascular unit. Blood 109:3270–3278
80. Sashindranath M, Sales E, Daglas M et al (2012) The tissue-type plasminogen activator-plasminogen activator inhibitor 1 complex promotes neurovascular injury in brain trauma: evidence from mice and humans. Brain 135:3251–3264
81. Lee SR, Guo SZ, Scannevin RH et al (2007) Induction of matrix metalloproteinase, cytokines and chemokines in rat cortical astrocytes exposed to plasminogen activators. Neurosci Lett 417:1–5
82. Wang X, Lee SR, Arai K et al (2003) Lipoprotein receptor-mediated induction of matrix metalloproteinase by tissue plasminogen activator. Nat Med 9:1313–1317
83. Suzuki Y, Nagai N, Yamakawa K et al (2009) Tissue-type plasminogen activator (t-PA) induces stromelysin-1 (MMP-3) in endothelial cells through activation of lipoprotein receptor-related protein. Blood 114:3352–3358
84. Hu K, Yang J, Tanaka S et al (2006) Tissue-type plasminogen activator acts as a cytokine that triggers intracellular signal transduction and induces matrix metalloproteinase-9 gene expression. J Biol Chem 281:2120–2127
85. Fredriksson L, Li H, Fieber C et al (2004) Tissue plasminogen activator is a potent activator of PDGF-CC. EMBO J 23:3793–3802
86. Li X, Ponten A, Aase K et al (2000) PDGF-C is a new protease-activated ligand for the PDGF alpha-receptor. Nat Cell Biol 2:302–309
87. Siao CJ, Tsirka SE (2002) Tissue plasminogen activator mediates microglial activation via its finger domain through annexin II. J Neurosci 22:3352–3358
88. Reijerkerk A, Kooij G, van der Pol SM et al (2008) Tissue-type plasminogen activator is a regulator of monocyte diapedesis through the brain endothelial barrier. J Immunol 181: 3567–3574
89. Sheehan JJ, Zhou C, Gravanis I et al (2007) Proteolytic activation of monocyte chemoattractant protein-1 by plasmin underlies excitotoxic neurodegeneration in mice. J Neurosci 27:1738–1745
90. Stamatovic SM, Keep RF, Kunkel SL et al (2003) Potential role of MCP-1 in endothelial cell tight junction 'opening': signaling via Rho and Rho kinase. J Cell Sci 116:4615–4628
91. Yao Y, Tsirka SE (2011) Truncation of monocyte chemoattractant protein 1 by plasmin promotes blood–brain barrier disruption. J Cell Sci 124:1486–1495
92. Sashindranath M, Samson AL, Downes CE et al (2011) Compartment- and context-specific changes in tissue-type plasminogen activator (tPA) activity following brain injury and pharmacological stimulation. Lab Invest 91:1079–1091
93. Bukhari N, Torres L, Robinson JK et al (2011) Axonal regrowth after spinal cord injury via chondroitinase and the tissue plasminogen activator (tPA)/plasmin system. J Neurosci 31: 14931–14943
94. Abe Y, Nakamura H, Yoshino O et al (2003) Decreased neural damage after spinal cord injury in tPA-deficient mice. J Neurotrauma 20:43–57
95. Mori T, Wang X, Kline AE et al (2001) Reduced cortical injury and edema in tissue plasminogen activator knockout mice after brain trauma. Neuroreport 12:4117–4120
96. Srivastava RA, Bhasin N, Srivastava N (1996) Apolipoprotein E gene expression in various tissues of mouse and regulation by estrogen. Biochem Mol Biol Int 38:91–101

97. Iwata A, Browne KD, Chen XH et al (2005) Traumatic brain injury induces biphasic upregulation of ApoE and ApoJ protein in rats. J Neurosci Res 82:103–114
98. Eisenberg DT, Kuzawa CW, Hayes MG (2010) Worldwide allele frequencies of the human apolipoprotein E gene: climate, local adaptations, and evolutionary history. Am J Phys Anthropol 143:100–111
99. Müller K, Ingebrigtsen T, Wilsgaard T et al (2009) Prediction of time trends in recovery of cognitive function after mild head injury. Neurosurgery 64:698–704, discussion 704
100. Teasdale GM, Nicoll JA, Murray G et al (1997) Association of apolipoprotein E polymorphism with outcome after head injury. Lancet 350:1069–1071
101. Alexander S, Kerr ME, Kim Y et al (2007) Apolipoprotein E4 allele presence and functional outcome after severe traumatic brain injury. J Neurotrauma 24:790–797
102. Jha A, Lammertse DP, Coll JR et al (2008) Apolipoprotein E ε4 allele and outcomes of traumatic spinal cord injury. J Spinal Cord Med 31:171–176
103. Sun C, Ji G, Liu Q et al (2011) Apolipoprotein E epsilon 4 allele and outcomes of traumatic spinal cord injury in a Chinese Han population. Mol Biol Rep 38:4793–4796
104. Moran LM, Taylor HG, Ganesalingam K et al (2009) Apolipoprotein E4 as a predictor of outcomes in pediatric mild traumatic brain injury. J Neurotrauma 26:1489–1495
105. Chamelian L, Reis M, Feinstein A (2004) Six-month recovery from mild to moderate traumatic brain injury: the role of APOE-ε4 allele. Brain 127:2621–2628
106. Teasdale GM, Murray GD, Nicoll JA (2005) The association between *APOE* ε4, age and outcome after head injury: a prospective cohort study. Brain 128:2556–2561
107. Corder EH, Saunders AM, Strittmatter WJ et al (1993) Gene dose of apolipoprotein E type 4 allele and the risk of Alzheimer's disease in late onset families. Science 261:921–923
108. Roberts GW, Gentleman SM, Lynch A et al (1994) β amyloid protein deposition in the brain after severe head injury: implications for the pathogenesis of Alzheimer's disease. J Neurol Neurosurg Psychiatry 57:419–425
109. Ikonomovic MD, Uryu K, Abrahamson EE et al (2004) Alzheimer's pathology in human temporal cortex surgically excised after severe brain injury. Exp Neurol 190:192–203
110. Hartman RE, Laurer H, Longhi L et al (2002) Apolipoprotein E4 influences amyloid deposition but not cell loss after traumatic brain injury in a mouse model of Alzheimer's disease. J Neurosci 22:10083–10087
111. Seitz A, Kragol M, Aglow E et al (2003) Apolipoprotein E expression after spinal cord injury in the mouse. J Neurosci Res 71:417–426
112. Chen Y, Lomnitski L, Michaelson DM et al (1997) Motor and cognitive deficits in apolipoprotein E-deficient mice after closed head injury. Neuroscience 80:1255–1262
113. Lynch JR, Wang H, Mace B et al (2005) A novel therapeutic derived from apolipoprotein E reduces brain inflammation and improves outcome after closed head injury. Exp Neurol 192:109–116
114. Hoane MR, Kaufman N, Vitek MP et al (2009) COG1410 improves cognitive performance and reduces cortical neuronal loss in the traumatically injured brain. J Neurotrauma 26:121–129
115. Kaufman NA, Beare JE, Tan AA et al (2010) COG1410, an apolipoprotein E-based peptide, improves cognitive performance and reduces cortical loss following moderate fluid percussion injury in the rat. Behav Brain Res 214:395–401
116. Lynch JR, Pineda JA, Morgan D et al (2002) Apolipoprotein E affects the central nervous system response to injury and the development of cerebral edema. Ann Neurol 51:113–117
117. Lynch JR, Tang W, Wang H et al (2003) APOE genotype and an ApoE-mimetic peptide modify the systemic and central nervous system inflammatory response. J Biol Chem 278:48529–48533
118. Fullerton SM, Shirman GA, Strittmatter WJ et al (2001) Impairment of the blood-nerve and blood–brain barriers in apolipoprotein E knockout mice. Exp Neurol 169:13–22
119. Hafezi-Moghadam A, Thomas KL, Wagner DD (2007) ApoE deficiency leads to a progressive age-dependent blood–brain barrier leakage. Am J Physiol Cell Physiol 292:C1256–C1262

120. Nishitsuji K, Hosono T, Nakamura T et al (2011) Apolipoprotein E regulates the integrity of tight junctions in an isoform-dependent manner in an *in vitro* blood–brain barrier model. J Biol Chem 286:17536–17542
121. Suzuki T, Elias BC, Seth A et al (2009) PKCη regulates occludin phosphorylation and epithelial tight junction integrity. Proc Natl Acad Sci U S A 106:61–66
122. Bell RD, Winkler EA, Singh I et al (2012) Apolipoprotein E controls cerebrovascular integrity via cyclophilin A. Nature 485:512–516
123. Handschumacher RE, Harding MW, Rice J et al (1984) Cyclophilin: a specific cytosolic binding protein for cyclosporin A. Science 226:544–547
124. Jin ZG, Lungu AO, Xie L et al (2004) Cyclophilin A is a proinflammatory cytokine that activates endothelial cells. Arterioscler Thromb Vasc Biol 24:1186–1191
125. Heine SJ, Olive D, Gao JL et al (2011) Cyclophilin A cooperates with MIP-2 to augment neutrophil migration. J Inflamm Res 4:93–104
126. Pineau I, Lacroix S (2009) Endogenous signals initiating inflammation in the injured nervous system. Glia 57:351–361
127. Au AK, Aneja RK, Bell MJ et al (2012) Cerebrospinal fluid levels of high-mobility group box 1 and cytochrome C predict outcome after pediatric traumatic brain injury. J Neurotrauma 29: 2013–2021
128. Wang H, Bloom O, Zhang M et al (1999) HMG-1 as a late mediator of endotoxin lethality in mice. Science 285:248–251
129. Nagyőszi P, Wilhelm I, Farkas AE et al (2010) Expression and regulation of toll-like receptors in cerebral endothelial cells. Neurochem Int 57:556–564
130. Farina C, Aloisi F, Meinl E (2007) Astrocytes are active players in cerebral innate immunity. Trends Immunol 28:138–145
131. Sumi N, Nishioku T, Takata F et al (2010) Lipopolysaccharide-activated microglia induce dysfunction of the blood–brain barrier in rat microvascular endothelial cells co-cultured with microglia. Cell Mol Neurobiol 30:247–253
132. Kilic U, Kilic E, Matter CM et al (2008) TLR-4 deficiency protects against focal cerebral ischemia and axotomy-induced neurodegeneration. Neurobiol Dis 31:33–40
133. Hyakkoku K, Hamanaka J, Tsuruma K et al (2010) Toll-like receptor 4 (TLR4), but not TLR3 or TLR9, knock-out mice have neuroprotective effects against focal cerebral ischemia. Neuroscience 171:258–267
134. Caso JR, Pradillo JM, Hurtado O et al (2007) Toll-like receptor 4 is involved in brain damage and inflammation after experimental stroke. Circulation 115:1599–1608
135. Yu ZQ, Zha JH (2012) Genetic ablation of toll-like receptor 2 reduces secondary brain injury caused by cortical contusion in mice. Ann Clin Lab Sci 42:26–33
136. Ziegler G, Harhausen D, Schepers C et al (2007) TLR2 has a detrimental role in mouse transient focal cerebral ischemia. Biochem Biophys Res Commun 359:574–579
137. Kigerl KA, Lai W, Rivest S et al (2007) Toll-like receptor (TLR)-2 and TLR-4 regulate inflammation, gliosis, and myelin sparing after spinal cord injury. J Neurochem 102:37–50
138. Stammers AT, Liu J, Kwon BK (2012) Expression of inflammatory cytokines following acute spinal cord injury in a rodent model. J Neurosci Res 90:782–790
139. Soares HD, Hicks RR, Smith D et al (1995) Inflammatory leukocytic recruitment and diffuse neuronal degeneration are separate pathological processes resulting from traumatic brain injury. J Neurosci 15:8223–8233
140. Royo NC, Wahl F, Stutzmann JM (1999) Kinetics of polymorphonuclear neutrophil infiltration after a traumatic brain injury in rat. Neuroreport 10:1363–1367
141. Beck KD, Nguyen HX, Galvan MD et al (2010) Quantitative analysis of cellular inflammation after traumatic spinal cord injury: evidence for a multiphasic inflammatory response in the acute to chronic environment. Brain 133:433–447
142. Szmydynger-Chodobska J, Fox LM, Lynch KM et al (2010) Vasopressin amplifies the production of proinflammatory mediators in traumatic brain injury. J Neurotrauma 27: 1449–1461

143. Semple BD, Bye N, Ziebell JM et al (2010) Deficiency of the chemokine receptor CXCR2 attenuates neutrophil infiltration and cortical damage following closed head injury. Neurobiol Dis 40:394–403
144. Kenne E, Erlandsson A, Lindbom L et al (2012) Neutrophil depletion reduces edema formation and tissue loss following traumatic brain injury in mice. J Neuroinflammation 9:17
145. Semple BD, Bye N, Rancan M et al (2010) Role of CCL2 (MCP-1) in traumatic brain injury (TBI): evidence from severe TBI patients and CCL2−/− mice. J Cereb Blood Flow Metab 30:769–782
146. Utagawa A, Bramlett HM, Daniels L et al (2008) Transient blockage of the CD11d/CD18 integrin reduces contusion volume and macrophage infiltration after traumatic brain injury in rats. Brain Res 1207:155–163
147. Bao F, Shultz SR, Hepburn JD et al (2012) A CD11d monoclonal antibody treatment reduces tissue injury and improves neurological outcome after fluid percussion brain injury in rats. J Neurotrauma 29:2375–2392
148. Popovich PG, Guan Z, Wei P et al (1999) Depletion of hematogenous macrophages promotes partial hindlimb recovery and neuroanatomical repair after experimental spinal cord injury. Exp Neurol 158:351–365
149. Lee SM, Rosen S, Weinstein P et al (2011) Prevention of both neutrophil and monocyte recruitment promotes recovery after spinal cord injury. J Neurotrauma 28:1893–1907
150. Gris D, Marsh DR, Oatway MA et al (2004) Transient blockade of the CD11d/CD18 integrin reduces secondary damage after spinal cord injury, improving sensory, autonomic, and motor function. J Neurosci 24:4043–4051
151. Geremia NM, Bao F, Rosenzweig TE et al (2012) CD11d antibody treatment improves recovery in spinal cord-injured mice. J Neurotrauma 29:539–550
152. Stirling DP, Liu S, Kubes P et al (2009) Depletion of Ly6G/Gr-1 leukocytes after spinal cord injury in mice alters wound healing and worsens neurological outcome. J Neurosci 29: 753–764
153. Martinez FO, Sica A, Mantovani A et al (2008) Macrophage activation and polarization. Front Biosci 13:453–461
154. Kigerl KA, Gensel JC, Ankeny DP et al (2009) Identification of two distinct macrophage subsets with divergent effects causing either neurotoxicity or regeneration in the injured mouse spinal cord. J Neurosci 29:13435–13444
155. Weckbach S, Neher M, Losacco JT et al (2012) Challenging the role of adaptive immunity in neurotrauma: *Rag1*$^{-/-}$ mice lacking mature B and T cells do not show neuroprotection after closed head injury. J Neurotrauma 29:1233–1242
156. Ankeny DP, Popovich PG (2009) Mechanisms and implications of adaptive immune responses after traumatic spinal cord injury. Neuroscience 158:1112–1121
157. Jones TB, Basso DM, Sodhi A et al (2002) Pathological CNS autoimmune disease triggered by traumatic spinal cord injury: implications for autoimmune vaccine therapy. J Neurosci 22:2690–2700
158. Hauben E, Nevo U, Yoles E et al (2000) Autoimmune T cells as potential neuroprotective therapy for spinal cord injury. Lancet 355:286–287
159. Hauben E, Butovsky O, Nevo U et al (2000) Passive or active immunization with myelin basic protein promotes recovery from spinal cord contusion. J Neurosci 20:6421–6430
160. Szmydynger-Chodobska J, Strazielle N, Zink BJ et al (2009) The role of the choroid plexus in neutrophil invasion after traumatic brain injury. J Cereb Blood Flow Metab 29:1503–1516
161. Szmydynger-Chodobska J, Strazielle N, Gandy JR et al (2012) Posttraumatic invasion of monocytes across the blood-cerebrospinal fluid barrier. J Cereb Blood Flow Metab 32: 93–104
162. McCormick BA, Hofman PM, Kim J et al (1995) Surface attachment of *Salmonella typhimurium* to intestinal epithelia imprints the subepithelial matrix with gradients chemotactic for neutrophils. J Cell Biol 131:1599–1608

163. McCormick BA, Parkos CA, Colgan SP et al (1998) Apical secretion of a pathogen-elicited epithelial chemoattractant activity in response to surface colonization of intestinal epithelia by *Salmonella typhimurium*. J Immunol 160:455–466
164. Dulin JN, Moore ML, Grill RJ (2013) The dual cyclooxygenase/5-lipoxygenase inhibitor licofelone attenuates p-glycoprotein-mediated drug resistance in the injured spinal cord. J Neurotrauma 30:211–226
165. Spudich A, Kilic E, Xing H et al (2006) Inhibition of multidrug resistance transporter-1 facilitates neuroprotective therapies after focal cerebral ischemia. Nat Neurosci 9:487–488
166. Pop V, Sorensen DW, Kamper JE et al (2013) Early brain injury alters the blood–brain barrier phenotype in parallel with β-amyloid and cognitive changes in adulthood. J Cereb Blood Flow Metab 33:205–214
167. Miller DS, Bauer B, Hartz AM (2008) Modulation of P-glycoprotein at the blood–brain barrier: opportunities to improve central nervous system pharmacotherapy. Pharmacol Rev 60:196–209
168. Potschka H (2010) Targeting regulation of ABC efflux transporters in brain diseases: a novel therapeutic approach. Pharmacol Ther 125:118–127
169. Simard JM, Tarasov KV, Gerzanich V (2007) Non-selective cation channels, transient receptor potential channels and ischemic stroke. Biochim Biophys Acta 1772:947–957
170. Woo SK, Kwon MS, Ivanov A et al (2013) The sulfonylurea receptor 1 (Sur1)-transient receptor potential melastatin 4 (Trpm4) channel. J Biol Chem 288:3655–3667
171. Sala-Rabanal M, Wang S, Nichols CG (2012) On potential interactions between non-selective cation channel TRPM4 and sulfonylurea receptor SUR1. J Biol Chem 287: 8746–8756
172. Simard JM, Chen M, Tarasov KV et al (2006) Newly expressed SUR1-regulated $NC_{Ca\text{-}ATP}$ channel mediates cerebral edema after ischemic stroke. Nat Med 12:433–440
173. Simard JM, Geng Z, Woo SK et al (2009) Glibenclamide reduces inflammation, vasogenic edema, and caspase-3 activation after subarachnoid hemorrhage. J Cereb Blood Flow Metab 29:317–330
174. Gerzanich V, Woo SK, Vennekens R et al (2009) *De novo* expression of Trpm4 initiates secondary hemorrhage in spinal cord injury. Nat Med 15:185–191
175. Simard JM, Yurovsky V, Tsymbalyuk N et al (2009) Protective effect of delayed treatment with low-dose glibenclamide in three models of ischemic stroke. Stroke 40:604–609
176. Patel AD, Gerzanich V, Geng Z et al (2010) Glibenclamide reduces hippocampal injury and preserves rapid spatial learning in a model of traumatic brain injury. J Neuropathol Exp Neurol 69:1177–1190
177. Ransohoff RM (2009) In the beginning. Nature 462:41–42
178. Guo S, Kim WJ, Lok J et al (2008) Neuroprotection via matrix-trophic coupling between cerebral endothelial cells and neurons. Proc Natl Acad Sci U S A 105:7582–7587
179. Sawada N, Kim HH, Moskowitz MA et al (2009) *Rac1* is a critical mediator of endothelium-derived neurotrophic activity. Sci Signal 2:ra10

Chapter 2
Mechanisms of Cerebral Edema Leading to Early Seizures After Traumatic Brain Injury

Philip H. Iffland II, Gerald A. Grant, and Damir Janigro

Abstract Epidemiological data support a link between traumatic brain injury (TBI) and seizures. TBI accounts for approximately 16 % of acute symptomatic seizures which usually occur in the first week after trauma. Children are at higher risk for posttraumatic seizures than adults and experience greater morbidity and mortality from cerebral edema (CE). CE is responsible for half of the mortality associated with TBI. A recent book chapter summarizes the most important features of posttraumatic seizure. In this chapter we will summarize features relevant to the link between cerebral edema, cerebrovascular events and seizures after TBI. In addition, we will discuss the potential autoimmune implications of TBI.

2.1 Introduction

Epidemiological data support a link between traumatic brain injury (TBI) and seizures. TBI accounts for approximately 16 % of acute symptomatic seizures which usually occur in the first week after trauma [1]. Children are at higher risk for posttraumatic seizures than adults and experience greater morbidity and mortality from cerebral edema (CE) [2]. CE is responsible for half of the mortality associated with TBI [3]. A recent book chapter summarizes the most important features of posttraumatic seizures [4]. In this chapter we will summarize features relevant to the link between cerebral edema, cerebrovascular events and seizures after TBI. In addition, we will discuss the potential autoimmune implications of TBI.

The mechanisms of early posttraumatic seizures and epilepsy following TBI are poorly understood. It is known that TBI causes both primary and secondary injury

D. Janigro (✉)
Cerebrovascular Research, Departments of Neurosurgery and Cellular and Molecular Medicine, Department of Molecular Medicine, Cleveland Clinic Foundation, NB-20 LRI, 9500 Euclid Avenue, Cleveland, OH 44195, USA
e-mail: janigrd@ccf.org

E.H. Lo et al. (eds.), *Vascular Mechanisms in CNS Trauma*, Springer Series in Translational Stroke Research 5, DOI 10.1007/978-1-4614-8690-9_2,

to the brain. Furthermore, the epileptogenic process may start with the initial insult to the brain which involves ionic, molecular, and cellular alterations resulting in cerebral edema and blood–brain barrier breakdown that may, or may not, lead to early seizures, late seizures, and/or epilepsy months to years following the initial trauma.

2.2 Posttraumatic Seizures

To study how TBI leads to changes in neuronal excitability, in human studies, one needs to focus on the fact that posttraumatic seizures refer only to seizures that occur *after* TBI and are *caused* by TBI. Exacerbation of preexisting seizures is not a good clinical example of post-TBI seizures. Another important preamble to this chapter is the acknowledgement that the temporal relationship between the traumatic event and seizures is a key factor in the underlying mechanisms of ictogenesis. Early posttraumatic seizures (hours after TBI; 2–5 % of all cases in mild TBI (mTBI); 10–15 % in severe TBI [2, 4, 5]) are likely different in mechanism from late seizures and are defined as occurring within 1 week of trauma.

Late seizures are most common after penetrating war-related events (53 % in Vietnam vets with penetrating TBI [5]). Additionally, 82 % of individuals who experience a late posttraumatic seizure will have another seizure within a year. This would suggest that patients be treated aggressively with anticonvulsant medication after a first unprovoked late seizure [6]. However, factors unrelated to TBI are at play in this population; these include infection, presence of foreign material in brain parenchyma, uncontrolled bleeding, etc. Recurrent seizures are chronic events that occur many months or years after TBI; whether these events are due to or consequence of late or early seizures remains unclear [7]. Use of acute prophylaxis to prevent conversion of early seizures into chronic epilepsy was suggested but the results were not conclusive [8].

Guidelines in the management of closed severe head injury recommend 1 week of anticonvulsants to prevent early seizures, though there is no effect on the risk of late seizures [9]. The prophylactic use of antiepileptic drugs should be short-lasting and therefore limited to the prevention of immediate and early seizures [10]. However, the routine use of antiepileptic drugs to prevent late posttraumatic seizures following severe TBI is not recommended.

2.3 The Blood–Brain Barrier

Prior to a discussion of edema etiology and its related pathologies, one must understand the homeostatic nature of cerebral ion gradients and how the blood–brain barrier (BBB) maintains optimal conditions in the brain. The BBB lines the cerebral microvasculature and is composed of, among other cellular components, differentiated endothelial cells and the tight junctions (TJs) that link them together [11]. Endothelial cells of the BBB are differentiated in that they display less pinocytic activity, lack fenestrations, and have an increased number of mitochondria compared to endothelial cells in the peripheral vasculature. Exposure to luminal flow is a key factor in endothelial cell differentiation [12]. This functional differentiation is most likely due to the tight regulation of transcellular tansport into the brain. While there are a number of molecules that can freely cross the BBB (e.g., ethanol), most substances, particularly those that are large and/or hydrophilic, must cross the BBB via primary or secondary active transport (transport requiring ATP) or by virtue of existing concentration gradients. Energy-dependent transport across the BBB provides a mechanism by which movement of substances into the brain can be regulated based on the requirements of the cerebral environment. For instance, glucose transporters are upregulated on the luminal side of the membrane when cerebral nutrient supply becomes low [13].

Of particular importance to the maintenance of ion homeostasis and integral to any discussion of the blood–brain barrier are tight junctions. These structures provide a means by which endothelial cells can be physically linked together creating a continuous impermeable barrier and forcing the movement of ions and macromolecules to occur across the endothelial membrane or, in instances of TBI, across a disrupted BBB. TJs are comprised of a number of proteins including the integral membrane proteins occludin and claudins −3, −5, and −12. These proteins serve to form the characteristic paracellular seal of the BBB. In the cytoplasmic compartment, occludin and claudins are linked to the zonula occludens (ZO) family of adaptor proteins. ZO−1, −2, and −3 bind to claudins and ZO-1 binds to occludin. Adaptor proteins are bound to secondary adaptor proteins that anchor the junctional complex to the cytoskeleton. In addition to TJs, a secondary barrier, the adherens junction, is located below the TJ in the paracellular space. Adherens junctions serve to further limit vascular permeability [14]. There are a number of detailed reviews regarding the cellular and molecular biology of tight junctions and related BBB junctional complexes [15, 16].

2.4 Ion and Water Homeostasis

The BBB maintains ion gradients that are specific to the cerebral environment. Of particular importance to the discussion that follows are those that, when increased due to disruption of the BBB, lead to neuronal excitability. BBB disruption and the accompanying neuronal hyperexcitability promote seizures [17]. Further, directly increasing potassium levels in the brain has also been shown to cause seizures [18]. A summary of the concentration gradients that exist on either side of the BBB and their effect on neuronal excitability are shown in Fig. 2.1b. What follows is a brief discussion of the regulation of ions and molecules relevant to TBI and seizure in the context of brain edema.

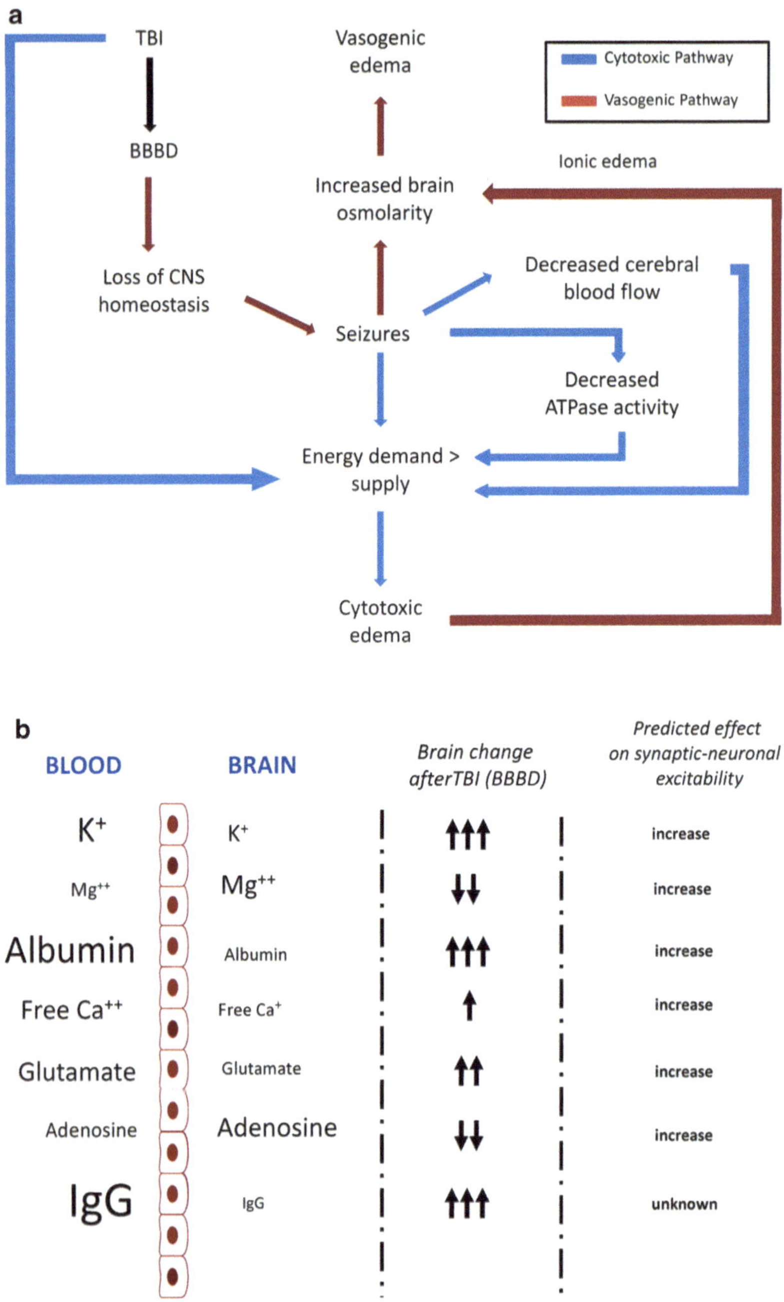

Fig. 2.1 (**a**) Mechanisms of homeostatic failure in the CNS. See text for details. (**b**) Quantitative gradients across the BBB and their predicted effect on neuronal excitability after TBI. The font sizes on the left and rights side of the idealized BBB are roughly proportional to their trans-BBB concentrations under homeostatic conditions. The brain concentration changes indicated by arrows are a semiquantitative means of showing what is expected after TBI induced BBB disruption

Potassium (K^+) regulation is key to the proper function of all excitable cells, including neurons [19]. K^+ in the brain extracellular space (ECS) is maintained at approximately 3 mM irrespective of serum potassium levels, though serum potassium is typically around 5 mM. Intracellular K^+ levels are kept high compared to the extracellular compartment (approximately 6 mM). There are therefore two distinct K^+ gradients: the gradient across cell membranes and the gradient across the BBB. This dramatic difference in concentration allows for the rapid repolarization required of CNS neurons after depolarization. However, neurons are unable to bring about such rapid changes in K^+ concentration alone. K^+ homeostasis is achieved via a glial buffering system in addition to the energy-dependent neuronal mechanism [20].

Neurons contribute to K^+ homeostasis through the Na^+/K^+ ATPase antiporter and sodium-potassium-chloride ($Na^+/K^+/Cl^-$) transporters. The Na^+/K^+ ATPase shuttles three Na^+ from the intracellular compartment and imports two K^+ into the cell per cycle of the pump. The $Na^+/K^+/Cl^-$ transporter moves ions in a 1:1:2 ratio, respectively, and will move ions in or out of the cell in order to maintain electroneutrality [21]. Glia, and in particular astroglia, play the largest role in K^+ buffering by two mechanisms: K^+ uptake and spatial buffering. In the K^+ uptake mechanism, astroglia remove K^+ in the ECS through a glial-specific Na^+/K^+ ATPase better suited for K^+ buffering than the neuronal isotype [22, 23]. Astrocytes also possess a $Na^+/K^+/Cl^-$ cotransporter (NKCC1) that has recently been implicated in potassium buffering [24]. Using these channels, astrocytes are able to temporarily sequester excess extracellular K^+ and release it back into the ECS when K^+ levels drop.

The second mechanism, spatial buffering, is a means by which astrocytes can remove K^+ from areas of high concentration and release it in areas of comparably lower concentration. This is achieved by a syncytium of astroglia connected by gap junctions that directly link neighboring cells [25]. In areas of high K^+ concentration, K^+ enters the cell primarily through potassium inward rectifier (Kir) channels, specifically Kir 4.1 [26]. Kir 4.1 channels localize at astrocyte foot processes [27]. These channels move K^+ into astrocytes and are unique in having a higher conductance at negative membrane potentials [28]. K^+ entry triggers a depolarization that travels through the astroglial network resulting in net outward movement of K^+. In this way, astrocytes are able to spread K^+ across a large area while causing only a transient increase in local intracellular K^+ concentrations [20].

The movement of water across the BBB and within the brain parenchyma follows osmotic gradients. In the brain, water moves through aquaporin channels, specifically aquaporin 1 and aquaporin 4 (AQP4). For the purposes of this discussion we will focus on the role of AQP4 as it is the primary water channel expressed on astrocytes and co-localizes with Kir 4.1 channels [29].

Fig. 2.1 (continued) (BBBD). Additionally, the predicted effect of altered BBB permeability on neuronal excitability is shown

Astrocytes express high levels of AQP4 on end-feet processes surrounding barrier capillaries; their proximity suggests a role in the regulation of water movement into and out of the brain parenchyma [29]. However, AQP4 can be found on end processes of astrocytes at synapses, suggesting an additional role in neuronal water uptake. It is interesting to note that immunocytochemical experiments do not detect the presence of the same aquaporin on neurons, indicating that astrocytes may be responsible for water homeostasis in the brain. AQP4 co-localization with Kir 4.1 channels suggests the coupling of K^+ and water movement within the brain as well as across the BBB [30]. The juxtaposition of these channels has led to the hypothesis that the astroglial syncytium may, in addition to the spatial buffering of potassium, redistribute water throughout the cortex and, in particular, perivascular areas thought to be sinks for excess water and K^+ [31]. Experiments have shown that an increase in extracellular K^+ leads to a decrease in ECS suggesting a link between K^+ and water movement [32]. Additionally, it has been demonstrated that Kir 4.1 channels and AQP4 channels may be associated, extracellularly, by the dystrophin-glycoprotein complex [33, 34].

Glutamate is an excitatory neurotransmitter whose concentration is tightly controlled across the BBB and in the ECS of the brain [35]. Concentrations of glutamate are considerably higher in blood than in the brain. Extracellular cerebral glutamate levels are maintained at approximately 1 μM while plasma glutamate levels are approximately 50 μM [36]. Similar to K^+, extracellular glutamate concentrations are regulated by astrocytes. Released at the synaptic cleft, excess glutamate is taken up by astrocytic processes that surround the synaptic cleft and are enriched with high affinity glutamate transporter 1 (GLT1) and glutamate aspartate transporter 1 (GLAST) receptors [37, 38]. Both transporters require the co-transport of Na^+ with H^+ and the movement of one K^+ and one bicarbonate (HCO_3^-) or OH^- out of the cell. Once in the cytoplasmic compartment of astroglia, glutamine synthase catalyzes the condensation of glutamate with ammonia to form glutamine. Glutamine is then released by the astrocyte and taken up by neurons that convert it back to glutamate.

2.5 Cerebral Edema

In general, CE is defined as an increase in total cerebral water volume leading to an increase in brain tissue volume and intracranial pressure (ICP) [39]. However, advances in basic and clinical science have shown this seemingly straightforward pathology to be a multifaceted process involving dramatic changes to intra/extracellular ion and water balance as well as changes in vascular permeability. Two broad categories of edema have been characterized and termed "cytotoxic" and "vasogenic" to refer to cellular swelling and increased vascular permeability, respectively. It must be noted that these two classifications of edema refer to events that rarely appear independently and are only useful to describe varying stages of a complex process. Additionally, there are a number of subclasses of cytotoxic and vasogenic edemas that appear as either part of the edema process or occur under specific pathological circumstances. In clinical situations, patients will present with varying degrees of the different types of edema depending on the time course and severity of the injury [40].

2.6 Vasogenic Edema

As mentioned previously, the BBB is essential to maintain appropriate ion, protein, and water levels in the brain. After TBI the integrity of the BBB is compromised due to mechanical disruption of the endothelial cells and their associated TJs. It has been widely stated that BBB disruption alone will lead to an increase in water entry into the brain [41, 42]. However, the total osmolarity of blood and cerebrospinal fluid is equal (289 mOsm/L) [43] which could not produce the dramatic movement of water into the brain that is characteristic of vasogenic edema. There are two plausible mechanisms by which water moves into the brain parenchyma after BBB disruption. (1) K^+ moves down its concentration gradient from blood into the brain resulting in a K^+ concentration that is sufficient to depolarize neurons, trigger action potentials, and drive repolarization further elevating cerebral K^+ levels. This high K^+ may lead to disruption of the osmotic homeostasis between brain and blood causing water to move into the brain. (2) Cellular damage from traumatic insult results in intracellular protein release into the brain parenchyma. As extracellular protein levels are normally kept low in the brain, the addition of such a large amount of protein would perturb the osmotic balance between the brain and blood resulting in water influx into the brain.

Mechanical injury to the BBB is not the only cause of increased vascular permeability leading to vasogenic edema. A number of molecules released after TBI have been found to play an active role in either BBB disruption and/or increases in vascular permeability. Matrix metalloproteinases (MMP) are a family of endopeptidases that have been shown to degrade the TJs of the BBB after TBI. Specifically, MMP-2 and −9 have been associated with the degradation of Z0-1, claudin-5, and occludin [44–46]. MMP-2, -3, and −9 expression is increased after TBI. Increased MMP-9 activity, in particular, has been observed in areas of BBB disruption and edema. MMP-9 knockout mice have supported this hypothesis, showing a decrease in BBB breakdown, edema, and inflammation [47].

Two members of the kinin family of proteins, bradykinin and tachykinin have been linked to increased BBB disruption after TBI [48, 49]. Bradykinin acts through two G-protein-coupled receptors, B_1 and B_2, linked to phospholipase C. After TBI, both receptors are highly upregulated for approximately 24 h, but it appears that binding of bradykinin to the B_2 receptor most dramatically affects edema. B_2 receptor knockout mice show less edema after TBI [50]. Further, administration of B_2 receptor antagonists to mice after TBI has been shown to reduce ICP [51]. It should be noted, however, that the same result was not observed in human clinical trials [52].

Tachykinins are neuropeptides that have been linked to neurogenic inflammation [49]. A specific tachykinin, substance P, is a mediator of vascular permeability and has been linked to increased BBB permeability after TBI [53]. In the brain, substance P binds to a G-protein-coupled receptor, neurokinin-1, that acts through phospholipase C. Human studies have shown an increase in immunoreactivity to substance P after TBI and elevated immunoreactivity to substance P in cortical microvasculature [54]. Substance P is co-released from neurons with calcitonin-gene-related peptide (CGRP), a vasodilator known to enhance edema in the presence of molecules similar to substance P [55]. While the link between substance P and vasogenic edema has not been directly established, evidence does suggest a potential role for substance P in the process.

Irrespective of the mediators or mechanism, the result of increased vascular permeability associated with BBB disruption is the paracellular leakage of protein- and ion-rich fluid into the brain. This can lead to a number of complications. (1) The increase in ICP from fluid accumulation. Eventually ICP will cause the ICP to become greater than that of vascular pressure causing blood vessels to collapse and nutrient flow to stop [56]. (2) Excess extracellular ions and neurotransmitter will disrupt the delicate neuronal and glial homeostatic mechanisms which may result in seizure. (3) Immunoglobulins, immune cells, and inflammatory mediators normally kept out of the immunologically privileged brain now have access to nervous tissue [57]. Conversely, proteins normally sequestered in the brain will then have access to peripheral circulation and tissues [58]. (4) The BBB disruption following TBI may prohibit adequate treatment of elevated ICPs with osmotic agents (e.g., mannitol or hypertonic saline) as the gradient which would normally drive water out of the brain might be impaired. There are some recent preclinical studies indicating that modulation of the BBB using small inhibitory RNA directed against claudin-5 may markedly improve the outcome of patients with cerebral edema [59].

2.7 Cytotoxic Edema

Ischemic, hypoxic, and impact injuries that coincide with TBI have been shown to induce the initial signs (cellular swelling) of cytotoxic edema in as little as 30 min [60]. Cytotoxic edema is characterized by changes in osmotic balance between the intracellular compartment and the ECS. This osmotic perturbation leads to an increase in cell volume and a 16 % [61] decrease in the volume of the ECS. However, this process does not directly lead to swelling of the brain, rather a net movement of water from the ECS to the intracellular compartment. Swelling of the brain may occur as a consequence of the ion gradient setup between the ECS and cerebral microvasculature in the absence of BBB disruption (Fig. 2.2). This gradient, caused by depletion of Na^+, water, and Cl^-, promotes the movement of ions and water across the BBB into the ECS leading to an increase in ICP [62]. This secondary movement of ions has been termed "ionic edema."

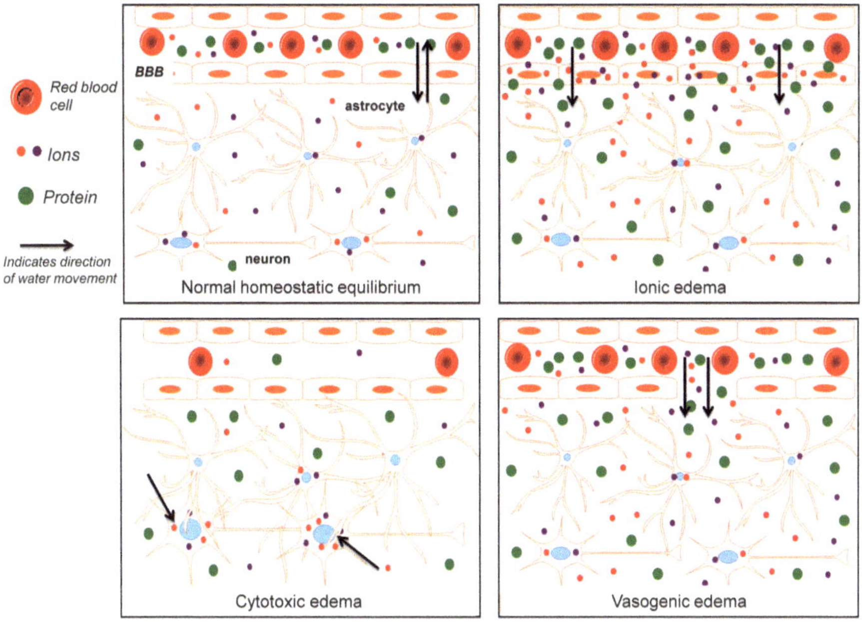

Fig. 2.2 Altered homeostasis leads to different types of edema. Note that water movements are a consequence of increased brain osmolarity due to either increased concentration of ions (BBB leakage) or loss of metabolic substrates (anoxia/ischemia)

A number of interacting mechanisms can produce cytotoxic edema after traumatic insult. However, it is chiefly the lack of nutrient (i.e., oxygen and glucose) supply to the brain and the accompanying loss of adenosine triphosphate (ATP) production that leads to a failure of neurons and glia to maintain proper ion gradients. Without ATP, the Na^+/K^+ ATPase antiporter shuts down. This results in an inadequate mechanism to rectify the passive efflux of K^+ though potassium leak channels and passive influx of Na^+ down its concentration gradient through α-amino-3-hydroxy-5-methyl-4-isoxazolepropionic acid (AMPA) receptors. Following Na^+, Cl^- enters the cell via chloride channels (e.g., $Cl^-/HCO3^-$ antiporter). As the intracellular compartment has now become hyperosmotic, water flows into the cell through aquaporins causing an increase in cell volume and decrease in ECS. Swelling can lead to lysis of the cell that further exacerbates the ionic disruption and leads to necrosis [63].

Astrocytes are particularly sensitive to changes in the cerebral environment and experience greater swelling than neurons after TBI [64]. High levels of extracellular K^+ cause glutamate transporters (GLT-1 and GLAST) to reverse direction and pump K^+ into the cell and glutamate, H^+ and Na^+ out. Additionally, rat models suggest that GLT-1 and GLAST-1 are down regulated after TBI [65]. In similar fashion, high extracellular K^+ leads to upregulation of Kir 4.1 channels leading to a number of downstream effects [66]. (1) High levels of glutamate in the ECS triggers overstimulation of glutamate receptors on neurons leading to excitotoxicity

[67]. (2) Calcium entry leads to the activation of a number of enzymes (e.g., phospholipases) that cause neuronal damage or death [68]. (3) The substantial influx of K^+ via the Kir 4.1 channel on astrocytes depolarizes the cell causing entry of water [69]. The resultant cell swelling can lead to lysis of the astrocyte, which releases more glutamate and K^+ into the ECS serving to drive the cytotoxic cycle [63].

Despite the destructive environment cytotoxic edema creates for neurons and glia, the movement of water from the ECS to the intracellular compartment does not directly lead to an increase in ICP. The extracellular fluid has merely shifted into the intracellular compartment. This water uptake into cells is mediated by aquaporins. AQP4 null mice show decreased susceptibility to cytotoxic edema compared to wild-type mice [70]. However, due to ECS water loss, the fluid in the ECS has become hypertonic compared to blood. The hypertonicity of the ECS will cause the net movement of water across the BBB into the brain parenchyma [62] causing the increase in ICP.

As mentioned previously, cytotoxic and vasogenic edema rarely appear alone. The presence of both creates a cycle that serves to enhance and spread edema from the site of injury into previously unaffected areas of the brain. Cytotoxic edema, which usually appears first after TBI, will eventually cause the movement of water out of the vasculature and into the brain. Excess water increases ICP, which lowers cerebral perfusion pressure and potentiates greater cytotoxic edema. Cytotoxic edema coupled with BBB disruption (vasogenic edema) causes an influx of an even greater amount of water leading to a spreading of the edema and further decreasing perfusion leading to more cytotoxic edema.

2.8 Early Seizure After TBI

TBI produces a number of changes that lead to early seizure after TBI. Acceleration and shearing forces rupture blood vessels, sever nerve fibers and lyse cells resulting in the dysregulation of normal homeostatic processes that manifests as edema and leads to seizure. Leakage of the BBB, irrespective of mechanism, results in an increase in extracellular potassium and an increase in extracellular glutamate [16]. The increase in glutamate will cause neurons to depolarize causing a further increase in extracellular potassium.

The combined effect of BBB leakage and depolarization will lead to extracellular K^+ levels higher than the glial buffering mechanisms can compensate for. High extracellular K^+ has been shown to increase neuronal hyperexcitability by increasing the membrane potential of neurons (i.e., bringing them closer to threshold) and potentiating the influx of Na^+ [71]. As one astrocyte can regulate the environment of multiple neurons, loss of a single cell can have widespread effects. Due to this large area of dysregulation at and around the site of injury, neurons can depolarize in a synchronous manner leading to seizure.

2.9 Autoimmune Implications of BBB Disruption After TBI

One of the obstacles in clinical research on the BBB was the lack of easy to adopt and sensitive measures of its integrity. A recently developed blood test allows for the measurement of BBB function by detection of serum "reporters" of BBB function; thus, a sudden opening of the cerebrovasculature causes a rapid elevation of serum S100B level [72–75]. A number of studies have shown that S100B increase is associated with TBI. A prospective, multicenter study showed that patients with mTBI with a serum S100B measurement of 0.12 ng/mL or lower did not have intracerebral lesions and did not require cranial computer tomography [76–79]. An ironic twist in the saga related to markers of BBB dysfunction is the fact that the use of S100B has not only helped diagnose BBBD but also made it possible to elucidate one of the mechanisms that can lead to autoimmune CNS diseases. In fact, studies have shown that S100B is a powerful autoantigen upon release in systemic circulation (see below) [58].

Evidence suggests that TBI with accompanying BBB disruption leads to the formation of autoantibodies against specific neuronal antigens. The brain, similar to other barrier organs, is an immunologically privileged site [80, 81]. Accordingly, breaking the BBB and allowing what the immune system perceives as foreign antigen into the systemic environment could lead to the formation of an autoimmune response. One known example of this is sympathetic opthalmia. After traumatic insult to the eye, proteins from the eye are exposed to the immune system. As the immune system is not tolerized to eye specific proteins, an immune response can be initiated. Blindness may result in the eye affected by trauma *and* the unaffected eye [82]. A similar mechanism may underlie autoimmune diseases of the CNS.

There are a number of proposed mechanisms for the formation of anti-CNS antibodies. After injury, BBB leakage causes potential CNS antigens to enter the systemic circulation where they enter immune organs and activate autoreactive B cells. These B cells differentiate into plasma cells that begin producing the autoantibody [83]. The proinflammatory and/or necrotic events that often persist after TBI lead to drainage of potential antigens into lymphoid follicles near the area of injury. This leads to a localized production of autoantibodies [84]. Additionally, previous infection may have lead to the formation of memory B cells producing antibodies that cross react with CNS proteins. Once the BBB is breached, these antibodies are able to enter the brain and bind their target. This process has been termed "molecular mimicry" [85, 86].

TBI-associated autoantibodies have been detected against myelin-associated glycoproteins, gangliosides, and ß-tubulin III [87–89]. While no direct evidence exists to support the development of chronic autoimmune disease after TBI, it is of interest to note that a recent study has shown that the development of an autoimmune response due to frequent "openings" of the BBB and extravasation of CNS antigen is sufficient to trigger a B cell response leading to serum autoantibodies

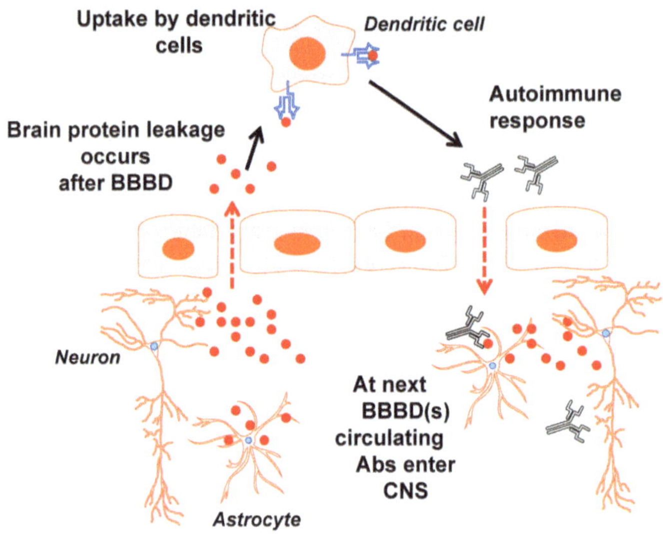

Fig. 2.3 Autoimmunity and TBI: example from football-related sub-concussive injuries [58]. Data have shown that sub-concussive head hits are accompanied by BBBD as measured by serum S100B. After release in systemic circulation, S100B is taken up by dendritic cells [92]; this leads to production of autoantibodies which may become pathogenic upon entry into the CNS

[58]. Interestingly, the authors also reported that the presence of these autoantibodies correlated with imaging and behavioral changes consistent with long-term sequelae of mTBI. Whether these autoantibodies gain entry into the CNS is yet to be demonstrated but results in epileptics show that this is a viable hypothesis (see below).

A number of anti-CNS autoantibodies are also linked to seizure (e.g., Rasmussen's encephalitis) [90]. A recently published article demonstrated that BBB disruption with seizure results in the accumulation of immunoglobulin in the cytoplasm and nuclei of neurons [91]. Whether or not this is a pathological autoimmune response, a neuroprotective response or some unknown phenomenon has yet to be determined. Animal studies and clinical data [92] have shown that a protein used as a marker of BBB disruption, astrocytic S100B, is perceived as nonself by the immune system after extravasation in blood. This protein triggers production of anti-S100B autoantibodies which may assume pathological significance. The sequelae of events linking blood–brain barrier disruption to immune dysregulation are shown in Fig. 2.3.

2.10 Conclusions

The discovery that non-neuronal mechanisms and systemic events are involved in the pathogenesis of CNS diseases is not novel. However, there is increasing evidence that this acquired knowledge may have a significant impact on how we treat and diagnose clinical consequences of, for example, TBI. This is in particular true for seizures occurring as a consequence of TBI. A multidisciplinary (e.g., immunology-neurology), and multimodal (laboratory-preclinical-clinical) approach is necessary to bring full fruition to this research and to distribute clinical dividends of translational research in the field.

Acknowledgements Supported by NIH R01NS43284, R41MH093302, R21NS077236, R42MH093302, and R21HD057256 awarded to DJ.

References

1. Hauser WA (2008) Epidemiology of acute symptomatic seizures. In: Engel G, Pedley TA (eds) Epilepsy: a comprehensive textbook. Lippincott Williams and Wilkins, Philadelphia, PA, pp 71–75
2. Hauser WA, Annegers JF, Kurland LT (1993) Incidence of epilepsy and unprovoked seizures in Rochester, Minnesota: 1935–1984. Epilepsia 34:453–468
3. Marmarou A (2003) Pathophysiology of traumatic brain edema: current concepts. Acta Neurochir Suppl 86:7–10
4. Langendorf FG, Pedley TA, Temkin NR (2008) Posttraumatic seizures. In: Engel G, Pedley TA (eds) Epilepsy: a comprehensive textbook. Lippincott Williams and Wilkins, Philadelphia, PA, pp 2537–2542
5. Salazar AM, Jabbari B, Vance SC, Grafman J, Amin D, Dillon JD (1985) Epilepsy after penetrating head-injury.1. Clinical correlates—a report of the Vietnam Head-Injury Study. Neurology 35:1406–1414
6. Haltiner AM, Temkin NR, Dikmen SS (1997) Risk of seizure recurrence after the first late posttraumatic seizure. Arch Phys Med Rehabil 78:835–840
7. Hesdorffer DC, Benn EK, Cascino GD, Hauser WA (2009) Is a first acute symptomatic seizure epilepsy? Mortality and risk for recurrent seizure. Epilepsia 50:1102–1108
8. Pieracci FM, Moore EE, Beauchamp K, Tebockhorst S, Barnett CC, Bensard DD, Burlew CC, Biffl WL, Stoval RT, Johnson JL (2012) A cost-minimization analysis of phenytoin versus levetiracetam for early seizure pharmacoprophylaxis after traumatic brain injury. J Trauma Acute Care Surg 72:276–281
9. Bullock MR, Povlishock JT (2007) Guidelines for the management of severe traumatic brain injury. Editor's commentary. J Neurotrauma 24(Suppl 1):2
10. Temkin NR, Dikmen SS, Anderson GD, Wilensky AJ, Holmes MD, Cohen W, Newell DW, Nelson P, Awan A, Winn HR (1999) Valproate therapy for prevention of posttraumatic seizures: a randomized trial. J Neurosurg 91:593–600
11. Daneman R (2012) The blood–brain barrier in health and disease. Ann Neurol 72:648–672
12. Cucullo L, Hossain M, Puvenna V, Marchi N, Janigro D (2011) The role of shear stress in blood–brain barrier endothelial physiology. BMC Neurosci 12:40
13. Fung C, Evans E, Shin D, Shin BC, Zhao Y, Sankar R, Chaudhuri G, Devaskar SU (2010) Hypoxic-ischemic brain injury exacerbates neuronal apoptosis and precipitates spontaneous seizures in glucose transporter isoform 3 heterozygous null mice. J Neurosci Res 88:3386–3398
14. Correale J, Villa A (2009) Cellular elements of the blood–brain barrier. Neurochem Res 34:2067–2077
15. Rubin LL, Staddon JM (1999) The cell biology of the blood–brain barrier. Annu Rev Neurosci 22:11–28

16. Janigro D (2012) Are you in or out? Leukocyte, ion, and neurotransmitter permeability across the epileptic blood–brain barrier. Epilepsia 53(Suppl 1):26–34
17. Seiffert E, Dreier JP, Ivens S, Bechmann I, Tomkins O, Heinemann U, Friedman A (2004) Lasting blood–brain barrier disruption induces epileptic focus in the rat somatosensory cortex. J Neurosci 24:7829–7836
18. Janigro D, Gasparini S, D'Ambrosio R, McKhann G, DiFrancesco D (1997) Reduction of K+ uptake in glia prevents long-term depression maintenance and causes epileptiform activity. J Neurosci 17:2813–2824
19. Misonou H (2010) Homeostatic regulation of neuronal excitability by K(+) channels in normal and diseased brains. Neuroscientist 16:51–64
20. Kofuji P, Newman EA (2004) Potassium buffering in the central nervous system. Neuroscience 129:1045–1056
21. Altamirano AA, Russell JM (1987) Coupled Na/K/Cl efflux. "Reverse" unidirectional fluxes in squid giant axons. J Gen Physiol 89:669–686
22. Sweadner KJ (1979) Two molecular forms of (Na+ + K+)-stimulated ATPase in brain. Separation, and difference in affinity for strophanthidin. J Biol Chem 254:6060–6067
23. Gloor SM (1997) Relevance of Na, K-ATPase to local extracellular potassium homeostasis and modulation of synaptic transmission. FEBS Lett 412:1–4
24. Kahle KT, Simard JM, Staley KJ, Nahed BV, Jones PS, Sun D (2009) Molecular mechanisms of ischemic cerebral edema: role of electroneutral ion transport. Physiology (Bethesda) 24:257–265
25. Pannasch U, Vargova L, Reingruber J, Ezan P, Holcman D, Giaume C, Sykova E, Rouach N (2011) Astroglial networks scale synaptic activity and plasticity. Proc Natl Acad Sci U S A 108:8467–8472
26. Ohno Y, Hibino H, Lossin C, Inanobe A, Kurachi Y (2007) Inhibition of astroglial Kir4.1 channels by selective serotonin reuptake inhibitors. Brain Res 1178:44–51
27. Higashi K, Fujita A, Inanobe A, Tanemoto M, Doi K, Kubo T, Kurachi Y (2001) An inwardly rectifying K(+) channel, Kir4.1, expressed in astrocytes surrounds synapses and blood vessels in brain. Am J Physiol Cell Physiol 281:C922–C931
28. Olsen ML, Sontheimer H (2008) Functional implications for Kir4.1 channels in glial biology: from K+ buffering to cell differentiation. J Neurochem 107:589–601
29. Medici V, Frassoni C, Tassi L, Spreafico R, Garbelli R (2011) Aquaporin 4 expression in control and epileptic human cerebral cortex. Brain Res 1367:330–339
30. Nielsen S, Nagelhus EA, Amiry-Moghaddam M, Bourque C, Agre P, Ottersen OP (1997) Specialized membrane domains for water transport in glial cells: high-resolution immunogold cytochemistry of aquaporin-4 in rat brain. J Neurosci 17:171–180
31. Rash JE, Yasumura T, Hudson CS, Agre P, Nielsen S (1998) Direct immunogold labeling of aquaporin-4 in square arrays of astrocyte and ependymocyte plasma membranes in rat brain and spinal cord. Proc Natl Acad Sci U S A 95:11981–11986
32. Dietzel I, Heinemann U, Hofmeier G, Lux HD (1980) Transient changes in the size of the extracellular space in the sensorimotor cortex of cats in relation to stimulus-induced changes in potassium concentration. Exp Brain Res 40:432–439
33. Ahn AH, Kunkel LM (1995) Syntrophin binds to an alternatively spliced exon of dystrophin. J Cell Biol 128:363–371
34. Connors NC, Kofuji P (2002) Dystrophin Dp71 is critical for the clustered localization of potassium channels in retinal glial cells. J Neurosci 22:4321–4327
35. O'Kane RL, Martinez-Lopez I, DeJoseph MR, Vina JR, Hawkins RA (1999) Na(+)-dependent glutamate transporters (EAAT1, EAAT2, and EAAT3) of the blood–brain barrier. A mechanism for glutamate removal. J Biol Chem 274:31891–31895
36. Hawkins RA (2009) The blood–brain barrier and glutamate. Am J Clin Nutr 90:867S–874S
37. Tanaka K, Watase K, Manabe T, Yamada K, Watanabe M, Takahashi K, Iwama H, Nishikawa T, Ichihara N, Kikuchi T, Okuyama S, Kawashima N, Hori S, Takimoto M,

Wada K (1997) Epilepsy and exacerbation of brain injury in mice lacking the glutamate transporter GLT-1. Science 276:1699–1702
38. Storck T, Schulte S, Hofmann K, Stoffel W (1992) Structure, expression, and functional analysis of a Na(+)-dependent glutamate/aspartate transporter from rat brain. Proc Natl Acad Sci U S A 89:10955–10959
39. Juge O (1982) Physiopathology of brain edema. Schweiz Med Wochenschr 112:956–959
40. Nag S, Manias JL, Stewart DJ (2009) Pathology and new players in the pathogenesis of brain edema. Acta Neuropathol 118:197–217
41. Donkin JJ, Vink R (2010) Mechanisms of cerebral edema in traumatic brain injury: therapeutic developments. Curr Opin Neurol 23:293–299
42. Unterberg AW, Stover J, Kress B, Kiening KL (2004) Edema and brain trauma. Neuroscience 129:1021–1029
43. Fishman RA (1992) Physiology of the cerebrospinal fluid. In: Mills LM (ed) Cerebrospinal fluid in diseases of the nervous system. W.B. Sauners Company, Philadelphia, pp 23–42
44. Asahi M, Wang X, Mori T, Sumii T, Jung JC, Moskowitz MA, Fini ME, Lo EH (2001) Effects of matrix metalloproteinase-9 gene knock-out on the proteolysis of blood–brain barrier and white matter components after cerebral ischemia. J Neurosci 21:7724–7732
45. Liu W, Hendren J, Qin XJ, Shen J, Liu KJ (2009) Normobaric hyperoxia attenuates early blood–brain barrier disruption by inhibiting MMP-9-mediated occludin degradation in focal cerebral ischemia. J Neurochem 108:811–820
46. Liu J, Jin X, Liu KJ, Liu W (2012) Matrix metalloproteinase-2-mediated occludin degradation and caveolin-1-mediated claudin-5 redistribution contribute to blood–brain barrier damage in early ischemic stroke stage. J Neurosci 32:3044–3057
47. Shigemori Y, Katayama Y, Mori T, Maeda T, Kawamata T (2006) Matrix metalloproteinase-9 is associated with blood–brain barrier opening and brain edema formation after cortical contusion in rats. Acta Neurochir Suppl 96:130–133
48. Plesnila N, Schulz J, Stoffel M, Eriskat J, Pruneau D, Baethmann A (2001) Role of bradykinin B2 receptors in the formation of vasogenic brain edema in rats. J Neurotrauma 18:1049–1058
49. Geppetti P, Bertrand C, Ricciardolo FL, Nadel JA (1995) New aspects on the role of kinins in neurogenic inflammation. Can J Physiol Pharmacol 73:843–847
50. Trabold R, Eros C, Zweckberger K, Relton J, Beck H, Nussberger J, Muller-Esterl W, Bader M, Whalley E, Plesnila N (2010) The role of bradykinin B(1) and B(2) receptors for secondary brain damage after traumatic brain injury in mice. J Cereb Blood Flow Metab 30:130–139
51. Zweckberger K, Plesnila N (2009) Anatibant, a selective non-peptide bradykinin B2 receptor antagonist, reduces intracranial hypertension and histopathological damage after experimental traumatic brain injury. Neurosci Lett 454:115–117
52. Shakur H, Andrews P, Asser T, Balica L, Boeriu C, Quintero JD, Dewan Y, Druwe P, Fletcher O, Frost C, Hartzenberg B, Mantilla JM, Murillo-Cabezas F, Pachl J, Ravi RR, Ratsep I, Sampaio C, Singh M, Svoboda P, Roberts I (2009) The BRAIN TRIAL: a randomised, placebo controlled trial of a Bradykinin B2 receptor antagonist (Anatibant) in patients with traumatic brain injury. Trials 10:109
53. Donkin JJ, Nimmo AJ, Cernak I, Blumbergs PC, Vink R (2009) Substance P is associated with the development of brain edema and functional deficits after traumatic brain injury. J Cereb Blood Flow Metab 29:1388–1398
54. Zacest AC, Vink R, Manavis J, Sarvestani GT, Blumbergs PC (2010) Substance P immunoreactivity increases following human traumatic brain injury. Acta Neurochir Suppl 106:211–216
55. Brain SD, Williams TJ (1985) Inflammatory oedema induced by synergism between calcitonin gene-related peptide (CGRP) and mediators of increased vascular permeability. Br J Pharmacol 86:855–860
56. Figaji AA, Zwane E, Fieggen AG, Argent AC, Le Roux PD, Siesjo P, Peter JC (2009) Pressure autoregulation, intracranial pressure, and brain tissue oxygenation in children with severe traumatic brain injury. J Neurosurg Pediatr 4:420–428

57. Soares HD, Hicks RR, Smith D, McIntosh TK (1995) Inflammatory leukocytic recruitment and diffuse neuronal degeneration are separate pathological processes resulting from traumatic brain injury. J Neurosci 15:8223–8233
58. Marchi N, Bazarian JJ, Puvenna V, Janigro M, Ghosh C, Zhong J, Zhu T, Blackman E, Stewart D, Ellis J, Butler R, Janigro D (2013) Consequences of repeated blood–brain barrier disruption in football players. PLoS One 8(3):e56805
59. Campbell M, Hanrahan F, Gobbo OL, Kelly ME, Kiang AS, Humphries MM, Nguyen AT, Ozaki E, Keaney J, Blau CW, Kerskens CM, Cahalan SD, Callanan JJ, Wallace E, Grant GA, Doherty CP, Humphries P (2012) Targeted suppression of claudin-5 decreases cerebral oedema and improves cognitive outcome following traumatic brain injury. Nat Commun 3:849
60. Garcia JH, Liu KF, Yoshida Y, Chen S, Lian J (1994) Brain microvessels: factors altering their patency after the occlusion of a middle cerebral artery (Wistar rat). Am J Pathol 145:728–740
61. Liang D, Bhatta S, Gerzanich V, Simard JM (2007) Cytotoxic edema: mechanisms of pathological cell swelling. Neurosurg Focus 22:E2
62. Simard JM, Kent TA, Chen M, Tarasov KV, Gerzanich V (2007) Brain oedema in focal ischaemia: molecular pathophysiology and theoretical implications. Lancet Neurol 6:258–268
63. Rumpel H, Nedelcu J, Aguzzi A, Martin E (1997) Late glial swelling after acute cerebral hypoxia-ischemia in the neonatal rat: a combined magnetic resonance and histochemical study. Pediatr Res 42:54–59
64. Su G, Kintner DB, Flagella M, Shull GE, Sun D (2002) Astrocytes from Na(+)-K(+)-Cl(−) cotransporter-null mice exhibit absence of swelling and decrease in EAA release. Am J Physiol Cell Physiol 282:C1147–C1160
65. Rao VL, Baskaya MK, Dogan A, Rothstein JD, Dempsey RJ (1998) Traumatic brain injury down-regulates glial glutamate transporter (GLT-1 and GLAST) proteins in rat brain. J Neurochem 70:2020–2027
66. Neusch C, Weishaupt JH, Bahr M (2003) Kir channels in the CNS: emerging new roles and implications for neurological diseases. Cell Tissue Res 311:131–138
67. Obrenovitch TP, Urenjak J (1997) Is high extracellular glutamate the key to excitotoxicity in traumatic brain injury? J Neurotrauma 14:677–698
68. Kristian T, Siesjo BK (1998) Calcium in ischemic cell death. Stroke 29:705–718
69. Kauppinen RA, Enkvist K, Holopainen I, Akerman KE (1988) Glucose deprivation depolarizes plasma membrane of cultured astrocytes and collapses transmembrane potassium and glutamate gradients. Neuroscience 26:283–289
70. Manley GT, Fujimura M, Ma T, Noshita N, Filiz F, Bollen AW, Chan P, Verkman AS (2000) Aquaporin-4 deletion in mice reduces brain edema after acute water intoxication and ischemic stroke. Nat Med 6:159–163
71. Baraban SC, Schwartzkroin PA (1998) Effects of hyposmolar solutions on membrane currents of hippocampal interneurons and mossy cells in vitro. J Neurophysiol 79:1108–1112
72. Fazio V, Bhudia SK, Marchi N, Aumayr B, Janigro D (2004) Peripheral detection of S100beta during cardiothoracic surgery: what are we really measuring? Ann Thorac Surg 78:46–52
73. Kanner AA, Marchi N, Fazio V, Mayberg MR, Koltz MT, Siomin V, Stevens GH, Masaryk T, Ayumar B, Vogelbaum MA, Barnett GH, Janigro D (2003) Serum S100beta: a noninvasive marker of blood–brain barrier function and brain lesions. Cancer 97:2806–2813
74. Vogelbaum MA, Masaryk T, Mazzone P, Mekhail T, Fazio V, McCartney S, Marchi N, Kanner A, Janigro D (2005) S100beta as a predictor of brain metastases: brain versus cerebrovascular damage. Cancer 104:817–824
75. Kapural M, Krizanac-Bengez L, Barnett G, Perl J, Masaryk T, apollo D, Rasmussen P, Mayberg MR, Janigro D (2002) Serum S-100beta as a possible marker of blood–brain barrier disruption. Brain Res 940:102–104
76. Begaz T, Kyriacou DN, Segal J, Bazarian JJ (2006) Serum biochemical markers for post-concussion syndrome in patients with mild traumatic brain injury. J Neurotrauma 23:1201–1210

77. Blyth B, Farahvar A, He H, Nayak A, Yang C, Shaw G, Bazarian JJ (2011) Elevated serum ubiquitin carboxy-terminal hydrolase L1 is associated with abnormal blood brain barrier function after traumatic brain injury. J Neurotrauma 28:2453–2462
78. Blyth BJ, Farhavar A, Gee C, Hawthorn B, He H, Nayak A, Stocklein V, Bazarian JJ (2009) Validation of serum markers for blood–brain barrier disruption in traumatic brain injury. J Neurotrauma 26:1497–1507
79. Biberthaler P, Mussack T, Wiedemann E, Kanz KG, Koelsch M, Gippner-Steppert C, Jochum M (2001) Evaluation of S-100b as a specific marker for neuronal damage due to minor head trauma. World J Surg 25:93–97
80. Bechmann I, Galea I, Perry VH (2007) What is the blood–brain barrier (not)? Trends Immunol 28:5–11
81. Galea I, Bechmann I, Perry VH (2007) What is immune privilege (not)? Trends Immunol 28:12–18
82. Chang GC, Young LH (2011) Sympathetic ophthalmia. Semin Ophthalmol 26:316–320
83. Gold M, Pul R, Bach JP, Stangel M, Dodel R (2012) Pathogenic and physiological autoantibodies in the central nervous system. Immunol Rev 248:68–86
84. Popovich PG, Stokes BT, Whitacre CC (1996) Concept of autoimmunity following spinal cord injury: possible roles for T lymphocytes in the traumatized central nervous system. J Neurosci Res 45:349–363
85. Fujinami RS, Oldstone MB (1985) Amino acid homology between the encephalitogenic site of myelin basic protein and virus: mechanism for autoimmunity. Science 230:1043–1045
86. Lang HL, Jacobsen H, Ikemizu S, Andersson C, Harlos K, Madsen L, Hjorth P, Sondergaard L, Svejgaard A, Wucherpfennig K, Stuart DI, Bell JI, Jones EY, Fugger L (2002) A functional and structural basis for TCR cross-reactivity in multiple sclerosis. Nat Immunol 3:940–943
87. Prochazka M, Voltnerova M, Stefan J (1971) Studies of immunologic reactions after brain injury. II. Antibodies against brain tissue lipids after blunt head injury in man. Int Surg 55:322–326
88. Lopez-Escribano H, Minambres E, Labrador M, Bartolome MJ, Lopez-Hoyos M (2002) Induction of cell death by sera from patients with acute brain injury as a mechanism of production of autoantibodies. Arthritis Rheum 46:3290–3300
89. Skoda D, Kranda K, Bojar M, Glosova L, Baurle J, Kenney J, Romportl D, Pelichovska M, Cvachovec K (2006) Antibody formation against beta-tubulin class III in response to brain trauma. Brain Res Bull 68:213–216
90. Granata T, Gobbi G, Spreafico R, Vigevano F, Capovilla G, Ragona F, Freri E, Chiapparini L, Bernasconi P, Giordano L, Bertani G, Casazza M, Dalla BB, Fusco L (2003) Rasmussen's encephalitis: early characteristics allow diagnosis. Neurology 60:422–425
91. Michalak Z, Lebrun A, Di Miceli M, Rousset MC, Crespel A, Coubes P, Henshall DC, Lerner-Natoli M, Rigau V (2012) IgG leakage may contribute to neuronal dysfunction in drug-refractory epilepsies with blood–brain barrier disruption. J Neuropathol Exp Neurol 71:826–838
92. Bargestock E, Puvenna V, Iffland P, Falcone T, Hossain M, Vetter S, Man S, Dickstein L, Carvalho-Tavares J, Marchi N, Janigro D (2013) Is peripheral immunity regulated by blood–brain barrier permeability changes? BMC Neurosci

Chapter 3
Human Cerebral Blood Flow and Traumatic Brain Injury

David A. Hovda and Thomas C. Glenn

Abstract The measurement of cerebral blood flow (CBF) in the clinical setting first began in earnest during the 1940s by Kety and Schmidt. By utilizing an inert diffusible gas (nitrous oxide), measuring it in arterial and jugular venous blood, and applying the Fick principle of blood flow determination they were able to calculate a global rate of CBF of mL/100 g/min. Kety–Schmidt-based CBF measurements evolved over the next 2 decades by the substitution of radioactive gases for nitrous oxide.

3.1 History of Cerebral Blood Flow

The measurement of cerebral blood flow (CBF) in the clinical setting first began in earnest during the 1940s by Kety and Schmidt [1]. By utilizing an inert diffusible gas (nitrous oxide), measuring it in arterial and jugular venous blood, and applying the Fick principle of blood flow determination they were able to calculate a global rate of CBF of mL/100 g/min. Additionally, other constituents within the blood such as oxygen and glucose could be measured and cerebral metabolic rates calculated by utilizing arterial and jugular venous differences. Kety–Schmidt-based CBF measurements evolved over the next 2 decades by the substitution of radioactive gases for nitrous oxide [2, 3].

One limitation of the Kety–Schmidt-based techniques was the global nature of CBF measurement and lack of three-dimensional tomographic reconstruction of CBF. During the 1970s and 1980s several new tomographic techniques were developed, which included stable xenon computed tomography (XeCT), positron emission tomography (PET), single photon emission computed tomography

---

D.A. Hovda (✉)
Departments of Neurosurgery, David Geffen School of Medicine at UCLA, 10833 Le Conte Avenue, 18-266 Semel, Box 957039, Los Angeles, CA, USA
e-mail: dhovda@mednet.ucla.edu

E.H. Lo et al. (eds.), *Vascular Mechanisms in CNS Trauma*, Springer Series in Translational Stroke Research 5, DOI 10.1007/978-1-4614-8690-9_3,

(SPECT), and CT perfusion. Thus, these techniques could provide quantitative CBF to morphological images such as CT and magnetic resonance imaging (MRI). More recent CBF measurement techniques include MR perfusion, thermal diffusion, laser Doppler flowmetry, and the indirect techniques of transcranial Doppler ultrasonography (TCD), and near-infrared spectroscopy.

Pediatric CBF can be variable based on age or stage of development. Suzuki [4] studied normal children with 133Xe-CBF and described a pattern of lowest CBF at birth followed by an increase of CBF to above 100 mL/100 g/min through age 4 followed by a gradual decline with age that plateaued to adult levels of approximately 50 mL/100 g/min during postadolescence. Using arterial spin labeling MRI, Biagi et al. [5] showed that CBF was highest at ages less than 10 years old and then decreased until a nadir between 30 and 40 years old.

3.2 Cerebral Blood Flow in Traumatic Brain Injury

Initial clinical studies of posttraumatic CBF in patients with severe TBI showed marked differences in the level of CBF [6]. Numerous investigators have described the dynamic nature of posttraumatic CBF. Some studies have described the phasic nature of CBF characterized by an acute phase of hypoperfusion relative to control subjects, followed by a period of hyperemia, which is then followed by another phase of reduced flow [7–10]. A similar pattern has been described for pediatric TBI patients [11]. It is during the third low-flow phase that posttraumatic vasospasm can occur [12, 13].

Numerous outcome studies, which have focused on CBF, have shown that the duration and degree of reduced CBF is a predictor of poor outcome [8, 14–17]. For example, Robertson et al. [15] reported increased mortality and poorer neurological outcome in patients categorized as reduced CBF compared to patients with normal or elevated flow. A recent pediatric study reported that early low CBF was associated with poor outcome [11]. However, hyperemic CBF-associated intracranial hypertension has also been linked to poor outcome [8]. A plausible explanation for posttraumatic hyperemia is an uncoupling of metabolism and blood flow.

Pediatric hyperemia has been suggested as a cause for diffuse cerebral swelling following TBI [18–20]. However, the definition of hyperemia and interpretation of results in pediatric patients is problematic due to the paucity of reliable reference values [21, 22]. As was shown above, CBF is highly variable in children.

Although oligemic CBF has been well described in many studies, few of the clinical studies have observed CBF in the ischemic range of less than 20 mL/100 g/min. Yet in the 1970s, postmortem evidence for ischemic CBF in as many as 55 % of patients with TBI was shown by Graham and Hume Adams [23]. Thus, the concept that TBI is the result of ischemic CBF is controversial. To fully understand ischemia, CBF must be studied in conjunction with cerebral metabolism [24, 25].

Early studies of cerebral oxygen metabolism and CBF using 133Xenon indicated that the arterial–jugular venous difference for oxygen ($AVDO_2$) was rarely in the ischemic range of 9 vol.%, even when CBF was reduced [25]. Consistent with these findings is the widely observed TBI-associated depression of cerebral metabolic rate for oxygen ($CMRO_2 = AVDO_2 \times CBF$). Hyperemic flow in the setting of low $CMRO_2$ was named “luxury perfusion” by Lassen [24]. Criticisms of studies showing no ischemia were that the timing of the initial CBF measurement was not early enough to capture ischemia [26]. CBF studies within the first 4–6 h post-injury showed a 33 % incidence of ischemia based on CBF and $AVDO_2$ [26]. Additionally, Hlatky et al. [27] reported a 12 % incidence of CBF below 18 mL/100 g/min in 77 patients using XeCT measurements. Clearly, it is difficult to determine if either the early or delayed ischemic events are responsible for those who died as a result of their TBI.

More recent studies have combined PET imaging with cerebral metabolism. As was described above, several studies showed that even when reduced levels of CBF were observed, the incidence of ischemia was rare. Diringer et al. [28] conducted a combined PET CBF, CMR02, and oxygen extraction fraction (OEF) studies in TBI patients before and during periods of brief hyperventilation. The investigators stated that CBF was in normal range while $CMRO_2$ was reduced. Hyperventilation did reduce CBF and increase oxygen extraction fraction without changing $CMRO_2$. Thus, the important distinction between ischemia from conditions such as stroke and TBI suggested by these authors is that low $CMRO_2$ coupled with low CBF may be the consequence of low substrate supply (ischemic conditions) in contrast to a decreased $CMRO_2$ which is caused by TBI (excitotoxic, ionic, and mitochondrial effects) followed by a secondary coupled decrease in CBF.

In another PET-based TBI study, Vespa et al. [29] reported a low incidence of cerebral ischemia and a mean ischemic brain volume of only 1.5 cm^3. Additionally, cerebral microdialysis results in this study revealed an increased lactate-to-pyruvate ratio corresponding to nonischemic reduction of $CMRO_2$. However, Coles et al. [30, 31] reported larger ischemic burden volumes of 67 cm^3. Thus, the degree of ischemic tissue within the brain appears to be variable.

Posttraumatic cerebral vasospasm is a secondary injury that has the potential to reduce regional CBF. Posttraumatic cerebral vasospasm has been confirmed by TCD and angiography [13, 32–34]. Using both 133Xe-CBF and TCD-based criteria to determine vasospasm, Oertel et al. [13] observed that vasospasm occurred in a third of all severely injured TBI patients. Moreover, Lee et al. [12] reported that the multimodal approach of concurrent TCD and CBF measurements could better define hemodynamically significant (TCD velocity/CBF = Spasm Index) vasospasm. Additionally, the investigators demonstrated that hemodynamically significant, a Spasm Index greater than 3.4, was associated with poor outcome.

3.3 Cerebral Autoregulation

Cerebral vasculature has the ability to respond to physiological changes such as a drop in mean arterial pressure (MAP) through a process called autoregulation. Autoregulation, which is a combination of intrinsic myogenic, neurogenic, and metabolic factors, is a hemostatic process that maintains CBF as a result of a variety of stimuli be they physical, chemical, or metabolic [35, 36]. Posttraumatic cerebral autoregulation has been widely studied in adults and children, and numerous reviews have been published [35–37]. Some of the first studies of posttraumatic autoregulation showed that impaired autoregulation occurred in both low- and high-flow states [6, 38]. By manipulating MAP these investigators found abnormal autoregulation in over half of the patients studied. Generally, impaired pressure autoregulation has been observed within the acute phase after injury and may begin to normalize after 5 days [39, 40]. The deleterious consequences of impaired autoregulation include low or ischemic flow at low arterial pressures and hyperemia at high pressure, which may lead to increased cerebral blood volume, intracranial pressure (ICP), and cerebral edema [41].

The advent of TCD has significantly developed the area of bedside autoregulation studies in the intensive care unit. Two methods of study exist; static or dynamic. The static method involves concurrent measurements of physiological factors such as MAP, CBF (velocity in the case of TCD), and ICP. Dynamic autoregulation studies measure physiological factors and then alter specific variables such as MAP, carbon dioxide levels (CO_2), or cerebral metabolism by means of metabolic agents such as anesthetics. From these techniques, indices of autoregulation can be calculated. Two well-described indices are the PRx and Mx. The PRx is the correlation coefficient between the slow waves in ICP and MAP, while the MRx is the Pearson correlation between cerebral perfusion pressure (CPP) and blood flow velocity measured by TCD [36, 42]. When either the PRx or Mx is positive, autoregulation is dysfunctional, thus showing pressure passive effects on ICP and blood flow, respectively.

In addition to pressure autoregulation, cerebrovascular reactivity to carbon dioxide (CO_2) has been studied. In contrast to pressure autoregulation, CO_2 cerebrovascular reactivity is relatively more robust. For example, Cold et al. [43] reported that CO_2 cerebral vascular reactivity were lower primarily in patients who were more deeply in a comatose state. Using the regional Xe-CT method, Marion et al. [10] reported that global or hemispheric measures of CO_2 cerebrovascular reactivity did not account for regions of disturbed vasoreactivity. Lee et al. [40] showed with TCD that CO_2 reactivity was below normal in 55 % of patients in the acute period (post-injury day 4 and below) while improving to only 25 % abnormal after day 4.

Prediction of clinical outcome, based on autoregulation studies, has shown varied results. Early studies reported that autoregulation was unrelated to outcome in patients who underwent 133Xe-CBF studies both before and after MAP was increased (Cold and Jensen [39]; Overgaard and Tweed [38]). However, other investigators have shown that indices of impaired pressure autoregulation are predictors of poor outcome [42, 44]. Studies conducted in pediatric patients also associated impaired CO_2 vasoreactivity [11] and autoregulation [45] with poor outcome.

3.4 Blood–Brain Barrier

The blood–brain barrier (BBB) is a complex of brain capillary endothelial, astrocytic glial cells, and other brain cellular components which selectively regulate the passage of molecules into and out of the brain parenchyma [46–48]. With the exception of the choroid plexus, there are few fenestrations and a high presence of continuous tight junctions between the cells. Following TBI, the cascade of events, including glutamate release and ionic disturbances, lead to a catabolic process which initiate BBB breakdown [49–51]. Furthermore, immune system activation, neutrophil recruitment, macrophage infiltration, activation of matrix metalloproteinase, microglial activation, and other inflammatory processes lead to reactive oxygen species (ROS) generation which can impact BBB permeability [51–55]. One possible consequence of disrupted BBB function is the genesis of posttraumatic vasogenic cerebral edema [53, 56–59]. Vasogenic edema can cause an increase in ICP, thus CBF and autoregulation may be negatively impacted [59, 60].

3.5 Summary

Traumatic brain injury in humans results in dynamic changes in CBF. Periods of oligemic perfusion are frequently followed by hyperemia and then a return to reduced or oligemic levels. During this acute period, a definitive consensus on the existence of cerebral ischemia seems elusive. Additionally, uncoupling of CBF and cerebral metabolism can occur as well as impaired autoregulation and vasoreactivity. Normal pediatric CBF varies with age and must be taken into consideration when studying this group of patients. Furthermore, the BBB can be undergoing dramatic posttraumatic changes which will affect cerebral metabolism, edema formation, and inflammation. Thus, CBF and its influence on brain function and recovery remains an area of intense interest and importance.

References

1. Kety S, Schmidt CF (1948) The nitrous oxide method for the quantitative determination of cerebral blood flow in man: theory, procedure and normal values. J Clin Invest 27(4):476–483. doi:10.1172/JCI101994
2. Lassen NA, Ingvar DH (1961) The blood flow of the cerebral cortex determined by radioactive krypton85. Experientia 17(1):42–43
3. Obrist WD, Thompson HK Jr, King CH (1967) Determination of regional cerebral blood flow by inhalation of 133-Xenon. Circ Res 20:124–135, PMID:6017706
4. Suzuki K (1990) The changes of regional cerebral blood flow with advancing age in normal children. Nagoya Med J 34:159–170
5. Biagi L, Abbruzzese A, Bianchi MC, Alsop DC, Del Guerra A, Tosetti M (2007) Age dependence of cerebral perfusion assessed by magnetic resonance continuous arterial spin labeling. J Magn Reson Imaging 25(4):696–702, PMID:17279531
6. Bruce DA, Langfitt TW, Miller JD, Schutz H, Vapalahti MP, Stanek A et al (1973) Regional cerebral blood flow, intracranial pressure, and brain metabolism in comatose patients. J Neurosurg 38(2):131–144, PubMed PMID: 4694215
7. Martin NA, Patwardhan RV, Alexander MJ, Africk CZ, Lee JH, Shalmon E et al (1997) Characterization of cerebral hemodynamic phases following severe head trauma: hypoperfusion, hyperemia, and vasospasm. J Neurosurg 87(1):9–19, PubMed PMID: 9202259
8. Kelly DF, Martin NA, Kordestani R, Counelis G, Hovda DA, Bergsneider M et al (1997) Cerebral blood flow as a predictor of outcome following traumatic brain injury. J Neurosurg 86 (4):633–641, PubMed PMID: 9120627
9. Bouma GJ, Muizelaar JP, Choi SC, Newlon PG, Young HF (1991) Cerebral circulation and metabolism after severe traumatic brain injury: the elusive role of ischemia. J Neurosurg 75 (5):685–693, PubMed PMID: 1919689
10. Marion DW, Darby J, Yonas H (1991) Acute regional cerebral blood flow changes caused by severe head injuries. J Neurosurg 74(3):407–414, PubMed PMID: 1899694
11. Adelson PD, Srinivas R, Chang Y, Bell M, Kochanek PM (2011) Cerebrovascular response in children following severe traumatic brain injury. Childs Nerv Syst 27(9):1465–1476, PubMed PMID: 21559825
12. Lee JH, Martin NA, Alsina G, McArthur DL, Zaucha K, Hovda DA et al (1997) Hemodynamically significant cerebral vasospasm and outcome after head injury: a prospective study. J Neurosurg 87(2):221–233, PubMed PMID: 9254085
13. Oertel M, Boscardin WJ, Obrist WD, Glenn TC, McArthur DL, Gravori T et al (2005) Posttraumatic vasospasm: the epidemiology, severity, and time course of an underestimated phenomenon: a prospective study performed in 299 patients. J Neurosurg 103(5):812–824, PubMed PMID: 16304984
14. Langfitt TW, Obrist WD, Gennarelli T, O'Connor J, Weeme CA (1977) Correlation of cerebral blood flow with outcome in head injured patients. Ann Surg 86:411–414, PMID:17020478
15. Robertson CS, Contant CF, Gokaslan ZL, Narayan RK, Grossman RG (1992) Cerebral blood flow, arteriovenous oxygen difference, and outcome in head injured patients. J Neurol Neurosurg Psychiatry 55:594–603, PMID:1640238
16. Glenn TC, Kelly DF, Boscardin WJ, McArthur DL, Vespa P, Oertel M, Hovda DA, Bergsneider M, Hillered L, Martin NA (2003) Energy dysfunction as a predictor of outcome after moderate or severe head injury: indices of oxygen, glucose, and lactate metabolism. J Cereb Blood Flow Metab 23:1239–1250, PMID:14526234
17. Obrist WD, Gennarelli TA, Segawa H, Dolinskas CA, Langfitt TW (1979) Relation of cerebral blood flow to neurological status and outcome in head-injured patients. J Neurosurg 51 (3):292–300, PubMed PMID: 469577
18. Bruce DA, Alavi A, Bilaniuk L, Dolinskas C, Obrist W, Uzzell B (1981) Diffuse cerebral swelling following head injuries in children: the syndrome of "malignant brain edema". J Neurosurg 54(2):170–178, PubMed PMID: 7452330

19. Zimmerman RA, Bilaniuk LT, Gennarelli T (1978) Computed tomography of shearing injuries of the cerebral white matter. Radiology 127:393–396, PMID:622489
20. Berger MS, Pitts LH, Lovely M, Edwards MSB, Bartkowski HM (1985) Outcome from severe head injury in children and adolescents. J Neurosurg 62(2):194–199, PubMed PMID: 3968558
21. Skippen P, Seear M, Poskitt K, Cochrane D, Annich G, Handel J (1997) Effect of hyperventilation on regional cerebral blood flow in head-injured children. Crit Care Med 25 (8):1402–1409, PMID:9267957
22. Zwienenberg M, Muizelaar JP (1999) Severe pediatric head injury: the role of hyperemia revisted. J Neurotrauma 16(10):937–943, PMID:10547102
23. Graham DI, Hume AJ (1971) Ischæmic brain damage in fatal head injuries. The Lancet 297 (7693):265–266, PMID:4100017
24. Lassen NA (1966) The luxury-perfusion syndrome and its possible relation to acute metabolic acidosis localised within the brain. The Lancet 288(7473):1113–1115, PMID:4162534
25. Obrist WD, Langfitt TW, Jaggi JL, Cruz J, Gennarelli TA (1984) Cerebral blood flow and metabolism in comatose patients with acute head injury. J Neurosurg 61(2):241–253, PubMed PMID: 6737048
26. Bouma GJ, Muizelaar JP, Stringer WA, Choi SC, Fatouros P, Young HF (1992) Ultra-early evaluation of regional cerebral blood flow in severely head-injured patients using xenon-enhanced computerized tomography. J Neurosurg 77(3):360–368, PubMed PMID: 1506882
27. Hlatky R, Contant C, Diaz-Marchan P, Valadka A, Robertson C (2004) Significance of a reduced cerebral blood flow during the first 12 hours after traumatic brain injury. Neurocrit Care 1(1):69–83, PMID:16174900
28. Diringer MN, Yundt K, Videen TO, Adams RE, Zazulia AR, Deibert E et al (2000) No reduction in cerebral metabolism as a result of early moderate hyperventilation following severe traumatic brain injury. J Neurosurg 92(1):7–13, PubMed PMID: 10616076
29. Vespa P, Bergsneider M, Hattori N, Wu HM, Huang SC, Martin NA et al (2005) Metabolic crisis without brain ischemia is common after traumatic brain injury: a combined microdialysis and positron emission tomography study. J Cereb Blood Flow Metab 25(6):763–774, PubMed PMID: 15716852
30. Coles JP, Steiner LA, Johnston AJ, Fryer TD, Coleman MR, Smieleweski P et al (2004) Does induced hypertension reduce cerebral ischaemia within the traumatized human brain? Brain 127(Pt 11):2479–2490, PubMed PMID: 15456706
31. Coles JP, Fryer TD, Smielewski P, Chatfield DA, Steiner LA, Johnston AJ et al (2004) Incidence and mechanisms of cerebral ischemia in early clinical head injury. J Cereb Blood Flow Metab 24(2):202–211, PubMed PMID: 14747747
32. Suwanwela C, Suwanwela N (1972) Intracranial arterial narrowing and spasm in acute head injury. J Neurosurg 36(3):314–323, PubMed PMID: 5059970
33. Marshall LF, Bruce DA, Bruno L, Langfitt TW (1978) Vertebrobasilar spasm: a significant cause of neurological deficit in head injury. J Neurosurg 48(4):560–564, PubMed PMID: 632879
34. Martin NA, Doberstein C, Zane C, Caron MJ, Thomas K, Becker DP (1992) Posttraumatic cerebral arterial spasm: transcranial Doppler ultrasound, cerebral blood flow, and angiographic findings. J Neurosurg 77(4):575–583, PubMed PMID: 1527618
35. Golding EM, Robertson CS, Bryan RM (1999) The consequences of traumatic brain injury on cerebral blood flow and autoregulation: a review. Clin Exp Hypertens 21(4):299–332, PMID:10369378
36. Rangel-Castilla L, Gasco J, Nauta HJ, Okonkwo DO, Robertson CS (2008) Cerebral pressure autoregulation in traumatic brain injury. Neurosurg Focus 25(4):E7, PubMed PMID: 18828705
37. Armstead WM, Vavilala MS (2013) Age and sex differences in cerebral blood flow and autoregulation after pediatric traumatic brain injury. In: Kreipke CW, Rafols JA (eds) Cerebral blood flow, metabolism, and head trauma: The pathotrajectory of traumatic brain injury. Springer, New York, pp 135–153

38. Overgaard J, Tweed WA (1974) Cerebral circulation after head injury. J Neurosurg 41 (5):531–541, PubMed PMID: 4418221
39. Cold GE, Jensen FT (1978) Cerebral autoregulation in unconscious patients with brain injury. Acta Anaesthesiol Scand 22(3):270–280, PMID:27938
40. Lee JH, Kelly DF, Oertel M, McArthur DL, Glenn TC, Vespa P et al (2001) Carbon dioxide reactivity, pressure autoregulation, and metabolic suppression reactivity after head injury: a transcranial Doppler study. J Neurosurg 95(2):222–232, PubMed PMID: 11780891
41. Czosnyka M, Smielewski P, Piechnik S, Steiner LA, Pickard JD (2001) Cerebral autoregulation following head injury. J Neurosurg 95(5):756–763, PubMed PMID: 11702864
42. Budohoski KP, Czosnyka M, de Riva N, Smielewski P, Pickard JD, Menon DK et al (2012) The relationship between cerebral blood flow autoregulation and cerebrovascular pressure reactivity after traumatic brain injury. Neurosurgery 71(3):652–660, PubMed PMID: 22653390
43. Cold GE, Jensen FT, Malmros R (1977) The cerebrovascular CO_2 reactivity during the acute phase of brain injury. Acta Anaesthesiol Scand 21(3):222–231, PMID:17991
44. Czosnyka M, Smielewski P, Kirkpatrick P, Menon DK, Pickard JD (1996) Monitoring of cerebral autoregulation in head-injured patients. Stroke 27(10):1829–1834, PMID:8841340
45. Vavilala MS, Lee LA, Boddu K et al (2004) Cerebral autoregulation in pediatric traumatic brain injury. Pediatr Crit Care Med 5:257–263, PMID:15115564
46. Abbott NJ, Ronnback L, Hansson E (2006) Astrocyte-endothelial interactions at the blood–brain barrier. Nature Rev Neurosci 7(1):41–53, PubMed PMID: 16371949
47. Oldendorf WH (1977) The blood–brain barrier. Exp Eye Res 1(0):177–190
48. Paolinelli R, Corada M, Orsenigo F, Dejana E (2011) The molecular basis of the blood brain barrier differentiation and maintenance. Is it still a mystery? Pharm Res 63(3):165–171
49. Baethmann A, Maier-Hauff K, Kempski O, Unterberg A, Wahl M, Schürer L (1988) Mediators of brain edema and secondary brain damage. Crit Care Med 16(10):972–978, PMID:2901938
50. Werner C, Engelhard K (2007) Pathophysiology of traumatic brain injury. Br J Anaesth 99 (1):4–9, PubMed PMID: 17573392
51. Das M, Mohapatra S, Mohapatra S (2012) New perspectives on central and peripheral immune responses to acute traumatic brain injury. J Neuroinflammation 9(1):236, PubMed PMID: doi:10.1186/1742-2094-9-236
52. Pun PB, Lu J, Moochhala S (2009) Involvement of ROS in BBB dysfunction. Free Radic Res 43(4):348–364, PubMed PMID: 19241241
53. Graves J, Betrus C (2013) Situating cerebral blood flow in the pathotrajectory of head trauma. In: Kreipke CW, Rafols JA (eds) Cerebral blood flow, metabolism and head trauma. Springer, New York, pp 29–51
54. Chodobski A, Zink BJ, Szmydynger-Chodobska J (2011) Blood–brain barrier pathophysiology in traumatic brain injury. Transl Stroke Res 2(4):492–516, PubMed PMID: 22299022. Pubmed Central PMCID: 3268209
55. Cervós-Navarro J, Lafuente JV (1991) Traumatic brain injuries: structural changes. J Neurol Sci 103(Suppl):3–14, PMID:1940963
56. Unterberg AW, Stover J, Kress B, Kiening KL (2004) Edema and brain trauma. Neurosci 129 (4):1021–1029, PubMed PMID: 15561417
57. Donkin JJ, Vink R (2010) Mechanisms of cerebral edema in traumatic brain injury: therapeutic developments. Curr Opin Neurol 23(3):293–299, PubMed PMID: 20168229
58. Graves J, Betrus C, Rafols JA (2013) Situating cerebral blood flow in the pathotrajectory of head trauma. In: Kreipke CW, Rafols JA (eds) Cerebral blood flow, metabolism, and head trauma: The pathotrajectory of traumatic brain injury. Springer, New York, pp 29–51
59. Schoknecht K, Shalev H (2012) Blood–brain barrier dysfunction in brain diseases: clinical experience. Epilepsia 53(Suppl 6):7–13, PubMed PMID: 23134490
60. Greve MW, Zink BJ (2009) Pathophysiology of traumatic brain injury. Mt Sinai J Med 76 (2):97–104, PMID:19306379

Chapter 4
Gliovascular Targets in Traumatic CNS Injury

Arjun Khanna, Brian P. Walcott, Kristopher T. Kahle, Volodymyr Gerzanich, and J. Marc Simard

Abstract The modern understanding of the cellular and molecular mechanisms responsible for cerebral edema and microvascular failure has given rise to the notion that the blood–brain barrier and its support cells, including endothelium, astrocytes, microglia, and pericytes, comprise a functional ensemble, termed the gliovascular unit, that plays a crucial role in secondary injury responses following traumatic injury to the central nervous system. This chapter provides an overview of the gliovascular unit as it relates to the pathological states encountered following trauma. In addition, this chapter briefly reviews a number of gliovascular molecules that have been identified as playing important roles in the response to trauma and that hold promise as therapeutic targets.

4.1 Introduction

The blood–brain barrier (BBB) restricts paracellular diffusion of compounds across the endothelium, thereby providing the specialized environment crucial for proper neural function, and protecting neural cells from potentially injurious circulating substances and pathogens. The BBB is formed by brain endothelial cells that are joined by tight junctions [1–3]. However, the integrity and normal functioning of the BBB requires more than endothelium—it requires complex interactions with other cells. Astrocytes, which maintain close contact with the microvascular endothelial basement membrane via foot processes, are critical for maintaining the BBB phenotype of brain endothelial cells and for normal functioning of the BBB [4–6]. In addition, microglia [7] and pericytes [8, 9] play important roles in BBB function. The complex functional interrelationships between these various cells and neurons is coordinated by paracrine signaling [4]. These anatomical and

J.M. Simard (✉)
Department of Neurosurgery, University of Maryland School of Medicine, 22 S. Greene Street, Suite S12D, Baltimore, MD, USA
e-mail: msimard@smail.umaryland.edu

E.H. Lo et al. (eds.), *Vascular Mechanisms in CNS Trauma*, Springer Series in Translational Stroke Research 5, DOI 10.1007/978-1-4614-8690-9_4,

functional interrelationships have given rise to the concept that the actual *functional* unit of the BBB encompasses the larger ensemble of cells types—the "gliovascular unit" [4, 5]. The term gliovascular unit emphasizes the important role of glial and other cells in the control of the local vasculature and the larger role of these cells in maintaining normal function of the BBB.

The gliovascular unit is disrupted following traumatic injury, and this disruption can have important effects on secondary injury progression, recovery, and functional outcome [10, 11]. Following injury, glial cells become activated and orchestrate numerous local responses, many of which can be detrimental. Injury to the endothelium can lead to a rearrangement of the cytoskeleton and cell retraction, with subsequent loss of integrity of tight junctions and the formation of intercellular gaps [12–14]. This allows for the paracellular flow of protein-rich plasma directly from the cerebral circulation, resulting in the pathological entity termed vasogenic cerebral edema. Additionally, traumatic insults leading to oncotic dysfunction or death of endothelial cells in the microvasculature can result in capillary fragmentation and secondary hemorrhage [15]. This phenomenon is frequently observed clinically in central nervous system (CNS) trauma involving both the brain and spinal cord and is characterized by a progressive enlargement of the lesion with fragmentation of capillaries, hemorrhagic expansion, and tissue necrosis, resulting in severe neurological dysfunction. In addition, the BBB plays a crucial role in regulating the innate inflammatory response to traumatic injury which, if unrestrained, can lead to bystander injury of otherwise uninjured neural cells. In combination, these phenomena—edema, secondary hemorrhage, and innate inflammation—are major determinants of secondary injury in the days and weeks following primary trauma. A better understanding of the gliovascular unit that regulates these processes, and the identification of potential targets in gliovascular signaling that are activated by trauma, promises to advance therapies to limit secondary injury.

Herein, we discuss several molecular targets that have been found to play important roles in gliovascular function following traumatic brain and spinal cord injury (TBI and SCI, respectively). Although the majority of the studies reviewed have been preclinical, a limited number of human investigations are underway that underscore the increasing emphasis on manipulating gliovascular molecular targets to improve outcomes following trauma.

4.1.1 Membrane Transporters

4.1.1.1 Aquaporin-4

Aquaporins are a class of water channels that serve as key regulators of transmembrane water conductance in cell types throughout the body, including those in the gliovascular unit. Of the six aquaporins identified in the rodent brain, aquaporin-4 (AQP-4) is the most abundant. It is expressed on capillary-facing astrocyte foot processes and in ependymal cells. The distribution of AQP-4 on fluid interfaces of the CNS correlates with its significant role in regulating the influx of cellular water associated with edema formation.

Following focal cortical contusion in rats, AQP-4 mRNA levels rise near the site of the injury, correlating with the edema that forms as a result of the trauma [16]. mRNA levels rise in the immediate area of injury, but fall in areas adjacent to the injury [16]. It is hypothesized that this pattern of up- and downregulation of AQP-4 may act to protect against the propagation of edema outside of the site of primary injury [16].

AQP-4 facilitates cytotoxic edema by allowing for transcellular water conductance, particularly in astrocytes. AQP-4-null mice appear to have less cytotoxic edema but more vasogenic edema following freeze-injury, suggesting different roles of AQP-4 in these pathological processes [17]. Reductions in overall edema following CNS trauma have been reported with both inhibition and over-expression of AQP-4. For example, inhibition of AQP-4 channel function with an anti-AQP-4 antibody significantly reduces edema formation in a rat model of TBI [18]. Conversely, sulforaphane enhances AQP-4 expression in the lesion site and penumbral region, which also results in less brain edema following injury [19]. These results suggest that AQP-4 is spatially regulated in a complex way surrounding the injury site.

In SCI, there is an early decrease in levels of AQP-4 that is followed by persistent upregulation [20].

At present there is no specific pharmacological way to inhibit AQP-4, but levels of this channel can be downregulated by manipulating upstream transcriptional mechanisms induced by trauma. The pro-inflammatory cytokine IL-1β, acting via nuclear factor κB (NF-κB), is a positive regulator of AQP-4 [21] and thus inhibiting NF-κB is expected to reduce AQP-4 expression. Similarly, inhibition of hypoxia-inducible factor 1α (Hif-1α) with 2-methoxyestradiol suppresses the expression of AQP-4 [18].

4.1.1.2 Sur1-Trpm4 Channel

The sulfonylurea receptor 1-transient receptor potential melastatin 4 (Sur1-Trpm4) channel recently was characterized molecularly in SCI [22]. Previously known as the Sur1-regulated $NC_{Ca\text{-}ATP}$ channel, it is not constitutively expressed but is transcriptionally upregulated de novo in neurons, astrocytes, oligodendrocytes, and microvascular endothelium within several hours of traumatic injury to the brain or spinal cord. In the context of ATP depletion inside the cell, the Sur1-Trpm4 channel opens, allowing the influx of monovalent cations which, if unchecked, results in cell depolarization, oncotic cell swelling (cytotoxic edema), and necrotic cell death [23].

De novo expression and activation of the channel in microvascular endothelium can lead to microvascular failure. In SCI, the Sur1-Trpm4 channel has been shown to play an obligatory role in the development of "progressive hemorrhagic necrosis," the autodestructive process initiated by trauma that is responsible for secondary hemorrhage and lesion expansion [23, 24]. In TBI, a similar pathological process ensues, which is termed "hemorrhagic progression of contusion" [15, 25]. In both cases, channel expression in microvascular endothelium is followed by necrotic death of endothelial cells, which leads to capillary fragmentation, extravasation of blood, and formation of petechial hemorrhages. Petechial hemorrhages coalesce to form larger hemorrhages surrounding the original site of the trauma, expanding the primary hemorrhagic contusion twofold.

Capillary fragmentation, secondary hemorrhage, and lesion expansion in TBI and in SCI have been shown to be blocked by pharmacological inhibition of Sur1 with glibenclamide or repaglinide, pharmacological inhibition of Trpm4 with flufenamic acid or riluzole, as well as by gene suppression of *Abcc8* (Sur1) or *Trpm4* [23–28]. As with AQP-4 (see above), NF-κB and Hif-1α are positive regulators of Sur1-Trpm4 and thus inhibiting these transcription factors reduces Sur1-Trpm4 expression.

A prospective, multicenter, placebo-controlled, double-blind, phase IIa trial of RP-1127 (glibenclamide for injection) is underway to test its effect in patients with moderate-to-severe TBI to reduce cerebral edema and hemorrhage (ClinicalTrials.gov Identifier: NCT01454154).

4.1.1.3 EAATs

Extracellular glutamate concentration in the CNS is typically regulated through reuptake of glutamate by excitatory amino acid transporters (EAATs) in the surrounding glia. Of the five different EAAT types that are known, EAAT-1 and EAAT-2 are expressed by glial cells, particularly astrocytes. Following traumatic injury, there is a significant increase in extracellular glutamate levels that is caused at least in part by a rapid decrease in EAAT-1 and EAAT-2 expression in the lesion and surrounding tissue, due to loss of EAAT-expressing glial cells and disrupted protein synthesis [29, 30]. Glutamate levels have been shown to be elevated 1.5–2-fold after injury, a condition that persists for weeks. This increase is detectable in perilesional areas remote from the primary site and represents a significant cause of excitotoxicity contributing to secondary neuronal damage.

It is hypothesized that, although early hyperactivity at the NMDA receptor causes excitotoxicity, glutamate signaling begins to facilitate recovery several hours after injury and becomes an important part of repair [30, 31]. Modulating the expression of EAAT-1 and EAAT-2 has been explored with some success as an alternative method of alleviating early excitotoxicity without NMDA receptor blockade. In animal models of spinal cord and traumatic brain injury, upregulation of EAAT-1 and EAAT-2 by pituitary andenylate cyclase-activating polypeptide (PACAP) or citicoline reduced apoptosis, edema, and axonal damage and improved neurological outcomes [32, 33]. Other compounds, including MS-153 [34], dibutyryl cyclic AMP [35], arundic acid [36], and β-lactam antibiotics [37], also are capable of modulating EAAT-1 and EAAT-2 expression levels in vivo.

4.1.1.4 P2X Receptors

Purinergic P2X4 and P2X7 receptors play a large role in initiating an inflammatory response to extracellular ATP that is released from the intracellular compartment by necrosis or mechanical rupture of nearby cells. Upon the detection of high extracellular ATP, P2X4- or P2X7-expressing neural and glial cells synthesize and release the potent pro-inflammatory cytokine IL-1β which, in turn, causes tissue edema, secondary CNS injury, and is associated with poorer outcomes. Expression of P2X4 increases in injured tissue after SCI, TBI, and ischemia. P2X7 expression, on the other hand, does not change significantly following injury. It co-localizes with AQP-4 to the foot processes of astrocytes in the gliovascular unit, suggesting a critical role for the astrocyte in the detection and response to extracellular ATP.

P2X4 knockout mice have an inhibited inflammatory response following SCI, reduced infiltration of neutrophils and macrophages to the injury site, and significantly improved neuromotor outcomes [38]. Similarly, P2X7 knockout mice have significantly attenuated expression of IL-1β, less edema, and improved neurobehavioral outcomes following TBI [39]. Pharmacological blockade of P2X7 with brilliant blue G (BBG) achieves similar results. There is no change in basal expression of IL-1β following BBG administration without TBI. Given their important effects on IL-1β expression and resulting edema, P2X4 and P2X7 may be attractive targets for reducing secondary injury from edema following injury.

4.1.1.5 Connexin43

The extracellular ATP that activates P2X7 receptors is in part released by surrounding astrocytes via connexin43 (Cx43) hemichannels. Although Cx43 hemichannels usually form gap junctions by docking to Cx43 hemichannels on neighboring cells, unopposed hemichannels are a potential pathway for the release of cytosolic compounds, including glutamate and ATP. It has been observed that Cx43-null mice have significantly less post-trauma ATP release, exhibit smaller lesions, and have improved motor recovery [40]. Attenuation of post-trauma ATP release by perilesional astrocytes by blockade of Cx43 is therefore a potential target for reducing the extracellular ATP-induced inflammatory response.

4.1.2 Other Transmembrane Proteins

4.1.2.1 EGFR/MAPK

The epidermal growth factor receptor (EGFR) is a potent regulator of microglial cell activation. EGFR ligands like epidermal growth factor (EGF) and TNFα transactivate mitogen-activated protein kinase (MAPK) and its associated signaling pathways. There exist three subtypes of MAPKs, namely, extracellular signal-regulated protein kinase (ERK), c-Jun NH_2-terminal kinase (JNK), and p38, which exert somewhat different cellular effects upon activation. MAPK activation generally produces other cytokines, such as IL-1β, TNFα, and IL-6, and so contributes to posttraumatic edema. EGFR is expressed in all members of the neurovascular unit, including microglial cells.

Consistent with the fact that EGFR/MAPK signaling mediates microglial activation and neuroinflammation, blockade of EGFR significantly reduces microglial activation and brain edema and improves functional outcomes compared to controls as early as 7 days after SCI [41]. Interestingly, MAPK signaling through the JNK pathway is stimulated by hyperthermia, which often accompanies injury. JNK signaling can be attenuated with hypothermia, suggesting a therapeutic role for hypothermia in CNS trauma [42]. However, the clinical evidence for this practice does not reveal any benefit in humans [43].

4.1.2.2 GPR17

A P2Y-like receptor, G-protein receptor 17 (GPR17), can be activated by uracil nucleotides as well as cysteinyl-leukotrienes. Cysteinyl-leukotrienes are arachidonic acid (AA)-derived, pro-inflammatory molecules, whose levels increase following CNS injury [44–46]. Interestingly, GPR17 is expressed in adult, but not fetal, neuroprogenitor cells, suggesting that GPR17 may be implicated in post-injury repair and that cells in the neurovascular unit may have differential responses to GPR17 ligands depending on their pre-injury expression of GPR17 [47]. Ependymal cells, which are now known to be true neural stem cells, constitutively express GPR17, as do some neurons and oligodendrocytes; astrocytes do not express any GPR17.

There is a biphasic GPR17-mediated response to trauma. Upon injury-induced release of uracil nucleotides and cysteinyl-leukotrienes, an initial increase of expression of GPR17 by neurons in the lesion site is quickly followed by cell death in the first, acute phase of the GPR17-mediated response. Next, proliferating GPR17-positive microglia/macrophages infiltrate the lesion site, and GPR17-positive ependymal cells begin to proliferate and express GFAP, which may indicate de-differentiation into progenitor cells. Thus, GPR17 is involved in an acute, cell-death response and a slower, progressive infiltration of cells for tissue repair. In vivo knockdown of GPR17 by an antisense oligonucleotide reduces initial tissue damage and improves histological and motor deficits following SCI [45]. Pharmacological inhibitors of GPR17 signaling, such as Montelukast (a cysteinyl-leukotriene receptor antagonist) and Zileuton (a 5-lipooxygenase inhibitor), have also shown benefit in SCI recovery [48].

4.1.2.3 PV-1

Plasmalemmal vesicle protein-1 (PV-1) encodes a transmembrane protein that is a major component in stomatal diaphragms of caveolae found in fenestrated microvascular endothelial cells. It is expressed in organs throughout the body but is normally silenced in the mature endothelium of the BBB, where transendothelial transport is limited. In pathological breakdown of the BBB, including SCI, PV-1 levels rise in activated endothelial cells and is associated with progressive inflammatory cell invasion [49]. PV-1 is considered a marker for BBB pathology, but given its impact on endothelial permeability, there is interest in targeting PV-1 to reduce BBB breakdown following CNS injury and subsequently reduce vasogenic edema. There is in vitro evidence that PV-1 targeted siRNA blocks Matrigel-induced tubulogenesis and inhibits cell migration induced by angiogenic growth factors in primary human endothelial cells [50], but this strategy has yet to be explored in CNS injury.

4.1.2.4 ADAM

A disintegrin and metalloprotease domain family (ADAMs) are transmembrane proteins with disintegrin domains that bind integrins and metalloprotease domains that are mainly responsible for activating cleavage-dependent transmembrane proteins. Among these cleavage-dependent proteins are factors that promote angiogenesis, such as Notch, Tie1, and TNFα, and factors that mediate inflammation, such as ICAM. There are a number of different ADAMs that are suspected of playing various roles in developing vasculature and pathological angiogenesis. Of particular interest is ADAM8, which has almost no CNS expression at baseline but is highly increased following SCI in mice [51]. Although there is no acute ADAM8 response to injury, expression begins increasing in the lesion epicenter about 3 days following contusion and steadily increases thereafter. ADAM8 localizes on the luminal side of blood vessels within endothelial cells that are actively undergoing angiogenesis. Although it is likely that ADAM8 plays an important role in modulating the inflammatory and angiogenic responses to injury, ADAM8 inhibition in CNS injury has yet to be thoroughly explored.

4.1.3 Intracellular Signaling Molecules

4.1.3.1 Calponin

Calponin (CP) is a regulator of contraction in vascular smooth muscle and may also play a role in vasospasm following TBI. In its non-phosphorylated form, CP inhibits smooth muscle contractility by binding to actin and inhibiting actomyosin cross-bridging. Phosphorylation of CP causes it to dissociate from the actomyosin complex, which disinhibits contraction. In vitro, CP can inhibit magnesium adenosine triphosphatase, which may also contribute to sustained contractility in smooth muscle.

CP is upregulated in smooth muscle cells within 4 h of injury, when it migrates from a diffuse cytosolic distribution toward localization at the cell membrane, suggesting phosphorylation and disinhibition of contraction [52]. Presence of phosphorylated CP seems to temporally correspond with vasospasm around the primary lesion site in CNS trauma, which can persist for up to 48 h after injury. Interestingly, ET-1 might be a regulator of CP, as ET-1 induces smooth muscle contractility in vitro that is accompanied by an increase in total CP. It has been hypothesized that ET-1 signaling through ETr activates protein kinase A (PKA), which may phosphorylate CP and cause its dissociation from actomyosin and migration to the cell membrane [53].

4.1.3.2 Akt and mTOR

Serine-threonine kinase, Akt, and mammalian target of rapamycin (mTOR) signaling play important roles in rodent TBI, but their exact role in the functioning of the gliovascular unit is still unclear. In models of stroke [54], nerve transection [55], and spinal cord injury [56], Akt seems to be neuroprotective. After injury, it is activated and localizes with uninjured neurons, where it may protect from further damage and promote regrowth.

Both Akt and mTOR signaling pathways are activated after controlled cortical impact in neurons, microglia, and astrocytes of mice [57]. Although inhibition of either Akt or mTOR has no detectable effect on recovery, simultaneous inhibition of both Akt and mTOR significantly decreased acute tissue damage in the lesion and significantly improved motor and cognitive outcomes [57]. Interestingly, simultaneous Akt and mTOR inhibition does not change lesion size or inflammation, and so likely affects recovery through some other mechanism that may be related to the unexpected increase in p-GSK3β phosphorylation seen when both Akt and mTOR are simultaneously inhibited [57]. It is suspected that Akt and mTOR inhibition may be beneficial in glial cells, where inhibition reduces gliosis. However, Akt inhibition may be harmful in neurons, where it may promote survival. Continued preclinical investigation is necessary to further elucidate the potential for manipulation of this signaling pathway.

4.1.4 Extracellular/Paracrine Ligands and Cytokines

4.1.4.1 Nitric Oxide/Endothelin-1

Hemorrhage at the primary lesion site stimulates vasoconstriction in the vessels of the surrounding penumbra. In this region, perfusion decreases by 50 % within 2 h of experimental TBI in rats and reaches a minimum at about 3 h, after which it begins to trend toward normalization [58]. Vasoconstriction is due in part to an increase in expression of the potent vasoconstrictor, endothelin-1 (ET-1), plus a reduction in the vasodilator, nitric oxide (NO), in the vascular endothelium after TBI. The balance between ET-1 and NO is thought to be an important mechanism by which perfusion through microvasculature is autoregulated in the CNS [58].

NO diffuses rapidly through membranes and promotes the formation of cyclic GMP. NO/cyclic guanosine monophosphate (cGMP) signaling causes dilation of the microvasculature and, at low concentrations, promotes angiogenesis. A reduction in NO synthase activity following trauma leads to decreased release of NO, decreased cGMP signaling in the microvascular endothelium, and contributes to vasoconstriction. Sildenafil, a cGMP phosphodiesterase-5 inhibitor, lowers the rate of degradation of cGMP and thereby increases NO/cGMP signaling. Following contusive SCI, mice treated with sildenafil show improved perfusion through the lesion epicenter, but do not show improved neuromotor outcomes [59]. While it is not clear whether improving perfusion through the primary lesion site via NO/cGMP signaling is capable of improving functional recovery, it may be an effective way to facilitate drug delivery to injured tissues in the salvageable penumbra.

In contrast to NO, ET-1 is a powerful vasoconstrictor that is associated with vasospasm, hypoperfusion, and oxidative damage following CNS trauma. Although there is evidence that ET-1 has a basal level of constitutive activity, levels of ET-1 and its receptors, ETrA and ETrB, increase in endothelial cells, pericytes, and astrocytes following injury for more than 24 h [58, 60]. ETrA is thought to stimulate pericyte contraction via signaling through the PLCβ/IP_3/Ca^{2+} pathway. Antagonism of ETrA, but not ETrB, lessens TBI-induced lesion hypoperfusion in rat [61], as does blockade of ET-1 by antisense oligodeoxynucleotides administered before injury [62]. Although blockade of ET-1 action at ETrA improves perfusion, more study is needed to determine if this will result in improved outcomes, something that has not been observed with the use of endothelin antagonists in the context of subarachnoid hemorrhage [63].

4.1.4.2 VEGF

Vascular endothelial growth factors are a family of secreted glycoproteins that play a critical role in angiogenesis. Of the five members of the VEGF family, VEGFA, VEGFB, VEGFC, VEGFD, and VEGFE, VEGFA has clear significance in the gliovascular response to CNS trauma. VEGFA has five isoforms ($VEGFA_{121}$, $VEGFA_{145}$, $VEGFA_{165}$, $VEGFA_{189}$, and $VEGFA_{206}$) that differ in their degree of binding to the extracellular matrix (ECM) and their resulting endogenous bioavailability. Of these, only $VEGFA_{121}$, which is not sequestered by the ECM and is freely diffusible, shows response to traumatic injury [64], although invading neutrophils also secrete VEGFA that may be of a different isoform, since some of it deposits in the ECM [65].

Primarily in astrocytes, but also in vascular endothelial cells and microglia, levels of $VEGFA_{121}$ acutely rise in the 3–4 h following injury and return to baseline levels after about 6 days [66]. Although VEGFA stimulates angiogenesis, it also causes increased BBB permeability through a number of mechanisms. For one, it redistributes and downregulates tight junction proteins, including occludin and CLN-5 [67]. It also causes tyrosine phosphorylation of adherens junction proteins, including VE-cadherin, β-catenin, plakoglobin, and p120 [68], and activates the PI3K/Akt signaling cascade in the vascular endothelium [69]. Given intravenously, VEGFA increases BBB permeability in experimental models [67] and therefore is unlikely to be considered as a therapeutic agent following traumatic injury. However, VEGFA delivered intraventricularly reduces traumatic lesion size, improves functional outcomes, and significantly increases the number of proliferating cells in the perilesional area [69]. This suggests the possibility of a differential effect of exogenous VEGFA on the apical and basal sides of the brain endothelium. Currently, approved indications for VEGF manipulation in the CNS of humans are limited to neoplastic processes.

4.1.4.3 Angiopoietin

Along with VEGF, angiopoietin-1 (Ang-1) is critical for maintaining vascular integrity. While VEGF promotes initial angiogenesis, Ang-1 plays an important role in maintaining the integrity of mature vasculature by suppressing protein leakage, inhibiting vascular inflammation, and preventing endothelial cell death. Ang-1 is a ligand for the Tie-2 receptor that promotes vascular integrity by promoting endothelial cell survival via the PI3k/Akt pathway, inhibits the breakdown of adhesions and tight junctions between endothelial cells, and modulates vasoconstriction through eNOS. Both Ang-1 and Tie-2 are expressed in all brain endothelial cells. Angiopoietin-2 is a natural competitive antagonist of Ang-1 at Tie-2. Its expression is low in the normal brain. Immediately following CNS injury, Ang-1 and Tie-2 levels fall and Ang-2 levels rise, coincident with acute post traumatic BBB breakdown and subsequent vasogenic edema [70]. After about 2 days, the levels of these proteins begin to return to normal. Because Ang-1 signaling through Tie-2 is associated with BBB integrity, it is a potential therapeutic target in the acute phase of CNS injury to protect against vasogenic edema development subsequent tissue damage.

4.1.4.4 Thrombospondin-1

Thrombospondin-1 (TSP-1) is a potent inhibitor of developmental and adaptive angiogenesis and is overexpressed almost 60-fold in spinal cord microvascular endothelial cells following contusive injury [71]. TSP-1 inhibits endothelial cell proliferation through many mechanisms, including activation of a CD36-Fyn-Caspase-3-p38 MAPK cascade that results in apoptosis. Its receptor, CD47, is also targeted by the signal regulatory protein α. Activation of CD47 allows neutrophil diapedesis across the endothelium to sites of injury, and it antagonizes vasodilation and angiogenesis by inhibiting synthesis of cGMP. Although TSP-1-null mice show improved vascular profiles in the primary lesion, they do not have improved microvascular perfusion or functional outcomes. CD47-null mice, however, show significantly improved perfusion, better locomotor recovery, and greater white matter sparing. This suggests that a combination of TSP-1 and signal regulatory protein α effect inhibition may be required for improved recovery from SCI [72].

4.1.4.5 Erythropoeitin

Erythropoeitin (EPO) is a cytokine that has the potential to improve recovery from CNS injury, largely by reducing both cytotoxic and vasogenic edema that contributes to secondary injury. Recombinant human EPO (rhEPO) reduces glutamate toxicity and oxidative stress in astrocytes, reduces VEGF-induced permeability of the BBB, and increases AQP-4 expression, leading to decreased vasogenic edema in animal models [73]. A single administration of rhEPO following injury also reduces gliosis and associated scarring [73]. Although the precise mechanisms behind these effects remain elusive, rhEPO is nevertheless of interest in ameliorating edema in secondary CNS injury.

4.1.4.6 Polyunsaturated Fatty Acids

Following injury, polyunsaturated fatty acids (PUFAs) play important roles in orchestrating the onset and resolution of the inflammatory response. Arachadonic acid is an ω-6 PUFA that gives rise to pro-inflammatory mediators such as prostaglandins, thromboxanes, and leukotrienes that contribute to edema by enhancing vascular permeability, promote the infiltration of leukocytes, and produce inflammatory cytokines like TNF-α and IL-1β. Metabolites of docosahexaenoic acid (DHA), an ω-3 PUFA, have both anti-inflammatory and anti-oxidant properties. Inhibition of AA metabolism (by selective inhibition of cyclooxygenases and lipoxygenases) has been shown to improve functional recovery following SCI [74]. Similarly, intravenous and dietary DHA following SCI in rats lessened tissue loss and improved functional outcomes [75]. Fenretinide is an analogue of retinoic acid that when taken orally simultaneously raises DHA while reducing AA in mice with SCI, which resulted in less production of pro-inflammatory molecules that significantly reduced tissue damage and improved locomotor recovery [76].

4.1.4.7 RANTES/CCL5

Regulated and normal T-cell expressed and secreted (RANTES) and chemokine (C-C motif) ligand 5 (CCL5), a potent chemokine, are strongly expressed in spinal cord microvascular endothelial cells [71]. Over-expression lasts for up to 21 days after injury and stimulates T-cell activation and infiltration into damaged tissue, resulting in neuroinflammation [77]. Increased expression of RANTES/CCL5 and other inflammatory mediators, such as IL-3, IL-6, IL-11, and interferon-β1, in the microvascular endothelium highlights an important mechanism by which vascular damage to the gliovascular unit may be caused by an unrestrained inflammatory response. Reduction of robust RANTES production in reactive astrocytes may contribute to its neuroprotection and potential application in SCI [78].

4.1.4.8 Interleukin-16

Interleukin-16 (IL-16) is a pro-inflammatory cytokine that induces lymphocyte migration, promotes the expression of other inflammatory cytokines including IL-1β and IL-6, increases activity of TNF-α, and modulates apoptosis. IL-16 becomes highly expressed by microglia/macrophages and perivascular cells around 3 days after SCI in rats [79]. Activated microglia/macrophages contribute the most to the post-injury increase in IL-16, but astrocytes, neurons, and granulocytes also upregulate and secrete this cytokine. Foamy macrophages expressing IL-16 persist for more than 30 days after injury, indicating that long-term post-injury modification of the environment by IL-16 may contribute to secondary injury. Dexamethasone can inhibit expression of IL-16 and does indeed reduce the numbers of IL-16-positive cells in the first 3 days of injury [80], although there are a number of inflammatory compounds besides IL-16 that also are modulated by dexamethasone [81].

4.1.5 Transcription Factors

4.1.5.1 NF-κB

In the CNS, astrocytes and endothelial cells, but not neurons, are the primary cell type subject to nuclear factor κB (NF-κB) regulation [82]. In models of brain and spinal cord injury, NF-κB and its downstream targets are highly expressed following trauma. Although it is a key regulator of neuroinflammation in secondary injury processes, downstream targets of NF-κB also serve neuroprotective roles, such as the synthesis of nerve growth factor, brain-derived neurotrophic factor, and calbindin.

Despite this mix of beneficial and detrimental effects, selective suppression of NF-κB in astrocytes improves outcomes following SCI in mice. In mice with astrocyte-specific NF-κB knockdown, functional recovery is improved, white matter is better preserved, and lesion volume is reduced following contusion [83]. The astrocytes in these mice also produce less pro-inflammatory cytokines and glial scarring. Conversely, in mice deficient in brain IκBα, which ordinarily sequesters NF-κB in the cytoplasm and prevents its migration into the nucleus, IκBα deficiency leads to elevated basal neuroinflammation, resulting in a failure to mount proper inflammatory responses following TBI and worsened brain damage [82]. In both of these studies with knockout mice, normal brain function was seemingly unaffected outside of the injury site, suggesting that the effects of NF-κB are relevant primarily in injured tissues.

These data suggest that the net effect of astrocyte NF-κB expression, which constitutes a large majority of the total expression of NF-κB following injury, serves to worsen secondary injury in SCI. Inhibition of NF-κB or its targets in astrocytes has the potential to protect against NF-κB-mediated secondary injury.

4.1.5.2 FOXO3a

Forkhead Class box O3a (FOXO3a) is a member of the forkhead transcription factor of the forkhead box class O subfamily, which is downstream of the PI3K/Akt pathway. Phosphorylation by activated Akt causes FOXO3a release of DNA and translocation into the cytoplasm, thereby decreasing the expression of FOXO3a target genes. Activity of FOXO3a and expression of its targets, which include p27kip1 and Bim, is associated with regulation of the cell cycle at the G1/S transition. Cell cycle activation has been found to contribute to post-mitotic cell death, glial cell activation, and scarring following SCI [84], while inhibition has been found to reduce glial proliferation and scar formation [85], indicating that regulation of cell cycle progression is critical in the recovery process. FOXO3a levels decrease significantly about 3 days after SCI, particularly in astrocytes [86]. Inhibition of PI3K ablates this effect. Modulating FOXO3a levels in the gliovascular unit following injury is a possible method to improve cell cycle regulation, but current understanding of the effect of FOXO3a phosphorylation on neurofunctional outcomes is limited.

4.1.5.3 Stat3 and Socs3

Reactive astrocytes have conflicting effects on recovery from CNS trauma. In the acute phase, astrocyte migration within the lesion site, compaction of inflammatory cells, and maintenance of BBB integrity minimize lesion size and preserve prenumbral tissue. In the chronic phase, the resulting glial scar prevents axonal regrowth. Signal transducer and activator of transcription 3 (Stat3) is an important mediator of activation in the acute astrocyte response. Knockdown of Stat3 in mice inhibits astrocyte activation and migration, which results in wider infiltration of inflammatory cells around the lesion site, neural disruption, and poorer motor outcomes, whereas knockdown of the protein suppressor of cytokine signaling 3 (Socs3), the negative regulator of Stat3, improves outcomes [87]. These regulators of astrocyte activity are potential targets for intervention, particularly in the acute response, when astrocyte activation appears to be beneficial.

4.2 Conclusion

Multiple molecular pathways mediate various types of secondary injury and various stages of secondary injury following TBI and SCI. However, one common denominator in many acute CNS injury pathways involves microvascular failure, resulting in cerebral edema, secondary hemorrhage, and inflammation, that is brought about by gliovascular dysregulation.

All of the molecular mechanisms discussed here must be considered to be adaptive, including not only those that are evidently beneficial but also those that are seemingly harmful, for if they were not adaptive, they would not have advanced evolutionarily. However, an adaptive response, if unrestrained or ill-timed, can become maladaptive, leading to morbidity. In this light, it is perhaps not surprising that several gliovascular targets have been shown to have seemingly ambiguous roles, at times protective, at other times detrimental. It is clear that a thorough understanding of the specific role of each molecule at each stage of the response after injury is mandatory to prevent inadvertent inhibition of a protective or otherwise desirable effect.

Among the many targets briefly reviewed above, arguably the most promising ones are those for which a specific small molecule inhibitor is available. Inhibition at a molecular level, e.g., using mRNA interference, is still not practical in acute injury because, with the exception of endothelium, it suffers from the problem that it is difficult to deliver agent to the cells involved. Similarly, inhibition of transcription factors typically is not desirable because of too many potential off-target effects. At present, small molecule pharmacological inhibitors remain the most promising agents for inhibiting gliovascular targets, especially if their specificity is such as to minimize undesirable off-target effects.

Progress in understanding the molecular mechanisms of gliovascular regulation and dysregulation continues to generate a rich, ever enlarging repertoire of potential therapeutic targets in CNS trauma that warrant further investigation.

References

1. Abbott NJ, Patabendige AA, Dolman DE, Yusof SR, Begley DJ (2010) Structure and function of the blood–brain barrier. Neurobiol Dis 37(1):13–25
2. Luissint AC, Artus C, Glacial F, Ganeshamoorthy K, Couraud PO (2012) Tight junctions at the blood brain barrier: physiological architecture and disease-associated dysregulation. Fluids Barriers CNS 9(1):23

3. Liu WY, Wang ZB, Zhang LC, Wei X, Li L (2012) Tight junction in blood–brain barrier: an overview of structure, regulation, and regulator substances. CNS Neurosci Ther 18 (8):609–615
4. Abbott NJ, Ronnback L, Hansson E (2006) Astrocyte-endothelial interactions at the blood–brain barrier. Nat Rev Neurosci 7(1):41–53
5. Wolburg H, Noell S, Mack A, Wolburg-Buchholz K, Fallier-Becker P (2009) Brain endothelial cells and the glio-vascular complex. Cell Tissue Res 335(1):75–96
6. Nag S (2011) Morphology and properties of astrocytes. Methods Mol Biol 686:69–100
7. Lassmann H, Zimprich F, Vass K, Hickey WF (1991) Microglial cells are a component of the perivascular glia limitans. J Neurosci Res 28(2):236–243
8. Liu S, Agalliu D, Yu C, Fisher M (2012) The role of pericytes in blood–brain barrier function and stroke. Curr Pharm Des 18(25):3653–3662
9. Sa-Pereira I, Brites D, Brito MA (2012) Neurovascular unit: a focus on pericytes. Mol Neurobiol 45(2):327–347
10. Nag S, Kapadia A, Stewart DJ (2011) Review: molecular pathogenesis of blood–brain barrier breakdown in acute brain injury. Neuropathol Appl Neurobiol 37(1):3–23
11. Shlosberg D, Benifla M, Kaufer D, Friedman A (2010) Blood–brain barrier breakdown as a therapeutic target in traumatic brain injury. Nat Rev Neurol 6(7):393–403
12. Blum MS, Toninelli E, Anderson JM, Balda MS, Zhou J, O'Donnell L et al (1997) Cytoskeletal rearrangement mediates human microvascular endothelial tight junction modulation by cytokines. Am J Physiol 273(1 Pt 2):H286–H294
13. Fanning AS, Little BP, Rahner C, Utepbergenov D, Walther Z, Anderson JM (2007) The unique-5 and −6 motifs of ZO-1 regulate tight junction strand localization and scaffolding properties. Mol Biol Cell 18(3):721–731
14. Wojciak-Stothard B, Entwistle A, Garg R, Ridley AJ (1998) Regulation of TNF-alpha-induced reorganization of the actin cytoskeleton and cell-cell junctions by Rho, Rac, and Cdc42 in human endothelial cells. J Cell Physiol 176(1):150–165
15. Kurland D, Hong C, Aarabi B, Gerzanich V, Simard JM (2012) Hemorrhagic progression of a contusion after traumatic brain injury: a review. J Neurotrauma 29(1):19–31
16. Sun MC, Honey CR, Berk C, Wong NL, Tsui JK (2003) Regulation of aquaporin-4 in a traumatic brain injury model in rats. J Neurosurg 98(3):565–569
17. Manley GT, Binder DK, Papadopoulos MC, Verkman AS (2004) New insights into water transport and edema in the central nervous system from phenotype analysis of aquaporin-4 null mice. Neuroscience 129(4):983–991
18. Higashida T, Kreipke CW, Rafols JA, Peng C, Schafer S, Schafer P et al (2011) The role of hypoxia-inducible factor-1alpha, aquaporin-4, and matrix metalloproteinase-9 in blood–brain barrier disruption and brain edema after traumatic brain injury. J Neurosurg 114(1):92–101
19. Zhao J, Moore AN, Clifton GL, Dash PK (2005) Sulforaphane enhances aquaporin-4 expression and decreases cerebral edema following traumatic brain injury. J Neurosci Res 82 (4):499–506
20. Nesic O, Lee J, Ye Z, Unabia GC, Rafati D, Hulsebosch CE et al (2006) Acute and chronic changes in aquaporin 4 expression after spinal cord injury. Neuroscience 143(3):779–792
21. Ito H, Yamamoto N, Arima H, Hirate H, Morishima T, Umenishi F et al (2006) Interleukin-1beta induces the expression of aquaporin-4 through a nuclear factor-kappaB pathway in rat astrocytes. J Neurochem 99(1):107–118
22. Woo SK, Kwon MS, Ivanov A, Gerzanich V, Simard JM (2013) The sulfonylurea receptor 1 (sur1)-transient receptor potential melastatin 4 (trpm4) channel. J Biol Chem 288 (5):3655–3667
23. Simard JM, Tsymbalyuk O, Ivanov A, Ivanova S, Bhatta S, Geng Z et al (2007) Endothelial sulfonylurea receptor 1-regulated NC Ca-ATP channels mediate progressive hemorrhagic necrosis following spinal cord injury. J Clin Invest 117(8):2105–2113
24. Simard JM, Woo SK, Norenberg MD, Tosun C, Chen Z, Ivanova S et al (2010) Brief suppression of Abcc8 prevents autodestruction of spinal cord after trauma. Sci Transl Med 2 (28):28ra29

25. Simard JM, Kilbourne M, Tsymbalyuk O, Tosun C, Caridi J, Ivanova S et al (2009) Key role of sulfonylurea receptor 1 in progressive secondary hemorrhage after brain contusion. J Neurotrauma 26(12):2257–2267
26. Gerzanich V, Woo SK, Vennekens R, Tsymbalyuk O, Ivanova S, Ivanov A et al (2009) De novo expression of Trpm4 initiates secondary hemorrhage in spinal cord injury. Nat Med 15 (2):185–191
27. Simard JM, Popovich PG, Tsymbalyuk O, Gerzanich V (2012) Spinal cord injury with unilateral versus bilateral primary hemorrhage—effects of glibenclamide. Exp Neurol 233 (2):829–835
28. Simard JM, Tsymbalyuk O, Keledjian K, Ivanov A, Ivanova S, Gerzanich V (2012) Comparative effects of glibenclamide and riluzole in a rat model of severe cervical spinal cord injury. Exp Neurol 233(1):566–574
29. van Landeghem FK, Weiss T, Oehmichen M, von DA (2006) Decreased expression of glutamate transporters in astrocytes after human traumatic brain injury. J Neurotrauma 23 (10):1518–1528
30. Beschorner R, Dietz K, Schauer N, Mittelbronn M, Schluesener HJ, Trautmann K et al (2007) Expression of EAAT1 reflects a possible neuroprotective function of reactive astrocytes and activated microglia following human traumatic brain injury. Histol Histopathol 22(5):515–526
31. Arvidsson A, Kokaia Z, Lindvall O (2001) N-methyl-D-aspartate receptor-mediated increase of neurogenesis in adult rat dentate gyrus following stroke. Eur J Neurosci 14(1):10–18
32. Cakir E, Usul H, Peksoylu B, Sayin OC, Alver A, Topbas M et al (2005) Effects of citicoline on experimental spinal cord injury. J Clin Neurosci 12(8):923–926
33. Chen WH, Tzeng SF (2005) Pituitary adenylate cyclase-activating polypeptide prevents cell death in the spinal cord with traumatic injury. Neurosci Lett 384(1–2):117–121
34. Shimada F, Shiga Y, Morikawa M, Kawazura H, Morikawa O, Matsuoka T et al (1999) The neuroprotective agent MS-153 stimulates glutamate uptake. Eur J Pharmacol 386 (2–3):263–270
35. Rozyczka J, Figiel M, Engele J (2004) Endothelins negatively regulate glial glutamate transporter expression. Brain Pathol 14(4):406–414
36. Mori T, Tateishi N, Kagamiishi Y, Shimoda T, Satoh S, Ono S et al (2004) Attenuation of a delayed increase in the extracellular glutamate level in the peri-infarct area following focal cerebral ischemia by a novel agent ONO-2506. Neurochem Int 45(2–3):381–387
37. Rothstein JD, Patel S, Regan MR, Haenggeli C, Huang YH, Bergles DE et al (2005) Beta-lactam antibiotics offer neuroprotection by increasing glutamate transporter expression. Nature 433(7021):73–77
38. de Rivero Vaccari JP, Bastien D, Yurcisin G, Pineau I, Dietrich WD, De KY et al (2012) P2X4 receptors influence inflammasome activation after spinal cord injury. J Neurosci 32 (9):3058–3066
39. Kimbler DE, Shields J, Yanasak N, Vender JR, Dhandapani KM (2012) Activation of P2X7 promotes cerebral edema and neurological injury after traumatic brain injury in mice. PLoS One 7(7):e41229
40. Huang C, Han X, Li X, Lam E, Peng W, Lou N et al (2012) Critical role of connexin 43 in secondary expansion of traumatic spinal cord injury. J Neurosci 32(10):3333–3338
41. Qu WS, Tian DS, Guo ZB, Fang J, Zhang Q, Yu ZY et al (2012) Inhibition of EGFR/MAPK signaling reduces microglial inflammatory response and the associated secondary damage in rats after spinal cord injury. J Neuroinflammation 9:178
42. Huang T, Solano J, He D, Loutfi M, Dietrich WD, Kuluz JW (2009) Traumatic injury activates MAP kinases in astrocytes: mechanisms of hypothermia and hyperthermia. J Neurotrauma 26 (9):1535–1545
43. Clifton GL, Miller ER, Choi SC, Levin HS, McCauley S, Smith KR Jr et al (2001) Lack of effect of induction of hypothermia after acute brain injury. N Engl J Med 344(8):556–563
44. Lecca D, Trincavelli ML, Gelosa P, Sironi L, Ciana P, Fumagalli M et al (2008) The recently identified P2Y-like receptor GPR17 is a sensor of brain damage and a new target for brain repair. PLoS One 3(10):e3579

45. Ceruti S, Villa G, Genovese T, Mazzon E, Longhi R, Rosa P et al (2009) The P2Y-like receptor GPR17 as a sensor of damage and a new potential target in spinal cord injury. Brain 132 (Pt 8):2206–2218
46. Zhao B, Zhao CZ, Zhang XY, Huang XQ, Shi WZ, Fang SH et al (2012) The new P2Y-like receptor G protein-coupled receptor 17 mediates acute neuronal injury and late microgliosis after focal cerebral ischemia in rats. Neuroscience 202:42–57
47. Maisel M, Herr A, Milosevic J, Hermann A, Habisch HJ, Schwarz S et al (2007) Transcription profiling of adult and fetal human neuroprogenitors identifies divergent paths to maintain the neuroprogenitor cell state. Stem Cells 25(5):1231–1240
48. Genovese T, Rossi A, Mazzon E, Di PR, Muia C, Caminiti R et al (2008) Effects of zileuton and montelukast in mouse experimental spinal cord injury. Br J Pharmacol 153(3):568–582
49. Mozer AB, Whittemore SR, Benton RL (2010) Spinal microvascular expression of PV-1 is associated with inflammation, perivascular astrocyte loss, and diminished EC glucose transport potential in acute SCI. Curr Neurovasc Res 7(3):238–250
50. Carson-Walter EB, Hampton J, Shue E, Geynisman DM, Pillai PK, Sathanoori R et al (2005) Plasmalemmal vesicle associated protein-1 is a novel marker implicated in brain tumor angiogenesis. Clin Cancer Res 11(21):7643–7650
51. Mahoney ET, Benton RL, Maddie MA, Whittemore SR, Hagg T (2009) ADAM8 is selectively up-regulated in endothelial cells and is associated with angiogenesis after spinal cord injury in adult mice. J Comp Neurol 512(2):243–255
52. Kreipke CW, Morgan NC, Petrov T, Rafols JA (2006) Calponin and caldesmon cellular domains in reacting microvessels following traumatic brain injury. Microvasc Res 71 (3):197–204
53. Kreipke CW, Morgan R, Roberts G, Bagchi M, Rafols JA (2007) Calponin phosphorylation in cerebral cortex microvessels mediates sustained vasoconstriction after brain trauma. Neurol Res 29(4):369–374
54. Carloni S, Girelli S, Scopa C, Buonocore G, Longini M, Balduini W (2010) Activation of autophagy and Akt/CREB signaling play an equivalent role in the neuroprotective effect of rapamycin in neonatal hypoxia-ischemia. Autophagy 6(3):366–377
55. Namikawa K, Honma M, Abe K, Takeda M, Mansur K, Obata T et al (2000) Akt/protein kinase B prevents injury-induced motoneuron death and accelerates axonal regeneration. J Neurosci 20(8):2875–2886
56. Hu LY, Sun ZG, Wen YM, Cheng GZ, Wang SL, Zhao HB et al (2010) ATP-mediated protein kinase B Akt/mammalian target of rapamycin mTOR/p70 ribosomal S6 protein p70S6 kinase signaling pathway activation promotes improvement of locomotor function after spinal cord injury in rats. Neuroscience 169(3):1046–1062
57. Park J, Zhang J, Qiu J, Zhu X, Degterev A, Lo EH et al (2012) Combination therapy targeting Akt and mammalian target of rapamycin improves functional outcome after controlled cortical impact in mice. J Cereb Blood Flow Metab 32(2):330–340
58. Petrov T, Rafols JA (2001) Acute alterations of endothelin-1 and iNOS expression and control of the brain microcirculation after head trauma. Neurol Res 23(2–3):139–143
59. Myers SA, DeVries WH, Gruenthal MJ, Andres KR, Hagg T, Whittemore SR (2012) Sildenafil improves epicenter vascular perfusion but not hindlimb functional recovery after contusive spinal cord injury in mice. J Neurotrauma 29(3):528–538
60. Dore-Duffy P, Wang S, Mehedi A, Katyshev V, Cleary K, Tapper A et al (2011) Pericyte-mediated vasoconstriction underlies TBI-induced hypoperfusion. Neurol Res 33(2):176–186
61. Kreipke CW, Schafer PC, Rossi NF, Rafols JA (2010) Differential effects of endothelin receptor A and B antagonism on cerebral hypoperfusion following traumatic brain injury. Neurol Res 32(2):209–214
62. Petrov T (2009) Amelioration of hypoperfusion after traumatic brain injury by in vivo endothelin-1 knockout. Can J Physiol Pharmacol 87(5):379–386
63. Macdonald RL (2012) Endothelin antagonists in subarachnoid hemorrhage: what next? Crit Care 16(6):171

64. Dore-Duffy P, Wang X, Mehedi A, Kreipke CW, Rafols JA (2007) Differential expression of capillary VEGF isoforms following traumatic brain injury. Neurol Res 29(4):395–403
65. Chodobski A, Zink BJ, Szmydynger-Chodobska J (2011) Blood–brain barrier pathophysiology in traumatic brain injury. Transl Stroke Res 2(4):492–516
66. Nag S, Eskandarian MR, Davis J, Eubanks JH (2002) Differential expression of vascular endothelial growth factor-A (VEGF-A) and VEGF-B after brain injury. J Neuropathol Exp Neurol 61(9):778–788
67. Argaw AT, Gurfein BT, Zhang Y, Zameer A, John GR (2009) VEGF-mediated disruption of endothelial CLN-5 promotes blood–brain barrier breakdown. Proc Natl Acad Sci U S A 106 (6):1977–1982
68. Esser S, Lampugnani MG, Corada M, Dejana E, Risau W (1998) Vascular endothelial growth factor induces VE-cadherin tyrosine phosphorylation in endothelial cells. J Cell Sci 111 (Pt 13):1853–1865
69. Thau-Zuchman O, Shohami E, Alexandrovich AG, Leker RR (2010) Vascular endothelial growth factor increases neurogenesis after traumatic brain injury. J Cereb Blood Flow Metab 30(5):1008–1016
70. Nourhaghighi N, Teichert-Kuliszewska K, Davis J, Stewart DJ, Nag S (2003) Altered expression of angiopoietins during blood–brain barrier breakdown and angiogenesis. Lab Invest 83 (8):1211–1222
71. Benton RL, Maddie MA, Worth CA, Mahoney ET, Hagg T, Whittemore SR (2008) Transcriptomic screening of microvascular endothelial cells implicates novel molecular regulators of vascular dysfunction after spinal cord injury. J Cereb Blood Flow Metab 28 (11):1771–1785
72. Myers SA, DeVries WH, Andres KR, Gruenthal MJ, Benton RL, Hoying JB et al (2011) CD47 knockout mice exhibit improved recovery from spinal cord injury. Neurobiol Dis 42(1):21–34
73. Vitellaro-Zuccarello L, Mazzetti S, Madaschi L, Bosisio P, Fontana E, Gorio A et al (2008) Chronic erythropoietin-mediated effects on the expression of astrocyte markers in a rat model of contusive spinal cord injury. Neuroscience 151(2):452–466
74. Genovese T, Mazzon E, Rossi A, Di PR, Cannavo G, Muia C et al (2005) Involvement of 5-lipoxygenase in spinal cord injury. J Neuroimmunol 166(1–2):55–64
75. Huang WL, King VR, Curran OE, Dyall SC, Ward RE, Lal N et al (2007) A combination of intravenous and dietary docosahexaenoic acid significantly improves outcome after spinal cord injury. Brain 130(Pt 11):3004–3019
76. Lopez-Vales R, Redensek A, Skinner TA, Rathore KI, Ghasemlou N, Wojewodka G et al (2010) Fenretinide promotes functional recovery and tissue protection after spinal cord contusion injury in mice. J Neurosci 30(9):3220–3226
77. Jones TB, Hart RP, Popovich PG (2005) Molecular control of physiological and pathological T-cell recruitment after mouse spinal cord injury. J Neurosci 25(28):6576–6583
78. Lin MS, Sun YY, Chiu WT, Hung CC, Chang CY, Shie FS et al (2011) Curcumin attenuates the expression and secretion of RANTES after spinal cord injury in vivo and lipopolysaccharide induced astrocyte reactivation in vitro. J Neurotrauma 28(7):1259–1269
79. Mueller CA, Schluesener HJ, Conrad S, Pietsch T, Schwab JM (2006) Spinal cord injury-induced expression of the immune-regulatory chemokine interleukin-16 caused by activated microglia/macrophages and CD8+ cells. J Neurosurg Spine 4(3):233–240
80. Zhang Z, Fauser U, Schluesener HJ (2008) Early attenuation of lesional interleukin-16 up-regulation by dexamethasone and FTY720 in experimental traumatic brain injury. Neuropathol Appl Neurobiol 34(3):330–339
81. Zhang ZY, Zhang Z, Fauser U, Artelt M, Burnet M, Schluesener HJ (2007) Dexamethasone transiently attenuates up-regulation of endostatin/collagen XVIII following traumatic brain injury. Neuroscience 147(3):720–726
82. Lian H, Shim DJ, Gaddam SS, Rodriguez-Rivera J, Bitner BR, Pautler RG et al (2012) IkappaBalpha deficiency in brain leads to elevated basal neuroinflammation and attenuated response following traumatic brain injury: implications for functional recovery. Mol Neurodegener 7:47

83. Brambilla R, Bracchi-Ricard V, Hu WH, Frydel B, Bramwell A, Karmally S et al (2005) Inhibition of astroglial nuclear factor kappaB reduces inflammation and improves functional recovery after spinal cord injury. J Exp Med 202(1):145–156
84. Becker EB, Bonni A (2004) Cell cycle regulation of neuronal apoptosis in development and disease. Prog Neurobiol 72(1):1–25
85. Di Giovanni S, Movsesyan V, Ahmed F, Cernak I, Schinelli S, Stoica B et al (2005) Cell cycle inhibition provides neuroprotection and reduces glial proliferation and scar formation after traumatic brain injury. Proc Natl Acad Sci U S A 102(23):8333–8338
86. Zhang S, Huan W, Wei H, Shi J, Fan J, Zhao J et al (2013) FOXO3a/p27kip1 expression and essential role after acute spinal cord injury in adult rat. J Cell Biochem 114(2):354–365
87. Okada S, Nakamura M, Katoh H, Miyao T, Shimazaki T, Ishii K et al (2006) Conditional ablation of Stat3 or Socs3 discloses a dual role for reactive astrocytes after spinal cord injury. Nat Med 12(7):829–834

Chapter 5
Neurovascular Responses to Traumatic Brain Injury

Josephine Lok, Ken Arai, Shu-zhen Guo, Wendy Leung, Takakuni Maki, Deepti Navaratna, Klaus van Leyen, Changhong Xing, Limin Wu, Natan Noviski, and Eng H. Lo

Abstract The coordinated action of cells within the neurovascular unit is critical for proper brain functioning. Traumatic brain injury causes cell injury and death, resulting in disruptions of the intricate interactions among the surviving cells. Neurons, oligodendrocytes, astrocytes, microglia, endothelial cells, and pericytes exhibit specific responses to injury. An examination of these responses is important in understanding the pathology after brain trauma, and will be reviewed in this chapter.

5.1 Introduction

The neurovascular unit consists of the neuron and the astrocytes, oligodendrocytes, endothelial cells, pericytes, and microglia that interact with the neuron [1–4]. Traumatic brain injury elicits many cell-type specific responses as well as complex interactions between various cellular pathways. An understanding of the intricate connections within the neurovascular unit is essential in the development of effective neuroprotective strategies. In this chapter the responses of the different cell types to brain trauma will be reviewed.

J. Lok (✉)
Neuroprotection Research Laboratory and Department of Pediatrics, Division of Pediatric Critical Care Medicine, Massachusetts General Hospital and Harvard Medical School, Boston, MA, USA
e-mail: jlok1@partners.org

E.H. Lo et al. (eds.), *Vascular Mechanisms in CNS Trauma*, Springer Series in Translational Stroke Research 5, DOI 10.1007/978-1-4614-8690-9_5,

5.2 Neurons

While the neuron is classically seen as the most important type of neural cell, it is now clear that the complete ensemble of cells in the brain, centered around the neurovascular unit, is essential for functional activity. Nonetheless, the neuron remains a major focus of brain injury studies, as neuron survival is fundamental to brain function. Neuronal injury determines immediate and long-term cognitive function. This is the case even in patients without direct severe brain injury. In athletes with repeated concussions and in military personnel exposed to repeat blast injury, brain pathology reveals increased cytoplasmic aggregates of proteins associated with neuronal degeneration, such as transactive response-DNA binding protein (TDP-43), tau, and β-amyloid [5].

5.2.1 Neuronal Cell Death Pathways After Brain Trauma

Neuronal death occurs through many different pathways, based on reports using different models of trauma. Fas-receptor-related caspase activation has been shown in mice after controlled cortical impact (CCI), as well as in humans after brain trauma [6]. The pro-apoptotic protein Bid contributes to early cell death after CCI in mice [7]. Early cytochrome c release suggests the importance of mitochondrial damage, which may be related to necrosis or apoptosis [8]. Rats subjected to fluid percussion injury show an inflammatory response, characterized by activation of the inflammasome [9]. Stretch injury in cultured cortical neurons leads to secretion of caspase-1, a pro-inflammatory agent, and activation of caspase-3, an executor of apoptotic processes. Lipid peroxidation and related oxidative stress appear to be important mechanisms, although this has been studied mainly in CSF or brain homogenates, and may thus not be limited to neurons [9, 10]. Cell death by necroptosis may be important, and the inhibitor necrostatin-1 is protective in the mouse CCI model [11]. In a mouse model of closed-head injury, the autophagic marker beclin-1 is increased in neurons and activated astrocytes [12]. Treatment with rapamycin, an activator of autophagy, reduced injury, suggesting that autophagy may be a protective mechanism [13]. Excitotoxicity has been considered to be an important component of traumatic injury to the brain [14]. Disappointingly, NMDA receptor antagonists, which were tested in clinical trials in the 1990s/early 2000s, did not show efficacy in the treatment of traumatic brain injury [15, 16].

5.2.2 Traumatic Penumbra

Neuronal pathology is not confined only to the area of direct impact. In focal brain injury, the peri-contusional area features ischemia, which leads to secondary damage. Reduced cerebral blood flow was found in the peri-contusional area in mice by ^{14}C-iodoantipyrine autoradiography [17], and clinical studies have suggested that intravascular coagulation may contribute to the reduction in blood flow [18, 19]. Analogous to the ischemic penumbra in stroke, it is likely that a traumatic penumbra exists [19–21], representing a site with potentially salvageable cells. Diffuse axonal injury has been observed in the peri-contusional area in humans after severe head injury. Presumably related to shear stress, it also compromises surviving neurons in the penumbra [22].

5.2.3 Selective Neuronal Vulnerability in Brain Trauma

Neurons from different brain regions appear to differ in their susceptibility to injury following brain trauma. In different types of experimental trauma models, the hippocampus, especially the CA3 region and the dentate gyrus, appear to be particularly susceptible to delayed neuronal damage [23–25], Etiologies proposed for this increased vulnerability include a higher concentration of NMDA receptors in hippocampal neurons [26]. Consistent with this hypothesis, hippocampal neurons exhibit higher intracellular calcium levels and increased cell death in an in vitro mechanical stretch model, compared to cultured cortical neurons, suggesting increased excitotoxic cell death [26]. However, the NMDA receptor antagonist MK-801 was not protective in this study or in clinical trials [15, 16, 26]. Other observations include a higher level of glia and macrophages in the hippocampus of rats after fluid percussion injury [27]. Using another approach with gene expression analysis, FluoroJade-positive and negative hippocampal CA3 neurons were isolated by laser capture microdissection and the transcriptional profile was examined. Compared to uninjured neurons, the transcriptional profile in FluoroJade-positive neurons was found to be different, with mRNAs encoding neuroprotective molecules such as glutathione peroxidase 1, heme oxygenase 1, and brain-derived neurotrophic factor (BDNF) expressed at lower levels in the injured neurons [28].

In addition to hippocampal neurons, Purkinje cells of the cerebellum are also very vulnerable to delayed cell death following TBI [29]. AMPA receptor-mediated calcium overload was found to be a possible factor in an in vitro stretch injury model [30]. Several studies have examined whether myelination state affects neuronal sensitivity to brain trauma. In a fluid percussion model in rats, unmyelinated axons in the corpus callosum appear to be especially susceptible [31]. In an in vitro correlate, myelinated axons are more resistant to axonal stretch injury in an oligodendrocyte–neuron co-culture system [32].

The fact that myelinated axons are more resilient suggests that oligodendrocytes have a protective effect on neurons under TBI conditions [32]. Similarly, endothelial-neuron co-cultures suggest that endothelial-secreted BDNF enhances survival of neurons in an in vitro model of hypoxia and re-oxygenation [33]. These examples of the interdependency between neurons and other types of cells highlight the need for neuroprotective strategies that target the many cells that interact with the neurons.

5.3 Oligodendrocytes

Oligodendrocytes constitute one of the major glial cell types in the white matter of the brain. Oligodendrocytes produce a lipid-rich membrane called myelin, which wraps around the axons of neurons and facilitates fast saltatory nerve impulse conduction [34].

5.3.1 *Physiological Functions of Oligodendrocytes*

Most myelination occurs during early postnatal life, but some myelination continues at least into late adolescence, contributing to the maturation of the functional circuits. In selected regions of the CNS, myelination actually increases throughout adult life [35], thereby conferring some degree of plasticity to the adult neural circuitry. For instance, successful learning of juggling is associated with an increase in fractional anisotropy in the white matter underlying the intraparietal sulcus, suggesting an increase in myelination even in adult brain [36]. Increases in myelination have been detected in the adult animals under environmental enrichment [37]. Adult oligodendrocyte precursor cells (OPCs) are abundant in both grey and white matter areas, comprising 5–8 % of all the cells in the adult brain, and can be stimulated to proliferate, migrate and differentiate to provide new oligodendrocytes [38].

5.3.2 *Interdependency of Oligodendrocytes with Neurons and Endothelial Cells*

Oligodendrocyte–neuron interactions are important in normal brain function. Oligodendrocytes signal to neurons and metabolically support axons via myelin–axon interactions [39, 40]. Acting through the monocarboxylate transporter 1 (MCT1), oligodendrocytes supply lactate, which is integral for axonal energy support [41]. In addition, oligodendrocyte-derived trophic factors, such as insulin-like growth factor (IGF-1) and glial cell-derived neurotrophic factor (GDNF), promote neuron survival, and axon outgrowth in vitro [42]. On the other hand, neurons can enhance or inhibit oligodendrocyte differentiation and maturation through axon-generated signals [40, 43–45]. Furthermore, there is evidence that axonal electrical activity affects myelination. During development, OPCs can generate postsynaptic potentials in response to synaptic input [46]; and in the mature mouse, OPCs from the subventricular zone (SVZ) receive synaptic input during remyelination [46].

Endothelial–oligodendrocyte interactions in the “oligovascular niche” may contribute to ongoing angiogenesis and oligodendrogenesis in adult white matter after brain injury [47, 48]. Cerebral endothelium enhances OPC proliferation and migration in cell culture [49]. Conversely, matrix metalloproteinase (MMP)-9 from oligodendrocytes may promote vascular remodeling after white matter injury [47]. Hence, strategies targeting oligovascular signaling may help promote remyelination along with vascular remodeling after white matter injury in the context of brain trauma.

5.3.3 Pathophysiology of Oligodendrocytes After Brain Trauma

White matter disruption due to loss of oligodendrocytes occurs across the spectrum of mild to severe TBI, both in animal models and in humans [48, 50–54]. Significant white matter atrophy occurs months to years after injury in a rat TBI model [52, 54, 55]. Recent studies have reported that activation of microglia, macrophages, and astrocytes, along with an increase in the number of OPCs, are found in white matter lesions after fluid percussion injury in a rat TBI model [50]. In human studies, atrophy of white matter tracts and subsequent demyelination [52, 56–59] is often seen along with chronic reductions in cerebral blood flow in white matter regions [60, 61]; whereas cognitive recovery is associated with improvements in white matter blood flow [62]. Most often, the degree of atrophy has a direct correlation with the degree of neuropsychological impairment, such as cognitive dysfunction [52, 54, 56–59].

Recent progress in MRI technique, including diffusion tensor imaging (DTI) and DTI tractography, has resulted in improved detection of subtle white matter injury which was not visualized with prior imaging techniques. For instance, it is now possible to detect diffuse axonal injury after mild TBI, even when the key identifying feature of microbleeds is not present [63–67]. Recent clinical studies using DTI have shown a relationship between white matter injury and the extent of impairment in functional connectivity within important brain networks [63–67]. Tractography measurements predicted learning and memory performance in patients during the acute phase of TBI, as well as their processing speed and executive function during the chronic phase [65].

Cell survival, inflammatory, and hemodynamic processes in white matter play a pivotal role in progressive pathophysiology of TBI. The loss or dysfunction of oligodendrocytes could also have devastating effects on the function of neurons and of axonal connectivity. On the other hand, endogenous regeneration of injured white matter can be induced after TBI [50]. Therapies to restore oligodendrocyte and myelin integrity and to promote oligodendrogenesis should be an important consideration in the therapeutic approach to TBI.

5.4 Astrocytes

Astrocytes play an essential role in neurovascular functioning. They constitute nearly half of all brain cells and outnumber neurons in the human brain. Astrocyte interaction with other cells is facilitated by its extensive network of fine membranous processes that ensheath synapses and microvessel. Astrocyte regulation of extracellular ionic balance is essential in maintaining an environment for normal brain function. Additionally, astrocytes play many other pivotal roles in NVU regulation, which will be discussed in this section.

5.4.1 Physiological Functions of Astrocytes

First, astrocytes affect the formation, function, and elimination of neuronal synapses [68]. Astrocytes secrete thrombospondins (TSP) [69], cholesterol [70], and glypicans 4 and 6 [71], which promote synaptic formation, presynaptic function, and postsynaptic function. When co-cultured with astrocytes, retinal ganglion cells form functional synapses whose activity increases by nearby100-fold [72]. Conversely, astrocytes are affected by neurons through their response to neurotransmitters, which activate signaling cascades in astrocytes. This process often results in a feedback loop in which astrocyte release of chemicals, such as ATP, in turn modulating neuronal activity.

Second, astrocytes couple the level of synaptic activity with cerebral blood flow through their effect on cerebral vascular tone. This process is accomplished by coordinated interactions between astrocytes, neurons, and endothelial cells, and facilitated by the numerous fine astrocytic processes which are closely associated with both blood vessels and synapses [73, 74]. In response to increases or decreases in synaptic activity, astrocytes modulate regional vascular tone to regulate the amount of regional blood flow, matching enhanced delivery of oxygen and glucose to metabolic needs in the active brain region.

Third, astrocytes constitute a key element of the blood–brain barrier (BBB). Scar-forming reactive astrocytes help seal injuries to the BBB [75]. Conversely, vascular endothelial growth factor (VEGF)-A, derived from reactive astrocytes, enhances pathological BBB breakdown accompanied with lymphocyte infiltration, tissue damage, and clinical deficit [76].

Fourth, astrocytes are highly secretory cells that release various substances including soluble trophic factors, that act on other types of cells. For example, astrocytes secrete a host of trophic factors which target oligodendrocyte lineage cells, including platelet-derived growth factor (PDGF), fibroblast growth factor-2 (FGF-2), leukemia inhibitory factor (LIF), ciliary neurotrophic factor (CNTF), IGF-1, BDNF, bone morphogenic proteins (BMPs), and ECM-related molecules [77, 78]. Astrocytes also have trophic effects on neurons [79] and on endothelial progenitor cells (EPCs), actions which may mediate neurovascular remodeling after stroke [80].

Fifth, astrocytes communicate with neighboring cells through gap junction channels that are regulated by extra- and intracellular signals [81]. For instance, connexin-43 (CX43) and CX30 hemichannels mediate astrocyte-endothelial and astrocyte-neuronal cell–cell transfer of nutrients, metabolite, secondary messengers and ions [82].

5.4.2 *Pathophysiology of Astrocytes After Brain Trauma*

Astrocytes play significant roles in brain function after brain trauma. Following experimental TBI in rodents [83–85] or neurotrauma in humans [86], astrocytes undergo a phenotypic change called "reactive astrocytosis." The morphology of these reactive astrocytes consists of cellular hypertrophy and hyperplasia, cytoplasmic enlargement, elongation of cytoplasmic processes, and increased expression of glial fibrillary acidic protein (GFAP) [87, 88].

Reactive astrocytosis after brain injury is generally regarded as detrimental. S100β, a marker of reactive astrocytosis, is elevated in the serum [89, 90] and CSF [91] of TBI patients and correlate with clinical outcome. Higher serum concentrations of GFAP are also associated with worse clinical outcome in patients with significant TBI [92]. Reactive astrocytes may be harmful by producing pro-inflammatory cytokines [93] and by inhibiting axon regeneration [94]. Down-regulation or dysfunction of the astrocytic glutamate transporters GLT-1 and GLAST is seen following CCI in rats [95] and in humans following severe brain trauma [96] and may exacerbate neuronal excitotoxicty following head trauma. Suppression of reactive astrocytes by simvastatin attenuated brain injury in a rat traumatic injury model [97], suggesting that reactive astrocytes may be a therapeutic target after brain trauma. However, reactive astrocytes can be beneficial under some conditions. For instance, they may upregulate synaptogenesis-inducing genes, such as TSP [69], and secrete trophic factors, including GDNF, BDNF, IGF1, and VEGF [98]. Additionally, reactive astrocytes may release tPA, which can enhance neuronal dendrite formation [99].

5.4.3 *Dichotomous Effects of Reactive Astrocytosis*

The divergent effects of reactive astrocytes after brain trauma are illustrated by experiments involving transgenic mice that lack reactive astrocytes. They have reduced glial scar formation after forebrain stab injury or spinal cord injury (SCI), but they exhibit increased inflammation and prolonged leukocyte infiltration [75, 100]. Likewise, conditional astrocyte ablation exacerbated neural tissue damage and increased the inflammatory response following moderate CCI in mice [101, 102].

It is intriguing that reactive astrocytosis may lead to beneficial or harmful results. Astrocytes appear to be a heterogenous group [68, 103, 104] in terms of morphology, antigenic phonotype, location, translational profile, electrophysiological properties and Ca^{2+} activity [105]. After brain injury, gene profiles of reactive astrocytes are differentially expressed in different brain regions, or at different time points in the same brain region. Whether each astrocyte can transform its phenotype or whether there are inherently different astrocyte populations in specific brain regions is an area of investigation [103]. Furthermore, alterations in gene expression of reactive astrocytes appears to differ by injury type, as was shown by the astrocyte transcriptome of mice after middle cerebral artery occlusion (MCAO) and systemic LPS injection [106]. Hence, the diversity of astrocytes may contribute to their dual roles in NVU functioning after brain injury.

5.5 Microglia

Microglia are resident immune cells of the CNS and serve as sensors and effectors of the immune system in the normal and pathologic brain [107, 108]. Microglia constantly monitor the microenvironment and respond to many types of pathologic events with typical macrophagic functions, such as phagocytosis, secretion of proinflammatory cytokines, and antigen presentation.

5.5.1 *Physiological Functions of Microglia*

Microglia are immunological sentinels of the brain and even in the "resting" state in the healthy CNS, they are not functionally silent [107, 109]. Microglial processes are highly dynamic in the intact cortex [110, 111], in which microglial processes constantly survey and respond to the functional status of synapses [112]. Furthermore, it appears that microglia contribute to the subsequent increased turnover of synaptic connections, based on the finding that some synapses in ischemic areas disappear after prolonged microglial contact [112]. Recent research has revealed many additional microglial functions beyond that of the microglia "immune network" [113]. Microglia may interact with axons of visual neurons, with a phenotype that is influenced by the postnatal visual experience [114]. Microglia are implicated in retinal blood vessel formation in humans and rodents [115]. Microglia act as phagocytes for the removal of dying cells during the process of programmed cell death [116]. In the adult brain, microglia participate in synapse remodeling and neurogenesis [117–120].

5.5.2 *Microglial Activation*

Microglial reactivity is a transition from the highly active surveillance state towards an even more reactive state, in which microglia respond to a pathological event with morphological and functional changes. Any insult to the CNS, including infection, trauma, or metabolic dysfunction, may cause microglial activation. Upon activation, microglia develop thicker processes, produce cytokines, chemokines, and growth factors, and generate reactive oxygen and nitrogen species. They may also increase the expression of immunomodulatory surface markers and acquire antigen-presenting ability [121, 122]. Activated microglia can exhibit phenotypic and functional diversity depending on the nature, strength, and duration of the stimulus [121, 123]. At least two activated phenotypes, "classically activated" (also called M1) or an "alternatively activated" (also called M2), have been identified [124–127]. M1 microglia are pro-inflammatory and release TNFα, IL-1β, nitric oxide, and reactive oxygen species (ROS). M2 microglia, in comparison to M1 microglia, have decreased production of nitric oxide and increased production of anti-inflammatory cytokines and neurotrophic factors such as GDNF, BDNF, bFGF, IGF-1, TGF-β, and VEGF [118, 128–131].

Whether microglial activation is neurotoxic is a long-standing debate. On the one hand, microglial over-activation or dysfunction may exacerbate a preexisting neuropathology or cause neurodegenerative diseases [132]. On the other hand, activated microglia can support neuronal survival [133, 134] by release of neurotrophic and anti-inflammatory molecules, clearance of toxic products or invading pathogens, as well as by guidance of stem cells to inflammatory lesion sites to promote neurogenesis [119, 120, 135].

5.5.3 *Pathophysiology of Microglia After Brain Trauma*

In rat models of experimental SCI, activated microglia appear between 12 and 24 h post-injury, with maximal infiltration 4–8 days post-injury [136, 137]. There is a second peak at 60 days and continued elevation through 180 days after SCI [138]. In animal models of experimental brain trauma, microglia react within minutes, and microglial processes rapidly converge at the injury site. Activated microglia surround the lesion and remain activated for weeks and months after the initial injury [139–143].

In humans with brain trauma from traffic accidents and falls, microglial activation has been reported as early as 72 h after injury [144], and can remain elevated for months after blunt injury [145, 146]. Sites of activation often coincide with those of neuronal degeneration and axonal abnormality [147]. Increased microglial activation has been reported as long as 17 years after moderate to severe brain trauma in a patient study [148]. This finding suggests that microglia may perpetuate a chronic inflammatory cycle after TBI. It also highlights the possibility that the brain's response to trauma may evolve over years or even decades [148].

5.5.4 *Dichotomous Effects of Microglial Activation*

Microglial activation may have harmful and beneficial effects. It is widely accepted that neuro-inflammation after TBI plays opposing roles. Many pathological changes of TBI are mediated through an inflammatory cascade characterized by activation of microglia [149] and an increase of pro-inflammatory cytokines [150, 151], both of which can exacerbate brain injury and increases the risk of delayed dementia [152]. A recent study showed that microglia alone, without the inflammatory effects of circulating monocytes or macrophages, can impair functional recovery in the injured spinal cord [153]. Activated microglia may also induce extensive retraction of dystrophic axons through direct physical interactions [154]. In humans, the long-term microglial activation and chronic inflammation may cause post-traumatic neurodegeneration that may underlie the cognitive decline seen in many long term survivors of TBI [146].

However, not all microglial activation is deleterious. There is evidence that microglia is involved in focal neurogenesis in the dentate gyrus of the hippocampus after TBI [155]. Interestingly, inflammation-activated microglia can attenuate neurogenesis, while microglia activated by certain T cell cytokines can promote neurogenesis [156]. Similarly, microglia may be protective after SCI, in which the M2 phenotype of activated microglia expresses growth factors that promote neuronal recovery and limit inflammation-mediated injury [157]. Thus, activated microglia may be important in promoting functional recovery after CNS trauma.

5.6 Endothelial Cells

Endothelial cells line the entire vasculature of the brain. The endothelium is extensive in its physical dimensions and in its interactions with other cells in the brain. Endothelial responses to traumatic brain injury, ranging from altered signaling pathways to physical breakdown of blood vessels, have far-reaching consequences for brain function and neurological outcome.

5.6.1 *Pathophysiology of Endothelial Cells After Brain Trauma*

Dysfunctional endothelial signaling can lead to vasospasm, vasoconstriction, microthrombi formation, oxidative stress, and amplification of inflammatory cascades [158–165]. Physical breakdown of blood vessels impose the burdens of ischemia, hemorrhage, and the toxic effects of extravasated iron and blood stream components [166–169]. Leukocyte–endothelium interactions are important after brain trauma and constitute an area of active research. Increased leukocyte adhesion to endothelial cells and migration into the brain parenchyma have been observed after brain trauma and may exacerbate pathological processes, but may not contribute directly to expansion of the primary lesion [170–175].

Endothelial dysfunction is detrimental to brain homeostasis and function, and pathological events in the microvasculature have great potential to contribute to secondary brain injury and impact neurological outcome. Of the myriad endothelial reactions to brain trauma, compromised BBB function and progressive secondary hemorrhage (PSH) are two of the most significant, and will be discussed in further detail in this section. Although some variations exist based on the modality of injury—concussion, contusion, blast injury, shear injury, and mixed models—the major elements of the pathways will be highlighted.

5.6.2 Compromised Blood–Brain Barrier Function

The blood–brain barrier (BBB), also known as the blood–brain interface, is a selective diffusion barrier, formed by endothelial cells which are closely apposed by astocytic end feet and pericyte processes [98]. The endothelial cells of the central nervous system are connected with specialized junctional complexes, generating a higher transendothelial resistance (~1,500 Ω/cm^2) than peripheral endothelial cells [176]. Additional transport proteins and solute carriers located on endothelial surfaces mediate the regulated transfer of substances in and out of the brain [177]. The maintenance of the barrier properties is crucial to proper neuronal functioning. In clinical series, patients who have disrupted BBB appear to have an increased risk of developing acute seizures, delayed epilepsy, cognitive impairment, and Alzheimer's disease [178–181].

The BBB is especially vulnerable to downstream effects of ROS which are generated by many pathological processes after TBI. These processes include excitotoxicity, neutrophil recruitment, mitochondrial energetic failure, and macrophage/microglial activation [182]. BBB breakdown after experimental brain injury is typically biphasic in nature, reaching a maximum within a few hours and subsequently declines, with a delayed smaller peak 3–7 days following injury [183–185]. Compromise of BBB integrity disrupts the precise ionic gradients necessary for proper neuronal function. Additionally, cerebral edema increases the total intracranial volume, which may increase intracranial pressure, leading to compromised perfusion, ischemia, brain herniation, and death. The increase in BBB permeability is often associated with a local immune response, including the activation of astrocytes and microglia [186, 187]. Some reports suggest that BBB opening alone, even without traumatic damage, may induce astrocyte activation [188]. In addition, activation and upregulation of MMPs degrade the endothelial and further increase vascular permeability [189].

5.6.3 Cerebral Edema Formation—Etiologic Considerations

A number of different pathways have been described in association with the increase in endothelial permeability and the development of cerebral edema. Several groups have examined the permeability increases provoked by IL-1β as it is an inflammatory cytokine which is upregulated and plays an important role after brain trauma. The pathophysiological processes include signaling through various isoforms of protein C kinase, resulting in phosphorylation of ZO-1 and a decrease in transendothelial electrical resistance [190]. Induction of tissue factor activity was reported to facilitate IL-1β effects on endothelial permeability [191]. In an in vitro model of hypoxia and re-perfusion, conditions which are often present in the brain after trauma, endothelial barrier function was found to be highly dependent on actin dynamics, mediated by complex interactions involving several Rho-GTPases such as Rho A and Rac1[192]. In another in vitro model of inflammation-induced permeability, VEGF-mediated disruption of claudin-5 was reported [193]. Data from tPA knockout mice subjected to CCI suggest that tPA may amplify edema and cortical brain damage [194].

A novel mechanism has been proposed by the Simard, based on research involving two ion transport proteins expressed in brain endothelial cells—NKCC1 (the Na^+–K^+–$2Cl^-$ cotransporter), and SUR1/TRPM4 (the SUR1-regulated $NC_{Ca\text{-}ATP}$ channel) [169, 195]. NKCC1, an intrinsic membrane protein which is constitutively expressed on the luminal side of brain endothelial cells, transports chloride ions, along with sodium and potassium ions, across plasma membranes. Brain trauma and the concomitant ischemia increases the expression and activity of NKCC1, which leads to excess sodium transport into the cells, resulting in cell swelling and increased capillary permeability [195–198]. The excess sodium inside endothelial cells is pumped out into the extracellular space by the Na^+–K^+ ATPase, which is expressed on the abluminal membrane, contributing to brain edema formation. NKCC1 activity is inhibited by a low dose of the diuretic bumetanide, which significantly reduces cerebral edema and neuronal injury following traumatic and ischemic brain injury [199, 200].

The SUR1-regulated $NC_{Ca\text{-}ATP}$ channel, also called SUR1/TRPM4, is another ion channel implicated in the development of cerebral edema [169, 195]. It consists of two subunits—SUR1, the regulatory subunit; and TRPM4, the pore-forming subunit. Expressed mostly during conditions of hypoxia or injury [201, 202], including brain trauma [195, 203] and intracranial hemorrhage [200], it is activated by Ca^{2+} and inhibited by intracellular ATP and transports inorganic monovalent cations across plasma membranes. Under conditions of ATP depletion, the SUR1/TRPM4 channel opens, depolarizing the cell and allowing excess sodium influx. These events contribute to cytoskeletal rearrangement, tight junction disruption, cell retraction, and the development of intercellular gaps, which promote the formation of vasogenic edema. The cell may also undergo oncotic swelling, leading to cytotoxic edema and necrotic cell death.

Because of their opposite relationships to ATP availability—NKCC1 function requires the presence of ATP and SUR1/TRPM4 requires the depletion of ATP—at least one of these proteins is likely to be active as ATP concentrations fluctuate in the aftermath of brain trauma [169]. Bumetanide and glibenclamide, two FDA-approved drugs in clinical use, can inhibit NKCC1 and SUR/TMRP4, respectively, and were found to decrease cerebral edema in in vivo TBI experiments [195, 200]. A recent analysis of patients who were taking a sulfonylurea drug (such as glibenclamide) for diabetes mellitus showed that a greater percentage of the patients (36 % vs. 7 %) who were on sulfonylureas had improved neurological status at discharge, compared to patients with similar characteristics but not taking a sulfonylurea drug [204].

Aquaporin-4 (APQ4), a major water channel located predominantly in the perivascular end feet of astrocytes, is also believed to be a participant in the tapestry of pathways involved in cerebral edema formation [205]. However, the relationship between APQ4 expression and edema formation is complex. In a focal cortical contusion model in rats, AQP4 was upregulated at the site of traumatic brain injury, but down-regulated adjacent to the site of injury [206]. In another rat CCI model, AQP4 expression was decreased in both hemispheres—maximally in the injured hemisphere at 48 h after injury, coinciding with edema development [207]. Primary cultures of astrocytes from AQP4-null mice had greatly reduced osmotic water permeability compared with wild-type astrocytes, consistent with the role of AQP4 as a principal water channel in these cells. Curiously, the absence of AQP-4 ameliorated cytotoxic brain edema, but worsened vasogenic brain edema [208]. In yet another series of experiments, inhibition of AQP-4 significantly decreased brain edema but did not affect BBB permeability [209]. In recent studies with a closed-skull model of TBI in mice, AQP4 expression was found to be generally increased; but the most prominent effect was the loss of polarized localization at end foot processes of reactive astrocytes. This alteration in AQP4 peaked at 7 days after injury, when cerebral edema and ICP had largely normalized. These temporal profiles suggest that changes in AQP4 expression and localization may be compensatory mechanisms rather than causative factors in the formation and resolution of cerebral edema [210]. The divergent data summarized here are likely manifestations of multiple complex actions of AQP-4 in the brain after trauma.

Experiments that incorporate the effects of secondary insults—hypoxia and hypotension—suggest that secondary insults blunt AQP4 upregulation after trauma [211]. However, in similar experiments which examined early cytotoxic edema after brain trauma and secondary insult or focal ischemia, an antagonist to the vasopressin V1A receptor reduced AQP4 upregulation and attenuated brain edema formation [212]. Since V1A receptors regulate AQP4-dependent brain water movement, the modulation of V1A receptor function may offer new insights into the role of APQ4 and vasopressin in brain edema formation.

The degree of cerebral edema after TBI is highly correlated with neurological outcome after brain trauma. There is also increasing evidence that microvascular injury may result in long-term abnormalities in the BBB and may contribute to chronic pathology, including post-trauma epilepsy and Alzheimer's disease

[178–181]. To investigate the delayed complications of BBB damage, Pop et al. [213] evaluated the effect of juvenile TBI (jTBI) on cognitive decline over time. 17-day-old rats were subjected to CCI. Sixty days after injury, jTBI mice had decreased P-glycoprotein (P-gp) on cortical blood vessels, indicating alterations in BBB properties. They also exhibited higher levels of endogenous β-amyloid in several brain regions compared to sham, along with impaired cognitive performance, raising the possibility that there is an association between BBB dysfunction and chronic cognitive decline.

5.6.4 Hemorrhagic Progression of a Contusion

Cerebral contusions are commonly seen in patients with severe brain injuries, and a significant proportion of these patients develop a progression in the size of the contusion [214]. Often associated with multiple micro-hemorrhages from ruptures of capillaries, the microvessels may undergo further breakdown in the hours after the initial trauma, resulting in expansion of the contusion or formation of additional non-contiguous hemorrhagic lesions [215]. Termed "hemorrhagic progression of a contusion (HPC)" or "PSH" by the Simard laboratory [169, 201], it is detrimental to brain tissue on several fronts. The capillary breakdown leads to decreased perfusion and tissue ischemia, and the extravasated heme and iron is extremely toxic to the surrounding cells and to the myelin of white matter tracts [166–168].

Historically, HPC has been attributed to continued extravasation of blood from damaged microvessels, along with exacerbation by a coagulopathy. A novel mechanism is being investigated, again involving the SUR1/TRPM4 channel [169]. Proposed to be a central agent in the development of cerebral edema, SUR1/TRPM4 appears to be integral to HPC as well. As previously described, opening of SUR1/TRPM4 allows excessive sodium influx, leading to oncotic cell swelling [195]. In the extreme case of critical ATP deletion, SUR1/TRPM4 opening is sustained, and oncotic swelling progresses to endothelial cell death; capillary fragmentation, and PSH. The microenvironment in the contused brain region promotes SUR1 activity because the kinetic energy of the impact activates mechano-sensitive transcriptional regulators of SUR1 [216, 217]. Two such transcription factors are protein 1 (Sp1) and nuclear factor-jB (NF-jB), both of which are involved in the transcriptional regulation of Sur1 [218–222]. Once activated, these proteins rapidly undergo nuclear translocation and lead to increased SUR1 expression in the microvessels. Transcriptional upregulation of SUR1 followed by opening of the channel after TBI sets the stage for progression of secondary bleeding and has been linked to oncotic cell swelling and death of neurons, astrocytes, and endothelial cells [202, 203, 223]. Treatment with the Sur1 inhibitor glibenclamide prevents the progression of hemorrhage and limits the final lesion volume [200, 217].

BBB dysfunction and progressive hemorrhage are two events at the endothelial level that significantly affect neurological outcome. Anatomical considerations and the temporal progression of these events add to their potential as targets for neuroprotective interventions—anatomically, the vasculature is more accessible than other brain compartments; temporally, these complications often occur with some time delay after the initial brain trauma. The incorporation of these therapeutic targets should be important considerations in the development of neuroprotective strategies.

5.7 Pericytes

Pericytes are cells that surround the endothelial cell layers of the capillary network in the brain. They have multiple functions, including the regulation of BBB integrity, regulation of cerebral blood flow, clearance of cellular debris, and being a source of pluripotent stem cell [224–228]. The pericyte density varies among different organs and vascular beds. Pericyte coverage in CNS is higher than in other organs, with approximately 30 % coverage of the endothelial abluminal surface [229], suggesting that pericyte functions are especially important in the brain vasculature. In this section we will focus on the key actions of pericytes on the cerebral microvasculature.

5.7.1 Physiological Functions of Pericytes

First, pericytes engage in functional coupling with endothelial cells. This is facilitated by elongated pericyte processes which ensheath the capillary wall [229]. Interdigitations of pericyte and endothelial cell membrane make direct contact containing cell–cell junction proteins, such as N-cadherin and connexin [227, 229]. Several transduction cascades, including PDGF-B, transforming growth factor-β (TGF-β), Notch, sphingosine-1 phosphate and angiopoietin signaling are involved [224, 226].

Second, pericytes have angiogenic actions [230]. Pericytes express MMPs, which enhance extracellular matrix degradation early in angiogenesis, thereby promoting endothelial migration and facilitating the release of matrix-sequestered angiogenic factors. Furthermore, pericytes directly contribute to the synthesis of extracellular matrix proteins, including laminin, nidogen, and fibronectin [231]. They also secrete tissue inhibitor of metalloproteinase 3 (TIMP3), a potent inhibitor of several MMPs, which inhibits degradation of basement membrane proteins during the vessel stabilization phase [231]. Pericyte-derived VEGF-A may also stimulate endothelial survival, proliferation, and sprout formation [232].

Third, pericytes play essential roles in maintaining BBB integrity. BBB forms early in embryogenesis, during a time period that coincides with initial pericyte recruitment, preceding astrocyte generation [233]. However, the role of pericytes in BBB function extends beyond the perinatal period to the adult and aging brain [225, 226]. Pericyte-deficient mice demonstrate both BBB breakdown and reductions in brain microcirculation, which lead to neuroinflammation and neurodegeneration in the adult and aging brain [226]. Loss of brain pericytes and the resulting BBB breakdown have been shown to impair neurovascular function through leakage and deposition of vasculotoxic or neurotoxic macromolecules, such as fibrin, thrombin, plasmin, and hemosiderin [226, 231].

5.7.2 Pathophysiology of Pericytes After Brain Trauma

The multiple functions of pericytes all come into play after brain trauma [234, 235]. Studies focusing on the role of pericyte in human trauma have been sparse; however, emergent studies demonstrate that pericytes play multiple functions in restoration of homeostasis in the neurovascular unit after trauma.

Pericytes are instrumental in the regulation of blood flow after TBI. Pericytes have been reported to be involved in endothelin-1-mediated hypoperfusion after TBI. Increased pericyte expression of alpha-SMA and endothelin-1 correlate with reductions in both arteriolar and capillary diameter after trauma [236]. ETrB immunolabeling was elevated in pericytes 24 h post-trauma, suggesting that pericytes could be involved in microvascular autoregulation and reduction of blood flow [237].

Pericytes are involved in adaptive cerebrovascular responses after TBI. The detachment and migration of brain pericytes after trauma is thought to trigger BBB disruption and edema formation, and to affect angiogenesis during the repair phase. In a model of TBI in the rat, pericytes close to the impact area started migrating as early as an hour after injury [238]. Approximately 40 % of brain microvascular pericytes are reported to migrate out to perivascular locations, often to adjacent neuropils. Coincident ultrastructural changes in the neurovascular unit are seen, such as thinning of abluminal basal laminar surfaces and increased protease activities at the leading tip of the migrating pericyte. A marked increase in the expression of urokinase plasminogen activator receptor and/or MMPs at the leading tip of the migratory pericyte has been reported [238, 239]. The migration is thought to be crucial for pericyte retention in the tissue as most cerebrovascular-embedded pericytes perish through apoptotic mechanisms following trauma [238]. However, in human brain edema, degenerative changes in pericytes have been reported without evidence of pericytic migration to perivascular spaces [235].

The presence or loss of pericytes from cerebrovascular plexus after trauma is thought to modulate VEGF expression which affects post-traumatic angiogenic responses [240]. Heat shock response in a TBI model, as demonstrated by a ubiquitous increase in HSP-70 staining of brain capillaries 48 h post-trauma, was preceded by a marked pericyte cell death [241]. Pericyte activity after trauma may contribute to BBB disruption. In several models of focal brain compression and ischemia, extensive morphological changes in the basal lamina, such as basement membrane thickening, are attributed to the proteolytic activities of the migrating pericyte [234, 235, 238–240]. In human traumatic brain edema, resident pericytes are reported to play a direct role in brain barrier dysfunction and edema resolution by phagocytic activities. Ultrastructural modifications, such as increased vacuolar and vesicular transport, transient transpericytal channels, and morphological variations point to altered pericyte function in the BBB [242].

Several submicroscopic changes have been documented in cortical capillary pericytes in human perifocal brain edema. Phagocytic pericytes were observed with ingested erythrocytes [243]. Hypertrophic pericytes with ruptured basement membrane were found as well. Resident-degenerated pericytes were found with lacunar enlargement of endoplasmic reticulum, vacuolization, and micropinocytotic vesicles orientated to multivesicular bodies, suggesting that pericytes may have phagocytic functions [242].

Apart from the well-characterized function of cerebrovascular regulation, brain pericytes are likely to have pleiotropic functions in repair and regeneration after brain trauma. Pericyte plasticity after trauma offers opportunities for therapeutic intervention. The current scientific evidence on pericytes is housed in different animal models and different injury paradigms. Research efforts to clarify the roles played by pericytes in acute brain trauma will add a great deal to the understanding of brain pathophysiology after trauma.

5.8 Conclusion

Each of the different types of cells in the neurovascular unit exhibits unique responses to brain trauma. Additionally, there are many intricate interactions between these cells, often with dichotomous manifestations at different time points. An understanding of these complex interactions within the neurovascular unit is a starting point for addressing the pathophysiology after brain trauma.

Acknowledgement *Conflict of interests*: None. Funding: R37NS037074-13 (EHL), RO1NS076694-02(EHL), P01NS055104-05 (EHL), R01NS0800991-01(KA), RO1NS049430-07 (KVL), RO1NS069939-02 (KVL), K08N5057339-04 (JL).

References

1. Hawkins BT, Davis TP (2005) The blood–brain barrier/neurovascular unit in health and disease. Pharmacol Rev 57(2):173–185
2. Iadecola C (2004) Neurovascular regulation in the normal brain and in Alzheimer's disease. Nat Rev Neurosci 5(5):347–360
3. Lo EH, Broderick JP, Moskowitz MA (2004) tPA and proteolysis in the neurovascular unit. Stroke 35(2):354–356
4. Park JA, Choi KS, Kim SY, Kim KW (2003) Coordinated interaction of the vascular and nervous systems: from molecule- to cell-based approaches. Biochem Biophys Res Commun 311(2):247–253

5. Blennow K, Hardy J, Zetterberg H (2012) The neuropathology and neurobiology of traumatic brain injury. Neuron 76(5):886–899
6. Qiu J, Whalen MJ, Lowenstein P, Fiskum G, Fahy B, Darwish R et al (2002) Upregulation of the Fas receptor death-inducing signaling complex after traumatic brain injury in mice and humans. J Neurosci 22(9):3504–3511
7. Bermpohl D, You Z, Korsmeyer SJ, Moskowitz MA, Whalen MJ (2006) Traumatic brain injury in mice deficient in Bid: effects on histopathology and functional outcome. J Cereb Blood Flow Metab 26(5):625–633
8. Lewen A, Fujimura M, Sugawara T, Matz P, Copin JC, Chan PH (2001) Oxidative stress-dependent release of mitochondrial cytochrome c after traumatic brain injury. J Cereb Blood Flow Metab 21(8):914–920
9. Tomura S, de Rivero Vaccari JP, Keane RW, Bramlett HM, Dietrich WD (2012) Effects of therapeutic hypothermia on inflammasome signaling after traumatic brain injury. J Cereb Blood Flow Metab 32(10):1939–1947
10. Ji J, Kline AE, Amoscato A, Samhan-Arias AK, Sparvero LJ, Tyurin VA et al (2012) Lipidomics identifies cardiolipin oxidation as a mitochondrial target for redox therapy of brain injury. Nat Neurosci 15(10):1407–1413
11. You Z, Savitz SI, Yang J, Degterev A, Yuan J, Cuny GD et al (2008) Necrostatin-1 reduces histopathology and improves functional outcome after controlled cortical impact in mice. J Cereb Blood Flow Metab 28(9):1564–1573
12. Diskin T, Tal-Or P, Erlich S, Mizrachy L, Alexandrovich A, Shohami E et al (2005) Closed head injury induces upregulation of Beclin 1 at the cortical site of injury. J Neurotrauma 22 (7):750–762
13. Erlich S, Alexandrovich A, Shohami E, Pinkas-Kramarski R (2007) Rapamycin is a neuroprotective treatment for traumatic brain injury. Neurobiol Dis 26(1):86–93
14. Lee JM, Zipfel GJ, Choi DW (1999) The changing landscape of ischaemic brain injury mechanisms. Nature 399(6738 Suppl):A7–A14
15. Ikonomidou C, Turski L (2002) Why did NMDA receptor antagonists fail clinical trials for stroke and traumatic brain injury? Lancet Neurol 1(6):383–386
16. Morris GF, Bullock R, Marshall SB, Marmarou A, Maas A, Marshall LF (1999) Failure of the competitive N-methyl-D-aspartate antagonist Selfotel (CGS 19755) in the treatment of severe head injury: results of two phase III clinical trials. The Selfotel Investigators. J Neurosurg 91 (5):737–743
17. Engel DC, Mies G, Terpolilli NA, Trabold R, Loch A, De Zeeuw CI et al (2008) Changes of cerebral blood flow during the secondary expansion of a cortical contusion assessed by 14C-iodoantipyrine autoradiography in mice using a non-invasive protocol. J Neurotrauma 25(7):739–753
18. Schroder ML, Muizelaar JP, Fatouros PP, Kuta AJ, Choi SC (1998) Regional cerebral blood volume after severe head injury in patients with regional cerebral ischemia. Neurosurgery 42 (6):1276–1280; discussion 80–1
19. von Oettingen G, Bergholt B, Gyldensted C, Astrup J (2002) Blood flow and ischemia within traumatic cerebral contusions. Neurosurgery 50(4):781–788; discussion 8–90
20. Terpolilli NA, Kim SW, Thal SC, Kuebler WM, Plesnila N (2013) Inhaled nitric oxide reduces secondary brain damage after traumatic brain injury in mice. J Cereb Blood Flow Metab 33(2):311–318
21. Bouma GJ, Muizelaar JP, Stringer WA, Choi SC, Fatouros P, Young HF (1992) Ultra-early evaluation of regional cerebral blood flow in severely head-injured patients using xenon-enhanced computerized tomography. J Neurosurg 77(3):360–368
22. Sahuquillo J, Poca MA, Amoros S (2001) Current aspects of pathophysiology and cell dysfunction after severe head injury. Curr Pharm Des 7(15):1475–1503
23. Nawashiro H, Shima K, Chigasaki H (1995) Selective vulnerability of hippocampal CA3 neurons to hypoxia after mild concussion in the rat. Neurol Res 17(6):455–460

24. Dietrich WD, Alonso O, Halley M (1994) Early microvascular and neuronal consequences of traumatic brain injury: a light and electron microscopic study in rats. J Neurotrauma 11 (3):289–301
25. Anderson KJ, Miller KM, Fugaccia I, Scheff SW (2005) Regional distribution of fluoro-jade B staining in the hippocampus following traumatic brain injury. Exp Neurol 193(1):125–130
26. Geddes DM, LaPlaca MC, Cargill RS II (2003) Susceptibility of hippocampal neurons to mechanically induced injury. Exp Neurol 184(1):420–427
27. Toulmond S, Duval D, Serrano A, Scatton B, Benavides J (1993) Biochemical and histological alterations induced by fluid percussion brain injury in the rat. Brain Res 620(1):24–31
28. Shimamura M, Garcia JM, Prough DS, Hellmich HL (2004) Laser capture microdissection and analysis of amplified antisense RNA from distinct cell populations of the young and aged rat brain: effect of traumatic brain injury on hippocampal gene expression. Brain Res Mol Brain Res 122(1):47–61
29. Igarashi T, Potts MB, Noble-Haeusslein LJ (2007) Injury severity determines Purkinje cell loss and microglial activation in the cerebellum after cortical contusion injury. Exp Neurol 203(1):258–268
30. Bell JD, Ai J, Chen Y, Baker AJ (2007) Mild in vitro trauma induces rapid Glur2 endocytosis, robustly augments calcium permeability and enhances susceptibility to secondary excitotoxic insult in cultured Purkinje cells. Brain 130(Pt 10):2528–2542
31. Reeves TM, Smith TL, Williamson JC, Phillips LL (2012) Unmyelinated axons show selective rostrocaudal pathology in the corpus callosum after traumatic brain injury. J Neuropathol Exp Neurol 71(3):198–210
32. Staal JA, Vickers JC (2011) Selective vulnerability of non-myelinated axons to stretch injury in an in vitro co-culture system. J Neurotrauma 28(5):841–847
33. Guo S, Kim WJ, Lok J, Lee SR, Besancon E, Luo BH et al (2008) Neuroprotection via matrix-trophic coupling between cerebral endothelial cells and neurons. Proc Natl Acad Sci U S A 105(21):7582–7587
34. Baumann N, Pham-Dinh D (2001) Biology of oligodendrocyte and myelin in the mammalian central nervous system. Physiol Rev 81(2):871–927
35. Paus T, Zijdenbos A, Worsley K, Collins DL, Blumenthal J, Giedd JN et al (1999) Structural maturation of neural pathways in children and adolescents: in vivo study. Science 283 (5409):1908–1911
36. Scholz J, Klein MC, Behrens TE, Johansen-Berg H (2009) Training induces changes in white-matter architecture. Nat Neurosci 12(11):1370–1371
37. Juraska JM, Kopcik JR (1988) Sex and environmental influences on the size and ultrastructure of the rat corpus callosum. Brain Res 450(1–2):1–8
38. Levine JM, Reynolds R, Fawcett JW (2001) The oligodendrocyte precursor cell in health and disease. Trends Neurosci 24(1):39–47
39. Franklin RJ, Ffrench-Constant C (2008) Remyelination in the CNS: from biology to therapy. Nat Rev Neurosc 9(11):839–855
40. Emery B (2010) Regulation of oligodendrocyte differentiation and myelination. Science 330 (6005):779–782
41. Lee Y, Morrison BM, Li Y, Lengacher S, Farah MH, Hoffman PN et al (2012) Oligodendroglia metabolically support axons and contribute to neurodegeneration. Nature 487 (7408):443–448
42. Wilkins A, Majed H, Layfield R, Compston A, Chandran S (2003) Oligodendrocytes promote neuronal survival and axonal length by distinct intracellular mechanisms: a novel role for oligodendrocyte-derived glial cell line-derived neurotrophic factor. J Neurosci 23 (12):4967–4974
43. Charles P, Hernandez MP, Stankoff B, Aigrot MS, Colin C, Rougon G et al (2000) Negative regulation of central nervous system myelination by polysialylated-neural cell adhesion molecule. Proc Natl Acad Sci U S A 97(13):7585–7590

44. Mi S, Miller RH, Lee X, Scott ML, Shulag-Morskaya S, Shao Z et al (2005) LINGO-1 negatively regulates myelination by oligodendrocytes. Nat Neurosci 8(6):745–751
45. Wang S, Sdrulla AD, diSibio G, Bush G, Nofziger D, Hicks C et al (1998) Notch receptor activation inhibits oligodendrocyte differentiation. Neuron 21(1):63–75
46. Etxeberria A, Mangin JM, Aguirre A, Gallo V (2010) Adult-born SVZ progenitors receive transient synapses during remyelination in corpus callosum. Nat Neurosci 13(3):287–289
47. Arai K, Lo EH (2009) An oligovascular niche: cerebral endothelial cells promote the survival and proliferation of oligodendrocyte precursor cells. J Neurosci 29(14):4351–4355
48. Arai K, Lo EH (2009) Oligovascular signaling in white matter stroke. Biol Pharm Bull 32 (10):1639–1644
49. Pham LD, Hayakawa K, Seo JH, Nguyen MN, Som AT, Lee BJ et al (2012) Crosstalk between oligodendrocytes and cerebral endothelium contributes to vascular remodeling after white matter injury. Glia 60(6):875–881
50. Bramlett HM, Dietrich WD (2002) Quantitative structural changes in white and gray matter 1 year following traumatic brain injury in rats. Acta Neuropathol 103(6):607–614
51. Flygt J, Djupsjo A, Lenne F, Marklund N (2013) Myelin loss and oligodendrocyte pathology in white matter tracts following traumatic brain injury in the rat. Eur J Neurosci 38:2153–2165
52. Lotocki G, de Rivero VJ, Alonso O, Molano JS, Nixon R, Dietrich WD et al (2011) Oligodendrocyte vulnerability following traumatic brain injury in rats: effect of moderate hypothermia. Ther Hypothermia Temp Manag 1(1):43–51
53. Lotocki G, de Rivero Vaccari JP, Alonso O, Molano JS, Nixon R, Safavi P et al (2011) Oligodendrocyte vulnerability following traumatic brain injury in rats. Neurosci Lett 499 (3):143–148
54. Saatman KE, Duhaime AC, Bullock R, Maas AI, Valadka A, Manley GT et al (2008) Classification of traumatic brain injury for targeted therapies. J Neurotrauma 25(7):719–738
55. Davenport ND, Lim KO, Armstrong MT, Sponheim SR (2012) Diffuse and spatially variable white matter disruptions are associated with blast-related mild traumatic brain injury. Neuroimage 59(3):2017–2024
56. Anderson CV, Bigler ED (1995) Ventricular dilation, cortical atrophy, and neuropsychological outcome following traumatic brain injury. J Neuropsychiatry Clin Neurosci 7(1):42–48
57. Cullum CM, Bigler ED (1986) Ventricle size, cortical atrophy and the relationship with neuropsychological status in closed head injury: a quantitative analysis. J Clin Exp Neuropsychol 8(4):437–452
58. Kinnunen KM, Greenwood R, Powell JH, Leech R, Hawkins PC, Bonnelle V et al (2011) White matter damage and cognitive impairment after traumatic brain injury. Brain 134 (Pt 2):449–463
59. Anderson CV, Bigler ED (1994) The role of caudate nucleus and corpus callosum atrophy in trauma-induced anterior horn dilation. Brain Inj 8(6):565–569
60. Barclay L, Zemcov A, Reichert W, Blass JP (1985) Cerebral blood flow decrements in chronic head injury syndrome. Biol Psychiatry 20(2):146–157
61. Bramlett HM, Dietrich WD, Green EJ, Busto R (1997) Chronic histopathological consequences of fluid-percussion brain injury in rats: effects of post-traumatic hypothermia. Acta Neuropathol 93(2):190–199
62. Terayama Y, Meyer JS, Kawamura J, Weathers S (1991) Role of thalamus and white matter in cognitive outcome after head injury. J Cereb Blood Flow Metab 11(5):852–860
63. Bramlett HM, Dietrich WD (2007) Progressive damage after brain and spinal cord injury: pathomechanisms and treatment strategies. Prog Brain Res 161:125–141
64. Moretti L, Cristofori I, Weaver SM, Chau A, Portelli JN, Grafman J (2012) Cognitive decline in older adults with a history of traumatic brain injury. Lancet Neurol 11(12):1103–1112
65. Palacios EM, Sala-Llonch R, Junque C, Roig T, Tormos JM, Bargallo N et al (2012) White matter integrity related to functional working memory networks in traumatic brain injury. Neurology 78(12):852–860

66. Wang JY, Bakhadirov K, Abdi H, Devous MD Sr, Marquez de la Plata CD, Moore C et al (2011) Longitudinal changes of structural connectivity in traumatic axonal injury. Neurology 77(9):818–826
67. Reider-Groswasser I, Cohen M, Costeff H, Groswasser Z (1993) Late CT findings in brain trauma: relationship to cognitive and behavioral sequelae and to vocational outcome. AJR Am J Roentgenol 160(1):147–152
68. Barres BA (2008) The mystery and magic of glia: a perspective on their roles in health and disease. Neuron 60(3):430–440
69. Christopherson KS, Ullian EM, Stokes CC, Mullowney CE, Hell JW, Agah A et al (2005) Thrombospondins are astrocyte-secreted proteins that promote CNS synaptogenesis. Cell 120 (3):421–433
70. Mauch DH, Nagler K, Schumacher S, Goritz C, Muller EC, Otto A et al (2001) CNS synaptogenesis promoted by glia-derived cholesterol. Science 294(5545):1354–1357
71. Allen NJ, Bennett ML, Foo LC, Wang GX, Chakraborty C, Smith SJ et al (2012) Astrocyte glypicans 4 and 6 promote formation of excitatory synapses via GluA1 AMPA receptors. Nature 486(7403):410–414
72. Pfrieger FW, Barres BA (1997) Synaptic efficacy enhanced by glial cells in vitro. Science 277 (5332):1684–1687
73. Iadecola C, Nedergaard M (2007) Glial regulation of the cerebral microvasculature. Nat Neurosci 10(11):1369–1376
74. Eroglu C, Barres BA (2010) Regulation of synaptic connectivity by glia. Nature 468 (7321):223–231
75. Bush TG, Puvanachandra N, Horner CH, Polito A, Ostenfeld T, Svendsen CN et al (1999) Leukocyte infiltration, neuronal degeneration, and neurite outgrowth after ablation of scar-forming, reactive astrocytes in adult transgenic mice. Neuron 23(2):297–308
76. Argaw AT, Asp L, Zhang J, Navrazhina K, Pham T, Mariani JN et al (2012) Astrocyte-derived VEGF-A drives blood–brain barrier disruption in CNS inflammatory disease. J Clin Invest 122(7):2454–2468
77. Arai K, Lo EH (2010) Astrocytes protect oligodendrocyte precursor cells via MEK/ERK and PI3K/Akt signaling. J Neurosci Res 88(4):758–763
78. Moore CS, Abdullah SL, Brown A, Arulpragasam A, Crocker SJ (2011) How factors secreted from astrocytes impact myelin repair. J Neurosci Res 89(1):13–21
79. Ricci G, Volpi L, Pasquali L, Petrozzi L, Siciliano G (2009) Astrocyte-neuron interactions in neurological disorders. J Biol Phys 35(4):317–336
80. Hayakawa K, Pham LD, Katusic ZS, Arai K, Lo EH (2012) Astrocytic high-mobility group box 1 promotes endothelial progenitor cell-mediated neurovascular remodeling during stroke recovery. Proc Natl Acad Sci U S A 109(19):7505–7510
81. Giaume C, Koulakoff A, Roux L, Holcman D, Rouach N (2010) Astroglial networks: a step further in neuroglial and gliovascular interactions. Nat Rev Neurosci 11(2):87–99
82. Chew SS, Johnson CS, Green CR, Danesh-Meyer HV (2010) Role of connexin43 in central nervous system injury. Exp Neurol 225(2):250–261
83. Cortez SC, McIntosh TK, Noble LJ (1989) Experimental fluid percussion brain injury: vascular disruption and neuronal and glial alterations. Brain Res 482(2):271–282
84. Hill SJ, Barbarese E, McIntosh TK (1996) Regional heterogeneity in the response of astrocytes following traumatic brain injury in the adult rat. J Neuropathol Exp Neurol 55 (12):1221–1229
85. Dietrich WD, Truettner J, Zhao W, Alonso OF, Busto R, Ginsberg MD (1999) Sequential changes in glial fibrillary acidic protein and gene expression following parasagittal fluid-percussion brain injury in rats. J Neurotrauma 16(7):567–581
86. Castejon OJ (1998) Morphological astrocytic changes in complicated human brain trauma. A light and electron microscopic study. Brain Inj 12(5):409–427; discussion 7
87. Baldwin SA, Scheff SW (1996) Intermediate filament change in astrocytes following mild cortical contusion. Glia 16(3):266–275

88. Amaducci L, Forno KI, Eng LF (1981) Glial fibrillary acidic protein in cryogenic lesions of the rat brain. Neurosci Lett 21(1):27–32
89. Pelinka LE, Toegel E, Mauritz W, Redl H (2003) Serum S 100 B: a marker of brain damage in traumatic brain injury with and without multiple trauma. Shock 19(3):195–200
90. Pelinka LE, Kroepfl A, Leixnering M, Buchinger W, Raabe A, Redl H (2004) GFAP versus S100B in serum after traumatic brain injury: relationship to brain damage and outcome. J Neurotrauma 21(11):1553–1561
91. Hayakata T, Shiozaki T, Tasaki O, Ikegawa H, Inoue Y, Toshiyuki F et al (2004) Changes in CSF S100B and cytokine concentrations in early-phase severe traumatic brain injury. Shock 22(2):102–107
92. Pelinka LE, Kroepfl A, Schmidhammer R, Krenn M, Buchinger W, Redl H et al (2004) Glial fibrillary acidic protein in serum after traumatic brain injury and multiple trauma. J Trauma 57(5):1006–1012
93. Zhang D, Hu X, Qian L, O'Callaghan JP, Hong JS (2010) Astrogliosis in CNS pathologies: is there a role for microglia? Mol Neurobiol 41(2–3):232–241
94. Laird MD, Vender JR, Dhandapani KM (2008) Opposing roles for reactive astrocytes following traumatic brain injury. Neurosignals 16(2–3):154–164
95. Rao VL, Baskaya MK, Dogan A, Rothstein JD, Dempsey RJ (1998) Traumatic brain injury down-regulates glial glutamate transporter (GLT-1 and GLAST) proteins in rat brain. J Neurochem 70(5):2020–2027
96. Beschorner R, Dietz K, Schauer N, Mittelbronn M, Schluesener HJ, Trautmann K et al (2007) Expression of EAAT1 reflects a possible neuroprotective function of reactive astrocytes and activated microglia following human traumatic brain injury. Histol Histopathol 22 (5):515–526
97. Wu H, Mahmood A, Lu D, Jiang H, Xiong Y, Zhou D et al (2010) Attenuation of astrogliosis and modulation of endothelial growth factor receptor in lipid rafts by simvastatin after traumatic brain injury. J Neurosurg 113(3):591–597
98. Abbott NJ, Ronnback L, Hansson E (2006) Astrocyte-endothelial interactions at the blood–brain barrier. Nat Rev Neurosci 7(1):41–53
99. Xin H, Li Y, Shen LH, Liu X, Wang X, Zhang J et al (2010) Increasing tPA activity in astrocytes induced by multipotent mesenchymal stromal cells facilitate neurite outgrowth after stroke in the mouse. PLoS One 5(2):e9027
100. Faulkner JR, Herrmann JE, Woo MJ, Tansey KE, Doan NB, Sofroniew MV (2004) Reactive astrocytes protect tissue and preserve function after spinal cord injury. J Neurosci 24 (9):2143–2155
101. Cui W, Allen ND, Skynner M, Gusterson B, Clark AJ (2001) Inducible ablation of astrocytes shows that these cells are required for neuronal survival in the adult brain. Glia 34 (4):272–282
102. Myer DJ, Gurkoff GG, Lee SM, Hovda DA, Sofroniew MV (2006) Essential protective roles of reactive astrocytes in traumatic brain injury. Brain 129(Pt 10):2761–2772
103. Zhang Y, Barres BA (2010) Astrocyte heterogeneity: an underappreciated topic in neurobiology. Curr Opin Neurobiol 20(5):588–594
104. Cahoy JD, Emery B, Kaushal A, Foo LC, Zamanian JL, Christopherson KS et al (2008) A transcriptome database for astrocytes, neurons, and oligodendrocytes: a new resource for understanding brain development and function. J Neurosci 28(1):264–278
105. Takata N, Hirase H (2008) Cortical layer 1 and layer 2/3 astrocytes exhibit distinct calcium dynamics in vivo. PLoS One 3(6):e2525
106. Zamanian JL, Xu L, Foo LC, Nouri N, Zhou L, Giffard RG et al (2012) Genomic analysis of reactive astrogliosis. J Neurosci 32(18):6391–6410
107. Hanisch UK, Kettenmann H (2007) Microglia: active sensor and versatile effector cells in the normal and pathologic brain. Nat Neurosci 10(11):1387–1394
108. Kreutzberg GW (1996) Microglia: a sensor for pathological events in the CNS. Trends Neurosci 19(8):312–318

109. Kettenmann H, Verkhratsky A (2008) Neuroglia: the 150 years after. Trends Neurosci 31 (12):653–659
110. Davalos D, Grutzendler J, Yang G, Kim JV, Zuo Y, Jung S et al (2005) ATP mediates rapid microglial response to local brain injury in vivo. Nat Neurosci 8(6):752–758
111. Nimmerjahn A, Kirchhoff F, Helmchen F (2005) Resting microglial cells are highly dynamic surveillants of brain parenchyma in vivo. Science 308(5726):1314–1318
112. Wake H, Moorhouse AJ, Jinno S, Kohsaka S, Nabekura J (2009) Resting microglia directly monitor the functional state of synapses in vivo and determine the fate of ischemic terminals. J Neurosci 29(13):3974–3980
113. Graeber MB, Streit WJ (2010) Microglia: biology and pathology. Acta Neuropathol 119 (1):89–105
114. Rochefort N, Quenech'du N, Watroba L, Mallat M, Giaume C, Milleret C (2002) Microglia and astrocytes may participate in the shaping of visual callosal projections during postnatal development. J Physiol Paris 96(3–4):183–192
115. Checchin D, Sennlaub F, Levavasseur E, Leduc M, Chemtob S (2006) Potential role of microglia in retinal blood vessel formation. Invest Ophthalmol Vis Sci 47(8):3595–3602
116. Caldero J, Brunet N, Ciutat D, Hereu M, Esquerda JE (2009) Development of microglia in the chick embryo spinal cord: implications in the regulation of motoneuronal survival and death. J Neurosci Res 87(11):2447–2466
117. Ekdahl CT, Kokaia Z, Lindvall O (2009) Brain inflammation and adult neurogenesis: the dual role of microglia. Neuroscience 158(3):1021–1029
118. Battista D, Ferrari CC, Gage FH, Pitossi FJ (2006) Neurogenic niche modulation by activated microglia: transforming growth factor beta increases neurogenesis in the adult dentate gyrus. Eur J Neurosci 23(1):83–93
119. Ziv Y, Ron N, Butovsky O, Landa G, Sudai E, Greenberg N et al (2006) Immune cells contribute to the maintenance of neurogenesis and spatial learning abilities in adulthood. Nat Neurosci 9(2):268–275
120. Thored P, Heldmann U, Gomes-Leal W, Gisler R, Darsalia V, Taneera J et al (2009) Long-term accumulation of microglia with proneurogenic phenotype concomitant with persistent neurogenesis in adult subventricular zone after stroke. Glia 57(8):835–849
121. Ransohoff RM, Perry VH (2009) Microglial physiology: unique stimuli, specialized responses. Annu Rev Immunol 27:119–145
122. Ransohoff RM, Cardona AE (2010) The myeloid cells of the central nervous system parenchyma. Nature 468(7321):253–262
123. Polazzi E, Monti B (2010) Microglia and neuroprotection: from in vitro studies to therapeutic applications. Prog Neurobiol 92(3):293–315
124. Benoit M, Desnues B, Mege JL (2008) Macrophage polarization in bacterial infections. J Immunol 181(6):3733–3739
125. Geissmann F, Auffray C, Palframan R, Wirrig C, Ciocca A, Campisi L et al (2008) Blood monocytes: distinct subsets, how they relate to dendritic cells, and their possible roles in the regulation of T-cell responses. Immunol Cell Biol 86(5):398–408
126. Mantovani A, Sica A, Sozzani S, Allavena P, Vecchi A, Locati M (2004) The chemokine system in diverse forms of macrophage activation and polarization. Trends Immunol 25 (12):677–686
127. Michelucci A, Heurtaux T, Grandbarbe L, Morga E, Heuschling P (2009) Characterization of the microglial phenotype under specific pro-inflammatory and anti-inflammatory conditions: effects of oligomeric and fibrillar amyloid-beta. J Neuroimmunol 210(1–2):3–12
128. Narantuya D, Nagai A, Sheikh AM, Masuda J, Kobayashi S, Yamaguchi S et al (2010) Human microglia transplanted in rat focal ischemia brain induce neuroprotection and behavioral improvement. PLoS One 5(7):e11746
129. Merson TD, Binder MD, Kilpatrick TJ (2010) Role of cytokines as mediators and regulators of microglial activity in inflammatory demyelination of the CNS. Neuromolecular Med 12 (2):99–132

130. Lalancette-Hebert M, Gowing G, Simard A, Weng YC, Kriz J (2007) Selective ablation of proliferating microglial cells exacerbates ischemic injury in the brain. J Neurosci 27 (10):2596–2605
131. Kiefer R, Streit WJ, Toyka KV, Kreutzberg GW, Hartung HP (1995) Transforming growth factor-beta 1: a lesion-associated cytokine of the nervous system. Int J Dev Neurosci 13 (3–4):331–339
132. Streit WJ, Xue QS (2009) Life and death of microglia. J Neuroimmune Pharmacol 4 (4):371–379
133. Harry GJ, McPherson CA, Wine RN, Atkinson K, Lefebvre d'Hellencourt C (2004) Trimethyltin-induced neurogenesis in the murine hippocampus. Neurotox Res 5(8):623–627
134. Streit WJ (2002) Microglia as neuroprotective, immunocompetent cells of the CNS. Glia 40 (2):133–139
135. Walton NM, Sutter BM, Laywell ED, Levkoff LH, Kearns SM, Marshall GP II et al (2006) Microglia instruct subventricular zone neurogenesis. Glia 54(8):815–825
136. Carlson SL, Parrish ME, Springer JE, Doty K, Dossett L (1998) Acute inflammatory response in spinal cord following impact injury. Exp Neurol. [Research Support, Non-U.S. Gov't]. 151 (1):77–88
137. Popovich PG, Wei P, Stokes BT (1997) Cellular inflammatory response after spinal cord injury in Sprague–Dawley and Lewis rats. J Comp Neurol. [Research Support, Non-U.-S. Gov't, P.H.S.]. 377(3):443–464
138. Beck KD, Nguyen HX, Galvan MD, Salazar DL, Woodruff TM, Anderson AJ (2010) Quantitative analysis of cellular inflammation after traumatic spinal cord injury: evidence for a multiphasic inflammatory response in the acute to chronic environment. Brain. [Comparative Study Research Support, N.I.H., Extramural, Research Support, Non-U.S. Gov't, Research Support, U.S. Gov't, Non-P.H.S.]. 133(Pt 2):433–447
139. Csuka E, Hans VH, Ammann E, Trentz O, Kossmann T, Morganti-Kossmann MC (2000) Cell activation and inflammatory response following traumatic axonal injury in the rat. Neuroreport. [Research Support, Non-U.S. Gov't]. 11(11):2587–2590
140. Maeda J, Higuchi M, Inaji M, Ji B, Haneda E, Okauchi T, et al (2007) Phase-dependent roles of reactive microglia and astrocytes in nervous system injury as delineated by imaging of peripheral benzodiazepine receptor. Brain Res. [Research Support, Non-U.S. Gov't]. 1157:100–111
141. Raghavendra Rao VL, Dogan A, Bowen KK, Dempsey RJ (2000) Traumatic brain injury leads to increased expression of peripheral-type benzodiazepine receptors, neuronal death, and activation of astrocytes and microglia in rat thalamus. Exp Neurol. [Research Support, Non-U.S. Gov't, Research Support, U.S. Gov't, P.H.S.]. 161(1):102–114
142. Koshinaga M, Katayama Y, Fukushima M, Oshima H, Suma T, Takahata T (2000) Rapid and widespread microglial activation induced by traumatic brain injury in rat brain slices. J Neurotrauma. [In Vitro Research Support, Non-U.S. Gov't]. 17(3):185–192
143. Holmin S, Mathiesen T (1999) Long-term intracerebral inflammatory response after experimental focal brain injury in rat. Neuroreport. [Research Support, Non-U.S. Gov't]. 10 (9):1889–1891
144. Engel S, Schluesener H, Mittelbronn M, Seid K, Adjodah D, Wehner HD et al (2000) Dynamics of microglial activation after human traumatic brain injury are revealed by delayed expression of macrophage-related proteins MRP8 and MRP14. Acta Neuropathol 100 (3):313–322
145. Beschorner R, Nguyen TD, Gozalan F, Pedal I, Mattern R, Schluesener HJ, et al (2002) CD14 expression by activated parenchymal microglia/macrophages and infiltrating monocytes following human traumatic brain injury. Acta Neuropathol. [Research Support, Non-U.-S. Gov't]. 103(6):541–549
146. Gentleman SM, Leclercq PD, Moyes L, Graham DI, Smith C, Griffin WS, et al (2004) Long-term intracerebral inflammatory response after traumatic brain injury. Forensic Sci Int. [Research Support, N.I.H., Extramural]. 146(2–3):97–104

147. Maxwell WL, MacKinnon MA, Stewart JE, Graham DI (2010) Stereology of cerebral cortex after traumatic brain injury matched to the Glasgow outcome score. Brain. [Comparative Study]. 133(Pt 1):139–160
148. Ramlackhansingh AF, Brooks DJ, Greenwood RJ, Bose SK, Turkheimer FE, Kinnunen KM, et al (2011) Inflammation after trauma: microglial activation and traumatic brain injury. Ann Neurol. [Research Support, Non-U.S. Gov't]. 70(3):374–383
149. Morganti-Kossmann MC, Satgunaseelan L, Bye N, Kossmann T (2007) Modulation of immune response by head injury. Injury. [Research Support, Non-U.S. Gov't Review]. 38 (12):1392–1400
150. Dietrich WD, Chatzipanteli K, Vitarbo E, Wada K, Kinoshita K (2004) The role of inflammatory processes in the pathophysiology and treatment of brain and spinal cord trauma. Acta Neurochir Suppl. [Research Support, U.S. Gov't, P.H.S.]. 89:69–74
151. Morganti-Kossmann MC, Rancan M, Otto VI, Stahel PF, Kossmann T (2001) Role of cerebral inflammation after traumatic brain injury: a revisited concept. Shock. [Research Support, Non-U.S. Gov't Review]. 16(3):165–177
152. Zhang B, West EJ, Van KC, Gurkoff GG, Zhou J, Zhang XM, et al (2008) HDAC inhibitor increases histone H3 acetylation and reduces microglia inflammatory response following traumatic brain injury in rats. Brain Res. [Research Support, N.I.H., Extramural]. 1226:181–191
153. Shechter R, London A, Varol C, Raposo C, Cusimano M, Yovel G, et al (2009) Infiltrating blood-derived macrophages are vital cells playing an anti-inflammatory role in recovery from spinal cord injury in mice. PLoS Med. [Research Support, Non-U.S. Gov't]. 6(7):e1000113
154. Horn KP, Busch SA, Hawthorne AL, van Rooijen N, Silver J (2008) Another barrier to regeneration in the CNS: activated macrophages induce extensive retraction of dystrophic axons through direct physical interactions. J Neurosci. [Research Support, N.I.H., Extramural]. 28(38):9330–9341
155. Urrea C, Castellanos DA, Sagen J, Tsoulfas P, Bramlett HM, Dietrich WD (2007) Widespread cellular proliferation and focal neurogenesis after traumatic brain injury in the rat. Restor Neurol Neurosci. [Research Support, N.I.H., Extramural]. 25(1):65–76
156. Olah M, Ping G, De Haas AH, Brouwer N, Meerlo P, Van Der Zee EA et al (2009) Enhanced hippocampal neurogenesis in the absence of microglia T cell interaction and microglia activation in the murine running wheel model. Glia 57(10):1046–1061
157. Kigerl KA, Gensel JC, Ankeny DP, Alexander JK, Donnelly DJ, Popovich PG (2009) Identification of two distinct macrophage subsets with divergent effects causing either neurotoxicity or regeneration in the injured mouse spinal cord. J Neurosci. [Research Support, N.I.H., Extramural, Research Support, Non-U.S. Gov't]. 29(43):13435–13444
158. Lenzlinger PM, Marx A, Trentz O, Kossmann T, Morganti-Kossmann MC (2002) Prolonged intrathecal release of soluble Fas following severe traumatic brain injury in humans. J Neuroimmunol 122(1–2):167–174
159. Morganti-Kossmann MC, Rancan M, Stahel PF, Kossmann T (2002) Inflammatory response in acute traumatic brain injury: a double-edged sword. Curr Opin Crit Care 8(2):101–105
160. Hong Y, Yan W, Chen S, Sun CR, Zhang JM (2010) The role of Nrf2 signaling in the regulation of antioxidants and detoxifying enzymes after traumatic brain injury in rats and mice. Acta Pharmacol Sin 31(11):1421–1430
161. Armin SS, Colohan AR, Zhang JH (2006) Traumatic subarachnoid hemorrhage: our current understanding and its evolution over the past half century. Neurol Res 28(4):445–452
162. Armin SS, Colohan AR, Zhang JH (2008) Vasospasm in traumatic brain injury. Acta Neurochir Suppl 104(13):421–425
163. Schwarzmaier SM, Kim SW, Trabold R, Plesnila N (2010) Temporal profile of thrombogenesis in the cerebral microcirculation after traumatic brain injury in mice. J Neurotrauma 27(1):121–130
164. Zink BJ, Szmydynger-Chodobska J, Chodobski A (2010) Emerging concepts in the pathophysiology of traumatic brain injury. Psychiatr Clin North Am 33(4):741–756

165. Chodobski A, Zink BJ, Szmydynger-Chodobska J (2011) Blood–brain barrier pathophysiology in traumatic brain injury. Transl Stroke Res 2(4):492–516
166. Chang EF, Claus CP, Vreman HJ, Wong RJ, Noble-Haeusslein LJ (2005) Heme regulation in traumatic brain injury: relevance to the adult and developing brain. J Cereb Blood Flow Metab 25(11):1401–1417
167. Wagner KR, Sharp FR, Ardizzone TD, Lu A, Clark JF (2003) Heme and iron metabolism: role in cerebral hemorrhage. J Cereb Blood Flow Metab 23(6):629–652
168. Aoyama N, Lee SM, Moro N, Hovda DA, Sutton RL (2008) Duration of ATP reduction affects extent of CA1 cell death in rat models of fluid percussion injury combined with secondary ischemia. Brain Res 1230:310–319
169. Kurland D, Hong C, Aarabi B, Gerzanich V, Simard JM (2012) Hemorrhagic progression of a contusion after traumatic brain injury: a review. J Neurotrauma 29(1):19–31
170. Whalen MJ, Carlos TM, Dixon CE, Robichaud P, Clark RS, Marion DW et al (2000) Reduced brain edema after traumatic brain injury in mice deficient in P-selectin and intercellular adhesion molecule-1. J Leukoc Biol 67(2):160–168
171. Whalen MJ, Carlos TM, Dixon CE, Schiding JK, Clark RS, Baum E et al (1999) Effect of traumatic brain injury in mice deficient in intercellular adhesion molecule-1: assessment of histopathologic and functional outcome. J Neurotrauma 16(4):299–309
172. McKeating EG, Andrews PJ, Mascia L (1998) Leukocyte adhesion molecule profiles and outcome after traumatic brain injury. Acta Neurochir Suppl 71:200–202
173. Hartl R, Medary MB, Ruge M, Arfors KE, Ghajar J (1997) Early white blood cell dynamics after traumatic brain injury: effects on the cerebral microcirculation. J Cereb Blood Flow Metab 17(11):1210–1220
174. Schoettle RJ, Kochanek PM, Magargee MJ, Uhl MW, Nemoto EM (1990) Early polymorphonuclear leukocyte accumulation correlates with the development of posttraumatic cerebral edema in rats. J Neurotrauma 7(4):207–217
175. Szmydynger-Chodobska J, Strazielle N, Zink BJ, Ghersi-Egea JF, Chodobski A (2009) The role of the choroid plexus in neutrophil invasion after traumatic brain injury. J Cereb Blood Flow Metab 29(9):1503–1516
176. Butt AM, Jones HC, Abbott NJ (1990) Electrical resistance across the blood–brain barrier in anaesthetized rats: a developmental study. J Physiol 429:47–62
177. Abbott NJ, Patabendige AA, Dolman DE, Yusof SR, Begley DJ (2010) Structure and function of the blood–brain barrier. Neurobiol Dis 37(1):13–25
178. Zlokovic BV (2008) The blood–brain barrier in health and chronic neurodegenerative disorders. Neuron 57(2):178–201
179. Shlosberg D, Benifla M, Kaufer D, Friedman A (2010) Blood–brain barrier breakdown as a therapeutic target in traumatic brain injury. Nat Rev Neurol 6(7):393–403
180. Rosenberg GA (2012) Neurological diseases in relation to the blood–brain barrier. J Cereb Blood Flow Metab 32(7):1139–1151
181. Tomkins O, Feintuch A, Benifla M, Cohen A, Friedman A, Shelef I (2011) Blood–brain barrier breakdown following traumatic brain injury: a possible role in posttraumatic epilepsy. Cardiovasc Psychiatry Neurol 2011:765923
182. Pun PB, Lu J, Moochhala S (2009) Involvement of ROS in BBB dysfunction. Free Radic Res 43(4):348–364
183. Adelson PD, Whalen MJ, Kochanek PM, Robichaud P, Carlos TM (1998) Blood brain barrier permeability and acute inflammation in two models of traumatic brain injury in the immature rat: a preliminary report. Acta Neurochir Suppl 71:104–106
184. Baskaya MK, Rao AM, Dogan A, Donaldson D, Dempsey RJ (1997) The biphasic opening of the blood–brain barrier in the cortex and hippocampus after traumatic brain injury in rats. Neurosci Lett 226(1):33–36
185. Shapira Y, Setton D, Artru AA, Shohami E (1993) Blood–brain barrier permeability, cerebral edema, and neurologic function after closed head injury in rats. Anesth Analg 77(1):141–148

186. Aihara N, Hall JJ, Pitts LH, Fukuda K, Noble LJ (1995) Altered immunoexpression of microglia and macrophages after mild head injury. J Neurotrauma 12(1):53–63
187. Stamatovic SM, Dimitrijevic OB, Keep RF, Andjelkovic AV (2006) Inflammation and brain edema: new insights into the role of chemokines and their receptors. Acta Neurochir Suppl 96:444–450
188. Seiffert E, Dreier JP, Ivens S, Bechmann I, Tomkins O, Heinemann U et al (2004) Lasting blood–brain barrier disruption induces epileptic focus in the rat somatosensory cortex. J Neurosci 24(36):7829–7836
189. Suehiro E, Fujisawa H, Akimura T, Ishihara H, Kajiwara K, Kato S et al (2004) Increased matrix metalloproteinase-9 in blood in association with activation of interleukin-6 after traumatic brain injury: influence of hypothermic therapy. J Neurotrauma 21(12):1706–1711
190. Rigor RR, Beard RS Jr, Litovka OP, Yuan SY (2012) Interleukin-1beta-induced barrier dysfunction is signaled through PKC-theta in human brain microvascular endothelium. Am J Physiol Cell Physiol 302(10):C1513–C1522
191. Puhlmann M, Weinreich DM, Farma JM, Carroll NM, Turner EM, Alexander HR Jr (2005) Interleukin-1beta induced vascular permeability is dependent on induction of endothelial tissue factor (TF) activity. J Transl Med 3:37
192. Aslam M, Schluter KD, Rohrbach S, Rafiq A, Nazli S, Piper HM et al (2013) Hypoxia-reoxygenation-induced endothelial barrier failure: role of RhoA, Rac1 and myosin light chain kinase. J Physiol 591(Pt 2):461–473
193. Argaw AT, Gurfein BT, Zhang Y, Zameer A, John GR (2009) VEGF-mediated disruption of endothelial CLN-5 promotes blood–brain barrier breakdown. Proc Natl Acad Sci U S A 106 (6):1977–1982
194. Mori T, Wang X, Kline AE, Siao CJ, Dixon CE, Tsirka SE et al (2001) Reduced cortical injury and edema in tissue plasminogen activator knockout mice after brain trauma. Neuroreport 12(18):4117–4120
195. Simard JM, Kahle KT, Gerzanich V (2010) Molecular mechanisms of microvascular failure in central nervous system injury—synergistic roles of NKCC1 and SUR1/TRPM4. J Neurosurg 113(3):622–629
196. Foroutan S, Brillault J, Forbush B, O'Donnell ME (2005) Moderate-to-severe ischemic conditions increase activity and phosphorylation of the cerebral microvascular endothelial cell Na+–K+–Cl- cotransporter. Am J Physiol Cell Physiol 289(6):C1492–C1501
197. O'Donnell ME, Lam TI, Tran L, Anderson SE (2004) The role of the blood–brain barrier Na-K-2Cl cotransporter in stroke. Adv Exp Med Biol 559:67–75
198. O'Donnell ME, Lam TI, Tran LQ, Foroutan S, Anderson SE (2006) Estradiol reduces activity of the blood–brain barrier Na-K-Cl cotransporter and decreases edema formation in permanent middle cerebral artery occlusion. J Cereb Blood Flow Metab 26(10):1234–1249
199. O'Donnell ME, Tran L, Lam TI, Liu XB, Anderson SE (2004) Bumetanide inhibition of the blood–brain barrier Na-K-Cl cotransporter reduces edema formation in the rat middle cerebral artery occlusion model of stroke. J Cereb Blood Flow Metab 24(9):1046–1056
200. Simard JM, Geng Z, Woo SK, Ivanova S, Tosun C, Melnichenko L et al (2009) Glibenclamide reduces inflammation, vasogenic edema, and caspase-3 activation after subarachnoid hemorrhage. J Cereb Blood Flow Metab 29(2):317–330
201. Simard JM, Tsymbalyuk O, Ivanov A, Ivanova S, Bhatta S, Geng Z et al (2007) Endothelial sulfonylurea receptor 1-regulated NC Ca-ATP channels mediate progressive hemorrhagic necrosis following spinal cord injury. J Clin Invest 117(8):2105–2113
202. Simard JM, Chen M, Tarasov KV, Bhatta S, Ivanova S, Melnitchenko L et al (2006) Newly expressed SUR1-regulated NC(Ca-ATP) channel mediates cerebral edema after ischemic stroke. Nat Med 12(4):433–440
203. Chen M, Dong Y, Simard JM (2003) Functional coupling between sulfonylurea receptor type 1 and a nonselective cation channel in reactive astrocytes from adult rat brain. J Neurosci 23 (24):8568–8577

204. Kunte H, Schmidt S, Eliasziw M, del Zoppo GJ, Simard JM, Masuhr F et al (2007) Sulfonylureas improve outcome in patients with type 2 diabetes and acute ischemic stroke. Stroke 38(9):2526–2530
205. Fukuda AM, Badaut J (2012) Aquaporin 4: a player in cerebral edema and neuroinflammation. J Neuroinflammation 9:279
206. Sun MC, Honey CR, Berk C, Wong NL, Tsui JK (2003) Regulation of aquaporin-4 in a traumatic brain injury model in rats. J Neurosurg 98(3):565–569
207. Kiening KL, van Landeghem FK, Schreiber S, Thomale UW, von Deimling A, Unterberg AW et al (2002) Decreased hemispheric Aquaporin-4 is linked to evolving brain edema following controlled cortical impact injury in rats. Neurosci Lett 324(2):105–108
208. Manley GT, Binder DK, Papadopoulos MC, Verkman AS (2004) New insights into water transport and edema in the central nervous system from phenotype analysis of aquaporin-4 null mice. Neuroscience 129(4):983–991
209. Higashida T, Kreipke CW, Rafols JA, Peng C, Schafer S, Schafer P et al (2011) The role of hypoxia-inducible factor-1alpha, aquaporin-4, and matrix metalloproteinase-9 in blood–brain barrier disruption and brain edema after traumatic brain injury. J Neurosurg 114(1):92–101
210. Ren Z, Iliff JJ, Yang L, Yang J, Chen X, Chen MJ et al (2013) 'Hit & Run' model of closed-skull traumatic brain injury (TBI) reveals complex patterns of post-traumatic AQP4 dysregulation. J Cereb Blood Flow Metab 33:834–845
211. Taya K, Marmarou CR, Okuno K, Prieto R, Marmarou A (2010) Effect of secondary insults upon aquaporin-4 water channels following experimental cortical contusion in rats. J Neurotrauma 27(1):229–239
212. Kleindienst A, Dunbar JG, Glisson R, Marmarou A (2013) The role of vasopressin V1A receptors in cytotoxic brain edema formation following brain injury. Acta Neurochir (Wien) 155(1):151–164
213. Pop V, Sorensen DW, Kamper JE, Ajao DO, Murphy MP, Head E et al (2013) Early brain injury alters the blood–brain barrier phenotype in parallel with beta-amyloid and cognitive changes in adulthood. J Cereb Blood Flow Metab 33(2):205–214
214. Alahmadi H, Vachhrajani S, Cusimano MD (2010) The natural history of brain contusion: an analysis of radiological and clinical progression. J Neurosurg 112(5):1139–1145
215. Khoshyomn S, Tranmer BI (2004) Diagnosis and management of pediatric closed head injury. Semin Pediatr Surg 13(2):80–86
216. Simard JM, Kilbourne M, Tsymbalyuk O, Tosun C, Caridi J, Ivanova S et al (2009) Key role of sulfonylurea receptor 1 in progressive secondary hemorrhage after brain contusion. J Neurotrauma 26(12):2257–2267
217. Patel AD, Gerzanich V, Geng Z, Simard JM (2010) Glibenclamide reduces hippocampal injury and preserves rapid spatial learning in a model of traumatic brain injury. J Neuropathol Exp Neurol 69(12):1177–1190
218. Abumiya T, Sasaguri T, Taba Y, Miwa Y, Miyagi M (2002) Shear stress induces expression of vascular endothelial growth factor receptor Flk-1/KDR through the CT-rich Sp1 binding site. Arterioscler Thromb Vasc Biol 22(6):907–913
219. Davis ME, Grumbach IM, Fukai T, Cutchins A, Harrison DG (2004) Shear stress regulates endothelial nitric-oxide synthase promoter activity through nuclear factor kappaB binding. J Biol Chem 279(1):163–168
220. Korenaga R, Yamamoto K, Ohura N, Sokabe T, Kamiya A, Ando J (2001) Sp1-mediated downregulation of P2X4 receptor gene transcription in endothelial cells exposed to shear stress. Am J Physiol Heart Circ Physiol 280(5):H2214–H2221
221. Yun S, Dardik A, Haga M, Yamashita A, Yamaguchi S, Koh Y et al (2002) Transcription factor Sp1 phosphorylation induced by shear stress inhibits membrane type 1-matrix metalloproteinase expression in endothelium. J Biol Chem 277(38):34808–34814
222. Verstraeten SV, Mackenzie GG, Oteiza PI (2010) The plasma membrane plays a central role in cells response to mechanical stress. Biochim Biophys Acta 1798(9):1739–1749

223. Chen M, Simard JM (2001) Cell swelling and a nonselective cation channel regulated by internal Ca2+ and ATP in native reactive astrocytes from adult rat brain. J Neurosci 21 (17):6512–6521
224. Peppiatt CM, Howarth C, Mobbs P, Attwell D (2006) Bidirectional control of CNS capillary diameter by pericytes. Nature 443(7112):700–704
225. Armulik A, Genove G, Mae M, Nisancioglu MH, Wallgard E, Niaudet C et al (2010) Pericytes regulate the blood–brain barrier. Nature 468(7323):557–561
226. Bell RD, Winkler EA, Sagare AP, Singh I, LaRue B, Deane R et al (2010) Pericytes control key neurovascular functions and neuronal phenotype in the adult brain and during brain aging. Neuron 68(3):409–427
227. Sa-Pereira I, Brites D, Brito MA (2012) Neurovascular unit: a focus on pericytes. Mol Neurobiol 45(2):327–347
228. Sims DE (1986) The pericyte—a review. Tissue Cell 18(2):153–174
229. Mathiisen TM, Lehre KP, Danbolt NC, Ottersen OP (2010) The perivascular astroglial sheath provides a complete covering of the brain microvessels: an electron microscopic 3D reconstruction. Glia 58(9):1094–1103
230. Gaengel K, Genove G, Armulik A, Betsholtz C (2009) Endothelial-mural cell signaling in vascular development and angiogenesis. Arterioscler Thromb Vasc Biol 29(5):630–638
231. Winkler EA, Bell RD, Zlokovic BV (2011) Central nervous system pericytes in health and disease. Nat Neurosci 14(11):1398–1405
232. Darland DC, Massingham LJ, Smith SR, Piek E, Saint-Geniez M, D'Amore PA (2003) Pericyte production of cell-associated VEGF is differentiation-dependent and is associated with endothelial survival. Dev Biol 264(1):275–288
233. Daneman R, Zhou L, Kebede AA, Barres BA (2010) Pericytes are required for blood–brain barrier integrity during embryogenesis. Nature 468(7323):562–566
234. Fisher M (2009) Pericyte signaling in the neurovascular unit. Stroke 40(3 Suppl):S13–S15
235. Dore-Duffy P (2008) Pericytes: pluripotent cells of the blood brain barrier. Curr Pharm Des 14(16):1581–1593
236. Dore-Duffy P, Wang S, Mehedi A, Katyshev V, Cleary K, Tapper A et al (2011) Pericyte-mediated vasoconstriction underlies TBI-induced hypoperfusion. Neurol Res 33(2):176–186
237. Kallakuri S, Kreipke CW, Rossi N, Rafols JA, Petrov T (2007) Spatial alterations in endothelin receptor expression are temporally associated with the altered microcirculation after brain trauma. Neurol Res 29(4):362–368
238. Dore-Duffy P, Owen C, Balabanov R, Murphy S, Beaumont T, Rafols JA (2000) Pericyte migration from the vascular wall in response to traumatic brain injury. Microvasc Res 60 (1):55–69
239. Takata F, Dohgu S, Matsumoto J, Takahashi H, Machida T, Wakigawa T et al (2011) Brain pericytes among cells constituting the blood–brain barrier are highly sensitive to tumor necrosis factor-alpha, releasing matrix metalloproteinase-9 and migrating in vitro. J Neuroinflammation 8:106
240. Dore-Duffy P, Wang X, Mehedi A, Kreipke CW, Rafols JA (2007) Differential expression of capillary VEGF isoforms following traumatic brain injury. Neurol Res 29(4):395–403
241. DeGracia DJ, Kreipke CW, Kayali FM, Rafols JA (2007) Brain endothelial HSP-70 stress response coincides with endothelial and pericyte death after brain trauma. Neurol Res 29 (4):356–361
242. Castejon OJ (2011) Ultrastructural pathology of cortical capillary pericytes in human traumatic brain oedema. Folia Neuropathol 49(3):162–173
243. Castejon OJ (1984) Submicroscopic changes of cortical capillary pericytes in human perifocal brain edema. J Submicrosc Cytol 16(3):601–618

Chapter 6
The Effects of Intravascular Coagulation and Microthrombosis on Cerebral Perfusion After Brain Trauma

Monisha A. Kumar, Douglas H. Smith, and Sherman C. Stein

Abstract Derangements in coagulation occur frequently after traumatic brain injury (TBI) and are associated with an increased risk of mortality or poor outcome. The coagulopathy after TBI is likely a variant of disseminated intravascular coagulation (DIC), composed of both hypocoagulable and hypercoagulable states with resultant hemorrhagic and thrombotic phenotypes. Much attention has been paid to the hemorrhagic phenotype of intravascular coagulation (IC) due to its association with progression of hemorrhagic injury. However, the coagulopathy after TBI also results in thrombosis, which may be responsible for compromised cerebral perfusion and thromboembolic phenomena as well as progression of injury. This chapter reviews the literature on intravascular coagulation after TBI, examines the effects of intravascular coagulation on microthrombosis, cerebral blood flow (CBF), and progression of injury, and reviews putative pathogenetic mechanisms.

6.1 Introduction

Derangements in coagulation occur frequently after traumatic brain injury (TBI) and are associated with an increased risk of mortality or poor outcome. The coagulopathy after TBI is likely a variant of disseminated intravascular coagulation (DIC), composed of both hypocoagulable and hypercoagulable states with resultant hemorrhagic and thrombotic phenotypes. Much attention has been paid to the hemorrhagic phenotype of intravascular coagulation (IC) due to its association with progression of hemorrhagic injury. However, the coagulopathy after TBI also results in thrombosis, which may be responsible for compromised cerebral perfusion and thromboembolic phenomena as well as progression of injury. This chapter reviews the literature on intravascular coagulation after TBI, examines the

D.H. Smith (✉)
Department of Neurosurgery, Perelman School of Medicine, University of Pennsylvania, 105 Hayden Hall/3320 Smith Walk, Philadelphia, PA 19104, USA
e-mail: smithdou@mail.med.upenn.edu

E.H. Lo et al. (eds.), *Vascular Mechanisms in CNS Trauma*, Springer Series in Translational Stroke Research 5, DOI 10.1007/978-1-4614-8690-9_6,

effects of intravascular coagulation on microthrombosis, cerebral blood flow (CBF), and progression of injury, and reviews putative pathogenetic mechanisms.

6.2 Intravascular Coagulation in TBI

Abnormalities of coagulation are commonly observed after TBI. The reported incidence varies widely in the literature, from 10 to 87.5 % [1–17]. The wide range primarily represents the lack of a standard definition of coagulopathy. Furthermore, studies have enrolled diverse patient populations, sampled blood at different time points, and used various coagulation assays; these factors have contributed to the diversity in reported incidence. Even when the same coagulation marker is studied, different cutoff values or sensitivity levels are employed, thereby limiting generalizability. A meta-analysis of 34 studies reported an overall incidence of coagulopathy of 32.7 % in patients with mostly moderate–severe TBI; the odds ratio of mortality was 9.0 [95 % CI (7.3, 11.6)] and of poor outcome was 36.3 [95 % CI (18.7, 70.5)] [14]. Despite the range in incidence, it is clear that the presence of coagulopathy after TBI significantly increases mortality [18–22] and worsens outcome [12, 23].

The association between IC and TBI was proposed in a study published in the New England Journal of Medicine in 1974 [24], although prior case reports had suggested it [25, 26]. Goodnight et al. compared hemostatic markers in two groups of head injured patients, one with evidence of parenchymal damage and the other without (e.g., skull fractures) [24]. Rates of defibrination, defined as hypofibrinogenemia, elevated fibrin degradation products, and low levels of coagulation factors or platelets, and DIC, defined as a positive protamine sulfate test, were assessed. Nine of 13 patients with parenchymal brain injury demonstrated defibrination, whereas none of the patients without gross evidence of brain injury did. Furthermore, 12/13 patients with brain tissue damage had evidence of DIC, whereas only 3/13 patients with skull fractures did. The patients with brain injury and resultant coagulopathy fared poorly; therefore, the authors advocated early empiric blood component replacement, including cryoprecipitate, FFP, and platelets. Although this was the first study to link intravascular coagulation to poor outcome after brain injury, this study had significant limitations. A small sample size, disparate levels of injury severity, and need for surgical intervention confounded the effect of DIC on outcome. Interestingly, the coagulation abnormalities noted in the study normalized within hours. The transient nature of the observed derangements may possibly explain the varied and incongruent results of prior studies.

Subsequent studies of coagulopathy after TBI have suffered from the lack of a standard definition of DIC. No single clinical sign or laboratory test possesses sufficient diagnostic accuracy to confirm the diagnosis of DIC [27]. The diagnosis of DIC relies on a combination of clinical findings and laboratory tests, including ones for molecular markers of thrombin generation and fibrin turnover [28]. Often, these tests are only available in the research setting and are not routinely available.

Studies addressing DIC after TBI have therefore focused on routine and available tests of coagulation. The most consistently deranged coagulation abnormality is the prothrombin time (PT). The PT represents the time of activation of the extrinsic, or tissue factor, pathway based on the cascade model of hemostasis. The International Mission on Prognosis and Analysis of Clinical Trials in TBI (IMPACT) study, a pooled analysis of six randomized clinical trials of admission biomarkers in TBI patients, demonstrated a positive linear relationship between PT prolongation and mortality [18].

Other indices of coagulation, namely, the PTT and platelet count, are less often deranged after TBI. Prolongation of the PTT is less commonly observed than PT prolongation [3, 5, 16]; however, when present, it may correlate more strongly with mortality [29, 30]. Thrombocytopenia complicates TBI less frequently than either PT or PTT elevation [5, 13, 16, 18]. Older studies with small cohorts of patients reported a normal platelet count on admission, with a nadir occurring at 48–72 h, followed by normalization or thrombocytosis at 1 week post-injury [15, 17]. The IMPACT study demonstrated a significant inverse linear relationship between platelet count and mortality; thrombocytopenia was associated with a 35 % increased risk of mortality at 6 months [18].

Although routinely available, serum coagulation tests, such as the PT, PTT, and platelet count, demonstrate poor sensitivity to the observed derangements in coagulation observed after TBI when compared to other assessments of coagulation [15, 30]. Furthermore, these routine coagulation tests are not sensitive to hypercoagulability [17]. Assessment of hypercoagulability and fibrinolysis using the D-dimer, fibrin degradation products [4, 19, 31–33], soluble fibrin levels [17], prothrombin fragment 1 + 2 [8], antithrombin [17], and thrombin–antithrombin complex [8] have provided useful information regarding coagulopathy after TBI.

In 2001, the International Society of Thrombosis and Haemostasis (ISTH) created a scoring system for overt DIC which was later validated in critically ill patients [34, 35]. The score is calculated from a five-step diagnostic algorithm. Diagnosis of DIC requires a *conditio sine qua non* or presence of an underlying disorder known to be associated with DIC, of which neurotrauma is one. In 2011, Sun et al. performed a prospective study of 242 TBI patients using this definition to accurately determine the incidence of DIC and to determine whether DIC was associated with functional outcome, defined by 3-month GOS [36]. They diagnosed overt DIC in one-third of TBI patients; this was evident within 6 h of injury. Higher DIC scores correlated with increased progression of hemorrhagic injury, higher mortality, and longer ICU and hospital stays.

Although the exact mechanisms that account for intravascular coagulation after TBI have not been fully elucidated, IC is generally attributed to the high concentration of tissue factor found in the central nervous system [37–40]. Tissue factor is an integral transmembrane protein found abundantly in the central nervous system. The extracellular domain contains two fibronectin type III domains with two potential disulfide bonds and can form a complex with factor VIIa. The tissue factor/FVIIa complex initiates the extrinsic pathway of the coagulation system.

Tissue factor is not uniformly distributed in the body. High levels are found in highly vascularized organs such as the brain, lung, and placenta [37]. Tissue factor has been identified in the adventitia of cerebral blood vessels as well as in astrocytes, where it is expressed constitutively [37]. However, TF is not normally exposed to circulating blood as it contained within the subendothelium. Monocytes and endothelial cells only express tissue factor after activation by direct injury or inflammation.

Control systems, including tissue factor pathway inhibitor (TFPI), the protein C system, antithrombin, and glycosaminoglycans are often overwhelmed by the massive release of TF [14]. When severe, this results in deposition of thrombi in the microvasculature, consumption of clotting factors, and activation of fibrinolysis, with both bleeding and thrombosis potentially occurring [41–43]. These derangements in hemostasis can lead to necrosis and hemorrhage in various organs resulting in multiorgan failure and delayed ICH [44–47]. The degree of brain injury appears to correlate with the severity of the coagulopathy [5, 48–51].

Whether intravascular coagulation after head injury is similar to other forms of DIC remains unknown. Coagulopathy, using a simple but unvalidated DIC scoring system, was studied in a cohort of 149 trauma patients with and without brain injury [20]. DIC scores were calculated on the basis of five admission laboratory parameters: the PT, PTT, platelet count, fibrinogen level, and D-dimer. Patients with TBI had significantly higher PT, PTT, and D-dimer levels, and coagulopathy was associated with a ninefold increase in the odds of death among patients with TBI. There was a dose-dependent relationship between DIC score and mortality. However, although two-thirds of brain-injured patients had evidence of DIC and fibrinolysis, only one-third sustained critical deficiencies of coagulation factors. The lack of consumption of coagulation factors is unusual in the classical definition of DIC. Furthermore, the reduction in coagulation factors was not accompanied by a concomitant fall in platelet counts, punctuating the difference between DIC after TBI and other forms of DIC. Also, multiorgan failure from microvascular occlusion, commonly observed in classical DIC, was not observed. An experimental study of TBI demonstrated that urokinase-type plasminogen activator knockout mice fared worse than wild-type mice, suggesting that fibrinolytic activity may be protective after brain injury [52]. The question remains whether the IC after TBI is the same as DIC observed in association with systemic illness.

One of the differences between intravascular coagulation after TBI and systemic DIC may be the fact that it is a localized, and not truly disseminated, process. Higher levels of procoagulant substances are found in jugular venous blood than in central venous blood in patients after TBI. In a study of severe TBI patients, global coagulation variables from blood samples obtained from jugular venous blood were compared to samples taken from the superior vena cava [8]. Patients in the head trauma group ($n = 20$) showed increased concentrations of thrombin–antithrombin complex ($p < 0.025$) and D-dimer ($p < 0.005$) in cerebrovenous and central venous blood, as well as higher concentrations of prothrombin fragment F1 + 2 in cerebrovenous blood ($p < 0.025$) when compared to the control group. The observed cerebrovenous—central venous gradient supports the contention that IC after TBI may be a local, and not widespread, phenomenon.

Coagulopathy is common after TBI and likely results from IC. The diagnosis of IC after TBI has been hampered by the lack of a standard definition of DIC. The transient nature of the IC coupled with a region-specific effect further complicates the diagnosis. Whether the IC observed after TBI is the same as classical definitions of DIC remains to be determined. IC after TBI may represent one end of a spectrum of systemic DIC. Regardless, it seems evident that coagulopathy is associated with worse outcome in this population.

6.2.1 *Microthrombosis*

The etiology of secondary cerebral ischemia after TBI has been attributed to many factors including: intracranial hypertension, cerebral hypoxia, and arterial hypotension [36, 53–55]. In 1979, Graham and Adams demonstrated pathological evidence of cerebral ischemia in over 90 % of patients who died from TBI [56]. However, despite improvements in intensive care management of these secondary injuries, the incidence of autopsy-proven cerebral ischemia has not decreased over recent decades [57]. This suggests a pathophysiological mechanism distinct from those previously mentioned.

Microthrombi are observed in histological samples of human patients [50, 58] and in experimental models [51, 55, 59–62] (Fig. 6.1). Autopsy studies confirm a higher concentration of microthrombi in specimens from patients with TBI than in those from patients with non-neurological demise [58]. Microthrombi are not only identified more frequently in the ipsilateral injured brain but also in the contralateral hemisphere when compared to controls [51, 58] (Fig. 6.2). The prevalence of microthrombi increases days after injury and younger patients have a higher concentration of microthrombi [58]. The timing of microthrombus formation coincides with an increased concentration of fibrin degradation products, suggesting a local hyperfibrinolytic state.

The temporal and spatial distribution of microthrombi may be indicative of primary and secondary injury and may vary by mechanism of injury [51, 61]. Microthrombi are not only found in contused tissue and traumatic penumbra but also in the hippocampus, striatum, and corpus callosum of both the lesioned and contralateral hemispheres in experimental models [51, 61]. In lateral fluid percussion models, microthrombi are found in high concentration in the perilesional tissue, in moderate concentration within the contusion, and diffusely in the ipsilateral hemisphere. In models of diffuse axonal injury (DAI), microthrombi are noted diffusely and in scattered clusters. Although there are a considerable number of microthrombi present 1 h post-trauma, the prevalence increases significantly over the first few days [51, 61].

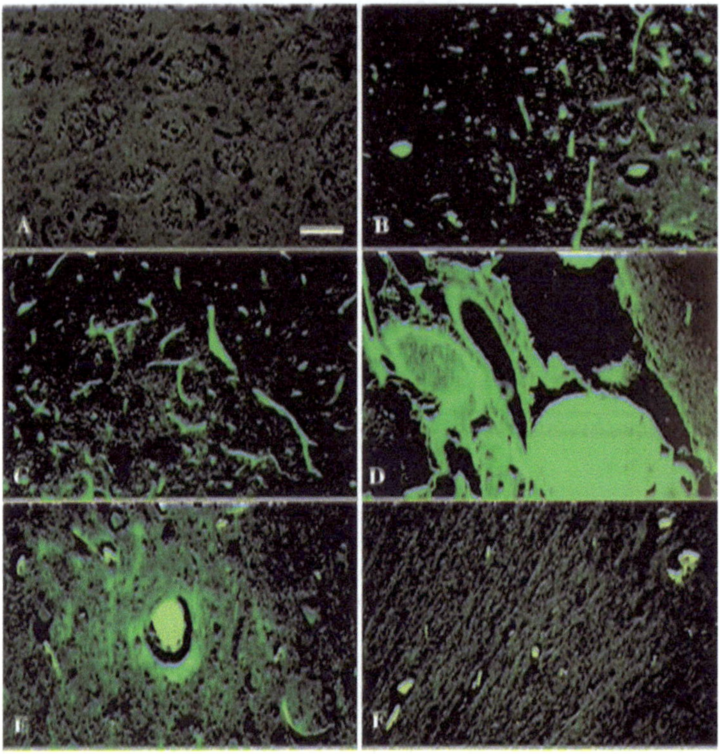

Fig. 6.1 Photomicrographs showing immunofluorescence staining of representative brain slices in laboratory animals. (**a**) Tissue obtained in an uninjured control rat; no fluorescence is seen. (**b**) Tissue obtained in a rat 24 h after lateral FPI; fluorescence shows intravascular thrombi. (**c**, **d**) Tissue obtained in a rat 48 h post-injury (**c**), and in a rat 1 h post-injury (**d**). Large vessels are seen in a cortical sulcus bordering the contusion. (**e**, **f**) Tissue obtained in a pig 6 h after DAI. Bar = 30 μm. Permission from © American Association of Neurological Surgeons

The temporal distribution of microthrombi, however, differs by location. Immediately after injury, many clots are identified in the injured area [51, 61]. Clot formation peaks between 1 h and 1 day and returns to baseline by approximately 2 weeks. The immediate thrombotic response within the lesioned tissue may represent an acute response to hemorrhage incurred by the primary traumatic insult. A large thrombus burden is also identified in the lesion boundary zone and in region CA3 of the hippocampus [61]. However, the appearance of microthrombi in these locations is delayed when compared to those contained within the injured region. The delayed thrombus formation suggests that secondary intravascular coagulation may be occurring in these areas. Outside the perilesional area, microthrombus burden is similar in temporal and spatial relation between the ipsilateral and contralateral hemisphere [61].

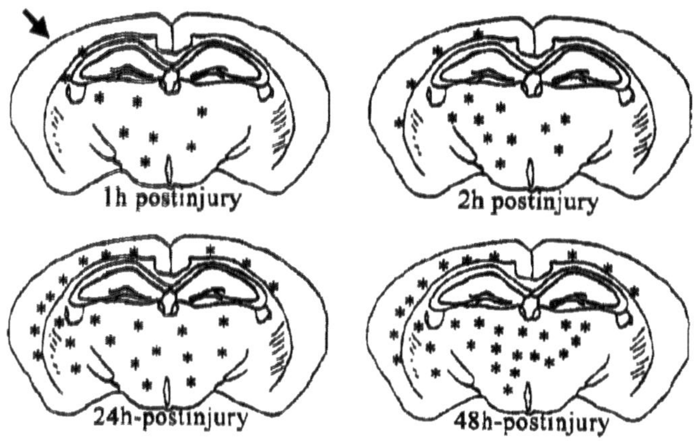

Fig. 6.2 Schematic representation of IC distribution in rats with lateral FPI at various times post-injury. Each *asterisk* represents a cluster of intravascular thrombi. The densities and distributions are averaged for the animals in each time group. *Arrow* designates the region of contusion

In animal pathology studies, microthrombus formation is linked to coagulopathy. Animal models of cortical contusion and DAI demonstrate abundant fibrin thrombi in small- and medium-sized arterioles and venules (10–600 μm) [51]. The number of intravascular microthrombi per square inch correlates linearly with the burden of coagulopathy defined by the number of abnormal serum coagulation tests, including the PT, PTT, and platelet count. These findings serve to correlate coagulation abnormalities to microthrombi deposition, thus serving as a possible causal link between intravascular thrombosis and traumatic coagulopathy.

6.2.2 *Effect of Microthrombosis on CBF*

Although the aforementioned pathology studies link microthrombi to neuronal injury and intravascular coagulation, the mechanism by which this occurs remains unclear. Murine models reveal information on leukocyte–endothelial interactions that may be central to microthrombus formation [63]. Intravital microscopy demonstrates that immediately after TBI, leukocytes start to roll on the venular, but not arteriolar, endothelium. The leukocyte–venular endothelial interactions increase over the first few hours after injury. Although there is no significant adhesion of leukocytes to the vascular endothelium, intravascular aggregates and microthrombi are identified almost exclusively in the injured subjects.

Intravascular aggregates, which are round structures composed of leukocytes and platelets that measure 15–25 μm in size, are observed primarily in post-capillary venules. Of note, these aggregates are identified during times of endothelial stress and result from an interaction of P-selectin on activated platelets and its ligand P-selectin glycoprotein ligand-1 (PSGL-1) on leukocytes [64], but have not been identified in other brain injury models, such as cerebral ischemia [65]. Since these aggregates are found predominantly in the venous microcirculation, it is unlikely that they are involved in arteriolar or capillary occlusion in the traumatic penumbra. However, it is possible that they are related to microthrombi identified in the pulmonary circulation after isolated TBI [31].

In contrast to aggregates, microthrombi, defined as oblong structures >25 μm in size, are seen in both arterioles and venules. Thirty minutes after injury, microthrombi are visualized in 15 % of investigated venules and 25 % of investigated arterioles [63]. Although more microthrombi are noted in arterioles at early time points, subsequent time points reveal reversal in the prevalence with microthrombi in up to 77 % of venules and 40 % of arterioles [63]. Microthrombosis results in demonstrable reductions in microcirculatory blood flow velocities, predominantly in the venules but also in the arterioles. Few venules demonstrate compromised flow velocities at 30-min post-TBI; however, by 2 h post-injury, 70 % of venules reveal no flow [63]. Thirty-five percent of arterioles display impaired flow velocities at 30 min post-injury. By 90 min post-injury, all arterioles with any amount of compromised flow demonstrate complete cessation of flow, indicating that 33 % of all arterioles in the traumatic penumbra have microvascular occlusion [63]. These data document that the first step of blood flow reduction after TBI results from platelets, adherent to the endothelial wall, that result in thrombus formation and ultimately, complete luminal occlusion and microvascular stasis. These data support the notion that microthrombi are responsible for compromised perfusion in the traumatic penumbra after TBI.

Microthrombosis in the traumatic penumbra precedes regional blood flow changes and thus may mediate secondary neuronal injury [60]. As mentioned previously, microthrombosis is evident immediately after injury, primarily within the core of the contusion. A number of microthrombi are identified in the pericontusion region at 6 h after injury, although no obvious tissue damage is yet evident [60]. Twenty-four hours after the injury, microthrombosis and edema formation are visualized in the periphery [60]. Regional CBF compromise follows formation of microthrombi, suggesting that secondary ischemic injury after TBI is mediated by microthrombus formation and resultant blood flow reduction. Microvascular thrombosis is an early and major pathological event following contusion that extends beyond the contusion and occurs prior to tissue changes, such as edema formation.

6.2.3 Cerebral Blood Flow

The changes in CBF observed after TBI are complex. After traumatic insult, CBF may be severely decreased or increased, depending on the type and severity of injury, and the timing and location of measurements [63, 66].

Within minutes of traumatic injury, there is no demonstrable perfusion at the core of a contusion [67, 68]. However, changes of CBF inside cortical contusions seem to depend heavily on injury severity. If the impact is strong enough to irreversibly damage the affected tissue immediately, then the low CBF in the core of the contusion likely represents direct tissue damage rather than secondary tissue ischemia [68]. If the initial impact is insufficient to obstruct rCBF, then contused tissue may survive and show normal or even elevated blood flow values [69, 70].

There may also be transient compromise of blood flow globally, not simply in the territory of the contusion. The global reduction in CBF can be quite significant with reduction of up to 50 % in the whole brain [67], but it depends considerably on the severity of injury [71]. Regional CBF compromise likely lasts more than 8 h [71] and is restored to baseline values by 24 h post-injury [72].

Profound reductions in CBF are observed in the peri-contusional region. Within the first few hours of injury, CBF is reduced to ischemic levels in the traumatic penumbra [67, 73]; however, if this region survives, long-term CBF may return to pre-morbid values [74, 75]. Diffuse injury generally leads to less severe reductions in blood flow and occasionally increased blood flow in the period immediately after injury [67, 68].

Perilesional ischemia is often due to systemic hypotension or intracranial hypertension when cerebral perfusion pressure (CPP) is compromised; however, ischemia in the traumatic penumbra also occurs when the CPP is normal. The mechanism of oligemia and ischemia in this latter case has been attributed to disturbances in the microvascular circulation from vasoconstriction, microvascular compression, or microthrombosis [73, 74, 76].

Vasoconstriction and vasocompression are unlikely to be the mechanisms of early reduction in CBF as vessel diameter is often increased, not decreased, after TBI [63]. There is an immediate and sustained vasodilation in the cerebral resistance vessels after injury. Arteriolar dilatation decreases slightly after 30 min, but remains dilated by almost 20 % compared to baseline [63]. Therefore, neither narrowing of the vessel lumen by vasoconstriction nor extrinsic compression by swollen astrocyte foot processes appears to be the primary mechanisms responsible for compromised CBF after TBI.

6.2.4 Secondary Contusion Expansion

Following TBI, many patients deteriorate after hospital admission due to the progression of intracranial lesions [77–79]. Contusions are the most frequent focal abnormalities in TBI patients and are associated with significant morbidity and mortality [80, 81]. Secondary contusion expansion (SCE) may occur from compromised perfusion.

Although the volume of contused tissue correlates strongly with the volume of peri-contusional ischemia, it remains unclear whether microthrombus-related perilesional ischemia is a significant factor in SCE [67]. In a study by Engel et al., ischemic volume defined by microthrombus formation did not correlate with SCE over the first 24 h after injury [67]. Therefore, microthrombus-induced peri-contusional ischemia may not be the only factor responsible for SCE.

6.3 Pathophysiological Mechanisms

Massive tissue factor release has long been heralded as the cause of intravascular coagulation. Recently, other mechanisms have been suggested as the cause of microvascular thrombosis. These include: microparticle upregulation, platelet hyperactivity, and altered protein C homeostasis.

6.3.1 Microparticle Upregulation

The mechanism which tissue integrates factor into the membrane surface of activated platelets remains unclear. Platelets express little, if any, tissue factor [37]. Microparticles (MPs), or anucleoid membrane vesicles, may serve as the conduit for tissue factor. In response to injury, cells of varied lineage, including platelets, erythrocytes, monocytes/macrophages, and endothelial cells, shed membrane fragments through microvesiculation [82, 83] or apoptosis [84–86]. Calcium influx activates the protease calpain, which cleaves cytoskeletal proteins to disrupt the membrane-cytoskeleton assembly and produce MPs [86].

After proinflammatory or apoptotic stimulation, there is an egress of anionic phospholipids, which are typically located on the inner leaflet of the lipid bilayer, to the outer leaflet of the plasma membrane on the cell surface. In contrast to anionic phospholipids, neutral phospholipids such as phosphatidylcholine and sphingomyelin are normally located on the outer leaflet. This results in an imbalance between the internal and external leaflets, with an advantage to the external membrane that results in membrane budding and MP release. Therefore, MPs are portions of bilayered membranes with an antigen distribution indicative of the cell type of origin. MPs are detectable markers of cell damage, even when their cells of origin have been cleared or sequestered.

MPs are enriched in lipid microdomains, which are the site of concentration of tissue factor, sphingolipids, and adhesion molecules. The adhesion molecules may theoretically be responsible for uniting tissue factor with cells expressing its appropriate receptor or ligand. For example, PSGL-1 expressed on monocytes could bring tissue factor to activated platelets and endothelial cells that express P-selectin [87–89]. After infusion into mice before vascular injury, fluorescent-labeled tissue factor-bearing MPs from mouse monocytes have been shown to accumulate along the leading edge of thrombus during formation. No significant accumulation was observed in P-selectin-deficient mice [90].

The anionic phospholipid, phosphatidylserine (PS), confers the significant procoagulant potential of MPs. PS links platelet activation to thrombin generation by providing a catalytic surface on which the enzyme complexes of the coagulation cascade can assemble. This provides an alternate locus for the generation of thrombin. Termed "blood-borne tissue factor" PS-bearing MPs allow an alternate procoagulant surface distinct from injured endothelium [130]. Furthermore, TF-bearing MPs may be recruited to trigger and amplify coagulation at sites of vascular injury [37, 89]. PS also increases the catalytic efficiency of the TF/FVIIa complex. PS is abundantly detected on the surface of MPs from ATP-stimulated microglial cells and injured neurons [91].

Once considered debris, circulating MPs are now felt to be involved in cell–cell cross talk [92]. MPs may fuse with other cells through endocytosis, conferring the ability to initiate coagulation to target cells that do not normally do so, such as neurons. The process of MP–cell fusion may also allow other cellular interactions, such as those between neurons and leukocytes or platelets, to occur [87]. Neurons may produce more tissue factor-bearing MPs as they are highly sensitive to apoptosis. In vitro, the concentration of MPs correlates with degree of apoptosis present in tissue [93].

Platelet-derived MPs are PS enriched and have 50–100-fold higher procoagulant properties than do platelets themselves [94]. Patients with head injury have a higher percentage of platelet-derived MPs at admission [13]. Similarly, the percentage of platelet-derived MPs is higher upon presentation in non-survivors compared to survivors [13, 92].

6.3.2 Platelet Hyperactivity

The composition of microthrombi consists of platelets, von Willebrand factor (vWF), and fibrin. vWF is synthesized by endothelial cells and megakaryocytes and stored in Weibel–Palade bodies and α-granules, respectively [95]. The stored forms of vWF are ultra large (ULVWF) and prothrombotic [96, 97]. These prothrombotic ULVWF multimers are rapidly cleaved by ADAMTS-13 (A Disintegrin and Metalloprotease with ThromboSpondin type 1 repeats) into smaller multimers that are active in hemostasis, but no longer prothrombotic. In the absence of this proteolytic cleavage, ULVWF can form rope-like strings that tether platelets and leukocytes to endothelium [87, 98–102]. ADAMTS-13 is upregulated after experimental spinal cord injury and is expressed by microglia and cultured astrocytes, but not neurons [103]. The secretion of vWF may be significantly increased in patients with severe TBI [104, 105].

vWF is one of the strongest stimuli for adhesion and activation of platelets [63]. This protein facilitates adhesion of platelets to injured endothelium via GP Ib, GP IIb/IIIa, and subendothelial collagen and incites formation of the platelet plug [106]. Activation of adherent platelets leads to upregulation of $\alpha_{IIb}\beta_{III}$ integrins (glycoproteins IIb/IIIa) on the platelet membrane. This causes a conformational change, which allows fibrinogen binding, and thus forms platelet–platelet aggregates, ultimately resulting in thrombus formation [107]. The temporal profile of vWF expression after experimental brain injury matches the profile of delayed thrombus formation [61]. Immunohistochemical staining corroborates vWF involvement in the formation of delayed microthrombi in the traumatic penumbra and CA3 regions of the hippocampus [61].

Platelet hyperactivity from excessive receptor activation on the membrane surface may also promote thrombosis [13, 108]. Platelet factor-4 elevation, consistent with increased activation of platelets, can persist in the systemic circulation for up to 8 days after experimental injury [61].

Brain injury induces endothelial damage and release of platelet activating factor (PAF), which induces further endothelial and tissue injury. PAF is a biologically active phospholipid that participates in diverse inflammatory conditions such as allergic reactions, shock, and ischemia–reperfusion [109]. It was originally identified as a molecule that induced aggregation and serotonin release in rabbit platelets [110, 111]. Various cell types including endothelial cells, platelets, neutrophils, basophils, neurons, and glia release PAF. PAF production increases significantly during post-ischemia–reperfusion [112] and is a potent vasoconstrictor when applied to cerebral arterioles [113]. Additionally, it induces cytotoxicity via blood–brain barrier breakdown, neutrophil adhesion, and thrombosis formation [114, 115]. PAF acts via a G-protein-coupled transmembrane receptor and can modify blood flow and cerebral metabolism [116]. It is released from neural cells during times of cerebral ischemia [112, 117, 118] and tissue hypoxia [119]; PAF levels have been demonstrated to increase 20-fold from baseline in experimental models of spinal cord ischemia [120].

PAF antagonism can improve outcomes in experimental TBI. Microthrombus formation is attenuated and rCBF is preserved when subjects are treated with PAF antagonists [60]. Administration of etizolam immediately after injury significantly reduces the volume of contusional necrosis compared to control subjects [60].

Although PAF is a potent platelet agonist, it should be noted that the mechanism by which it appears to improve outcome after experimental TBI may not be through platelet mechanisms. PAF enhances BBB permeability possibly mediated through ICAM-1 upregulation [109]. This may result in edema formation and release of prothrombotic mediators into the systemic circulation [87]. PAF antagonists have been shown to attenuate cytotoxic edema and post-ischemic hyperperfusion [121, 122]. Whether the benefit conferred by PAF antagonists is due to mitigation of direct neurotoxicity or microvascular thrombosis remains unknown. However, pre-morbid use of aspirin or clopidogrel does not appear to confer neuroprotection after TBI [123].

6.3.3 Activated Protein C

A maladaptive protein C response to trauma-induced shock may result in both hypercoagulability and an increased bleeding tendency [40, 123, 124]. Protein C is a serine protease with two primary functions, namely, coagulation and inflammation. Protein C is activated through a reaction with a four-part structure composed of thrombin, thrombomodulin, and the endothelial protein C receptor (EPRC) [125]. Activated protein C (aPC) causes coagulopathy by directly inactivating factors Va and VIIIa and indirectly by inhibiting plasminogen activator inhibitor-1 (PAI-1).

Normally, traumatic injury leads to extrinsic pathway activation, resulting in thrombin generation, fibrinogen cleavage, and clot formation. A localized prothrombotic response to trauma seems an appropriate response to repair injured vessels. However, tissue hypoperfusion from trauma-induced shock results instead in an anticoagulant response through expression of thrombomodulin from the vascular endothelium [40, 124]. The thrombin–thrombomodulin complex shunts thrombin away from fibrinogen cleavage and toward protein C activation, which results in inhibition of factors V and VIII [124]. This may represent the mechanism for increased bleeding tendency. Activated protein C also consumes PAI-1, which allows rampant action of tissue plasminogen activator and thus promotes fibrinolysis. Subsequently, the post-traumatic inflammatory response may lead to depletion of activated protein C, causing inhibition of fibrinolysis, thereby promoting a hypercoagulable state. The aPC theory may explain the observed initial hypocoagulable phase and later hypercoagulable phase of TBI-related coagulopathy. In support of this theory, prolongation of the PT has been identified as a risk factor for the subsequent development of venous thromboembolism in TBI patients [126]. Other inflammatory mediators, e.g., cytokines and complement, may also contribute to the pathogenesis of the acute coagulopathy after TBI [127].

6.4 Monitoring

Accurate laboratory assessment of hemostasis is vital to identification, monitoring, and treatment of coagulopathy after TBI. However, there are significant limitations to traditional serum studies in characterizing TBI-AC. The coagulation profile, including the PT/INR and PTT, measures the in vitro clotting time of the extrinsic and intrinsic clotting, but this may not accurately reflect in vivo coagulation. The platelet count provides information about the number of platelets, but does not provide information about platelet function or state of activation. Fibrin degradation assays lack standardization across laboratories and may not be specific to intravascular coagulation, thus limiting their use. Most importantly, the traditional markers of coagulation fail to account for the contributions of cellular elements to coagulation and hemostasis.

A rapid point-of-care test, thromboelastography (TEG), uses whole blood to better account for the effects of platelets, tissue factor-bearing cells, and erythrocytes on coagulopathy. TEG has been proposed as an alternative to traditional laboratory tests of thrombosis and hemostasis in this patient population [30]. This technology has been validated in cardiac [128], liver transplant [129], and trauma surgery [33]. Whole blood is activated by kaolin or tissue factor

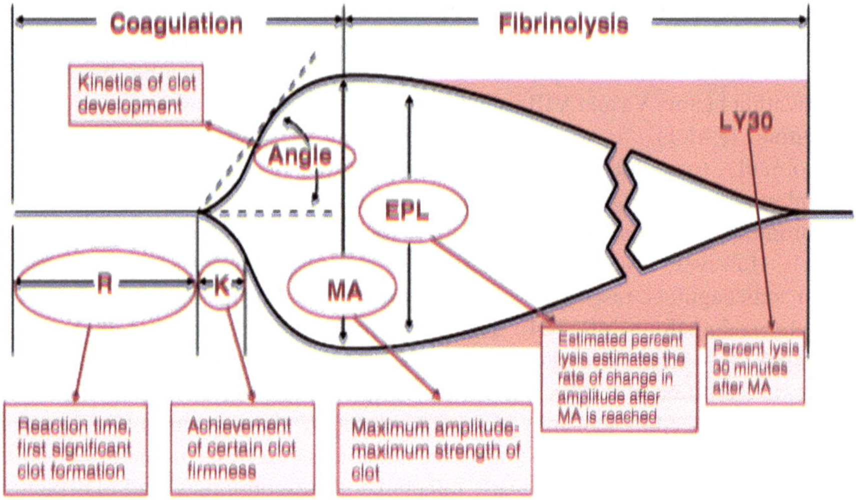

Fig. 6.3 Example of a TEG tracing

resulting in clot formation, and the viscoelastic properties of clot formation are displayed graphically. The tracing reveals objective measurements of clot formation including: (1) the R time, which represents the time to clot formation; (2) the alpha angle, which reflects the rate of fibrin cross-linking and fibrinogen function; and (3) the maximum amplitude (MA), which is a measure of clot strength and thus an indicator platelet function (Fig. 6.3).

In a study of 69 patients with TBI, 9 % presented with evidence of hypocoagulability, based on a prolonged R time indicative of enzymatic dysfunction. Both the mortality rate and ICU length of stay were significantly higher in the subjects with hypocoagulability [30]. Interestingly, prolongation of the R time correlated more strongly with mortality, ICU length of stay, hospital length of stay and better than any traditional serum measure of coagulation, including the PT, PTT, platelet count, or fibrinogen.

A variety of POC testing (POCT) devices are available for platelet function. These devices have been developed primarily to measure the effect of antiplatelet medications and not necessarily to diagnose platelet function defects. The Platelet Function Analyzer 100 (PFA-100, Siemens Corporation, NY) is an optical detection device that measures closure time (CT), R time to the cessation of the blood flow through a channel coated with collagen and a platelet activator, either adenosine diphosphate (ADP) or epinephrine. Only 800 μL of citrated whole blood is required for the analysis.

Platelet function and activation were studied in a cohort of 100 trauma patients, using the platelet function analyzer, the PFA-100 [13]. Although platelet counts were not significantly different between groups, platelet function in the first 24 h was significantly decreased in patients with TBI compared to those without TBI. Thereafter, platelet function was not significantly different between the two groups.

Platelet activation was measured by three different parameters: expression of the activated conformation of GPIIb/IIIA on activated platelets, expression of the P-selectin (CD62P) on the platelet surface, and concentration of platelet-derived microparticles (MPs). All markers of platelet activation were significantly higher at all timepoints (admission, 24-, 48-, and 72-h) in trauma patients compared to controls. The duration of increased platelet activation was longer than the duration of decreased platelet function in those patients with TBI.

6.5 Conclusions

Secondary ischemic injury occurs commonly after TBI. When CPP is preserved, the mechanism of ischemia may be intravascular coagulation. Intravascular coagulation is a common and early form of ischemic injury in experimental models and may predispose to compromised blood flow and tissue injury in the traumatic penumbra. Intravascular coagulation results in microthrombus formation, compromised tissue perfusion, and progression of traumatic injury. DIC has long been thought to be the cause, but microparticle upregulation, platelet hyperactivity, and altered protein C homeostasis may play a significant role. Advanced monitoring may be indicated for accurate diagnosis as standard laboratory tests may not be sensitive to the abnormalities of coagulation. Widespread thrombin generation may have implications beyond hemostasis in TBI; through protease activated receptor (PAR) activity, thrombin may be involved in lowering the threshold of hippocampal neurons for generating epileptic seizures. Further study into the relationship between coagulation, inflammation, and neuronal transmission is sure to yield greater insight and novel therapeutic targets to improve the outcome after TBI.

References

1. Sawada Y et al (1984) Lack of correlation between delayed traumatic intracerebral haematoma and disseminated intravascular coagulation. J Neurol Neurosurg Psychiatry 47 (10):1125–1127
2. Touho H et al (1986) Relationship between abnormalities of coagulation and fibrinolysis and postoperative intracranial hemorrhage in head injury. Neurosurgery 19(4):523–531
3. Cortiana M et al (1986) Coagulation abnormalities in patients with head injury. J Neurosurg Sci 30(3):133–138
4. Olson JD et al (1989) The incidence and significance of hemostatic abnormalities in patients with head injuries. Neurosurgery 24(6):825–832
5. Stein SC et al (1992) Delayed brain injury after head trauma: significance of coagulopathy. Neurosurgery 30(2):160–165
6. Sorensen JV et al (1993) Haemostatic activation in patients with head injury with and without simultaneous multiple trauma. Scand J Clin Lab Invest 53(7):659–665
7. Selladurai BM et al (1997) Coagulopathy in acute head injury—a study of its role as a prognostic indicator. Br J Neurosurg 11(5):398–404

8. Scherer RU, Spangenberg P (1998) Procoagulant activity in patients with isolated severe head trauma. Crit Care Med 26(1):149–156
9. Piek J et al (1992) Extracranial complications of severe head injury. J Neurosurg 77 (6):901–907
10. Murshid WR, Gader AG (2002) The coagulopathy in acute head injury: comparison of cerebral versus peripheral measurements of haemostatic activation markers. Br J Neurosurg 16(4):362–369
11. May AK et al (1997) Coagulopathy in severe closed head injury: is empiric therapy warranted? Am Surg 63(3):233–236, discussion 236–237
12. Kushimoto S et al (2001) Implications of excessive fibrinolysis and alpha(2)-plasmin inhibitor deficiency in patients with severe head injury. Neurosurgery 49(5):1084–1089, discussion 1089–1090
13. Jacoby RC et al (2001) Platelet activation and function after trauma. J Trauma 51(4):639–647
14. Harhangi BS et al (2008) Coagulation disorders after traumatic brain injury. Acta Neurochir (Wien) 150(2):165–175, discussion 175
15. Gando S, Nanzaki S, Kemmotsu O (1999) Coagulofibrinolytic changes after isolated head injury are not different from those in trauma patients without head injury. J Trauma 46 (6):1070–1076, discussion 1076–1077
16. Lustenberger T et al (2010) Time course of coagulopathy in isolated severe traumatic brain injury. Injury 41(9):924–928
17. Bredbacka S, Edner G (1994) Soluble fibrin and D-dimer as detectors of hypercoagulability in patients with isolated brain trauma. J Neurosurg Anesthesiol 6(2):75–82
18. Van Beek JG et al (2007) Prognostic value of admission laboratory parameters in traumatic brain injury: results from the IMPACT study. J Neurotrauma 24(2):315–328
19. Kumura E et al (1987) Coagulation disorders following acute head injury. Acta Neurochir (Wien) 85(1–2):23–28
20. Hulka F, Mullins RJ, Frank EH (1996) Blunt brain injury activates the coagulation process. Arch Surg 131(9):923–927, discussion 927–928
21. Carrick MM et al (2005) Subsequent development of thrombocytopenia and coagulopathy in moderate and severe head injury: support for serial laboratory examination. J Trauma 58 (4):725–729, discussion 729–730
22. Brohi K et al (2003) Acute traumatic coagulopathy. J Trauma 54(6):1127–1130
23. Tan JE et al (2004) Patients who talk and deteriorate: a new look at an old problem. Ann Acad Med Singapore 33(4):489–493
24. Goodnight SH et al (1974) Defibrination after brain-tissue destruction: a serious complication of head injury. N Engl J Med 290(19):1043–1047
25. McGehee WG, Rapaport SI (1968) Systemic hemostatic failure in the severely injured patient. Surg Clin North Am 48(6):1247–1256
26. Druskin MS, Drijansky R (1972) Afibrinogenemia with severe head trauma. JAMA 219 (6):755–756
27. Levi M, de Jonge E, Meijers J (2002) The diagnosis of disseminated intravascular coagulation. Blood Rev 16(4):217–223
28. Boisclair MD, Ireland H, Lane DA (1990) Assessment of hypercoagulable states by measurement of activation fragments and peptides. Blood Rev 4(1):25–40
29. MacLeod JB et al (2003) Early coagulopathy predicts mortality in trauma. J Trauma 55 (1):39–44
30. Kunio NR et al (2012) Thrombelastography-identified coagulopathy is associated with increased morbidity and mortality after traumatic brain injury. Am J Surg 203(5):584–588
31. Kaufmann H, Milkowitz K (1994) [Results of surgical treatment of Stilling-Turk-Duane retraction syndrome]. Klin Monbl Augenheilkd 204(2):90–97
32. Ueda S et al (1985) Correlation between plasma fibrin-fibrinogen degradation product values and CT findings in head injury. J Neurol Neurosurg Psychiatry 48(1):58–60

33. Kaufmann CR et al (1997) Usefulness of thrombelastography in assessment of trauma patient coagulation. J Trauma 42(4):716–720, discussion 720–722
34. Taylor FB Jr et al (2001) Towards definition, clinical and laboratory criteria, and a scoring system for disseminated intravascular coagulation. Thromb Haemost 86(5):1327–1330
35. Voves C, Wuillemin WA, Zeerleder S (2006) International Society on Thrombosis and Haemostasis score for overt disseminated intravascular coagulation predicts organ dysfunction and fatality in sepsis patients. Blood Coagul Fibrinolysis 17(6):445–451
36. Sun Y et al (2011) Validating the incidence of coagulopathy and disseminated intravascular coagulation in patients with traumatic brain injury—analysis of 242 cases. Br J Neurosurg 25 (3):363–368
37. Osterud B, Bjorklid E (2006) Sources of tissue factor. Semin Thromb Hemost 32(1):11–23
38. Lustenberger T et al (2010) Early coagulopathy after isolated severe traumatic brain injury: relationship with hypoperfusion challenged. J Trauma 69(6):1410–1414
39. Halpern CH et al (2008) Traumatic coagulopathy: the effect of brain injury. J Neurotrauma 25 (8):997–1001
40. Cohen MJ et al (2007) Early coagulopathy after traumatic brain injury: the role of hypoperfusion and the protein C pathway. J Trauma 63(6):1254–1261, discussion 1261–1262
41. Bredbacka S et al (1992) Disseminated intravascular coagulation in neurosurgical patients: diagnosis by new laboratory methods. J Neurosurg Anesthesiol 4(2):128–133
42. Kuo JR, Chou TJ, Chio CC (2004) Coagulopathy as a parameter to predict the outcome in head injury patients—analysis of 61 cases. J Clin Neurosci 11(7):710–714
43. Nekludov M et al (2007) Coagulation abnormalities associated with severe isolated traumatic brain injury: cerebral arterio-venous differences in coagulation and inflammatory markers. J Neurotrauma 24(1):174–180
44. Kaufman HH (1984) Delayed traumatic intracerebral hematomas. Neurosurgery 14 (6):784–785
45. Levi M (2005) Pathogenesis and treatment of DIC. Thromb Res 115(Suppl 1):54–55
46. Levi M, Ten Cate H (1999) Disseminated intravascular coagulation. N Engl J Med 341 (8):586–592
47. Roberts HR, Monroe DM III, Hoffman M (2004) Safety profile of recombinant factor VIIa. Semin Hematol 41(1 Suppl 1):101–108
48. Stein SC, Spettell CM (1995) Delayed and progressive brain injury in children and adolescents with head trauma. Pediatr Neurosurg 23(6):299–304
49. Stein SC et al (1993) Delayed and progressive brain injury in closed-head trauma: radiological demonstration. Neurosurgery 32(1):25–30, discussion 30–31
50. Stein SC et al (2004) Association between intravascular microthrombosis and cerebral ischemia in traumatic brain injury. Neurosurgery 54(3):687–691, discussion 691
51. Stein SC et al (2002) Intravascular coagulation: a major secondary insult in nonfatal traumatic brain injury. J Neurosurg 97(6):1373–1377
52. Morales D et al (2006) Impaired fibrinolysis and traumatic brain injury in mice. J Neurotrauma 23(6):976–984
53. Arvigo F et al (1985) Cerebral blood flow in minor cerebral contusion. Surg Neurol 24 (2):211–217
54. Dickman CA et al (1991) Continuous regional cerebral blood flow monitoring in acute craniocerebral trauma. Neurosurgery 28(3):467–472
55. Hekmatpanah J, Hekmatpanah CR (1985) Microvascular alterations following cerebral contusion in rats. Light, scanning, and electron microscope study. J Neurosurg 62(6):888–897
56. Graham DI, Adams JH, Doyle D (1978) Ischaemic brain damage in fatal non-missile head injuries. J Neurol Sci 39(2–3):213–234
57. Graham DI et al (1989) Ischaemic brain damage is still common in fatal non-missile head injury. J Neurol Neurosurg Psychiatry 52(3):346–350
58. Lafuente JV, Cervos-Navarro J (1999) Craniocerebral trauma induces hemorheological disturbances. J Neurotrauma 16(5):425–430

59. van der Sande JJ, Emeis JJ, Lindeman J (1981) Intravascular coagulation: a common phenomenon in minor experimental head injury. J Neurosurg 54(1):21–25
60. Maeda T et al (1997) Hemodynamic depression and microthrombosis in the peripheral areas of cortical contusion in the rat: role of platelet activating factor. Acta Neurochir Suppl 70:102–105
61. Lu D et al (2004) Delayed thrombosis after traumatic brain injury in rats. J Neurotrauma 21 (12):1756–1766
62. Dietrich WD et al (1994) Photothrombotic infarction triggers multiple episodes of cortical spreading depression in distant brain regions. J Cereb Blood Flow Metab 14(1):20–28
63. Schwarzmaier SM et al (2010) Temporal profile of thrombogenesis in the cerebral microcirculation after traumatic brain injury in mice. J Neurotrauma 27(1):121–130
64. Lehr HA et al (1994) P-selectin mediates the interaction of circulating leukocytes with platelets and microvascular endothelium in response to oxidized lipoprotein in vivo. Lab Invest 71(3):380–386
65. Kataoka H, Kim SW, Plesnila N (2004) Leukocyte-endothelium interactions during permanent focal cerebral ischemia in mice. J Cereb Blood Flow Metab 24(6):668–676
66. Werner C, Engelhard K (2007) Pathophysiology of traumatic brain injury. Br J Anaesth 99 (1):4–9
67. Engel DC et al (2008) Changes of cerebral blood flow during the secondary expansion of a cortical contusion assessed by 14C-iodoantipyrine autoradiography in mice using a non-invasive protocol. J Neurotrauma 25(7):739–753
68. Zweckberger K et al (2006) Effect of early and delayed decompressive craniectomy on secondary brain damage after controlled cortical impact in mice. J Neurotrauma 23 (7):1083–1093
69. Ginsberg MD et al (1997) Uncoupling of local cerebral glucose metabolism and blood flow after acute fluid-percussion injury in rats. Am J Physiol 272(6 Pt 2):H2859–H2868
70. Dietrich WD et al (1998) Posttraumatic cerebral ischemia after fluid percussion brain injury: an autoradiographic and histopathological study in rats. Neurosurgery 43(3):585–593, discussion 593–594
71. Cherian L et al (1994) Lateral cortical impact injury in rats: cerebrovascular effects of varying depth of cortical deformation and impact velocity. J Neurotrauma 11(5):573–585
72. Kochanek PM et al (1995) Severe controlled cortical impact in rats: assessment of cerebral edema, blood flow, and contusion volume. J Neurotrauma 12(6):1015–1025
73. Bryan RM Jr, Cherian L, Robertson C (1995) Regional cerebral blood flow after controlled cortical impact injury in rats. Anesth Analg 80(4):687–695
74. Plesnila N et al (2003) Relative cerebral blood flow during the secondary expansion of a cortical lesion in rats. Neurosci Lett 345(2):85–88
75. Bouma GJ et al (1991) Cerebral circulation and metabolism after severe traumatic brain injury: the elusive role of ischemia. J Neurosurg 75(5):685–693
76. von Oettingen G et al (2002) Blood flow and ischemia within traumatic cerebral contusions. Neurosurgery 50(4):781–788, discussion 788–790
77. Oertel M et al (2002) Can hyperventilation improve cerebral microcirculation in patients with high ICP? Acta Neurochir Suppl 81:71–72
78. Tian HL et al (2010) D-dimer as a predictor of progressive hemorrhagic injury in patients with traumatic brain injury: analysis of 194 cases. Neurosurg Rev 33(3):359–365, discussion 365–366
79. Servadei F et al (1995) Evolving brain lesions in the first 12 hours after head injury: analysis of 37 comatose patients. Neurosurgery 37(5):899–906, discussion 906–907
80. Maas AI et al (2007) Prognosis and clinical trial design in traumatic brain injury: the IMPACT study. J Neurotrauma 24(2):232–238
81. Leitgeb J et al (2007) Severe traumatic brain injury in Austria V: CT findings and surgical management. Wien Klin Wochenschr 119(1–2):56–63

82. Siljander P et al (2001) Platelet adhesion enhances the glycoprotein VI-dependent procoagulant response: involvement of p38 MAP kinase and calpain. Arterioscler Thromb Vasc Biol 21(4):618–627
83. Heemskerk JW et al (1997) Collagen but not fibrinogen surfaces induce bleb formation, exposure of phosphatidylserine, and procoagulant activity of adherent platelets: evidence for regulation by protein tyrosine kinase-dependent Ca2+ responses. Blood 90(7):2615–2625
84. Dale GL, Remenyi G, Friese P (2005) Quantitation of microparticles released from coated-platelets. J Thromb Haemost 3(9):2081–2088
85. Brown SB et al (2000) Constitutive death of platelets leading to scavenger receptor-mediated phagocytosis. A caspase-independent cell clearance program. J Biol Chem 275 (8):5987–5996
86. Fox JE et al (1990) Role of the membrane skeleton in preventing the shedding of procoagulant-rich microvesicles from the platelet plasma membrane. J Cell Biol 111 (2):483–493
87. Zhang J et al (2012) Traumatic brain injury-associated coagulopathy. J Neurotrauma 29 (17):2597–2605
88. Bevilacqua MP et al (1986) Regulation of the fibrinolytic system of cultured human vascular endothelium by interleukin 1. J Clin Invest 78(2):587–591
89. Giesen PL et al (1999) Blood-borne tissue factor: another view of thrombosis. Proc Natl Acad Sci U S A 96(5):2311–2315
90. Falati S et al (2003) Accumulation of tissue factor into developing thrombi in vivo is dependent upon microparticle P-selectin glycoprotein ligand 1 and platelet P-selectin. J Exp Med 197(11):1585–1598
91. Bianco F et al (2005) Astrocyte-derived ATP induces vesicle shedding and IL-1 beta release from microglia. J Immunol 174(11):7268–7277
92. Morel N et al (2008) Generation of procoagulant microparticles in cerebrospinal fluid and peripheral blood after traumatic brain injury. J Trauma 64(3):698–704
93. Aupeix K et al (1997) The significance of shed membrane particles during programmed cell death in vitro, and in vivo, in HIV-1 infection. J Clin Invest 99(7):1546–1554
94. Sinauridze EI et al (2007) Platelet microparticle membranes have 50- to 100-fold higher specific procoagulant activity than activated platelets. Thromb Haemost 97(3):425–434
95. Ruggeri ZM (2003) Von Willebrand factor. Curr Opin Hematol 10(2):142–149
96. Sporn LA, Marder VJ, Wagner DD (1986) Inducible secretion of large, biologically potent von Willebrand factor multimers. Cell 46(2):185–190
97. Brouland JP et al (1999) In vivo regulation of von Willebrand factor synthesis: von Willebrand factor production in endothelial cells after lung transplantation between normal pigs and von Willebrand factor-deficient pigs. Arterioscler Thromb Vasc Biol 19(12):3055–3062
98. Furlan M (1996) Von Willebrand factor: molecular size and functional activity. Ann Hematol 72(6):341–348
99. Tsai HM (1996) Physiologic cleavage of von Willebrand factor by a plasma protease is dependent on its conformation and requires calcium ion. Blood 87(10):4235–4244
100. Levy GG et al (2001) Mutations in a member of the ADAMTS gene family cause thrombotic thrombocytopenic purpura. Nature 413(6855):488–494
101. Dong JF et al (2002) ADAMTS-13 rapidly cleaves newly secreted ultralarge von Willebrand factor multimers on the endothelial surface under flowing conditions. Blood 100 (12):4033–4039
102. Bernardo A et al (2005) Platelets adhered to endothelial cell-bound ultra-large von Willebrand factor strings support leukocyte tethering and rolling under high shear stress. J Thromb Haemost 3(3):562–570
103. Tauchi R et al (2012) ADAMTS-13 is produced by glial cells and upregulated after spinal cord injury. Neurosci Lett 517(1):1–6
104. De Oliveira CO et al (2007) Plasma von Willebrand factor levels correlate with clinical outcome of severe traumatic brain injury. J Neurotrauma 24(8):1331–1338

105. Yokota H et al (2002) Cerebral endothelial injury in severe head injury: the significance of measurements of serum thrombomodulin and the von Willebrand factor. J Neurotrauma 19 (9):1007–1015
106. de Groot PG (2002) The role of von Willebrand factor in platelet function. Semin Thromb Hemost 28(2):133–138
107. Rivera J et al (2009) Platelet receptors and signaling in the dynamics of thrombus formation. Haematologica 94(5):700–711
108. Dietrich GV et al (1996) Platelet function and adrenoceptors during and after induced hypotension using nitroprusside. Anesthesiology 85(6):1334–1340
109. Fang W et al (2011) Platelet activating factor induces blood brain barrier permeability alteration in vitro. J Neuroimmunol 230(1–2):42–47
110. Henson PM (1970) Release of vasoactive amines from rabbit platelets induced by sensitized mononuclear leukocytes and antigen. J Exp Med 131(2):287–306
111. Benveniste J, Henson PM, Cochrane CG (1972) Leukocyte-dependent histamine release from rabbit platelets. The role of IgE, basophils, and a platelet-activating factor. J Exp Med 136 (6):1356–1377
112. Nishida K, Markey SP (1996) Platelet-activating factor in brain regions after transient ischemia in gerbils. Stroke 27(3):514–518, discussion 518–519
113. Armstead WM et al (1988) Platelet activating factor: a potent constrictor of cerebral arterioles in newborn pigs. Circ Res 62(1):1–7
114. Frerichs KU et al (1990) Platelet-activating factor and progressive brain damage following focal brain injury. J Neurosurg 73(2):223–233
115. Kochanek PM et al (1988) Cerebrovascular and cerebrometabolic effects of intracarotid infused platelet-activating factor in rats. J Cereb Blood Flow Metab 8(4):546–551
116. Feuerstein G, Yue TL, Lysko PG (1990) Platelet-activating factor. A putative mediator in central nervous system injury? Stroke 21(11 Suppl):III90–III94
117. Satoh K et al (1992) Increased levels of blood platelet-activating factor (PAF) and PAF-like lipids in patients with ischemic stroke. Acta Neurol Scand 85(2):122–127
118. Pettigrew LC et al (1995) Delayed elevation of platelet activating factor in ischemic hippocampus. Brain Res 691(1–2):243–247
119. Rink C, Khanna S (2011) Significance of brain tissue oxygenation and the arachidonic acid cascade in stroke. Antioxid Redox Signal 14(10):1889–1903
120. Lindsberg PJ et al (1990) Evidence for platelet-activating factor as a novel mediator in experimental stroke in rabbits. Stroke 21(10):1452–1457
121. Faden AI, Tzendzalian PA (1992) Platelet-activating factor antagonists limit glycine changes and behavioral deficits after brain trauma. Am J Physiol 263(4 Pt 2):R909–R914
122. Tokutomi T et al (1994) Effect of platelet-activating factor antagonist on brain injury in rats. Acta Neurochir Suppl (Wien) 60:508–510
123. Wong DK, Lurie F, Wong LL (2008) The effects of clopidogrel on elderly traumatic brain injured patients. J Trauma 65(6):1303–1308
124. Brohi K et al (2007) Acute traumatic coagulopathy: initiated by hypoperfusion: modulated through the protein C pathway? Ann Surg 245(5):812–818
125. Esmon CT (2003) The protein C pathway. Chest 124(3 Suppl):26S–32S
126. Knudson MM et al (1992) Thromboembolism following multiple trauma. J Trauma 32 (1):2–11
127. Laroche M et al (2012) Coagulopathy after traumatic brain injury. Neurosurgery 70 (6):1334–1345
128. Spiess BD (1995) Thromboelastography and cardiopulmonary bypass. Semin Thromb Hemost 21(Suppl 4):27–33
129. McNicol PL et al (1994) Patterns of coagulopathy during liver transplantation: experience with the first 75 cases using thrombelastography. Anaesth Intensive Care 22(6):659–665
130. Cimmino G, Golino P, Badimon JJ (2011) Pathophysiological role of blood-borne tissue factor: should the old paradigm be revistited. Intern Emerg Med 6:29–34

Chapter 7
Barriers to Drug Delivery for Brain Trauma

F. Anthony Willyerd, Philip E. Empey, Patrick M. Kochanek, and Robert S.B. Clark

We can't go over it. We can't go under it. Oh no! We've got to go through it!
—Michael Rosen, *We're Going on a Bear Hunt*
Got to go through it to get to it
—Graham Central Station, *Release Yourself*

Abstract One of the most opposing barriers—both literally and figuratively—challenging the translational advancement of therapeutics targeting brain trauma involves *effective* delivery of potentially neuroprotective agents to the damaged brain. Aspects distinguishing delivery of drugs to the injured brain relative to other tissues include: (1) several unique physical features of the blood–brain barrier (BBB) vs. other blood–tissue barriers; (2) added diffusion distance associated with astrocyte swelling (in addition to interstitial edema) away from the therapeutic target; and (3) membrane transporters at both the BBB and blood–CSF barriers (BCSFB), such as ATP-binding cassette (ABC) transporters, organic anion transporters (OAT), and organic anion transporting peptides (OATP) that have the capacity to move drug substrates out of the brain, actively reducing brain bioavailability. While these "barriers" represent unique challenges to the development of efficacious neuroprotective agents for the treatment of traumatic brain injury (TBI), recent pharmacological advancements provide an optimistic outlook for designing strategies that impact outcome for victims of TBI in the near future.

7.1 Introduction

One of the most opposing barriers—both literally and figuratively—challenging the translational advancement of therapeutics targeting brain trauma involves effective delivery of potentially neuroprotective agents to the damaged brain. Further

R.S.B. Clark (✉)
Department of Pediatric Critical Care Medicine, Safar Center for Resuscitation Research, University of Pittsburgh School of Medicine, 3550 Terrace Street, Pittsburgh, PA15261, USA
e-mail: clarkrs@upmc.edu

E.H. Lo et al. (eds.), *Vascular Mechanisms in CNS Trauma*, Springer Series in Translational Stroke Research 5, DOI 10.1007/978-1-4614-8690-9_7,

challenging is maintenance of therapeutic levels of compounds that do manage to enter the brain (i.e. bioavailability). Aspects distinguishing delivery of drugs to, and maintenance of drug levels in, the injured brain relative to other tissues include: 1) unique physical features of the blood-brain barrier (BBB); 2) added diffusion distance away from the therapeutic target (most often neurons); and 3) membrane transporters at both the BBB and blood-CSF barriers (BCSFB), such as ATP-binding cassette (ABC) transporters, organic anion transporters (OAT), and organic anion transporting peptides (OATP) that have the capacity to move drug substrates out of the brain, actively reducing brain bioavailability and therapeutic concentration. These "barriers" represent unique challenges to the development of neuroprotective agents for the treatment of traumatic brain injury (TBI). This Chapter discusses these unique barriers and recent pharmacological advancements for surmounting them, in order to impact outcome for victims of TBI in the near future.

7.2 "Barriers" to Effective Systemically Administered CNS Drug Delivery

A typical model for drug delivery in the uninjured CNS is depicted in Fig. 7.1. Compartment 1 represents the systemic circulation after administration of drug into the blood stream via oral, mucosal, intravenous (IV), or intramuscular delivery. Drug delivery into compartment 2 (the brain) is driven by diffusion and hydrostatic pressure with the rate and extent of distribution affected by tissue perfusion (cerebral blood flow—CBF) and drug permeability. k_{12} represents the influx rate and in the normal (uninjured) brain, is affected by the size, charge, and lipophilicity of the drug, the unique physical and biochemical properties of the BBB relative to other blood–tissue barriers, and the diffusion distance between the brain-side of the BBB and the therapeutic target (typically neurons). k_{21} represents the efflux rate of drug from the brain back into the systemic circulation, either via diffusion or active membrane transport back across the BBB or from CSF to blood via the BCSFB. k_{21} is affected by drug-substrate specificity for membrane transporters on endothelium, epithelium, ependymal cells, as well as astrocytes and neurons themselves. k_{10} is affected by hepatic metabolism and/or renal excretion (as well as metabolism in other tissues, such as plasma), and like k_{21}, is also impacted by the presence of membrane transporters in the liver, gut, and kidney.

TBI adds a significant layer of complexity upon this simplistic model, shown in Fig. 7.2. For example, the degree of BBB disruption, local CBF to the injured brain, astrocyte swelling, and extracellular edema all further impact k_{12}. Furthermore, little is known regarding (any) dynamic changes of ABC drug transporters (*aka* multidrug resistance [MDR] and multidrug resistance-associated protein [MRP]) and organic anion transporters (e.g., OAT and OATP) after TBI in the CNS—and liver and/or kidney for that matter—which would further impact k_{21} and k_{10}. These complexities are further discussed below.

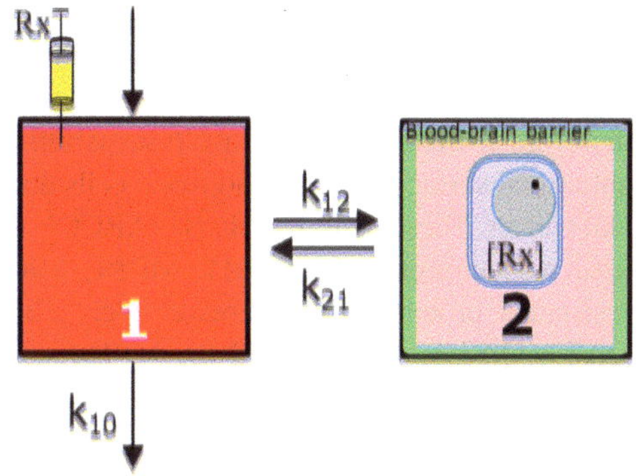

Fig. 7.1 Conventional model for drug delivery in the uninjured CNS. *k* represents rate of drug moving between or from compartment 1 (blood) and compartment 2 (brain)

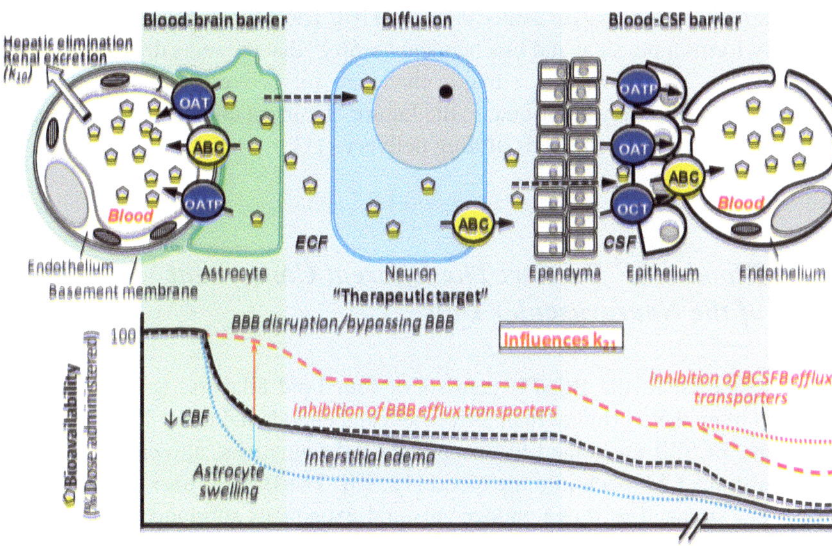

Fig. 7.2 Modified model for drug delivery in the injured CNS. In addition to the barriers to drug delivery in the normal brain, the BBB and BCSFB, factors unique to CNS trauma influence drug influx and efflux, and therefore, bioavailability. Factors influencing rate of drug entering the brain (k_{12}) include BBB disruption and hyperemia (increases); and hypoperfusion, astrocyte swelling, and interstitial edema (decreases). Factors influencing rate of drug exiting the brain (k_{21}) include efflux transporters at the BBB and BCSFB, and bulk flow of interstitial fluid and increased CSF flux (increases); and impaired CSF flow as seen with obstructive hydrocephalus (decreases)

7.2.1 *Blood–Brain Barrier: The Afferent Component of the Neurovascular Unit*

In contrast to endothelial cells in non-CNS tissues, the endothelial cells that form the capillaries in the brain generally do not have gap junctions. Each cell is connected by tight junctions (zona occludens) that are impermeable to hydrophilic substances.

Surrounding these endothelial cells is a basement membrane and a cellular barrier composed of pericytes and astrocyte foot processes [1, 2]. In addition to this mechanical barrier (basement membrane sandwiched between four lipid bilayers), there are a multitude of efflux transporters that are arranged on the luminal and abluminal surfaces of the endothelial cells that form an active and effective biochemical barrier. These include ABCB1, ABCC1, ABCC2, ABCC4, ABCC5, and ABCG2 (*aka* p-glycoprotein or MDR1, MRP1, MRP2, MRP4, MRP5, and breast cancer resistance protein [BRCP], respectively), and organic anion transporters such as OAT3, OATP1, and OATP2 [3, 4]. These membrane transporters rapidly extrude endogenous and exogenous (drug) substrates from the brain interstitium back into the blood [5–9]. As if this were not sufficient, neurons, glia, and pericytes also appear to have an array of transporters on their plasma membranes, which impact drug efflux and bioavailability [6]. The BBB is so effective that even the most lipophilic drugs that are larger than 500 Da are denied entrance [10]. The imposing nature of the BBB essentially forms a physical and biochemical "sentry" that prevents drug distribution into the brain. This represents a major challenge for delivery of neuroprotective agents after TBI and may contribute to the failure of current clinical trials in humans that utilize conventional methods of drug delivery [11].

7.2.2 *Blood–CSF Barrier: The Efferent Component of the Neurovascular Unit*

Much like the BBB, the BCSFB consists of four lipid bilayers, with the BCSFB composed of epithelial (rather than astrocytes) and endothelial cells. The endothelial cells at the BCSFB are joined by incomplete tight junctions, allowing passage of substrates in the CSF back into the blood [12, 13]. The BCSFB also contains many transporters arranged to move endogenous and exogenous substances via the CSF through the ventricular system and back to the systemic circulation at the arachnoid-blood intersection and choroid plexus. The endothelial cells of the choroid plexus maintain a 100- to 150-mL fluid volume that is completely replaced every 4–5 h in the adult human [14]. Thus, CSF flow facilitates convective elimination of drugs and other substances. This contributes to a limited exposure time of drugs that have crossed the BBB, diminishing local drug concentration, working in concert with the BBB to further reduce brain bioavailability.

7.3 Aspects of Drug Delivery Unique to CNS Trauma

Disruption of the BBB following TBI has been well established [15] and appears to be biphasic in nature [16, 17]. While on the surface it seems that trauma may make it easier for delivery of drugs via mechanical disruption of the BBB, evidence from

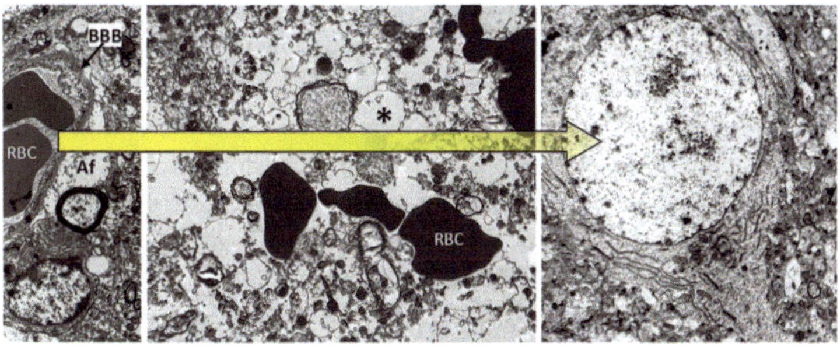

Fig. 7.3 Drug concentration is expected to diminish with distance from the blood vessel to the therapeutic target due to the BBB, astrocyte swelling, and tissue edema. *Af* astrocytes foot process, *BBB* blood–brain barrier, *RBC* red blood cell; * = contusion core with edema and hemorrhage

animal models evaluating therapeutic windows [18] and clinical trials [11], would suggest otherwise. Experimental models of TBI demonstrate that the therapeutic window is typically narrow, measured in minutes rather than hours in most cases [19], suggesting that most of the BBB that is primarily disrupted early after TBI reconstitutes relatively quickly [20], at least in terms of permeability to typical pharmacological agents. Furthermore, the site of damage where BBB disruption occurs may represent irreversibly damaged tissue, thus delivering drug to "dead brain." Additionally, regions of secondary injury, a common therapeutic target after TBI, do not seem to have a proximally disturbed BBB [21]. The duration and location of BBB disruption after TBI make drug delivery and efficacy at the very least unpredictable.

If a drug crosses a disrupted BBB, or a BBB-permeable drug enters into the brain, obstacles must still be overcome before reaching its therapeutic target. The drug must diffuse down a concentration gradient to the target site, often the injured neuron. Studies have shown intercapillary distances to be ~40 μm in the brain [14]. This distance increases with interstitial edema and astrocyte swelling, interstitial hemorrhage, and contusion, all common pathological events after TBI (Fig. 7.3). Further, lipid membrane permeable agents may sequester in cells such as microglia, astrocytes, extravasated erythrocytes from hemorrhage, or infiltrated circulating leukocytes before reaching their therapeutic target. As the concentration of drug likely decreases significantly for each unit distance it must transverse [22], the effective concentration at the target site may be considerably lower in traumatized CNS tissue. It is logical to infer that increasing systemic concentration may overcome these barriers and increased diffusion distance; however, this would come at the expense of increasing systemic toxicity. The capacity for drugs to penetrate traumatically injured myelinated axons, an only recently appreciated aspect of TBI [23], adds yet another layer of complexity. Diffusion distance also plays a meaningful role in treatment of other diseases such as cancer, and disease processes such as abscesses.

Alterations in global and regional CBF after TBI also influence delivery of therapeutics. Hypoperfusion related to primary damage or reduced cerebral perfusion pressure as a consequence of intracranial hypertension or hypotension would reduce diffusion into the brain. The ability of hypoperfusion to limit drug delivery after TBI may be particularly important, given that CBF reductions are usually greatest early after injury—the time when BBB disruption is generally maximal. In contrast, hyperemia would favor diffusion into the brain. Finally, edema at and around the primary injury site can prevent diffusion of drug to the target via increased interstitial pressure and bulk flow of interstitial fluid away from the core of the contusion [24]. The opposite circumstance, impaired CSF flow as can be seen in obstructive hydrocephalus, would be expected to reduce drug efflux to a degree.

7.4 Surmounting the Barriers: Getting Drugs into the CNS after Trauma

Several novel and a few not so novel techniques can be utilized to enhance systemically administered drug delivery to the CNS and improve drug bioavailability after TBI. These techniques include chemical drug modifications to increase BBB permeability, linking drugs ("payload") to endogenous or exogenous carrier molecules, encapsulating the drug in a BBB-penetrating vehicle, inhibiting membrane transporters at the BBB and/or BCSFB using currently available therapeutics, and of course disrupting the BBB concurrently with drug administration or bypassing the BBB altogether by direct administration.

7.4.1 Drug Modifications

Since hydrophilic and many lipophilic drugs do not make it past the BBB [10], pharmaceutical approaches have been used to modify the structure of potentially neuroprotective drugs to enhance delivery into the brain to achieve therapeutic concentrations. These modifications include addition of chemical or biological moieties to the drug to facilitate entry into the brain and, in some cases, localized targeting of the drug, while maintaining its therapeutic efficacy.

One of the simplest modifications involves chemical alterations to increase the lipophilicity of the drug. For example, amidation, esterfication, or pegylation of a drug may increase its lipophilicity and if small enough, should allow entry to the CNS from the blood. One example of this is based on the antioxidant cysteine donor *N*-acetylcysteine (NAC). NAC is currently used clinically to prevent hepatic necrosis after acetaminophen overdose; however, it is notoriously impermeable to the intact BBB. Modification via addition of an ethyl ester and glutamate residue to cysteine results in the BBB-permeable drug, γ-glutamylcysteine ethyl ester (GCEE). GCEE serves as a precursor for replenishing reduced glutathione (GSH) and has been shown to reduce histological damage and improve functional outcome after TBI in mice [25].

Another approach is to utilize a prodrug that is either converted to its (more) active form or has its active site exposed by an enzymatic or chemical reaction (e.g., via hydrolysis or cytochrome P450) once it enters the brain [26]. Other classes of prodrugs include therapeutic agents coupled with fatty acids, phospholipids, or glycerides. An example includes utilizing brain cytochrome P450 enzymes to convert drug compounds to more active metabolites [27]. More sophisticated systems are also being developed, including "lock in systems," that require multiple transformations to release the active drug [28]. These therapeutics use a penetration-enhancing modifier that allows for passage into the CNS and a targetor moiety directing the drug to its target. The shedding of the transport modifier exposes the active drug and locks it into the CNS [29].

7.4.2 *Biologic Modifications*

Another avenue of delivering therapeutics to the CNS from the blood involves taking advantage of biologic systems at the BBB instead of circumventing them. These include prompting transcytosis of a drug and/or using carrier- or receptor-mediated transport to move the drug through the BBB. Several types of proteins can move across the BBB [30]. The best explored proteins include albumin and antibodies. One example reported is the coupling of tumor necrosis factor (TNF) receptor to a monoclonal antibody against the transferrin receptor, to facilitate BBB-penetration, inhibit TNF and improve neurological outcome in a mouse model of Parkinson's disease [31].

Another way to gain access to the CNS is to create a drug that has a similar structure to a nutrient or molecule that has an uptake transporter in the endothelial cell membrane. Then allow the carrier protein to shuttle the drug into the CNS. Many transporters exist for essential amino acids, sugars, minerals, and other necessary molecules. A well known example of a drug that uses this method of drug delivery is dopamine for Parkinson's disease. A large neutral amino acid form, levodopa, is carried across the endothelium via the large neutral amino acid transporter, LNAT [28]. Similarly, one can couple therapeutics to relevant substrates that undergo receptor-mediated transport across the BBB, such as insulin, transferrin, leptin, and acetylated low density lipoprotein [32, 33].

Finally, small peptides (e.g., the protein transduction domain derived from human immunodeficiency virus TAT protein) that cross the BBB can also be utilized to deliver therapeutics to the CNS. An example of this strategy includes linking doxorubicin with a small peptide vector and found increased CNS penetration [34]. This method is also being utilized in neurodiagnostic studies [14].

7.4.3 "Payload" Delivery Systems

The "classic" payload delivery method involves encapsulation of the therapeutic in liposomes to form a vesicle with a lipid bilayer surrounding drug. The liposome fuses with the cell membrane and releases its contents ("payload"). The use of traditional liposomes for CNS indications is limited by promiscuous fusing of the liposomes to all cell membranes and rapid elimination from the circulation by the reticuloendothelial system. As such, liposome-encapsulated drugs such as NAC have been shown to be efficacious in models of liver or lung injury, but there have been no reports evaluating efficacy in CNS diseases [35]. To address these limitations liposomes have been conjugated with antibodies to allow specific targeting of the liposome to certain cell types [36, 37], and attaching polymers to the liposome can help reduce their clearance [38, 39].

A more contemporary payload delivery method of drug delivery to the CNS takes advantage of non-biologically active, membrane-penetrating moieties, combining them with small molecule therapeutics to form BBB-penetrating, membrane-permeable compounds. One such delivery system involves the antibiotic Gramicidin S. Conjugating nitroxides to a hemi-Gramicidin S moiety has yielded a class of powerful antioxidants [40] that not only penetrate the BBB and cell membranes but also target the mitochondria, one of the cell's major sources of free radicals. One of these compounds, XJB-5-131, demonstrated therapeutic concentrations in brain and brain mitochondria after systemic delivery and had powerful neuroprotective effects in a model of TBI in immature rats [41].

7.4.4 Maintaining Brain Bioavailability by Inhibiting Efflux Transporters

Numerous efflux transporters exist at the BBB and BCSFB, as previously mentioned, adding a second line of defense in the event that xenobiotics breach the BBB and enter the brain. Thus, even if temporary BBB disruption occurs after TBI allowing penetration of drug if delivered during this window, many xenobiotic substrates are actively eliminated out of the brain. The list of xenobiotics with clinical relevance is strikingly long [42], given the variety of membrane efflux transporters in the brain (Table 7.1).

Blocking the active efflux of a drug is another way to increase and maintain drug levels in the CNS. Drug development in the 1900s successfully focused on the kidney and renal elimination, and many effective drugs were developed, many still used today as antibiotic adjuvants. Historically speaking, the most prominent of these membrane transport inhibitors was benemid (later developed into probenecid), synthesized during World War II due to the short supply of penicillin desperately needed to treat wounded soldiers [43]. Probenecid blocks efflux of penicillin by blocking the action of organic anion transporters in nephrons [44]. This same

Table 7.1 CNS membrane efflux transporters, clinically relevant substrates, and inhibitors (in addition to cited sources includes verification from the UCSF/FDA transportal http://bts.ucsf.edu/fdatransportal/transporters/)

Transporter	Other names	Substrates	Inhibitors
ABCB1	MDR1, PgP	Organic cations and weak organic bases with hydrophobic regions (about half of commonly prescribed): e.g., Aldosterone; amiodarone; amitriptyline; anthracyclines; atorvastatin; beta blockers; cetirizine; clarithromycin; colchicine; cyclosporin A; daunorubicin; dexamethasone; digoxin; diltiazem; doxorubicin; erythromycin; etoposide; fexofenadine; fluoxetine; glucocorticoids; HIV protease inhibitors; indinavir; itraconazole; ivermectin; levofloxacin; loperamide; loratadine; losartan; lovastatin; methadone; methotrexate; morphine; nicardipine; nifedipine; ondansetron; paciltaxel; paroxetine; phenobarbital; phenytoin; progesterone; ranitidine; rapimycin; rifampin; sertraline; spironolactone; St. Johns wort; tacrolimus; tetracycline; vinblastine; vincristine	Cyclosporin A; elacridar; ketoconazole; mefipristone; quinidine; tariquidar; verapamil
ABCC1	MRP1	Mostly organic anions with hydrophobic regions: e.g., Daunorubicin; dehydroepiandrosterone 3-sulfate; etoposide; estradiol 17-beta glucuronide; folate; fluorescein; glutathione; glucuronide conjugates; leukotriene C4; methotrexate; p-aminohippurate; vincristine	Benzbronarone; probenecid; sulfinpyrazone
ABCC2	MRP2	Mostly organic anions with hydrophobic regions: e.g., Ampicillin; cisplatin; estradiol 17-beta glucuronide; etoposide; glutathione conjugates; irinotecan; leukotriene C4; methotrexate; mitoxantrone; pravastatin; temocaprilate; valsartan	Benzbronarone; cyclosporin A; probenecid; sulfinpyrazone
ABCC4	MRP4	Cyclic nucelotides and derivatives, some organic anions and weak organic acids: e.g., Adefovir, cAMP; cGMP; dehydroepiandrosterone 3-sulfate; estradiol; folic acid, methotrexate; prostaglandin; tenofovir; topotecan	Celecoxib, diclofenac
ABCC5	MRP5	Mostly cyclic nucelotides and derivatives: e.g., cAMP; cGMP; folic acid, methotrexate	

(continued)

Table 7.1 (continued)

Transporter	Other names	Substrates	Inhibitors
ABCG2	BCRP, BCRP1	Mostly organic cations and weak organic bases with hydrophobic regions, some anionic drugs: e.g., Azathioprine; bisantrene; cimetidine; ciprofloxacin; daunrobicin; dipyridamole; doxorubicin; etoposide; flavonoids; glucuronide conjugates; imatinib; methotrexate, mitoxantrone; ofloxacin; porphyrins, prazosin; statins, sulfate conjugates; topotecan	Estrone, 17β-oestradiol; elacridar; fumitremorgin C
OAT3		Mostly organic anions: e.g., estrone-3-sulfate, nonsteroidal anti-inflammatory drugs, cefaclor, ceftizoxime, furosemide, bumetanide	Probenecid
OATP1		Mostly weak organic acids: e.g., aldosterone; cholate; cortisole; digoxin; estrone-3-sulfate, dehydroepiadrosterone 3-sulfate; estradiol; fexofenadine; leukotriene C4; levofloxacin; methotrexate; statins	Ritonavir; lopinavir; rifampicin

ABC ATP binding cassette, *BCRP* breast cancer resistance protein, *MDR* multidrug resistance protein, *MRP* multidrug resistance-associated protein, *OAT* organic anion transporter, *OATP* organic anion transporting peptides, *PgP* p-glycoprotein

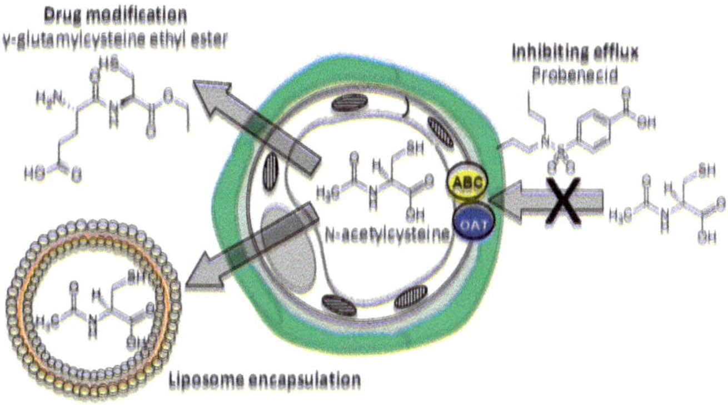

Fig. 7.4 Three means for delivering the BBB-impermeable drug, NAC, to the CNS

pharmacological effect could be applied to active efflux transporters in the brain. Challenges to this approach includes achieving high enough unbound inhibitor concentrations to block transporter proteins at the BBB and minimizing systemic toxicity resulting from collateral decreased systemic drug clearance [45, 46]. Lessons learned from failed pharmaceutical industry drug development programs that have tried to achieve targeted p-glycoprotein inhibition to reverse cancer chemoresistance over the past 2 decades suggest that systemic pharmacokinetic interactions and systemic toxicities are of real concern. However, applications using probenecid as an "adjuvant" to maintain NAC (and GSH) concentrations in the brain via inhibition of other transporters such as ABCCs and/or OATs seems promising. A Phase 1 clinical to examine the effect of the combination of probenecid and NAC on oxidative stress in children with TBI is ongoing (*ClinicalTrials.gov* NCT01322009). Figure 7.4 summarizes various means for increasing NAC concentrations in the CNS.

7.4.5 Disruption of the Blood–Brain Barrier

A pharmacologically invasive method for delivery of drugs to the CNS is by disrupting the BBB. This would ideally be a transient event, which causes reversible disruption and minimal collateral damage, to avoid serious adverse events such as toxin accumulation, seizures, and/or hemorrhage. Three methods have been utilized to disrupt the BBB to allow drug delivery: osmotic disruption, vasoactive disruption, and alkylglycerol disruption.

Osmotic disruption entails use of a rapid intra-carotid infusion of high doses of a hyperosmotic solution to create a hypertonic environment in proximity to the endothelial cells of the BBB. This draws water out of the cells, shrinking them and creating openings in the normally tight junctions for drug to pass through [47]. The most common agent for hyperosmolar therapy has been mannitol.

A few studies suggest that this method can lead to altered gene expression and aberrant glucose uptake [48]. However, this method does allow passage of the drug into the CNS. Osmotic disruption may have a place for enhancing drug delivery to brain in the treatment of chronic diseases such as brain tumors; however, this approach would be contraindicated in the setting of TBI given the importance of brain edema and raised ICP in that condition. It should be recognized that intravenous administration of osmolar agents such as mannitol and hypertonic saline, used in a conventional manner, do not disrupt the BBB. Vasoactive disruption involves infusion of a vasoactive substance such as histamine, bradykinin, or other similar substance. This leads to activation of endothelial receptors which allow permeability of tight junctions, thereby allowing drug entry [49]. Another approach to BBB disruption involves alkylglycerols to open the way for drug delivery. Some studies report that use of alkylglycerols causes a reversible and concentration-dependant permeability of drugs to the BBB [50]. The exact mechanism of action through this pathway is not well understood, but it seems that there is temporary breakdown of tight junction allowing drug to cross unhindered [50].

7.4.6 Going Around the Blood–Brain Barrier

So far, this chapter has discussed techniques to overcome the BBB using drug modifications, chemical and biological vehicles and carrier systems, and pharmacological disruption. And, while minimally invasive methods are always most desirable, in diseases such as TBI, the benefit/risk ratio may favor direct (invasive) administration in the event a powerful neuroprotective drug is discovered. There are several methods of direct CNS delivery that have shown promise. These tools have both benefits and drawbacks. They are invasive methods and carry all the associated risks. Further, once directly delivered, the therapeutics may also have to contend with the aforementioned barriers of diffusion and convection.

Open or imaging-assisted placement of microspheres, lipid-based polymers impregnated with drug, is one such drug delivery method that can be employed. Microspheres can be placed in an area at or near the target site of the drug and allow the drug to leech out and affect its target. Using advanced radiologic techniques, the drug can be placed in a precise location [51]. This method presently finds use in cancer and Parkinson's disease treatment; however, the technology could be used in other disease processes where a constant amount of drug can be delivered at the desired region of injured brain, for example, after a decompressive craniotomy is performed in a patient with severe TBI.

Like drug-impregnated microspheres, novel biodegradable membranes can be suffused with drug and placed at target regions in the CNS. The benefits are similar to microsphere placement but the amount of drug can be higher. In addition, the polymer can be degraded with no residue, toxin, or foreign substance remaining in the subject [52]. The drawbacks include size, which could limit where it can be placed, and diffusion. If the membrane is not directly adjacent to the target it may not reach the intended area in any significant quantity.

Nanoparticles and nanogels are small (10–1,000 nm)-sized units with a core shell structure (nanocapsule) or continuous matrix (nanosphere) that house drug and can be modified like liposomes to decrease elimination and target-specific CNS areas or cells [53]. Nanogels are a newer class of carrier system that involves nanoparticles cross-linked in a hydrogel. There is some concern that nanoparticles can cause increased reactive oxygen species leading to unintended collateral damage [53].

Tissue implants are strictly that; implantation of a biological tissue into the brain that secretes a therapeutic substance. This method has been used in Parkinson's treatment but the tissue typically does not survive for long periods of time [28]. Recent studies have been performed using embryonic neural grafts and gene therapy in attempts to improve duration of survival of the biologic tissue [54].

Microchip technology has advanced to the point of allowing hundreds of drug-filled wells to be delivered by a single biocompatible microchip [55]. The release of drug can be controlled through the chip which is implanted at or near the area of need. Many obstacles remain; however, it appears that this drug delivery technology is on the horizon in the future.

There are several types of pumps/catheters available which serve to provide a continuous or bolus dose of drug directly into the CSF or target brain area. Some of the pumps are currently in use in brain tumor trials but they seem to have several drawbacks, including incidents of increase intracranial pressure, hemorrhage, CSF leaks, and infection [28]. The continuous delivery method appears to be better tolerated but promising animal studies have not shown the same benefit in clinical trials [56]. It is difficult to know if the failure is a function of the machine or the drug, or both. As mentioned, diffusion to the target site may be effected by flow of CSF [28]. Baclofen is a common CNS drug that employs a pump/catheter method of delivery to the intrathecal space, bypassing the BBB. Baclofen pumps are used for treatment of severe spasticity; however, this approach involves surgery and includes a risk of infection, pump failure, and other complications [57]. It is not outside the realm of possibility to believe that chronic neuroprotectant drugs or other symptom-relieving drugs may be delivered via intrathecal pumps for victims of TBI in the near future.

A final method of bypassing the BBB is via the intranasal route. By this means of administration, drug enters the CNS via the olfactory neurons, which then bypasses the BBB by traveling along the olfactory and trigeminal nerves to the brain and brain stem [58]. Uptake still depends on molecular weight and lipophilicity. Other problems arise in delivering through the intranasal route if there is mucosal swelling, mucous production, or obstruction. Many biotechnology companies are exploring this route of administration for multiple drug uses.

7.5 Conclusion

A developing understanding of barriers to CNS drug delivery in both the uninjured and injured brain—and methods to overcome those barriers—raise hope for future development of neuroprotective strategies for the treatment of CNS trauma. It is possible that the failure of many promising drugs to translate from animal models to humans was at least in-part due to an incomplete understanding of the drug's brain bioavailability, and not to its pharmacological actions, or worse yet, to untreatable disease. A clearer understanding of brain pharmacokinetics in the injured brain, and optimization of CNS drug delivery, appears warranted.

References

1. Armulik A et al (2010) Pericytes regulate the blood–brain barrier. Nature 468(7323):557–561
2. Abbott NJ et al (2010) Structure and function of the blood–brain barrier. Neurobiol Dis 37(1): 13–25
3. Redzic Z (2011) Molecular biology of the blood–brain and the blood-cerebrospinal fluid barriers: similarities and differences. Fluids Barriers CNS 8(1):3
4. Miller DS (2010) Regulation of P-glycoprotein and other ABC drug transporters at the blood–brain barrier. Trends Pharmacol Sci 31(6):246–254
5. Begley DJ (2004) ABC transporters and the blood–brain barrier. Curr Pharm Des 10(12): 1295–1312
6. Hartz AM, Bauer B (2011) ABC transporters in the CNS—an inventory. Curr Pharm Biotechnol 12(4):656–673
7. Hartz AM, Bauer B (2010) Regulation of ABC transporters at the blood–brain barrier: new targets for CNS therapy. Mol Interv 10(5):293–304
8. Shen S, Zhang W (2010) ABC transporters and drug efflux at the blood–brain barrier. Rev Neurosci 21(1):29–53
9. Sun H et al (2003) Drug efflux transporters in the CNS. Adv Drug Deliv Rev 55(1):83–105
10. Pardridge WM (2003) Blood–brain barrier drug targeting: the future of brain drug development. Mol Interv 3(2):90–105, 51
11. Narayan RK et al (2002) Clinical trials in head injury. J Neurotrauma 19(5):503–557
12. Graff CL, Pollack GM (2004) Drug transport at the blood–brain barrier and the choroid plexus. Curr Drug Metab 5(1):95–108
13. Johanson CE, Stopa EG, McMillan PN (2011) The blood-cerebrospinal fluid barrier: structure and functional significance. Methods Mol Biol 686:101–131
14. Pardridge WM (2012) Drug transport across the blood–brain barrier. J Cereb Blood Flow Metab 32(11):1959–1972
15. Rosenberg GA (2012) Neurological diseases in relation to the blood–brain barrier. J Cereb Blood Flow Metab 32(7):1139–1151
16. Kuroiwa T et al (1985) The biphasic opening of the blood–brain barrier to proteins following temporary middle cerebral artery occlusion. Acta Neuropathol 68(2):122–129
17. Baskaya MK et al (1997) The biphasic opening of the blood–brain barrier in the cortex and hippocampus after traumatic brain injury in rats. Neurosci Lett 226(1):33–36
18. Marklund N, Hillered L (2011) Animal modelling of traumatic brain injury in preclinical drug development: where do we go from here? Br J Pharmacol 164(4):1207–1229
19. Marklund N et al (2006) Evaluation of pharmacological treatment strategies in traumatic brain injury. Curr Pharm Des 12(13):1645–1680

20. Shlosberg D et al (2010) Blood–brain barrier breakdown as a therapeutic target in traumatic brain injury. Nat Rev Neurol 6(7):393–403
21. Lo EH et al (2001) Drug delivery to damaged brain. Brain Res Brain Res Rev 38(1–2): 140–148
22. Pardridge WM (2010) Biopharmaceutical drug targeting to the brain. J Drug Target 18(3): 157–167
23. Smith DH, Hicks R, Povlishock JT (2013) Therapy development for diffuse axonal injury. J Neurotrauma 30(5):307–323
24. Reulen HJ et al (1978) Clearance of edema fluid into cerebrospinal fluid. A mechanism for resolution of vasogenic brain edema. J Neurosurg 48(5):754–764
25. Lai Y et al (2008) Autophagy is increased after traumatic brain injury in mice and is partially inhibited by the antioxidant gamma-glutamylcysteinyl ethyl ester. J Cereb Blood Flow Metab 28(3):540–550
26. Huttunen KM, Raunio H, Rautio J (2011) Prodrugs–from serendipity to rational design. Pharmacol Rev 63(3):750–771
27. Tekes K et al (2011) Prodrugs and active metabolites among antidepressive compounds. Neuropsychopharmacol Hung 13(2):103–110
28. Patel MM et al (2009) Getting into the brain: approaches to enhance brain drug delivery. CNS Drugs 23(1):35–58
29. Bodor N, Buchwald P (1997) Drug targeting via retrometabolic approaches. Pharmacol Ther 76(1–3):1–27
30. Herve F, Ghinea N, Scherrmann JM (2008) CNS delivery via adsorptive transcytosis. AAPS J 10(3):455–472
31. Zhou QH et al (2011) Neuroprotection with a brain-penetrating biologic tumor necrosis factor inhibitor. J Pharmacol Exp Ther 339(2):618–623
32. Jones AR, Shusta EV (2007) Blood–brain barrier transport of therapeutics via receptor-mediation. Pharm Res 24(9):1759–1771
33. Dehouck B et al (1997) A new function for the LDL receptor: transcytosis of LDL across the blood–brain barrier. J Cell Biol 138(4):877–889
34. Rousselle C et al (2000) New advances in the transport of doxorubicin through the blood–brain barrier by a peptide vector-mediated strategy. Mol Pharmacol 57(4):679–686
35. Suntres ZE (2011) Liposomal antioxidants for protection against oxidant-induced damage. J Toxicol 2011:152474
36. Huwyler J et al (2002) By-passing of P-glycoprotein using immunoliposomes. J Drug Target 10(1):73–79
37. Boado RJ, Pardridge WM (2011) The Trojan horse liposome technology for nonviral gene transfer across the blood–brain barrier. J Drug Deliv 2011:296151
38. Voinea M, Simionescu M (2002) Designing of 'intelligent' liposomes for efficient delivery of drugs. J Cell Mol Med 6(4):465–474
39. Schmidt J et al (2003) Drug targeting by long-circulating liposomal glucocorticosteroids increases therapeutic efficacy in a model of multiple sclerosis. Brain 126(Pt 8):1895–1904
40. Wipf P et al (2005) Mitochondrial targeting of selective electron scavengers: synthesis and biological analysis of hemigramicidin-TEMPO conjugates. J Am Chem Soc 127(36): 12460–12461
41. Ji J et al (2012) Lipidomics identifies cardiolipin oxidation as a mitochondrial target for redox therapy of brain injury. Nat Neurosci 15(10):1407–1413
42. Giacomini KM et al (2010) Membrane transporters in drug development. Nat Rev Drug Discov 9(3):215–236
43. Burnell JM, Kirby WM (1951) Effectiveness of a new compound, benemid, in elevating serum penicillin concentrations. J Clin Invest 30(7):697–700
44. Robbins N et al (2012) The history and future of probenecid. Cardiovasc Toxicol 12(1):1–9
45. Coley HM (2010) Overcoming multidrug resistance in cancer: clinical studies of p-glycoprotein inhibitors. Methods Mol Biol 596:341–358

46. Kalvass JC, et al (2013) Why clinical inhibition of efflux transport at the blood–brain barrier is unlikely: the ITC evidence-based position. Clin Pharmacol Ther 94:80–94
47. Siegal T et al (2000) In vivo assessment of the window of barrier opening after osmotic blood–brain barrier disruption in humans. J Neurosurg 92(4):599–605
48. Arima H et al (2003) Hyperosmolar mannitol simulates expression of aquaporins 4 and 9 through a p38 mitogen-activated protein kinase-dependent pathway in rat astrocytes. J Biol Chem 278(45):44525–44534
49. Matsukado K et al (1996) Enhanced tumor uptake of carboplatin and survival in glioma-bearing rats by intracarotid infusion of bradykinin analog, RMP-7. Neurosurgery 39(1): 125–133, discussion 133–4
50. Erdlenbruch B et al (2003) Alkylglycerol opening of the blood–brain barrier to small and large fluorescence markers in normal and C6 glioma-bearing rats and isolated rat brain capillaries. Br J Pharmacol 140(7):1201–1210
51. Benoit JP et al (2000) Development of microspheres for neurological disorders: from basics to clinical applications. J Control Release 65(1–2):285–296
52. Halliday AJ et al (2012) Novel methods of antiepileptic drug delivery—polymer-based implants. Adv Drug Deliv Rev 64(10):953–964
53. Wong HL, Wu XY, Bendayan R (2012) Nanotechnological advances for the delivery of CNS therapeutics. Adv Drug Deliv Rev 64(7):686–700
54. Leigh K, Elisevich K, Rogers KA (1994) Vascularization and microvascular permeability in solid versus cell-suspension embryonic neural grafts. J Neurosurg 81(2):272–283
55. Farra R et al (2012) First-in-human testing of a wirelessly controlled drug delivery microchip. Sci Transl Med 4(122):122ra21
56. Salvatore MF et al (2006) Point source concentration of GDNF may explain failure of phase II clinical trial. Exp Neurol 202(2):497–505
57. Varhabhatla NC, Zuo Z (2012) Rising complication rates after intrathecal catheter and pump placement in the pediatric population: analysis of national data between 1997 and 2006. Pain Physician 15(1):65–74
58. Heydel JM et al (2010) UDP-glucuronosyltransferases (UGTs) in neuro-olfactory tissues: expression, regulation, and function. Drug Metab Rev 42(1):74–97

Chapter 8
Angiogenesis and Functional Recovery After Traumatic Brain Injury

Yanlu Zhang, Ye Xiong, Asim Mahmood, Zheng Gang Zhang, and Michael Chopp

Abstract Brain injuries caused by trauma remain a major cause of death and serious long-term disability worldwide, especially in children and young adults. However, nearly all Phase III traumatic brain injury (TBI) clinical trials have failed to provide safe and effective treatment for improving functional recovery after TBI. This review discusses recent promising preclinical and clinical data indicating that TBI promotes angiogenesis (formation of new blood vessels from preexisting endothelial cells), which couples with neurogenesis (generation of new neurons) and oligodendrogenesis (generation of new oligodendrocytes), in concert, contributing to spontaneous functional recovery. Selected cell-based and pharmacological therapies that can amplify these endogenous neurorestorative effects to enhance cognitive and neurological functional recovery after TBI are discussed. Perspectives for further investigation of angiogenesis after TBI and associated therapeutic treatments are provided.

8.1 Introduction

Traumatic brain injury (TBI) is a leading cause of mortality and morbidity worldwide. TBI survivors often suffer cognitive deficits and sensorimotor dysfunctions [1]. Many therapeutic strategies have shown promise in the laboratory setting [2–4] but failed in human clinical trial [5, 6]. Thus, it is imperative to develop therapies for TBI to reduce neurological deficits.

M. Chopp (✉)
Department of Neurology, Henry Ford Hospital, Education and Research Building, Room 3056, 2799 West Grand Boulevard, Detroit, MI 48202, USA

Department of Physics, Henry Ford Hospital, Detroit, MI, USA

Oakland University, Rochester, MI, USA
e-mail: mchopp1@hfhs.org

E.H. Lo et al. (eds.), *Vascular Mechanisms in CNS Trauma*, Springer Series in Translational Stroke Research 5, DOI 10.1007/978-1-4614-8690-9_8,

Emerging data from preclinical TBI studies indicate that angiogenesis plays an important role in mediating brain repair by coupling with neurogenesis and oligodendrogenesis and that cell-based and pharmacological therapies targeting amplification of angiogenesis and white matter remodeling substantially improve sensorimotor functions and reduce cognitive impairments. In this chapter, we will review TBI-induced angiogenesis and the coupling of angiogenesis with neurogenesis, oligodendrogenesis, and white matter remodeling. We will then highlight therapies that amplify these events, leading to improvement in neurological outcomes after TBI.

8.2 Angiogenesis After TBI

The endothelial cells (ECs) of cerebral capillaries, unlike those from non-cerebral capillaries, are linked by complex tight junctions that along with astrocyte end-feet, microglial cells, and pericytes form the blood–brain barrier (BBB) [7]. Under physiological conditions, the cerebral ECs are relatively quiescent with a turnover rate of approximately 3 years in the adult rodent [8]. Angiogenesis is the sprouting of new capillaries from preexisting vessels, involving the proliferation and migration of ECs, formation, branching, and anastomosis of tubes [9, 10]. TBI induces angiogenesis at an early stage after injury. After TBI, using immunohistochemistry with antibodies against bromodeoxyuridine (BrdU) and measurement of capillary density, newly formed vessels are found and capillary density increases [11]. To monitor development of angiogenesis after TBI noninvasively and longitudinally, magnetic resonance imaging (MRI) indices including cerebral blood volume (CBV), cerebral blood flow (CBF), blood-to-brain transfer constant (Ki) marked with extrinsic-contrast agents, such as gadolinium DTPA (diethylene triamine pentaacetic acid), and T1- or T2-weighted imaging have been used. Hyperpermeabilities on the Ki map in the injured brain indicate vascular leakage. New vessels are permeable at the early phase of angiogenesis, and become less leaky as they mature [12–14]. The feature of a transient increase in vascular permeability is used to detect formation of new blood vessels [13]. Newly generated leaky cerebral vessels with immature BBB which are present in the lesion boundary zone 2 days after TBI are detected by Ki maps [15, 16]. Angiogenic areas identified on the Ki map become apparent 3–4 weeks after TBI [15]. These vessels appear less leaky 6 weeks after TBI and may contribute to an increase in CBF [15]. Furthermore, as confirmed with endothelial barrier antigen (EBA) immunoreactivity, the angiogenic area on the Ki map identifies enlarged thin-walled vessels [15]. The correlation between increase of CBF and enhancement of vessel density indicates that TBI induces functional new vessels in the lesion boundary zone. Vessels with BrdU-positive ECs are detected in the ipsilateral dentate gyrus (DG) in the rats after TBI, indicating that TBI induces angiogenesis in the DG [17]. Additionally, elevated CBV is reported starting at day 1 after injury and lasting for 2 weeks in the ipsilateral DG after TBI [18]. Elevated CBV in the DG after TBI suggests that newly generated vessels by TBI-induced angiogenesis present to be functional.

Vascular endothelial growth factor (VEGF) and its receptors initiate the formation of immature vessels, while angiopoietins (Ang 1, Ang 2) and their receptor Tie2 are essentially involved in maturation, stabilization, and remodeling of vessels [19]. VEGFR1 and VEGFR2 mRNA and protein are present in vessels adjacent to the lesion from 1 day after injury [20]. VEGF and VEGFR2 assessed by Western blot analysis also increase in the ipsilateral hippocampus after TBI [21]. With double immunofluorescent staining of endothelium and VEGFR2, it is revealed that increased VEGFR2 is expressed in the endothelium [11]. Although there is no direct evidence that TBI induces angiogenesis in human brain in response to TBI, serum and intracerebral extracellular fluid levels of angiogenic factors, such as VEGF and Ang-1, peak at 14 days post-trauma and subsequently decline [22, 23]. Furthermore, in a 21-day clinical study, serum VEGF level significantly increases during the entire period while there is no difference of serum Ang-1 level between severe TBI patients and control subjects during the first 4 days after TBI but then Ang-1 increases after 4 days [23]. However, VEGF and Ang-1 are expressed not only within vasculature but also in large numbers of platelets [24–26]. Therefore, in vitro, increased level of VEGF and Ang-1 in serum collected from patients may also be caused by platelet clotting in serum tubes [27]. For both healthy controls and patients in this 21-day clinical study [23], the average level of Ang-1 in serum on day 1 is tenfold higher than the levels reported from other studies [24, 28]. Therefore, the levels of VEGF and Ang-1 could result from an artifact caused by platelet activation after TBI [27]. In uninjured rats, none or weak immunoreactivity with matrix metalloproteinases (MMPs), such as MMP2 and MMP9, is detected in cortical capillaries [29]. As measured using gelatin zymography, MMP9 is elevated from 3 h after TBI, reaches a maximum at 24 h, and persists to 2 weeks, while MMP2 is increased from 1 day and persists to 2 weeks [30–32]. Likewise in humans, as measured by ELISA, serum MMP9 is significantly increased during the follow-up period after TBI [33]. Robust MMP2 and MMP9 immunoreactivities are found to colocalize to the vessels adjacent to the lesion site, and particularly in the immature ECs [29]. In vitro studies show that ECs secrete MMP2 and MMP9 before and after tumor necrosis factor (TNF)-α stimulated injury [34]. Increased expression of VEGFR2 and MMP9 is detected in the basement membrane of the new capillaries 2 days after TBI [29]. These results show that TBI also induces the expression of EC MMPs which are involved in angiogenesis [29]. In vitro, exogenous MMP2 increases EC tube formation while addition of MMP inhibitors or synthetic MMP agonists decreases EC tube formation [35].

Collectively, these data indicate that angiogenic factors and MMPs play important roles in TBI-induced angiogenesis.

8.3 Angiogenesis Couples with Neurogenesis and Oligodendrogenesis

In addition to providing nutritive blood flow, cerebral ECs regulate biological activity of neural progenitor cells [36].

Cerebral ECs activated by cerebral ischemia secrete VEGF that acts on neural stem cells (NSCs) and consequently leads to augmentation of newborn neurons [37]. Blockage of VEGFR2 not only reduces EBA-immunoreactive vascular density (an indicator of angiogenesis) but also reduces the number of newborn neurons in the DG in rats after TBI, which is associated with decreased neurological function recovery [21]. These data indicate that angiogenesis cooperates with neurogenesis and is involved in the recovery of neurological function after brain injury.

Neurogenesis occurs in the subventricular zone of the lateral ventricle and the subgranular zones of the DG in mammalian adult brains under normal conditions [38, 39] and pathological situations including TBI [40–44]. In the normal hippocampus, newborn neurons are detected after 1 h of [^{3}H]thymidine injection and continually generated from the DG subgranular zone, with a significant increase after 2 weeks, and these newborn cells migrate laterally into the granule cell layer, projecting axons to the CA3 region of the hippocampus within a 4-week study [45]. Following TBI, by immunofluorescent double-labeling of the proliferation cell marker BrdU and the mature neuronal marker NeuN or the astrocytic marker GFAP, there is a significant peak period of cell proliferation at 2 days post-injury in the DG both in injured juvenile and adult rats compared to shams [44], and the majority of BrdU+ cells which survive for 10 weeks become dentate granule neurons [46]. Injured animals display significant cognitive deficits at 11–15 days post-moderate injury, while there is no significant difference of cognitive deficits at days 56–60 between injured and sham animals, which shows cognitive recovery over time following TBI [46]. Therefore, injury-induced limited endogenous neurogenesis may partially contribute to spontaneous cognitive functional recovery after TBI.

In the central nervous system (CNS), the neurovascular unit (NVU) comprises ECs, pericytes, neurons and glial cells, as well as growth factors and extracellular matrix proteins close to the endothelium [47, 48]. New blood vessels in peri infarct cortex are closely associated with new neurons identified by BrdU+/doublecortin+ (DCX, a marker of migrating neuroblasts) after ischemic injury, which indicates that neurogenesis coexists with angiogenesis in peri-infarct cortex [49]. Correlation analysis shows that the cognitive function outcome is significantly correlated with the number of the newborn neurons generated in the DG [50] and also with the increased number of vessels in ipsilateral cortex, DG, and CA3 region examined 35 days after TBI [51]. This evidence indicates that angiogenesis is coupled with neurogenesis and may improve neurological functional outcome after TBI spontaneously or with therapy [15, 16, 52, 53].

Oligodendrocytes (OLGs) are the major cell type in the white matter in the CNS, maintaining the integrity of the white matter in the adult brain. Mature OLGs generate myelin which forms sheaths for axons in the adult mammalian CNS but are unable to proliferate in response to injury [54]. However, there are plentiful oligodendrocyte progenitor cells (OPCs) in the white matter of normal CNS with functions including proliferation and maturation to remyelinate the demyelinated axons [55, 56]. Oligodendrogenesis occurs after injury in the lesion area and corpus callosum where OPCs proliferate, mature, migrate, and are regulated by the factors secreted from ECs such as fibroblast growth factor (FGF), VEGF, brain-derived neurotrophic factor (BDNF), and MMPs. TBI alone significantly increases the number of OPCs in the ipsilateral cortex and hippocampus (CA3, DG) compared to sham controls, suggesting that oligodendrogenesis may partially be responsible for spontaneous functional recovery presumably myelinating axons [57, 58]. Cognitive function recovery is significantly and positively correlated with both angiogenesis and neurogenesis in the hippocampus region after TBI [51]. After TBI, myelin content, as measured by staining with myelin-specific stain Luxol fast blue, is reduced in many white matter regions [59]. Therefore, OPCs may play an important role in remyelination in the injured brain even though the axonal regeneration is limited in adult brain after injury [60].

An interaction between OLGs and cerebral ECs has been investigated. In OLGs culture system derived from 1- to 2-day SD rats, MMP9 is not detected in OLGs under normal conditions but is secreted after stimulation by IL-β [61]. U0126, a MEK inhibitor, is able to block MMP9 secretion in IL-β-treated OLGs culture system, indicating that the MEK/ERK signaling pathway regulates OLGs to secrete MMP9 under stimulated situation [61]. Furthermore, the inhibition of MMP9 decreases newborn ECs and EC density [61]. Therefore, MMP9 released from OLGs after injury plays a critical role in white matter remodeling.

In vitro, coculture of OPCs with cerebral ECs promotes OPC survival and proliferation via the Akt and Scr signaling pathways [62]. In addition, VEGF, involved in angiogenesis, is primarily released by EC and its receptor, VEGFR2, is expressed in OPCs [63]. In endothelial-conditioned media, VEGF promotes OPC migration at 24 h after incubation and its effect can be inhibited by anti-VEGFR2 antibody. Therefore, these findings provide a novel concept of the oligovascular niche, with trophic factors secreted from ECs, activated through the Akt and Src signaling pathways to regulate OPC function including proliferation and migration. Furthermore, sensorimotor function is significantly correlated with the axonal density of ipsilateral hemisphere after TBI [64]. Taken together, these data indicate that angiogenesis coupling with oligodendrogenesis may contribute to white matter remodeling including axons and synapses after brain injury, which improves sensorimotor function recovery.

8.4 Enhancement of TBI-Induced Angiogenesis by Cell-Based Therapies

8.4.1 Bone Marrow Stromal Cells

Marrow stromal cells (MSCs) are extracted from bone marrow and include mesenchymal stem or progenitor cells [65–68], which can replicate and differentiate to other cells including neural cells [69–72]. The restorative therapy of MSCs has been performed intravenously or through direct implantation. Intravenous administration of MSCs after TBI significantly enhanced improvement in functional outcome [73]. In vitro, TBI-conditioned cultured hMSCs increase BDNF, NGF, VEGF, and hepatocyte growth factor (HGF) [74]. Furthermore, in vivo, MSCs also induce intrinsic parenchymal cells to produce the above growth factors after TBI [73]. Rats treated with MSCs cultured with BDNF and NGF have more engrafted cells than the group treated with MSCs cultured without these factors, and more robust motor function recovery is detected in the MSC groups cultured with neurotrophic factors [75]. These data suggest that motor function recovery after TBI is accomplished by transplantation of MSCs and enhanced by additional neurotrophic factors. To investigate the changes of the vascular system after TBI with acute treatment of human MSCs (hMSCs), MRI T2 maps are used to monitor and quantify the volumetric changes in the lesion area, while CBF (measured by perfusion-weighted MRI) and Ki (extrinsic-contrast agents) are used to monitor hemodynamic alteration and the BBB permeability, respectively [15]. After TBI, Ki-detected angiogenesis occurs significantly earlier in the MSC-treated group compared to the control group and the angiogenic area on Ki map is confirmed histologically by enlarged thin-walled vessels [15]. Furthermore, compared to control subjects, this early angiogenesis is not only associated with a significantly higher vessel density in the lesion boundary region of cell-treated animals but also associated with improved behavioral status after MSC treatment [15]. Pre-labeled MSCs can be tracked in the brain using MRI and verified by immunostaining [76–78]. After hMSCs are injected intravenously and detected to migrate into brain around the injury site [79], they promote cell proliferation in the subventricular zone, hippocampus, and boundary zone of injury, and some of these newly generated cells expressed the neuronal markers (Tuj1 for immature neurons, DCX for migrating neuroblasts, NeuN for mature neurons) with improved cognitive function recovery [80]. Compared to hMSCs alone injected intracerebrally or intravenously 1 week after TBI, hMSC-impregnated scaffolds transplanted into the lesion cavity 1 week after TBI significantly increase the number of hMSCs which migrate from lesion cavity to the boundary zone and also increase the vascular density in the boundary zone and hippocampus after TBI, and enhance cognitive and sensorimotor function [81]. Thus, scaffolds impregnated with MSCs provide a promising therapy option to tissue repair and functional recovery after TBI. With enhanced neurological and cognitive function recovery,

MRI shows that white matter reorganization is located in the extended area of the corpus callosum where labeled hMSCs are co-localized [78]. hMSCs secrete angiogenic factors such as VEGFA, FGF1, and MMP9 after TBI associated with enhanced neurologic and cognitive function recovery [82], which may lead to the restructuring of axons and myelin after TBI to reorganize the white matter through oligodendrogenesis.

In summary, MSC treatment amplifies neurogenesis, angiogenesis, and oligodendrogenesis after TBI.

8.4.2 Neural Stem Cells

Stem cells are able to self-renew and differentiate into multiple cell types. NSCs can differentiate into neurons, astrocytes, and OLGs [83, 84]. NSCs with their intrinsic ability of regeneration have been used in the treatment of many neurological diseases in animal models including TBI [85–87]. After transplantation into corpus callosum of brain-injured animals 2 days after TBI, some of NSCs which are derived from the neonatal murine cerebellar external germinal layer express NG2 (marker for OPCs) and migrate to the injury area 2 weeks after transplantation [88]. In a subacute therapy (1 week after injury), NSCs injected in the striatum remain in the brain and improve motor recovery on a rotarod test at 14 days after cell placement [89]. There are two possible strategies for NSC treatment of TBI: transplantation of exogenous NSCs and stimulation of endogenous NSCs [90]. Local or systemic administration of pre-differentiated human fetal neural progenitor cells improves long-term motor and sensory function recovery, decreases trauma lesion size, and increases neuronal survival in the border zone of the lesion, which are likely to be attributable to transiently increased angiogenesis and reduced astrogliosis in the border zone instead of cell replacement from donor cell transplantation [91].

8.5 Augmentation of TBI-Induced Angiogenesis by Pharmacological Therapies

8.5.1 Erythropoietin

Erythropoietin (EPO) and its receptor (EPOR) are essential for erythropoiesis and EPO has been widely used in the clinic for treatment of anemia since it regulates the maturation, differentiation, and survival of hematopoietic progenitor cells [4]. In normal adult brains, low levels of EPO and EPO receptors are detected, while after injury, increased levels of EPO and EPO receptors are found in neurons, NPCs, glial cells, and ECs [92]. EPO treatment (24 h after TBI) increases expression of

VEGF and phosphorylation of VEGFR2 as well as results in a significant increase of newborn neurons and vascular density in the cortex, DG, and CA3. However, after blockage of VEGFR2 with SU5416, newborn neurons and vascular density are all significantly decreased and functional recovery in EPO-treated TBI rats is abolished [21]. Therefore, EPO therapy improves sensorimotor and cognitive functional recovery after TBI by promoting neurogenesis and angiogenesis through upregulating VEGF/VEGFR2 expression in the brain [21, 43, 93]. Previous studies show that cognitive function recovery is mediated by neurogenesis coupled with angiogenesis in the hippocampus [42, 43, 51] while sensorimotor function recovery is associated with brain angiogenesis and spinal cord axon remodeling [94]. However, in animals null for the EPOR gene in neural cells, EPO treatment still significantly reduces cell loss in the hippocampus compared with saline controls, as well as improves sensorimotor and cognitive function after TBI, which suggests that therapeutic benefits of EPO may be mediated through its vascular protection but not via neural EPOR [95]. Carbamylated erythropoietin (CEPO), a modified erythropoietin molecule that does not affect hematocrit, is as effective as EPO in terms of reducing hippocampal cell loss, enhancing angiogenesis and neurogenesis in the injured cortex and hippocampus, and improving sensorimotor functional recovery and spatial learning in rats after TBI [53, 96]. EPO and its derivatives have a potential value in TBI therapy.

8.5.2 *Statins*

Statins lower cholesterol levels and also have neuroprotective and neurorestorative effects including angiogenesis, neurogenesis, and synaptogenesis and improve function recovery in rats after TBI [97, 98]. To measure the effect of atorvastatin on improvement of microvascular integrity and cognitive function recovery after TBI, animals are perfused with FITC–dextran to track the vascular changes, and the water maze test is performed to investigate the spatial learning on injured rats [97]. Compared with saline treatment group, atorvastatin treatment significantly improves spatial learning, increases the vessel-to-tissue ratio and vascular length on days 8 and 15 after TBI in both hippocampal CA3 region and boundary zone of injured area, and also augments vascular diameter on day 8 after TBI in the boundary zone of contusion [97]. Simvastatin upregulates VEGFR2 expression after TBI, increases the BrdU+ ECs in the lesion boundary zone and hippocampus with improved functional recovery in rats, and enhances in vitro capillary-like tube formation after oxygen glucose deprivation (OGD), indicating that simvastatin-enhanced angiogenesis may be related to activation of the VEGFR-2/Akt/eNOS signaling pathway [16]. A recent study shows that increasing circulating EPCs with atorvastatin treatment may contribute to the observed increase in brain angiogenesis and improved functional outcome after TBI [99].

8.5.3 Thymosin Beta 4

Tβ4 is a multifunctional regenerative small peptide comprising 43 amino acids and its major function is G-actin-sequestering [100]. Tβ4 is involved in many cellular procedures including cell proliferation, mobility, antiapoptosis, anti-inflammation, and promotion of wound healing [101–104]. Tβ4 is a novel therapeutic choice for CNS trauma, which promotes endogenous neurorestorative processes in animal models of TBI [52, 57]. Tβ4 is evaluated to be safe in clinical treatment of acute myocardial infarction [105]. Early treatment (6 h after TBI) shows that Tβ4 significantly improves spatial learning and sensorimotor functional recovery, and promotes neurogenesis in the DG [52]. Late treatment (24 h after TBI) indicates that Tβ4 significantly increases the vascular density in the injured cortex, CA3, and DG of the ipsilateral hemisphere, and enhances neurogenesis in the injured cortex and hippocampus, along with increased generation of mature OLGs in the CA3 region, which are associated with improved sensorimotor and cognitive functional recovery after TBI [57]. The mechanisms underlying the beneficial effects of Tβ4 remain unknown. However, a recent study shows that Tβ4 is able to induce endothelial progenitor cell migration via the phosphatidylinositol 3 kinase/Akt/endothelial nitric oxide synthase (eNOS) signal transduction pathway, which may mediate angiogenesis [106]. Tβ4 treatment induces OLG differentiation by inducing p38MAPK with parallel inactivation of ERK1 and JNK1, thus preventing the accumulation of phosphorylated c-Jun [107]. Therefore, Tβ4 treatment-induced angiogenesis, neurogenesis, and oligodendrogenesis, in concert, may contribute to functional recovery in rats after brain injury.

8.6 Other Growth Factors

VEGF is an important regulator of angiogenesis. VEGF is neuroprotective in several models of experimental brain injury [108–112]. The expression of VEGF and VEGFR2 is increased in rodents subjected to TBI [20, 113], and inhibition of VEGF expression after injury decreases newborn neurons and newly generated vessels with aggravated function outcome [21], suggesting that VEGF-induced angiogenesis and neurogenesis promote neurological and cognitive function recovery [21]. Hepatocyte growth factor (HGF) is an important molecule for tissue repair [114]. Enhancement of vascular pixel intensity and GAP-43-positive cells (a crucial component of the axon and presynaptic terminal) is detected at the ischemic boundary zone with HGF treatment [115], indicating that HGF is involved in angiogenesis and synaptogenesis after injury. HGF is also known to induce angiogenesis in cooperation with VEGF [116]. Basic fibroblast growth factor (FGF2) is a potent angiogenic agent present in neurons and glia, vascular basement membrane of blood vessels, and in the ependymal cells of the ventricles [117]. After TBI, FGF2 treatment significantly decreases lesion size, increases the number of blood vessels in the cortex around the lesion, and improves sensorimotor function recovery [110]. Granulocyte-colony stimulating factor (G-CSF), a hematopoietic growth factor, significantly increases 3 h after TBI and peaks at 8 h [118]. In the ischemic hemisphere post-stroke, G-CSF treatment increases endothelial proliferation, vascular density, expression of eNOS and angiopoietin-2, and decreases BBB disruption and function deficits [119]. Most of these growth factors, with large molecular-weighted and hydrophilic proteins, have a limited access to the CNS after systemic administration, principally due to poor BBB permeability. Cerebrolysin is a mixture of low-molecular-weight neuropeptides derived from purified brain proteins by standardized enzymatic proteolysis, with proposed neuroprotective and neurotrophic properties similar to naturally occurring growth and neurotrophic factors [120]. Direct and indirect evidences indicate that low-molecular-weight Cerebrolysin, which contains many neurotrophic factor-like peptides, is able to cross the BBB [121]. Early intervention with Cerebrolysin reduces BBB permeability changes, attenuates brain pathology and brain edema, and mitigates functional deficits [120]. Recent data show that Cerebrolysin enhances neurogenesis in the ischemic brain and improves functional outcome after stroke [122]. Taken together, these data suggest that Cerebrolysin has potential therapeutic value in TBI.

8.7 Perspectives

There is evidence for a prominent role of angiogenesis in the recovery of neurological function post-TBI. It is well known that TBI induces angiogenesis, particularly in the injury boundary zone. An angiogenic environment is essential for tissue repair and functional recovery after injury. The contribution of endogenous angiogenesis, however, may not be sufficient to support the degree of neuroplasticity required for functional recovery after TBI. The therapeutic approaches that enhance brain remodeling via angiogenesis are promising. In addition to cell-based therapy including MSCs and NSCs, many promising drugs such as EPO, CEPO, Tβ4, statins, and growth and neurotrophic factors, all of which amplify endogenous angiogenesis, have been evaluated in TBI. Enhanced angiogenesis, coupled with neurogenesis, oligodendrogenesis, and white matter remodeling, contributes to improvement of functional recovery induced by these treatments. The therapeutic window for stimulation of angiogenesis, neurogenesis, and white matter remodeling after TBI has not been ascertained. Further investigation of angiogenesis and its correlation between neurogenesis and white matter remodeling is also warranted to better understand mechanisms underlying functional recovery after TBI and to develop effective therapeutic treatment for improving outcomes in patients with the CNS injury.

Acknowledgments This work was supported by National Institutes of Health grants RO1 NS062002 (Y.X.), RO1AG037506 (M.C.).

References

1. Davis AE (2000) Mechanisms of traumatic brain injury: biomechanical, structural and cellular considerations. Crit Care Nurs Q 23:1–13
2. Kwon BK, Okon E, Hillyer J et al (2011) A systematic review of non-invasive pharmacologic neuroprotective treatments for acute spinal cord injury. J Neurotrauma 28:1545–1588
3. Kwon BK, Okon EB, Plunet W et al (2011) A systematic review of directly applied biologic therapies for acute spinal cord injury. J Neurotrauma 28:1589–1610
4. Xiong Y, Mahmood A, Chopp M (2010) Neurorestorative treatments for traumatic brain injury. Discov Med 10:434–442
5. Hawryluk GW, Rowland J, Kwon BK et al (2008) Protection and repair of the injured spinal cord: a review of completed, ongoing, and planned clinical trials for acute spinal cord injury. Neurosurg Focus 25:E14
6. Narayan RK, Michel ME, Ansell B et al (2002) Clinical trials in head injury. J Neurotrauma 19:503–557
7. Correale J, Villa A (2009) Cellular elements of the blood–brain barrier. Neurochem Res 34: 2067–2077
8. Polverini PJ (2002) Angiogenesis in health and disease: insights into basic mechanisms and therapeutic opportunities. J Dent Educ 66:962–975
9. Risau W (1997) Mechanisms of angiogenesis. Nature 386:671–674
10. Carmeliet P (2000) VEGF gene therapy: stimulating angiogenesis or angioma-genesis? Nat Med 6:1102–1103
11. Morgan R, Kreipke CW, Roberts G et al (2007) Neovascularization following traumatic brain injury: possible evidence for both angiogenesis and vasculogenesis. Neurol Res 29:375–381
12. Zhang ZG, Zhang L, Jiang Q et al (2000) VEGF enhances angiogenesis and promotes blood–brain barrier leakage in the ischemic brain. J Clin Invest 106:829–838
13. Jiang Q, Zhang ZG, Ding GL et al (2005) Investigation of neural progenitor cell induced angiogenesis after embolic stroke in rat using MRI. Neuroimage 28:698–707
14. Palmer TD, Schwartz PH, Taupin P et al (2001) Cell culture. Progenitor cells from human brain after death. Nature 411:42–43
15. Li L, Chopp M, Ding GL et al (2012) MRI measurement of angiogenesis and the therapeutic effect of acute marrow stromal cell administration on traumatic brain injury. J Cereb Blood Flow Metab 32:2023–2032
16. Wu H, Jiang H, Lu D et al (2011) Induction of angiogenesis and modulation of vascular endothelial growth factor receptor-2 by simvastatin after traumatic brain injury. Neurosurgery 68:1363–1371, discussion 1371
17. Lu D, Qu C, Goussev A et al (2007) Statins increase neurogenesis in the dentate gyrus, reduce delayed neuronal death in the hippocampal CA3 region, and improve spatial learning in rat after traumatic brain injury. J Neurotrauma 24:1132–1146
18. Immonen R, Heikkinen T, Tahtivaara L et al (2010) Cerebral blood volume alterations in the perilesional areas in the rat brain after traumatic brain injury–comparison with behavioral outcome. J Cereb Blood Flow Metab 30:1318–1328
19. Yancopoulos GD, Klagsbrun M, Folkman J (1998) Vasculogenesis, angiogenesis, and growth factors: ephrins enter the fray at the border. Cell 93:661–664
20. Skold MK, von Gertten C, Sandberg-Nordqvist AC et al (2005) VEGF and VEGF receptor expression after experimental brain contusion in rat. J Neurotrauma 22:353–367
21. Xiong Y, Zhang Y, Mahmood A et al (2011) Erythropoietin mediates neurobehavioral recovery and neurovascular remodeling following traumatic brain injury in rats by increasing expression of vascular endothelial growth factor. Transl Stroke Res 2:619–632
22. Mellergard P, Sjogren F, Hillman J (2010) Release of VEGF and FGF in the extracellular space following severe subarachnoidal haemorrhage or traumatic head injury in humans. Br J Neurosurg 24:261–267

23. Gong D, Zhang S, Liu L et al (2011) Dynamic changes of vascular endothelial growth factor and angiopoietin-1 in association with circulating endothelial progenitor cells after severe traumatic brain injury. J Trauma 70:1480–1484
24. Lukasz A, Hellpap J, Horn R et al (2008) Circulating angiopoietin-1 and angiopoietin-2 in critically ill patients: development and clinical application of two new immunoassays. Crit Care 12:R94
25. Nadar SK, Blann A, Beevers DG et al (2005) Abnormal angiopoietins 1&2, angiopoietin receptor Tie-2 and vascular endothelial growth factor levels in hypertension: relationship to target organ damage [a sub-study of the Anglo-Scandinavian Cardiac Outcomes Trial (ASCOT)]. J Intern Med 258:336–343
26. Kusumanto YH, Dam WA, Hospers GA et al (2003) Platelets and granulocytes, in particular the neutrophils, form important compartments for circulating vascular endothelial growth factor. Angiogenesis 6:283–287
27. Padberg JS, Wiesinger A, Kumpers P (2011) Platelet activation accounts for excessive angiopoietin-1 levels in patients' sera. J Trauma 71:1480–1481
28. Kumpers P, Nickel N, Lukasz A et al (2010) Circulating angiopoietins in idiopathic pulmonary arterial hypertension. Eur Heart J 31:2291–2300
29. Frontczak-Baniewicz M, Walski M, Madejska G et al (2009) MMP2 and MMP9 in immature endothelial cells following surgical injury of rat cerebral cortex—a preliminary study. Folia Neuropathol 47:338–346
30. Wang X, Jung J, Asahi M et al (2000) Effects of matrix metalloproteinase-9 gene knock-out on morphological and motor outcomes after traumatic brain injury. J Neurosci 20:7037–7042
31. Yamaguchi M, Jadhav V, Obenaus A et al (2007) Matrix metalloproteinase inhibition attenuates brain edema in an in vivo model of surgically-induced brain injury. Neurosurgery 61:1067–1075, discussion 1075–1066
32. Shigemori Y, Katayama Y, Mori T et al (2006) Matrix metalloproteinase-9 is associated with blood–brain barrier opening and brain edema formation after cortical contusion in rats. Acta Neurochir Suppl 96:130–133
33. Gong D, Hao M, Liu L et al (2012) Prognostic relevance of circulating endothelial progenitor cells for severe traumatic brain injury. Brain Inj 26:291–297
34. Anderson DE, Hinds MT (2012) Extracellular matrix production and regulation in micropatterned endothelial cells. Biochem Biophys Res Commun 427:159–164
35. Schnaper HW, Grant DS, Stetler-Stevenson WG et al (1993) Type IV collagenase(s) and TIMPs modulate endothelial cell morphogenesis in vitro. J Cell Physiol 156:235–246
36. Zhang ZG, Chopp M (2009) Neurorestorative therapies for stroke: underlying mechanisms and translation to the clinic. Lancet Neurol 8:491–500
37. Teng H, Zhang ZG, Wang L et al (2008) Coupling of angiogenesis and neurogenesis in cultured endothelial cells and neural progenitor cells after stroke. J Cereb Blood Flow Metab 28:764–771
38. Gage FH (2002) Neurogenesis in the adult brain. J Neurosci 22:612–613
39. Zhao C, Deng W, Gage FH (2008) Mechanisms and functional implications of adult neurogenesis. Cell 132:645–660
40. Richardson RM, Sun D, Bullock MR (2007) Neurogenesis after traumatic brain injury. Neurosurg Clin N Am 18:169–181, xi
41. Sun D, Bullock MR, McGinn MJ et al (2009) Basic fibroblast growth factor-enhanced neurogenesis contributes to cognitive recovery in rats following traumatic brain injury. Exp Neurol 216:56–65
42. Xiong Y, Lu D, Qu C et al (2008) Effects of erythropoietin on reducing brain damage and improving functional outcome after traumatic brain injury in mice. J Neurosurg 109:510–521
43. Xiong Y, Mahmood A, Meng Y et al (2010) Delayed administration of erythropoietin reducing hippocampal cell loss, enhancing angiogenesis and neurogenesis, and improving functional outcome following traumatic brain injury in rats: comparison of treatment with single and triple dose. J Neurosurg 113:598–608

44. Sun D, Colello RJ, Daugherty WP et al (2005) Cell proliferation and neuronal differentiation in the dentate gyrus in juvenile and adult rats following traumatic brain injury. J Neurotrauma 22:95–105
45. Cameron HA, Woolley CS, McEwen BS et al (1993) Differentiation of newly born neurons and glia in the dentate gyrus of the adult rat. Neuroscience 56:337–344
46. Sun D, McGinn MJ, Zhou Z et al (2007) Anatomical integration of newly generated dentate granule neurons following traumatic brain injury in adult rats and its association to cognitive recovery. Exp Neurol 204:264–272
47. Lok J, Gupta P, Guo S et al (2007) Cell-cell signaling in the neurovascular unit. Neurochem Res 32:2032–2045
48. Guo S, Lo EH (2009) Dysfunctional cell-cell signaling in the neurovascular unit as a paradigm for central nervous system disease. Stroke 40:S4–S7
49. Ohab JJ, Fleming S, Blesch A et al (2006) A neurovascular niche for neurogenesis after stroke. J Neurosci 26:13007–13016
50. Zhang Y, Chopp M, Mahmood A et al (2012) Impact of inhibition of erythropoietin treatment-mediated neurogenesis in the dentate gyrus of the hippocampus on restoration of spatial learning after traumatic brain injury. Exp Neurol 235:336–344
51. Meng Y, Xiong Y, Mahmood A et al (2011) Dose-dependent neurorestorative effects of delayed treatment of traumatic brain injury with recombinant human erythropoietin in rats. J Neurosurg 115:550–560
52. Xiong Y, Zhang Y, Mahmood A et al (2012) Neuroprotective and neurorestorative effects of thymosin beta4 treatment initiated 6 hours after traumatic brain injury in rats. J Neurosurg 116:1081–1092
53. Xiong Y, Mahmood A, Zhang Y et al (2011) Effects of posttraumatic carbamylated erythropoietin therapy on reducing lesion volume and hippocampal cell loss, enhancing angiogenesis and neurogenesis, and improving functional outcome in rats following traumatic brain injury. J Neurosurg 114:549–559
54. McTigue DM, Tripathi RB (2008) The life, death, and replacement of oligodendrocytes in the adult CNS. J Neurochem 107:1–19
55. Carroll WM, Jennings AR (1994) Early recruitment of oligodendrocyte precursors in CNS demyelination. Brain 117(Pt 3):563–578
56. Gensert JM, Goldman JE (1997) Endogenous progenitors remyelinate demyelinated axons in the adult CNS. Neuron 19:197–203
57. Xiong Y, Mahmood A, Meng Y et al (2011) Treatment of traumatic brain injury with thymosin beta(4) in rats. J Neurosurg 114:102–115
58. Oshima T, Lee S, Sato A et al (2009) TNF-alpha contributes to axonal sprouting and functional recovery following traumatic brain injury. Brain Res 1290:102–110
59. Abdel Baki SG, Schwab B, Haber M et al (2010) Minocycline synergizes with N-acetylcysteine and improves cognition and memory following traumatic brain injury in rats. PLoS One 5:e12490
60. Huebner EA, Strittmatter SM (2009) Axon regeneration in the peripheral and central nervous systems. Results Probl Cell Differ 48:339–351
61. Pham LD, Hayakawa K, Seo JH et al (2012) Crosstalk between oligodendrocytes and cerebral endothelium contributes to vascular remodeling after white matter injury. Glia 60:875–881
62. Arai K, Lo EH (2009) An oligovascular niche: cerebral endothelial cells promote the survival and proliferation of oligodendrocyte precursor cells. J Neurosci 29:4351–4355
63. Hayakawa K, Seo JH, Pham LD et al (2012) Cerebral endothelial derived vascular endothelial growth factor promotes the migration but not the proliferation of oligodendrocyte precursor cells in vitro. Neurosci Lett 513:42–46
64. Wu H, Mahmood A, Qu C et al (2012) Simvastatin attenuates axonal injury after experimental traumatic brain injury and promotes neurite outgrowth of primary cortical neurons. Brain Res 1486:121–130

65. Friedenstein AJ, Gorskaja JF, Kulagina NN (1976) Fibroblast precursors in normal and irradiated mouse hematopoietic organs. Exp Hematol 4:267–274
66. Javazon EH, Colter DC, Schwarz EJ et al (2001) Rat marrow stromal cells are more sensitive to plating density and expand more rapidly from single-cell-derived colonies than human marrow stromal cells. Stem Cells 19:219–225
67. Lennon DP, Haynesworth SE, Young RG et al (1995) A chemically defined medium supports in vitro proliferation and maintains the osteochondral potential of rat marrow-derived mesenchymal stem cells. Exp Cell Res 219:211–222
68. Dennis JE, Merriam A, Awadallah A et al (1999) A quadripotential mesenchymal progenitor cell isolated from the marrow of an adult mouse. J Bone Miner Res 14:700–709
69. Pittenger MF, Mackay AM, Beck SC et al (1999) Multilineage potential of adult human mesenchymal stem cells. Science 284:143–147
70. Azizi SA, Stokes D, Augelli BJ et al (1998) Engraftment and migration of human bone marrow stromal cells implanted in the brains of albino rats–similarities to astrocyte grafts. Proc Natl Acad Sci U S A 95:3908–3913
71. Woodbury D, Schwarz EJ, Prockop DJ et al (2000) Adult rat and human bone marrow stromal cells differentiate into neurons. J Neurosci Res 61:364–370
72. Sanchez-Ramos J, Song S, Cardozo-Pelaez F et al (2000) Adult bone marrow stromal cells differentiate into neural cells in vitro. Exp Neurol 164:247–256
73. Mahmood A, Lu D, Chopp M (2004) Intravenous administration of marrow stromal cells (MSCs) increases the expression of growth factors in rat brain after traumatic brain injury. J Neurotrauma 21:33–39
74. Chen X, Katakowski M, Li Y et al (2002) Human bone marrow stromal cell cultures conditioned by traumatic brain tissue extracts: growth factor production. J Neurosci Res 69:687–691
75. Mahmood A, Lu D, Wang L et al (2002) Intracerebral transplantation of marrow stromal cells cultured with neurotrophic factors promotes functional recovery in adult rats subjected to traumatic brain injury. J Neurotrauma 19:1609–1617
76. Lu D, Mahmood A, Wang L et al (2001) Adult bone marrow stromal cells administered intravenously to rats after traumatic brain injury migrate into brain and improve neurological outcome. Neuroreport 12:559–563
77. Mahmood A, Lu D, Wang L et al (2001) Treatment of traumatic brain injury in female rats with intravenous administration of bone marrow stromal cells. Neurosurgery 49:1196–1203, discussion 1203–1194
78. Jiang Q, Qu C, Chopp M et al (2011) MRI evaluation of axonal reorganization after bone marrow stromal cell treatment of traumatic brain injury. NMR Biomed 24:1119–1128
79. Mahmood A, Lu D, Lu M et al (2003) Treatment of traumatic brain injury in adult rats with intravenous administration of human bone marrow stromal cells. Neurosurgery 53:697–702, discussion 702–693
80. Mahmood A, Lu D, Chopp M (2004) Marrow stromal cell transplantation after traumatic brain injury promotes cellular proliferation within the brain. Neurosurgery 55:1185–1193
81. Qu C, Xiong Y, Mahmood A et al (2009) Treatment of traumatic brain injury in mice with bone marrow stromal cell-impregnated collagen scaffolds. J Neurosurg 111:658–665
82. Qu C, Mahmood A, Liu XS et al (2011) The treatment of TBI with human marrow stromal cells impregnated into collagen scaffold: functional outcome and gene expression profile. Brain Res 1371:129–139
83. Davis AA, Temple S (1994) A self-renewing multipotential stem cell in embryonic rat cerebral cortex. Nature 372:263–266
84. Alvarez-Buylla A, Lois C (1995) Neuronal stem cells in the brain of adult vertebrates. Stem Cells 13:263–272
85. Hong SQ, Zhang HT, You J et al (2011) Comparison of transdifferentiated and untransdifferentiated human umbilical mesenchymal stem cells in rats after traumatic brain injury. Neurochem Res 36:2391–2400

86. Wallenquist U, Brannvall K, Clausen F et al (2009) Grafted neural progenitors migrate and form neurons after experimental traumatic brain injury. Restor Neurol Neurosci 27:323–334
87. Gao J, Prough DS, McAdoo DJ et al (2006) Transplantation of primed human fetal neural stem cells improves cognitive function in rats after traumatic brain injury. Exp Neurol 201:281–292
88. Boockvar JA, Schouten J, Royo N et al (2005) Experimental traumatic brain injury modulates the survival, migration, and terminal phenotype of transplanted epidermal growth factor receptor-activated neural stem cells. Neurosurgery 56:163–171, discussion 171
89. Harting MT, Sloan LE, Jimenez F et al (2009) Subacute neural stem cell therapy for traumatic brain injury. J Surg Res 153:188–194
90. Shear DA, Tate CC, Tate MC et al (2011) Stem cell survival and functional outcome after traumatic brain injury is dependent on transplant timing and location. Restor Neurol Neurosci 29:215–225
91. Skardelly M, Gaber K, Burdack S et al (2011) Long-term benefit of human fetal neuronal progenitor cell transplantation in a clinically adapted model after traumatic brain injury. J Neurotrauma 28:401–414
92. Grasso G, Sfacteria A, Cerami A et al (2004) Erythropoietin as a tissue-protective cytokine in brain injury: what do we know and where do we go? Neuroscientist 10:93–98
93. Wang L, Chopp M, Gregg SR et al (2008) Neural progenitor cells treated with EPO induce angiogenesis through the production of VEGF. J Cereb Blood Flow Metab 28:1361–1368
94. Zhang Y, Xiong Y, Mahmood A et al (2010) Sprouting of corticospinal tract axons from the contralateral hemisphere into the denervated side of the spinal cord is associated with functional recovery in adult rat after traumatic brain injury and erythropoietin treatment. Brain Res 1353:249–257
95. Xiong Y, Mahmood A, Qu C et al (2010) Erythropoietin improves histological and functional outcomes after traumatic brain injury in mice in the absence of the neural erythropoietin receptor. J Neurotrauma 27:205–215
96. Mahmood A, Lu D, Qu C et al (2007) Treatment of traumatic brain injury in rats with erythropoietin and carbamylated erythropoietin. J Neurosurg 107:392–397
97. Lu D, Mahmood A, Goussev A et al (2004) Atorvastatin reduction of intravascular thrombosis, increase in cerebral microvascular patency and integrity, and enhancement of spatial learning in rats subjected to traumatic brain injury. J Neurosurg 101:813–821
98. Lu D, Goussev A, Chen J et al (2004) Atorvastatin reduces neurological deficit and increases synaptogenesis, angiogenesis, and neuronal survival in rats subjected to traumatic brain injury. J Neurotrauma 21:21–32
99. Wang B, Sun L, Tian Y et al (2012) Effects of atorvastatin in the regulation of circulating EPCs and angiogenesis in traumatic brain injury in rats. J Neurol Sci 319:117–123
100. Yarmola EG, Klimenko ES, Fujita G et al (2007) Thymosin beta4: actin regulation and more. Ann N Y Acad Sci 1112:76–85
101. Malinda KM, Sidhu GS, Mani H et al (1999) Thymosin beta4 accelerates wound healing. J Invest Dermatol 113:364–368
102. Huff T, Muller CS, Otto AM et al (2001) beta-Thymosins, small acidic peptides with multiple functions. Int J Biochem Cell Biol 33:205–220
103. Smart N, Dube KN, Riley PR (2013) Epicardial progenitor cells in cardiac regeneration and neovascularisation. Vascul Pharmacol 58(3):164–173. doi:10.1016/j.vph.2012.08.001
104. Wei C, Kumar S, Kim IK et al (2012) Thymosin beta 4 protects cardiomyocytes from oxidative stress by targeting anti-oxidative enzymes and anti-apoptotic genes. PLoS One 7: e42586
105. Crockford D (2007) Development of thymosin beta4 for treatment of patients with ischemic heart disease. Ann N Y Acad Sci 1112:385–395
106. Qiu FY, Song XX, Zheng H et al (2009) Thymosin beta4 induces endothelial progenitor cell migration via PI3K/Akt/eNOS signal transduction pathway. J Cardiovasc Pharmacol 53: 209–214

107. Santra M, Chopp M, Zhang ZG et al (2012) Thymosin beta 4 mediates oligodendrocyte differentiation by upregulating p38 MAPK. Glia 60:1826–1838
108. Thau-Zuchman O, Shohami E, Alexandrovich AG et al (2010) Vascular endothelial growth factor increases neurogenesis after traumatic brain injury. J Cereb Blood Flow Metab 30: 1008–1016
109. Thau-Zuchman O, Shohami E, Alexandrovich AG et al (2012) Subacute treatment with vascular endothelial growth factor after traumatic brain injury increases angiogenesis and gliogenesis. Neuroscience 202:334–341
110. Thau-Zuchman O, Shohami E, Alexandrovich AG et al (2012) Combination of vascular endothelial and fibroblast growth factor 2 for induction of neurogenesis and angiogenesis after traumatic brain injury. J Mol Neurosci 47:166–172
111. Siddiq I, Park E, Liu E et al (2012) Treatment of traumatic brain injury using zinc-finger protein gene therapy targeting VEGF-A. J Neurotrauma 29:2647–2659
112. Ma Y, Liu W, Wang Y et al (2011) VEGF protects rat cortical neurons from mechanical trauma injury induced apoptosis via the MEK/ERK pathway. Brain Res Bull 86:441–446
113. Dore-Duffy P, Wang X, Mehedi A et al (2007) Differential expression of capillary VEGF isoforms following traumatic brain injury. Neurol Res 29:395–403
114. Mizuno S, Matsumoto K, Kurosawa T et al (2000) Reciprocal balance of hepatocyte growth factor and transforming growth factor-beta 1 in renal fibrosis in mice. Kidney Int 57:937–948
115. Shang J, Deguchi K, Ohta Y et al (2011) Strong neurogenesis, angiogenesis, synaptogenesis, and antifibrosis of hepatocyte growth factor in rats brain after transient middle cerebral artery occlusion. J Neurosci Res 89:86–95
116. Takeuchi S, Wang W, Li Q et al (2012) Dual inhibition of Met kinase and angiogenesis to overcome HGF-induced EGFR-TKI resistance in EGFR mutant lung cancer. Am J Pathol 181:1034–1043
117. Cuevas P, Gimenez-Gallego G, Martinez-Murillo R et al (1991) Immunohistochemical localization of basic fibroblast growth factor in ependymal cells of the rat lateral and third ventricles. Acta Anat (Basel) 141:307–310
118. Takamiya M, Fujita S, Saigusa K et al (2007) Simultaneous detections of 27 cytokines during cerebral wound healing by multiplexed bead-based immunoassay for wound age estimation. J Neurotrauma 24:1833–1844
119. Lee ST, Chu K, Jung KH et al (2005) Granulocyte colony-stimulating factor enhances angiogenesis after focal cerebral ischemia. Brain Res 1058:120–128
120. Sharma HS, Zimmermann-Meinzingen S, Johanson CE (2010) Cerebrolysin reduces blood-cerebrospinal fluid barrier permeability change, brain pathology, and functional deficits following traumatic brain injury in the rat. Ann N Y Acad Sci 1199:125–137
121. Anton Alvarez X, Fuentes P (2011) Cerebrolysin in Alzheimer's disease. Drugs Today (Barc) 47:487–513
122. Zhang C, Chopp M, Cui Y et al (2010) Cerebrolysin enhances neurogenesis in the ischemic brain and improves functional outcome after stroke. J Neurosci Res 88:3275–3281

Chapter 9
Vascular Mechanisms in Spinal Cord Injury

Theo Hagg

Abstract This review provides an introduction to the field of vascular dysfunction as a cause of secondary degeneration following traumatic spinal cord injury. Major breakthroughs have been made by using endothelial cell-selective treatments to show that endothelial cell survival and function is key to protecting the spinal cord tissue. Other vascular treatment strategies involve the reduction of detrimental leakage and improving microvascular perfusion and function. The toxicity of the acutely developing hemorrhage most likely is a major problem but has not been solved so far. The role of angiogenesis in spinal cord injury remains unclear. The advances made between late 2011 and the end of 2012 in understanding molecular mechanisms and opportunities for therapeutic intervention receive particular attention. Lastly, future directions are discussed, including the need for microvascular diagnostic tools and challenges of clinical translation.

Abbreviations

BSCB	Blood–spinal cord-barrier
CHOP	CCAAT enhancer binding protein (C/EBP) homologous protein
FGF2	Fibroblast growth factor 2 (basic)
JNK	c-Jun N-terminal kinase
PDGF	Platelet-derived growth factor
Sur1	Sulfonylurea receptor 1
TRPM4	Transient receptor potential cation channel, subfamily M, member 4
VEGF	Vascular endothelial growth factor

T. Hagg (✉)
Kentucky Spinal Cord Injury Research Center, Department of Neurological Surgery, University of Louisville, 511 S. Floyd Street, MDR Building 616, Louisville, KY 40292, USA
e-mail: theo.hagg@louisville.edu

E.H. Lo et al. (eds.), *Vascular Mechanisms in CNS Trauma*, Springer Series in Translational Stroke Research 5, DOI 10.1007/978-1-4614-8690-9_9,

9.1 Historical Perspective

Traumatic spinal cord injury consists of an acute primary injury by which some neural tissue is irreparably lost. This is followed by secondary degeneration of the spinal cord tissue at the injury site over the first hours which progresses into the penumbra over the first 2 weeks and sometimes longer [1–5] (Fig. 9.1). The ongoing nature of the secondary injury processes provides opportunities for developing treatments to improve recovery of function. However, no successful FDA-approved neuroprotective treatment for human spinal cord injury exists yet. Experimental treatments during the acute phase are primarily aimed at rescuing neurons and white matter [6, 7]. They have included the reduction of inflammation [8, 9], sodium channel permeability [10], and free radical and lipid peroxidation damage [1]. Vascular dysfunction as a major contributor to degeneration after contusive spinal cord injury was discovered more than 100 years ago in dogs [11, 12]. One of the first studies to focus on the microvasculature showed very early pathological changes after a thoracic contusion in monkeys, including hemorrhage and ischemic endothelial injury [13]. The roles of vascular damage and vasospasm in the central hemorrhagic necrosis were also recognized [14–16]. Thus, it was proposed that "the initial vascular damage and subsequent reparative changes within the spinal cord appear to adequately explain the cavitation observed" [17]. Moreover, using perfusion measurements with colloidal carbon and microCT analyses in cats it was concluded that "The hypoperfusion of the white matter found in this study suggests that ischemia plays a role in paraplegia resulting from experimental compression injury of the spinal cord" [18]. One of the earliest pharmacological studies following spinal cord injury was aimed at improving vascular perfusion, showing the best effects with antioxidants in cats with a contusive injury [19]. Many more details of the developing vascular pathology have come to light and many other treatment strategies have been found over the years. The number of publications dealing with endothelial cells in the context of spinal cord injury has gained momentum over the past 20 years (Fig. 9.2). Many of these advances have been described in our last comprehensive reviews of 2011 [20, 21]. This review will summarize the current understanding of the mechanisms and therapies for microvascular dysfunction, each followed by an update of studies performed between 2011 and late 2012. The reader is also referred to other comprehensive reviews dealing specifically with vascular mechanisms of spinal cord injury [22–25].

9.2 Mechanisms of Microvascular Dysfunction

Contusive and compressive acute spinal cord injuries cause rapid local hemorrhage and disruption of normal blood flow due to the direct physical disruption of the microvessels [26–28]. The injury and hemorrhage also lead to vasoconstriction

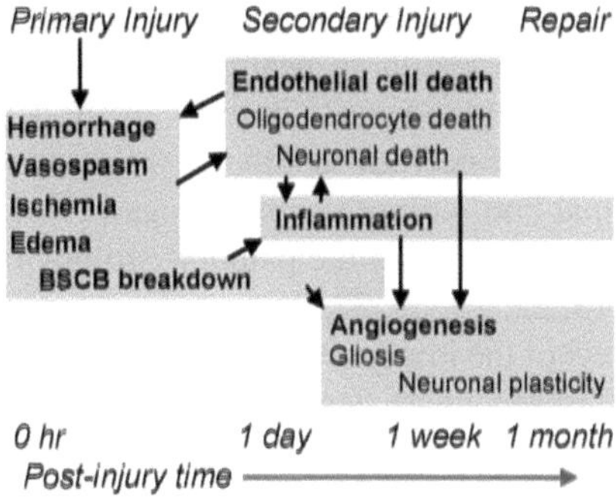

Fig. 9.1 Sequence of events following acute traumatic spinal cord injury. Shown here is a simplified schematic of the consequences of spinal cord injury. The high energy shear and blunt contusive forces of the primary injury cause an evolving hemorrhage, vasospasm, and ischemia, the latter can be exacerbated by continuing compression. The ensuing blood–spinal cord-barrier (BSCB) breakdown contributes to dysregulated water homeostasis with edema which can cause additional compression. The ischemic condition and the toxicity of the hemorrhage cause early endothelial and other cell death at the epicenter which leads to additional hemorrhage and a progressive secondary degeneration. Inflammation is initiated during the first day and can last for months, and has largely detrimental effects further exacerbating the secondary tissue loss. A reparative phase follows, which is accompanied by angiogenesis and gliosis reestablishing, among others, the disrupted BSCB. Neuronal plasticity also contributes to functional recovery. All events associated directly with endothelial cells are shown in *bold* and highlight the central role of vascular dysfunction in the pathogenesis of tissue loss following spinal cord injury

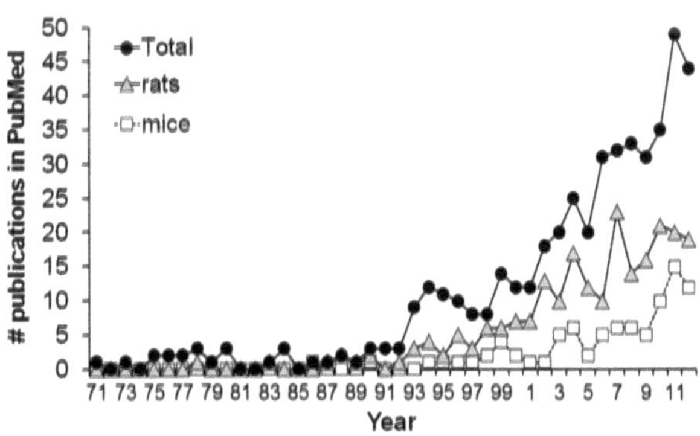

Fig. 9.2 Publications on endothelial cells and spinal cord injury continue to gain momentum. The study of the vasculature in the context of spinal cord injury continues to gain momentum as shown by the "Total" number of publications per year retrieved from Pubmed using the keywords "endothelial" and "spinal cord injury." This is in part due to the basic science studies as identified by using the additional keywords "rats" or "mice." Note that the use of mice to study the vasculature following spinal cord injury lags by approximately 10 years but it is steadily increasing. The increase in endothelial cell studies as a percentage of the total number of spinal cord injury studies is also increasing (data not shown), possibly reflecting the recognition of the crucial role microvascular pathology plays

causing reduced or absent perfusion of many surviving microvessels [14, 23, 29]. This ischemic condition results in death of spinal cord cells and a necrotic injury epicenter. Moreover, blood itself is toxic to the CNS cells [30, 31] and removal of the hemorrhagic necrotic debris in the lesion epicenter resulted in improvement after spinal cord compression in dogs [11]. Hemorrhage and vasoconstriction are considered central to the pathophysiology of spinal cord injury in animals and humans [15, 22, 23, 27, 32, 33].

The detrimental effect of the injury on endothelial cells has also been recognized as a central component of secondary degeneration. Thus, the normal attachment of endothelial cells to their basement membrane is disrupted and within the injury epicenter they undergo degenerative changes as early as 15 min [22, 34–37]. Many of the endothelial cells die over the next hours to 48 h, primarily by necrosis and an oncotic form of cell death caused by de novo expression of nonselective ion channels, and the surviving blood vessels show increased permeability [27, 38, 39] (Figs. 9.3 and 9.4). This pathology is caused in part by loss of survival signaling from basement membrane-integrin binding and decreased expression of survival-promoting growth factors such as angiopoietin-1 [40], as well as reactive oxygen species and lipid peroxidation [1, 41].

The health of the normal CNS is dependent on restricted entry of peripheral substances, which is accomplished by the blood–brain-barrier and blood–spinal cord-barrier (BSCB), which have unique properties, including adherens and tight junctions between the endothelial cells [42–44]. Spinal cord injury results in lengthy increases in permeability [26, 34] and BSCB breakdown [45, 46] in part by leakage through impaired or absent tight junctions and expression of plasmalemmal vesicle-associated protein-1 of damaged and (fenestrated) new vessels [23, 37, 46, 47]. This can be seen in dynamic MRI in injured rats and mice [48]. The extracellular matrix and perivascular cells such as pericytes and astrocytes also contribute to the barrier, including regulation of leukocyte entry into the CNS and astrocytes in particular are important for recovery [49–51].

Surviving endothelial cells at the injury site initiate the growth of new blood vessels or angiogenesis over the first 2 weeks after spinal cord injury in rats and mice, although it is transient in rats [35–38, 46, 52]. Whether or not spontaneous angiogenesis is detrimental or beneficial has not been conclusively shown. The mechanisms leading to angiogenesis following spinal cord injury most likely include increased expression of angiopoietin-2 and vascular endothelial growth factor (VEGF), although they may also contribute to increased permeability [53]. The $\alpha1\beta1$ integrin [54] and a disintegrin and metalloprotease 8 [55] also may play a role.

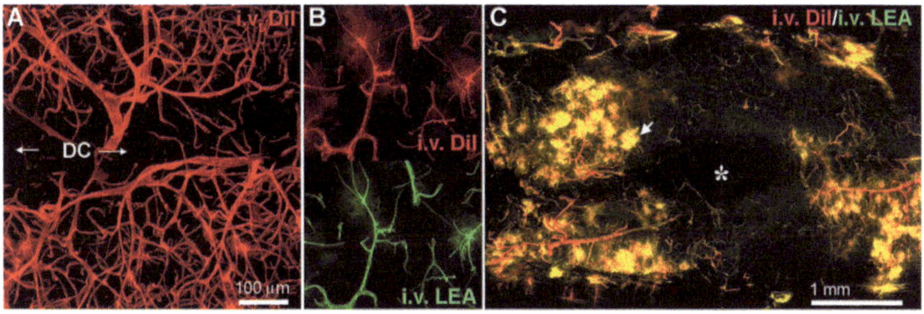

Fig. 9.3 Spinal cord injury causes early endothelial cell death and blood–spinal cord-barrier breakdown. (**a**) Normal microvasculature can be seen in a confocal image of a horizontal 100 μm thick slice of a mouse injected intravenously with the fluorescent dye DiI. Note the sparse perfusion of the dorsal column (DC) white matter where axons are located compared to the gray matter where the spinal cord neurons are located. (**b**) Intravenous injection with a fluorescent LEA lectin, which binds to endothelial cell surfaces, can identify perfused blood vessels with an endothelial lining. (**c**) One day following contusive spinal cord injury in adult mice the vasculature was labeled with intravenous injection of DiI and LEA and a thick horizontal section of the spinal cord imaged by confocal microscopy. Rostral is to the left, caudal to the right. The epicenter is devoid of perfused blood vessels (*asterisk*), representing endothelial cell death and loss of local perfusion creating an ischemic injury. The penumbra contains many leaky blood vessels leading to extravasation of fluorescent vascular labeling (for example, *arrow*)

9.3 Mechanisms Update: Endothelial Cell Death and Dysfunction

One of the earliest defensive cellular responses to injury-induced protein misfolding is found in the endoplasmic reticulum stress response, but continued signaling results in cell death. Expression of two of these signaling proteins, activating transcription factor 4 and CCAAT enhancer binding protein (C/EBP) homologous protein (CHOP or DNA-damage inducible transcript 3), is upregulated in microvessels that are not perfused at 24 h following spinal cord contusion in mice [56]. Moreover, CHOP−/− mice have more blood vessels at 3 days post-injury and better white matter sparing and locomotor function over 6 weeks. This identifies endoplasmic reticulum stress response pathways as additional vascular targets for potential pharmacological inhibition which is expected to lead to better outcomes.

9.4 Mechanisms Update: Hemorrhage-Induced Degeneration

Progressive hemorrhage during the acute phase is one of the main causes of the initial damage to the spinal cord tissue. A recent replication study identified a major role of the method used to contuse the cervical spinal cord where a unilateral hemorrhagic injury with less spread occurs in a more lateral impact than after a

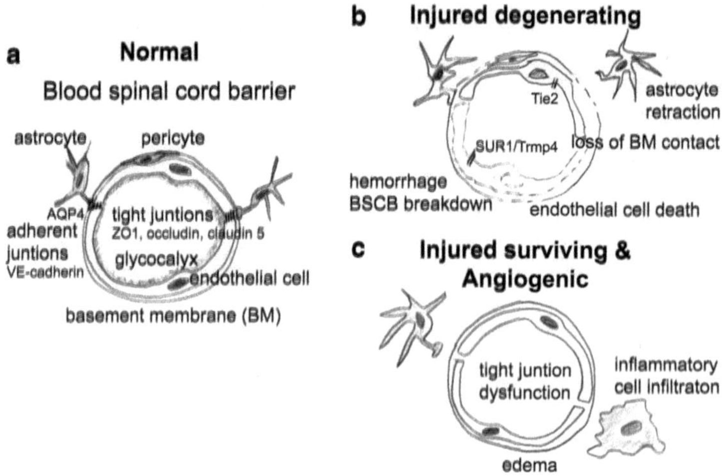

Fig. 9.4 Mechanisms leading to endothelial cell dysfunction after spinal cord injury. (**a**) The normal blood–spinal cord-barrier is maintained by tight and adherent junctions between the endothelial cells, as well as by a glycocalyx on the luminal side of the microvessels. This barrier regulates transport into the CNS, restricting entry of potentially harmful molecules and cells. Astrocytes contribute to the barrier by their endfeet positioned over the abluminal side of the endothelial cells, and their expression of aquaporin-4 (AQP4) maintains water balance thus preventing edema in the nervous tissue. The basement membrane provides structural support as well as survival-promoting signaling to endothelial cells. Pericytes are also found around microvessels but their normal function is not well-defined. (**b**) Endothelial cells die early following spinal cord injury in part by their de novo expression of nonselective ion channels (Sur1/Trpm4) leading to influx of sodium and oncotic death, as well as by loss of basement membrane contact and Tie2 signaling. Torn microvessels and those with dying endothelial cells cause an evolving hemorrhage which is toxic to the nervous tissue. Retraction of astrocyte endfeet exacerbates the pathological permeability caused by the blood–spinal cord-barrier breakdown. (**c**) Surviving endothelial cells and those produced by the angiogenic response in the injury epicenter and penumbra are functionally immature, leading to a protracted pathological permeability. Inflammatory cells exit into the spinal cord through the activated endothelial cell layer. Over time the barrier can be recreated by the formation of junctions and astroglial apposition, and the influx of new inflammatory cells diminishes

more medially placed impact which results in a more bilateral hemorrhage and lesion pattern [57]. This might have implications for diagnostics of human spinal cord injury. Perhaps a vector analysis of the primary injury could predict the ensuing hemorrhage and tissue injury.

9.5 Mechanisms Update: Blood–Spinal Cord-Barrier

BSCB dysfunction is known to be promoted by local inflammation. Expression of the histone H3K27me3 demethylase Jmjd3 is acutely upregulated in blood vessels of the spinal cord after contusion in rats and drives the expression of the pro-inflammatory cytokine IL-6 [58]. This can be replicated by oxygen glucose deprivation/reperfusion injury in cultured endothelial cells and blocked by siRNA. Thus, the ischemic conditions in the spinal cord in vivo might contribute to the initiation of inflammatory responses. It will be important to test whether inhibition of de novo expression of specific demethylases such as Jmjd3 would be beneficial. Interestingly, peripherally nerve injury causes leakage and inflammation in the spinal cord possibly by circulating IL1beta and MCP-1 in the cord [59]. The vascular dysfunction may be related to the development of pain as also suggested by the microglial responses. Importantly, the anti-inflammatory cytokines TGF-beta1 and IL-10 reduced the increase in permeability and spinal cord inflammation. It will be important to determine whether and by which mechanisms peripheral injury and inflammation affect outcomes after acute spinal cord injury.

The compromised BSCB is eventually repaired primarily by astrocytes which need to migrate and regrow their endfeet. The adhesion molecule tenascin-C is probably involved in this process as tenascin-C−/− mice have worse vascular leakage 14–21 days after injury after spinal cord compression injuries [60]. This correlates with worse functional outcomes. On the other hand, CHL1 may play an opposing role as CHL1−/− mice have less vascular leakage and improved recovery. These types of data will help to define mechanisms that are involved in vascular repair which will be a valuable addition to the acute vascular protection approaches.

The destabilizing role of angiopoietin-2 on endothelial cell adhesion is well-known and angiopoietin-2 is involved in adaptive angiogenesis and pathological permeability. Its expression following spinal cord contusion is increased over a long 10-week period in rats and seen in perivascular cells and astrocytes [61]. This temporal profile does not correlate with BSCB dysfunction, as suggested by the authors, or angiogenesis. The effect of angiopoietin-2 is context-dependent and it therefore may also play other roles. In fact, the protein expression levels measured by Western blot correlated with the locomotor function suggesting that it has a protective effect [61]. Protein levels of angiopoietin-2 in the CSF and serum of 15 humans with acute spinal cord injury is also increased over at least 5 days [24], whereas angiopoietin-1 peaked at 12 h and the potent angiogenic factor angiogenin did not change. It is possible that angiogenin levels would change at later times. Angiogenesis seems to occur in humans following spinal cord injury as more blood vessels are found at 7–10 days after injury [33]. The expression levels of the growth factors were not predictive of the neurological outcome at 6 months and 1 year although the sample of patients was too small to exclude these growth factors as potential biomarkers [24].

9.6 Mechanisms Update: Angiogenesis

The origin of angiogenesis into the injury site is thought to be from the surviving endothelial cells. However, mobilized and circulating bone marrow-derived endothelial progenitor cells detected 3 days following spinal cord injury may also contribute to neovascularization as shown with intravenous infusion of endothelial progenitors from transgenic mice with a Tie2 reporter [62]. Treatment with these cells also enhanced locomotor function after 2 weeks which would be consistent with an angiogenic response. The cell treatment also increased astrogliosis, which is perhaps associated with vascular repair. It remains to be tested whether the functional recovery was vascular and/or astrocyte-related, or neither. The astroglial response occurs via Jagged1-dependent Notch signaling, as shown by infusion of endothelial progenitors from jagged−/− mice into wildtype mice [63]. Jagged1+/+ endothelial progenitors caused morphologically abnormal vessel stabilization. Even so, locomotor function was improved but not after treatment with jagged−/− cells. One potential confound of this study is the use of nude mice which have an impaired immune response with greatly reduced numbers of T cells. In short, like immune cells, bone marrow-derived endothelial cells may have beneficial and detrimental effects and it will be important to understand the mechanisms of each of these effects.

9.7 Vascular-Selective Treatments

Given the central role of microvascular dysfunction after spinal cord injury it is not surprising that preservation of endothelial cells, blood flow, and reduced pathological permeability at the injury site would lead to better tissue preservation and functional outcomes. As also reviewed elsewhere in more detail [21] several different molecular mechanisms of the microvasculature have been targeted with pharmacological drugs. Reduced local blood perfusion has been improved following compressive injuries with the central vasodilator nimodipine and vasopressors in rats, although functional improvements were modest at best [64–66]. Reduction of oxidative damage has also led to improved blood flow and BSCB [19, 67, 68]. Matrix metalloproteinase (MMP) 9 plays a detrimental role following spinal cord injury and metalloprotease inhibition with GM6001 reduces leakiness and improves white matter sparing and locomotor function in mice with a contusive injury [69]. Intrathecal injection of SB-3CT, a selective MMP2/MMP9 inhibitor, also reduces permeability and cell death in rats with a spinal cord injury [70]. Spinal cord injury induces protective heme oxygenase in endothelial cells and its induction before the injury with hemin treatments improves outcomes [71].

Recently, more direct evidence for the crucial role of endothelial cell survival for better outcomes following spinal cord injury has emerged. Nonselective sulfonylurea receptor 1 (Sur1)-regulated NC(Ca-ATP) cation channels comprise the regulatory Sur1 (or Abcc8) subunit and a pore-forming transient receptor potential cation channel, subfamily M, member 4 (TRPM4) subunit. Both are expressed de novo in endothelial cells following spinal cord injury and contribute to their death possibly by persistent sodium influx [39, 72, 73]. Inhibition of Sur1 with glibenclamide or repaglinide, or genetic knock-down with antisense [39, 73], or genetic knock-down of TRPM4 [72] leads to improved endothelial cell survival and functional outcomes in cervical injuries in mice and rats. We have also found direct evidence for the crucial role of endothelial cell survival in mice treated with angiopoietin-1, which activates the endothelial-selective Tie2 receptor, leading to improved permeability, white matter sparing, and function [74]. Interestingly, the vascular protective effect could be enhanced by a peptide that stimulates the αvβ3 integrin, which is also involved in endothelial survival. Angiopoietin-1 has also been combined with VEGF to improve permeability and tissue sparing following spinal cord injury in rats [75]. Lastly, intrathecal infusion of an inhibitor of protein tyrosine phosphatases, which would lead to enhanced neurotrophic signaling, also rescues endothelial cells, white matter, and improves function [76].

We have promoted the intravenous administration route for vascular protective agents as a clinically relevant approach, because of its ease, which enables rapid treatment, and because it results in immediate bioavailability. Time is of the essence after acute neurological conditions such as spinal cord injury. Besides better metabolic support of the tissue, improved microvascular perfusion is expected to improve distribution of intravenously delivered therapeutics. Targeting the endothelial cell from the intravenous luminal side also circumvents the issue of low CNS bioavailability of most drugs.

The role of angiogenesis following spinal cord injury is still unclear. Several treatments have shown correlative effects of increased vascularity and tissue sparing and functional outcomes. VEGF is known to promote angiogenesis and improved function was seen in rats treated with VEGF over or into the spinal cord, although the vascularity was not assessed [53, 77]. Induction of VEGF with an engineered transcription factor or delivery by neural stem cells causes increased vascularity and improves outcomes after spinal cord injury [78, 79]. Treatment with an inhibitor of semaphorin 3A, which is an antagonist of the VEGF receptor, leads to increased vascularity and improved outcomes after spinal cord transection in rats, perhaps also related to axonal regeneration [80]. It should be noted that others have found that VEGF is ineffective [81] or even detrimental [82] following spinal cord injury suggesting that the effects of increased VEGF are very much context-dependent. Hepatocyte growth factor is also known as an angiogenic factor and improves vascularity and functional outcomes following spinal cord injury [83]. Intravenous treatment with endothelial cell precursors increases vascularity and function in injured rats [84]. Lastly, intravenous treatment with an αvβ3 integrin agonist peptide stimulates angiogenesis in mice following spinal cord injury and the number of microvessels correlates with the functional improvement [74]. In short, all studies so far only show circumstantial evidence for angiogenesis and for its relationship to improvements. Experimental approaches to provide more definitive evidence might be very challenging.

9.8 Treatment Update: Endothelial Cell Survival and Function

As discussed, newly upregulated Sur1 in endothelial cells in the epicenter after spinal cord injury contributes to their death perhaps due to pathological influx of ions, including sodium, whereas systemic treatment with the inhibitor glibenclamide reduces this [85]. Riluzole also blocks sodium channels, perhaps indirectly reducing excitotoxicity via glutamate receptors, and is being tested in phase I clinical trials for acute human spinal cord injury (ClinicalTrials.gov identifier NCT00876889; [86]). In a comparison study to glibenclamide, riluzole was found to also block Sur1-regulated NC(Ca-ATP) channels in endothelial cells and to also reduce endothelial cell fragmentation, hemorrhage, tissue loss, and functional deficits [87]. This suggests that some of the effects of riluzole in other spinal cord injury models are due to vascular protection. If so, this would be consistent with a central role of microvascular function in traumatic and other neurological disorders. Glibenclamide acts by inactivating the Sur1 regulatory part and riluzole inhibits the putative pore-forming subunit suggesting that a combination might even be better in reducing endothelial cell death.

Imatinib (Gleevac) is a tyrosine kinase inhibitor with selectivity for *bcr-abl* and can reduce hemorrhage, edema, and inflammation. After moderately severe spinal cord injury, a 5-day oral treatment with imatinib enhances tissue preservation, including vascular integrity, and provides a modest functional improvement in rats [88]. The BSCB function was improved as shown by protection by tight junctions and reduced leakiness most likely because of endothelial protection. The improvement in locomotor function was most evident from day 21 onwards which suggests that repair processes might have been involved. Therefore, it is of interest that the pericyte proliferation, as measured by proliferation markers, was reduced, because pericytes have now been shown as a source of the fibroblast scar following spinal cord injury [89].

9.9 Treatment Update: Hemorrhage

One of the major outcomes of Sur1 and TRPM4 inhibition following spinal cord injury is reduced hemorrhage. One of the few successful independent replication studies so far has shown that glibenclamide has the same improved outcomes in a unilateral cervical contusion model in rats [90]. This treatment reduced hemorrhage, although the fate of the endothelial cells was not analyzed as was done in the study that was being replicated [39]. As also noted in a commentary [57] the replication was not easy and depended on very specific injury parameters, with a unilateral impact applied from the dorsolateral side of the spinal cord. This replication is an important step forward and clinically relevant because glibenclamide is already FDA-approved as a drug for diabetes. The therapeutic effects might be due to penumbral effects on the gray matter rather than the injury core and the white matter. This is important in the cervical level where improved motor neuron survival is expected to substantially improve hand and arm function, as well as vital respiratory functions. Glibenclamide is effective in more medial unilateral injuries, although less so, suggesting that the benefit is inversely correlated to the primary hemorrhage [91]. It remains to be seen whether this relationship also exists in humans and how patients could be selected for drugs such as these which are perhaps only effective in certain types of injuries.

The toxic effects of blood theoretically would make it a good therapeutic target if toxin-reducing agents could reach the breakdown products of blood in the parenchyma rapidly enough. Ferrous iron is a catalyst for the production of reactive oxygen and nitrogen species which injured cells. Systemic treatment with the iron chelator salicylaldehyde isonicotinoyl hydrazine in mice with a spinal cord contusion over 6 weeks reduces the iron deposits and improves locomotor function most evident after 21 days [92]. The latter finding suggests that iron deposits in the tissue or released by macrophages have long-lasting detrimental effects on the nervous system. In the same study, ceruloplasmin was identified as an endogenous detoxifying protein and the critical role of macrophages and astrocytes in processing iron. These findings provide a platform for investigating potential additional therapeutic strategies.

The anticoagulant heparin can reduce tissue ischemia by maintaining patency of the microcirculation but its safety after spinal cord injury is unknown. In rats with a mild cervical spinal cord injury, heparin treatment improved microvascular perfusion as measured by i.v. injection of India ink at 3 days and gray matter sparing, but had no effects on white matter sparing or functional outcomes [93]. Interestingly, despite increased subpial hemorrhage, intraparenchymal hemorrhage was much reduced in the heparin-treated rats, suggesting that the endothelial cells survived due to the mild injury combined with continued microvascular perfusion. Heparin treatment after more severe injuries did not worsen the hemorrhage and did not improve outcomes, perhaps because such injuries cause maximal hemorrhage and are associated with dying endothelial cells.

9.10 Treatment Update: Vasospasm

As described before, improvement of microvascular perfusion caused by the vasoconstriction after spinal cord injury is one goal that has been pursued. Treatment with the vasodilator sildenafil (Viagra), which inhibits cGMP phosphodiesterase 5, for 1 week following contusion in mice improved perfusion of blood vessels that were present in the injury core [94]. The vasodilatory effects could have been directly via action on the smooth muscles which expressed phosphodiesterase 5. Microvascular endothelial cells also expressed phosphodiesterase 5 but it remains to be determined what increased cGMP might do as they do not regulate microvascular diameter. Sildenafil had no significant effect on the injury penumbra or on angiogenesis, perhaps explaining why it did not affect the locomotor deficits.

Magnesium is used clinically as a vasodilator. Previous studies have shown that intravenous injections with magnesium can improve functional outcomes following spinal cord compression [95] or when combined with polyethylene glycol after contusion in rats [96, 97]. We have recently documented microvascular hypoperfusion during the first 48 h after a contusion in rats which can be resolved with continuous intravenous infusion of magnesium chloride [113]. However, this treatment did not rescue endothelial cells, or tissue, and did not affect locomotor function, suggesting that better microvascular perfusion needs to be accompanied by protection of endothelial cells.

9.11 Treatment Update: Blood–Spinal Cord-Barrier

Metalloproteases are known to contribute to BSCB breakdown. Systemic treatment with the antidepressant drug fluoxetine has been shown to have neuroprotective activity potentially by inhibiting MMP2, 9, and 12 expression after spinal cord contusion in mice [98]. Fluoxetine treatment reduced pathological leakiness and loss of tight junction molecules at 1 day post-injury. A 2-week treatment also resulted in marked improved of locomotor function already at 10 days, consistent with a neuroprotective effect. Fluoxetine also reduced inflammatory markers, suggesting that the effects could be due to its anti-inflammatory effects. Because fluoxetine is a serotonin reuptake inhibitor some of the functional locomotor effects could be related to increases in serotonin, although this was not tested.

The c-Jun N-terminal kinase (JNK mitogen-activated protein kinase 8)-c-Jun pathway was known to play a role in causing blood–brain-barrier dysfunction in other models. A single i.p. injection of a specific JNK inhibitor 6 h after a T8 dorsal hemisection in mice improves white matter sparing and locomotor function [99]. The functional difference was already present at 1 week consistent with a protective effect, which was modest in terms of functionality because the scores were in the top range of the test. At 24 h, JNK inhibitor-treated mice showed less hemorrhage and BSCB permeability. The mechanism underlying the improved permeability is not clear because of the low number of endothelial cells with activated c-Jun in the control-treated mice. Thus, JNK inhibition in other cells may contribute to the improvement or JNK has another target within endothelial cells, which would be important to define.

9.12 Treatment Update: Angiogenesis

Several groups have used growth factors to enhance angiogenesis following spinal cord injury. The angiogenic growth factors VEGF and fibroblast growth factor 2 basic (FGF2) have been combined in slow-release microspheres and when implanted as bridges into a T9-10 hemi-resection in rats promote angiogenesis as shown by infiltrated endothelial cells and 3D casts of microfil-filled blood vessels [100]. VEGF alone was also effective. Interestingly, as also shown by others, neurofilament-positive axons were seen alongside blood vessels, reiterating the pro-regenerative potential of angiogenesis. Whether this treatment would result in a mature microvascular system and better functional outcomes remains to be determined. A growth factor mixture of EGF, FGF2, and PDGF-AA has been infused at the injury site in a study aimed at increasing production of oligodendrocytes from endogenous neural precursors after spinal cord compression [101]. Digestion of CSPGs was aimed at improving cell migration. Four days following the injury, there was an increased number of new endothelial cells as shown by double-labeling for BrdU (indicating proliferation) and RECA1. Interestingly, the growth factors or ChABC or both increased the number of new endothelial cells, suggesting a role for CSPGs in reducing endothelial cell migration. FGF2 alone has been loaded in an intrathecal nanoparticle scaffold for localized and sustained delivery after a clip compression injury in rats [102]. Four weeks later more blood vessels were seen in the dorsal horns, which could result from an angiogenic response or protection of the endothelial cells early after the injury. The functional outcomes remain to be tested. The advantages of such delivery approaches are that they circumvent the BSCB and avoid systemic side effects. Another advantage is the absence of proliferative lesions seen with local infusions via catheters.

New blood vessels have been considered potentially detrimental after spinal cord injury because of their leakiness. Platelet-derived growth factor (PDGF), which plays a role in vascular maturation, has therefore been infused for 7 days over the injury site together with VEGF, or applied by a slow-release patch, and reduces secondary degeneration after spinal cord hemisection [103]. However, there was no significant effect of the VEGF/PDGF treatment on the density of blood vessels at 1 and 3 months compared to the control treatment, with both having more than normal numbers of mature microvessels identified as tubular structures stained for smooth muscle actin in the injury penumbra. However, whether these microvessels were leaky or not remains to be tested. Also, whether angiogenesis might have been affected is unknown as endothelial cell markers were not used. Importantly, infusion of VEGF or PDGF alone greatly exacerbated the secondary degeneration. However, whether or not this could be related to an angiogenic response and/or immature phenotype of the microvessels remains to be determined as no blood vessel analysis was performed.

Others have used cell grafts to enhance angiogenesis. In the most direct manner so far (see also Sect. 9.6) bone marrow-derived endothelial progenitor cells have been shown to increase the number of endothelial cells in the injury site following spinal cord injury and result in improved locomotor function [62, 63]. However, whether the functional recovery was due to angiogenesis or not remains to be tested. Endothelial cells from human umbilical veins transplanted together with neural stem cells following a clip compression injury in rats resulted in much more angiogenesis, and less hypoxia and better survival of the neural stem cells within the ischemic core [104]. Whether the cells inosculated with perfused microvasculature was not tested. Also, no functional analyses were performed, leaving the question whether angiogenesis is beneficial unanswered. Intravenous infusion of mesenchymal stem cells in rats with a compression injury led to tissue sparing and a substantial improvement in locomotor function at 4 weeks post-injury [105]. The density of blood vessels was higher than in non-treated injured rats but it is not clear whether this represents angiogenesis or endothelial cell preservation. The treatment induced neurotrophic factors such as NGF as shown by microarray at 3 days post-injury. The late recovery of locomotor function would be consistent with increased angiogenesis or enhanced plasticity. Surprisingly, no mesenchymal stem cells were found in the spinal cord at any time, which leaves the nature of the therapeutic effect unanswered. Lastly, bone marrow stromal cells grafted directly into the injury site after a contusion improved white matter sparing, modest functional recovery, and reduced allodynia in rats [106]. The cell graft also resulted in a greater density of blood vessels, most likely due to an enhanced angiogenic response because the density in epicenter white matter of grafted rats was more than twofold higher than in uninjured rats. Also, the graft was injected at 3 days when most endothelial cell loss would already have occurred, suggesting that the effect on the microvasculature is due to angiogenesis. It will be important to define what agents these bone marrow stromal cells produce that enhances angiogenesis.

9.13 Future Directions

Most of the experimental techniques that have been used to document changes in microvascular function in animals have relied on histological methods [107]. These techniques include various endothelial cell markers, intravenous labeling techniques, and permeability assays. A few genetic reporter mouse models have been used. Although 3D reconstructions of the vascular network have been performed by microCT of colloidal carbon or silicone rubber [18, 108], the resolution has not been sufficient for capillaries. Insight into the real-time changes, especially the 3D relationship between different cell types, would be important but is limited by current imaging technology. A few groups have started to apply multiphoton microscopy to assess the vasculature in the spinal cord [109, 110]. The expectation is that a combination of this technology with fluorescent antibodies for endothelial cell markers or functional assays will greatly enhance our understanding of the pathophysiology after spinal cord injury. Importing noninvasive imaging technology from other vascular fields will also be helpful, especially those that can be readily translated to the clinic.

Repetitive measures will be important as diagnostics and to assess the efficacy of treatments. Such tools would include blood flow measurements, angiography, PET scanning with endothelial cell ligands, and contrast MRI. Biomarkers for spinal cord injury will also be important for understanding human pathology and translation of treatments to the clinic. They have been studied in the CSF of rats with a cervical contusion [111] and in serum and CSF of humans with spinal cord injury [112]. So far, no endothelial cell markers have been identified although HO-1 was suggested as a potential marker which is interesting because of its relationship to detrimental hemorrhage. It should be noted that CNS endothelial cell markers will be most useful given the fact that many people with acute spinal cord injury also have other injuries which damage or affect blood vessels.

One area of investigation that will require much more effort is the documentation of the newly identified mechanisms of microvascular pathology in humans with spinal cord injury. The overall tissue changes seem to be largely similar in different species [20] but more detailed comparisons between the microvascular response of different species, including larger animals, and their similarity to humans remain to be performed. This is important because it will inform us better whether the vascular-selective experimental therapies under development might be relevant. Of course, any such treatment that would show efficacy in humans with spinal cord injury would guide such comparative studies and would further stimulate the field to focus attention on endothelial cells.

Collectively, the experimental work in animals suggests that the extent of therapeutic protection of microvascular has reached a ceiling where up to half of the blood vessels can be protected. A few areas of research might help to determine whether a greater effect can be expected. For example, hemorrhage is thought to play a crucial role in the degenerative events following spinal cord injury and treatments which would reduce the toxic effects are needed. Also, the finding that many microvessels are not perfused points to the need to better define the vascular dynamics and methods to improve blood flow at the epicenter and which would lead to better tissue preservation. Adaptive and potentially pathological angiogenesis is another area which needs more attention, and will require experimental approaches that go beyond the largely descriptive ones applied so far.

In conclusion, the understanding of vascular pathology and vascular-selective therapies following spinal cord injury is rapidly evolving. The crucial role of the microvascular pathology in the secondary degeneration has opened up a window of opportunity to develop intravenous treatments. However, large gaps remain in our knowledge and the road to clinical translation of the experimental findings may be a long one yet.

Acknowledgements Justin Geralds is thanked for his contribution to Fig. 9.3. Supported by NIH grant NS045734 and the Commonwealth of Kentucky Challenge for Excellence.

References

1. Hall ED, Springer JE (2004) Neuroprotection and acute spinal cord injury: a reappraisal. NeuroRx 1:80–100
2. Hagg T, Oudega M (2006) Degenerative and spontaneous regenerative processes after spinal cord injury. J Neurotrauma 23:264–280
3. Bramlett HM, Dietrich WD (2007) Progressive damage after brain and spinal cord injury: pathomechanisms and treatment strategies. Prog Brain Res 161:125–141
4. Alexander JK, Popovich PG (2009) Neuroinflammation in spinal cord injury: therapeutic targets for neuroprotection and regeneration. Prog Brain Res 175:125–137
5. Popovich P, McTigue D (2009) Damage control in the nervous system: beware the immune system in spinal cord injury. Nat Med 15:736–737
6. Kwon BK, Okon E, Hillyer J et al (2011) A systematic review of non-invasive pharmacologic neuroprotective treatments for acute spinal cord injury. J Neurotrauma 28:1545–1588
7. Kwon BK, Okon EB, Plunet W et al (2011) A systematic review of directly applied biologic therapies for acute spinal cord injury. J Neurotrauma 28:1589–1610
8. Popovich PG, Guan Z, Wei P, Huitinga I, van Rooijen N, Stokes BT (1999) Depletion of hematogenous macrophages promotes partial hindlimb recovery and neuroanatomical repair after experimental spinal cord injury. Exp Neurol 158:351–365
9. Weaver LC, Gris D, Saville LR et al (2005) Methylprednisolone causes minimal improvement after spinal cord injury in rats, contrasting with benefits of an anti-integrin treatment. J Neurotrauma 22:1375–1387
10. Schwartz G, Fehlings MG (2001) Evaluation of the neuroprotective effects of sodium channel blockers after spinal cord injury: improved behavioral and neuroanatomical recovery with riluzole. J Neurosurg 94:245–256
11. Allen AR (1911) Surgery of experimental lesion of spinal cord equivalent to crush injury of fracture dislocation of spinal column. A preliminary report. JAMA 57:878–880
12. Allen AR (1914) Remarks on the histopathological changes in the spinal cord due to impact. An experimental study. J Nerv Ment Dis 41:141–147
13. Dohrmann GJ, Wagner FC Jr, Bucy PC (1971) The microvasculature in transitory traumatic paraplegia. An electron microscopic study in the monkey. J Neurosurg 35:263–271
14. Dohrmann GJ, Allen WE (1975) Microcirculation of traumatized spinal cord. A correlation of microangiography and blood flow patterns in transitory and permanent paraplegia. J Trauma 15(11):1003–1013
15. Nelson E, Gertz SD, Rennels ML, Ducker TB, Blaumanis OR (1977) Spinal cord injury. The role of vascular damage in the pathogenesis of central hemorrhagic necrosis. Arch Neurol 34:332–333
16. Schneider RC, Crosby EC (1959) Vascular insufficiency of brain stem and spinal cord in spinal trauma. Neurology 9:643–656
17. Wagner FC Jr, Van Gilder JC, Dohrmann GJ (1977) The development of intramedullary cavitation following spinal cord injury: an experimental pathological study. Paraplegia 14:245–250
18. Means ED, Anderson DK, Nicolosi G, Gaudsmith J (1978) Microvascular perfusion experimental spinal cord injury. Surg Neurol 9:353–360
19. Hall ED, Wolf DL (1986) A pharmacological analysis of the pathophysiological mechanisms of posttraumatic spinal cord ischemia. J Neurosurg 64:951–961
20. Benton RL, Hagg T (2011) Vascular pathology as potential therapeutic target in SCI. Trans Stroke Res 2:556–572
21. Fassbender JM, Whittemore SR, Hagg T (2011) Targeting microvasculature for neuroprotection after SCI. Neurotherapeutics 8:240–251
22. Tator CH, Fehlings MG (1991) Review of the secondary injury theory of acute spinal cord trauma with emphasis on vascular mechanisms. J Neurosurg 75:15–26

23. Mautes AE, Weinzierl MR, Donovan F, Noble LJ (2000) Vascular events after spinal cord injury: contribution to secondary pathogenesis. Phys Ther 80:673–687
24. Ng MT, Stammers AT, Kwon BK (2011) Vascular disruption and the role of angiogenic proteins after spinal cord injury. Transl Stroke Res 2:474–491
25. Oudega M (2012) Molecular and cellular mechanisms underlying the role of blood vessels in spinal cord injury and repair. Cell Tissue Res 349:269–288
26. Beggs JL, Waggener JD (1979) The acute microvascular responses to spinal cord injury. Adv Neurol 22:179–189
27. Noble LJ, Wrathall JR (1989) Correlative analyses of lesion development and functional status after graded spinal cord contusive injuries in the rat. Exp Neurol 103:34–40
28. Tator CH (1991) Review of experimental spinal cord injury with emphasis on the local and systemic circulatory effects. Neurochirurgie 37:291–302
29. Anthes DL, Theriault E, Tator CH (1996) Ultrastructural evidence for arteriolar vasospasm after spinal cord trauma. Neurosurgery 39:804–814
30. Sadrzadeh SM, Anderson DK, Panter SS, Hallaway PE, Eaton JW (1987) Hemoglobin potentiates central nervous system damage. J Clin Invest 79:662–664
31. Matz PG, Fujimura M, Chan PH (2000) Subarachnoid hemolysate produces DNA fragmentation in a pattern similar to apoptosis in mouse brain. Brain Res 858:312–319
32. Dumont RJ, Okonkwo DO, Verma S et al (2001) Acute spinal cord injury, part I: pathophysiologic mechanisms. Clin Neuropharmacol 24:254–264
33. Norenberg MD, Smith J, Marcillo A (2004) The pathology of human spinal cord injury: defining the problems. J Neurotrauma 21:429–440
34. Noble LJ, Mautes AE, Hall JJ (1996) Characterization of the microvascular glycocalyx in normal and injured spinal cord in the rat. J Comp Neurol 376:542–556
35. Loy DN, Crawford CH, Darnall JB, Burke DA, Onifer SM, Whittemore SR (2002) Temporal progression of angiogenesis and basal lamina deposition after contusive spinal cord injury in the adult rat. J Comp Neurol 445:308–324
36. Casella GT, Marcillo A, Bunge MB, Wood PM (2002) New vascular tissue rapidly replaces neural parenchyma and vessels destroyed by a contusion injury to the rat spinal cord. Exp Neurol 173:63–76
37. Benton RL, Maddie MA, Minnillo DR, Hagg T, Whittemore SR (2008) Griffonia simplicifolia isolectin B4 identifies a specific subpopulation of angiogenic blood vessels following contusive spinal cord injury in the adult mouse. J Comp Neurol 507:1031–1052
38. Casella GT, Bunge MB, Wood PM (2006) Endothelial cell loss is not a major cause of neuronal and glial cell death following contusion injury of the spinal cord. Exp Neurol 202:8–20
39. Simard JM, Tsymbalyuk O, Ivanov A et al (2007) Endothelial sulfonylurea receptor 1-regulated NC Ca-ATP channels mediate progressive hemorrhagic necrosis following spinal cord injury. J Clin Invest 117:2105–2113
40. Ritz MF, Graumann U, Gutierrez B, Hausmann O (2010) Traumatic spinal cord injury alters angiogenic factors and TGF-beta1 that may affect vascular recovery. Curr Neurovasc Res 7 (4):301–310
41. Hall ED (1995) Inhibition of lipid peroxidation in central nervous system trauma and ischemia. J Neurol Sci 134 suppl:79–83
42. Engelhardt B, Sorokin L (2009) The blood–brain and the blood-cerebrospinal fluid barriers: function and dysfunction. Semin Immunopathol 31:497–511
43. Quaegebeur A, Lange C, Carmeliet P (2011) The neurovascular link in health and disease: molecular mechanisms and therapeutic implications. Neuron 71:406–424
44. Del Zoppo GJ (2013) Toward the neurovascular unit. A journey in clinical translation: 2012 Thomas Willis Lecture. Stroke 44(1):263–269
45. Popovich PG, Horner PJ, Mullin BB, Stokes BT (1996) A quantitative spatial analysis of the blood-spinal cord barrier. I. Permeability changes after experimental spinal contusion injury. Exp Neurol 142:258–275

46. Whetstone WD, Hsu JY, Eisenberg M, Werb Z, Noble-Haeusslein LJ (2003) Blood-spinal cord barrier after spinal cord injury: relation to revascularization and wound healing. J Neurosci Res 74:227–239
47. Mozer AB, Whittemore SR, Benton RL (2010) Spinal microvascular expression of PV-1 is associated with inflammation, perivascular astrocyte loss, and diminished EC glucose transport potential in acute SCI. Curr Neurovasc Res 7:238–250
48. Byrnes KR, Fricke ST, Faden AI (2010) Neuropathological differences between rats and mice after spinal cord injury. J Magn Reson Imaging 32:836–846
49. Faulkner JR, Herrmann JE, Woo MJ, Tansey KE, Doan NB, Sofroniew MV (2004) Reactive astrocytes protect tissue and preserve function after spinal cord injury. J Neurosci 24:2143–2155
50. Okada S, Nakamura M, Katoh H et al (2006) Conditional ablation of Stat3 or Socs3 discloses a dual role for reactive astrocytes after spinal cord injury. Nat Med 12:829–834
51. Voskuhl RR, Peterson RS, Song B et al (2009) Reactive astrocytes form scar-like perivascular barriers to leukocytes during adaptive immune inflammation of the CNS. J Neurosci 29:11511–11522
52. Griffiths IR, McCulloch M, Crawford RA (1978) Ultrastructural appearances of the spinal microvasculature between 12 hours and 5 days after impact injury. Acta Neuropathol 43:205–211
53. Patel CB, Cohen DM, Ahobila-Vajjula P, Sundberg LM, Chacko T, Narayana PA (2009) Effect of VEGF treatment on the blood-spinal cord barrier permeability in experimental spinal cord injury: dynamic contrast-enhanced magnetic resonance imaging. J Neurotrauma 26:1005–1016
54. Baker KA, Hagg T (2007) Developmental and injury-induced expression of alpha1beta1 and alpha6beta1 integrins in the rat spinal cord. Brain Res 1130:54–66
55. Mahoney ET, Benton RL, Maddie MA, Whittemore SR, Hagg T (2009) ADAM8 is selectively up-regulated in endothelial cells and is associated with angiogenesis after spinal cord injury in adult mice. J Comp Neurol 512:243–255
56. Fassbender JM, Saraswat-Ohri S, Myers SA, Gruenthal MJ, Benton RL, Whittemore SR (2012) Deletion of endoplasmic reticulum stress-induced chop protects microvasculature post-spinal cord injury. Curr Neurovasc Res 9(4):274–281
57. Simard JM, Gerzanich V (2012) When replication teaches more than the original experiment—the saga of the unknown unknown. Exp Neurol 233:623–624
58. Lee K, Na W, Lee JY et al (2012) Molecular mechanism of Jmjd3-mediated interleukin-6 gene regulation in endothelial cells underlying spinal cord injury. J Neurochem 122 (2):272–282
59. Echeverry S, Shi XQ, Rivest S, Zhang J (2011) Peripheral nerve injury alters blood-spinal cord barrier functional and molecular integrity through a selective inflammatory pathway. J Neurosci 31:10819–10828
60. Peter NR, Shah RT, Chen J, Irintchev A, Schachner M (2012) Adhesion molecules close homolog of L1 and tenascin-C affect blood-spinal cord barrier repair. Neuroreport 23:479–482
61. Durham-Lee JC, Wu Y, Mokkapati VU, Paulucci-Holthauzen AA, Nesic O (2012) Induction of angiopoietin-2 after spinal cord injury. Neuroscience 202:454–464
62. Kamei N, Kwon SM, Kawamoto A et al (2012) Contribution of bone marrow-derived endothelial progenitor cells to neovascularization and astrogliosis following spinal cord injury. J Neurosci Res 90:2281–2292
63. Kamei N, Kwon SM, Ishikawa M et al (2012) Endothelial progenitor cells promote astrogliosis following spinal cord injury through Jagged1-dependent Notch signaling. J Neurotrauma 29:1758–1769
64. Guha A, Tator CH, Smith CR, Piper I (1989) Improvement in post-traumatic spinal cord blood flow with a combination of a calcium channel blocker and a vasopressor. J Trauma 29:1440–1447

65. Fehlings MG, Tator CH, Linden RD (1989) The effect of nimodipine and dextran on axonal function and blood flow following experimental spinal cord injury. J Neurosurg 71:403–416
66. Ross IB, Tator CH, Theriault E (1993) Effect of nimodipine or methylprednisolone on recovery from acute experimental spinal cord injury in rats. Surg Neurol 40:461–470
67. Hall ED (1988) Effects of the 21-aminosteroid U74006F on posttraumatic spinal cord ischemia in cats. J Neurosurg 68:462–465
68. Hall ED, McCall JM, Means ED (1994) Therapeutic potential of the lazaroids (21-aminosteroids) in acute central nervous system trauma, ischemia and subarachnoid hemorrhage. Adv Pharmacol 28:221–268
69. Noble LJ, Donovan F, Igarashi T, Goussev S, Werb Z (2002) Matrix metalloproteinases limit functional recovery after spinal cord injury by modulation of early vascular events. J Neurosci 22:7526–7535
70. Yu F, Kamada H, Niizuma K, Endo H, Chan PH (2008) Induction of mmp-9 expression and endothelial injury by oxidative stress after spinal cord injury. J Neurotrauma 25:184–195
71. Yamauchi T, Lin Y, Sharp FR, Noble-Haeusslein LJ (2004) Hemin induces heme oxygenase-1 in spinal cord vasculature and attenuates barrier disruption and neutrophil infiltration in the injured murine spinal cord. J Neurotrauma 21:1017–1030
72. Gerzanich V, Woo SK, Vennekens R et al (2009) De novo expression of Trpm4 initiates secondary hemorrhage in spinal cord injury. Nat Med 15:185–191
73. Simard JM, Woo SK, Norenberg MD et al (2010) Brief suppression of Abcc8 prevents autodestruction of spinal cord after trauma. Sci Transl Med 2:28ra29
74. Han S, Arnold SA, Sithu SD et al (2010) Rescuing vasculature with intravenous angiopoietin-1 and alpha v beta 3 integrin peptide is protective after spinal cord injury. Brain 133:1026–1042
75. Herrera JJ, Sundberg LM, Zentilin L, Giacca M, Narayana PA (2010) Sustained expression of vascular endothelial growth factor and angiopoietin-1 improves blood spinal cord barrier integrity and functional recovery after spinal cord injury. J Neurotrauma 27(11):2067–2076
76. Nakashima S, Arnold SA, Mahoney ET et al (2008) Small-molecule protein tyrosine phosphatase inhibition as a neuroprotective treatment after spinal cord injury in adult rats. J Neurosci 28:7293–7303
77. Widenfalk J, Lipson A, Jubran M et al (2003) Vascular endothelial growth factor improves functional outcome and decreases secondary degeneration in experimental spinal cord contusion injury. Neuroscience 120:951–960
78. Kim HM, Hwang DH, Lee JE, Kim SU, Kim BG (2009) Ex vivo VEGF delivery by neural stem cells enhances proliferation of glial progenitors, angiogenesis, and tissue sparing after spinal cord injury. PLoS One 4:e4987
79. Liu Y, Figley S, Spratt SK et al (2010) An engineered transcription factor which activates VEGF-A enhances recovery after spinal cord injury. Neurobiol Dis 37:384–393
80. Kaneko S, Iwanami A, Nakamura M et al (2006) A selective Sema3A inhibitor enhances regenerative responses and functional recovery of the injured spinal cord. Nat Med 12:1380–1389
81. van Neerven S, Joosten EA, Brook GA et al (2010) Repetitive intrathecal VEGF(165) treatment has limited therapeutic effects after spinal cord injury in the rat. J Neurotrauma 27:1781–1791
82. Benton RL, Whittemore SR (2003) VEGF165 therapy exacerbates secondary damage following spinal cord injury. Neurochem Res 28:1693–1703
83. Kitamura K, Iwanami A, Nakamura M et al (2007) Hepatocyte growth factor promotes endogenous repair and functional recovery after spinal cord injury. J Neurosci Res 85:2332–2342
84. Sasaki H, Ishikawa M, Tanaka N et al (2009) Administration of human peripheral blood-derived CD133+ cells accelerates functional recovery in a rat spinal cord injury model. Spine (Phila Pa 1976) 34:249–254

85. Simard JM, Woo SK, Schwartzbauer GT, Gerzanich V (2012) Sulfonylurea receptor 1 in central nervous system injury: a focused review. J Cereb Blood Flow Metab 32:1699–1717
86. Fehlings MG, Wilson JR, Frankowski RF et al (2012) Riluzole for the treatment of acute traumatic spinal cord injury: rationale for and design of the NACTN Phase I clinical trial. J Neurosurg Spine 17:151–156
87. Simard JM, Tsymbalyuk O, Keledjian K, Ivanov A, Ivanova S, Gerzanich V (2012) Comparative effects of glibenclamide and riluzole in a rat model of severe cervical spinal cord injury. Exp Neurol 233:566–574
88. Abrams MB, Nilsson I, Lewandowski SA et al (2012) Imatinib enhances functional outcome after spinal cord injury. PLoS One 7:e38760
89. Goritz C, Dias DO, Tomilin N, Barbacid M, Shupliakov O, Frisen J (2011) A pericyte origin of spinal cord scar tissue. Science 333:238–242
90. Popovich PG, Lemeshow S, Gensel JC, Tovar CA (2012) Independent evaluation of the effects of glibenclamide on reducing progressive hemorrhagic necrosis after cervical spinal cord injury. Exp Neurol 233:615–622
91. Simard JM, Popovich PG, Tsymbalyuk O, Gerzanich V (2012) Spinal cord injury with unilateral versus bilateral primary hemorrhage—effects of glibenclamide. Exp Neurol 233:829–835
92. Rathore KI, Kerr BJ, Redensek A et al (2008) Ceruloplasmin protects injured spinal cord from iron-mediated oxidative damage. J Neurosci 28:12736–12747
93. Zheng Y, Zhang YP, Shields LB et al (2011) Effect of heparin following cervical spinal cord injuries in rats. Neurosurgery 69:930–941
94. Myers SA, DeVries WH, Gruenthal MJ, Andres KR, Hagg T, Whittemore SR (2012) Sildenafil improves epicenter vascular perfusion but not hindlimb functional recovery after contusive spinal cord injury in mice. J Neurotrauma 29:528–538
95. Ditor DS, John SM, Roy J, Marx JC, Kittmer C, Weaver LC (2007) Effects of polyethylene glycol and magnesium sulfate administration on clinically relevant neurological outcomes after spinal cord injury in the rat. J Neurosci Res 85:1458–1467
96. Kwon BK, Roy J, Lee JH et al (2009) Magnesium chloride in a polyethylene glycol formulation as a neuroprotective therapy for acute spinal cord injury: preclinical refinement and optimization. J Neurotrauma 26:1379–1393
97. Lee JH, Roy J, Sohn HM et al (2010) Magnesium in a polyethylene glycol formulation provides neuroprotection after unilateral cervical spinal cord injury. Spine (Phila Pa 1976) 35:2041–2048
98. Lee JY, Kim HS, Choi HY, Oh TH, Yune TY (2012) Fluoxetine inhibits matrix metalloprotease activation and prevents disruption of blood-spinal cord barrier after spinal cord injury. Brain 135:2375–2389
99. Repici M, Chen X, Morel MP et al (2012) Specific inhibition of the JNK pathway promotes locomotor recovery and neuroprotection after mouse spinal cord injury. Neurobiol Dis 46:710–721
100. De Laporte L, des Rieux A, Tuinstra HM et al (2011) Vascular endothelial growth factor and fibroblast growth factor 2 delivery from spinal cord bridges to enhance angiogenesis following injury. J Biomed Mater Res A 98:372–382
101. Karimi-Abdolrezaee S, Schut D, Wang J, Fehlings MG (2012) Chondroitinase and growth factors enhance activation and oligodendrocyte differentiation of endogenous neural precursor cells after spinal cord injury. PLoS One 7:e37589
102. Kang CE, Baumann MD, Tator CH, Shoichet MS (2013) Localized and sustained delivery of fibroblast growth factor-2 from a nanoparticle-hydrogel composite for treatment of spinal cord injury. Cells Tissues Organs 197(1):55–63
103. Lutton C, Young YW, Williams R, Meedeniya AC, Mackay-Sim A, Goss B (2012) Combined VEGF and PDGF treatment reduces secondary degeneration after spinal cord injury. J Neurotrauma 29:957–970

104. Oh J, Kim KN, Yoon DH, Han SR, Shin DA, Ha Y (2012) Rapid recovery of tissue hypoxia by cotransplantation of endothelial cells. Neuroreport 23:658–662
105. Quertainmont R, Cantinieaux D, Botman O, Sid S, Schoenen J, Franzen R (2012) Mesenchymal stem cell graft improves recovery after spinal cord injury in adult rats through neurotrophic and pro-angiogenic actions. PLoS One 7:e39500
106. Ritfeld GJ, Tewarie RN, Vajn K et al (2012) Bone marrow stromal cell-mediated tissue sparing enhances functional repair after spinal cord contusion in adult rats. Cell Transplant 7:1561–1575
107. Hagg T, Benton RL, Fassbender JM, Whittemore SR (2012) Assessing microvessels after spinal cord injury. In: Chen J, Xu XM, Xu ZC, Zhang JH (eds) Animal models of acute neurological injuries II: injury and mechanistic assessments. Humana Press, Inc., Totowa, NJ, pp 499–519
108. Hu JZ, Wu TD, Zhang T, Zhao YF, Pang J, Lu HB (2012) Three-dimensional alteration of microvasculature in a rat model of traumatic spinal cord injury. J Neurosci Methods 204:150–158
109. Dray C, Rougon G, Debarbieux F (2009) Quantitative analysis by in vivo imaging of the dynamics of vascular and axonal networks in injured mouse spinal cord. Proc Natl Acad Sci U S A 106:9459–9464
110. Davalos D, Akassoglou K (2012) In vivo imaging of the mouse spinal cord using two-photon microscopy. J Vis Exp 59:e2760
111. Lubieniecka JM, Streijger F, Lee JH et al (2011) Biomarkers for severity of spinal cord injury in the cerebrospinal fluid of rats. PLoS One 6:e19247
112. Kwon BK, Casha S, Hurlbert RJ, Yong VW (2011) Inflammatory and structural biomarkers in acute traumatic spinal cord injury. Clin Chem Lab Med 49:425–433
113. Muradov JM, Hagg T (2013) Intravenous infusion of magnesium chloride improves epicenter blood flow during the acute stage of contusive spinal cord injury in rats. J Neurotrauma 30:840–52

Chapter 10
Neurovascular Mechanisms of Ischemia Tolerance Against Brain Injury

Kunjan R. Dave, John W. Thompson, Jake T. Neumann, Miguel A. Perez-Pinzon, and Hung W. Lin

Abstract Traumatic brain injury (TBI) can result in secondary ischemia. This secondary ischemic insult is implicated in post-TBI pathophysiology. Pharmacological intervention to elevate cerebral blood flow can improve outcomes following TBI. The brain and other organ systems have an innate ability to induce protection against ischemic injury, limiting the severity of the ischemia-induced damage. This "self" protection can be initiated by exposing the brain to a stimulus before ischemia called "preconditioning," such as exposure to a mild episode(s) of ischemia, hypoxia, anesthesia, or pharmacologically induced mild cell stressors. Current efforts to reduce ischemia-induced brain damage have been the focus in determining the mechanisms of preconditioning-induced ischemia tolerance as findings may help lower cerebral ischemia-induced brain damage in at-risk patients including TBI patients. Different preconditioning paradigms have been shown to lower TBI-induced damage. Although not all of the mechanisms of preconditioning are confirmed in models of TBI, basic mechanisms of preconditioning applies here as ischemia is a major part of TBI. Ischemic preconditioning, in part, confers protection by modulating regulators of cerebral blood flow, increase angiogenesis, and prevent cerebral ischemia-induced increase in blood–brain barrier permeability. This chapter highlights preconditioning-induced changes in components of the neurovascular system involved in ischemia tolerance. Understanding of these pathways may aid in the development of novel therapies to protect the brain from TBI-induced secondary ischemic insult.

M.A. Perez-Pinzon (✉)
The Cerebral Vascular Disease Research Laboratories, Department of Neurology, Leonard M. Miller School of Medicine, University of Miami, Miami, FL 33136, USA
e-mail: perezpinzon@miami.edu

E.H. Lo et al. (eds.), *Vascular Mechanisms in CNS Trauma*, Springer Series in Translational Stroke Research 5, DOI 10.1007/978-1-4614-8690-9_10,

10.1 Introduction

Traumatic brain injury (TBI) is one of the leading causes of death and disability in the USA [1]. Motor vehicle accidents, sports injuries, and simple falls are leading causes of TBI [1]. TBI-induced secondary ischemic insult is associated with poor outcome following severe TBI [2]. To counteract TBI, ischemia tolerance, which is the brain's (along with other organs) innate ability to protect itself against ischemic injury, can help lower TBI-induced ischemic damage. Ischemia tolerance can protect the brain against ischemic injury via activation of endogenous cellular pathways prior to ischemia, resulting in the induction of ischemic tolerance. The process that induces ischemia tolerance is termed preconditioning. Numerous preconditioning stimuli, including hypoxia/anoxia [3, 4], anesthetics [5–7], cortical spreading depression [8, 9], metabolic inhibitors [10–12], and ischemia itself [13–16], have demonstrated to protect the brain against ischemic damage. Preconditioning of the brain can also lower TBI-induced damage [17–22]. Most of the protective pathways are commonly activated by various preconditioning stimulus. We, and others, have identified several cell death and survival pathways that are inhibited or activated, following the induction of ischemic tolerance. These pathways directly affect the parenchyma and the neurovascular unit, which is defined as "a conceptual framework that links microvessel and neuron function and their responses to injury, but also represents a structural arrangement, recognizing that microvessel components and neurons connect via common astrocytes" [23–25].

Dysfunction of the neurovascular unit (i.e., perturbations in cerebral blood flow (CBF) and damage to the blood–brain barrier (BBB)) plays a critical role in cerebral ischemic damage. Thus, the focus of this chapter will be to discuss the role of the neurovascular components involved (i.e., CBF, ischemia/reperfusion, angiogenesis, and BBB permeability) in mitigating cerebral ischemia-induced damage in the "preconditioned" brain.

10.2 Tolerance Against TBI

Ischemic preconditioning (IPC) is defined as a brief period of mild ischemic stress followed by a period of recovery that induces protective mechanisms against a subsequent brain insults such as TBI and severe ischemia. IPC, when given 48 h prior to TBI resulted in 79 % lower contusion volume as compared to TBI alone group [17]. Another preconditioning stimulus, heat acclimation also protects the brain against closed head injury [26]. Hyperbaric oxygen a preconditioning stimulus has been shown to attenuate TBI at high altitudes [19, 20]. Preconditioning by exposing mice to sublethal doses of *N*-methyl-D-aspartate can improve TBI-induced motor and cognitive deficits [21, 22]. Overall, these studies suggest that preconditioning by various stimuli can induce tolerance against TBI. It is suggestive that these preconditioning stimuli may confer protection by lowering secondary ischemic injury along with primary TBI.

10.3 Basic Mechanisms of Ischemia Tolerance

IPC is characterized by an early or immediate window of ischemia tolerance which occurs within minutes of the preconditioning insult (lasting for a few hours) and is followed approximately 24–48 h later by a second window of protection which persists for days to even weeks [14, 27, 28]. The early window of protection is characterized by a rapid change in cellular physiology through the post-translational modification of proteins and ion channel permeability, whereas the delayed window of protection is dependent upon changes in gene transcription leading to altered protein expression [29]. Some of the molecular changes following IPC include maintenance of the mitochondrial function [30–33], reduced reactive oxygen species (ROS) formation [33–35], and suppression of cellular death pathways [36–38].

In recent years, there have been significant gains in the understanding of the trigger mechanisms, cellular signaling mediator pathways, and effector proteins which mediate IPC-induced ischemic tolerance. Some of the known mechanisms that initiate an IPC response include tumor necrosis factor alpha (TNF-α) [39], adenosine [40, 41], mitochondrial ATP-sensitive K^+ channels [3, 42], and oxidative stress induced by ROS and reactive nitrogen species [35, 43, 44]. Numerous signaling pathways activated by IPC have also been identified such as mitogen-activated protein kinase family (c-Jun N-terminal kinases (JNK), extracellular signal-regulated kinases (ERK1/2), and p38) [45–47], protein kinase B (Akt) [48], nuclear factor kappa-light-chain-enhancer of activated B cells (NF-κB) [49], signal transducers and activator of transcription 3 (STAT3) [50], protein kinase C epsilon (PKCε) [45, 51], and sirtuin 1 (SIRT1) [52–55]. Of these effector proteins, PKCε appears to be a converging point for the activation of numerous cell survival signaling pathways following IPC exposure [49, 50, 56].

10.3.1 PKC Epsilon

PKCε is a member of the PKC family of serine/threonine kinases, which has a pivotal role in IPC-induced ischemic tolerance in both the heart and brain. This central role for PKCε has been shown through the pharmacological inhibition of PKCε (using an isozyme-specific peptide, εV1-2), which attenuated IPC-induced neuroprotection in both in vivo and in vitro models of cerebral ischemia [42, 51, 56]. Similarly, the pharmacological activation of PKCε, in the absence of IPC, has been demonstrated to be sufficient to induce ischemic tolerance [45, 51, 56]. The importance of PKCε in IPC-induced ischemic tolerance is further demonstrated in the heart, where the overexpression of PKCε was sufficient to protect against ischemic injury [57] and knockout of PKCε in mice prevented IPC-mediated ischemic protection [58].

Currently, the molecular mechanisms of PKCε-induced ischemic tolerance are not fully understood, but PKCε appears to be involved in the direct protection of the mitochondria. In the heart, the translocation of PKCε to the mitochondria, using a specifically designed peptide ψεHSP90, was sufficient to protect against ischemia/reperfusion injury [59]. In the brain, PKCε also translocates to the mitochondria following IPC exposure, where PKCε modulates mitochondrial function through the regulation of mitochondrial ROS production, increase in mitochondrial membrane potential, and oxygen consumption by specific respiratory chain complexes including complex IV [42, 60]. Numerous mitochondrial proteins have also been identified as targets for PKCε phosphorylation, these proteins include aldehyde dehydrogenase 2 (ALDH2) [61], cytochrome *c* oxidase subunit IV (COXIV) [60, 62, 63], proteins of the mitochondrial permeability transition pore [64], and the mitochondrial ATP-sensitive K^+ channel (mitoK^+_{ATP}) [42]. The mitoK^+_{ATP} is of particular interest as the opening of the channel during the initial phase of IPC is associated with the generation of mild levels of ROS formation which serve as a trigger for the initiation of IPC [42, 65–67]. It is of interest to note that ROS are known to activate PKCε [35] leading to lower ROS production resulting in mitochondrial protection. Severe ROS production induces endothelial cell dysfunction, cerebral circulation abnormalities and affects BBB permeability [68, 69]. PKCε activation-decreased ROS production may help preserve endothelial function and cerebral circulation following lethal cerebral ischemia.

10.3.2 SIRT1

SIRT1 is a member of the sirtuin family of nicotinamide adenine dinucleotide (NAD^+)-dependent deacetylases, which is activated during periods of energy deprivation [70]. A role for SIRT1 in IPC-induced ischemic tolerance was demonstrated through the activation of SIRT1 with resveratrol, which emulated IPC-induced neuroprotection in both in vitro and in vivo models of cerebral ischemia [52, 53]. SIRT1 is primarily localized to the nucleus where it regulates the expression of proteins associated with survival pathways against apoptosis, oxidative stress, and inflammation. SIRT1 regulates the cellular stress responses by altering gene expression through the deacetylation of histone and non-histone proteins, thereby activating or repressing their activities. Some of the transcription factors targeted by SIRT1 include TAF168 (TATA-box binding protein)-associated factor I, MEF2 (MADS box transcription enhancer factor 2), NF-κβ, the tumor suppressor p53, members of the FOXO transcription factor family, and others [70]. SIRT1 can also regulate mitochondrial function by activating the transcriptional co-activator of genes associated with energy metabolism, peroxisome proliferator-activated receptor γ co-activator 1α (PGC-1α) [71]. Additionally,

SIRT1 and PGC-1α are also present in the mitochondria, which may allow for a coordination of nuclear and mitochondrial gene expression [72]. Preconditioning-induced SIRT1 activation may participate in ischemia tolerance by affecting the abovementioned pathways.

Besides the activation of the previously mentioned-cellular stress response pathways within the ischemic tissue, preconditioning can also have effects on components of the neurovascular unit. SIRT1 is present in the endothelium and vascular smooth muscle cells. Inhibition of SIRT1 reduces nitric oxide-mediated endothelium-dependent vasodilation [73]. It is possible that preconditioning-induced SIRT1 activation may also participate in endothelium-dependent vasodilation which may help prevent cerebral ischemia-induced hypoperfusion and ultimately lower cerebral damage. PKCε, another anti-apoptotic pathway activated by IPC may also participate in protection of the neurovascular unit. PKCε also participates in protection against inflammatory cytokine-induced excessive BBB collapse [74]. Overall, the signaling pathways, namely SIRT1 and PKCε, activated by preconditioning may also affect the function of the neurovascular unit. These pathways/effects are highlighted in subsequent sections.

10.4 Preconditioning and Cerebral Blood Flow

The effect of different types of preconditioning stimuli on CBF has been studied in both early, as well as delayed preconditioning models. Numerous studies were performed to address how preconditioning affects (1) the neurovascular unit immediately prior to/or during lethal ischemia and (2) lethal cerebral ischemia-induced abnormalities in cerebral perfusion following an ischemic event.

10.4.1 Effects of Preconditioning on Neurovascular Unit Immediately Prior to/During Lethal Ischemia

Preconditioning induced by three 10-min intervals of transient ischemia (separated by 45 min) did not significantly affect regional CBF (via the [^{14}C]iodoantipyrine distribution method) in ischemic-tolerant regions 3 days after IPC induction, as well as lethal cerebral ischemia [75, 76]. Lipopolysaccharide (LPS, a preconditioning stimulus) preconditioning attenuates ischemic severity, however, local CBF (using [14C] iodoantipyrine) during ischemia was not significantly different between LPS or saline-treated rats [77]. On the contrary, regional CBF was found to be significantly higher in the frontoparietal region during lethal ischemia (middle cerebral artery occlusion, MCAO) of IPC-induced rats (by transient MCAO) without a reduction in cerebral damage [78, 79]. Additionally, induction of IPC (via global cerebral ischemia [four-vessel occlusion]) with subsequent lethal ischemia (via MCAO) 4 days later did not improve CBF; however, there was a significant reduction in cerebral ischemic damage [79]. Also, cortical spreading depression-induced preconditioning exerted neuroprotection against lethal cerebral ischemia, but had no effect on CBF during or immediately after lethal ischemia [80]. Furthermore, transient MCAO-induced preconditioning protected the brain against permanent MCAO without affecting CBF (via laser Doppler flowmetry) [81]. Altogether, the conclusion of these studies is not without controversy as the effects of preconditioning as a correlate of CBF have not been directly demonstrated. Furthermore, preconditioning-induced changes in CBF resulting in subsequent neuroprotection have not been readily shown and may be considered an epiphenomenon.

10.4.2 Preconditioning and Lethal Cerebral Ischemia

Preconditioning also prevents TBI and cerebral ischemia-induced abnormalities in CBF. For example, Hu et al. (2010) observed that hyperbaric oxygen preconditioning attenuated TBI at high altitudes. This suppression of TBI was accompanied by improved regional cerebral blood flow (rCBF) [20]. Using a gerbil model of transient cerebral ischemia, Nakamura et al. (2006) studied the effect of IPC on rCBF after global cerebral ischemia [82]. rCBF was monitored for 7 days following reperfusion using [^{14}C] iodoantipyrine autoradiography, where they observed that IPC (induced by 2 min of global cerebral ischemia) reduced post-ischemic hypoperfusion in the hippocampus (CA1, CA2, and dentate gyrus). This reduction in post-ischemic hypoperfusion was also observed throughout the brain including the ischemia vulnerable hippocampus. Similarly, IPC-induction reduced edema-corrected infarct volume by 49 % as compared to sham-operated rats using permanent MCAO in spontaneously hypertensive rats [83], while CBF was also restored in the penumbra in the ipsilateral cortex faster in IPC-treated rats suggesting that this timely reperfusion plays a crucial role in salvaging penumbral region leading to preconditioning-induced cerebral protection (Fig. 10.1) [83].

Administration of LPS (another preconditioning stimuli) in mice has been shown to prevent MCAO-induced dysfunction in cerebrovascular regulation [84]. LPS-mediated preconditioning stimuli in mice lacking iNOS or the nox2 subunit of the superoxide-producing enzyme nicotinamide adenine dinucleotide phosphate oxidase, failed to prevent MCAO-induced dysfunction in cerebrovasculature. This suggests that iNOS-derived NO and nox2-derived superoxides are key mediators of preserved post-cerebral ischemia neurovascular function in LPS-preconditioned mice (Fig. 10.1). Similar to LPS-induced preconditioning, an IPC-induced increase in endothelial nitric oxide synthase (eNOS) plus inducible nitric oxide synthase (iNOS) expression, and prostaglandin E_2 (PGE2) production have all been attributed to increased reperfusion following ischemia [83, 85–88] (Fig. 10.1). This timely restoration of CBF in the ischemic territory has been correlated with moderate to severe brain damage [89]; therefore, a reduction of post-ischemic

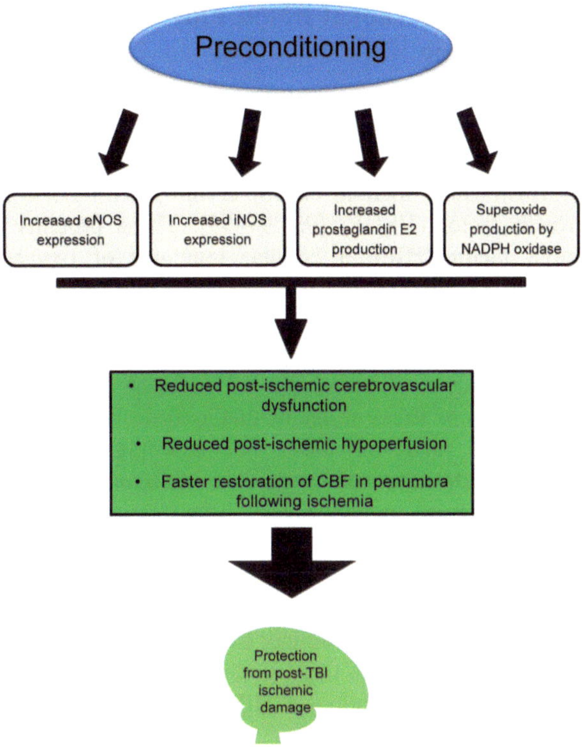

Fig. 10.1 Schematic representation of preconditioning-activated pathways that preserves cerebral blood flow (CBF) following traumatic brain injury (TBI)-induced cerebral ischemia. Lower degree of post-ischemic cerebrovascular dysfunction, lower post-ischemic hypoperfusion, and rapid restoration of post-ischemic reperfusion in penumbra are key neurovascular unit-dependent mechanisms that lead to protection from ischemic damage. These mechanisms also involve increased eNOS expression, increased iNOS expression, increased prostaglandin E2 production, and superoxide production by NADPH. *NOS* endothelial nitric oxide synthase, *iNOS* inducible nitric oxide synthase, *NADPH* Nicotinamide adenine dinucleotide phosphate

hypoperfusion (faster recovery of CBF) in preconditioned animals may contribute to lower cerebral ischemic damage (Fig. 10.1). Relative deficiency of NO is reported to occur several hours after TBI [90]. Post-TBI administration of L-arginine improved CBF and lowered contusion volume potentially by the restoration of NO levels in injured brain [90]. Overall, these studies indicate that preconditioning alone has beneficial improvement on the CBF abnormalities following lethal ischemia; however, the overall effect of preconditioning on CBF remains unclear. These studies also indicate that these beneficial effects are independent of any preconditioning stimuli such as ischemic or LPS-induced preconditioning. Improved CBF following TBI can help lower the brain damage.

10.5 Regulators of Cerebral Blood Flow Following Preconditioning

Numerous agents that improve CBF have an innate ability to afford neuroprotection [91–96]. In fact, several proteins/biochemical pathways that modulate CBF are activated after IPC, which include eNOS, erythropoietin (EPO), and vascular endothelial growth factor (VEGF) [93, 97–99]. These proteins are targets of hypoxia-inducible factor (HIF), a cellular regulator of oxygen homeostasis [100]. Under normoxia, HIF is targeted for proteasomal degradation through the hydroxylation of HIF-α by iron- and 2-oxoglutarate-dependent oxygenases; however, during ischemia, the reduction of oxygen levels reduces the hydroxylation and thus degradation of HIF-α, allowing for HIF-induced gene transcription. This suggests that conditions of stress or agents that activate HIF would increase these proteins/pathways and modulate CBF. For example, dimethyloxalylglycine (DMOG, a 2-oxoglutarate-dependent oxygenase inhibitor) administration prior to MCAO decreased infarct size, and improved CBF post-ischemia [94]. Additionally, desferroxamine (an iron chelator that increases HIF-1 activity) infusion has been shown to induce significant cerebral vasodilation and improve CBF in healthy humans and has been suggested as a potential neuroprotective agent [101]. Improvement in CBF following desferroxamine treatment may help lower cerebral ischemia-induced damage. Overall, these studies suggest that the pharmacological or preconditioning-induced activation of HIF-1 may be necessary for improved CBF and neuroprotection following IPC.

Enhanced VEGF (pro-angiogenic factor) production can increase microvascular density consequently improving cerebral perfusion [98]. Additionally, recombinant human EPO administered prior to permanent focal cerebral ischemia up-regulated EPO receptor levels in vascular endothelial cells resulting in increased CBF (due to increased angiogenesis) necessary for neuroprotection [93]. VEGF or EPO can also improve CBF through the activation of phosphoinositide 3-kinase-Akt pathway (a known regulator of eNOS activity), where an increase in activation of this pathway is neuroprotective [102]. Neuroprotection from increased eNOS activity has been shown to occur through NO's actions on soluble guanylate cyclase, which induces vasodilation [103]. In fact, eNOS is a requirement for preconditioning-induced protection, as eNOS knockout mice do not benefit from IPC [104, 105].

CBF modulation can also occur by glutamate activation of NMDA receptors in neurons or glutamate metabotropic receptors in astrocytes. Under normoxia in neurons, the presynaptic release of glutamate activates NMDA receptors and increases postsynaptic intracellular Ca^{2+} levels. This increase in Ca^{2+} can cause the activation of neuronal NOS on perivascular nerves resulting in the release of NO, and activation of guanylate cyclase in vascular smooth muscle to induce vasodilation and increase cerebral perfusion [106–111]. In astrocytes, presynaptic glutamate release activates metabotropic glutamate receptors to increase intracellular Ca^{2+} [112]. This increase in intracellular Ca^{2+} in astrocytes increases the production of arachidonic acid from the activation of phospholipase A_2, where the accumulation of arachidonic acid leads to the production of prostaglandins and epoxyeicosatrienoic acids to induce vasodilation [106, 107, 113–115]. Prostaglandin E_2 (PGE2) has been suggested to induce vasodilation through the activation of the EP4 prostaglandin receptor [116]. Stimulation of the EP4 receptor increases cyclic AMP activation of protein kinase A to decrease the phosphorylation of myosin light chain (PKA inhibits myosin light chain kinase and activates myosin light chain phosphatase [117, 118]), inducing vasodilation [119] through an increase in K^+ conductance, which hyperpolarizes smooth muscle cells and decreases intracellular Ca^{2+} [120]. On the contrary, stimulation of EP1 and EP3 receptors by PGE2 can produce vasoconstriction in porcine large cerebral arteries mediated by the phosphatidyl-inositol pathway on vascular smooth muscle [121].

While increases in CBF occur with neuronal or astrocyte glutamate stimulation, excessive glutamate during ischemia is cytotoxic. Therefore, NMDA receptor antagonists (e.g., MK-801 [91] and CGS-19755 [122]) have been widely studied as neuroprotective agents administered pre- and post-ischemia, where numerous agents have displayed neuroprotective properties and increased CBF in animal models. This increase in CBF has been suggested to be a result of hypercapnia [123] or preservation of pH [122]; however, these agents have not translated successfully into clinically based neuroprotection [124]. Additionally, IPC limits glutamate release following ischemia [125–127]; therefore, reduced glutamate release during lethal ischemia in the preconditioned brain may be, in part, responsible for preserved CBF following lethal ischemia. In summary, these studies indicate that preconditioning activates several pathways which in turn, in synergy, aid in the preservation of cerebral perfusion following lethal cerebral ischemia.

10.6 Preconditioning and Angiogenesis

Angiogenesis is defined as a process of growth of microvessel sprouts into the capillary blood vessel from existing vasculature [128]. This process is highly regulated by the recruit, proliferation, and alignment of vascular endothelial cells through actions such as excretion of angiogenic growth factors, matrix metalloproteinases, and remodeling of extracellular membranes and endothelial cells [128]. The consequences of angiogenesis can aid in improved blood perfusion in the ischemic region due to enhanced vascularization [95, 129]. The protective effects of preconditioning also implicate angiogenesis. Preconditioning with 4 weeks of bilateral common carotid artery (BCCA) ligation-induced cerebral hypoperfusion, has been suggested to increase matrix metalloproteinase-2 (MMP-2) activity (MMP-2 are involved in the degradation of the extracellular matrix (ECM), the basement membrane, and interstitial stroma resulting in protection of the brain against cerebral ischemia [130, 131]. Degradations of the ECM are essential for endothelial cell invasion into the ECM (a crucial process of angiogenesis) and are associated with adaptive arteriogenesis or angiogenesis (Fig. 10.2) [130, 131]. In a model of hypoxic preconditioning, increased expression of blood

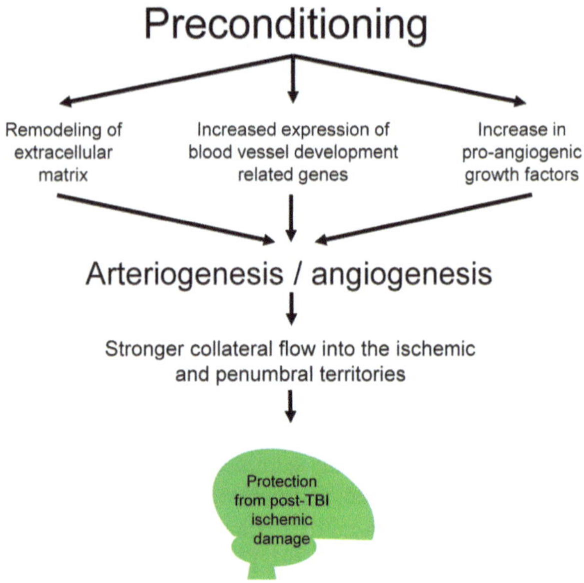

Fig. 10.2 Schematic representation of preconditioning-activated pathways that reduces traumatic brain injury (TBI)-induced cerebral ischemic damage by inducing angiogenesis. Preconditioning promotes arteriogenesis/angiogenesis providing better collateral flow into the ischemic and penumbral territories. Preconditioning induces angiogenesis by affecting remodeling of extracellular matrix, by increasing expression of blood vessel development-related genes, and by increasing levels of pro-angiogenic growth factors

vessel development-related genes (within 24 h of hypoxic preconditioning) were observed in neonatal rats (postnatal day 6) exposed to hypoxic preconditioning suggesting potential angiogenesis following hypoxic preconditioning [99]. Additionally, physical exercise has been shown to increase cerebral angiogenesis by promoting an increase in pro-angiogenic growth factors including VEGF and insulin-like growth factor (Fig. 10.2) [132–138]. Overall, hypoperfusion-, hypoxia-, and physical exercise-preconditioning increases factors responsible for the promotion of angiogenesis suggesting a "preconditioning-induced angiogenesis." Increased angiogenesis and/or vascular density may be of therapeutic benefit during TBI-induced cerebral ischemia through more robust collateral flow into the ischemic and penumbral territories [95, 129].

10.7 Ischemia Tolerance Following Controlled Restoration of CBF

The restoration of blood flow to ischemic tissue (reperfusion) has been established as a major contributing factor to ischemia-mediated injury. Reperfusion injury expands the infarct size greater than that observed during ischemia alone and is caused by the excessive generation of ROS (at levels that cause lipid, protein, and DNA oxidation), which impedes the normal cellular functions that lead to cell death [139]. This injury can also occur via more clinically relevant emission of micro emboli, which is independent of ROS production [140, 141]. For example, permanent occlusion of the middle cerebral artery (MCA) for 6 h generated a smaller infarct than transient MCA occlusion for 3 h following 3 h of reperfusion, when infarct volume was measured 6 h after the start of ischemia [142]. The detrimental effects of reperfusion is further emphasized in the study by Aronowski et al., in which the infarct size at 24 h of permanent common carotid artery (CCA) and MCA occlusion was smaller than 2–5 h of transit CCA and MCA occlusion in rats [143].

Numerous investigators have controlled the rate of reperfusion as an effort to decrease reperfusion injury. In contrast to IPC, which is administered prior to the ischemic event, ischemic postconditioning refers to a stimulus that is administered prior to the complete restoration of blood flow to the ischemic tissue. Ischemic postconditioning requires a series of brief ischemia/reperfusion events to induce tolerance. In fact, a key aspect of postconditioning-induced neuroprotection is the "postconditioning algorithm" which consists of the number of cycles and the duration of the re-occlusion/reperfusion episodes. For example, in the brain, Zhao et al. showed that three cycles of 30 s reperfusion and 10 s CCA occlusion was sufficient to reduce infarct size and cellular apoptosis 2 days following focal ischemia generated by permanent distal MCA occlusion plus transient bilateral CCA occlusion [144]. They also reported that controlled reperfusion paradigm was able to reduce infarct size by as much as 80 % and that the degree of protection was inversely proportional to the severity of cerebral ischemia. Using the same cerebral ischemia and postconditioning model described above, Gao et al. demonstrated that postconditioning improved motor asymmetry behavioral function 30 days following focal ischemia [145]. Other "postconditioning algorithms" have also been described with protective effects [146] similar to preconditioning and which appear to activate related protective pathways, since the combination of pre- and postconditioning did not increase ischemic protection in the brain [147]. Postconditioning-mediated neuroprotection is also observed when administered acute (immediate or soon after I/R) or chronically (many days after I/R) following the ischemic event [147, 148]. Controlled restoration of CBF by postconditioning can protect the brain from reperfusion injury as demonstrated in models of cerebral ischemia [146]. It remains to be determined if controlled reperfusion of the injured brain can help lower TBI.

10.8 Pre-/Postconditioning and the Blood–Brain Barrier

The BBB is considered an important and integral part of the neurovascular unit [149, 150]. Cerebral ischemia-induced breakdown of the BBB is one of the major events leading to injury [151]. The degradation of components from the ECM/basal lamina such as collagen IV, laminin, and fibronectin by proteases has been suggested to have an important role in post-ischemia breakdown of BBB [152–155]. Several groups have investigated the effect of pre- and postconditioning on post-TBI and cerebral ischemia BBB function. One month of heat acclimation resulted in attenuation in TBI-induced BBB permeability as measured by Evans blue extravasation (an index of BBB permeability) [26]. Ren et al. showed that delayed ischemic postconditioning (6 h post-lethal ischemia) reduced Evans blue extravasation when measured at 48 h post-lethal ischemia [156]. These results suggest that reduced BBB permeability in postconditioned animals possesses reduced cerebral ischemic damage. Similarly, in a rat model of exercise preconditioning, a reduction in post-lethal ischemic brain edema and Evans blue leakage were observed at 24 h after focal cerebral ischemia, along with an increase in pre- and post-cerebral ischemia collagen IV levels [153]. The same study also suggested a post-cerebral ischemia increase in matrix metalloproteinase-9 (MMP-9, a proteinase responsible for the disruption of the BBB via enhanced degradation of ECM/basal lamina, i.e., collagen IV) was blunted in exercise-preconditioned animals as compared to control [153]. This reduction in MMP-9 levels was associated with an increase in the levels of endogenous MMP-9 inhibitor, tissue inhibitor of metalloproteinase-1 (TIMP-1) suggesting that exercise preconditioning protects post-cerebral ischemia BBB integrity by reducing MMP-9 levels via TIMP-1 [153].

TNFα has also known to play a role in BBB permeability, as TNFα affects the organization of cytoskeleton, expression of tight junction protein, and production of serine proteases that affect BBB integrity [157–159]. Pretreatment with rapamycin (a preconditioning stimulus) blunted lethal cerebral ischemia-induced increase in TNFα [160]. An exercise preconditioning paradigm that protects against cerebral ischemic damage, increased TNFα levels following exercise, but prevented a lethal cerebral ischemia-induced increase in TNFα [161]. Another study showed that increased (pre-ischemia) TNFα levels through physical exercise decreased post-ischemic BBB dysfunction via the extracellular signal-regulated kinase 1 and 2 pathways [162]. Besides exercise preconditioning, LPS-preconditioning in mice also increased plasma TNFα levels; however, the lethal cerebral ischemia-induced increase in TNFα levels was blunted [163]. LPS preconditioning also increased the levels of neuronal soluble TNF-receptor 1 (s-TNFR1) following lethal cerebral ischemia. S-TNFR1 binds and inhibits the actions of TNFα suggesting that increased s-TNFR1 may neutralize the toxic effects of TNFα [163]. LPS-preconditioning also reduces post-ischemic neutrophil infiltration indirectly suggesting preserved BBB integrity [164]. Various preconditioning stimuli such as ischemic-, exercise- and LPS-preconditioning, heat acclimation protect BBB integrity following TBI and lethal cerebral ischemia by affecting TNFα-mediated BBB-damaging pathways.

10.9 Summary

Different preconditioning stimuli have been shown to protect the brain by affecting the parenchyma. While the current literature is limited, the data from these studies indicates that various preconditioning stimulus have a pronounced beneficial effect on cerebral circulation following ischemia. It is likely that the synergistic effect of preconditioning on the neurovascular unit and parenchymal cell death/cell survival pathways confer protection against cerebral ischemia following TBI. Preconditioning itself prevents post-ischemia/reperfusion abnormalities in the brain by preserving CBF via increased brain perfusion maintaining adequate microcirculation and overall BBB health. Controlled reperfusion using postconditioning paradigms also lowers extent of injury potentially by lowering reperfusion injury. Preconditioning improves overall cerebral circulation ultimately leading to protection against subsequent ischemic insult. Further understanding into the mechanisms of preconditioning may aid in the development of novel therapies to protect the brain following TBI.

References

1. https://www.braintrauma.org/tbi-faqs/tbi-statistics/. Accessed 14 May 2013
2. Hlatky R, Valadka AB, Robertson CS (2003) Intracranial hypertension and cerebral ischemia after severe traumatic brain injury. Neurosurg Focus 14(4):e2
3. Perez-Pinzon MA, Born JG (1999) Rapid preconditioning neuroprotection following anoxia in hippocampal slices: role of the K+ ATP channel and protein kinase C. Neuroscience 89:453–459
4. Schurr A, Reid KH, Tseng MT, West C, Rigor BM (1986) Adaptation of adult brain tissue to anoxia and hypoxia in vitro. Brain Res 374:244–248
5. De Hert SG, Turani F, Mathur S, Stowe DF (2005) Cardioprotection with volatile anesthetics: mechanisms and clinical implications. Anesth Analg 100:1584–1593. doi:10.1213/01.ANE.0000153483.61170.0C
6. Freed RS, Freed SA (1990) Ghost illness of children in north India. Med Anthropol 12:401–417. doi:10.1080/01459740.1990.9966034
7. Widimsky J, Stolz I (1977) The adaptation of cardiovascular system to exercise and training in healthy subjects and in heart disease. Acta Univ Carol Med Monogr 82:1–26
8. Horiguchi T, Snipes JA, Kis B, Shimizu K, Busija DW (2005) The role of nitric oxide in the development of cortical spreading depression-induced tolerance to transient focal cerebral ischemia in rats. Brain Res 1039:84–89. doi:10.1016/j.brainres.2005.01.047
9. Kawahara N, Ruetzler CA, Klatzo I (1995) Protective effect of spreading depression against neuronal damage following cardiac arrest cerebral ischaemia. Neurol Res 17:9–16
10. Hellweg R, von Arnim CA, Buchner M, Huber R, Riepe MW (2003) Neuroprotection and neuronal dysfunction upon repetitive inhibition of oxidative phosphorylation. Exp Neurol 183:346–354
11. Riepe MW, Esclaire F, Kasischke K, Schreiber S, Nakase H, Kempski O, Ludolph AC, Dirnagl U, Hugon J (1997) Increased hypoxic tolerance by chemical inhibition of oxidative phosphorylation: "chemical preconditioning". J Cereb Blood Flow Metab 17:257–264. doi:10.1097/00004647-199703000-00002
12. Wiegand F, Liao W, Busch C, Castell S, Knapp F, Lindauer U, Megow D, Meisel A, Redetzky A, Ruscher K, Trendelenburg G, Victorov I, Riepe M, Diener HC, Dirnagl U (1999) Respiratory chain inhibition induces tolerance to focal cerebral ischemia. J Cereb Blood Flow Metab 19:1229–1237. doi:10.1097/00004647-199911000-00007
13. Kirino T, Tsujita Y, Tamura A (1991) Induced tolerance to ischemia in gerbil hippocampal neurons. J Cereb Blood Flow Metab 11:299–307. doi:10.1038/jcbfm.1991.62
14. Stagliano NE, Perez-Pinzon MA, Moskowitz MA, Huang PL (1999) Focal ischemic preconditioning induces rapid tolerance to middle cerebral artery occlusion in mice. J Cereb Blood Flow Metab 19:757–761. doi:10.1097/00004647-199907000-00005

15. Truettner J, Busto R, Zhao W, Ginsberg MD, Perez-Pinzon MA (2002) Effect of ischemic preconditioning on the expression of putative neuroprotective genes in the rat brain. Brain Res Mol Brain Res 103:106–115
16. Xu GP, Dave KR, Vivero R, Schmidt-Kastner R, Sick TJ, Perez-Pinzon MA (2002) Improvement in neuronal survival after ischemic preconditioning in hippocampal slice cultures. Brain Res 952:153–158
17. Perez-Pinzon MA, Alonso O, Kraydieh S, Dietrich WD (1999) Induction of tolerance against traumatic brain injury by ischemic preconditioning. Neuroreport 10:2951–2954
18. Shein NA, Horowitz M, Shohami E (2007) Heat acclimation: a unique model of physiologically mediated global preconditioning against traumatic brain injury. Prog Brain Res 161:353–363. doi:10.1016/S0079-6123(06)61025-X
19. Hu SL, Hu R, Li F, Liu Z, Xia YZ, Cui GY, Feng H (2008) Hyperbaric oxygen preconditioning protects against traumatic brain injury at high altitude. Acta Neurochir Suppl 105:191–196
20. Hu S, Li F, Luo H, Xia Y, Zhang J, Hu R, Cui G, Meng H, Feng H (2010) Amelioration of rCBF and PbtO2 following TBI at high altitude by hyperbaric oxygen pre-conditioning. Neurol Res 32:173–178. doi:10.1179/174313209X414524
21. Costa T, Constantino LC, Mendonca BP, Pereira JG, Herculano B, Tasca CI, Boeck CR (2010) N-methyl-D-aspartate preconditioning improves short-term motor deficits outcome after mild traumatic brain injury in mice. J Neurosci Res 88:1329–1337. doi:10.1002/jnr.22300
22. Moojen VK, Damiani-Neves M, Bavaresco DV, Pescador BB, Comim CM, Quevedo J, Boeck CR (2012) NMDA preconditioning prevents object recognition memory impairment and increases brain viability in mice exposed to traumatic brain injury. Brain Res 1466:82–90. doi:10.1016/j.brainres.2012.05.041
23. del Zoppo GJ (2012) Aging and the neurovascular unit. Ann N Y Acad Sci 1268:127–133. doi:10.1111/j.1749-6632.2012.06686.x
24. del Zoppo GJ (2006) Stroke and neurovascular protection. N Engl J Med 354:553–555. doi:10.1056/NEJMp058312
25. del Zoppo GJ (2010) The neurovascular unit, matrix proteases, and innate inflammation. Ann N Y Acad Sci 1207:46–49. doi:10.1111/j.1749-6632.2010.05760.x
26. Shohami E, Novikov M, Horowitz M (1994) Long term exposure to heat reduces edema formation after closed head injury in the rat. Acta Neurochir Suppl 60:443–445
27. Kitagawa K, Matsumoto M, Tagaya M, Hata R, Ueda H, Niinobe M, Handa N, Fukunaga R, Kimura K, Mikoshiba K et al (1990) 'Ischemic tolerance' phenomenon found in the brain. Brain Res 528:21–24
28. Perez-Pinzon MA, Xu GP, Dietrich WD, Rosenthal M, Sick TJ (1997) Rapid preconditioning protects rats against ischemic neuronal damage after 3 but not 7 days of reperfusion following global cerebral ischemia. J Cereb Blood Flow Metab 17:175–182. doi:10.1097/00004647-199702000-00007
29. Iadecola C, Anrather J (2011) Stroke research at a crossroad: asking the brain for directions. Nat Neurosci 14:1363–1368. doi:10.1038/nn.2953
30. Cabrera JA, Ziemba EA, Colbert R, Anderson LB, Sluiter W, Duncker DJ, Butterick TA, Sikora J, Ward HB, Kelly RF, McFalls EO (2012) Altered expression of mitochondrial electron transport chain proteins and improved myocardial energetic state during late ischemic preconditioning. Am J Physiol Heart Circ Physiol 302:H1974–H1982. doi:10.1152/ajpheart.00372.2011
31. Dave KR, Saul I, Busto R, Ginsberg MD, Sick TJ, Perez-Pinzon MA (2001) Ischemic preconditioning preserves mitochondrial function after global cerebral ischemia in rat hippocampus. J Cereb Blood Flow Metab 21:1401–1410. doi:10.1097/00004647-200112000-00004
32. Kurian GA, Berenshtein E, Saada A, Chevion M (2012) Rat cardiac mitochondrial sub-populations show distinct features of oxidative phosphorylation during ischemia,

reperfusion and ischemic preconditioning. Cell Physiol Biochem 30:83–94. doi:10.1159/000339043
33. Quarrie R, Lee DS, Steinbaugh G, Cramer B, Erdahl W, Pfeiffer DR, Zweier JL, Crestanello JA (2012) Ischemic preconditioning preserves mitochondrial membrane potential and limits reactive oxygen species production. J Surg Res 178:8–17. doi:10.1016/j.jss.2012.05.090
34. Glantz L, Avramovich A, Trembovler V, Gurvitz V, Kohen R, Eidelman LA, Shohami E (2005) Ischemic preconditioning increases antioxidants in the brain and peripheral organs after cerebral ischemia. Exp Neurol 192:117–124. doi:10.1016/j.expneurol.2004.11.012
35. Perez-Pinzon MA, Dave KR, Raval AP (2005) Role of reactive oxygen species and protein kinase C in ischemic tolerance in the brain. Antioxid Redox Signal 7:1150–1157. doi:10.1089/ars.2005.7.1150
36. Ding ZM, Wu B, Zhang WQ, Lu XJ, Lin YC, Geng YJ, Miao YF (2012) Neuroprotective effects of ischemic preconditioning and postconditioning on global brain ischemia in rats through the same effect on inhibition of apoptosis. Int J Mol Sci 13:6089–6101. doi:10.3390/ijms13056089
37. El-Achkar TM (2012) Modulation of apoptosis by ischemic preconditioning: an emerging role for miR-21. Kidney Int 82:1149–1151. doi:10.1038/ki.2012.305
38. Lin WY, Chang YC, Ho CJ, Huang CC (2013) Ischemic preconditioning reduces neurovascular damage after hypoxia-ischemia via the cellular inhibitor of apoptosis 1 in neonatal brain. Stroke 44:162–169. doi:10.1161/STROKEAHA.112.677617
39. Lin HY, Huang CC, Chang KF (2009) Lipopolysaccharide preconditioning reduces neuroinflammation against hypoxic ischemia and provides long-term outcome of neuroprotection in neonatal rat. Pediatr Res 66:254–259. doi:10.1203/PDR.0b013e3181b0d336
40. Perez-Pinzon MA, Mumford PL, Rosenthal M, Sick TJ (1996) Anoxic preconditioning in hippocampal slices: role of adenosine. Neuroscience 75:687–694
41. Zhou AM, Li WB, Li QJ, Liu HQ, Feng RF, Zhao HG (2004) A short cerebral ischemic preconditioning up-regulates adenosine receptors in the hippocampal CA1 region of rats. Neurosci Res 48:397–404. doi:10.1016/j.neures.2003.12.010
42. Raval AP, Dave KR, DeFazio RA, Perez-Pinzon MA (2007) EpsilonPKC phosphorylates the mitochondrial K(+) (ATP) channel during induction of ischemic preconditioning in the rat hippocampus. Brain Res 1184:345–353. doi:10.1016/j.brainres.2007.09.073
43. Gaspar T, Snipes JA, Busija AR, Kis B, Domoki F, Bari F, Busija DW (2008) ROS-independent preconditioning in neurons via activation of mitoK(ATP) channels by BMS-191095. J Cereb Blood Flow Metab 28:1090–1103. doi:10.1038/sj.jcbfm.9600611
44. Wang Q, Sun AY, Simonyi A, Kalogeris TJ, Miller DK, Sun GY, Korthuis RJ (2007) Ethanol preconditioning protects against ischemia/reperfusion-induced brain damage: role of NADPH oxidase-derived ROS. Free Radic Biol Med 43:1048–1060. doi:10.1016/j.freeradbiomed.2007.06.018
45. Lange-Asschenfeldt C, Raval AP, Dave KR, Mochly-Rosen D, Sick TJ, Perez-Pinzon MA (2004) Epsilon protein kinase C mediated ischemic tolerance requires activation of the extracellular regulated kinase pathway in the organotypic hippocampal slice. J Cereb Blood Flow Metab 24:636–645. doi:10.1097/01.WCB.0000121235.42748.BF
46. Zhang J, Bian HJ, Li XX, Liu XB, Sun JP, Li N, Zhang Y, Ji XP (2010) ERK-MAPK signaling opposes rho-kinase to reduce cardiomyocyte apoptosis in heart ischemic preconditioning. Mol Med 16:307–315. doi:10.2119/molmed.2009.00121
47. Zhang QG, Wang RM, Han D, Yang LC, Li J, Brann DW (2009) Preconditioning neuroprotection in global cerebral ischemia involves NMDA receptor-mediated ERK-JNK3 crosstalk. Neurosci Res 63:205–212
48. Bhuiyan MI, Jung SY, Kim HJ, Lee YS, Jin C (2011) Major role of the PI3K/Akt pathway in ischemic tolerance induced by sublethal oxygen-glucose deprivation in cortical neurons in vitro. Arch Pharm Res 34:1023–1034. doi:10.1007/s12272-011-0620-3

49. Kim EJ, Raval AP, Hirsch N, Perez-Pinzon MA (2010) Ischemic preconditioning mediates cyclooxygenase-2 expression via nuclear factor-kappa B activation in mixed cortical neuronal cultures. Transl Stroke Res 1:40–47
50. Kim EJ, Raval AP, Perez-Pinzon MA (2008) Preconditioning mediated by sublethal oxygen-glucose deprivation-induced cyclooxygenase-2 expression via the signal transducers and activators of transcription 3 phosphorylation. J Cereb Blood Flow Metab 28:1329–1340. doi:10.1038/jcbfm.2008.26
51. Raval AP, Dave KR, Mochly-Rosen D, Sick TJ, Perez-Pinzon MA (2003) Epsilon PKC is required for the induction of tolerance by ischemic and NMDA-mediated preconditioning in the organotypic hippocampal slice. J Neurosci 23:384–391
52. Della-Morte D, Dave KR, DeFazio RA, Bao YC, Raval AP, Perez-Pinzon MA (2009) Resveratrol pretreatment protects rat brain from cerebral ischemic damage via a sirtuin 1-uncoupling protein 2 pathway. Neuroscience 159:993–1002. doi:10.1016/j.neuroscience.2009.01.017
53. Raval AP, Dave KR, Perez-Pinzon MA (2006) Resveratrol mimics ischemic preconditioning in the brain. J Cereb Blood Flow Metab 26:1141–1147. doi:10.1038/sj.jcbfm.9600262
54. Wurdak H, Zhu S, Min KH, Aimone L, Lairson LL, Watson J, Chopiuk G, Demas J, Charette B, Halder R, Weerapana E, Cravatt BF, Cline HT, Peters EC, Zhang J, Walker JR, Wu C, Chang J, Tuntland T, Cho CY, Schultz PG (2010) A small molecule accelerates neuronal differentiation in the adult rat. Proc Natl Acad Sci U S A 107:16542–16547. doi:10.1073/pnas.1010300107
55. Kitagawa K (2012) Ischemic tolerance in the brain: endogenous adaptive machinery against ischemic stress. J Neurosci Res 90:1043–1054. doi:10.1002/jnr.23005
56. Kim E, Raval AP, Defazio RA, Perez-Pinzon MA (2007) Ischemic preconditioning via epsilon protein kinase C activation requires cyclooxygenase-2 activation in vitro. Neuroscience 145:931–941. doi:10.1016/j.neuroscience.2006.12.063
57. Cross HR, Murphy E, Bolli R, Ping P, Steenbergen C (2002) Expression of activated PKC epsilon (PKC epsilon) protects the ischemic heart, without attenuating ischemic H(+) production. J Mol Cell Cardiol 34:361–367. doi:10.1006/jmcc.2001.1518
58. Saurin AT, Pennington DJ, Raat NJ, Latchman DS, Owen MJ, Marber MS (2002) Targeted disruption of the protein kinase C epsilon gene abolishes the infarct size reduction that follows ischaemic preconditioning of isolated buffer-perfused mouse hearts. Cardiovasc Res 55:672–680
59. Budas GR, Churchill EN, Disatnik MH, Sun L, Mochly-Rosen D (2010) Mitochondrial import of PKCepsilon is mediated by HSP90: a role in cardioprotection from ischaemia and reperfusion injury. Cardiovasc Res 88:83–92. doi:10.1093/cvr/cvq154
60. Dave KR, DeFazio RA, Raval AP, Torraco A, Saul I, Barrientos A, Perez-Pinzon MA (2008) Ischemic preconditioning targets the respiration of synaptic mitochondria via protein kinase C epsilon. J Neurosci 28:4172–4182. doi:10.1523/JNEUROSCI.5471-07.2008
61. Churchill EN, Disatnik MH, Mochly-Rosen D (2009) Time-dependent and ethanol-induced cardiac protection from ischemia mediated by mitochondrial translocation of varepsilonPKC and activation of aldehyde dehydrogenase 2. J Mol Cell Cardiol 46:278–284. doi:10.1016/j.yjmcc.2008.09.713
62. Ogbi M, Chew CS, Pohl J, Stuchlik O, Ogbi S, Johnson JA (2004) Cytochrome c oxidase subunit IV as a marker of protein kinase Cepsilon function in neonatal cardiac myocytes: implications for cytochrome c oxidase activity. Biochem J 382:923–932. doi:10.1042/BJ20040468
63. Ogbi M, Johnson JA (2006) Protein kinase Cepsilon interacts with cytochrome c oxidase subunit IV and enhances cytochrome c oxidase activity in neonatal cardiac myocyte preconditioning. Biochem J 393:191–199. doi:10.1042/BJ20050757
64. Baines CP, Song CX, Zheng YT, Wang GW, Zhang J, Wang OL, Guo Y, Bolli R, Cardwell EM, Ping P (2003) Protein kinase Cepsilon interacts with and inhibits the permeability

transition pore in cardiac mitochondria. Circ Res 92:873–880. doi:10.1161/01.RES. 0000069215.36389.8D
65. Carroll R, Gant VA, Yellon DM (2001) Mitochondrial K(ATP) channel opening protects a human atrial-derived cell line by a mechanism involving free radical generation. Cardiovasc Res 51:691–700
66. Forbes RA, Steenbergen C, Murphy E (2001) Diazoxide-induced cardioprotection requires signaling through a redox-sensitive mechanism. Circ Res 88:802–809
67. Obata T, Yamanaka Y (2000) Block of cardiac ATP-sensitive K(+) channels reduces hydroxyl radicals in the rat myocardium. Arch Biochem Biophys 378:195–200. doi:10.1006/abbi.2000.1830
68. Chrissobolis S, Faraci FM (2008) The role of oxidative stress and NADPH oxidase in cerebrovascular disease. Trends Mol Med 14:495–502. doi:10.1016/j.molmed.2008.09.003
69. Chrissobolis S, Miller AA, Drummond GR, Kemp-Harper BK, Sobey CG (2011) Oxidative stress and endothelial dysfunction in cerebrovascular disease. Front Biosci (Landmark Ed) 16:1733–1745
70. Morris KC, Lin HW, Thompson JW, Perez-Pinzon MA (2011) Pathways for ischemic cytoprotection: role of sirtuins in caloric restriction, resveratrol, and ischemic preconditioning. J Cereb Blood Flow Metab 31:1003–1019. doi:10.1038/jcbfm.2010.229
71. Nemoto S, Fergusson MM, Finkel T (2005) SIRT1 functionally interacts with the metabolic regulator and transcriptional coactivator PGC-1{alpha}. J Biol Chem 280:16456–16460. doi:10.1074/jbc.M501485200
72. Aquilano K, Vigilanza P, Baldelli S, Pagliei B, Rotilio G, Ciriolo MR (2010) Peroxisome proliferator-activated receptor gamma co-activator 1alpha (PGC-1alpha) and sirtuin 1 (SIRT1) reside in mitochondria: possible direct function in mitochondrial biogenesis. J Biol Chem 285:21590–21599. doi:10.1074/jbc.M109.070169
73. Tajbakhsh N, Sokoya EM (2012) Regulation of cerebral vascular function by sirtuin 1. Microcirculation 19:336–342. doi:10.1111/j.1549-8719.2012.00167.x
74. Sonobe Y, Takeuchi H, Kataoka K, Li H, Jin S, Mimuro M, Hashizume Y, Sano Y, Kanda T, Mizuno T, Suzumura A (2009) Interleukin-25 expressed by brain capillary endothelial cells maintains blood–brain barrier function in a protein kinase Cepsilon-dependent manner. J Biol Chem 284:31834–31842. doi:10.1074/jbc.M109.025940
75. Chen J, Graham SH, Zhu RL, Simon RP (1996) Stress proteins and tolerance to focal cerebral ischemia. J Cereb Blood Flow Metab 16:566–577. doi:10.1097/00004647-199607000-00006
76. Sakurada O, Kennedy C, Jehle J, Brown JD, Carbin GL, Sokoloff L (1978) Measurement of local cerebral blood flow with iodo [14C] antipyrine. Am J Physiol 234:H59–H66
77. Dawson DA, Furuya K, Gotoh J, Nakao Y, Hallenbeck JM (1999) Cerebrovascular hemodynamics and ischemic tolerance: lipopolysaccharide-induced resistance to focal cerebral ischemia is not due to changes in severity of the initial ischemic insult, but is associated with preservation of microvascular perfusion. J Cereb Blood Flow Metab 19:616–623. doi:10.1097/00004647-199906000-00004
78. von Kummer R, von Kries F, Herold S (1986) Hydrogen clearance method for determining local cerebral blood flow. II. Effect of heterogeneity in cerebral blood flow. J Cereb Blood Flow Metab 6:492–498. doi:10.1038/jcbfm.1986.84
79. Matsushima K, Hakim AM (1995) Transient forebrain ischemia protects against subsequent focal cerebral ischemia without changing cerebral perfusion. Stroke 26:1047–1052
80. Matsushima K, Hogan MJ, Hakim AM (1996) Cortical spreading depression protects against subsequent focal cerebral ischemia in rats. J Cereb Blood Flow Metab 16:221–226. doi:10.1097/00004647-199603000-00006
81. Barone FC, White RF, Spera PA, Ellison J, Currie RW, Wang X, Feuerstein GZ (1998) Ischemic preconditioning and brain tolerance: temporal histological and functional outcomes, protein synthesis requirement, and interleukin-1 receptor antagonist and early gene expression. Stroke 29:1937–1950, discussion 1950–1931

82. Nakamura H, Katsumata T, Nishiyama Y, Otori T, Katsura K, Katayama Y (2006) Effect of ischemic preconditioning on cerebral blood flow after subsequent lethal ischemia in gerbils. Life Sci 78:1713–1719. doi:10.1016/j.lfs.2005.08.008
83. Zhao L, Nowak TS Jr (2006) CBF changes associated with focal ischemic preconditioning in the spontaneously hypertensive rat. J Cereb Blood Flow Metab 26:1128–1140. doi:10.1038/sj.jcbfm.9600269
84. Kunz A, Park L, Abe T, Gallo EF, Anrather J, Zhou P, Iadecola C (2007) Neurovascular protection by ischemic tolerance: role of nitric oxide and reactive oxygen species. J Neurosci 27:7083–7093. doi:10.1523/JNEUROSCI.1645-07.2007
85. Cho S, Park EM, Zhou P, Frys K, Ross ME, Iadecola C (2005) Obligatory role of inducible nitric oxide synthase in ischemic preconditioning. J Cereb Blood Flow Metab 25:493–501. doi:10.1038/sj.jcbfm.9600058
86. Jefayri MK, Grace PA, Mathie RT (2000) Attenuation of reperfusion injury by renal ischaemic preconditioning: the role of nitric oxide. BJU Int 85:1007–1013
87. Kempski O, Shohami E, von Lubitz D, Hallenbeck JM, Feuerstein G (1987) Postischemic production of eicosanoids in gerbil brain. Stroke 18:111–119
88. Puisieux F, Deplanque D, Pu Q, Souil E, Bastide M, Bordet R (2000) Differential role of nitric oxide pathway and heat shock protein in preconditioning and lipopolysaccharide-induced brain ischemic tolerance. Eur J Pharmacol 389:71–78
89. Pulsinelli WA, Levy DE, Duffy TE (1982) Regional cerebral blood flow and glucose metabolism following transient forebrain ischemia. Ann Neurol 11:499–502. doi:10.1002/ana.410110510
90. Cherian L, Hlatky R, Robertson CS (2004) Nitric oxide in traumatic brain injury. Brain Pathol 14:195–201
91. Buchan AM, Slivka A, Xue D (1992) The effect of the NMDA receptor antagonist MK-801 on cerebral blood flow and infarct volume in experimental focal stroke. Brain Res 574:171–177
92. Hoyte LC, Papadakis M, Barber PA, Buchan AM (2006) Improved regional cerebral blood flow is important for the protection seen in a mouse model of late phase ischemic preconditioning. Brain Res 1121:231–237. doi:10.1016/j.brainres.2006.08.107
93. Li Y, Lu Z, Keogh CL, Yu SP, Wei L (2007) Erythropoietin-induced neurovascular protection, angiogenesis, and cerebral blood flow restoration after focal ischemia in mice. J Cereb Blood Flow Metab 27:1043–1054. doi:10.1038/sj.jcbfm.9600417
94. Nagel S, Papadakis M, Chen R, Hoyte LC, Brooks KJ, Gallichan D, Sibson NR, Pugh C, Buchan AM (2011) Neuroprotection by dimethyloxalylglycine following permanent and transient focal cerebral ischemia in rats. J Cereb Blood Flow Metab 31:132–143. doi:10.1038/jcbfm.2010.60
95. Sugiyama Y, Yagita Y, Oyama N, Terasaki Y, Omura-Matsuoka E, Sasaki T, Kitagawa K (2011) Granulocyte colony-stimulating factor enhances arteriogenesis and ameliorates cerebral damage in a mouse model of ischemic stroke. Stroke 42:770–775. doi:10.1161/STROKEAHA.110.597799
96. Zwagerman N, Sprague S, Davis MD, Daniels B, Goel G, Ding Y (2010) Pre-ischemic exercise preserves cerebral blood flow during reperfusion in stroke. Neurol Res 32:523–529. doi:10.1179/016164109X12581096796431
97. Gu GJ, Li YP, Peng ZY, Xu JJ, Kang ZM, Xu WG, Tao HY, Ostrowski RP, Zhang JH, Sun XJ (2008) Mechanism of ischemic tolerance induced by hyperbaric oxygen preconditioning involves upregulation of hypoxia-inducible factor-1alpha and erythropoietin in rats. J Appl Physiol 104:1185–1191. doi:10.1152/japplphysiol.00323.2007
98. Gustavsson M, Mallard C, Vannucci SJ, Wilson MA, Johnston MV, Hagberg H (2007) Vascular response to hypoxic preconditioning in the immature brain. J Cereb Blood Flow Metab 27:928–938. doi:10.1038/sj.jcbfm.9600408

99. Gustavsson M, Wilson MA, Mallard C, Rousset C, Johnston MV, Hagberg H (2007) Global gene expression in the developing rat brain after hypoxic preconditioning: involvement of apoptotic mechanisms? Pediatr Res 61:444–450. doi:10.1203/pdr.0b013e3180332be4
100. Semenza GL (1999) Regulation of mammalian O2 homeostasis by hypoxia-inducible factor 1. Annu Rev Cell Dev Biol 15:551–578. doi:10.1146/annurev.cellbio.15.1.551
101. Sorond FA, Shaffer ML, Kung AL, Lipsitz LA (2009) Desferroxamine infusion increases cerebral blood flow: a potential association with hypoxia-inducible factor-1. Clin Sci 116:771–779. doi:10.1042/CS20080320
102. Hashiguchi A, Yano S, Morioka M, Hamada J, Ushio Y, Takeuchi Y, Fukunaga K (2004) Up-regulation of endothelial nitric oxide synthase via phosphatidylinositol 3-kinase pathway contributes to ischemic tolerance in the CA1 subfield of gerbil hippocampus. J Cereb Blood Flow Metab 24:271–279. doi:10.1097/01.WCB.0000110539.96047.FC
103. Willmot M, Gray L, Gibson C, Murphy S, Bath PM (2005) A systematic review of nitric oxide donors and L-arginine in experimental stroke; effects on infarct size and cerebral blood flow. Nitric Oxide 12:141–149. doi:10.1016/j.niox.2005.01.003
104. Atochin DN, Clark J, Demchenko IT, Moskowitz MA, Huang PL (2003) Rapid cerebral ischemic preconditioning in mice deficient in endothelial and neuronal nitric oxide synthases. Stroke 34:1299–1303. doi:10.1161/01.STR.0000066870.70976.57
105. Huang Z, Huang PL, Ma J, Meng W, Ayata C, Fishman MC, Moskowitz MA (1996) Enlarged infarcts in endothelial nitric oxide synthase knockout mice are attenuated by nitro-L-arginine. J Cereb Blood Flow Metab 16:981–987. doi:10.1097/00004647-199609000-00023
106. Attwell D, Buchan AM, Charpak S, Lauritzen M, Macvicar BA, Newman EA (2010) Glial and neuronal control of brain blood flow. Nature 468:232–243. doi:10.1038/nature09613
107. Busija DW, Bari F, Domoki F, Louis T (2007) Mechanisms involved in the cerebrovascular dilator effects of N-methyl-D-aspartate in cerebral cortex. Brain Res Rev 56:89–100. doi:10.1016/j.brainresrev.2007.05.011
108. Garthwaite J, Charles SL, Chess-Williams R (1988) Endothelium-derived relaxing factor release on activation of NMDA receptors suggests role as intercellular messenger in the brain. Nature 336:385–388. doi:10.1038/336385a0
109. Chen FY, Lee TJ (1993) Role of nitric oxide in neurogenic vasodilation of porcine cerebral artery. J Pharmacol Exp Ther 265:339–345
110. Ishine T, Yu JG, Asada Y, Lee TJ (1999) Nitric oxide is the predominant mediator for neurogenic vasodilation in porcine pial veins. J Pharmacol Exp Ther 289:398–404
111. Lee TJ, Sarwinski S, Ishine T, Lai CC, Chen FY (1996) Inhibition of cerebral neurogenic vasodilation by L-glutamine and nitric oxide synthase inhibitors and its reversal by L-citrulline. J Pharmacol Exp Ther 276:353–358
112. Porter JT, McCarthy KD (1996) Hippocampal astrocytes in situ respond to glutamate released from synaptic terminals. J Neurosci 16:5073–5081
113. Liu X, Li C, Gebremedhin D, Hwang SH, Hammock BD, Falck JR, Roman RJ, Harder DR, Koehler RC (2011) Epoxyeicosatrienoic acid-dependent cerebral vasodilation evoked by metabotropic glutamate receptor activation in vivo. Am J Physiol Heart Circ Physiol 301: H373–H381. doi:10.1152/ajpheart.00745.2010
114. Gordon GR, Choi HB, Rungta RL, Ellis-Davies GC, MacVicar BA (2008) Brain metabolism dictates the polarity of astrocyte control over arterioles. Nature 456:745–749. doi:10.1038/nature07525
115. Wendling WW, Chen D, Daniels FB, Monteforte MR, Fischer MB, Harakal C, Carlsson C (1996) The effects of N-methyl-D-aspartate agonists and antagonists on isolated bovine cerebral arteries. Anesth Analg 82:264–268
116. Davis RJ, Murdoch CE, Ali M, Purbrick S, Ravid R, Baxter GS, Tilford N, Sheldrick RL, Clark KL, Coleman RA (2004) EP4 prostanoid receptor-mediated vasodilatation of human middle cerebral arteries. Br J Pharmacol 141:580–585. doi:10.1038/sj.bjp.0705645

117. Conti MA, Adelstein RS (1981) The relationship between calmodulin binding and phosphorylation of smooth muscle myosin kinase by the catalytic subunit of 3′:5′ cAMP-dependent protein kinase. J Biol Chem 256:3178–3181
118. Nakamura K, Koga Y, Sakai H, Homma K, Ikebe M (2007) cGMP-dependent relaxation of smooth muscle is coupled with the change in the phosphorylation of myosin phosphatase. Circ Res 101:712–722. doi:10.1161/CIRCRESAHA.107.153981
119. Takata F, Dohgu S, Nishioku T, Takahashi H, Harada E, Makino I, Nakashima M, Yamauchi A, Kataoka Y (2009) Adrenomedullin-induced relaxation of rat brain pericytes is related to the reduced phosphorylation of myosin light chain through the cAMP/PKA signaling pathway. Neurosci Lett 449:71–75. doi:10.1016/j.neulet.2008.10.082
120. Serebryakov V, Zakharenko S, Snetkov V, Takeda K (1994) Effects of prostaglandins E1 and E2 on cultured smooth muscle cells and strips of rat aorta. Prostaglandins 47:353–365
121. Jadhav V, Jabre A, Lin SZ, Lee TJ (2004) EP1- and EP3-receptors mediate prostaglandin E2-induced constriction of porcine large cerebral arteries. J Cereb Blood Flow Metab 24:1305–1316. doi:10.1097/01.WCB.0000139446.61789.14
122. Takizawa S, Hogan M, Hakim AM (1991) The effects of a competitive NMDA receptor antagonist (CGS-19755) on cerebral blood flow and pH in focal ischemia. J Cereb Blood Flow Metab 11:786–793. doi:10.1038/jcbfm.1991.136
123. Nehls DG, Park CK, MacCormack AG, McCulloch J (1990) The effects of N-methyl-D-aspartate receptor blockade with MK-801 upon the relationship between cerebral blood flow and glucose utilisation. Brain Res 511:271–279
124. Ginsberg MD (2008) Neuroprotection for ischemic stroke: past, present and future. Neuropharmacology 55:363–389. doi:10.1016/j.neuropharm.2007.12.007
125. Dave KR, Lange-Asschenfeldt C, Raval AP, Prado R, Busto R, Saul I, Perez-Pinzon MA (2005) Ischemic preconditioning ameliorates excitotoxicity by shifting glutamate/gamma-aminobutyric acid release and biosynthesis. J Neurosci Res 82:665–673. doi:10.1002/jnr.20674
126. Douen AG, Akiyama K, Hogan MJ, Wang F, Dong L, Chow AK, Hakim A (2000) Preconditioning with cortical spreading depression decreases intraischemic cerebral glutamate levels and down-regulates excitatory amino acid transporters EAAT1 and EAAT2 from rat cerebal cortex plasma membranes. J Neurochem 75:812–818
127. Johns L, Sinclair AJ, Davies JA (2000) Hypoxia/hypoglycemia-induced amino acid release is decreased in vitro by preconditioning. Biochem Biophys Res Commun 276:134–136. doi:10.1006/bbrc.2000.3443
128. Folkman J (2007) Angiogenesis: an organizing principle for drug discovery? Nat Rev Drug Discov 6:273–286. doi:10.1038/nrd2115
129. Lay CC, Davis MF, Chen-Bee CH, Frostig RD (2010) Mild sensory stimulation completely protects the adult rodent cortex from ischemic stroke. PLoS One 5:e11270. doi:10.1371/journal.pone.0011270
130. Heissig B, Hattori K, Friedrich M, Rafii S, Werb Z (2003) Angiogenesis: vascular remodeling of the extracellular matrix involves metalloproteinases. Curr Opin Hematol 10:136–141
131. Choi SA, Kim EH, Lee JY, Nam HS, Kim SH, Kim GW, Lee BI, Heo JH (2007) Preconditioning with chronic cerebral hypoperfusion reduces a focal cerebral ischemic injury and increases apurinic/apyrimidinic endonuclease/redox factor-1 and matrix metalloproteinase-2 expression. Curr Neurovasc Res 4:89–97
132. Bullitt E, Rahman FN, Smith JK, Kim E, Zeng D, Katz LM, Marks BL (2009) The effect of exercise on the cerebral vasculature of healthy aged subjects as visualized by MR angiography. AJNR Am J Neuroradiol 30:1857–1863. doi:10.3174/ajnr.A1695
133. Isaacs KR, Anderson BJ, Alcantara AA, Black JE, Greenough WT (1992) Exercise and the brain: angiogenesis in the adult rat cerebellum after vigorous physical activity and motor skill learning. J Cereb Blood Flow Metab 12:110–119. doi:10.1038/jcbfm.1992.14

134. Swain RA, Harris AB, Wiener EC, Dutka MV, Morris HD, Theien BE, Konda S, Engberg K, Lauterbur PC, Greenough WT (2003) Prolonged exercise induces angiogenesis and increases cerebral blood volume in primary motor cortex of the rat. Neuroscience 117:1037–1046
135. Zhang F, Wu Y, Jia J (2011) Exercise preconditioning and brain ischemic tolerance. Neuroscience 177:170–176. doi:10.1016/j.neuroscience.2011.01.018
136. Carro E, Nunez A, Busiguina S, Torres-Aleman I (2000) Circulating insulin-like growth factor I mediates effects of exercise on the brain. J Neurosci 20:2926–2933
137. Cotman CW, Berchtold NC, Christie LA (2007) Exercise builds brain health: key roles of growth factor cascades and inflammation. Trends Neurosci 30:464–472. doi:10.1016/j.tins.2007.06.011
138. Tang K, Xia FC, Wagner PD, Breen EC (2010) Exercise-induced VEGF transcriptional activation in brain, lung and skeletal muscle. Respir Physiol Neurobiol 170:16–22. doi:10.1016/j.resp.2009.10.007
139. Zhao Y, Zhao B (2010) Protective effect of natural antioxidants on heart against ischemia-reperfusion damage. Curr Pharm Biotechnol 11:868–874
140. Watson BD, Prado R, Mirzabeigi M, Veloso A, Morales A (2003) A tissue plasminogen activator (reteplase) augments the efficacies of UV laser-facilitated dethrombosis in recanalizing aged platelet and platelet-rich occlusive thrombi in rat middle cerebral artery. J Cereb Blood Flow Metab 23:279
141. Watson BD, Prado R, Veloso A, Brunschwig JP, Dietrich WD (2002) Cerebral blood flow restoration and reperfusion injury after ultraviolet laser-facilitated middle cerebral artery recanalization in rat thrombotic stroke. Stroke 33:428–434
142. Yang GY, Betz AL (1994) Reperfusion-induced injury to the blood–brain barrier after middle cerebral artery occlusion in rats. Stroke 25:1658–1664, discussion 1664–1655
143. Aronowski J, Strong R, Grotta JC (1997) Reperfusion injury: demonstration of brain damage produced by reperfusion after transient focal ischemia in rats. J Cereb Blood Flow Metab 17:1048–1056. doi:10.1097/00004647-199710000-00006
144. Zhao H, Sapolsky RM, Steinberg GK (2006) Interrupting reperfusion as a stroke therapy: ischemic postconditioning reduces infarct size after focal ischemia in rats. J Cereb Blood Flow Metab 26:1114–1121. doi:10.1038/sj.jcbfm.9600348
145. Gao X, Zhang H, Takahashi T, Hsieh J, Liao J, Steinberg GK, Zhao H (2008) The Akt signaling pathway contributes to postconditioning's protection against stroke; the protection is associated with the MAPK and PKC pathways. J Neurochem 105:943–955. doi:10.1111/j.1471-4159.2008.05218.x
146. Zhao H (2009) Ischemic postconditioning as a novel avenue to protect against brain injury after stroke. J Cereb Blood Flow Metab 29:873–885. doi:10.1038/jcbfm.2009.13
147. Gao X, Ren C, Zhao H (2008) Protective effects of ischemic postconditioning compared with gradual reperfusion or preconditioning. J Neurosci Res 86:2505–2511. doi:10.1002/jnr.21703
148. Bonini PA, Banfi G, Murone M (1990) Enhanced chemiluminescence in the measurement of proteins and haptens: evaluation of choriogonadotropin (hCG) and free thyroxin. J Biolumin Chemilumin 5:193–195. doi:10.1002/bio.1170050309
149. Xing C, Hayakawa K, Lok J, Arai K, Lo EH (2012) Injury and repair in the neurovascular unit. Neurol Res 34:325–330. doi:10.1179/1743132812Y.0000000019
150. Zlokovic BV (2008) The blood–brain barrier in health and chronic neurodegenerative disorders. Neuron 57:178–201. doi:10.1016/j.neuron.2008.01.003
151. Lipton P (1999) Ischemic cell death in brain neurons. Physiol Rev 79:1431–1568
152. del Zoppo GJ, von Kummer R, Hamann GF (1998) Ischaemic damage of brain microvessels: inherent risks for thrombolytic treatment in stroke. J Neurol Neurosurg Psychiatry 65:1–9
153. Guo M, Cox B, Mahale S, Davis W, Carranza A, Hayes K, Sprague S, Jimenez D, Ding Y (2008) Pre-ischemic exercise reduces matrix metalloproteinase-9 expression and ameliorates blood–brain barrier dysfunction in stroke. Neuroscience 151:340–351. doi:10.1016/j.neuroscience.2007.10.006

154. Hamann GF, Burggraf D, Martens HK, Liebetrau M, Jager G, Wunderlich N, DeGeorgia M, Krieger DW (2004) Mild to moderate hypothermia prevents microvascular basal lamina antigen loss in experimental focal cerebral ischemia. Stroke 35:764–769. doi:10.1161/01.STR.0000116866.60794.21
155. Mun-Bryce S, Rosenberg GA (1998) Matrix metalloproteinases in cerebrovascular disease. J Cereb Blood Flow Metab 18:1163–1172. doi:10.1097/00004647-199811000-00001
156. Ren C, Gao X, Niu G, Yan Z, Chen X, Zhao H (2008) Delayed postconditioning protects against focal ischemic brain injury in rats. PLoS One 3:e3851. doi:10.1371/journal.pone.0003851
157. Pan W, Kastin AJ (2007) Tumor necrosis factor and stroke: role of the blood–brain barrier. Prog Neurobiol 83:363–374. doi:10.1016/j.pneurobio.2007.07.008
158. Watters O, O'Connor JJ (2011) A role for tumor necrosis factor-alpha in ischemia and ischemic preconditioning. J Neuroinflammation 8:87. doi:10.1186/1742-2094-8-87
159. Franzen B, Duvefelt K, Jonsson C, Engelhardt B, Ottervald J, Wickman M, Yang Y, Schuppe-Koistinen I (2003) Gene and protein expression profiling of human cerebral endothelial cells activated with tumor necrosis factor-alpha. Brain Res Mol Brain Res 115:130–146
160. Yin L, Ye S, Chen Z, Zeng Y (2012) Rapamycin preconditioning attenuates transient focal cerebral ischemia/reperfusion injury in mice. Int J Neurosci 122:748–756. doi:10.3109/00207454.2012.721827
161. Ding YH, Young CN, Luan X, Li J, Rafols JA, Clark JC, McAllister JP 2nd, Ding Y (2005) Exercise preconditioning ameliorates inflammatory injury in ischemic rats during reperfusion. Acta Neuropathol 109:237–246. doi:10.1007/s00401-004-0943-y
162. Guo M, Lin V, Davis W, Huang T, Carranza A, Sprague S, Reyes R, Jimenez D, Ding Y (2008) Preischemic induction of TNF-alpha by physical exercise reduces blood–brain barrier dysfunction in stroke. J Cereb Blood Flow Metab 28:1422–1430. doi:10.1038/jcbfm.2008.29
163. Rosenzweig HL, Minami M, Lessov NS, Coste SC, Stevens SL, Henshall DC, Meller R, Simon RP, Stenzel-Poore MP (2007) Endotoxin preconditioning protects against the cytotoxic effects of TNFalpha after stroke: a novel role for TNFalpha in LPS-ischemic tolerance. J Cereb Blood Flow Metab 27:1663–1674. doi:10.1038/sj.jcbfm.9600464
164. Rosenzweig HL, Lessov NS, Henshall DC, Minami M, Simon RP, Stenzel-Poore MP (2004) Endotoxin preconditioning prevents cellular inflammatory response during ischemic neuroprotection in mice. Stroke 35:2576–2581. doi:10.1161/01.STR.0000143450.04438.ae

Chapter 11
Stem Cells for Neurovascular Repair in CNS Trauma

Mibel M. Pabón, Travis Dailey, Naoki Tajiri, Kazutaka Shinozuka, Hiroto Ishikawa, Sandra Acosta, Yuji Kaneko, and Cesar V. Borlongan

Abstract Stem cells exert therapeutic effects for central nervous system (CNS) trauma. Accumulating evidence reveals that stem cell-based therapies for CNS trauma can be achieved via transplantation of exogenous stem cells or stimulation of endogenous stem cells from the neurogenic niches of subventricular zone and subgranular zone, or recruited from the bone marrow through peripheral circulation. In this chapter, we review the different sources of stem cells that have been tested in animal models of CNS trauma, highlighting the research progress on stem cell-based therapeutics in stroke and their extension to traumatic brain injury (TBI). In addition, we discuss specific mechanisms of action, in particular neurovascular repair by endothelial progenitor cells, as key translational research for advancing the clinical applications of stem cells for CNS trauma.

11.1 Introduction

Traumatic brain injury (TBI) is the third leading cause of death and the leading cause of long-term disability in the United States [1]. In 2000, the direct and indirect costs of stroke in the United States were estimated to be $76.5 billion [2]. The mean lifetime cost of TBI to a single patient in the United States is estimated at $196,460; this includes inpatient care, rehabilitation, and follow-up care necessary for lasting deficits [3]. Approximately 1.7 million people sustain a TBI annually each year [1]. The numbers of affected individuals, the costs necessary to facilitate their care, and rehabilitation coupled with the lack of therapies indicate that TBI represents a current significant unmet medical need.

C.V. Borlongan (✉)
Department of Neurosurgery and Brain Repair, University of South Florida College of Medicine, 12901 Bruce B. Downs Boulevard, Tampa, FL 33612, USA
e-mail: cborlong@health.usf.edu

E.H. Lo et al. (eds.), *Vascular Mechanisms in CNS Trauma*, Springer Series in Translational Stroke Research 5, DOI 10.1007/978-1-4614-8690-9_11,

The current therapy for TBI is limited, with decompressive craniectomy to relieve intracranial pressure as a treatment of choice for TBI patients [4–8] and thereafter patients relegated mostly to rehabilitation therapy [9–12] either via aerobic exercise and cognitive rehabilitation to improve learning and memory [13]. Recent clinical trials have targeted the acute phase of injury using neuroprotective drugs [14–18] and have also tested prophylactic treatments such as hypothermia to lessen TBI injury [19]. An opportunity exists for treatment regimens designed to abrogate the secondary cell death associated with TBI. Along this line of investigations, stem cell therapy may prove beneficial in treating TBI-secondary cell death beyond the acute phase of injury. Hematopoietic stem cells and mesenchymal stem cells have been used for many years to treat disorders with some observed degree of benefit for neurological disorders such as stroke and TBI [19].

To this end, we advance the approach that cell therapy can abrogate the blood–brain barrier (BBB) breakdown associated with TBI, and such BBB repair should directly benefit TBI in view of the BBB damage inherent in the disease itself. Our group has examined the BBB destruction accompanying the progressive pathology of TBI after the initial injury. The interaction between endothelial cells, pericytes, astrocytes, neurons, and smooth muscle cells create the neurovascular unit, which plays an important role as a barrier between the CNS and the blood stream [20]. The BBB not only serves as a barrier but it also transports nutrients back and forth crossing the endothelial layer; it also inactivates molecules that threaten to cross the barrier [20]. TBI is characterized mainly by the primary injury that results from blunt force to the brain matter, but we cannot neglect the fact that TBI is also associated with secondary events. These secondary cell death events, including BBB breakdown, occur with some delay and if left untreated can exacerbate the primary injury and can be detrimental with long-lasting adverse effects. Clinical and experimental animal models demonstrate that there are some behavioral changes that occur in the first days and changes in cerebral blood flow [21, 22]. As seen in stroke, TBI is also associated with pro-inflammatory components in the neurovasculature that accumulate within the brain. First, leukocytes accumulate within the first couple of hour after TBI followed by surplus in pro-inflammatory cytokines and oxidative stress [23]. After the initial injury in TBI, there is disruption of tight junctions, channels, pericytes, and astrocytic foot processes within the neurovasculature. For many years, it was believed that the opening of the BBB occur transiently after TBI but now we recognize that it is an event that occurs within the first day post-injury and that BBB permeability persists over time due to the progressive tight junctions alterations [24]. Targeting neurovascular repair has been explored recently but the results have been very limited in animal models. Further research needs to be performed focusing on the repair of neurovasculature in TBI. More recently, the inflammatory response has been implicated in the integrity of BBB after TBI, characterized by an increase in production of pro-inflammatory cytokines eventually leading to an increase in influx of inflammatory cells from blood to brain [25]. Altogether, these studies suggest that neurovascular alterations accompany TBI pathology, and that finding a strategy geared towards neurovascular repair is likely to augment the progressive nature of the disease's secondary cell death. This chapter discusses the preclinical basis for testing stem cell therapy in CNS trauma. Because of significant research strides achieved in stem cell-based therapeutics in stroke, and with overlapping etiologies and pathologies between stroke and TBI, we provide relevant insights in stroke that may prove critical to extending the safety and efficacy of stem cell therapy for TBI. We outline below the potential of cell-based therapy in view of the current treatments for TBI. In particular, the wider therapeutic window for stem cell transplantation, which may allow neuroregeneration at the chronic stage of the disease as opposed to the acute phase targeted by neuroprotection, makes cell-based therapy an appealing strategy. Finally, we address the gap in knowledge concerning mechanisms underlying the therapeutic benefit of stem cells in CNS trauma. Here, we highlight the underexplored concept of neurovascular repair as a major mode of action of cell therapy, and emphasize the major role of endothelial progenitor cells (EPC) as an effective cell source for transplantation. Our strategy is to exploit this neurovascular repair mechanism via EPC transplantation as a standalone or as an adjunct therapy for augmenting existing treatments for TBI.

11.2 Stem Cell Sources

We attempt here to provide an overview of stem cell sources that have been investigated in stroke, then subsequently discuss stem cells that have been tested in TBI. Several sources of stem cells have been demonstrated as safe and effective in animal models of stroke. In a historical order, the major types of cells transplanted in stroke include fetal-derived cells, neuroteratocarcinoma cells (NT2N), xenogenic pig-derived cells, embryonic stem (ES) cells, adult stem cells (bone marrow, human umbilical cord, placenta, amnion fluid, menstrual blood), and induced pluripotent stem cells (iPS). Due to ethical and logistical concerns, the use of adult stem cells has flourished over the last decade, which was further aided by a moratorium for using federal funds on ES research. Interestingly, the ongoing FDA-approved stem cell clinical trials in stroke use adult stem cells. It is likely that adult stem cells may also be the frontrunner for cell therapy in TBI. For this chapter, we highlight the potential of adult bone marrow-derived EPC in neurovascular repair for stroke and TBI.

11.3 Stem Cell Therapy for CNS Trauma

Cell transplantation therapies and stem cell treatments have emerged as potential treatments for numerous diseases and medical conditions, including stroke. One approach using stem cells involved the direct transplantation of neural stem cells (NSCs) into the damaged region of the brain. NSCs transplanted following transient global ischemia differentiated into neurons and improved spatial recognition in rats [26]. Post-mitotic neuron-like cells (NT2N) cells, derived from a human embryonal carcinoma cell line, migrated over long distances after implantation into brains of immunocompetent newborn mice and differentiated into neuron- and oligodendrocyte-like cells [27]. NT2N cells promoted functional recovery following focal cerebral ischemia after direct transplantation [28]. Similarly, MHP36 cells, a stem cell line derived from mouse neuroepithelium, improved functional outcome in rats after global ischemia [29] and also following focal cerebral ischemia or stroke [30]. NCSs grafted into brain developed morphological and electrophysiological characteristics of neurons [31].

Other direct transplantation experiments in the brain have utilized cells derived from bone marrow. Bone marrow stromal cells (MSCs), when injected into the lateral ventricle of the brain, migrated, and differentiated into astrocytes [32]. Fresh bone marrow transplanted directly into the ischemic boundary zone of rat brain improved functional recovery from middle cerebral artery occlusion [33]. Similarly, MSCs implanted into the striatum of mice after stroke, improved functional recovery [34]. MSCs differentiated into presumptive neurons in culture [35] and assumed functional neuronal characteristics in embryonic rats [36]. Intracerebral grafts of mouse bone marrow also facilitated restoration of cerebral blood flow and BBB after stroke in rats [37]. Indirect transplant methods, via intravenous or intra-arterial injection, also have been shown to afford positive effects. Following bone marrow transplantation with tagged donor cells, tagged bone marrow stem cells were shown to differentiate into microglia and astrocyte-like cells [38]. Intra-carotid administration of MSCs following middle cerebral artery occlusion in a rat model improved functional outcome [39]. Similarly, intravenous administration of umbilical cord blood cells ameliorated functional deficits after stroke in rats [40]. Rats, which had received tagged bone marrow cell transplantation, showed the tagged cells as putative neurons and endothelial cells following middle cerebral artery occlusion and reperfusion [41]. It has also been reported that intravenous administration of cord blood cells was more effective than intra-striatal administration in producing functional benefit following stroke in rats [42]. Intravenous administration of MSCs has also been found to induce angiogenesis in the ischemic boundary zone following stroke in rats [43].

Along this vein, stem cell therapy appears to be a promising treatment for TBI. We summarize below recent studies on cell-based therapeutics for TBI (Table 11.1). Here, we acknowledge that the field of stem cell therapy for TBI remains in its infancy. Stem cell sources range from cells derived from embryonic, fetal, and adult tissue sources (e.g., umbilical cord, placenta, amnion, bone marrow) [44–64]. Functional readouts have been mostly generated during short-term post-transplantation, thereby necessitating long-term investigations for monitoring of stable and robust benefits, as well as assessing any adverse effects over time in order to reveal both efficacy and safety profiles of stem cell therapy for TBI.

Table 11.1 Recent stem cell-based therapies for TBI

Author	Stem cell type	TBI animal model	Outcomes/results	References
Ma et al. 2011	Neural stem cells (NSCs) genetically modified to encode BDNF gene (BDNF/NSCs)	Controlled cortical impact (CCI) Rat	Enhanced neurite growth and upregulated synaptic proteins in BDNF/NSCs-transplanted TBI rats. Over expression of BDNF-mediated motor behavior improvement in transplanted TBI rats	[44]
Mahmood et al. 2006	Bone marrow stromal cells	Controlled cortical impact (CCI) Rat	Transplanted BMSCs were present in the injured brain 3 months after TBI and functional outcome was significantly improved	[45]
Qu et al. 2008	Marrow stromal cells (MSCs)	Controlled cortical impact (CCI) Mouse	Significant neurological improvements as revealed by morris water maze and foot fault test in MSC-treated TBI mice	[46]
Lu et al. 2007	Human marrow stromal cells (hMSCs)	Controlled cortical impact (CCI) Rat	hMSCs improved spatial learning and sensorimotor function, accompanied by reduced lesion volume in TBI animals	[47]
Harting et al. 2010	Mesenchymal stem cell (MSC)	Unilateral controlled cortical impact (CCI) Rat	Intravenous-delivered MSCs were identified in the lungs 48 h post-infusion; therefore, there was no functional improvement seen	[48]
Riess et al. 2002	Neural stem cells (NSC)	Controlled cortical impact (CCI) Mouse	TBI animals that received NSC transplant improved motor function with graft survival after 13 weeks post-transplantation	[49]
Hattiangady et al. 2012	Neural stem cell (NSC)	Unilateral partial hippocampal injury Rat	TBI animals that received SVZ-NSC grafts after injury exhibited improved mood and memory function as compared to control. The cells derived from grafts exhibited migration, survival, and neuronal differentiation	[50]

(continued)

Table 11.1 (continued)

Author	Stem cell type	TBI animal model	Outcomes/results	References
Nichols et al. 2013	Human peripheral blood derived (HPBD) MSC HPBD CD133+, ATP-binding cassette sub-family G member 2 (ABCG2)+, C-X-C chemokine receptor type 4 (CXCR4)+ MSCs combined with trans-retinoic acid (RA) mixture	Fluid percussion injury Rat	CD133+ ABCG2+ CXCR4+ MSCs expressed neuronal lineage markers and survived for 1 and 3 month post-transplantation with the potential to reduce cognitive impairment seen in TBI	[51]
Yan et al. 2013	Human amnion-derived mesenchymal stem cells (AMSC)	Controlled TBI impact model a weight-drop device Rat	Transplanted TBI rats demonstrated significant increase in neurological function, brain morphology, and increase in expression of neurotrophic and growth factors, thereafter stimulating endogenous growth factors and promoting neurorehabilitation	[52]
Wallenquist et al. 2012	Neural stem and progenitor cells (NSPC)	Controlled cortical impact (CCI) Mouse	Ibuprofen down-regulated TBI-induced inflammatory response. Interestingly transplanted neuroblast were found near the impacted area and ipsilateral hippocampus suggesting that ibuprofen anti-inflammatory properties is crucial for the survival and differentiation of the grafts	[53]
Shear et al. 2011	Neural stem cells (NSCs)	Controlled cortical impact (CCI) Mouse	NSCs are optimal when used 2–7 days post-TBI. The transplant location plays a key role in cell survival, differentiation, migration, and functional efficacy. NSC also stimulate protective and neurotrophic factors rather than replacing neuronal or glial cells	[54]

(continued)

Table 11.1 (continued)

Author	Stem cell type	TBI animal model	Outcomes/results	References
Lee et al. 2013	Neural stem cells (NSCs)	Corticectomy Rat	NSCs improved behaviors and motor evoked potentials when combined with rehabilitation therapy. The groups that received combination therapy and only rehabilitation demonstrated a prolonged effect in expression of the endogenous NSCs	[55]
de Freitas et al. 2012	Bone marrow-derived mesenchymal stem cells (MSCs) or bone marrow mononuclear cells (BMMCs)	Ablation by aspiration Rat	Bone marrow mononuclear (BMMC) cells are more efficient and accessible than MSCs	[56]
Chuang et al. 2012	Secretome from human mesenchymal stem cells	Fluid percussion injury Rat	MSC-derived secretome attenuated motor deficits seen after TBI injury. Markers for apoptosis and neuronal cell loss were also decreased in the secretome-treated animals. Conversely, secretome increased the levels of VEGF positive cells	[57]
Walker et al. 2012	Bone marrow-derived mesenchymal stromal cells (MSCs)	TBI model Rat	Neuroprotection produced by MSC via enhanced M2 cell activation of anti-inflammatory response, thereby reducing exacerbated inflammatory reaction associated TBI	[58]
Tu et al. 2012	Mesenchymal stem cells from umbilical cord (UCSMCs) and temperature-sensitive UCSMCs (tsUCSMCs)	Fluid percussion injury Rat	The combination of hypothermia with UCSMCs, or tsUCSMCs is beneficial in improving motor and cognitive function when used together rather than stem cell therapy alone after TBI injury	[59]

(continued)

Table 11.1 (continued)

Author	Stem cell type	TBI animal model	Outcomes/results	References
Antonucci et al. 2012	Amniotic Fluid-derived Stem cells (AFS)	Controlled cortical impact (CCI) and fluid percussion injury Rat and mouse	AFS cells are good transplant donor cells due to high renewable capacity and have a capacity to effectively differentiate to multiple lineages	[60]
Joo et al. 2012	Neural stem cells (NSCs)	Focused brain irradiation Mouse	NSC supplementation enhanced endogenous neurotrophic factors and was able to differentiate into astrocytes and neurons which migrated to the irradiated areas of the brain	[61]
Shi et al. 2012	Human umbilical cord mesenchymal stem cells (hUC-MSCs) In vitro BDNF blended chitosan scaffolds on neural stem cell	TBI model Rat	BDNF is beneficial in promoting neuronal differentiation of NSC	[62]
Yang et al. 2011	Schwann cells differentiated from adipose-derived stem cells (ADSC-SCs)	Contusion brain injury Rat	Transplantation of ADSC-SCs into rats with contused brain promoted locomotor function and reduced reactive gliosis compared to undifferentiated ADSCs	[63]
Skardelly et al. 2011	Human fetal neural progenitor cell (hfNPC)	Controlled cortical impact (CCI) Rat	MRI analysis showed a smaller lesion size in animals that received the transplants as compared to non-transplanted animals. Histological analysis demonstrated increased levels of angiogenic markers and reduced astroglial reaction at 4 weeks after transplantation	[64]

11.4 Mechanistic Interpretation of Therapeutic Benefit Involving Stem Cells

It is unclear what brings about the purported benefit from stem cell transplantation. One possibility is the transformation of the transplanted cells into neurons [65]. There appears to be a positive relationship between the degree of behavioral improvement and the number of transplanted cells that stain positive for neuron-specific markers [26]. However, transplanted cells often do not develop normal processes, and thus the benefit may not be mediated only by neuronal circuitry [66].

A second hypothesis that is not mutually exclusive is that the transplanted cells may also assist via differentiation into neuroectodermal-derived cell types other than neurons. MSCs migrate and transform into astrocytes [32]. Hematopoietic cells can differentiate into microglia and macroglia [38]. Bone marrow-derived stem cells may also assist in blood vessel regeneration following brain tissue damage in several ways. The stromal cell-derived factor-1 (SDF-1)/CXCR4 system assists in integration of cells into injured tissue by promoting the adhesion of CXCR4-positive cells onto vascular endothelium [67]. SDF-1 also augments vasculogenesis and neo-vasculogenesis of ischemic tissue by recruitment of EPC [68]. Bone marrow is a source of these endothelial progenitors [69]. Adult bone marrow-derived cells have been shown to participate in angiogenesis by the formation of periendothelial vascular cells [70]. Intravenous administration of MSCs induced angiogenesis in the ischemic boundary zone after stroke [43]. We also observed that crude bone marrow is a source of endothelial cells after experimental stroke [41]. Interestingly TBI induces cell proliferation in the hippocampus and the subventricular zone differentiated into mature neuronal cells 10 days post-TBI [71, 72]. The vasculature in the CNS becomes activated after injury and initiates a self-repair mechanism to combat the compromised site through the activation and mobilization of EPC from bone marrow and peripheral blood. Angiogenesis is believed to be a neuroprotective factor that can rescue nerve cells from secondary cell death injury [73]. Vascular endothelial growth factor (VEGF) induces angiogenesis and mobilizes EPC in diseases associated with blood vessel disorders such as stroke and TBI [74], thereby serves as a crucial growth factor in the creation of new vascular cells for BBB repair. VEGF also stimulates and supports preexisting endothelium-derived angiogenic cells which in synergy all these components are key players for brain repair [75].

Trophic factors produced by the transplanted cells could be a factor. Via this mechanism, bone marrow grafts may assist in restoring brain blood flow and also repairing the BBB [37]. Trophic factors from MSCs may play a role in brain repair itself. Recent evidence suggests that intravenous administration of MSCs increases the expression of nerve growth factor and brain-derived neurotrophic factor following TBI [76]. Understanding the exact mechanism(s) responsible for the therapeutic benefit seen following stem cell transplantation in the CNS is now at a critical junction in view of the planned FDA allowance for limited clinical trials of bone marrow-derived multipotent adult progenitor cells in acute ischemic stroke

[77]. Similarly, insights into the mechanism of action mediating stem cell therapeutic benefits in TBI will aid in optimization of cell dose, route of delivery, and timing of initiation of cell transplantation for clinical applications.

In accordance with the STAIR (Stroke Therapy Academic Industry Roundtable) and STEPS (Stem cell Therapeutics as an Emerging Paradigm for Stroke) criteria, investigations of the mechanism of action mediating experimental therapeutics in stroke are vital for extending their potential clinical utility [78, 79]. A similar call for strict translational guidelines has been advanced for TBI [80], and a set of consensus recommendations has been published to provide standards and best practices for future investigations in testing novel therapeutics in TBI animal models [81].

11.5 BBB Breakdown in CNS Trauma

We again draw from our observations of BBB breakdown in stroke as we advance the hypothesis of BBB compromise in TBI. A closely associated cell death cascade involved in stroke pathogenesis is impairment of the BBB, which further exacerbates brain damage. The central nervous system (CNS) is an immunologically privileged zone, protected from entry of immune cells and serum proteins by the BBB (as well as by the blood–spinal cord barrier and blood–cerebrospinal fluid barrier, but we will focus here on BBB). These CNS barriers control cerebral/spinal cord homeostasis by selective transport of molecules and cells [69–76, 82, 83]. This control is possible due to the unique structure of the microvasculature—in particular capillaries formed by endothelial cells which are connected via adherens and tight junctions [84–86]. Functional integrity of all BBB elements is critical for protection of the CNS from harmful blood substances. Impairment of this cellular machinery may cause BBB breakdown, leading to edema in many cases of brain diseases or injuries, including stroke. Degradation of the extracellular matrix may be concomitant with BBB disruption and tissue softening, leading to more pronounced brain swelling and to severe cerebral edema in stroke patients [87] and other brain disorders such as Alzheimer's disease [88] and multiple sclerosis [89, 90]. Examination of BBB status in stroke reveals evidence of the barrier's altered permeability. Whereas the first phase of stroke is characterized by a surge in tissue Na+ and water content concomitant with an increased pinocytosis and Na+, K+ ATPase activity across the endothelium, the second stage of stroke ensues with BBB breakdown that is associated with infarction of both the parenchyma and the vasculature itself [91]. At this second stage, tissue Na+ level still remains, but the extravasation of serum proteases stands as a likely exacerbating factor [92]. Accumulating evidence implicates serum proteases in degradation of the extracellular matrix metalloproteinases (MMPs), which in turn aggravate BBB disruption and softening of the tissue, eventually manifesting into a well-defined form of brain swelling [91–93]. Part of the reason for the tPA's limited time window is that the surge in production of free radicals associated with delayed reperfusion brings a second wave of oxidative and nitrative stress that increases the risk of brain hemorrhage and edema [94]. With delayed reperfusion, there is a surge in production of superoxide, NO, and peroxynitrate. Formation of these radicals in the vicinity of blood vessels plays an important role in reperfusion-induced injury. These radicals activate MMPs, which degrade collagen and laminin in the basal lamina, disrupting the integrity of the basement membrane and increasing BBB permeability. Oxidative and nitrative stress also triggers recruitment and migration of neutrophils and other leukocytes to the cerebral vasculature, which release enzymes that further increase basal lamina degradation and vascular permeability. These BBB pathological events can lead to parenchymal hemorrhage, vasogenic brain edema, and neutrophil infiltration into the brain [95]. In the clinic, significant brain edema, such as that seen in malignant MCA infarction, develops in a delayed fashion after large hemispheric strokes and accounts for a high mortality rate (80 % in the case of malignant MCA infarction) [96]. The primary BBB function is controlling CNS homeostasis by selective transport. Substances with molecular weights higher than 400 Da generally cannot cross the BBB by free diffusion. Some molecules cross the barriers via endothelial carrier-mediated or receptor-mediated transporters, see review [69, 70, 82, 97]. It is possible that barrier disruption or dysfunction occurs in stroke, altering CNS homeostasis and allowing entry of harmful molecules from the periphery to the brain [98–100]. Among these injurious molecules are immune/inflammatory factors, such as monocyte/macrophage cells, activated microglia, and reactive astrocytes possibly secreting pro-inflammatory cytokines, which have been detected in stroke patients and animal models [101–103]. Although additional studies are warranted to confirm the BBB status in stroke patients, the above results taken together imply that BBB dysfunction may contribute to stroke pathology. Thus, there could be an impaired endothelium-mediated mechanism in stroke leading to barrier dysfunction.

In the TBI field, 16 patients demonstrated regions with enhanced signals within the brain showing BBB leakage in the cortical regions of at least 15 of the patients [104]. The disrupted BBB regions were surrounding old contusions which suggest that a local trauma had occurred. Models of TBI have been helpful in identifying what occurs to the BBB after impact. After stress from impact the vasculature is a primary target of the injury, leading to leakage of blood-borne proteins and the extravasation of red blood cells [104, 105]. Isolated petechial hemorrhages have also been identified contralaterally to the injury [105]. Along with the extravasation of red blood cells there has been disrupted endothelial lining and endothelial vacuolation, increase intracranial pressure leading to altered cerebral flow and poor neurological outcome due to increase levels of lactate overall causing brain damage and functional deficits [105].

As noted above, VEGF is enhanced after injury and stimulates angiogenesis as well therefore is beneficial for newborn neuronal cells and endogenous neurogenesis. In a closed head mice injury model, VEGF was seen to decrease the lesion volume caused by TBI and it also increased the amount of BrdU positive cells, demonstrating an increase in neurogenesis and gliogenesis after TBI [106]. This study indicates the vascular repair may be beneficial for TBI. Accordingly, we discuss below the potential of stem cell therapy for BBB repair.

11.6 EPC Therapy for BBB Repair in CNS Trauma

In discussing neurovascular repair for TBI, we build upon the more mature field of BBB repair in stroke. EPC, initially described by Asahara et al. [107] are immature endothelial cells that circulate in peripheral blood. In their pioneering study, transplanted EPC, isolated from human blood, were found in the endothelium of newly formed vessels in ischemic regions, indicating that a discrete cell population within the human blood participates in the formation of new vessels after ischemia. Griese et al. also found that grafted EPC populated the endothelium in animals with experimentally induced endothelial damage [108], further advancing the notion that EPC contribute to the repair of damaged endothelium. The dogma that existed until recently is that neovascularization, or formation of new blood vessels, results exclusively from proliferation and migration of preexisting endothelial cells, a process referred as to angiogenesis [109]. Furthermore, vasculogenesis or vascularization, defined as in situ differentiation of vascular endothelial cells from endothelial precursor cells, was thought to occur only in the embryo during vascular development. However, recent evidence has now established that circulating bone marrow-derived EPCs are capable of homing to neovascularization sites, proliferating, and differentiating into endothelial cells [110, 111]. EPCs have been identified mainly in the mononuclear cell fraction of peripheral blood, leukapheresis products, and in umbilical cord blood [107, 112], but can also be harvested from bone marrow. Over the last few years, EPCs have been studied as biomarkers to assess the risk of cardiovascular disease in human subjects. For example, a low EPC count predicts severe functional impairments in several cardiovascular pathologies such as diabetes [113], hypercholesterolemia [114], hypertension [115, 116], scleroderma [117, 118], aging [116, 119], cigarettes smoking [116, 120, 121], and coronary artery disease [84]. In addition, EPCs have been examined as potent donor graft cells for transplantation therapy.

Transplantation of EPCs into ischemic tissues has emerged as a promising approach in the treatment of diseases with blood vessels disorders [122–124]. In mouse models of ischemic injury, EPCs injection led to improved neovascularization in hind limb ischemia [122–124]. Based largely on these laboratory findings suggesting angiogenic and vasculogenic potential of EPCs, clinical studies have been initiated to reveal whether patients with lower EPC numbers are at higher risk for atherosclerotic events, and whether patients with ischemic events may benefit from EPC administration [125].

Clinical studies to date suggest the therapeutic potential of EPC transplantation, although this assumption should be approached with much caution due to being open label trials, observational and/or anecdotal accounts, and limited number of patients. Ex vivo expanded EPC, isolated from peripheral blood mononuclear cells, can incorporate into the foci of myocardial neovascularization [126, 127], and intracoronary infusion of peripheral blood or bone marrow-derived progenitors in patients with acute myocardial infarction was associated with significant benefits in post-infarction remodeling [128–135]. Still in observational studies in patients with myocardial infarction, higher numbers of EPC correlate with better prognosis, more myocardial salvage [136], viability and perfusion [137], and more collaterals in the ischemic zone [138]. Randomized clinical trials on autologous bone marrow-derived cells are mixed; whereas transplanted coronary artery disease patients display improved left ventricular function at least in the short term [139], transplanted patients with chronic ischemic heart failure exhibit modest to no effects on change in left ventricular function [140].

Similar randomized trials of autologous bone marrow-derived cells have been carried out in patients with peripheral artery disease and showed improved endothelium-dependent vasodilation [141], ankle brachial index, rest pain, and pain-free walking time [142], but the degree of functional recovery was not as robust as seen in animal models. Clearly, these results are obtained from autologous bone marrow-derived cells, which are heterogenous with scarce number of EPCs, thus may not closely approximate EPC endpoints. For clinical application of EPC in neurovascular disease, the available studies are much more limited with only three observational studies in patients with stroke. In 25 patients with an ischemic stroke, CD34+ cells peaked 7 days after stroke but generally reverted to baseline after 30 days [143]. Interestingly, higher CD34+ cell levels at 30 days related to higher numbers of infarcts on magnetic resonance imaging and also to cerebrovascular function as measured with positron emission tomography scanning (cerebral metabolic rate of oxygen and cerebral blood flow). On the other hand, decreased numbers of clusters of rapidly adhering cells were seen after stroke and in "stable cerebrovascular disease," compared to controls free of vascular disease [144]. Higher age and the presence of cerebrovascular disease in general independently related to lower EPC numbers. The discrepancies in the results of these studies may be due to mismatched controls for age of patients and/or the lack of methodological design for testing specific hypotheses on the causal role of EPC in cerebrovascular disease [144]. Although the primary mitigating mechanisms underlying stroke pathogenesis and its abrogation by cell therapy are still uncertain, there is substantial evidence implicating immunological attack upon the brain and/or its vasculature; widespread inflammatory reactions in stroke may trigger a cascade of events which alter the integrity of the BBB, resulting in migration of leukocytes into the CNS. Leukocyte transmigration across the BBB during stroke immune/inflammatory processes could influence inter-endothelial junctional complex function leading to vascular endothelium damage and BBB breakdown. Equally a key component to our mechanism-based thesis is that disruption or dysfunction of the BBB, preceding entry of harmful substances into the brain parenchyma, could be a key initial factor in stroke pathogenesis. Thus, restoration of barrier integrity may have a critical role in preventing stroke progression. Our studies have begun to address these questions, particularly, whether endothelial cell replacement can restore structural and functional properties of the BBB after stroke. Results of this study will provide the basis for pursuing cell therapy both for non-tPA and tPA-treated ischemic stroke patients, as well as for patients with neurodegenerative disorders characterized by BBB dysfunction.

As we extend EPC therapy for BBB repair in TBI, we apply similar concepts of transplanting exogenous EPC for TBI, stimulating endogenous EPC in TBI, and augmenting blood flow, angiogenesis and/or vasculogenesis in TBI using drugs. A soluble factor known as tissue inhibitor of matrix metalloproteinase-3 (TIMP3) is produced by MSCs and has been demonstrated to mediate the beneficial effects of MSCs on endothelial function including the structural and functional restoration of a compromised BBB caused by TBI [145]. Following transplantation of MSCs, TIMP3 upregulated and attenuated the TBI-associated BBB permeability after TBI; blocking TIMP3 expression led to a compromised BBB [145]. Repairing the BBB by the transplantation of exogenous EPCs has also been explored, taking advantage of EPCs' capacity to migrate to the site of injury and contribute to the regeneration of vascular tissue by releasing angiogenic factors and creating structural components of capillaries. In addition, BBB repair can benefit from transplantation of stem cell progenitors and growth factors that are released by the grafted cells to the host microenvironment facilitating BBB repair and maintenance [146].

11.7 Conclusion

The recognition that BBB breakdown closely accompanies CNS trauma warrants therapies designed to arrest this BBB dysfunction. Currently, much of the therapy implemented for CNS trauma does not consider the capacity of BBB damage after injury. It is our contention that if EPC transplantation promotes restoration of the vascular endothelium, the clinical effects could be far reaching and substantially help a large population of patients that may be excluded from the current therapeutic window of neuroprotection for TBI. Although a plethora of accumulating stem cell research is quickly translating into clinical trials, it is important to gain insights into the mechanisms of action, which will aid in optimizing the safety and efficacy of these stem cells in CNS trauma. TBI is a public health problem that afflicts children and adults, and in the last decade is rampant to our military soldiers. Almost half a million of visits yearly to the emergency wards are related to TBI. The need for better understanding of cell death pathways associated with TBI, especially secondary cell loss, is crucial to developing an effective treatment. Here, we advance the notion that a treatment regimen directed at attenuating TBI deficits should consider the pivotal role of BBB repair in order to maintain CNS homeostasis and enhance neuronal regeneration. Structurally and functionally restoring the BBB in an acute, sub-acute, and even chronic phases of injury setting may afford therapeutic benefits against TBI. A regenerative mechanism involving the repair of the damaged BBB by EPC is key to the successful outcome of cell therapy in CNS trauma. Cell therapy tailored at EPC recruitment and/or directed secretion of EPC-soluble factors into the traumatized brain stands as a potent strategy for BBB repair in TBI.

Disclosures/Conflict of Interests: CVB is supported by NIH NINDS 5U01NS055914-04 and NIH NINDS R01NS071956-01, James and Esther King Foundation for Biomedical Research Program, and receives research grant support for his projects on bone marrow stem cell therapy for stroke from SanBio Inc., Celgene Cellular Therapeutics, KMPHC and NeuralStem Inc.

References

1. Faul M, Xu L, Wald MM, Coronado VG (2010) Traumatic brain injury in the United States: emergency department visits, hospitalizations, and deaths. Centers for Disease Control and Prevention, National Center for Injury Prevention and Control, Atlanta, GA
2. Centers for Disease Control and Prevention (CDC), National Center for Injury Prevention and Control (2003) Report to Congress on mild traumatic brain injury in the United States: steps to prevent a serious public health problem. Centers for Disease Control and Prevention, Atlanta, GA
3. Brooks A, Lindstrom J, McCray J et al (1995) Cost of medical care for a population-based sample of persons surviving traumatic brain injury. J Head Trauma Rehabil 10:1–13
4. Oladunjoye AO, Schrot RJ, Zwienenberg-Lee M, Muizelaar JP, Shahlaie K (2013) Decompressive craniectomy using gelatin film and future bone flap replacement. J Neurosurg 118(4):776–782
5. Swadron SP, LeRoux P, Smith WS, Weingart SD (2012) Emergency neurological life support: traumatic brain injury. Neurocrit Care 17:S112–S121
6. Farahvar A, Gerber LM, Chiu YL, Carney N, Hartl R, Ghajar J (2012) Increased mortality in patients with severe traumatic brain injury treated without intracranial pressure monitoring. J Neurosurg 117:729–734
7. Bor-Seng-Shu E, Figueiredo EG, Amorim RL, Teixeira MJ, Valbuza JS, de Oliveira MM, Panerai RB (2012) Decompressive craniectomy: a meta-analysis of influences on intracranial pressure and cerebral perfusion pressure in the treatment of traumatic brain injury. J Neurosurg 117:589–596
8. Bor-Seng-Shu E, Figueiredo EG, Fonoff ET, Fujimoto Y, Panerai RB, Teixeira MJ (2013) Decompressive craniectomy and head injury: brain morphometry, ICP, cerebral hemodynamics, cerebral microvascular reactivity, and neurochemistry. Neurosurg Rev 36 (3):361–370
9. Brasure M, Lamberty GJ, Sayer NA, Nelson NW, MacDonald R, Ouellette J, Tacklind J, Grove M, Rutks IR, Butler ME, Kane RL, Wilt TJ (2012) Multidisciplinary postacute rehabilitation for moderate to severe traumatic brain injury in adults [internet]. Agency for Healthcare Research and Quality (US), Rockville, MD
10. Brasure M, Lamberty GJ, Sayer NA, Nelson NW, Macdonald R, Ouellette J, Wilt TJ (2013) Participation after multidisciplinary rehabilitation for moderate to severe traumatic brain injury in adults: a systematic review. Arch Phys Med Rehabil 94(7):1398–1420
11. Krawczyk DC, Marquez de la Plata C, Schauer GF, Vas AK, Keebler M, Tuthill S, Gardner C, Jantz T, Yu W, Chapman SB (2013) Evaluating the effectiveness of reasoning training in military and civilian chronic traumatic brain injury patients: study protocol. Trials 14:29
12. Brown JM, Deriso DM, Tansey KE (2012) From contemporary rehabilitation to restorative neurology. Clin Neurol Neurosurg 114:471–474

13. Ploughman M (2008) Exercise is brain food: the effects of physical activity on cognitive function. Dev Neurorehabil 11:236–240
14. Zafonte RD, Bagiella E, Ansel BM, Novack TA, Friedewald WT, Hesdorffer DC, Timmons SD, Jallo J, Eisenberg H, Hart T, Ricker JH, Diaz-Arrastia R, Merchant RE, Temkin NR, Melton S, Dikmen SS (2012) Effect of citicoline on functional and cognitive status among patients with traumatic brain injury: citicoline brain injury treatment trial (COBRIT). JAMA 308:1993–2000
15. Sánchez-Aguilar M, Tapia-Pérez JH, Sánchez-Rodríguez JJ, Viñas-Ríos JM, Martínez-Pérez P, de la Cruz-Mendoza E, Sánchez-Reyna M, Torres-Corzo JG, Gordillo-Moscoso A (2013) Effect of rosuvastatin on cytokines after traumatic head injury. J Neurosurg 118:669–675
16. McConeghy KW, Hatton J, Hughes L, Cook AM (2012) A review of neuroprotection pharmacology and therapies in patients with acute traumatic brain injury. CNS Drugs 26:613–636
17. Stein DG (2013) A clinical/translational perspective: can a developmental hormone play a role in the treatment of traumatic brain injury? Horm Behav 63:291–300
18. Stein SC et al (2009) Erythrocyte-bound tissue plasminogen activator is neuroprotective in experimental traumatic brain injury. J Neurotrauma 26:1585–1592
19. Fox JL, Vu EN, Doyle-Waters M, Brubacher JR, Abu-Laban R, Hu Z (2010) Prophylactic hypothermia for traumatic brain injury: a quantitative systematic review. CJEM 12:355–364
20. Joan Abbott N, Patabendige AAK, Dolman DEM, Yusof SR, Begley DJ (2010) Structure and function of the blood–brain barrier. Neurobiol Dis 37:13–25
21. Gaetz M (2004) The neurophysiology of brain injury. Clin Neurophysiol 115(1):4–18
22. Zweckberger K et al (2006) Effect of early and delayed decompressive craniectomy on secondary brain damage after controlled cortical impact in mice. J Neurotrauma 23:1083–1093
23. Rhodes J (2011) Peripheral immune cells in the pathology of traumatic brain injury? Curr Opin Crit Care 17:122–130
24. Beaumont A et al (2006) Bolus tracer delivery measured by MRI confirms edema without blood–brain barrier permeability in diffuse traumatic brain injury. Acta Neurochir Suppl 96:171–174
25. Soares HD, Hicks RR, Smith D, McIntosh TK (1995) Inflammatory leukocytic recruitment and diffuse neuronal degeneration are separate pathological processes resulting from traumatic brain injury. J Neurosci 15:8223–8233
26. Toda H, Takahashi J, Iwakami N (2001) Grafting neural cells improved the impaired spatial recognition in ischemic rats. Neurosci Lett 31:9–12
27. Ferrari A, Ehler E, Nitsch RM, Gotz J (2000) Immature human NT2 cells grafted into mouse brain differentiate into neuronal and glial cell types. FEBS Lett 486:121–125
28. Borlongan CV, Tajima Y, Trojanowski JQ, Lee VM, Sanberg PR (1998) Transplantation of cryopreserved human embryonal carcinoma-derived neurons (NT2N cells) promotes functional recovery in ischemic rats. Exp Neurol 149:310–321
29. Veizovic T, Beech JS, Stroemer PR, Watson WP, Hodges H (2001) Resolution of stroke deficits following contralateral grafts of conditionally immortal neuroepithelial stem cells. Stroke 32:1012–1019
30. Modo M, Stroemer RP, Tang E, Veizovic T, Sowniski P, Hodges H (2000) Neurological sequelae and long-term behavioural assessment of rats with transient middle cerebral artery occlusion. J Neurosci Methods 104:99–109
31. Englund U, Bjorklund A, Wictorin K, Lindvall O, Kokaia M (2002) Grafted neural stem cells develop into functional pyramidal neurons and integrate into host cortical circuitry. Proc Natl Acad Sci U S A 99:17089–17094
32. Kopen GC, Prockop DJ, Phinney DG (1999) Marrow stromal cells migrate through out forebrain and cerebellum, and they differentiate into astrocytes after injection into neonatal mouse brains. Proc Natl Acad Sci U S A 96:10711–10716

33. Chen J, Li Y, Chopp M (2000) Intracerebral transplantation of bone marrow with BDNF after MCAo in rat. Neuropharmacology 39:711–716
34. Li Y, Chopp M, Chen J, Wang L, Gautam SC, Xu XY, Zhang Z (2000) Intrastriatal transplantation of bone marrow nonhematopoietic cells improves functional recovery after stroke in adult mice. J Cereb Blood Flow Metab 20:1311–1319
35. Woodbury D, Schwarz EJ, Prockop DJ, Black IB (2000) Adult rat and human bone marrow stromal cells differentiate into neurons. J Neurosci Res 61:364–370
36. Munoz-Elias G, Marcus AJ, Coyne M, Woodbury D, Black IB (2004) Adult bone marrow stromal cells in the embryonic brain: engraftment, migration, differentiation, and long-term survival. J Neurosci 24:4585–4595
37. Borlongan CV, Hadman M, Davis C, Sanberg PR (2004) CNS entry of peripherally injected umbilical cord blood cells is not required for neuroprotection in stroke. Stroke 35:2385–2389
38. Eglitis MA, Mezey E (1997) Hematopoietic cell differentiate into both microglia and macroglia in the brains of adult mice. Proc Natl Acad Sci U S A 94:4080–4085
39. Li Y, Chen J, Wang L, Lu M, Chopp M (2001) Treatment of stroke in rat with intracarotid administration of marrow stromal cells. Neurology 56:1666–1672
40. Chen J, Sanberg PR, Li Y, Wang L, Lu M, Willing AE, Sanchez-Ramos J, Chopp M (2001) Intravenous administration of human umbilical cord blood reduces behavioral deficits after stroke in rats. Stroke 32:2682–2688
41. Hess DC, Hill WD, Martin-Studdard A, Carroll J, Brailer J, Carothers J (2002) Bone marrow as a source of endothelial cells and NeuN-expressing cells after stroke. Stroke 33:1362–1368
42. Willing AE, Milliken M, Poulos S, Zigova T, Song S, Davis CD et al (2003) Intravenous versus intrastriatal cord blood administration in a rodent model of stroke. J Neurosci Res 73:296–307
43. Chen J, Zhang ZG, Li Y, Wang L, Xu YX, Gautam SC et al (2003) Intravenous administration of human bone marrow stromal cells induces angiogenesis in the ischemic boundary zone after stroke in rats. Circ Res 92:692–699
44. Ma H et al (2012) Neural stem cells over-expressing brain-derived neurotrophic factor (BDNF) stimulate synaptic protein expression and promote functional recovery following transplantation in rat model of traumatic brain injury. Neurochem Res 37:69–83
45. Mahmood A et al (2006) Long-term recovery after bone marrow stromal cell treatment of traumatic brain injury in rats. J Neurosurg 104:272–277
46. Qu C et al (2008) Treatment of traumatic brain injury in mice with marrow stromal cells. Brain Res 1208:234–239
47. Lu D et al (2007) Collagen scaffolds populated with human marrow stromal cells reduce lesion volume and improve functional outcome after traumatic brain injury. Neurosurgery 61:596–602
48. Harting MT et al (2009) Intravenous mesenchymal stem cell therapy for traumatic brain injury. J Neurosurg 110:1189–1197
49. Riess P et al (2002) Transplanted neural stem cells survive, differentiate, and improve neurological motor function after experimental traumatic brain injury. Neurosurgery 51:1043–1054
50. Hattiangady B, Shetty AK (2012) Neural stem cell grafting counteracts hippocampal injury-mediated impairments in mood, memory, and neurogenesis. Stem Cells Transl Med 1:696–708
51. Nichols JE et al (2013) Neurogenic and neuro-protective potential of a novel subpopulation of peripheral blood-derived CD133+ ABCG2 + CXCR4+ mesenchymal stem cells: development of autologous cell-based therapeutics for traumatic brain injury. Stem Cell Res Ther 4:3
52. Yan ZJ et al (2013) Neural stem-like cells derived from human amnion tissue are effective in treating traumatic brain injury in rat. Neurochem Res 38(5):1022–1033

53. Wallenquist U et al (2012) Ibuprofen attenuates the inflammatory response and allows formation of migratory neuroblasts from grafted stem cells after traumatic brain injury. Restor Neurol Neurosci 30:9–19
54. Shear DA et al (2011) Stem cell survival and functional outcome after traumatic brain injury is dependent on transplant timing and location. Restor Neurol Neurosci 29:215–225
55. Lee DH et al (2013) Functional recovery after injury of motor cortex in rats: effects of rehabilitation and stem cell transplantation in a traumatic brain injury model of cortical resection. Childs Nerv Syst 29:403–411
56. Giraldi-Guimaraes A et al (2012) Bone marrow mononuclear cells and mannose receptor expression in focal cortical ischemia. Brain Res 1452:173–184
57. Chuang TJ et al (2012) Effects of secretome obtained from normoxia-preconditioned human mesenchymal stem cells in traumatic brain injury rats. J Trauma Acute Care Surg 73:1161–1167
58. Walker PA et al (2012) Bone marrow-derived stromal cell therapy for traumatic brain injury is neuroprotective via stimulation of non-neurologic organ systems. Surgery 152:790–793
59. Tu Y et al (2012) Combination of temperature-sensitive stem cells and mild hypothermia: a new potential therapy for severe traumatic brain injury. J Neurotrauma 29:2393–2403
60. Antonucci I et al (2012) Amniotic fluid stem cells: a promising therapeutic resource for cell-based regenerative therapy. Curr Pharm Des 18:1846–1863
61. Joo KM et al (2012) Trans-differentiation of neural stem cells: a therapeutic mechanism against the radiation induced brain damage. PLoS One 7:e25936
62. Shi W et al (2012) BDNF blended chitosan scaffolds for human umbilical cord MSC transplants in traumatic brain injury therapy. Biomaterials 33:3119–3126
63. Yang L et al (2011) Transplantation of Schwann cells differentiated from adipose-derived stem cells modifies reactive gliosis after contusion brain injury in rats. J Int Med Res 39:1344–1357
64. Skardelly M et al (2011) Long-term benefit of human fetal neuronal progenitor cell transplantation in a clinically adapted model after traumatic brain injury. J Neurotrauma 28:401–414
65. Reiss P, Zhang C, Saatman KE (2002) Transplanted neural cells survive, differentiate, and improve neurological motor function after experimental traumatic brain injury. Neurosurgery 51:1043–1052
66. Zhao LR, Duan WM, Reyes M, Keene CD, Verfaillie CM, Low WC (2002) Human bone marrow cells exhibit neural phenotypes and ameliorate neurological deficits after grafting into the ischemic brain of rats. Exp Neurol 174:11–20
67. Peled A, Kollet O, Ponomaryov T, Petit I, Frantza S, Grabovsky V et al (2000) The chemokine SDF-1 activated the integrins LFA-1, VLA-4, and VLA-5 on immature CD34+ cells: role in transendothelial/stromal migration and engraftment of NOD/SCID mice. Blood 95:3289–3296
68. Yamaguichi J, Kusano K, Masuo O, Kawamoto A, Silver M, Murasawa S et al (2003) Stromal cell-derived factor-1 effects on ex vivo expanded endothelial progenitor cell recruitment for ischemic neovascularization. Circulation 107:1322–1328
69. Reyes M, Dudek A, Jahagirdar B, Koodie L, Marker PH, Verfaillie CM (2002) Origin of endothelial progenitors in human postnatal bone marrow. J Clin Invest 109:337–346
70. Rajantie L, Llmonen M, Alminaite A, Ozer U, Alitalo K, Salven P (2004) Adult bone marrow-derived cells recruited during angiogenesis comprise precursors for periendothelial vascular mural cells. Blood 104:2084–2086
71. Lu D, Mahmood A, Qu C, Goussev A, Schallert T, Chopp M (2005) Erythropoietin enhances neurogenesis and restores spatial memory in rats after traumatic brain injury. J Neurotrauma 22:1011–1017
72. Sun D, Mcginn MJ, Zhou Z, Harvey HB, Bullock MR, Colello RJ (2007) Anatomical integration of newly generated dentate granule neurons following traumatic brain injury in adult rats and its association to cognitive recovery. Exp Neurol 204:264–272

73. Guo X, Liu L, Zhang M, Bergeron A, Cui Z, Dong JF, Zhang J (2009) Correlation of CD34+ cells with tissue angiogenesis after traumatic brain injury in a rat model. J Neurotrauma 26:1337–1344
74. Madeddu P (2005) Therapeutic angiogenesis and vasculogenesis for tissue regeneration. Exp Physiol 90:315–326
75. Besler C, Doerries C, Giannotti G, Lüscher TF, Landmesser U (2008) Pharmacological approaches to improve endothelial repair mechanisms. Expert Rev Cardiovasc Ther 6:1071–1082
76. Mahmoud A, Lu D, Chopp M (2004) Intravenous administration of marrow stromal cells (MSCs) increases the expression of growth factors after traumatic brain injury. J Neurotrauma 21:33–39
77. Suarez-Monteagudo C, Hernandez-Ramirez P, Alvarez-Gonzalez L, Garcia-Maeso I, de la Cuetara-Bernal K, Castillo-Diaz L et al (2009) Autologous bone marrow stem cell neurotransplantation in stroke patients. An open study. Restor Neurol Neurosci 27:151–161
78. Borlongan CV (2009) Cell therapy for stroke: remaining issues to address before embarking on clinical trials. Stroke 40:146–148
79. Feuerstein GZ, Zaleska MM, Krams M, Wang X, Day M, Rutkowski JL et al (2008) Missing steps in the STAIR case: a translational medicine perspective on the development of NXY-059 for treatment of acute ischemic stroke. J Cereb Blood Flow Metab 28:217–219
80. Santiago LA, Oh BC, Dash PK, Holcomb JB, Wade CE (2012) A clinical comparison of penetrating and blunt traumatic brain injuries. Brain Inj 26:107–125
81. Manley GT, Diaz-Arrastia R, Brophy M, Engel D, Goodman C, Gwinn K, Veenstra TD, Ling G, Ottens AK, Tortella F, Hayes RL (2010) Common data elements for traumatic brain injury: recommendations from the biospecimens and biomarkers working group. Arch Phys Med Rehabil 91:1667–1672
82. Keene CD, Ortiz-Gonzalez XR, Jiang Y, Largaespada DA, Verfaillie CM, Low WC (2003) Neural differentiation and incorporation of bone marrow-derived multipotent adult progenitor cells after single cell transplantation into blastocyst stage mouse embryos. Cell Transplant 2:201–213
83. Di Nicola M, Carlo-Stella C, Magni M, Milanesi M, Longoni PD, Matteucci P et al (2002) Human bone marrow stromal cells suppress T lymphocyte proliferation induced by cellular or nonspecific mitogenic stimuli. Blood 99:3838–3843
84. Jorgensen C, Djouad F, Apparailly F (2003) Engineering mesenchymal stem cells for immunotherapy. Gene Ther 10:928–931
85. Le Blanc K, Tammik L, Sundberg B, Haynesworth SE et al (2003) Mesenchymal stem cells inhibit and stimulate mixed lymphocyte cultures and mitogenic responses independently of the MHC. Scand J Immunol 57:11–20
86. McIntosh K, Bartholomew A (2000) Stromal cell modulation of the immune system: a potential role for mesenchymal stem cells. Graft 3:324–328
87. Aggarwal S, Pittenger MF (2005) Human mesenchymal stem cells modulate allogeneic immune cell responses. Blood 105:1815–1822
88. Tse WT, Pendleton JD, Beyer WM, Egalka MC, Guinan EC (2003) Suppression of allogeneic T-cell proliferation by human marrow stromal cells: implications for transplantation. Transplantation 75:389–397
89. Le Blanc K, Rasmusson I, Sundberg B, Gotherstrom C, Hassan M, Uzunel M, Ringden O (2004) Treatment of severe acute graft-versus-host disease with third party haploidentical mesenchymal stem cells. Lancet 363:1411–1412
90. Reyes M, Verfaillie CM (2001) Characterization of multipotent adult progenitor cells, a subpopulation of mesenchymal stem cells. Ann N Y Acad Sci 938:231–233, discussion; 233–5
91. Ayata C, Ropper AH (2002) Ischaemic brain edema. J Clin Neurosci 9:113–124

92. Busch E, Kruger K, Fritze K, Allegrini PR, Hoehn-Berlage M, Hossmann KA (1997) Blood–brain barrier disturbances after rt-PA treatment of thromboembolic stroke in the rat. Acta Neurochir Suppl 70:206–208
93. Kaur J, Zhao Z, Klein GM, Lo EH, Buchan AM (2004) The neurotoxicity of tissue plasminogen activator? J Cereb Blood Flow Metab 24:945–963
94. De Brouns R, Deyn PP (2009) The complexity of neurobiological processes in acute ischemic stroke. Clin Neurol Neurosurg 111:483–495
95. Aoki T, Sumii T, Mori T, Wang X, Lo EH (2002) Blood–brain barrier disruption and matrix metalloproteinase-9 expression during reperfusion injury: mechanical versus embolic focal ischemia in spontaneously hypertensive rats. Stroke 33:2711–2717
96. Subramaniam S, Hill MD (2009) Decompressive hemicraniectomy for malignant middle cerebral artery infarction: an update. Neurologist 15:178–184
97. Chang CF, Lin SZ, Chiang YH, Morales M, Chou J, Lein P et al (2003) Intravenous administration of bone morphogenetic protein-7 after ischemia improves motor function in stroke rats. Stroke 34:558–564
98. Castillo J, Alvarez-Sabin J, Martinez-Vila E, Montaner J, Sobrino T, Vivancos J (2009) MITICO study investigators. Inflammation markers and prediction of post-stroke vascular disease recurrence: the MITICO study. J Neurol 256:217–224
99. Vila N, Castillo J, Davalos A, Chamorro A (2000) Proinflammatory cytokines and early neurological worsening in ischemic stroke. Stroke 31:2325–2329
100. Castillo J, Leira R (2002) Predictors of deteriorating cerebral infarct: role of inflammatory mechanisms. Would its early treatment be useful? Cerebrovasc Dis 1:40–48
101. Doyle KP, Simon RP, Stenzel-Poore MP (2008) Mechanisms of ischemic brain damage. Neuropharmacology 55:310–318
102. Wang GJ, Deng HY, Maier CM, Sun GH, Yenari MA (2002) Mild hypothermia reduces ICAM-1 expression, neutrophil infiltration and microglia/monocyte accumulation following experimental stroke. Neuroscience 114:1081–1090
103. Tang Y, Xu H, Du X, Lit L, Walker W, Lu A et al (2006) Gene expression in blood changes rapidly in neutrophils and monocytes after ischemic stroke in humans: a microarray study. J Cereb Blood Flow Metab 26:1089–1102
104. Tomkins O, Feintuch A, Benifla M, Cohen A, Friedman A, Shelef I (2011) Blood–brain barrier breakdown following traumatic brain injury: a possible role in posttraumatic epilepsy. Cardiovasc Psychiatry Neurol 2011:765923
105. Dietrich WD, Alonso O, Halley M (1994) Early microvascular and neuronal consequences of traumatic brain injury: a light and electron microscopic study in rats. J Neurotrauma 11:289–301
106. Thau-Zuchman O, Shohami E, Alexandrovich AG, Leker RR (2010) Vascular endothelial growth factor increases neurogenesis after traumatic brain injury. J Cereb Blood Flow Metab 30:1008–1016
107. Asahara T, Murohara T, Sullivan A, Silver M, van der Zee R, Li T et al (1997) Isolation of putative progenitor endothelial cells for angiogenesis. Science 275:964–967
108. Griese DP, Ehsan A, Melo LG, Kong D, Zhang L, Mann MJ et al (2003) Isolation and transplantation of autologous circulating endothelial cells into denuded vessels and prosthetic grafts: implications for cell-based vascular therapy. Circulation 108:2710–2715
109. Carmeliet P (2005) Angiogenesis in life, disease and medicine. Nature 438:932–936
110. Masuda H, Asahara T (2003) Post-natal endothelial progenitor cells for neovascularization in tissue regeneration. Cardiovasc Res 58:390–398
111. Shi Q, Rafii S, Wu MH, Wijelath ES, Yu C, Ishida A et al (1998) Evidence for circulating bone marrow-derived endothelial cells. Blood 92:362–367
112. Bompais H, Chagraoui J, Canron X, Crisan M, Liu XH, Anjo A et al (2004) Human endothelial cells derived from circulating progenitors display specific functional properties compared with mature vessel wall endothelial cells. Blood 103:2577–2584

113. Fadini GP (2008) An underlying principle for the study of circulating progenitor cells in diabetes and its complications. Diabetologia 51:1091–1094
114. Chen JZ, Zhang FR, Tao QM, Wang XX, Zhu JH (2004) Number and activity of endothelial progenitor cells from peripheral blood in patients with hypercholesterolaemia. Clin Sci 107:273–280
115. Pirro M, Schillaci G, Menecali C, Bagaglia F, Paltriccia R, Vaudo G et al (2007) Reduced number of circulating endothelial progenitors and HOXA9 expression in CD34+ cells of hypertensive patients. J Hypertens 25:2093–2099
116. Umemura T, Soga J, Hidaka T, Takemoto H, Nakamura S, Jitsuiki D et al (2008) Aging and hypertension are independent risk factors for reduced number of circulating endothelial progenitor cells. Am J Hypertens 21:1203–1209
117. Del Papa N, Quirici N, Soligo D, Scavullo C, Cortiana M, Borsotti C et al (2006) Bone marrow endothelial progenitors are defective in systemic sclerosis. Arthritis Rheum 54:2605–2615
118. Kuwana M, Okazaki Y, Yasuoka H, Kawakami Y, Ikeda Y (2004) Defective vasculogenesis in systemic sclerosis. Lancet 364:603–610
119. Heiss C, Keymel S, Niesler U, Ziemann J, Kelm M, Kalka C (2005) Impaired progenitor cell activity in age-related endothelial dysfunction. J Am Coll Cardiol 45:1441–1448
120. Kondo T, Hayashi M, Takeshita K, Numaguchi Y, Kobayashi K, Iino S et al (2004) Smoking cessation rapidly increases circulating progenitor cells in peripheral blood in chronic smokers. Arterioscler Thromb Vasc Biol 24:1442–1447
121. Michaud SE, Dussault S, Haddad P, Groleau J, Rivard A (2006) Circulating endothelial progenitor cells from healthy smokers exhibit impaired functional activities. Atherosclerosis 187:423–432
122. Botta R, Gao E, Stassi G, Bonci D, Pelosi E, Zwas D et al (2004) Heart infarct in NOD-SCID mice: therapeutic vasculogenesis by transplantation of human CD34+ cells and low dose CD34 + KDR + cells. FASEB J 18:1392–1394
123. Kawamoto A, Gwon HC, Iwaguro H, Yamaguchi JI, Uchida S, Masuda H et al (2001) Therapeutic potential of ex vivo expanded endothelial progenitor cells for myocardial ischemia. Circulation 103:634–637
124. Madeddu P, Emanueli C, Pelosi E, Salis MB, Cerio AM, Bonanno G et al (2004) Transplantation of low dose CD34 + KDR + cells promotes vascular and muscular regeneration in ischemic limbs. FASEB J 18:1737–1739
125. van Rouhl RP, Oostenbrugge RJ, Damoiseaux J, Cohen Tervaert JW, Lodder J (2008) Endothelial progenitor cell research in stroke: a potential shift in pathophysiological and therapeutical concepts. Stroke 39:2158–2165
126. Erbs S, Linke A, Adams V, Lenk K, Thiele H, Diederich KW, Emmrich F et al (2005) Transplantation of blood-derived progenitor cells after recanalization of chronic coronary artery occlusion: first randomized and placebo-controlled study. Circ Res 97:756–762
127. Li ZQ, Zhang M, Jing YZ, Zhang WW, Liu Y, Cui LJ et al (2007) The clinical study of autologous peripheral blood stem cell transplantation by intracoronary infusion in patients with acute myocardial infarction (AMI). Int J Cardiol 115:52–56
128. Fernandez-Aviles F, San Roman JA, Garcia-Frade J, Fernandez ME, de la Penarrubia MJ, Fuente L et al (2004) Experimental and clinical regenerative capability of human bone marrow cells after myocardial infarction. Circ Res 95:742–748
129. Meluzin J, Janousek S, Mayer J, Groch L, Hornacek I, Hlinomaz O et al (2008) Three-, 6-, and 12-month results of autologous transplantation of mononuclear bone marrow cells in patients with acute myocardial infarction. Int J Cardiol 128:185–192
130. Meluzin J, Mayer J, Groch L, Janousek S, Hornacek I, Hlinomaz O et al (2006) Autologous transplantation of mononuclear bone marrow cells in patients with acute myocardial infarction: the effect of the dose of transplanted cells on myocardial function. Am Heart J 152:975

131. Mocini D, Staibano M, Mele L, Giannantoni P, Menichella G, Colivicchi F et al (2006) Autologous bone marrow mononuclear cell transplantation in patients undergoing coronary artery bypass grafting. Am Heart J 151:192–197
132. Perin EC, Dohmann HF, Borojevic R, Silva SA, Sousa AL, Mesquita CT et al (2003) Transendocardial, autologous bone marrow cell transplantation for severe, chronic ischemic heart failure. Circulation 107:2294–2302
133. Perin EC, Dohmann HF, Borojevic R, Silva SA, Sousa AL, Silva GV et al (2004) Improved exercise capacity and ischemia 6 and 12 months after transendocardial injection of autologous bone marrow mononuclear cells for ischemic cardiomyopathy. Circulation 110:213–218
134. Strauer BE, Brehm M, Zeus T, Bartsch T, Schannwell C, Antke C et al (2005) Regeneration of human infarcted heart muscle by intracoronary autologous bone marrow cell transplantation in chronic coronary artery disease: the IACT study. J Am Coll Cardiol 46:1651–1658
135. Strauer BE, Brehm M, Zeus T, Kostering M, Hernandez A, Sorg RV et al (2002) Repair of infarcted myocardium by autologous intracoronary mononuclear bone marrow cell transplantation in humans. Circulation 106:1913–1918
136. Numaguchi Y, Sone T, Okumura K, Ishii M, Morita Y, Kubota R et al (2006) The impact of the capability of circulating progenitor cell to differentiate on myocardial salvage in patients with primary acute myocardial infarction. Circulation 114:114–119
137. Dobert N, Britten M, Assmus B, Berner U, Menzel C, Lehmann R et al (2004) Transplantation of progenitor cells after reperfused acute myocardial infarction: evaluation of perfusion and myocardial viability with fdg-pet and thallium spect. Eur J Nucl Med Mol Imaging 31:1146–1151
138. Lev EI, Kleiman NS, Birnbaum Y, Harris D, Korbling M, Estrov Z (2005) Circulating endothelial progenitor cells and coronary collaterals in patients with non-st segment elevation myocardial infarction. J Vasc Res 42:408–414
139. Hristov M, Heussen N, Schober A, Weber C (2006) Intracoronary infusion of autologous bone marrow cells and left ventricular function after acute myocardial infarction: a meta-analysis. J Cell Mol Med 10:727–733
140. Dimmeler S, Zeiher AM, Schneider MD (2005) Unchain my heart: the scientific foundations of cardiac repair. J Clin Invest 115:572–583
141. Higashi Y, Kimura M, Hara K, Noma K, Jitsuiki D, Nakagawa K et al (2004) Autologous bone-marrow mononuclear cell implantation improves endothelium-dependent vasodilation in patients with limb ischemia. Circulation 109:1215–1218
142. Tateishi-Yuyama E, Matsubara H, Murohara T, Ikeda U, Shintani S, Masaki H, Amano K et al (2002) Therapeutic angiogenesis for patients with limb ischaemia by autologous transplantation of bone marrow cells: a pilot study and a randomised controlled trial. Lancet 360:427–435
143. Taguchi A, Matsuyama T, Moriwaki H, Hayashi T, Hayashida K, Nagatsuka K et al (2004) Circulating cd34-positive cells provide an index of cerebrovascular function. Circulation 109:2972–2975
144. Ghani U, Shuaib A, Salam A, Nasir A, Shuaib U, Jeerakathil T et al (2005) Endothelial progenitor cells during cerebrovascular disease. Stroke 36:151–153
145. Menge T, Zhao Y, Zhao J, Wataha K, Gerber M, Zhang J, Letourneau P, Redell J, Shen L, Wang J, Peng Z, Xue H, Kozar R, Cox CS, Khakoo AY, Holcomb JB, Dash PK, Pati S (2012) Mesenchymal stem cells regulate blood–brain barrier integrity through timp3 release after traumatic brain injury. Sci Transl Med 4:161ra150
146. Jujo K, Ii M, Losordo DW (2008) Endothelial progenitor cells in neovascularization of infarcted myocardium. J Mol Cell Cardiol 45:530–544

Chapter 12
Vascular Actions of Hypothermia in Brain Trauma

W. Dalton Dietrich and Helen M. Bramlett

Abstract Traumatic brain injury (TBI) is a serious condition that affects approximately 1.5 million people in the United States each year. Currently, there are no approved therapies to treat the devastating consequences of severe TBI. Therapeutic hypothermia has a history of showing efficacy both in animal models and clinical studies. Potential mechanisms by which hypothermia may improve traumatic outcome include targeting vascular alterations such as by reducing the incidence of blood–brain barrier (BBB) permeability. In addition, therapeutic hypothermia promotes normal vascular reactivity and reduces a variety of inflammatory processes that are activated by trauma-induced cerebrovascular damage. Ongoing research in the laboratory and clinic is demonstrating that therapeutic hypothermia may indeed protect specific populations of severe TBI patients by targeting vascular perturbations including altered BBB function.

12.1 Introduction

Therapeutic hypothermia is considered one of the most powerful neuroprotective strategies for a variety of neurological disorders including traumatic brain injury (TBI), for reviews see [1–3]. Previous experimental studies have reported that pre- or posttraumatic cooling reduces contusion volume, protects against neuronal and axonal damage, and in some cases improves functional outcome [4, 5]. The fact that temperature reductions protect against a spectrum of neuropathological events that occur under controlled experimental conditions is important when considering an

W.D. Dietrich (✉)
Department of Neurological Surgery and Miami Project to Cure Paralysis, University of Miami, Leonard M. Miller School of Medicine, Lois Pope LIFE Center, 1095 NW 14th Terrace, Suite 2-30 (R48), Miami, FL 33136-1060, USA
e-mail: ddietrich@miami.edu

E.H. Lo et al. (eds.), *Vascular Mechanisms in CNS Trauma*, Springer Series in Translational Stroke Research 5, DOI 10.1007/978-1-4614-8690-9_12,

experimental therapy for a very heterogeneous patient population [6–8]. Indeed, posttraumatic hypothermia has been shown to improve a spectrum of behavioral abnormalities including sensorimotor as well as cognitive dysfunction frequently observed in trauma animal models [6, 9, 10]. Early cooling can reduce mortality rates after severe TBI and limit a wide variety of comorbidities that are associated with clinical TBI [7, 8, 11].

Early studies showed that one major mechanism by which hypothermia may protect against ischemic and TBI is by reducing the degree of vascular permeability [1, 12, 13]. Like protection afforded by hypothermia against neuronal cell death, early cooling strategies following TBI decreases the extravasation of several vascular tracers across the endothelial barrier. In other studies, abnormalities in vascular reactivity to vasoactive substances commonly reported after TBI [14] are improved with early cooling which may be an important mechanism of reducing secondary injury [15–17]. More recently, the effects of hypothermia in TBI models have also been reported to include a reduction in patterns of reperfusion injury that could occur following various surgical interventions to reduce the detrimental consequences of evolving brain edema [18].

The purpose of this chapter is to summarize previous work in the area of therapeutic hypothermia targeting TBI. The effects of early cooling strategies on the vascular perturbations commonly associated with TBI will be reviewed and discussed. A rich experimental and clinical literature supports the fact that therapeutic hypothermia targets a variety of vascular alterations including abnormal permeability, vasoreactivity, cerebral blood flow as well as posttraumatic inflammatory cascades [6]. Importantly, therapeutic hypothermia has now been successfully translated to specific TBI patient populations where these vascular perturbations are observed.

12.2 Effects of TBI on Blood–Brain Barrier Dysfunction

TBI leads to a variety of blood–brain barrier (BBB) alterations that are felt to participate in some of the long-term consequences of neurotrauma [19–24]. Patterns of altered vascular permeability have been evaluated in models of both diffuse as well as focal TBI [25–27]. Various vascular tracers including Evans blue, albumin, and horseradish peroxidase (HRP) have demonstrated that specific vascular beds are highly vulnerable to different types of brain trauma [4, 27, 28]. Early electron microscopic studies by Povlishock et al. [28] showed that in models of more diffuse injury without overt neuropathology, increased vascular transport across intact endothelial barriers was demonstrated. In subsequent studies with more focal areas of cortical contusive injury with associated neuronal damage, more severe damage to vascular endothelial barriers was shown to lead to extensive HRP extravasation across injured endothelial cells as well as through tight junctional complexes [26, 29]. Increased vascular permeability can have detrimental consequences on normal brain function including producing alterations in ionic homeostasis, increasing brain water and causing hemodynamic perturbations [24]. Indeed, several studies have shown regional correlations between early vascular permeability changes and patterns of neuronal and glial damage or activation [4, 30]. Thus, altered BBB breakdown after TBI is considered an important therapeutic target for the acute and more chronic consequences of TBI [21, 23].

12.3 Therapeutic Hypothermia Reduces BBB Breakdown

In 1992, Jiang et al. [12] first reported that hypothermia significantly affected patterns of BBB permeability after fluid percussion (FP) brain injury in rats. In that study, pre- and post-cooling to 30 ° C had a dramatic effect on tracer extravasation compared to normothermic (37 °C) rats. Increased vascular permeability to an endogenous serum albumin (IgG) was seen throughout the dorsal cortical gray and white matter structures and within the underlying hippocampal regions in normothermic animals. In contrast, hypothermia introduced prior to injury significantly reduced albumin immunoreactivity which was confined to the gray–white interface (Fig. 12.1). Interestingly, pre-traumatic hypothermia also significantly reduced the acute hypertensive response to trauma that is normally seen in normothermic TBI animals. Thus, the dramatic reduction in vascular protein tracer leakage reported in this study could have been due in part related to hypothermia modulation of systemic blood pressure after TBI.

Using another model of TBI, Smith and Hall [13] also evaluated the effects of pre-traumatic hypothermia on BBB permeability following controlled cortical impact (CCI) injury. In that study, animals were subjected to variable periods of hypotension under either normothermic or mildly hypothermic conditions. In normothermic animals, extravasation of plasma protein-bound Evans Blue dye was seen within the injured cortex at 60 min post-injury. In contrast, when brain temperature was allowed to become hypothermic spontaneously during the study, reduced BBB permeability was demonstrated. Because previous studies from that laboratory had shown that oxygen radical formation and lipid peroxidation occurs at the site of cortical injury, the beneficial effects of spontaneous hypothermia on altered BBB function were suggested to involve temperature effects upon free radical-induced lipid peroxidation.

These early studies emphasized the early changes in BBB permeability seen after experimental TBI and the effects of temperature reductions on early permeability changes. More recently, other studies have now emphasized that some models of TBI can produce more long-term effects on vascular permeability that may also participate in the progressive damage and more chronic functional abnormalities observed in these trauma models [30–32]. For example, in the study by Habgood et al. [32] evidence for the leakage of small molecule tracers into the brain was seen up to 4 days after the injury. To determine whether posttraumatic hypothermia would have a significant effect on these more long-term vascular perturbations, Lotocki et al. [30] tested small molecular weight

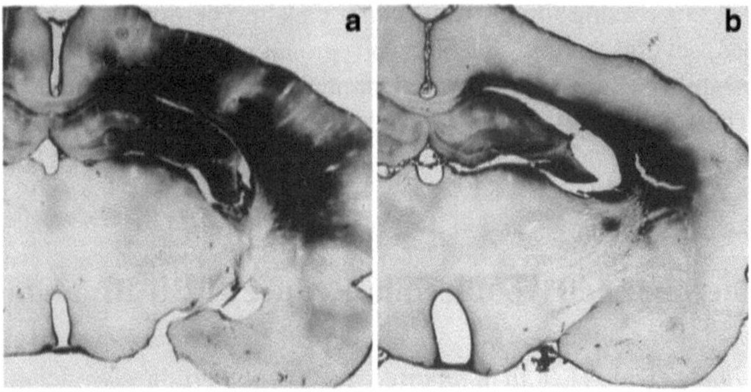

Fig. 12.1 (**a**, **b**) Comparison of blood–brain barrier (BBB) permeability in normothermic and hypothermic traumatic brain injury (TBI) rats. (**a**) Typical normothermic (37.5 °C) TBI rat: permeability changes were found throughout the cortical gray and white matter as well as in the underlying hippocampi. (**b**) Hypothermic (30 °C) TBI rat: this is a section from a rat which showed the greatest permeability. The permeability changes were greatly reduced compared to the typical normothermic rat. Permeability was confined to the gray–white interface with minimal involvement in the overlying cortical gray matter and reduced involvement in the hippocampi. Unstained, 50-μm-thick tissue sections, × 9. Reprinted from [12] with permission by Springer Publishing

tracers that remained visible up to several days after moderate FP brain injury in normothermic rats. Importantly, mild hypothermia (33 ° C) initiated 30 min after the traumatic insult significantly reduced the permeability of these tracers 3 days later (Fig. 12.2). It should be emphasized that in the clinic, hypothermia is used as an adjunctive therapy to reduce severe elevations in intracranial pressure in some severe TBI patients. Thus, the established effects of posttraumatic hypothermia on alterations in vascular permeability appear to be clinically relevant. The ability of an early cooling strategy to have profound effects on more long-lasting vascular perturbations indicates the importance of temperature management strategies in the days following a traumatic insult.

12.4 Effects of Therapeutic Hypothermia on Alterations in Vascular Reactivity

The normal complex cerebrovascular responses to altered blood pressure, dilation to vasoactive substances or hemodynamic changes are all critical in maintaining the normal homeostasis of the brain [33]. Under normal conditions, the topical application of specific vasodilators including acetycholine and adenosine led to significant vasodilation of pial vessels. After TBI, alterations in regional cerebral blood flow (rCBF) and cerebral capillary perfusion (CCP) are present that can be a consequence of altered vasomotor function [14, 17, 33–35]. In models of TBI where brain surface vessel responses can be precisely monitored using a pial window approach,

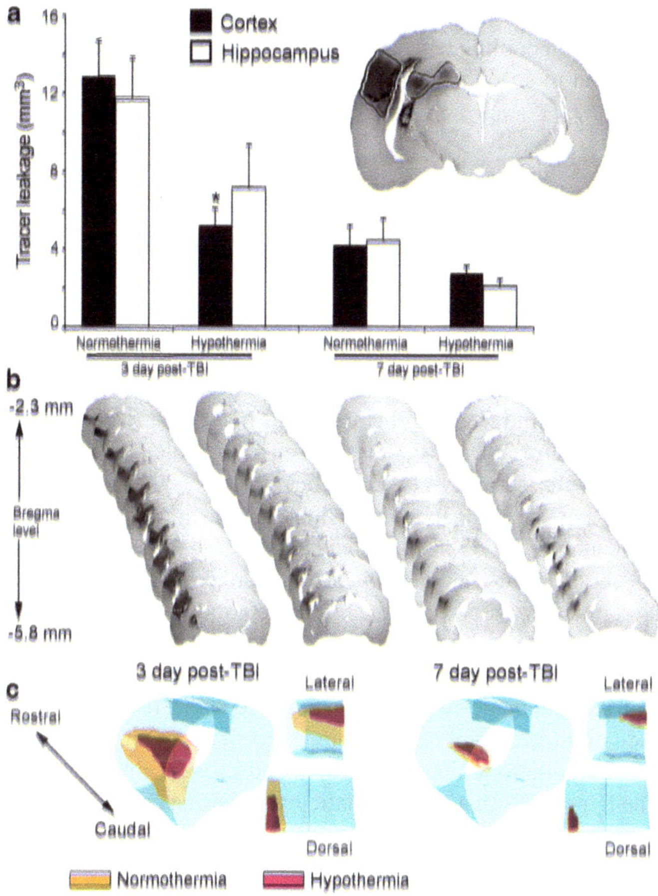

Fig. 12.2 Posttraumatic hypothermia reduces BBB permeability changes seen after moderate TBI. (**a**) At 3 and 7 days after TBI, posttraumatic hypothermia significantly reduces the volume of BDA-3 K protein extravasation in the ipsilateral rat cortex and hippocampus. (**b**) Representative serial brain sections stained with DAB showing BBB permeability alterations at 3 days (first and second column) and 7 days (third and fourth column) in normothermia (*left*) and hypothermia (*right*) animals. (**c**) 3-D reconstruction of serial sections shown in (**b**). *Red*: Hypothermia, *yellow*: normothermia. Data are presented as mean ± SEM, $*p < 0.05$, compared to normothermia for the respective time point and brain region. $N = 5$ per group. Reprinted from [30] with permission by Mary Ann Liebert Publishing

significant alterations in microvascular dilation are reported to be affected following brain injury [14, 15].

In addition to therapeutic hypothermia having significant effects on BBB permeability, other TBI studies have reported its effects on vascular reactivity [15, 17, 33]. Importantly, several investigations reported that TBI-induced long-term vascular dysfunction in terms of altered vascular reactivity to vasodilators was significantly improved with the use of delayed hypothermic treatment [36]. The ability of hypothermia to improve vascular reactivity function may be another important mechanism by which cooling can effect traumatic outcome by promoting impaired vessels to respond to secondary challenges that commonly occur in patients with severe TBI [37].

12.5 The Effect of Hypothermia on Hemorrhage and Inflammatory Responses

In addition to TBI altering the endothelial transport of various tracers across the vascular endothelium, TBI can also have more dramatic effects on endothelial integrity and lead to intracerebral hemorrhage. Indeed, in models of severe TBI, hemorrhagic contusions are commonly seen in vulnerable brain regions including gray and white matter structures [4, 26]. Thus, strategies that reduce the formation of intracerebral hemorrhages remain a clinical concern. In this regard, Kinoshita et al. [38] tested the effect of posttraumatic hypothermia on hemoglobin extravasation after FP brain injury in rats. In this moderate TBI model, a hemorrhagic contusion is commonly observed in the lateral cerebral cortex bordering underlying white matter tracts in normothermic rats. Hemoglobin levels in specific brain regions were quantified using a spectrophotometric hemoglobin assay. In that study, 3 h of posttraumatic hypothermia were reported to significantly decrease the magnitude of intracerebral hemoglobin levels in the traumatized hemisphere as compared to normothermic animals. These studies were important because they supported the hypothesis that posttraumatic hypothermia could minimize the more severe cerebrovascular events associate with TBI that frequently lead to severe brain swelling and high rates of mortality and morbidity. It also appears that a slow post-hypothermic rewarming procedure is an important variable in assuring maximal efficiency with hypothermia treatments [17, 39].

12.6 Hypothermia and Posttraumatic Inflammation

Vascular damage can be associated with brain edema function and multiple posttraumatic inflammatory events [19, 22, 40, 41]. Trauma-induced damage to vascular components can include endothelial cell damage and as previously described cerebral hemorrhage. As a consequence of TBI, a variety of endothelial and leukocyte adhesion molecules are upregulated, resulting in the enhanced recruitment and extravasation of circulating leukocytes into the brain parenchyma [42]. Circulating monocytes that gain access to the brain parenchyma can contribute to macrophage-induced neuronal degeneration. Together with endogenous astrocyte and microglial activation, these inflammatory cascades have been shown to participate in secondary injury mechanisms and are an important target for therapeutic drug development.

Previous studies from various laboratories have reported that posttraumatic cooling reduces the infiltration of circulating inflammatory cells [42–45]. Also, decreased levels of several proinflammatory cytokines and nitric oxide synthase (NOS) are reported after TBI with modest hypothermia treatment [2, 46, 47]. In a study by Lotocki et al. [30], posttraumatic hypothermia was shown to significantly reduce the extravasation of CD68 immunoreactive inflammatory cells. In that study, quantitative cell counts of CD68 immunoreactive cells indicated a significant decrease in numbers of infiltrating macrophages in hypothermic animals compared to normothermia in cortical and hippocampal regions. Interestingly, areas of increased CD68 cell counts were commonly seen in areas showing alterations in BBB permeability. The spatial relationships between BBB leakage and CD68-positive macrophage infiltration indicate a relationship between these early vascular perturbations and secondary inflammatory cascades.

In another study, Chatzipanteli et al. [42] reported the beneficial effects of early cooling after TBI on the temporal and regional profile of polymorphonuclear leukocyte (PMNL) infiltration. In that study, animals were allowed to survive up to 7 days after trauma and brains were dissected for biochemical analysis of myeloperoxidase (MPO) activity. In normothermic animals, significantly increased MPO activity was observed in several traumatized brain regions. In contrast, posttraumatic hypothermia significantly reduced MPO activity in these vulnerable brain regions. Because PMNL are some of the earliest inflammatory cells that accumulate after vascular injury, these results again indicate a relationship between trauma-induced vascular perturbations and early inflammatory responses.

In related biochemical and immunocytochemistry studies, posttraumatic hypothermia has been also reported to affect hydroxyl radicals, lipid peroxidation, and NOS activity after TBI [46, 48]. Because nitric oxide leading to oxidative stress is thought to participate in the pathogenesis of TBI [49–51] the observation that hypothermia significantly altered constitutive and inducible NOS activity provides another mechanism by which hypothermia can protect the vasculature and parenchyma after TBI. Together, these studies support the hypothesis that temperature-dependent alterations in vascular pathology including BBB disruption and inflammatory cell accumulation could participate in the beneficial effects of posttraumatic hypothermia on traumatic outcome.

12.7 Biochemical and Molecular Mechanisms of Hypothermic Protection

An important mechanism by which hypothermia may protect against trauma-induced vascular damage is by targeting excessive matrix metalloproteinase (MMP) activity. Previous investigations have stressed the importance of abnormal

MMP activation in models of brain injury including TBI [52, 53]. Pharmacological treatments or genetic models targeting specific various MMPs including MMP-3 and MMP-9 have been reported to improve outcome in models of stroke and Neurotrauma [53]. In a TBI study, Truettner et al. [54] tested whether early posttraumatic hypothermia would influence the upregulation of MMP-3 and MMP-9 after moderate FP brain injury. Although levels and activities of MMP-3 and MMP-9 were significantly increased in normothermic rats, posttraumatic hypothermia reduced MMP-9 activity within vulnerable cortical and subcortical areas. In contrast, hypothermic treatment did not affect the delayed activation of MMP-3. Similar investigations were conducted by Jia et al. [55] where again posttraumatic hypothermia was shown to attenuate both the mRNA and protein levels of MMP-9 compared to normothermic levels. Taken together, these studies emphasize the temperature sensitivity of abnormal MMP expression following TBI. Abnormal MMP activity has been implicated in the pathophysiology of cerebral ischemia [52]. Thus, a mechanism by which hypothermia protects against BBB dysfunction after TBI is by reducing the detrimental effects of trauma-induced increased MMP-9 activity on vascular function.

Subsequent studies by a variety of laboratories have also investigated molecular mechanisms by which posttraumatic hypothermia can affect inflammatory cascades after trauma [56, 57]. Several inflammatory genes that are reported to be upregulated after TBI have been shown to be altered by posttraumatic hypothermia. Therapeutic hypothermia has been shown to modify various signaling cascades including tumor necrosis factor as well as ligand–receptor interactions [3, 58]. More recently, abnormal inflammasome activation which is an important component of the early innate inflammatory response to injury has also been shown to be activated after TBI [59]. In studies where posttraumatic hypothermia was initiated after FP brain injury [60], significantly reduced expression of various inflammatory proteins including caspase 1, caspase 11, and the purinergic receptor P2X7 was reported. Finally, the possible modulation of specific microRNA patterns of expression after TBI has also been evaluated in terms of hypothermic consequences. In one study, Truettner et al. [57] using a microarray strategy evaluated the expression of 388 rat microRNAs after FP brain injury. Interestingly, posttraumatic hypothermia led to several of the microRNAs being elevated compared to normothermia while five were reduced by the treatment. These data indicate that in addition to a variety of cellular consequences of therapeutic hypothermia in various TBI models, temperature-sensitive molecular processes including alterations in microRNAs thought to be involved in basic cell processing events are altered with early cooling.

12.8 Clinical Studies

The use of therapeutic hypothermia in TBI patients has been utilized in a variety of clinical investigations with various consequences [7, 8]. Individual institutional studies testing the efficacy of moderate hypothermia in patients with severe TBI led to several positive reports. However, disappointing results resulted from several multicenter trials where large numbers of severe TBI patients were evaluated for the efficacy of therapeutic hypothermia [7].

Most recently, attention has been directed toward identifying a subpopulation of severe TBI patients that may be most sensitive to early hypothermic treatment [8]. In a recent study by Clifton et al. [8], for example, patients undergoing early decompressive surgery who were cooled prior to the surgical intervention appeared to be protected with early cooling. This preliminary result may indicate that early cooling may target some of the early pathophysiological events that are activated in severe TBI patients where early surgical procedures are warranted.

Evidence for reperfusion injury is seen in many models of brain injury [18]. The surgical or pharmacologically induced opening of previously occluded or compressed blood vessels leads to immediate reflow of blood into previously ischemic areas. This hemodynamic response can activate various detrimental consequences including free radical generation and inflammatory cascades that can damage vulnerable vascular beds and produce alterations in vascular permeability. Hypothermia introduced prior to reperfusion injury may significantly attenuate many of these harmful effects. In one recent experimental study that tested this hypothesis in a subdural hematoma model, early but not delayed cooling prior to decompression surgery significantly reduced early neuronal and glial damage [18]. In this regard, a new multicenter clinical trial (HOPES) is now being organized to test the hypothesis that early cooling prior to decompression surgery in severe TBI patients can have beneficial effects in terms of long-term outcome.

12.9 Summary

Therapeutic hypothermia has been successfully used in a variety of experimental and clinical situations to target the devastating consequences of TBI. Although many preclinical studies using a variety of experimental models showed efficacy with hypothermic treatment in terms of blocking early vascular changes and improving outcome, these findings have been slow to be successfully translated to the clinic. As investigators continue to evaluate and clarify the cellular and molecular mechanisms by which early cooling can protect against vascular perturbations including BBB damage, this critical knowledge will aid in the further development of therapeutic strategies to promote recovery after TBI including combination approaches. Recent biomarker studies of plasma and cerebral spinal fluid samples have already demonstrated the possibility of using altered BBB function as a mechanism to evaluate the biochemical status of the injured brain [61, 62]. It is clear that additional experimental investigations are required to understand mechanisms by which trauma can damage the neurovascular network so that more powerful strategies that may include therapeutic hypothermia can be used to protect and promote repair after TBI.

Acknowledgements The authors would like to thank Jeremy Lytle for his editorial assistance. This work was supported by grants NIH NS030291, NS042133, NS056072, and a Veterans Affairs grant BX000521.

References

1. Dietrich WD (1992) The importance of brain temperature in cerebral injury. J Neurotrauma 9 (Suppl 2):S475–S485
2. Dietrich WD, Busto R, Globus MY, Ginsberg MD (1996) Brain damage and temperature: cellular and molecular mechanisms. Adv Neurol 71:177–194
3. Yenari MA, Han HS (2012) Neuroprotective mechanisms of hypothermia in brain ischaemia. Nat Rev Neurosci 13(4):267–278
4. Dietrich WD, Alonso O, Busto R, Globus MY, Ginsberg MD (1994) Post-traumatic brain hypothermia reduces histopathological damage following concussive brain injury in the rat. Acta Neuropathol 87:250–258
5. Clifton GL, Jiang JY, Lyeth BG, Jenkins LW, Hamm RJ, Hayes RL (1991) Marked protection by moderate hypothermia after experimental traumatic brain injury. J Cereb Blood Flow Metab 11(1):114–121
6. Dietrich WD, Bramlett HM (2010) The evidence for hypothermia as a neuroprotectant in traumatic brain injury. Neurotherapeutics 7(1):43–50
7. Polderman KH (2008) Induced hypothermia and fever control for prevention and treatment of neurological injuries. Lancet 371(9628):1955–1969
8. Clifton GL, Coffey CS, Fourwinds S, Zygun D, Valadka A, Smith KR Jr (2012) Early induction of hypothermia for evacuated intracranial hematomas: a post hoc analysis of two clinical trials. J Neurosurg 117(4):714–720
9. Atkins CM, Truettner JS, Lotocki G, Sanchez-Molano J, Kang Y, Alonso OF, Sick TJ, Dietrich WD, Bramlett HM (2010) Post-traumatic seizure susceptibility is attenuated by hypothermia therapy. Eur J Neurosci 32(11):1912–1920
10. Bramlett HM, Green EJ, Dietrich WD, Busto R, Globus MY, Ginsberg MD (1995) Posttraumatic brain hypothermia provides protection form sensorimotor and cognitive behavioral deficits. J Neurotrauma 12:289–298
11. Urbano LA, Oddo M (2012) Therapeutic hypothermia for traumatic brain injury. Curr Neurol Neurosci Rep 12(5):580–591
12. Jiang JY, Lyeth BG, Kapasi MZ, Jenkins LW, Povlishock JT (1992) Moderate hypothermia reduces blood–brain barrier disruption following traumatic brain injury in the rat. Acta Neuropathol 84:495–500
13. Smith SL, Hall ED (1996) Mild pre- and posttraumatic hypothermia attenuates blood–brain barrier damage following controlled cortical impact injury in the rat. J Neurotrauma 13(1):1–9
14. Wei EP, Dietrich WD, Povlishock JT, Navari RM, Kontos HA (1980) Functional, morphological, and metabolic abnormalities of the central microcirculation after concussive brain injury in cats. Circ Res 46:37–47

15. Wei EP, Hamm RJ, Baranova AI, Povlishock JT (2009) The long-term microvascular and behavioral consequences of experimental traumatic brain injury after hypothermic intervention. J Neurotrauma 26(4):527–537
16. Oda Y, Gao G, Wei EP, Povlishock JT (2011) Combinational therapy using hypothermia and immunophilin ligand FK506 to target altered pial arteriolar reactivity, axonal damage, and blood–brain barrier dysfunction after traumatic brain injury in rat. J Cereb Blood Flow Metab 32(4):1143–1154
17. Suehiro E, Ueda Y, Wei EP, Kontos HA, Povlishock JT (2003) Postraumatic hypothermia followed by slow rewarming protects the cerebral microcirculation. J Neurotrauma 20:381–390
18. Yokobori S, Gajavelli S, Mondello S, Mo-Seaney J, Bramlett HM, Dietrich WD, Bullock MR (2012) Neuroprotective effect of preoperatively induced mild hypothermia as determined by biomarkers and histopathological estimation in a rat subdural hematoma decompression model. J Neurosurg 118:370–80
19. Beaumont A, Marmarou A, Hayasaki K, Barzo P, Fatouros P, Corwin F, Marmarou C, Dunbar J (2000) The permissive nature of blood brain barrier (BBB) opening in edema formation following traumatic brain injury. Acta Neurochir Suppl 76:125–129
20. Morganti-Kossmann MC, Bye N, Nguyen P, Kossmann T (2005) Influence of brain trauma on blood–brain barrier properties. In: De Vries E, Prat A (eds) The blood–brain barrier and its microenvironment. New York, Taylor & Francis, pp 457–480
21. Zlokovic BV (2008) The blood–brain barrier in health and chronic neurodegenerative disorders. Neuron 57(2):178–201
22. Shapira Y, Setton D, Artru AA, Shohami E (1993) Blood–brain barrier permeability, cerebral edema, and neurologic function after closed head injury in rats. Anesth Analg 77(1):141–148
23. Shlosberg D, Benifla M, Kaufer D, Friedman A (2010) Blood–brain barrier breakdown as a therapeutic target in traumatic brain injury. Nat Rev Neurol 6(7):393–403
24. Chodobski A, Zink BJ, Szmydynger-Chodobska J (2011) Blood–brain barrier pathophysiology in traumatic brain injury. Transl Stroke Res 2(4):492–516
25. Adelson PD, Whalen MJ, Kochanek PM, Robichaud P, Carlos TM (1998) Blood brain barrier permeability and acute inflammation in two models of traumatic brain injury in the immature rat: a preliminary report. Acta Neurochir Suppl 71:104–106
26. Cortez SC, McIntosh TK, Noble LJ (1989) Experimental fluid percussion brain injury: vascular disruption and neuronal and glial alterations. Brain Res 482(2):271–282
27. Hicks RR, Baldwin SA, Scheff SW (1997) Serum extravasation and cytoskeletal alterations following traumatic brain injury in rats. Comparison of lateral fluid percussion and cortical impact models. Mol Chem Neuropathol 32(1–3):1–16
28. Povlishock JT, Becker DP, Sullivan HG, Miller JD (1978) Vascular permeability alterations to horseradish peroxidase in experimental brain injury. Brain Res 22:223–239
29. Dietrich WD, Alonso O, Halley M (1994) Early microvascular and neuronal consequences of traumatic brain injury: a light and electron microscopic study in rats. J Neurotrauma 11:289–301
30. Lotocki G, de Rivero Vaccari JP, Perez ER, Sanchez-Molano J, Furones-Alonso O, Bramlett HM, Dietrich WD (2009) Alterations in blood–brain barrier permeability to large and small molecules and leukocyte accumulation after traumatic brain injury: effects of post-traumatic hypothermia. J Neurotrauma 26(7):1123–1134
31. Başkaya MK, Rao AM, Doğan A, Donaldson D, Dempsey RJ (1997) The biphasic opening of the blood–brain barrier in the cortex and hippocampus after traumatic brain injury in rats. Neurosci Lett 226(1):33–36
32. Habgood MD, Bye N, Dziegielewska KM, Ek CJ, Lane MA, Potter A et al (2007) Changes in blood–brain barrier permeability to large and small molecules following traumatic brain injury in mice. Eur J Neurosci 25:231–238

33. Ueda Y, Wei EP, Kontos HA, Suehiro E, Povlishock JT (2003) Effects of delayed, prolonged hypothermia on the pial vascular response after traumatic brain injury in rats. J Neurosurg 99:899–906
34. Golding EM, Robertson CS, Bryan RM Jr (1999) The consequences of traumatic brain injury on cerebral blood flow and autoregulation: a review. Clin Exp Hypertens 21:299–332
35. Lin Y, Pan Y, Wang M, Huang X, Yin Y, Wang Y, Jia F, Xiong W, Zhang N, Jiang JY (2012) Blood–brain barrier permeability is positively correlated with cerebral microvascular perfusion in the early fluid percussion-injured brain of the rat. Lab Invest 92(11):1623–1634
36. Fujita M, Oda Y, Wei EP, Povlishock JT (2011) The combination of either tempol or FK506 with delayed hypothermia: implications for traumatically induced microvascular and axonal protection. J Neurotrauma 28:1209–1218
37. Fujita M, Wei EP, Povlishock JT (2012) Effects of hypothermia on cerebral autoregulatory vascular responses in two rodent models of traumatic brain injury. J Neurotrauma 29 (7):1491–1498
38. Kinoshita K, Chatzipanteli K, Alonso OF, Howard M, Dietrich WD (2002) The effect of brain temperature on hemoglobin extravasation after traumatic brain injury. J Neurosurg 97:945–953
39. Povlishock JT, Wei EP (2009) Posthypothermic rewarming considerations following traumatic brain injury. J Neurotrauma 26(3):333–340
40. Hartl R, Medary M, Ruge M, Arfors KE, Ghajar J (1997) Blood–brain barrier breakdown occurs early after traumatic brain injury and is not related to white blood cell adherence. Acta Neurochir Suppl 70:240–242
41. Stahel PF, Shohami E, Younis FM, Kariya K, Otto VI, Lenzlinger PM et al (2000) Experimental closed head injury: analysis of neurological outcome, blood–brain barrier dysfunction, intra-cranial neutrophil infiltration, and neuronal cell death in mice deficient in genes for pro-inflammatory cytokines. J Cereb Blood Flow Metab 20:369–380
42. Chatzipanteli K, Alonso OF, Kraydieh S, Dietrich WD (2000) Importance of post-traumatic hypothermia and hyperthermia on the inflammatory response after fluid-percussion brain injury: biochemical and immunocytochemical studies. J Cereb Blood Flow Metab 20:531–542
43. Whalen MJ, Carlos TM, Clark RS, Marion DW, DeKosky MS, Heineman S (1997) The relationship between brain temperature and neutrophil accumulation after traumatic brain injury in rats. Acta Neurochir Suppl 70:260–261
44. Whalen MJ, Carlos TM, Clark RS, Marion DW, DeKosky ST, Heineman S et al (1997) The effect of brain temperature on acute inflammation after traumatic brain injury in rats. J Neurotrauma 14:561–572
45. Sutcliffe IT, Smith HA, Stanimirovic D, Hutchison SB (2001) Effects of moderate hypothermia on IL-1 beta-induced leukocyte rolling and adhesion in pial microcirculation of mice and on proinflammatory gene expression in human cerebral endothelial cells. J Cereb Blood Flow Metab 21:1310–1319
46. Chatzipanteli K, Wada K, Busto R, Dietrich WD (1999) Effects of moderate hypothermia on constitutive and inducible nitric oxide synthase activities after traumatic brain injury in the rat. J Neurochem 72(5):2047–2052
47. Kinoshita K, Chatzipanteli K, Vitarbo E, Truettner JS, Alonso OF, Dietrich WD (2002) Interleukin-1β messenger ribonucleic acid and protein levels after fluid-percussion brain injury in rats: importance of injury severity and brain temperature. Neurosurgery 51(1):195–203; discussion 203
48. Globus MY, Alonso O, Dietrich WD, Busto R, Ginsberg MD (1995) Glutamate release and free radical production following brain injury: effects of posttraumatic hypothermia. J Neurochem 65(4):1704–1711
49. Smith SL, Andrus PK, Zhang JR, Hall ED (1994) Direct measurement of hydroxyl radicals, lipid peroxidation, and blood–brain barrier disruption following unilateral cortical impact head injury in the rat. J Neurotrauma 11(4):393–404

50. Readnower RD, Chavko M, Adeeb S, Conroy MD, Pauly JR, McCarron RM, Sullivan PG (2010) Increase in blood–brain barrier permeability, oxidative stress, and activated microglia in a rat model of blast-induced traumatic brain injury. J Neurosci Res 88(16):3530–3539
51. Haorah J, Ramirez SH, Schall K, Smith D, Pandya R, Persidsky Y (2007) Oxidative stress activates protein tyrosine kinase and matrix metalloproteinases leading to blood–brain barrier dysfunction. J Neurochem 101(2):566–576
52. Lo EH, Wang X, Cuzner ML (2002) Extracellular proteolysis in brain injury and inflammation: role for plasminogen activators and matrix metalloproteinases. J Neurosci Res 69(1):1–9
53. Asahi M, Asahi K, Jung JC, del Zoppo GJ, Fini ME, Lo EH (2000) Role for matrix metalloproteinase 9 after focal cerebral ischemia: effects of gene knockout and enzyme inhibition with BB-94. J Cereb Blood Flow Metab 20(12):1681–1689
54. Truettner JS, Alonso OF, Dietrich WD (2005) Influence of therapeutic hypothermia on matrix metalloproteinase activity after traumatic brain injury in rats. J Cereb Blood Flow Metab 25:1505–1516
55. Jia F, Pan YH, Mao Q, Liang YM, Jiang JY (2010) Matrix metalloproteinase-9 expression and protein levels after fluid percussion injury in rats: the effect of injury severity and brain temperature. J Neurotrauma 27(6):1059–1068
56. Truettner JS, Suzuki T, Dietrich WD (2005) The effect of therapeutic hypothermia on the expression of inflammatory response genes following moderate traumatic brain injury in the rat. Brain Res Mol Brain Res 138(2):124–134
57. Truettner JS, Alonso OF, Bramlett HM, Dietrich WD (2011) Therapeutic hypothermia alters microRNA responses to traumatic brain injury in rats. J Cereb Blood Flow Metab 31 (9):1897–1907
58. Lotocki G, de Rivero Vaccari JP, Perez ER, Alonso OF, Curbelo K, Keane RW, Dietrich WD (2006) Therapeutic hypothermia modulates TNFR1 signaling in the traumatized brain via early transient activation of the JNK pathway and suppression of XIAP cleavage. Eur J Neurosci 24:2283–2290
59. De Rivero Vaccari JP, Lotocki G, Alonso OF, Bramlett HM, Dietrich WD, Keane RW (2009) Therapeutic neutralization of the NLRP1 inflammasome reduces the innate immune response and improves histopathology after traumatic brain injury. J Cereb Blood Flow Metab 29 (7):1251–1261
60. Tomura S, de Rivero Vaccari JP, Keane RW, Bramlett HM, Dietrich WD (2012) Effects of therapeutic hypothermia on inflammasome signaling after traumatic brain injury. J Cereb Blood Flow Metab 32(10):1939–1947
61. Kochanek PM, Bramlett H, Dietrich WD, Dixon CE, Hayes RL, Povlishock J, Tortella FC, Wang KK (2011) A novel multicenter preclinical drug screening and biomarker consortium for experimental traumatic brain injury: operation brain trauma therapy. J Trauma 71(1 Suppl): S15–S24
62. Adamczak S, Dale G, de Rivero Vaccari JP, Bullock MR, Dietrich WD, Keane RW (2012) Inflammasome proteins in cerebrospinal fluid of brain-injured patients as biomarkers of functional outcome. J Neurosurg 117(6):1119–1125

Part II
Experimental Models and Methods

Chapter 13
Vascular Responses in Rodent Models of Traumatic Brain Injury

Xiaoshu Wang, Zhanyang Yu, Zhengbu Liao, Qi Liu, MingMing Ning, Xiaochuan Sun, Josephine Lok, Eng H. Lo, and Xiaoying Wang

Abstract The heterogeneity of traumatic brain injury (TBI) is considered one of the most significant barriers to finding effective therapeutic interventions. Among multiple confirmed pathological events after TBI, vascular response or cerebrovascular pathophysiology is one of the most important pathophysiological components of TBI, but its role and molecular mechanisms remain largely unknown. This chapter reviews experimental studies of cerebrovascular pathophysiology, especially in rodent TBI models. Clinically translational advantages and limitations of each commonly used rodent TBI models in the study of vascular responses are also discussed.

13.1 Pathophysiology of TBI

13.1.1 General Pathophysiology of TBI

The heterogeneity of traumatic brain injury (TBI) is considered one of the most significant barriers to finding effective therapeutic interventions [1]. Classifications of TBI pathophysiology vary between researchers in use of the terms. Generally, the first stage of cerebral injury after TBI is characterized by direct tissue damage, impaired regulation of cerebral blood flow (CBF), and metabolism. The decreased CBF leads to accumulation of lactic acid due to anaerobic glycolysis and consequently inducing ATP-stores deplete and energy failure, increased membrane permeability, and consecutive edema formation. The second stage of the

X. Wang (✉)
Department of Neurosurgery, The First Affiliated Hospital of Chongqing Medical University, Chongqing, China

Departments of Neurology and Radiology, Massachusetts General Hospital, Harvard Medical School, Charlestown, MA, USA
e-mail: xswang789@gmail.com

E.H. Lo et al. (eds.), *Vascular Mechanisms in CNS Trauma*, Springer Series in Translational Stroke Research 5, DOI 10.1007/978-1-4614-8690-9_13,

pathophysiological cascade is characterized by terminal membrane depolarization along with excessive release of excitatory neurotransmitters and Ca^{2+}-influx. The Ca^{2+}-influx leads to self-digesting (catabolic) intracellular processes by activating lipid peroxidases, proteases, and phospholipases that result in increases of the intracellular concentration of free fatty acids and free radicals. Together, these events lead to membrane degradation of vascular and cellular structures and vascular dysfunction. Interactions between pathological factors or cascades may cause neurovascular unit decoupling and cerebrovascular dysfunction.

13.1.2 *Primary and Secondary Brain Damage after TBI*

TBI can result in the development of complex neurological deficits which is caused by both primary and secondary injury mechanisms. Primary injury events encompass the mechanical damage that occurs at the time of trauma to neurons, axons, glia, and blood vessels as a result of shearing, tearing, or stretching. In addition, secondary injury evolves over minutes to days and even months after the initial traumatic insult and results from delayed biochemical, metabolic, and cellular changes that are initiated by the primary event. These secondary injury cascades are thought to account for the development of many of the neurological deficits observed after TBI, and their delayed nature suggests that there is a therapeutic window for pharmacological treatment to prevent progressive tissue damage and improve outcome [2]. Intracranial hypertension, ischemic brain injury, cerebral edema, and other pathophysiologic squeal for cerebrovascular dysfunction, in most cases, would be more accurately described as pathophysiologic cascades or secondary insults [3].

13.2 Cerebrovascular Pathophysiology after TBI

Normal brain function depends on the exquisite balance between vascular, neuronal, and glial interactions. The complex interplay between different types of cells in the neurovascular unit dictates the final evolution of pathogenesis. Cerebrovascular function is recognized as a key part of the neurovascular unit, a major determinant of patient outcome after brain injury [4], and it has been hypothesized that therapies that do not address that dysfunction are unlikely to promote recovery [5]. Among multiple confirmed pathological events after TBI, the vascular response is one of the most important pathophysiological components. Tearing of blood vessels by primary mechanical damage results in dynamic vascular responses, including abnormalities of vasodilation and vasoconstriction, reduction of CBF, loss of integrity of blood–brain barrier (BBB), the vasogenic edema, inflammation, and microthrombosis. It is likely that some of the aspects of cerebral vascular dysfunction due to TBI contribute to the increased sensitivity of the injured brain to secondary hypoxia and hypotension and the increased mortality and morbidity that occurs due to secondary insults after TBI in humans [4, 6]. Normally, the high metabolic demand of the brain is met by maintenance of CBF even under altered systemic blood pressure, through a process termed cerebral autoregulation [7]. It is achieved by the ability of the brain vasculature to dilate during low blood pressure and constrict when increased blood pressure in order to maintain its perfusion within physiological range. Loss of autoregulation can expose the brain to hypoperfusion during low blood pressure (hypotension) and to excess blood flow and potentially bleeding during high blood pressure (hypertension) [8]. However, pathophysiologic classification schemes have not been commonly used in treatment trials. This may be due, in part, to challenges associated with capturing a spatio-temporal profile of the patient's injury, limited availability and usage of sophisticated monitoring techniques needed for measurement of physiologic parameters, and difficulties in distinguishing inevitable but progressive cell damage from potentially reversible injury cascades [1].

13.2.1 Reduction of Cerebral Blood Flow

The brain is unusual in that it is only able to withstand very short periods of lack of blood supply (ischemia). This is because neurons produce energy (ATP) almost entirely by oxidative metabolism of substrates including glucose and ketone bodies, with very limited capacity for anaerobic metabolism. Without oxygen, energy-dependent processes cease, leading to irreversible cellular injury if blood flow is not reestablished rapidly (3–8 min under most circumstances). Therefore, adequate CBF must be maintained to ensure a constant delivery of oxygen and substrates and to remove the waste products of metabolism. CBF is dependent on a number of factors that can broadly be divided into those affecting cerebral perfusion pressure (CPP) and those affecting the radius of cerebral blood vessels [9]. Ischemia, especially pericontusional ischemia in the traumatic penumbra, is one of the leading causes of secondary posttraumatic brain damage and leads to poor functional outcome [10]. Generally, CBF is markedly reduced immediately after and during the first few hours after trauma, especially in the pericontusional region [11]. The primary lesion evolving at the time of injury increases in size when CBF in the critically perfused tissue around the contusion ("traumatic penumbra") gradually decreases below the ischemic threshold giving rise to secondary injury [12].

Reduced CBF is a feature of many models of experimental TBI. Fluid percussion injury (FPI), the most commonly used model of experimental TBI, caused 40–50 % reductions in CBF that occurred within 15–30 min after injury and persisted for about 4 h [13]. Controlled cortical impact (CCI), another widely used model of TBI in rats, reduced cortical CBF by about 50–85 % [14, 15]. CBF reductions due to CCI persisted for at least 3 h after injury, with 85 % and 49 % reductions in a contused and contralateral cortex, respectively. At 24 h after trauma, a focal

CBF reduction persists in the vicinity of the injury site [15]. In a weight drop model of rats, reduction of CBF was detected up to 48 h after TBI in injured cortex area [16].

13.2.2 Cerebrovascular Autoregulation Impairment

The high metabolic demand of the brain is met by maintenance of CBF even under altered systemic blood pressure, in a process termed cerebral autoregulation [7]. It is achieved by the ability of the brain vasculature to dilate during low blood pressure and constrict when increased blood pressure in order to maintain its perfusion within physiological range. Loss of autoregulation can expose the brain to hypoperfusion during low blood pressure (hypotension) and to excess blood flow and potentially bleeding during high blood pressure (hypertension) [8]. After TBI, CBF autoregulation (i.e., cerebrovascular constriction or dilation in response to increases or decreases in CPP) is impaired or abolished in most patients. The extent and temporal profile of this autoregulatory failure does not always correlate with the severity of injury. Defective CBF autoregulation may be present immediately after trauma or may develop over time and is transient or persistent in nature irrespective of the presence of mild, moderate, or severe damage [17].

In a closed-head injury of weight drop model in rats, CBF was monitored using laser Doppler flowmetry along with monitoring of ICP and arterial blood pressure. Decrease in CBF and increase of ICP was observed as a result of loss of cerebral autoregulation in injured rats during first 4 h [18]. MRI and spin-labeled carotid artery water protons as an endogenous tracer was used to measure CBF and CO_2 reactivity after TBI with CCI device in rats. During normocarbia, CBF was reduced within a cortical region of interest (ROI, injured versus contralateral) after TBI. Within a contusion-enriched ROI, CBF was reduced after TBI. In the contusion-enriched ROI, only controls showed a significant reduction in CBF, suggesting blunted CO_2 reactivity in the sham and TBI group [19]. In a modified closed-head injury model of impact-acceleration using weight drop in rats [20, 21] TBI significantly reduced the ability of the cerebral vasculature to alter cerebral vascular resistance in order to maintain constant levels of CBF when arterial blood pressure is changed. This cerebral autoregulation was impaired 24 h after TBI [20]. In addition to the abolished CBF response to hyperventilation [19], vasodilatory responses to increases in $PaCO_2$ also were reduced or abolished by the CCI model of rats [17]. These studies have demonstrated that clinically observed impairments in blood pressure autoregulation and CO_2 reactivity are also detectable in experimental TBI of rats [17, 22]. Furthermore, there are some reports on a variety of other cerebral vascular compensatory mechanisms in experimental TBI of rats. For example, FPI in rats altered CBF–metabolism coupling—the normally close relationship between CBF and metabolic activity [23]. Posttraumatic cerebral vasospasm is an important secondary insult that occurs in more than one-third of patients with TBI and indicates severe damage to the brain [24]. For example, in a closed-head acceleration impact model, there was decreased vasoconstriction following TBI, which may be due to changes induced by endothelin-1 induced pericyte-mediated regulation of microvessel blood flow following TBI. Moreover, the results also suggested that endothelin receptor-A antagonists ameliorate trauma-induced hypoperfusion, in part, by inhibiting endothelin-1-mediated upregulation of alpha-smooth muscle actin in pericytes [25]. The posttraumatic status of vascular reactivity and autoregulation also has important implications with regard to the treatment of high ICP, in particular for the use of hyperventilation and pharmacologic management of blood pressure, which have been discussed in detail [26]. Clinical studies suggested TBI patient outcome was correlated with the degree of autoregulation [27]. TBI can be associated with an imbalance between cerebral oxygen delivery and cerebral oxygen consumption. Although this mismatch is induced by several different vascular and hemodynamic mechanisms as indicated earlier, the final common endpoint is brain tissue hypoxia. Dysfunction of cerebral autoregulation may be caused by multiple pathogenic factors and their dynamic interactions, but the pathophysiology and related therapy development remain largely lacking [4].

13.2.3 Other Cerebrovascular Responses After TBI

Besides the reduced CBF and cerebral autoregulation dysfunction, there are other pathophysiological mechanism related to vascular response after TBI, such as edema (cytotoxic and vasogenic edema), BBB dysfunction, vascular inflammation, microthrombosis, and vascular remodeling. There are some review articles which summarized these topics [28–33] and are discussed on other chapters of this book. In this chapter, only a brief introduction is provided below.

13.2.3.1 Brain Edema

The current classification of brain edema relates to the structural damage induced by the primary or secondary injury [28]. Vasogenic brain edema is caused by mechanical or autodigestive disruption or functional breakdown of the endothelial cell layer (an essential structure of the blood–brain barrier) of brain vessels. Disintegration of the cerebral vascular endothelial wall allows for uncontrolled ion and protein transfer from the intravascular to the extracellular (interstitial) brain compartments with ensuring water accumulation. Anatomically, this pathology increases the volume of the extracellular space. For many years, vasogenic brain edema was accepted as the prevalent edema type following TBI. The development of mechanical TBI models ("weight drop," "FPI," and "CCI injury") and the use of magnetic resonance imaging, however, revealed that "cytotoxic" edema is of decisive pathophysiological importance following TBI as it develops early and persists while BBB integrity is gradually restored. These findings suggest that cytotoxic and vasogenic brain edema are two entities which can be targeted simultaneously or according to their temporal prevalence [28, 34, 35].

13.2.3.2 Blood–Brain Barrier Damage

The majority of the models of experimental TBI produce some degree of increased BBB permeability. Both clinical and experimental investigations demonstrate that TBI results in BBB opening within minutes and persists for hours to days after injury, providing further evidence that the cerebral vasculature is functionally and morphologically damaged by TBI. Evidence from experimental animal models have demonstrated that BBB damage contributes to posttraumatic inflammation and edema formation [29].

13.2.3.3 Vascular Inflammation

TBI induces a complex array of immunological/inflammatory reperfusion injury. Both primary and secondary insults activate the release of cellular mediators including proinflammatory cytokines, prostaglandins, free radicals, and complement [36]. These processes induce chemokines and adhesion molecules and in turn mobilize immune and glial cells in a parallel and synergistic fashion. Brian tissue infiltration by leukocytes is facilitated via upregulation of cellular adhesion molecules such as P-selectin, intercellular adhesion molecules (ICAM-1), and vascular adhesion molecules (VCAM-1) [37]. In response to these inflammatory processes, injured and adjacent tissue will be eliminated and within hours, days, and weeks astrocytes produce microfilaments and neutropines ultimately to synthesize scar tissue [38]. The progression of tissue damage relates to direct release of neurotoxic mediators or indirectly to the release of nitric oxide and cytokines. The additional release of vasoconstrictors (prostaglandins and leucotrienes), the obliteration of microvasculature through adhesion of leucocytes and platelets, the BBB leakage and the edema formation further reduce tissue perfusion and consequently aggravate secondary brain damage [31].

13.2.3.4 Cerebrovascular Microthrombosis

Cerebrovascular damages leading to subsequent reductions in regional cerebral blood flow (rCBF) may play an important role in secondary cell damages following TBI. In a cortical contusion model of rats, histological examinations revealed microthrombosis formation in the contused area, extending from the center to the peripheral areas within 6 h post-injury. The rCBF decrease and the contusion necrosis volume were significantly attenuated by platelet activating factor antagonist etizolam administration, suggesting that platelet activating factor is involved in microthrombosis formation and hemodynamic depression, resulting in ischemic damages within areas surrounding the contusion [39]. In rats subjected to CCI, in vivo fluorescence microscopy determined rolling of leukocytes on the cerebrovascular endothelium of both in arterioles and venules, leukocyte-platelet aggregates in venules, and microthrombi occlusion in up to 70 % of venules and 33 % of arterioles, suggesting that immediate posttraumatic decrease in pericontusional blood flow is not caused only by arteriolar vasoconstriction, but by platelet activation and the subsequent formation of thrombi in the cerebral microcirculation [33].

13.2.3.5 Vascular Remodeling

The recovery of common functional disabilities after TBI can continue for months to years. A critical factor for recovery of tissue structure and function is the availability of oxygen and nutrients in the blood stream. Hemodynamic and cerebrovascular factors are crucially involved in secondary damage after TBI. In rat TBI model of LFP, there was widespread ipsilateral and contralateral hypoperfusion. Hemodynamic unrest may partly be explained by an increase in blood vessel density over a period of 2 weeks in the ipsilateral hippocampus and perilesional cortex [40]. Several studies have also suggested that enhanced neurovascularization, especially in the acute phase after stroke or TBI, improves motor recovery [41]. One study for chronic vascular response in an experimental CCI models of rats demonstrated remarkable reduction in CBF in the cortex and hippocampus but not in the thalamus at 1 year after TBI [42]. In a TBI rat model of LFP injury, animals were followed-up for 9 months and examined with MRI-CBF, histologic assessment of vascular density, and neurobehavioral analysis [43]. Experimental results showed that each of the investigated brain areas has a unique pattern of vascular abnormalities; chronic alterations in CBF could not be attributed to changes in vascular density; chronic hippocampal hypoperfusion and impaired performance in hippocampus-dependent memory task may predict a later increase in seizure susceptibility [43].

The adult central nervous system (CNS) vasculature is extremely stable under physiological conditions, but is activated after injury [44]. Adult vascular remodeling includes angiogenesis by mature endothelial cells (that is, the formation of new capillaries from preexisting vessels) and vasculogenesis (de novo formation of blood vessels when there are no preexisting ones) by endothelial progenitor cells (EPCs) [41]. There is a substantial increase in vasculogenesis following TBI [45]. Pharmacological agents such as erythropoietin (EPO), statins, bone marrow stromal cells (MSCs), and thymosin beta4 promote angiogenesis and improve functional recovery in rats after TBI. Neurorestorative treatments targeting vascular remodeling might be a novel and effective approach for TBI [41].

13.3 Rodent Models of TBI in Study of Cerebrovascular Responses

13.3.1 Commonly Used Rodent Models of TBI

In the past 3 decades, several animal models of TBI have been established using a variety of species including rodents, cats, pigs, dogs, and nonhuman primates. However, in the recent 20 years, the use of rodents dominated the TBI field and comprises today the most widely adopted approach for TBI research [46]. This is due to several reasons including the simplicity of carrying out the surgery, ease of utilizing larger groups, limited costs, and available transgenic animal for targeting specific genes or signaling pathways. Additionally, that rodent TBI models are commonly used because use of large animals are associated with controversial ethical issues, difficulty in providing surgical facilities and post-surgery care and lack of established behavioral testing [47]. Previous literature has provided excellent summaries of the TBI animal models [47–52]; more details will not be discussed in this chapter.

The establishment of animal models of TBI shares the ultimate goals of reproducing patterns of tissue damage observed in humans, reproducible and highly standardized to allow for the manipulation of individual variables, and to finally explore novel therapeutics for clinical translation. When confronted with the choice of the experimental model, it becomes clear that the ideal animal model does not exist. This limitation derives from the fact that most models mimic either focal or diffuse brain injury, whereas the clinical reality suggests that each patient has an individual form of TBI characterized by various combinations of focal and diffuse patterns of tissue damage. These heterogeneous TBI in humans are additionally complicated by the secondary insults, modalities of traumatic events, age, gender, and heterogeneity of medical treatments and preexisting conditions [49, 52]. There are three specific and most commonly used rodent models of TBI (see paper) [52], including FPI, CCI and closed-head weight drop impact (WDI) injury models [47, 50, 52].

13.3.2 Translational Advantages and Limitations of TBI Rodent Models

13.3.2.1 Fluid Percussion Injury Model

The FPI model of TBI is currently one of the most commonly used and well-characterized preclinical models of TBI by applying direct brain deformation [53, 54]. This model involves exposing intact dura by trephination craniotomy and performing impact of a rapid fluid bolus on dura surface by a falling pendulum.

FPI model can produce mostly focal damage to regions of cortex and hippocampus, but there is also a significant degree of axonal injury in the ipsilateral corpus callosum, capsula interna and externa, and thalamus making this model more clinically relevant [55]. The unilateral FPI may also induce contralateral damage [53, 54]. The immediate physiological responses to FPI comprise change in blood pressure [56], elevated ICP [57], reduced and fluctuating CBF [40], and compromised vascular autoregulation [58, 59]. The increased cerebral vascular resistance reflecting the enhanced vascular tone or vasospasm is also shown in FPI models, but usually in larger animal models such as porcine instead of rodent [60]. The patterns of BBB breakdown after TBI have also been well characterized in rodent model of FPI [61]. Alteration of BBB permeability and cerebral edema in rodents are frequently studied subjects in this model, with mechanisms of either inflammatory infiltration or structural loss of microvasculature [62, 63]. However, there are a few limitations of the FPI model, such as: (1) the pressure impact characteristics and the mechanism of tissue injury in FPI model are not very similar to those observed in clinical cases, making difficulties in comparison; (2) accurate biomechanical analyses in FPI models are limited; (3) FPI model may be technically challenging with differences in outcome among technicians and research centers even at similar setting of severity level [64], thus impairing the reliability and reproducibility of this model; (4) FPI model can cause relatively higher severity and morbidity due to disproportional involvement of brainstem and development of neurogenic pulmonary edema [47, 50, 52].

13.3.2.2 Controlled Cortical Impact Injury Model

CCI model is another widely used animal model of TBI. Its popularity in present days may partly come from the fact that the most common brain lesion following human TBI is cortical contusion, which is defined as a focal destruction of brain tissue with micro hemorrhage [65, 66]. CCI may have the best ability of controlling or adjusting severity of TBI among all the current models. Some models have an attached computer-based device to accurately control the impact velocity, impact depth, and the duration of compression, thus it offers potential advantages over FPI model. CCI is thought to produce mostly focal brain injury, including local cortical tissue loss, hippocampal and thalamic damage [67], and cortical spreading depression [68]. CCI-induced injury may also be widespread or even diffuse including diffuse axonal injury in the hippocampus and thalamus [69, 70]. As described earlier, the rat CCI model has been most commonly used the studying cerebrovascular responses after TBI, making this model ideal for studying focal cerebrovascular effects. In addition, CCI can produce specific, distinct, and long-lasting behavioral deficits; hence it is suitable for analyzing outcomes related to vascular events. CCI has some limitations including the following: (1) it relatively lacks power to mimic diffused secondary damage; (2) CCI causes a large amount of tissue loss in the cortex and even hippocampus, which is not comparable with the extent of brain injury in human survivors; (3) the craniotomy procedure performed in CCI can attenuate raised ICP by mimicking decompression surgery in clinical cases [71], this may compromise the modeling validity of CCI and leads to misinterpretation of research data; (4) CCI has less involvement of brainstem and mild diffused injury that could not produce long-lasting coma and carries much lower mortality than other models, which limited its clinical relevance [47, 50, 52].

13.3.2.3 Weight Drop Impact Acceleration Injury Model

WDI injury model is a classic model of TBI; it is probably the easiest, fastest, and cheapest one among all TBI models. The popularity of WDI led to many modifications to its original design to meet various requirements of TBI studies, but the core mechanism of impact device in this model never changed. As its name implies, WDI models employ a weight which is dropped through a guiding apparatus to impact either the closed skull (closed skull WDI), a metal plate attached to the vertex, or directly onto dura via a craniotomy (open skull WDI). For vascular responses, the closed skull WDI model has been widely used in both mice and rats [72, 73]; this is because the vast majority of clinical TBI cases is closed-head injury [74]. A typical representation of closed skull WDI is the Marmarou's weight drop model [75]. Marmarou's model has been used in the study of vascular response in a hemisphere scale or even bigger scope and a wide range of reduced CBF and elevated ICP have been exhibited due to impaired cerebral autoregulation during the acute phase following impact [76–78], but without marked blood pressure fluctuation [79]. WDI model shares most phenotypes of vascular responses with the FPI model, but its most important feature is the unique advantage of being exempt from craniotomy [80, 81]. The integrity of skull provides a better basis to simulate the pathophysiological condition of TBI. Closed WDI model also has limitations such as: (1) the biomechanics of impact in this model is not fully and strictly controlled, likely causing variation of severity among animals, especially compared with CCI; (2) WDI may have the highest mortality compared with FPI and CCI; (3) there is a possibility of "second hit" induced by the weight rebounding from the impact point, which may lead to failure of delivering precise load of injury [47, 50, 52].

13.4 Summary

The heterogeneity of TBI is considered one of the most significant barriers to finding effective therapeutic interventions. Among multiple confirmed pathological events after TBI, the vascular response is one of the most important pathophysiological components. It contributes to the increased sensitivity of the injured brain to secondary hypoxia and hypotension and the increased mortality and morbidity that occurs due to secondary insults after TBI in humans. A better understanding of the vascular responses after TBI may lead to new therapeutic strategy development. Rodent models of TBI are commonly used to study pathophysiology after

TBI. Translational advantages and limitations of each rodent TBI model should be considered for experimental investigations of cerebrovascular responses after TBI.

References

1. Saatman KE, Duhaime AC, Bullock R, Maas AI, Valadka A, Manley GT (2008) Classification of traumatic brain injury for targeted therapies. J Neurotrauma 25(7):719–738
2. Loane DJ, Faden AI (2010) Neuroprotection for traumatic brain injury: translational challenges and emerging therapeutic strategies. Trends Pharmacol Sci 31(12):596–604
3. Povlishock JT (2008) The classification of traumatic brain injury (tbi) for targeted therapies. J Neurotrauma 25(7):717–718
4. Adelson PD, Srinivas R, Chang Y, Bell M, Kochanek PM (2011) Cerebrovascular response in children following severe traumatic brain injury. Childs Nerv Syst 27(9):1465–1476
5. Lo EH (2008) Experimental models, neurovascular mechanisms and translational issues in stroke research. Br J Pharmacol 153(Suppl 1):S396–S405
6. Butcher I, Maas AI, Lu J, Marmarou A, Murray GD, Mushkudiani NA, McHugh GS, Steyerberg EW (2007) Prognostic value of admission blood pressure in traumatic brain injury: results from the impact study. J Neurotrauma 24(2):294–302
7. Rangel-Castilla L, Gasco J, Nauta HJ, Okonkwo DO, Robertson CS (2008) Cerebral pressure autoregulation in traumatic brain injury. Neurosurg Focus 25(4):E7
8. Bitner BR, Marcano DC, Berlin JM, Fabian RH, Cherian L, Culver JC, Dickinson ME, Robertson CS, Pautler RG, Kent TA, Tour JM (2012) Antioxidant carbon particles improve cerebrovascular dysfunction following traumatic brain injury. ACS Nano 6(9):8007–8014
9. Schmidt B, Czosnyka M, Raabe A, Yahya H, Schwarze JJ, Sackerer D, Sander D, Klingelhofer J (2003) Adaptive noninvasive assessment of intracranial pressure and cerebral autoregulation. Stroke 34(1):84–89
10. Bouma GJ, Muizelaar JP, Stringer WA, Choi SC, Fatouros P, Young HF (1992) Ultra-early evaluation of regional cerebral blood flow in severely head-injured patients using xenon-enhanced computerized tomography. J Neurosurg 77(3):360–368
11. Engel DC, Mies G, Terpolilli NA, Trabold R, Loch A, De Zeeuw CI, Weber JT, Maas AI, Plesnila N (2008) Changes of cerebral blood flow during the secondary expansion of a cortical contusion assessed by 14c-iodoantipyrine autoradiography in mice using a non-invasive protocol. J Neurotrauma 25(7):739–753
12. Menon DK (2003) Procrustes, the traumatic penumbra, and perfusion pressure targets in closed head injury. Anesthesiology 98(4):805–807
13. Yamakami I, McIntosh TK (1989) Effects of traumatic brain injury on regional cerebral blood flow in rats as measured with radiolabeled microspheres. J Cereb Blood Flow Metab 9 (1):117–124
14. Alessandri B, Schwandt E, Kamada Y, Nagata M, Heimann A, Kempski O (2012) The neuroprotective effect of lactate is not due to improved glutamate uptake after controlled cortical impact in rats. J Neurotrauma 29(12):2181–2191
15. Hendrich KS, Kochanek PM, Williams DS, Schiding JK, Marion DW, Ho C (1999) Early perfusion after controlled cortical impact in rats: quantification by arterial spin-labeled mri and the influence of spin–lattice relaxation time heterogeneity. Magn Reson Med 42(4):673–681
16. Shen Y, Kou Z, Kreipke CW, Petrov T, Hu J, Haacke EM (2007) In vivo measurement of tissue damage, oxygen saturation changes and blood flow changes after experimental traumatic brain injury in rats using susceptibility weighted imaging. Magn Reson Imaging 25 (2):219–227

17. Golding EM, Robertson CS, Bryan RM Jr (1999) The consequences of traumatic brain injury on cerebral blood flow and autoregulation: a review. Clin Exp Hypertens 21(4):299–332
18. Prat R, Markiv V, Dujovny M, Misra M (1998) Failure of cerebral autoregulation in an experimental diffuse brain injury model. Acta Neurochir Suppl 71:123–126
19. Forbes ML, Hendrich KS, Kochanek PM, Williams DS, Schiding JK, Wisniewski SR, Kelsey SF, DeKosky ST, Graham SH, Marion DW, Ho C (1997) Assessment of cerebral blood flow and co2 reactivity after controlled cortical impact by perfusion magnetic resonance imaging using arterial spin-labeling in rats. J Cereb Blood Flow Metab 17(8):865–874
20. Engelborghs K, Haseldonckx M, Van Reempts J, Van Rossem K, Wouters L, Borgers M, Verlooy J (2000) Impaired autoregulation of cerebral blood flow in an experimental model of traumatic brain injury. J Neurotrauma 17(8):667–677
21. Nawashiro H, Shima K, Chigasaki H (1995) Immediate cerebrovascular responses to closed head injury in the rat. J Neurotrauma 12(2):189–197
22. Ter Minassian A, Dube L, Guilleux AM, Wehrmann N, Ursino M, Beydon L (2002) Changes in intracranial pressure and cerebral autoregulation in patients with severe traumatic brain injury. Crit Care Med 30(7):1616–1622
23. Ginsberg MD, Zhao W, Alonso OF, Loor-Estades JY, Dietrich WD, Busto R (1997) Uncoupling of local cerebral glucose metabolism and blood flow after acute fluid-percussion injury in rats. Am J Physiol 272(6 Pt 2):H2859–H2868
24. Shahlaie K, Keachie K, Hutchins IM, Rudisill N, Madden LK, Smith KA, Ko KA, Latchaw RE, Muizelaar JP (2011) Risk factors for posttraumatic vasospasm. J Neurosurg 115(3):602–611
25. Dore-Duffy P, Wang S, Mehedi A, Katyshev V, Cleary K, Tapper A, Reynolds C, Ding Y, Zhan P, Rafols J, Kreipke CW (2011) Pericyte-mediated vasoconstriction underlies tbi-induced hypoperfusion. Neurol Res 33(2):176–186
26. Bouma GJ, Muizelaar JP (1992) Cerebral blood flow, cerebral blood volume, and cerebrovascular reactivity after severe head injury. J Neurotrauma 9(Suppl 1):S333–S348
27. Bowles AP, Pasierb L, Simunich T, Updyke M (2012) Implications of neurophysiological parameters in persons with severe brain injury with respect to improved patient outcomes: a retrospective review. Brain Inj 26(12):1415–1424
28. Donkin JJ, Vink R (2010) Mechanisms of cerebral edema in traumatic brain injury: therapeutic developments. Curr Opin Neurol 23(3):293–299
29. Chodobski A, Zink BJ, Szmydynger-Chodobska J (2011) Blood–brain barrier pathophysiology in traumatic brain injury. Transl Stroke Res 2(4):492–516
30. Shlosberg D, Benifla M, Kaufer D, Friedman A (2010) Blood–brain barrier breakdown as a therapeutic target in traumatic brain injury. Nat Rev Neurol 6(7):393–403
31. Ziebell JM, Morganti-Kossmann MC (2010) Involvement of pro- and anti-inflammatory cytokines and chemokines in the pathophysiology of traumatic brain injury. Neurotherapeutics 7(1):22–30
32. Stein SC, Graham DI, Chen XH, Smith DH (2004) Association between intravascular microthrombosis and cerebral ischemia in traumatic brain injury. Neurosurgery 54 (3):687–691, discussion 691
33. Schwarzmaier SM, Kim SW, Trabold R, Plesnila N (2010) Temporal profile of thrombogenesis in the cerebral microcirculation after traumatic brain injury in mice. J Neurotrauma 27(1):121–130
34. Marmarou A (2007) A review of progress in understanding the pathophysiology and treatment of brain edema. Neurosurg Focus 22(5):E1
35. Unterberg AW, Stover J, Kress B, Kiening KL (2004) Edema and brain trauma. Neuroscience 129(4):1021–1029
36. Abdul-Muneer PM, Schuetz H, Wang F, Skotak M, Jones J, Gorantla S, Zimmerman MC, Chandra N, Haorah J (2013) Induction of oxidative and nitrosative damage leads to cerebrovascular inflammation in an animal model of mild traumatic brain injury induced by primary blast. Free Radic Biol Med 60:282–291

37. Carlos TM, Clark RS, Franicola-Higgins D, Schiding JK, Kochanek PM (1997) Expression of endothelial adhesion molecules and recruitment of neutrophils after traumatic brain injury in rats. J Leukoc Biol 61(3):279–285
38. Di Giovanni S, Movsesyan V, Ahmed F, Cernak I, Schinelli S, Stoica B, Faden AI (2005) Cell cycle inhibition provides neuroprotection and reduces glial proliferation and scar formation after traumatic brain injury. Proc Natl Acad Sci U S A 102(23):8333–8338
39. Maeda T, Katayama Y, Kawamata T, Aoyama N, Mori T (1997) Hemodynamic depression and microthrombosis in the peripheral areas of cortical contusion in the rat: role of platelet activating factor. Acta Neurochir Suppl 70:102–105
40. Hayward NM, Tuunanen PI, Immonen R, Ndode-Ekane XE, Pitkanen A, Grohn O (2011) Magnetic resonance imaging of regional hemodynamic and cerebrovascular recovery after lateral fluid-percussion brain injury in rats. J Cereb Blood Flow Metab 31(1):166–177
41. Xiong Y, Mahmood A, Chopp M (2010) Angiogenesis, neurogenesis and brain recovery of function following injury. Curr Opin Investig Drugs 11(3):298–308
42. Kochanek PM, Hendrich KS, Dixon CE, Schiding JK, Williams DS, Ho C (2002) Cerebral blood flow at one year after controlled cortical impact in rats: assessment by magnetic resonance imaging. J Neurotrauma 19(9):1029–1037
43. Hayward NM, Immonen R, Tuunanen PI, Ndode-Ekane XE, Grohn O, Pitkanen A (2010) Association of chronic vascular changes with functional outcome after traumatic brain injury in rats. J Neurotrauma 27(12):2203–2219
44. Greenberg DA, Jin K (2005) From angiogenesis to neuropathology. Nature 438 (7070):954–959
45. Morgan R, Kreipke CW, Roberts G, Bagchi M, Rafols JA (2007) Neovascularization following traumatic brain injury: possible evidence for both angiogenesis and vasculogenesis. Neurol Res 29(4):375–381
46. Cernak I (2005) Animal models of head trauma. NeuroRx 2(3):410–422
47. Morganti-Kossmann MC, Yan E, Bye N (2010) Animal models of traumatic brain injury: is there an optimal model to reproduce human brain injury in the laboratory? Injury 41(Suppl 1): S10–S13
48. Duhaime AC (2006) Large animal models of traumatic injury to the immature brain. Dev Neurosci 28(4–5):380–387
49. Finnie J (2001) Animal models of traumatic brain injury: a review. Aust Vet J 79(9):628–633
50. O'Connor WT, Smyth A, Gilchrist MD (2011) Animal models of traumatic brain injury: a critical evaluation. Pharmacol Ther 130(2):106–113
51. Werner C, Engelhard K (2007) Pathophysiology of traumatic brain injury. Br J Anaesth 99 (1):4–9
52. Xiong Y, Mahmood A, Chopp M (2013) Animal models of traumatic brain injury. Nat Rev Neurosci 14(2):128–142
53. Thompson HJ, Lifshitz J, Marklund N, Grady MS, Graham DI, Hovda DA, McIntosh TK (2005) Lateral fluid percussion brain injury: a 15-year review and evaluation. J Neurotrauma 22(1):42–75
54. Kabadi SV, Hilton GD, Stoica BA, Zapple DN, Faden AI (2010) Fluid-percussion-induced traumatic brain injury model in rats. Nat Protoc 5(9):1552–1563
55. Graham DI, Raghupathi R, Saatman KE, Meaney D, McIntosh TK (2000) Tissue tears in the white matter after lateral fluid percussion brain injury in the rat: relevance to human brain injury. Acta Neuropathol 99(2):117–124
56. Dixon CE, Lighthall JW, Anderson TE (1988) Physiologic, histopathologic, and cineradiographic characterization of a new fluid-percussion model of experimental brain injury in the rat. J Neurotrauma 5(2):91–104
57. Prins ML, Lee SM, Cheng CL, Becker DP, Hovda DA (1996) Fluid percussion brain injury in the developing and adult rat: a comparative study of mortality, morphology, intracranial pressure and mean arterial blood pressure. Brain Res Dev Brain Res 95(2):272–282

58. Fujita M, Wei EP, Povlishock JT (2012) Effects of hypothermia on cerebral autoregulatory vascular responses in two rodent models of traumatic brain injury. J Neurotrauma 29 (7):1491–1498
59. Armstead WM, Kiessling JW, Kofke WA, Vavilala MS (2010) Impaired cerebral blood flow autoregulation during posttraumatic arterial hypotension after fluid percussion brain injury is prevented by phenylephrine in female but exacerbated in male piglets by extracellular signal-related kinase mitogen-activated protein kinase upregulation. Crit Care Med 38(9):1868–1874
60. Armstead WM, Kiessling JW, Cines DB, Higazi AA (2011) Glucagon protects against impaired nmda-mediated cerebrovasodilation and cerebral autoregulation during hypotension after brain injury by activating camp protein kinase a and inhibiting upregulation of tpa. J Neurotrauma 28(3):451–457
61. Schmidt RH, Grady MS (1993) Regional patterns of blood–brain barrier breakdown following central and lateral fluid percussion injury in rodents. J Neurotrauma 10(4):415–430
62. Beziaud T, Ru Chen X, El Shafey N, Frechou M, Teng F, Palmier B, Beray-Berthat V, Soustrat M, Margaill I, Plotkine M, Marchand-Leroux C et al (2011) Simvastatin in traumatic brain injury: effect on brain edema mechanisms. Crit Care Med 39(10):2300–2307
63. Lotocki G, de Rivero Vaccari JP, Perez ER, Sanchez-Molano J, Furones-Alonso O, Bramlett HM, Dietrich WD (2009) Alterations in blood–brain barrier permeability to large and small molecules and leukocyte accumulation after traumatic brain injury: effects of post-traumatic hypothermia. J Neurotrauma 26(7):1123–1134
64. Floyd CL, Golden KM, Black RT, Hamm RJ, Lyeth BG (2002) Craniectomy position affects morris water maze performance and hippocampal cell loss after parasagittal fluid percussion. J Neurotrauma 19(3):303–316
65. Smith DH, Soares HD, Pierce JS, Perlman KG, Saatman KE, Meaney DF, Dixon CE, McIntosh TK (1995) A model of parasagittal controlled cortical impact in the mouse: cognitive and histopathologic effects. J Neurotrauma 12(2):169–178
66. Dixon CE, Kochanek PM, Yan HQ, Schiding JK, Griffith RG, Baum E, Marion DW, DeKosky ST (1999) One-year study of spatial memory performance, brain morphology, and cholinergic markers after moderate controlled cortical impact in rats. J Neurotrauma 16(2):109–122
67. Saatman KE, Feeko KJ, Pape RL, Raghupathi R (2006) Differential behavioral and histopathological responses to graded cortical impact injury in mice. J Neurotrauma 23(8):1241–1253
68. von Baumgarten L, Trabold R, Thal S, Back T, Plesnila N (2008) Role of cortical spreading depressions for secondary brain damage after traumatic brain injury in mice. J Cereb Blood Flow Metab 28(7):1353–1360
69. Hall ED, Bryant YD, Cho W, Sullivan PG (2008) Evolution of post-traumatic neurodegeneration after controlled cortical impact traumatic brain injury in mice and rats as assessed by the de olmos silver and fluorojade staining methods. J Neurotrauma 25(3):235–247
70. Soblosky JS, Matthews MA, Davidson JF, Tabor SL, Carey ME (1996) Traumatic brain injury of the forelimb and hindlimb sensorimotor areas in the rat: physiological, histological and behavioral correlates. Behav Brain Res 79(1–2):79–92
71. Zweckberger K, Stoffel M, Baethmann A, Plesnila N (2003) Effect of decompression craniotomy on increase of contusion volume and functional outcome after controlled cortical impact in mice. J Neurotrauma 20(12):1307–1314
72. Chen Y, Constantini S, Trembovler V, Weinstock M, Shohami E (1996) An experimental model of closed head injury in mice: pathophysiology, histopathology, and cognitive deficits. J Neurotrauma 13(10):557–568
73. Henninger N, Dutzmann S, Sicard KM, Kollmar R, Bardutzky J, Schwab S (2005) Impaired spatial learning in a novel rat model of mild cerebral concussion injury. Expe Neurol 195 (2):447–457
74. Flierl MA, Stahel PF, Beauchamp KM, Morgan SJ, Smith WR, Shohami E (2009) Mouse closed head injury model induced by a weight-drop device. Nat Protoc 4(9):1328–1337

75. Marmarou A, Foda MA, van den Brink W, Campbell J, Kita H, Demetriadou K (1994) A new model of diffuse brain injury in rats. Part i: pathophysiology and biomechanics. J Neurosurg 80 (2):291–300
76. Roof RL, Hall ED (2000) Estrogen-related gender difference in survival rate and cortical blood flow after impact-acceleration head injury in rats. J Neurotrauma 17(12):1155–1169
77. Engelborghs K, Verlooy J, Van Reempts J, Van Deuren B, Van de Ven M, Borgers M (1998) Temporal changes in intracranial pressure in a modified experimental model of closed head injury. J Neurosurg 89(5):796–806
78. Prat R, Markiv V, Dujovny M, Misra M (1998) Failure of cerebral autoregulation in an experimental diffuse brain injury model. Acta Neurochir Suppl 71:123–126
79. Nawashiro H, Shima K, Chigasaki H (1995) Immediate cerebrovascular responses to closed head injury in the rat. J Neurotrauma 12(2):189–197
80. Heath DL, Vink R (1995) Impact acceleration-induced severe diffuse axonal injury in rats: characterization of phosphate metabolism and neurologic outcome. J Neurotrauma 12 (6):1027–1034
81. Vink R, O'Connor CA, Nimmo AJ, Heath DL (2003) Magnesium attenuates persistent functional deficits following diffuse traumatic brain injury in rats. Neurosci Lett 336(1):41–44

Chapter 14
SAH Models: Review, New Modification, and Prospective

Sheng Chen, Damon Klebe, Alexander Vakhmyanin, Mutsumi Fujii, and John H. Zhang

Abstract Subarachnoid hemorrhage (SAH) is a devastating type of hemorrhagic stroke. It is characterized by spontaneous or traumatic bleeding in the subarachnoid space and is associated with a high rate of morbidity and mortality. A reproducible animal model of SAH that mimics the acute and delayed brain injury history after SAH will be an invaluable tool for exploring the underlying mechanisms of SAH-induced brain injury and evaluating potential therapeutic interventions. At present, a number of models have been developed, mainly the double injection model and the endovascular puncture model. While different species have been studied, rodents have become the most popular and widely utilized animal subjects. In this summary, we will explore in detail the various models and animal species. We will also introduce the emerging modified model, which was recently developed within the past 5 years, and discuss the prospective study.

14.1 Introduction

When healthy persons are suddenly seized with thunder pains in the head, quickly laid down speechless, and breathing with stridor, they die in 7 days. It was the earliest known description of subarachnoid hemorrhage (SAH) from Hippocrates. A World Health Organization report found the annual incidence of aneurismal SAH ranges from 2.0 to 22.5 cases per 100,000 population in age-adjusted adults [1]. Although SAH accounts for only 5 % of all strokes, its burden on society is relevant due to the young age at which it occurs, high mortality and disability, and poor outcomes [2]. Despite obliterating the offending aneurysm and removing the risk of rebleeding, the median mortality rate of SAH in epidemiological studies from the United States is 32 %, although this statistics does not fully account for

J.H. Zhang (✉)
Department of Physiology and Pharmacology, School of Medicine, Loma Linda University, Loma Linda, CA 92354, USA
e-mail: johnzhang3910@yahoo.com

E.H. Lo et al. (eds.), *Vascular Mechanisms in CNS Trauma*, Springer Series in Translational Stroke Research 5, DOI 10.1007/978-1-4614-8690-9_14,

prehospital deaths from SAH [3]. The hope of improving SAH patient outcomes has not been realized. Hence, further studies on the pathophysiology of SAH remain highly valued.

Recently, the theory that cerebral vasospasm is the only cause of brain injury in SAH patients is being increasingly questioned, and hypotheses of other mechanisms contributing to early or delayed brain damage are being discussed [4]. Delayed vasospasm was regarded as the single most important, treatable cause of mortality and morbidity after SAH. However, disappointing results from an endothelin receptor antagonist treatment, clazosentan, were observed in CONSCIOUS-2 trials [5]. Clazosentan succeeded in steering patients out of vasospasm in a randomized, double-blind, placebo-controlled, phase III trial (CONSCIOUS-3), but failed to reduce mortality [6]. Currently, more attention has been focusing on global cerebral injury within 72 h post-ictus, which was termed early brain injury (EBI) [7]. Multifactorial pathophysiological processes are considered to be the important triggers in SAH-evoked EBI, which involve acute elevated intracranial pressure (ICP), reduction in global cerebral blood flow (CBF), brain edema, oxidative stress, inflammation, and apoptosis [8]. Hence, it is very important to design and utilize a suitable model according to the intended research purpose.

Ideally, studies on the pathogenesis and treatment of SAH should be using human cerebral arteries. However, using live human vessels is clearly not possible in in vivo experimental studies. In addition, only a little information can be obtained from postmortem evaluation of arteries taken from humans that died from SAH [9]. Moreover, there are no naturally occurring animal models of SAH. Hence, animal models of SAH in various species have been developed for this purpose. The first SAH model was invented in 1928. They infused blood into the subarachnoid space of a dog to simulate the lesion occurring in humans, where a small amount of blood escapes into the subarachnoid space. After its introduction, many techniques have been used to deliberately produce SAH in animals that recapitulate this vexing problem in humans [10]. So far, several animal species and SAH models have been developed to better understand SAH pathophysiology, including monkeys, baboons, pigs, rabbits, dogs, cats, rats, and mice [11]. Among those species, the rat is excellent for mimicking SAH because it is relatively inexpensive, easy to manipulate in a laboratory setting, and has been extensively studied. However, no animal model perfectly mimics the phenomenon of EBI and cerebral vasospasm in humans after SAH.

In this review, we will describe both the existing and emerging models and compare their advantages and disadvantages in order to help you choose the most suitable models for your aim.

14.2 In Vivo Models

In vivo experimental models have been used to investigate diverse aspects of SAH, including its natural history, pathophysiology, diagnosis, and treatment. More than 50 different experimental models have been identified with the aim of reproducing the clinical sequelae after SAH. These models used two major techniques to simulate SAH: (1) an artery was punctured, allowing blood to escape and collect around the artery and neighboring vessels; (2) an artery was surgically exposed and autologous blood, obtained from another site, was placed around the artery.

14.2.1 Vessel Dissection and Puncture Model

SAH is often elicited by aneurysmal rupture in humans. To closely mimic the human situation after SAH, many different bleeding techniques have been proposed and applied in several species, including mice, rats, rabbits, dogs, primates, cats, pigs, and goats [12].

The vessel avulsion in dogs was the first model used to characterize the biphasic, vasospastic response of large cerebral arteries after SAH [13]. The authors left a suture in the anterior cerebral artery (ACA) by a 4–0 silk loose ligature. Then pulling the suture in the external ear canal produced the ACA rupture. More importantly, they observed a biphasic phenomenon of SAH-induced vasospasm using this experimental technique. The acute phase occurred within minutes of the rupture and lasted no more than 1 h. The chronic phase occurred 3–24 h after the catastrophic hemorrhage and lasted for days. One clear advantage of this model is the biphasic phenomenon occurred repeatedly within the animals as they were awake, which mimics the clinical situation of sudden unconsciousness seen in humans. However, neurological function in this model was highly variable. A similar technique was performed to observe a biphasic vasospasm by Nagai and colleagues [14]. Willis circle was explored after a left subtemporal craniotomy. A threaded needle (~0.33 mm) was left in the posterior communicating artery (PCoA) to induce SAH. Their results suggested rapidly increased ICP seemed to be a crucial factor in producing diffuse early spasm. Asano et al. also performed the internal carotid artery (ICA) avulsion model by withdrawing a needle that was previously inserted into the ICA through a small craniectomy in the lateral base of the skull in dogs [15]. They observed the biphasic ICP pattern, which resulted in reactive hyperemia after the first ICP peak and failure to recover adequate CBF and EEG after the second, prolonged ICP elevation. Thus, they suggest an important role for the no-reflow phenomenon in the acute stage of SAH.

With the development of endoscopic methods, the preferred species in SAH studies changed from dog to murine. Murine became the most commonly used animals for SAH induction. Murine are relatively cheap and convenient for housing care and surgery. They also provoke less opposition from animal rights groups.

Several early models that focused on vasospasm involved basilar artery puncture through the midline portion of clivus. In 1979, Barry et al. first tried to induce SAH by developing a lesion around the wall of the basilar artery with a stereotactically controlled tungsten needle, thereby emulating the natural occurrence of SAH. They controlled bleeding under an operating microscope, which is helpful in making a consistent and reproducible model [16]. Most importantly, this study determined the suitability of the rat for SAH models, although there were some deficiencies. The basilar artery puncture method of experimental SAH in rats holds promise for studying the chronic changes in CBF in the telencephalon, but it failed to produce chronic cerebral vasospasm and delayed cerebral ischemia [17]. It is widely accepted that the models utilizing transclival puncture of the basilar artery produce vasospasm 24 and 48 h after surgery and return to baseline after 72 h. However, the model presented several limitations, including the need for craniotomy and arachnoidal dissection or surgical placement of an infusion catheter, small and slow elevations of ICP, and limited blood distribution. To overcome these disadvantages, one new rat SAH model was developed, the endovascular suture technique [18, 19]. The endovascular perforation results in a very rapid and reproducible elevation of ICP, which is detectable from the beginning to 72 h after SAH [20, 21]. The endovascular perforation model exhibits pathological characteristics, such as rupturing of the main cerebral artery with arterial bleeding, and the model produces more profound pathophysiological and histological changes to the brain and cerebral vessels. The surgical procedure involves perforating the ICA near its intracranial bifurcation between the ACA and the middle cerebral artery (MCA) by means of a sharp ended 3–0 or 4–0 monofilament suture inserted through the stump of the external carotid artery (ECA). The suture passes through the common carotid artery (CCA) bifurcation and into the ICA. The filament is advanced distally into the intracranial ICA until resistance is felt and then pushed 2–5 mm farther to perforate the artery. Subsequently, the suture is withdrawn, producing SAH. The endovascular perforation model is able to represent clinical situations because it mimics intracranial aneurysmal rupture with a considerably high mortality ranging between 30 and 50 % after 24 h [22]. The best advantage of this puncture model is it does not require a craniotomy. Recently, the puncture SAH model is believed to better mimic the pathophysiological events after aneurysm rupture with a mortality rate similar to the clinical catastrophe seen in humans (40–50 %). Hence, of all the vessel puncture or avulsion models, the endovascular perforation model is very popular and most widely applied, especially in translational studies, although uncontrollable blood distribution is a well-known weakness of this model. In this approach, the types and sizes of the filament used to perforate a cerebral artery seems to be correlated with the amount of extravasated blood [23–25]. Some researchers tried to control the bleeding by keeping the suture in the artery, but they failed because ischemic development complicated the model.

Numerous transgenic mice have been generated, which are strong research tools. The transgenic mice are able to isolate a single gene product by either nullifying or augmenting gene expression. The transgenic models not only have the potential to provide insights into disease mechanisms but also are useful tools for testing new treatments for SAH. The inherent technical difficulty, however, in establishing a simple, reliable, and reproducible model in mice is the small size of the animal. With the development of microsurgical technology, the puncture technique was also adapted to mice. A mouse SAH model was produced by endovascular arterial rupture in the bifurcation of the ACA and MCA [26]. Recently, the therapeutic effect of some drugs against EBI was examined based on the mouse endovascular puncture model [27, 28]. Overcoming the apparent technical impediments in establishing a mouse model of SAH would be considerably valuable. Increased efforts should been directed toward developing a simple, reliable, and reproducible mouse model of SAH.

Non-human primates are very close in genome, anatomy, and physiology to humans, thus the models of SAH in non-human primates are believed to be the best candidates for replicating clinical SAH. The puncture monkey model through a small anterior craniotomy was first reported in 1968 [29]. A penetrating insult to the wall of a major cerebral vessel is sufficient to induce long-standing constriction in monkeys. Puncturing the ICA in primates can produce acute and delayed vasospasm [30]. However, a craniotomy is needed and delayed vasospasm was found in only 65 % of cases. More importantly, non-human primate models are usually time-consuming and demand complex surgical manipulation, such as anesthesia, angiography, and craniotomy.

Taken together, the endovascular perforation model (a non-craniotomy model) is believed to closely mimic the acute pathophysiological changes of an aneurysmal rupture in humans [25, 31] and was, therefore, selected for the investigation of EBI. However, the lack of control over the hemorrhage volume, resulting in a high morbidity and mortality rate, significantly impacts its use for therapeutic studies.

14.2.2 Blood Injection Model

The blood injection model, which directly injects blood into the subarachnoid space, is another widely used technique for inducing experimental SAH. This technique is able to elicit early and delayed vasospasm in a variety of species, although its presence depends on the site of injection.

In 1961, autologous blood was injected into the chiasmatic cistern using a small needle, which was the first to report a "one-hemorrhage" SAH experimental model [32]. It appears some animals had severe injuries with SAH symptoms. After 1 year, another new method was introduced to inject blood into the chiasmatic cistern using a needle passed under the zygomatic arch and directed toward the optic foramen. This and subsequent studies using the same model in dogs were used to analyze the effects of dehydrating agents in increased ICP [33, 34]. Also, primate models of SAH utilizing blood injection into the chiasmatic cistern through a transfrontal approach generally have high success rates of inducing cerebral vasospasm and producing high mortality rates, although scarce data have been reported on mortality and neurological scores [35]. Moreover, the prechiasmatic injection model in experimental rats was a suitable candidate to study the acute phase after SAH [36], because it produced a sudden increase in ICP to MABP levels, a significant decrease in CBF, and substantial neuronal death, even when observing the distribution of blood clots in the basal cisterns. Additionally, more than 90 % of SAH cases resulted from rupture of an aneurysm in the anterior circulation [37]. In some studies, however, the biphasic pattern of vasospasm was present only in small subgroups of animals after injecting fresh blood into the chiasmatic cistern through the optic canal following orbital exenteration, and the neurological deficits were none or only lasting a few hours [38].

To overcome the weakness of the chiasmatic cistern infusion model, a pioneering model was performed by means of blood injection into the cisterna magna in a small group of cynomolgus monkeys [39]. In this study, vasospasm seemed more pronounced with whole blood compared with serotonin, which corroborated with Echlin's opinion that "other factors, not serotonin, in blood cause vasospasm" [40]. The study also demonstrated that introduction of blood into the cisterna magna induced at least as much or more constriction in the MCA than other intracranial arteries. Another line of investigation utilized the injection of autologous erythrocyte hemolysate into the cortex of mice. The arteries in the subarachnoid space were not directly assessed for vasospasm in this study. However, it is noteworthy that the hemolysate did not reach the basal cisterns and the acute mortality rate was 16 % [41]. In addition, vein blood obtained from a dog leg was injected into the cisterna magna, the subarachnoid space over the cerebral hemispheres, and, occasionally, the cerebral ventricles. Dogs with injected blood exhibited an altered clinical course as SAH patients, and pathological changes were observed in stained sections of their brains. The single blood injection model had observed maximal narrowing on days 2–4, but the double injection model had observed maximal narrowing on days 7–10 [12]. To improve the success rate, repeated injections of blood into the basal cisterns in the head-down position after injection was stressed to induce cerebral vasospasm, worsen neurological status, and create dysfunctional autoregulation [42]. In order to increase the precision of blood injections, stereotactic methods and catheter implantations were encouraged [43]. The next major advancement in the development of an SAH model was created by Varsos and collaborators [44]. They performed double-injections of autologous blood into the cisterna magna 48 h apart via small-gauge needles in dogs. The "two-hemorrhage" model was certified for better simulating vasospasm and has been used most frequently in inducing vasospasm after SAH. This double-injection model resulted in cerebral vasospasm development in 100 % of animals. This slight modification is a great step in the field of vasospasm study. This model also demonstrated an acceptable mortality [45]. The simplicity makes it a widely applied technique. However, there are some considerable drawbacks: (1) after injection of blood into the cisterna magna, neurological functioning usually remains intact, and the mortality rate is surprisingly lower than in humans, (2) because leakage of a considerable amount of blood during injection and after removal of the needle leads to an insufficient quantity of subarachnoid blood around the circle of Willis, even with the use of hemostyptic materials, the possibility of excluding animals is high, (3) this model requires two large surgeries to expose the femoral artery and the atlanto-occipital membrane. The second blood injection 24 or 48 h after the initial injection, which is obscured by the discolored and nontransparent dura mater, increases the risk of severe injury to the brain stem or cerebellum [46].

In conclusion, the double hemorrhage model represents a well-established experimental model frequently employed for delayed cerebral vasospasm, mainly defined by proximal vessel narrowing and neurological deterioration observed in the clinic. The great advantage of the SAH injection model is it allows for close control of subarachnoid blood volumes [47] and has a low morbidity and mortality rate, which are criteria for an ideal animal model [45].

14.3 In Vitro Model

Generally, there are two kinds of compounds that could produce delayed vasospasm after SAH: (1) some compounds or their metabolites found in blood, such as hemoglobin (Hb) and bilirubin oxidation products (BOXes), and (2) compounds regulated or induced from production of blood around blood vessels directly or indirectly from blood itself, blood vessels, or brain, such as endothelin or nitric oxide [48]. Neuron, astrocyte, and vascular smooth muscle cells were grown in culture and exposed to those compounds to mimic an in vitro model of SAH [49]. Vascular smooth muscle was predominantly used from CSF in SAH patients to force generation and reserve oxidative stress for vasospasm, but it is histologically, metabolically, and functionally abnormal in vessel walls. However, there are no good alternatives for in vitro SAH modeling that equal in vivo models.

14.4 Emerging Modified Models in the Last 5 Years

A CCA-prechiasmal cistern shunt model was established in which the SAH severity could be controlled by changing the bleeding time [50]. However, this model has uncontrolled pressure, is time-consuming, and demands delicate surgical manipulation. Fortunately, a closed-cranium ICP-controlled rabbit SAH model was performed using an extra-intracranial blood shunt from the subclavian artery into the cerebromedullar cistern [51, 52]. The blood shunt rabbit SAH model can mimic both EBI and vasospasm, eliciting acute physiological derangements, provoking marked and consistent early damage to the brain 24 h post-surgery, and triggering a high degree of delayed cerebral vasospasm. These findings make this model a valid tool for investigating pathophysiological mechanisms and novel treatment modalities for SAH. An advantage of this model is the consistent ICP, which is a desirable characteristic in an experimental SAH model. The initial event of increased ICP is considered to be crucial for early physiological derangements that are responsible for acute brain injury in aneurysmal SAH.

Another modification rat SAH model was presented by percutaneous injection of autologous, non-heparinized blood into the intracisternal space [53]. Once anesthetized, rats were fixed in a prone position in a stereotaxic frame. After finding the projection of the occipital bone, the needle is advanced toward the foramen magnum until it punctures through the atlanto-occipital membrane and obtains cerebrospinal fluid. Autologous blood was withdrawn from the tail and injected intracisternally. The average time between obtaining the blood and the start of the injection was very short, so the model did not require treating the blood with heparin to avoid coagulation at the time of injection. It could reduce the effect of Heparin in SAH models, since Heparin can reduce neuroinflammation and neuronal apoptosis [54]. This model is simple to perform, minimally invasive, quick, reproducible, and has low mortality, which makes it suitable for future studies on vasospasm-related delayed SAH complications.

The classical double injection model requires twice large surgeries to expose both the femoral artery and the atlanto-occipital membrane, which induces severe damage. To overcome these weaknesses, Dusick et al. set out to refine a minimally invasive modification that prevents confounding effects of surgical procedures, reduces leakage of blood from the subarachnoid space, and minimizes risk of infection [55]. The rat was placed prone with the head downward to 90° in a stereotactic frame, so that the cisterna magna was aligned with the intermeatal line. An angiocatheter is advanced in a vertical trajectory level with the external auditory canals. The needle is slowly advanced to puncture the atlanto-occipital membrane and observed for CSF. A syringe withdraws a small amount of CSF and the blood is injected into the subarachnoid space. After 24 h, the procedure was repeated. This model was feasible and produced neurological deficits as previous double-injection models. It consequently appears to be an excellent platform for future laboratories to study neuroprotective treatments for delayed vasospasm and neurological dysfunction following SAH.

A modified SAH model was presented in mice [56, 57]. Briefly, under general anesthesia, animals were flipped into the prone position. A mechanical support was placed under the clavicles to position the head flexed at a 30° angle. Using a surgical microscope, the posterior cervical muscles were initially dissected through a suboccipital midline skin incision. The transparent atlanto-occipital membrane and an underlying intracisternal vein were exposed. Blood was allowed to transect from the vein to the subarachnoid space through a pair of microjewelers. This procedure has the advantage of requiring surgery at only one site. The model not only inherited the advantages of currently used mouse models of vasospasm but also streamlined the procedure without losing reliable vasospasm development.

To circumvent the disadvantages of directly injecting blood into the cisterna magna, especially those related to the blind nature of the second injection, this study, based on the method from Solomon et al. [58], employed the injection of autologous blood into the cisterna magna via a thin catheter, which was carefully implanted through a small burr hole in the parieto-occipital. The catheter was left in position with a tight-fitting, flexible inner wire, allowing for SAH induction that is several times safer. Autologous blood was aspirated from a tube in the left femoral artery. Consequently, one strength of this model is its low mortality rate, only 1.5 %. Furthermore, the authors observed a lower blood volume resulted in a lower mortality rate. This low-mortality, double hemorrhage model seemed to be suitable for studying both acute and chronic injury after SAH [46].

14.5 Future Direction of the SAH Model

A suitable animal model to optimize the pathophysiology of SAH in humans remains a challenge. The limitations of recent models must be carefully considered. We propose some aspects to improve the SAH model for successful translation of pathophysiological concepts and therapeutics from bench to bedside.

Previously, neurovascular unit was restricted to endothelial cells, neurons and glia within millimeters of the cerebral capillary microcirculation. We now highlighted the roles of vascular smooth muscle, endothelial cells, and perivascular innervation of cerebral arteries in the neurovascular unit to prevent and progress the stroke. We also proposed that the vascular neural network should be considered the key target for therapeutic intervention after cerebrovascular damage in future [59].

All recent models aimed to mimic EBI and delayed vasospasm. However, little literature focused on long-term studies. Since one-third of SAH patients will survive and suffer with long-term disabilities, we should pay more attention to long-term SAH studies. Thus, a suitable model for long-term SAH study should be developed.

SAH is well known for inducing hydrocephalus, seizures, and other complications. Most studies report an overall 20–30 % incidence of hydrocephalus after SAH [60]. Those complications are often associated with unfavorable outcomes [61]. However, animal models focusing on complications after SAH are very rare. Future studies should differentiate suitable models in detail that target those complications.

SAH is also associated with weakness, fatigue, sleep disturbances, anxiety, depression, posttraumatic stress disorder, as well as cognitive and physical impairment [62]. Symptoms of depression and/or posttraumatic stress disorder have been shown to be present up to 41 % of post-SAH patients. Those symptoms are usually associated with white matter injury. The mechanism of post-SAH white brain injury remains unclear, possibly due to the lack of experimental studies. Thus, there is great need for development of animal models for post-SAH white matter injury.

Genome-wide association studies have identified novel genetic factors that contribute to intracranial aneurysm susceptibility. Hence, it is possible to produce an aneurysm on cerebral vessels experimentally using genetic technology [63]. We can study aneurysmal SAH using transgenic mice, which is close to the whole natural history in SAH patients.

14.6 Conclusion

In conclusion, although there are no ideal models or animal species that can mimic the natural history in SAH patients, each model can be used to study certain aspects of the pathophysiological process behind SAH. At present, there are no good alternatives for in vitro SAH modeling that equal in vivo models. Furthermore, future models focusing on complications after SAH are needed to improve SAH research and develop better outcomes for SAH patients.

References

1. Ingall T, Asplund K, Mahonen M, Bonita R (2000) A multinational comparison of subarachnoid hemorrhage epidemiology in the WHO MONICA stroke study. Stroke 31:1054–1061
2. Venti M (2012) Subarachnoid and intraventricular hemorrhage. Front Neurol Neurosci 30:149–153
3. Connolly ES Jr, Rabinstein AA, Carhuapoma JR, Derdeyn CP, Dion J, Higashida RT, Hoh BL, Kirkness CJ, Naidech AM, Ogilvy CS, Patel AB, Thompson BG, Vespa P (2012) Guidelines for the management of aneurysmal subarachnoid hemorrhage: a guideline for healthcare professionals from the American Heart Association/American Stroke Association. Stroke 43:1711–1737
4. Broderick JP, Brott TG, Duldner JE, Tomsick T, Leach A (1994) Initial and recurrent bleeding are the major causes of death following subarachnoid hemorrhage. Stroke 25:1342–1347
5. Macdonald RL, Higashida RT, Keller E, Mayer SA, Molyneux A, Raabe A, Vajkoczy P, Wanke I, Bach D, Frey A, Marr A, Roux S, Kassell N (2011) Clazosentan, an endothelin receptor antagonist, in patients with aneurysmal subarachnoid haemorrhage undergoing surgical clipping: a randomised, double-blind, placebo-controlled phase 3 trial (CONSCIOUS-2). Lancet Neurol 10:618–625
6. Macdonald RL, Higashida RT, Keller E, Mayer SA, Molyneux A, Raabe A, Vajkoczy P, Wanke I, Bach D, Frey A, Nowbakht P, Roux S, Kassell N (2012) Randomized trial of clazosentan in patients with aneurysmal subarachnoid hemorrhage undergoing endovascular coiling. Stroke 43:1463–1469
7. Kusaka G, Ishikawa M, Nanda A, Granger DN, Zhang JH (2004) Signaling pathways for early brain injury after subarachnoid hemorrhage. J Cereb Blood Flow Metab 24:916–925
8. Cahill J, Calvert JW, Zhang JH (2006) Mechanisms of early brain injury after subarachnoid hemorrhage. J Cereb Blood Flow Metab 26:1341–1353
9. Crompton MR (1964) The pathogenesis of cerebral infarction following the rupture of cerebral berry aneurysms. Brain 87:491–510
10. Megyesi JF, Vollrath B, Cook DA, Findlay JM (2000) In vivo animal models of cerebral vasospasm: a review. Neurosurgery 46:448–460, discussion 460–441
11. Sehba FA, Bederson JB (2006) Mechanisms of acute brain injury after subarachnoid hemorrhage. Neurol Res 28:381–398
12. Marbacher S, Fandino J, Kitchen ND (2010) Standard intracranial in vivo animal models of delayed cerebral vasospasm. Br J Neurosurg 24:415–434
13. Brawley BW, Strandness DE Jr, Kelly WA (1968) The biphasic response of cerebral vasospasm in experimental subarachnoid hemorrhage. J Neurosurg 28:1–8
14. Nagai H, Suzuki Y, Sugiura M, Noda S, Mabe H (1974) Experimental cerebral vasospasm. 1: Factors contributing to early spasm. J Neurosurg 41:285–292

15. Asano T, Sano K (1977) Pathogenetic role of no-reflow phenomenon in experimental subarachnoid hemorrhage in dogs. J Neurosurg 46:454–466
16. Barry KJ, Gogjian MA, Stein BM (1979) Small animal model for investigation of subarachnoid hemorrhage and cerebral vasospasm. Stroke 10:538–541
17. Kader A, Krauss WE, Onesti ST, Elliott JP, Solomon RA (1990) Chronic cerebral blood flow changes following experimental subarachnoid hemorrhage in rats. Stroke 21:577–581
18. Bederson JB, Germano IM, Guarino L (1995) Cortical blood flow and cerebral perfusion pressure in a new noncraniotomy model of subarachnoid hemorrhage in the rat. Stroke 26:1086–1091, discussion 1091–1082
19. Veelken JA, Laing RJ, Jakubowski J (1995) The Sheffield model of subarachnoid hemorrhage in rats. Stroke 26:1279–1283, discussion 1284
20. Bederson JB, Levy AL, Ding WH, Kahn R, DiPerna CA, Jenkins AL III, Vallabhajosyula P (1998) Acute vasoconstriction after subarachnoid hemorrhage. Neurosurgery 42:352–360, discussion 360–352
21. Sugawara T, Ayer R, Jadhav V, Chen W, Tsubokawa T, Zhang JH (2008) Simvastatin attenuation of cerebral vasospasm after subarachnoid hemorrhage in rats via increased phosphorylation of Akt and endothelial nitric oxide synthase. J Neurosci Res 86:3635–3643
22. Prunell GF, Mathiesen T, Svendgaard NA (2004) Experimental subarachnoid hemorrhage: cerebral blood flow and brain metabolism during the acute phase in three different models in the rat. Neurosurgery 54:426–436, discussion 436–427
23. Alkan T, Tureyen K, Ulutas M, Kahveci N, Goren B, Korfali E, Ozluk K (2001) Acute and delayed vasoconstriction after subarachnoid hemorrhage: local cerebral blood flow, histopathology, and morphology in the rat basilar artery. Arch Physiol Biochem 109:145–153
24. Park IS, Meno JR, Witt CE, Suttle TK, Chowdhary A, Nguyen TS, Ngai AC, Britz GW (2008) Subarachnoid hemorrhage model in the rat: modification of the endovascular filament model. J Neurosci Methods 172:195–200
25. Schwartz AY, Masago A, Sehba FA, Bederson JB (2000) Experimental models of subarachnoid hemorrhage in the rat: a refinement of the endovascular filament model. J Neurosci Methods 96:161–167
26. Kamii H, Kato I, Kinouchi H, Chan PH, Epstein CJ, Akabane A, Okamoto H, Yoshimoto T (1999) Amelioration of vasospasm after subarachnoid hemorrhage in transgenic mice overexpressing CuZn-superoxide dismutase. Stroke 30:867–871, discussion 872
27. Altay O, Hasegawa Y, Sherchan P, Suzuki H, Khatibi NH, Tang J, Zhang JH (2012) Isoflurane delays the development of early brain injury after subarachnoid hemorrhage through sphingosine-related pathway activation in mice. Crit Care Med 40:1908–1913
28. Altay O, Suzuki H, Hasegawa Y, Caner B, Krafft PR, Fujii M, Tang J, Zhang JH (2012) Isoflurane attenuates blood–brain barrier disruption in ipsilateral hemisphere after subarachnoid hemorrhage in mice. Stroke 43:2513–2516
29. Simeone FA, Ryan KG, Cotter JR (1968) Prolonged experimental cerebral vasospasm. J Neurosurg 29:357–366
30. Simeone FA, Trepper PJ, Brown DJ (1972) Cerebral blood flow evaluation of prolonged experimental vasospasm. J Neurosurg 37:302–311
31. Schwartz AY, Sehba FA, Bederson JB (2000) Decreased nitric oxide availability contributes to acute cerebral ischemia after subarachnoid hemorrhage. Neurosurgery 47:208–214, discussion 214–205
32. Lougheed WM, Tom M (1961) A method of introducing blood into the subarachnoid space in the region of the circle of Willis in dogs. Can J Surg 4:329–337
33. McQueen JD, Jeanes LD (1964) Dehydration and rehydration of the brain with hypertonic urea and mannitol. J Neurosurg 21:118–128
34. McQueen JD, Jelsma LF (1967) Intracranial hypertension. Cerebrospinal fluid pressure rises following intracisternal infusions of blood components in dogs. Arch Neurol 16:501–508
35. Martins AN, Doyle TF, Newby N, Kobrine AI, Ramirez A (1975) The effect of a simulated subarachnoid hemorrhage on cerebral blood flow in the monkey. Stroke 6:664–672

36. Prunell GF, Mathiesen T, Svendgaard NA (2002) A new experimental model in rats for study of the pathophysiology of subarachnoid hemorrhage. Neuroreport 13:2553–2556
37. Kassell NF, Torner JC, Haley EC Jr, Jane JA, Adams HP, Kongable GL (1990) The international cooperative study on the timing of aneurysm surgery. Part 1: Overall management results. J Neurosurg 73:18–36
38. Peters ND, Di Chiro G (1976) A model for spasm of the anterior cerebral artery. Stroke 7:243–247
39. Chow RW, Newton TH, Smith MC, Adams JE (1968) Cerebral vasospasm induced by subarachnoid blood and serotonin. An angiographic study. Invest Radiol 3:402–407
40. Echlin FA (1965) Spasm of basilar and vertebral arteries caused by experimental subarachnoid hemorrhage. J Neurosurg 23:1–11
41. Matz PG, Fujimura M, Lewen A, Morita-Fujimura Y, Chan PH (2001) Increased cytochrome c-mediated DNA fragmentation and cell death in manganese-superoxide dismutase-deficient mice after exposure to subarachnoid hemolysate. Stroke 32:506–515
42. Peerless SJ, Fox AJ, Komatsu K, Hunter IG (1982) Angiographic study of vasospasm following subarachnoid hemorrhage in monkeys. Stroke 13:473–479
43. Delgado-Zygmunt TJ, Arbab MA, Shiokawa Y, Svendgaard NA (1992) A primate model for acute and late cerebral vasospasm: angiographic findings. Acta Neurochir 118:130–136
44. Varsos VG, Liszczak TM, Han DH, Kistler JP, Vielma J, Black PM, Heros RC, Zervas NT (1983) Delayed cerebral vasospasm is not reversible by aminophylline, nifedipine, or papaverine in a "two-hemorrhage" canine model. J Neurosurg 58:11–17
45. Prunell GF, Mathiesen T, Diemer NH, Svendgaard NA (2003) Experimental subarachnoid hemorrhage: subarachnoid blood volume, mortality rate, neuronal death, cerebral blood flow, and perfusion pressure in three different rat models. Neurosurgery 52:165–175, discussion 175–166
46. Lee JY, Huang DL, Keep R, Sagher O (2008) Characterization of an improved double hemorrhage rat model for the study of delayed cerebral vasospasm. J Neurosci Methods 168:358–366
47. Krafft PR, Bailey EL, Lekic T, Rolland WB, Altay O, Tang J, Wardlaw JM, Zhang JH, Sudlow CL (2012) Etiology of stroke and choice of models. Int J Stroke 7:398–406
48. Clark JF, Sharp FR (2006) Bilirubin oxidation products (boxes) and their role in cerebral vasospasm after subarachnoid hemorrhage. J Cereb Blood Flow Metab 26:1223–1233
49. Pyne GJ, Cadoux-Hudson TA, Clark JF (2001) Cerebrospinal fluid from subarachnoid haemorrhage patients causes excessive oxidative metabolism compared to vascular smooth muscle force generation. Acta Neurochir 143:59–62, discussion 62–53
50. Zhao W, Ujiie H, Tamano Y, Akimoto K, Hori T, Takakura K (1999) Sudden death in a rat subarachnoid hemorrhage model. Neurol Med Chir (Tokyo) 39:735–741, discussion 741–733
51. Marbacher S, Andereggen L, Neuschmelting V, Widmer HR, von Gunten M, Takala J, Jakob SM, Fandino J (2012) A new rabbit model for the study of early brain injury after subarachnoid hemorrhage. J Neurosci Methods 208:138–145
52. Marbacher S, Sherif C, Neuschmelting V, Schlappi JA, Takala J, Jakob SM, Fandino J (2010) Extra-intracranial blood shunt mimicking aneurysm rupture: Intracranial-pressure-controlled rabbit subarachnoid hemorrhage model. J Neurosci Methods 191:227–233
53. Munoz-Sanchez MA, Egea-Guerrero JJ, Revuelto-Rey J, Moreno-Valladares M, Murillo-Cabezas F (2012) A new percutaneous model of subarachnoid haemorrhage in rats. J Neurosci Methods 211:88–93
54. Simard JM, Tosun C, Ivanova S, Kurland DB, Hong C, Radecki L, Gisriel C, Mehta R, Schreibman D, Gerzanich V (2012) Heparin reduces neuroinflammation and transsynaptic neuronal apoptosis in a model of subarachnoid hemorrhage. Transl Stroke Res 3:155–165
55. Dusick JR, Evans BC, Laiwalla A, Krahl S, Gonzalez NR (2011) A minimally-invasive rat model of subarachnoid hemorrhage and delayed ischemic injury. Surg Neurol Int 2:99
56. Lin CL, Calisaneller T, Ukita N, Dumont AS, Kassell NF, Lee KS (2003) A murine model of subarachnoid hemorrhage-induced cerebral vasospasm. J Neurosci Methods 123:89–97

57. Altay T, Smithason S, Volokh N, Rasmussen PA, Ransohoff RM, Provencio JJ (2009) A novel method for subarachnoid hemorrhage to induce vasospasm in mice. J Neurosci Methods 183:136–140
58. Solomon RA, Antunes JL, Chen RY, Bland L, Chien S (1985) Decrease in cerebral blood flow in rats after experimental subarachnoid hemorrhage: a new animal model. Stroke 16:58–64
59. Zhang JH, Badaut J, Tang J, Obenaus A, Hartman R, Pearce WJ (2012) The vascular neural network-a new paradigm in stroke pathophysiology. Nat Rev Neurol 8:711–716
60. Germanwala AV, Huang J, Tamargo RJ (2010) Hydrocephalus after aneurysmal subarachnoid hemorrhage. Neurosurg Clin N Am 21:263–270
61. Wang YM, Lin YJ, Chuang MJ, Lee TH, Tsai NW, Cheng BC, Lin WC, Su BY, Yang TM, Chang WN, Huang CC, Kung CT, Lee LH, Wang HC, Lu CH (2012) Predictors and outcomes of shunt-dependent hydrocephalus in patients with aneurysmal sub-arachnoid hemorrhage. BMC Surg 12:12
62. Kagerbauer SM, Rothoerl RD, Brawanski A (2007) Pituitary dysfunction after aneurysmal subarachnoid hemorrhage. Neurol Res 29:283–288
63. Suda N, Moriyama K, Ganburged G (2013) Effect of angiotensin II receptor blocker on experimental periodontitis in a mouse model of Marfan syndrome. Infect Immun 81 (1):182–188

Chapter 15
Age and Sex Differences in Hemodynamics in a Large Animal Model of Brain Trauma

William M. Armstead and Monica S. Vavilala

Abstract Traumatic brain injury (TBI) is a global health concern and is the leading cause of traumatic morbidity and mortality in children. Despite a lower overall mortality than in adult traumatic brain injury, the cost to society from the sequelae of pediatric traumatic brain injury is very high. Predictors of poor outcome after traumatic brain injury include altered systemic and cerebral physiology, including altered cerebral hemodynamics. Impaired cerebral hemodynamics, including cerebral blood flow and autoregulation following TBI may adversely impact poor outcome and may be age and or sex dependent. Yet, there is a paucity of information regarding changes in cerebral blood flow and cerebral autoregulation after pediatric TBI by age and sex. In this chapter, we first discuss clinical observations of pediatric TBI and how these can be used to better mimic TBI in a basic science large animal model of brain injury, the pig. Mechanistic explanations for age- and sex-dependent differences of TBI will focus on the spasmogen endothelin-1, glutamate, and the signaling system mitogen-activated protein kinase. Treatment strategies used for improvement of cerebral hemodynamics after TBI will consider the role of pressor choice in outcome. These data advocate for the consideration of development of sex-based therapies for treatment of hemodynamic sequelae of pediatric TBI.

15.1 Introduction

Traumatic brain injury (TBI) is the leading form of pediatric trauma and accounts for 36 % of deaths in US children 1–14 years of age [1, 2]. Despite a lower overall mortality than in adult traumatic brain injury, the cost to society from the sequelae of pediatric traumatic brain injury is very high. Clinically, it is necessary to

W.M. Armstead (✉)
Anesthesiology and Critical Care, University of Pennsylvania, 3620 Hamilton Walk, JM3, Philadelphia, PA 19104, USA
e-mail: armsteaw@uphs.upenn.edu

E.H. Lo et al. (eds.), *Vascular Mechanisms in CNS Trauma*, Springer Series in Translational Stroke Research 5, DOI 10.1007/978-1-4614-8690-9_15,

understand the pathophysiologic responses in children after TBI in order to optimize mechanistically appropriate therapeutic modalities. In addition to the deleterious effects of early hypotension, low Glasgow Coma Scale score, coagulopathy, hyperglycemia, low compromised CBF, and impaired autoregulation are also associated with poor outcome after pediatric TBI [3–9]. Yet, little is still known about changes in cerebral blood flow (CBF) and cerebral autoregulation after TBI. Moreover, unlike the adult TBI literature, very little is known about the influence of age and sex on TBI and TBI-related outcomes. Small studies suggest that children less than 4 years and boys have worse outcomes compared to girls after TBI [1, 10]. Most cerebrovascular data in children are derived from critically ill neonates but the role of sex has not been typically examined. The paucity of information in these areas and on the effect of age and sex on the course of pediatric TBI and its outcomes leaves clinicians without a clear understanding of how to optimally manage hemodynamics [11]. However, basic science approaches enhance our understanding of the mechanisms underlying the cerebrovascular pathophysiology after TBI [11]. In this review, the current understanding of cerebrovascular hemodynamics, including cerebral autoregulation, in TBI will be discussed, focusing on the influences of age and sex using a large animal (pig) model of TBI.

15.1.1 Types and Models of Injury

Each of the models available for use in the immature can be categorized in terms of the model having characteristics of one of the two types of brain injury: focal and diffuse [12]. Focal models of TBI produce either a contusion or a localized regional area of injury. Clinically relevant focal brain injuries include cerebral contusion and laceration, as well as hemorrhages and hematomas. Diffuse brain injuries include cerebral concussion and prolonged traumatic coma caused by diffuse axonal injury. Some of the focal models of TBI include weight drop, controlled cortical impact, and subdural hematoma [13]. An example of a diffuse model of TBI is impact acceleration, while fluid percussion brain injury (FPI) has elements of both diffuse and focal injuries [13]. FPI, in fact, can be of two types: lateral and medial. A lateral injury primarily includes cerebral cortical damage whereas a medial injury involves the brainstem in injury [14, 15]. Diffuse TBI and diffuse cerebral swelling in general are more common in children than adults [12]. The lateral FPI technique has been suggested to be a good model for shaken impact syndrome [13], an example of child abuse.

15.1.2 The Role of Species in Model Choice

Many studies of head injury in immature animals to date have utilized rodent models. However, rodents have a paucity of white matter. Piglets, in contrast, provide many advantages in modeling the human brain. The overall shape, gyral pattern, and distribution of grey and white matter are similar in pigs and humans [16]. The growth pattern of the postnatal brain is similar to that of human infants [17]. The response of the piglet to hypoxia and ischemia appears parallel to that observed in human infants [18]. Additionally, CBF and metabolism are similar and the piglet matures in a manner similar to the human with respect to myelination and brain electrical activity [19]. Selective vulnerability in the white matter similar to that observed in the human infant has been demonstrated in piglets with a model of acute subdural hematoma [20]. Therefore, the gyrencephalic pig brain containing substantial white matter appears to be an excellent choice to model human TBI.

15.1.3 Clinical Observations Inform Basic Science Modeling of TBI

Much of the data estimating CBF in healthy children and in children with TBI have been obtained using transcranial Doppler (TCD) ultrasonography. TCD ultrasonography is the more commonly used tool to estimate CBF largely because it is a noninvasive imaging tool operable at the bedside. TCD ultrasonography measures cerebral blood flow velocity (CBFV) of the basal cerebral arteries. Although it is not a direct measure of CBF, changes in CBFV generally correlate well with changes in CBF [21, 22]. Compared to children without TBI, children with TBI have lower middle cerebral artery velocity [23] and cerebral hypoperfusion (CBF < 25 mL/100 g/min) is the dominant derangement [24]. Cerebral hypoperfusion is associated with cerebral ischemia and poor outcome [25–27]. Nonetheless, following severe pediatric TBI, CBF may also be normal or high [27] and may result in cerebral hyperemia and cerebral hemorrhage.

Cerebral autoregulation is a homeostatic process; arterioles dilate and constrict to maintain CBF nearly constant over a range of blood pressures. In healthy adults, changes in mean arterial pressure (MAP) between 60 and 160 mmHg or cerebral perfusion pressure (CPP) between 50 and 150 mmHg produce little or no change in CBF [28, 29]. Conversely, this adaptive mechanism maintains constant (adequate) CBF by decreasing cerebrovascular resistance or when MAP/CPP decreases. Beyond these limits of autoregulation, CBF depends on CPP; hypotension may result in cerebral ischemia, and hypertension may cause cerebral hyperemia. Hypotension after pediatric TBI is associated with poor outcome [7–9]. Cerebral autoregulation is impaired more often following severe compared to mild pediatric TBI [23, 30–32]. Additionally, severity of impaired cerebral autoregulation may be increased in children with blunt (intentional or abusive) TBI than in children with non-inflicted TBI [32]. Vavilala et al. show that moderate to severe pediatric TBI associated with hemispheric differences in cerebral autoregulation are common (40 %) after focal TBI. Young children with TBI and less than 4 years of age are more at risk of impaired cerebral autoregulation than older children and there is suggestion that after moderate-severe TBI, boys may also have more impairment of autoregulation compared to girls after adjusting for injury severity [10].

If cerebral autoregulation is impaired, lower blood pressure may result in diminished CPP and CBF. Decreased MAP causes cerebral vasodilation, increase in cerebral blood volume, and thus an increase in intracranial pressure (ICP). Increase in ICP further decreases CPP, leading to more cerebral vasodilation, resulting in a vicious cycle. Controversy exists regarding empirically increasing MAP to prevent cerebral ischemia in the presence of impaired cerebral autoregulation, since theoretically, augmenting MAP in the hyperemic brain could result in cerebral hemorrhage [33–35]. Impaired cerebral autoregulation has been associated with poor 6-month outcomes after pediatric TBI [32, 36], but there is controversy as to whether impairment is a marker of injury or whether it contributes to poor outcome. We examined the relationship between cerebral hemodynamic predictors during the first 72 h after injury, including cerebral autoregulation, and long-term outcome after severe (Glasgow coma Scale score < 9) pediatric TBI. Ten (28 %) of the 36 children examined had poor outcome. Independent risk factors for poor 6-month GOS were impaired cerebral autoregulation and hypotension (SBP $<$5th percentile). In this study, both impaired cerebral autoregulation and SBP $<$5th percentile were independent risk factors for poor 6-month GOS, suggesting a causal role for impaired autoregulation, independent of hypotension on outcome.

15.2 Pediatric Basic Science Models of TBI

Studies of TBI in models involving immature animals have revealed age-related differences in the mechanisms of injury in the brain [37, 38]. Although several rodent models of juvenile TBI have been described [38], all have the disadvantage of not permitting repeated measurements of systemic physiological variables and regional CBF because of the small size of the subjects. Additionally, rodents have a lissencephalic brain containing more grey than white matter. In contrast, piglets have a gyrencepahalic brain that contains substantial white matter similar to humans, which is more sensitive to ischemic damage than grey matter. A number of neuroprotectants identified in preclinical rodent stroke and TBI studies have yielded disappointing results when entered into clinical trials. We speculate that the reason for failure may rest on these drugs being primarily grey matter protective owing to the greater amount of grey compared to white matter in the rodent. Others have used pigs to study TBI. While most have investigated parameters other than cerebral hemodynamics as an index of outcome post TBI (e.g., lesion volume and edema) [39–41], one characterized age-related effects on CBF [42]. However, the injury model used was focal (cortical contusion), not diffuse or mixed focal/diffuse as in FPI, thought to be a good mimic of diffuse shaken impact and motor vehicle concussive injury [13].

15.3 Age- and Sex-Dependent Effects of TBI on Cerebral Hemodynamics in Piglets

We have endeavored to use clinical observations of TBI in children to inform the design of basic science studies [11]. First, we were interested in characterizing the influence of TBI on cerebral hemodynamics. Using newborn and juvenile pigs, which mimic young (<4 years) and older (≥4 years) children [17], the following were observed: (1) pial arteries constricted more and regional CBF fell and remained depressed longer in newborn versus juvenile pigs; (2) there were marked increases in ICP in the newborn, but modest increases in ICP in the juvenile; (3) there were differences in cerebral oxygenation, an index of metabolism: in the newborn saturation increased, followed by profound prolonged desaturation of hemoglobin for oxygen, while in the juvenile saturation increased modestly, followed by mild desaturation [43]. Since decreases in CBF were associated with decreased saturation of hemoglobin for oxygen, these data suggested that ischemia may well occur following TBI, particularly in the newborn pig [43]. Furthermore, systemic arterial pressure has been observed to increase in adults of other species [15, 44] and in juvenile pigs [43], whereas systemic arterial blood pressure decreased following FPI in the newborn pig [43]. Subsequent studies investigated the time duration for cerebral hemodynamic effects of FPI in these two age groups of pigs. For example, pial arteries remained constricted and CBF reduced for up to 3 days post-insult in the newborn; by 7 days pial artery diameter and CBF had returned to pre-injury values [45]. In contrast, pial artery diameter and CBF reductions had normalized within 8 h post-injury in the juvenile pig [45]. These studies using an equivalent degree of injury indicate that there are age-dependent differences in the cerebral hemodynamic effects of brain injury, with the newborn being affected to a greater degree. Similarly, differences in the effects of TBI on CBF and pial artery diameter were observed in male compared to female pigs. For example, pial artery diameter and CBF were reduced more in male compared to female newborn pigs after equivalent FPI [46, 47]. However, how the force of the brain insult acts once it enters the skull may well depend on differences in the composition and compliance of the newborn and juvenile brain. It is also unclear how developmental parameters such as brain water content or skull dimensions will affect the biomechanics of the brain insult delivered to these two age groups.

15.3.1 Mechanisms for Age- and Sex-Dependent Effects on Cerebral Hemodynamics After TBI in Pigs

15.3.1.1 Endothelin-1

Since ethical considerations constrain mechanistic studies in children with TBI, we have used an established porcine model of FPI that mimics TBI to corroborate clinical observations regarding cerebral hemodynamics and autoregulation after TBI [11]. In the context of the neurovascular unit, CBF is thought to contribute to neuronal cell integrity and function. Therefore, CBF often is an important therapeutic target in strategies for treatment of TBI. One of our early studies investigated the role of the cerebrovascular peptide spasmogen endothelin-1 (ET-1) [48–50]. We observed that a larger increase in the CSF ET-1 concentration was associated with a more robust decrease in CBF in the newborn and male compared to the juvenile and female pig after induction of a comparable level of FPI [50, 51]. Administration of the ET-1 antagonist BQ 123 largely prevented reductions in CBF, pial artery diameter, and impairment of autoregulation during hypotension in both ages and sexes of pigs [50, 51], along with loss of neuronal cells in the cerebral cortex [52], indicating the importance of upregulation of this peptide in outcome after TBI. Subsequent studies were designed to investigate mechanisms whereby ET-1 release might couple impaired cerebral hemodynamics to histopathology in the setting of TBI.

Relaxation of blood vessels can be mediated by several mechanisms, including cGMP, cAMP, and K^+ channels [48]. Membrane potential of vascular muscle is a major determinant of vascular tone and activity of K^+ channels is a major regulator of membrane potential [48]. Activation or opening of these channels increases K^+ efflux, producing hyperpolarization of vascular muscle. Membrane hyperpolarization closes voltage-dependent calcium channels and causes relaxation of vascular muscle. Direct measurements of membrane potential and K^+ current in vitro indicate that several types of K^+ channels are present in cerebral blood vessels. In addition, a number of pharmacological studies using activators and inhibitors have provided functional evidence that K^+ channels, especially ATP-sensitive (Katp) and calcium-sensitive (Kca) channels, regulate cerebrovascular tone [48]. Cromakalim and calcitonin gene-related peptide (CGRP) are examples of Katp channel agonists, while a Kca channel agonist is NS 1619. Vasodilation in response to these drugs can be used as an index of the intactness of K channel function after TBI and cerebral ischemia [53, 54]. Pial artery dilation in response to hypotension is due to activation of Katp and Kca channels [55], thereby giving functional significance to intactness of K channel function. Since pial artery dilation in response to Katp and Kca channel agonists is blunted more in the male than the female after FPI [56], such observations suggest that sex-dependent greater reductions in CBF and impairment of cerebral autoregulation in the male compared to the female may relate to more aggravated impairment of K channel function in that sex. Since an endogenous K channel agonist, adrenomedullin (ADM), is upregulated in the female but not the male pig after FPI, while exogenous ADM administration prevents sex-dependent impairment of autoregulation in both sexes, these data indicate that the presence of an endogenous neuropotectant in the female contributes to differential sex-dependent outcome post insult [46]. Mechanistically, ADM achieves cerebrohemodynamic protection through blockade of the upregulation of the ERK isoform of mitogen-activated protein kinase, a family of at least three kinases (ERK, p38, and JNK) that are critically important in hemodynamics after TBI [46, 47, 57, 58]. ET-1 contributes to blunted K channel agonist-mediated dilation after FPI via release of activated oxygen (O_2^-) which can then activate ERK MAPK [11, 51, 59]. Because more ERK MAPK is released after FPI in the male compared to the female [47, 51], there is correspondingly greater impairment of autoregulation and Katp and Kca channel agonist-mediated cerebrovasodilation post-injury in males compared to females [11].

15.4 Glutamate

Glutamate is known to bind to each of the three ionotropic receptor subtypes named after synthetic analogues: *N*-methyl-D-aspartate (NMDA), kainate, and α-amino-3-hydroxy-5-methyl-4-isoxazolepropionate (AMPA). Activation of NMDA receptors (R) contributes to excitotoxicity [60] but also elicits cerebrovasodilation and represents a mechanism by which local metabolism is coupled to CBF [48]. In healthy brain, tissue plasminogen activator (tPA) is critical for the full expression of the flow increase evoked by activation of the mouse whisker barrel cortex [61]. tPA promotes nitric oxide (NO) synthesis during NMDA receptor activation by modulating the phosphorylation state of neuronal nitric oxide synthase [61]. These findings suggest that tPA is a key mediator linking NMDA receptor activation to NO synthesis and functional hyperemia. Glutamatergic hyperactivity occurs in animal models of TBI, while NMDA antagonists are protective [62, 63]. Although CBF is thought to contribute to neurologic outcome, little attention has been given to the role of NMDA in this process. In our studies, NMDA-R-mediated vasodilation is reversed to vasoconstriction after FPI in the piglet [64].

Similarly, tPA upregulation not only enhances excitotoxic neuronal cell death through the NMDA-R [65, 66] but also contributes to impaired NMDA-R-mediated vasodilation, autoregulation during hypotension, and histopathology after FPI [58, 67–69] via ERK and JNK MAPK [70]. A potential explanation for the differential role of tPA in normal and injured brain could relate to increased superoxide production after FPI [71], which together with increased NO will generate excessive peroxynitrite. Once formed, peroxynitrite could impair cerebrovasodilator systems post injury. However, the severity of constriction observed with NMDA after FPI + tPA in the newborn pig is substantial and probably not the sole result of loss of a dilator, such as NO scavenging by superoxide, but also production of a vasoconstrictor. While the identity of this vasoconstrictor is not known with certainty, ET-1 may play a role since it has been found to be upregulated and contribute to impaired dilation induced by NMDA receptor activation after FPI [72]. While the NMDA antagonist MK 801 protects against cerebral dysregulation after FPI [73], its toxicity limits use in humans. We found that glucagon minimizes the glutamate surge after TBI in mice and pigs, prevents brain tissue damage, and preserves autoregulation during hypotension by blunting upregulation of tPA [67, 68].

Based on these studies, we posit that glutamate and tPA act in concert to induce neurotoxicity. In absence of tPA (tPA null mice), even high levels of CNS glutamate occurring after brain injury are weakly neurotoxic. In addition, exogenous tPA is not neurotoxic when glutamate levels are kept low. Based on this, we propose that tPA and glutamate create a vicious cycle wherein tPA increases the toxicity of glutamate by increasing the sensitivity of NMDA-Rs to tPA and glutamate increases the neurotoxicity of tPA by signal transduction through NMDA-Rs that have been activated by tPA [70]. Furthermore, neurotoxicity induced by tPA increases CSF levels of glutamate [68] and neurotoxicity induced by glutamate increases the levels of tPA [67, 68], which further exacerbates injury. Break of this vicious cycle through inhibition of interactions between tPA and NMDA is predicted to prevent disturbed cerebral autoregulation in the setting of TBI. Excessive tPA released after TBI may cause tissue damage either by over-activating NMDA-Rs or cause premature clot lysis leading to progressive hemorrhage. Use of a catalytically inactive tPA variant (tPA-S^{481}A) that competes with wild-type tPA for binding to NMDA-R through its receptor docking site but cannot activate it, prevents activation of ERK MAPK and thereby impairment of autoregulation after FPI [74]. Administration of tPA-S^{481}A at 30 min post-FPI also blocked neuronal cell necrosis in the CA1 and CA3 hippocampus. More recent studies showing similar protection of cerebral autoregulation and prevention of histopathology when administered at 3 h post insult suggest there is a realistic therapeutic window within which tPA-S^{481}A can improve outcome post TBI (unpublished observations). An alternative strategy that similarly prevented histopathology in CA1 and CA3 hippocampus post-FPI was the administration of the synthetic PAI-1 analogue, EEIIMD [69]. However, while peptide antagonists such as EEIIMD may be of benefit, they are rapidly degraded, suggesting that use of a tPA variant such as tPA-S^{481}A might have greater therapeutic utility.

15.4.1 Treatment Strategies Currently Used to Improve Cerebral Hemodynamic Outcome After TBI: Pressor Choice Influences Outcome

Hypotension and low cerebral perfusion pressure (CPP, mean arterial pressure [MAP] minus intracranial pressure [ICP]) are associated with low CBF, cerebral ischemia, and poor outcomes after pediatric TBI [75]. Cerebral autoregulation is often impaired after TBI and with concomitant hypotension and high ICP, lead to poor outcome [10]. Current 2012 Pediatric Guidelines recommend maintaining CPP above 40 mmHg, noting that an age-related continuum for the optimal CPP is between 40 and 65 mmHg [76–81]. Despite these therapeutic targets, there are no guidelines regarding how this should be achieved other than therapies to lower ICP by using mannitol or hypertonic saline [78], the latter which may be desirable because of the added benefit of increasing CPP beyond what would be expected due to the drop in ICP [80]. In addition to decreasing ICP, vasopressors to elevate mean arterial pressure (MAP) are commonly used to preserve CPP by normalizing BP in

TBI patients with hypotension. However, CPP-directed therapy has remained somewhat controversial because it has been observed to either have no effect or in fact worsen outcome [82]. Additionally, CPP has been considered to be a poor surrogate for CBF [81], since regional or local CBF may be markedly reduced even if CPP is normal [82]. Since ICP monitors are not universally used to guide CPP therapy, especially in young children, clinicians often rely on MAP to estimate CBF.

Three vasopressors commonly used to elevate MAP are phenylephrine (Phe), norepinephrine (NE), and dopamine (DA) [83], but the pressor of choice often may be Phe due to its longer duration of action and peak elevation of MAP [84]. However, when used in our piglet TBI model, we were surprised to observe that while Phe is protective of cerebral hemodynamics, particularly for autoregulation, in female piglets, it aggravates cerebrovascular dysregulation in male piglets post injury [85]. Because of this perplexing observation, we hypothesized that pressor choice may influence outcome. Indeed, we obtained provocative data indicating that this may, in fact, be the case in that another pressor, DA, produced the opposite outcome, equivalent cerebrohemodynamic protection in both male and female piglets in the setting of TBI [86].

Additional experiments were then designed to utilize the advantage of use of a basic science animal mimic of the clinical situation to ask mechanistically driven questions with the intent of understanding why two pressors could produce such divergent cerebrohemodynamic outcomes. While Phe blunted ET-1 and ERK MAPK upregulation in female piglets after FPI, there was an unanticipated and unwanted Phe-mediated aggravation of ET-1 and ERK MAPK upregulation in male piglets post injury [85]. The latter compounded the already greater release of ET-1 and ERK MAPK in males compared to females after FPI and appeared to contribute to the sex-dependent impairment of autoregulation [85]. The ET-1 antagonist BQ 123 blocked elevation of CSF ERK MAPK and the aggravation of such elevation by Phe after FPI [85]. Co-administered BQ 123 with Phe also prevented impairment of autoregulatory pial artery dilation during hypotension after FPI, supportive of the intermediary role for ET-1 in sex-dependent Phe-mediated hemodynamic dysregulation. Papaverine-induced pial artery dilation was unchanged after FPI and Phe in male and female piglets, indicating the specificity of the Phe effect. In contrast, DA completely blocked ET-1 and ERK MAPK upregulation equivalently in male and female piglets after FPI [86]. Taken together, choice of pressor to elevate MAP achieves differential cerebrohemodynamic outcome mechanistically due to sex-specific modulatory effects of the pressor on ET-1 upregulation, thereby influencing subsequent release of ERK MAPK. Figure 15.1 summarizes the sex-dependent differences in the roles of ET-1 and ERK MAPK in cerebral hemodynamics after TBI. In particular, this figure illustrates the sex-dependent differences in cerebral hemodynamics after TBI when treated with Phe and DA to normalize CPP. This cartoon figure crystallizes the idea that choice of pressor influences cerebrohemodynamic outcome after TBI.

As noted previously in this review, activation of Katp and Kca channels mediates autoregulation during hypotension [48, 55]. ET-1 contributes to blunted K

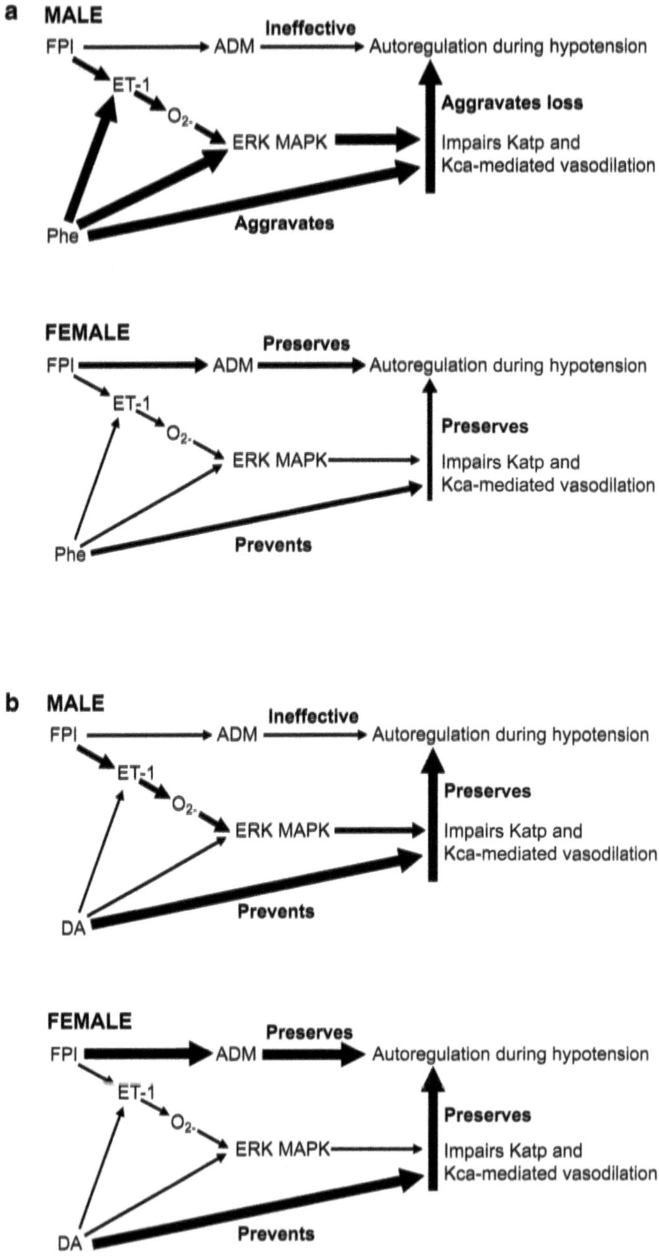

Fig. 15.1 Comparison of proposed mechanisms for Phe (**a**) and DA (**b**) in control of cerebral hemodynamics after FPI. *Arrow* thickness in proportion to probability of action

channel agonist-mediated dilation after FPI via release of activated oxygen (O_2^-) which can then activate ERK MAPK [11, 51, 59]. Because more ERK MAPK is released after FPI in the male compared to the female [47, 51], there is correspondingly greater impairment of autoregulation and Katp and Kca channel agonist-mediated cerebrovasodilation post-injury in males compared to females [11]. Systemic pressor support with Phe exacerbates dysregulation via aggravation of the sequential impairment of K channel-mediated cerebrovasodilation in males but abrogates such impairment and is protective in females after TBI in females but not males (Fig 15.1) [51, 56]. Whether DA protection of K channel agonist-mediated cerebrovasodilation after FPI can serve as an explanation for equivalent prevention of impairment of cerebral autoregulation in male and female pigs post-insult is an intriguing question for investigation in future studies. Nonetheless, data from these studies support the hypothesis that pressor choice is important in determining cerebrohemodynamic outcome after TBI. These data advocate for the consideration of development of sex-based therapies for treatment of hemodynamic sequelae of pediatric TBI (Fig 15.1).

However, several somewhat paradoxical observations were also noted in the above study. Release of a spasmogen like ET-1 that produces greater reductions in pial artery diameter in males than females should result in a larger decreased cerebral blood volume and decreased ICP in males compared to females. However, males in the current studies were observed to demonstrate higher ICP compared to females after FPI. Blunting abnormal arterial diameter increases during hypotension would also be expected to increase ICP through increased blood volume, but ICP was actually lower with the administration of DA. The results of these studies are also inconsistent with the observations in human studies where DA increased ICP [87]. We speculate that the relative constrictor/dilator ratio in the cerebral circulation, in fact, determines the ultimate disposition of ICP, and not just the mere presence of a spasmogen.

Others have investigated the cerebrohemodynamic effects of DA in the setting of TBI in basic science animal models as well as in patients. In a rodent impact acceleration model of rapidly increasing ICP, DA restored CBF through elevation of CPP [88]. However, despite the restoration of CBF, ICP and edema formation were not improved by DA [88]. In a second rodent TBI model, cortical contusion, DA also increased tissue water content [88]. In an adult swine model of FPI, CPP-directed therapy with combined Phe and DA improved brain oxygenation and maintained cerebrovascular CO_2 reactivity while decreasing brain edema [89]. When comparing DA to NE in head-injured patients, NE was observed to produce more significant and predictable increases in TCD flow velocity than DA [90], while DA increased ICP more for the same MAP compared to NE [87]. No significant differences between NE and DA on cerebral oxygenation or metabolism either at baseline or following a CPP intervention were observed in head-injured patients [91]. However, similar to the TCD flow velocity study, the response to a CPP intervention with DA on oxygenation seemed to be more variable than that with NE [91]. DA, then, could be viewed as a pressor whose use in CPP intervention could lead to an unpredictable outcome. When comparing the therapeutic use of

Phe, DA, and NE in head-injured patients, Phe was associated with higher MAP and CPP compared to DA and NE [83].

15.5 Conclusion

Pediatric TBI remains an important topic of research since TBI is the leading cause of morbidity and mortality in children. However, there is a paucity of data on the influence of age and sex on cerebrovascular pathophysiology in pediatric TBI. The basic science model has advanced our understanding of the mechanisms underlying the observations made in children with TBI. Increasing our understanding of the mechanisms governing cerebral autoregulation by age and sex is important to improving outcome in children with TBI. Using a bidirectional translational approach, our recent data advocate for the consideration of development of sex-based therapies for treatment of hemodynamic sequelae of pediatric TBI.

References

1. Langlois JA, Rutland-Brown W, Thomas KE (2005) The incidence of traumatic brain injury among children in the United States: differences by race. J Head Trauma Rehabil 20 (3):229–238
2. Hoyert DL, Heron MP, Murphy SL et al (2006) Deaths: final data for 2003. Natl Vital Stat Rep 54:1–120
3. Luerrsen TG, Klauber MR, Marshall LF (1988) Outcome from head injury related to patient's age: a longitudinal prospective study of adult and pediatric study of adult and pediatric head injury. J Neurosurg 68:409–416
4. Nakayama DK, Copes WS, Sacco WJ (1999) The effect of patient age upon survival in pediatric trauma. J Trauma 31:1521–1526
5. Vavilala MS, Dunbar PJ, Rivara FP et al (2001) Coagulopathy predicts poor outcome following head injury in children less than 16 years of age. J Neurosurg Anesthesiol 13:13–18
6. Jennett B, Teasdale G, Braakman R et al (1979) Prognosis of patients with severe head injury. Neurosurgery 4:283–289
7. Kokoska ER, Smith GS, Pittman T et al (1998) Early hypotension worsens neurological outcome in pediatric patients with moderately severe head trauma. J Pediatr Surg 33:333–338
8. Pigula FA, Wald SL, Shackford SR et al (1993) The effect of hypotension and hypoxia on children with severe head injuries. J Pediatr Surg 28:310–316
9. Vavilala MS, Bowen A, Lam AM et al (2003) Blood pressure and outcome after severe pediatric traumatic brain injury. J Trauma 55:1039–1044
10. Freeman SS, Udomphorn Y, Armstead WM, Fisk DM, Vavilala MS (2008) Young age as a risk factor for impaired cerebral autoregulation after moderate to severe pediatric traumatic brain injury. Anesthesiology 108(4):588–595
11. Armstead WM, Vavilala MS (2012) Age and sex differences in cerebral blood flow and autoregulation after pediatric traumatic brain injury. In: Kreipke CW, Rafols JA (eds) Cerebral blood flow, metabolism, and head trauma: the pathotrajectory of traumatic brain injury. Springer, New York, pp 135–154

12. Adelson PD (1999) Animal models of traumatic brain injury in the immature: a review. Exp Toxicol Pathol 51:130–136
13. Gennarelli TA (1994) Animate models of human head injury. J Neurotrauma 11:357–368
14. McIntosh TK, Noble L, Andrews B, Faden AI (1987) Traumatic brain injury in the cat: characterization of a midline fluid percussion model. Cent Nerv Syst Trauma 4:119–134
15. McIntosh TK, Vink R, Noble L, Yamakami I, Fernyak S, Soares H, Faden AI (1989) Traumatic brain injury in the rat: characterization of a lateral fluid percussion model. Neuroscience 28:233–244
16. Duhaime AC, Raghupathi R (1999) Age-specific therapy for traumatic injury of the immature brain: experimental approaches. Exp Toxicol Pathol 51:172–177
17. Dickerson JWT, Dobbing J (1967) Prenatal and postnatal growth and development of the central nervous system of the pig. Proc R Soc Lond B Biol Sci 166:384–395
18. Pampiglione G (1971) Some aspects of development of cerebral function in mammals. Proc R Soc Med 64:492–495
19. Buckley JM (1986) Maturation of circulatory system in three mammalian models of human development. Comp Biochem Physiol 83:1–7
20. Shaver E, Duhaime AC, Curtis M, Gennarelli LM, Barrett R (1996) Experimental acute subdural hematoma in infant piglets. Pediatr Neurosurg 25:123–129
21. Fisher AQ, Truemper EJ (1999) Applications in the neonate and child. In: Babikian VL, Weschler LR (eds) Transcranial Doppler ultrasonography, 2nd edn. Butterworth-Heinemann, Boston, pp 355–376
22. Giller CA, Hatab MR, Giller AM (1998) Estimation of vessel flow and diameter during cerebral vasospasm using transcranial Doppler indices. Neurosurgery 42:1076–1081
23. Vavilala MS, Lee LA, Boddu K et al (2004) Cerebral autoregulation in pediatric traumatic brain injury. Pediatr Crit Care Med 5:257–263
24. Adelson PD, Clyde B, Kochanek PM et al (1997) Cerebrovascular response in infants and young children following severe traumatic brain injury: a preliminary report. Pediatr Neurosurg 26:200–207
25. Coles JP, Fryer TD, Smielewski P et al (2004) Incidence and mechanisms of cerebral ischemia in early clinical head injury. J Cereb Blood Flow Metab 24:202–211
26. Skippen P, Seear M, Poskitt K et al (1997) Effect of hyperventilation on regional cerebral blood flow in head-injured children. Crit Care Med 25:1402–1409
27. Sharples PM, Stuart AG, Matthews DS et al (1995) Cerebral blood flow and metabolism in children with severe head injury. Part I: relation to age, Glasgow coma score, outcome, intracranial pressure, and time after injury. J Neurol Neurosurg Psychiatry 58:145–152
28. Paulson OB, Strandgaard S, Edvinsson L (1990) Cerebral autoregulation. Cerebrovasc Brain Metab Rev 2:161–191
29. Lassen NA (1959) Cerebral blood flow and oxygen consumption in man. Physiol Rev 39:183–238
30. Bouma GJ, Muizelaar JP, Fatouros P (1998) Pathogenesis of traumatic brain swelling: role of cerebral blood volume. Acta Neurochir Suppl 71:272–275
31. Stoyka WW, Schutz HH (1975) The cerebral response to sodium nitroprusside and trimethaphan controlled hypotension. Can Anaesth Soc J 22:275–283
32. Vavilala MS, Muangman S, Waitayawinyu P et al (2007) Neurointensive care; impaired cerebral autoregulation in infants and young children early after inflicted traumatic brain injury: a preliminary report. J Neurotrauma 24:87–96
33. Mandera M, Larysz D, Wojtacha M (2002) Changes in cerebral hemodynamics assessed by transcranial Doppler ultrasonography in children after head injury. Childs Nerv Syst 18:124–128
34. Bruce DA, Alavi A, Bilaniuk L et al (1981) Diffuse cerebral swelling following head injuries in children; the syndrome of malignant brain edema. J Neurosurg 54:170–178
35. Aldrich EF, Eisenberg HM, Saydjari C et al (1992) Diffuse brain swelling in severely head-injured children. A report from the NIH Traumatic Coma Data Bank. J Neurosurg 76:450–454

36. Vavilala MS, Muangman S, Tontisirin N et al (2006) Impaired cerebral autoregulation and 6-month outcome in children with severe traumatic brain injury: preliminary findings. Dev Neurosci 28:348–353
37. Babikian T, Prins ML, Cai Y, Barkhoudarian G, Hartonian I, Hovda DA, Giza CC (2010) Molecular and physiological responses to juvenile traumatic brain injury: focus on growth and metabolism. Dev Neurosci 32:431–441
38. Robertson CL, Scafidi S, McKenna MC, Fiskum G (2009) Mitochondrial mechanisms of cell death and neuroprotection in pediatric ischemic and traumatic brain injury. Exp Neurol 218:371–380
39. Grate LL, Golden JA, Hoopes PJ, Hunter JV, Duhaime AC (2003) Traumatic brain injury in piglets of different ages: techniques for lesion analysis using histology and magnetic resonance imaging. J Neurosci Methods 123:201–206
40. Duhaime AC, Margulies SS, Durham SR, O'Rouke MM, Golden JA, Marwaha S, Raghupathi R (2000) Maturation-dependent response of the piglet brain to scaled cortical impact. J Neurosurg 93:455–462
41. Duhaime AC, Hunter JV, Grate LL, Kim A, Golden J, Demidenko E, Harris C (2003) Magnetic resonance imaging studies of age-dependent responses to scaled focal brain injury in the piglet. J Neurosurg 99:542–548
42. Durham SR, Raghupathi R, Helfaer MA, Marwaha S, Duhaime AC (2000) Age-related differences in acute physiologic response to focal traumatic brain injury in piglets. Pediatr Neurosurg 33:76–82
43. Armstead WM, Kurth CD (1994) Different cerebral hemodynamic responses following fluid percussion brain injury in the newborn and juvenile pig. J Neurotrauma 11:487–497
44. Wei EP, Dietrich WD, Povlishock JT, Navari RM, Kontos HA (1980) Functional, morphological, and metabolic abnormalities of the cerebral microcircuit after concussive brain injury in cats. Circ Res 46:37–47
45. Armstead WM (2000) Age-dependent cerebral hemodynamic effects of traumatic brain injury in newborn and juvenile pigs. Microcirculation 7:225–235
46. Armstead WM, Vavilala MS (2007) Adrenomedullin reduces sex dependent loss of hypotensive cerebrovasodilation after newborn brain injury through activation of ATP dependent K channels. J Cereb Blood Flow Metab 27:1702–1709
47. Armstead WM, Kiessling JW, Bdeir K, Kofke WA, Vavilala MS (2010) Adrenomedullin prevents sex dependent impairment of autoregulation during hypotension after piglet brain injury through inhibition of ERK MAPK upregulation. J Neurotrauma 27:391–402
48. Faraci FM, Heistad DD (1998) Regulation of the cerebral circulation: role of endothelium and potassium channels. Physiol Rev 78:53–97
49. Armstead WM, Mirro R, Leffler CW, Busija DW (1989) Influence of endothelin on piglet cerebral microcirculation. Am J Physiol 257:H707–H710
50. Armstead WM (1999) Role of endothelin-1 in age dependent cerebrovascular hypotensive responses after brain injury. Am J Physiol 277:H1884–H1894
51. Armstead WM, Riley J, Vavilala MS (2012) TBI sex dependently upregulates ET-1 to impair autoregulation which is aggravated by phenylephrine in males but is abrogated in females. J Neurotrauma 29:1483–1490
52. Armstead WM, Raghupathi R (2011) Endothelin and the neurovascular unit in pediatric traumatic brain injury. Neurol Res 33:127–132
53. Armstead WM (1997) Brain injury impairs ATP-sensitive K^+ channel function in piglet cerebral arteries. Stroke 28:2273–2280
54. Bari F, Louis T, Meng W, Busija DW (1996) Global ischemia impairs ATP-sensitive K channel function in cerebral arterioles in piglets. Stroke 27:1874–1881
55. Armstead WM (1999) Hypotension dilates pial arteries by K_{ATP} and K_{ca} channel activation. Brain Res 816:158–164
56. Armstead WM, Kiessling JW, Riley J, Kofke WA, Vavilala MS (2011) Phenylephrine infusion prevents impairment of ATP and calcium sensitive K channel mediated cerebrovasodilation

after brain injury in female but aggravates impairment in male piglets through modulation of ERK MAPK upregulation. J Neurotrauma 28:105–111
57. Laher I, Zhang JH (2001) Protein kinase C and cerebral vasospasm. J Cereb Blood Flow Metab 21:887–906
58. Armstead WM, Cines DB, Bdeir K, Bdeir Y, Stein SC, Higazi AAR (2009) uPA modulates the age dependent effect of brain injury on cerebral hemodynamics through LRP and ERK MAPK. J Cereb Blood Flow Metab 29:524–533
59. Armstead WM (2005) Age and cerebral circulation. Pathophysiology 12:5–15
60. Choi DW (1992) Excitotoxic cell death. J Neurobiol 23:1261–1276
61. Park L, Gallo EF, Anrather J, Wang G, Norris EH, Paul J et al (2008) Key role of tissue plasminogen activator in neurovascular coupling. Proc Natl Acad Sci U S A 105:1073–1078
62. Katayama Y, Becker DP, Tamura T, Hovda DA (1990) Massive increases in extracellular potassium and the indiscriminate release of glutamate following concussive brain injury. J Neurosurg 73:889–900
63. Merchant RE, Bullock MR, Carmack CA, Shah AK, Wilner DK, Ko G, Williams SA (1999) A double blind, placebo controlled study of the safety, tolerability and pharmacokinetics of CP-101,606 in patients with a mild or moderate traumatic brain injury. Ann N Y Acad Sci 890:41–50
64. Armstead WM, Cines DB, Higazi AAR (2055) Plasminogen activators contribute to age dependent impairment of NMDA cerebraovasodilaton after brain injury. Brain Res Dev Brain Res 156:136–146
65. Nicole O, Docagne F, Ali C, Margail I, Carmeliet P, Mackenzie ET, Vivien D, Buisson A (2001) The proteolytic activity of tissue-plasminogen activator enhances NMDA receptor-mediated signaling. Nat Med 87:59–64
66. Wang YF, Tsirka SE, Strickland S, Stiege PE, Lipton SA (1998) Tissue plasminogen activator (tPA) increases neuronal damage after focal cerebral ischemia in wild-type and tPA deficient mice. Nat Med 4:228–231
67. Armstead WM, Kiessling JW, Cines DB, Higazi AAR (2011) Glucagon protects against impaired NMDA-mediated cerebrovasodilation and cerebral autoregulation during hypotension after brain injury by activating cAMP protein kinase a and inhibiting upregulation of tPA. J Neurotrauma 28:451–457
68. Fanne RA, Nassar T, Mazuz A, Waked O, Heyman SN, Hijazi N, Goelman G, Higazi AAR (2011) Neuroprotection by glucagons: role of gluconeogenesis. J Neurosurg 114:85–91
69. Armstead WM, Nassar T, Akkawi S, Smith DH, Chen XH, Cines DB, Higazi AA-R (2006) Neutralizing the neurotoxic effects of exogenous and endogenous tPA. Nat Neurosci 9:1150–1157
70. Armstead WM, Kiessling JW, Riley J, Cines DB, Higazi AAR (2011) tPA contributes to impaired NMDA cerebrovasodilation after traumatic brain injury through activation of JNK MAPK. Neurol Res 33:726–733
71. Kulkarni M, Armstead WM (2002) Relationship between NOC/oFQ, dynorphin and COX-activation in impaired NMDA cerebrovasodilation after brain injury. J Neurotrauma 19:965–973
72. Armstead WM (2001) Age dependent endothelin contribution to NOC/oFQ induced impairment of NMDA cerebrovasodilation after brain injury. Peptides 22:39–46
73. Armstead WM (2002) Age dependent NMDA contribution to impaired hypotensive cerebral hemodynamics following brain injury. Brain Res Dev Brain Res 139:19–28
74. Armstead WM, Riley J, Yarovoi S, Cines DB, Smith DH, Higazi AAR (2012) tPA-S481A prevents neurotoxicity of endogenous tPA in traumatic brain injury. J Neurotrauma 29:1794–1802
75. Català-Temprano A, Claret Teruel G, Cambra Lasaosa FJ et al (2007) Intracranial pressure and cerebral perfusion pressure as risk factors in children with traumatic brain injury. J Neurosurg 106(6):463–466

76. Carter BG, Butt W, Taylor A (2008) ICP and CPP: Excellent predictors of long term outcome in severely brain injured children. Childs Nerv Syst 24:245–251
77. Kochanek PM, Carney N, Adelson PD, Ashwal S, Bell MJ, Bratton S, Carson S, Chesnut RM, Ghajar J, Goldstein B, Grant GA, Kissoon N, Peterson K, Selden NR, Tasker RC, Tong KA, Vavilala MS, Wainwright MS, Warden CR (2012) Guidelines for the acute medical management of severe traumatic brain injury in infants, children, and adolescents-Second Edition. Pediatr Crit Care Med 13(Suppl 1):S24–S29
78. Bratton SL, Chestnut RM, Ghajar J, Brain Trauma Foundation, American Association of Neurological Surgeons, Congress of Neurological Surgeons, Joint Section on Neurotrauma and Critical Care, AANS/CNS et al (2008) Guidelines for the management of severe traumatic brain injury. J Neurotrauma 25(3):276–278
79. Khanna S, Davis D, Peterson B et al (2000) Use of hypertonic saline in the treatment of severe refractory posttraumatic intracranial hypertension in pediatric traumatic brain injury. Crit Care Med 28(4):1144–1151
80. Keenan HT, Nocera M, Bratton SL (2005) Frequency of intracranial pressure monitoring in infants and young toddlers with traumatic brain injury. Pediatr Crit Care Med 6(5):537–541
81. Cremer OL, van Dijk GW, van Wensen E et al (2005) Effect of intracranial pressure monitoring and targeted intensive care on functional outcome after severe head injury. Crit Care Med 33(10):2207–2213
82. Coles JP, Steiner A, Johnston AJ et al (2004) Does induced hypertension reduce cerebral ischemia within traumatized human brain? Brain 127:2479–2490
83. Sookplung P, Siriussawakul A, Malakouti A et al (2011) Vasopressor use and effect on blood pressure after severe adult traumatic brain injury. Neurocrit Care 15:46–54
84. Digennaro JL, Mack CD, Malakouti A et al (2011) Use and effect of vasopressors after pediatric traumatic brain injury. Dev Neurosci 32:420–430
85. Armstead WM, Kiessling JW, Kofke WA, Vavilala MS (2010) Impaired cerebral blood flow autoregulation during post traumatic arterial hypotension after fluid percussion brain injury is prevented by phenylephrine in female but exacerbated in male piglets by ERK MAPK upregulation. Crit Care Med 38:1868–1874
86. Armstead WM, Riley J, Vavilala MS (2013) Dopamine prevents impairment of autoregulation after TBI in the newborn pig through inhibition of upregulation of ET-1 and ERK MAPK. Pediatr Crit Care Med 14(2):e103–e111
87. Ract C, Vigue B (2001) Comparison of the cerebral effects of dopamine and norepinephrine in severely head-injured patients. Intensive Care Med 27:101–106
88. Beaumont A, Hayasaki K, Marmarou A et al (2001) Contrasting effects of dopamine therapy in experimental brain injury. J Neurotrauma 18:1359–1372
89. Malhotra AK, Schweitzer JB, Fabian TC et al (2003) Cerebral perfusion pressure directed therapy following traumatic brain injury and hypotension in swine. J Neurotrauma 20:827–839
90. Steiner LA, Johnston AJ, Czosnyka M et al (2004) Direct comparison of cerebrovascular effects of norepinephrine and dopamine in head injured patients. Crit Care Med 32:1049–1054
91. Johnston AJ, Steiner LA, Chatfield DA et al (2004) Effect of cerebral perfusion pressure augmentation with dopamine and norepinephrine on global and focal brain oxygenation after traumatic brain injury. Intensive Care Med 30:791–799

Chapter 16
Neutrophils as Determinants of Vascular Stability in the Injured Spinal Cord

Alpa Trivedi, Sang Mi Lee, Haoqian Zhang, and Linda J. Noble-Haeusslein

Abstract While a number of studies have examined the complex roles of leukocytes in the acute and chronically injured cord, few have specifically focused on neutrophils, where we have only recently begun to appreciate their involvement in both vascular pathogenesis and early wound healing. Here we address the mechanisms underlying neutrophil-mediated endothelial destabilization, their synergism with monocytes in modulating permeability, and their putative role as initiators of angiogenesis in the acutely injured spinal cord. Neutrophils contain a variety of bioactive molecules that are stored in granules. Studies have shown that certain of these molecules, and most notably proteases, contribute to endothelial destabilization as neutrophils degranulate during their transmigration across this front. Neutrophils have historically been regarded as detrimental to the acutely injured cord. However, there is growing evidence that this may be an oversimplified view as it fails to take into account their ability to release proteases that degrade the extracellular matrix, releasing latent growth factors that may in turn support early angiogenesis.

16.1 Introduction

This chapter focuses on how spinal cord injury-induced infiltration of neutrophils influences vascular integrity and their emerging role as initiators of angiogenesis during wound healing. Abnormal vascular permeability is evident as early as 35 min after spinal cord injury and persists thereafter for at least 28 days (Table 16.1). The early trafficking of neutrophils into the injured cord has been characterized by a number of studies [1–4] with general agreement that this trafficking begins within hours and peaks at approximately 24 h post-injury.

A. Trivedi (✉)
University of California, San Francisco, 513 Parnassus Avenue, HSE 722, San Francisco, CA 94143, USA
e-mail: alpa.mahuvakar@ucsf.edu

E.H. Lo et al. (eds.), *Vascular Mechanisms in CNS Trauma*, Springer Series in Translational Stroke Research 5, DOI 10.1007/978-1-4614-8690-9_16,

Table 16.1 Abnormal vascular permeability after spinal cord injury: time course studies

Species	Type of injury	Tracer	Time course for vascular leakage	References
Mouse	Contusion	Luciferase	35 min–14 days	[31]
Rat	Contusion	[^{14}C]-alpha-Aminoisobutyric acid	3–28 days	[83]
Rat	Transection	Horseradish peroxidase	15 min–24 h	[84, 85]
Rat	Contusion	Horseradish peroxidase	3 h–14 days	[86]
Rat	Contusion	Longitudinal dynamic contrast-enhanced MRI	3–56 days	[87]
Rat	Contusion	Horseradish peroxidase	1 h–2 days	[88]

MRI magnetic resonance imaging

There is, however, some disagreement with regard to when trafficking of neutrophils into the injured cord ceases. While trafficking has been reported to be resolved within the first week post-injury [2], others describe a more prolonged time course of infiltration corresponding to two distinct peaks—1 day post-injury followed by a second peak at 2 weeks [5]—that may persist for even months after injury [1]. It has been hypothesized that prolonged trafficking may be due to continuous expression of chemotactic factors that serve to stimulate recruitment of these cells to the injured cord [6].

While we have yet to fully appreciate what neutrophils may be doing in the more chronically injured cord, it is noteworthy that they are restricted to zones of tissue degeneration at 3–14 days and to areas of fibrosis at 28 days post-injury [5]. Importantly, neutrophil-mediated vascular interactions in the injured cord are likely influenced by both timing and context. Thus, while the temporal pattern of infiltration of neutrophils corresponds to early abnormal vascular permeability, it is conceivable that these leukocytes may be involved not only in vascular dysfunction but also angiogenesis, where newly forming immature vessels exhibit a leaky phenotype.

16.2 Overview of the Endothelial Cell

The endothelium is only one component of the neurovascular unit, which also includes pericytes, astrocytes, neurons, and the extracellular matrix [7]. Here we focus on the endothelial cell, and as a working definition, we will consider the endothelial cell proper that is flanked on the luminal and abluminal sides by its glycocalyx and the basal lamina, respectively.

Glycocalyx: For a more complete discussion of the endothelial glycocalyx, please refer to the review by Schmidt and colleagues [8]. In brief, this structure is <0.1 μm in thickness and is primarily composed of proteoglycans of the syndecan and glypican families that carry highly sulfated heparan, chondroitin and dermatan sulfates, as well as receptor-bound hyaluronan (Fig. 16.1). These glycosaminoglycans form a charged tight meshwork [9] that serves to regulate paracellular protein and fluid transit [10].

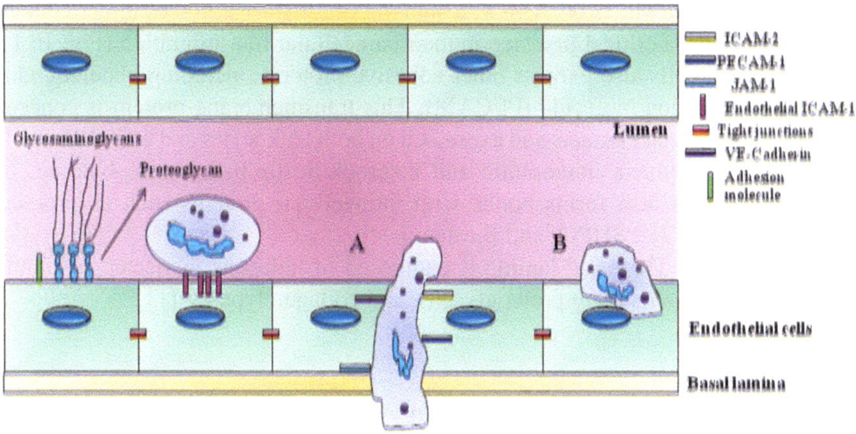

Fig. 16.1 Modes of neutrophil transmigration. Neutrophils tether, roll, adhere, crawl, and finally transmigrate. (*A*) Paracellular migration reflects migration between adjacent endothelial cells where neutrophil-derived proteases degrade the junctional complexes. (*B*) Transcellular migration occurs across the apical surface of the endothelial cell and is initiated by endothelial-neutrophil adhesion. These adhesion molecules are exposed as a result of loss of the glycocalyx. Neutrophils form actin-rich podosomes and are associated with invaginations of the endothelium. Movement through the cell is facilitated by formation of both pores and vesicles

The endothelial cell proper: Endothelial cells within the central nervous system (CNS) are characterized in part by two classes of proteins that join adjacent endothelial cells, namely, adherens junctions and tight junction. Adherens junctions (VE-cadherin, β-catenin, or γ-catenin) are intercellular junctional complexes that join adjacent endothelial cells by linking through homophilic adhesion or formation of multimeric complexes. VE-cadherin binds to β-catenin and γ-catenin which in turn binds to the intracellular proteins α-catenin, α-actinin, vinculin, and zonula occludens-1 (ZO-1). These interactions mediate anchorage to the actin cytoskeleton. Adherens junctions play an important role in contact inhibition during endothelial cell proliferation and paracellular (between cells) transmigration of leukocytes [11]. Extravasation of neutrophils has been shown to involve disorganization of the junctional components VE-cadherin and β-catenin [12].

Tight junctions are composed of both transmembrane (claudins, occludins, junctional adherens molecule (JAM) endothelial cell-specific adhesion molecule) and intracellular proteins (ZO-1 and cingulin) that seal the paracellular spaces between overlapping endothelial cells. These endothelial adhesion molecules are localized on the apical surface of the cell or within the junction. Endothelial tight junction proteins are involved in intercellular contacts and interactions with intracellular proteins, such as ZO-1, and the actin cytoskeleton and associated proteins, such as protein kinases, small GTPases, and heterotrimeric G-proteins [11]. Of the tight junction proteins, it is JAM-A, a subtype of JAM, that makes direct contact with neutrophils, a critical first step in their paracellular transmigration (Fig. 16.1).

Endothelial cells also express other adhesive proteins, including platelet endothelial cell adhesion molecule (PECAM). This transmembrane protein is concentrated at intercellular contacts and expressed on both leukocytes and endothelium. It promotes adhesion via homophilic and heterophilic ($\alpha_V\beta_3$-integrin) binding. In addition PECAM also forms bonds with intracellular proteins such as protein-tyrosine phosphatase, SHP-2, and β-catenin.

Basal lamina: The basal lamina is composed of nidogens, perlecan, laminins, fibronectin, heparan sulfate proteoglycans, and collagen type IV [11].

16.3 Neutrophil Transmigration

The glycocalyx: During inflammation, the glycocalyx changes its properties such that leukocyte rolling and adhesion may be facilitated. Cytokine-mediated activation of proteases, secreted by endothelial cells or neutrophils, partially degrades the glycocalyx thus facilitating the exposure of the selectin family of adherent molecules and hence tethered rolling of leukocytes. The importance of the glycocalyx is best illustrated in studies where this structure has been enzymatically removed. The end result is an abnormal increase in endothelial permeability and neutrophil adhesion to the endothelium (see review [8]).

There have been no studies to date that have addressed neutrophil/glycocalyx interactions in the injured spinal cord. Nevertheless, there has been a detailed analysis of the glycocalyx in response to spinal cord injury [13]. In these early studies, electron microscopy, combined with cationized ferritin, revealed this anionic charged interface is evenly distributed along the endothelial front. In response to spinal cord injury, there is a loss of these anionic properties that coincides with abnormal permeability to the protein tracer horseradish peroxidase. Such findings suggest that anionic sites contribute to vascular stabilization. Of note, the loss of the glycocalyx is transient with regeneration of this structure by 3 days post-injury. Such regeneration is likewise demonstrated in mice, infused with tumor necrosis factor-α (TNF-α), where the damaged glycocalyx is restored by 5–7 days [14].

The endothelial cell proper: The process of neutrophil transmigration is a regulated, complex, multistep process. Neutrophils interact with different parts of the endothelium leading to its activation (upregulation of adhesion molecules) and migration that begins with tethering, rolling, and finally adhesion (Fig. 16.1). Molecular players of these processes on both the neutrophils and endothelium are well defined [8, 15, 16]. After adhesion, neutrophils migrate along the endothelium and ultimately transmigrate. While transmigration may occur through the cytoplasm of the endothelial cell (transcellular route) in the CNS, the preferential route for neutrophils is between adjacent endothelial cells (paracellular route) (Fig. 16.1) [8].

For neutrophil transmigration to occur, there is a rapid transition from nonadherent to an adherent state. The first step is tethering and rolling of neutrophils along the endothelium. This involves upregulation of adhesion molecules on the endothelium. The rolling step involves L-selectin expressed on neutrophils and E- and P-selectins expressed on inflamed endothelium. Firm neutrophil adhesion to endothelial cells is mediated by the interaction of leukocyte integrins (CD11a, CD11b, CD11c/CD18) with endothelial intercellular adhesion molecule-1 (ICAM-1). During neutrophil transmigration via the paracellular route, ICAM-1 clusters beneath the adherent neutrophil along with JAM-A and PECAM-1 to facilitate adhesion and migration.

There is indirect evidence that disruption of tight junction proteins by transmigrating leukocytes contributes to destabilization of endothelial cells after spinal cord injury [17, 18]. This is inferred by the demonstration of the loss of transmembrane proteins occludin and claudin-5 as well as the intracellular tight junction component ZO-1 in spinal cord vasculature, which also appear abnormally leaky. While disruption of these proteins by leukocyte transmigration may be transient, in pathologic conditions activated neutrophils can facilitate abnormal endothelial permeability by virtue of releasing cytotoxic molecules during their transmigration [8].

Basal lamina: Constituents of the basal lamina serve as substrates for proteases, released from activated neutrophils, allowing these leukocytes to migrate across this interface [19]. Neutrophils adhere to those regions of the basal lamina that have reduced expression of matrix proteins such as laminin and collagen IV (low expression regions, LER). These regions coincide with endothelial cell junction and gaps between pericytes. Neutrophil transmigration is facilitated by their release of neutrophil elastase as well as other proteases, which collectively serve to increase the size of LER [15] and degrade junctional complexes [20–22].

16.4 Neutrophils, Their Granular Contents, and Cytokines

As noted in the preceding paragraphs, activated neutrophils interact sequentially with the glycocalyx, the endothelial cell proper, and the abluminal basal lamina. These collective interactions contribute to transient destabilization that is in part attributed to release of granular contents as neutrophils transmigrate across this interface.

Various effector molecules are presynthesized and stored in neutrophilic granules and released upon activation of neutrophils. Neutrophilic granules are divided into three groups, azurophilic, specific, and gelatinase, based on their granular contents (Table 16.2). Granules, derived from the fusion of vesicles that bud from the Golgi apparatus [23], vary in content depending on the maturity of the neutrophils. While Table 16.2 provides a comprehensive description of the contents of these granules, we will focus on those contents that have been linked to endothelial

Table 16.2 Content of neutrophilic granules[a,b]

Category of proteins	Azurophilic granules	Specific granules	Gelatinase granules	Secretory vesicles
Adhesion molecules	N/A	CD11b/CD18, CD66, CD67	CD11b/CD18, CD67	CD11b/CD18, CD67
Receptors	N/A	TNFR, uPAR	TNFR	LIR1-4, -6, -7, -9; CD35; CD16; C1q-R; IFN-αR1-αR2; IFN-γR1, -γR2; TNF-R1,-R2; IL-R; TGF-βR2; CXCR-1,-2,-4; CCR-1,-2,-3; TLR-1,-2-,4,-6,-8; CD14; MyD88; fMLPR; TREM1
Antibacterial	**Defensins**, bactericidal permeability-increasing protein, lysozyme, **myeloperoxidase**, Gp91phox/p22phox	Lysozyme, vit. B_{12}-binding protein, lactoferrin, haptoglobin, pentraxin 3, prodefensin, NGAL, hCAP18, Gp91phox/p22phox	Lysozyme, Gp91phox/p22phox	Gp91phox/p22phox
Proteases	**Cathepsins, elastase, proteinase 3**	Collagenase (MMP-8), gelatinase, plasminogen activator, cystatins	**Gelatinase**, arginase 1	
Other proteins with variable functions	Acid β-glycerophosphatase, **azurocidin**, β-glucuronidase, acid mucopolysaccharide, α1 antitrypsin, α-mannosidase, β-glycerophosphatase, heparin-binding protein, sialidase	β_2-microglobulin, histaminase, heparanase, sialidase, CRISP3	Acetyltransferase, β_2-microglobulin, CRISP3	Plasma proteins

[a] See reviews [19, 23, 89]

[b] Text in *bold* are discussed in the chapter

TNFR tumor necrosis factor receptor; *uPAR* urokinase receptor; *LIR1-4, -6, -7, and -9* leukocyte immunoglobulin-like receptors; *C1q-R* complement component 1q receptor; *CD93* IFN-αR1-αR2-γR1, and -γR2: interferons; *IL-R* interleukin receptor; *TGF-βR2* transforming growth factor beta receptor; *CXCR-1,-2,* and *-4* C-X-C chemokine receptors; *CCR-1,-2, -3* C-C chemokines; *TLR-1, -2-,4,-6, and -8* Toll-like receptors; *MyD88* myeloid differentiation primary response gene; *fMLPR N*-formyl-methionine-leucine-phenylalanine receptor; *TREM1* triggering receptor expressed on myeloid cells-1; *Gp91phox/p22phox* components of NADPH oxidase; *NGAL* neutrophil gelatinase-associated lipocalin; *hCAP18* cathelin/pro-bactenecin-like protein; *CRISP3* cysteine-rich secretory protein 3 destabilization and/or studied in the context of spinal cord injury, namely, myeloperoxidase (MPO), gelatinase, and neutrophil elastase.

Azurophilic granules are the largest granules and are also known as primary or peroxidase-positive granules based upon their content of MPO, an enzyme responsible for the oxidative burst. Neutrophil elastase, a potent proteolytic enzyme, is likewise located in azurophilic granules. Gelatinase granules contain the gelatinase, matrix metalloproteinase-9 (MMP-9), which is not complexed with its inhibitor and as such can be quickly activated upon its release [24].

When activated, neutrophils degranulate and release pro-inflammatory mediators that activate endothelial cells [25]. Cytokines such as TNF-α, interleukin (IL)-1β, and IL-17 activate endothelial cells leading to expression of adhesion molecules including P-selectins, E-selectins, and ICAMs. These activated endothelial cells then interact with neutrophils that then tether through L-selectin, roll, and finally transmigrate.

16.5 Neutrophil-Directed Destabilization of the Endothelium

Release of granular contents, including neutrophil elastase, MMP-9 and MPO, from activated neutrophils mediates abnormal vascular permeability.

Matrix metalloproteinases (MMPs): Of the proteases conveyed by infiltrating neutrophils, MMP-9 is a central player in vascular destabilization. Substrates for MMP-9 include constituents of the endothelial basal lamina (specifically laminin-5, fibronectin, heparan sulfate, and collagen type IV) [26] and tight junction proteins (occludin and claudin-5) [27].

Immunologically depleting neutrophils prior to spinal cord injury result in reduced MMP-9 activity in the injured rodent cord demonstrating that neutrophils are the principal source of this protease in the injured tissue [28, 29]. Similarly, neutrophils are thought to be a key source of MMP-9 in the injured human spinal cord [30].

There are several lines of evidence linking MMP-9 to vascular destabilization in the injured spinal cord. Abnormal vascular permeability to the protein luciferase corresponds to peak activity of MMP-9 in the injured cord [31, 32] (Table 16.1). Spinal cord-injured mice, deficient in MMP-9, or spinal cord-injured wild-type mice, treated with the broad-spectrum MMP inhibitor, GM6001, result in stabilization of the barrier [32] (Table 16.3). Similar to findings with GM6001, intrathecal administration of the selective gelatinase inhibitor SB-3CT results in reduced MMP-9 activity and stabilization of the barrier [33]. Administration of other drugs including fluoxetine, valproic acid, and atorvastatin likewise reduces abnormal vascular permeability in rodent models of spinal cord injury, as a result of diminished MMP-9 activity [34–36] (Table 16.3). This reduction in vascular permeability is at least in part correlated with attenuation in the degradation of

Table 16.3 Examples of strategies to reduce neutrophil infiltration into the injured cord

Targets	Strategy	Categories of targets	Type of injury	Species	Barrier disruption	Improved neurological outcome	References
MMP-9	KO			Mouse	↓	Yes	[32]
MMPs	Inhibitor- GM6001			Mouse	↓	Yes	[32]
MMP-9	Reduces MMP-9 activity-fluoxetine			Mouse	↓	Yes	[34]
MMP-9	Reduces MMP-9 activity-valproic acid	Enzyme	Contusion	Rat	↓	Yes	[35]
MMP-9	Reduces MMP-9 activity-atorvastatin			Rat	↓	N.D.	[36]
MMP-9	Inhibitor – SB-3CT		Compression	Rat	↓	N.D.	[33]
MPO	KO			Mouse	N.D.	Yes	[44]
Neutrophil elastase	Inhibitor- Eglin			Rat	N.D.	Yes	[39]
Neutrophil elastase	Inhibitor- L658, 758		Contusion	Rat	N.D.	Yes	[39]
Neutrophil elastase	Inhibitor-ONO-5046			Rat	N.D.	Yes	[41]
Neutrophil elastase	Inhibitor- gabexate mesilate		Compression	Rat	N.D.	Yes	[40]
GR-1	Anti-Ly6G/GR-1	Neutrophil depletion		Mouse	N.D.	Adverse outcome	[69]
GR-1	Anti-Ly6G-GR-1	Neutrophil depletion		Mouse	No effect	No change	[29]
GR-1 + circulating macrophages	Anti-Ly6-GR-1 + clodronate	Neutrophil and monocyte depletion	Contusion	Mouse	↓	Yes	[29]
CD11d/CD18 integrin	Anti-CD11d		Compression	Mouse	N.D.	Yes	[51, 55, 90, 91]
Integrin α4β1 integrin	Anti-α4, BIO5192	Adhesion		Rat	N.D.	Yes	[52, 54, 92]
ICAM-1	Anti-ICAM-1		Contusion	Rat	N.D.	Yes	[56]
ICAM-1	KO		Compression	Mouse	N.D.	No change	[57]

MMP-9 matrix metalloproteinase-9; *KO* knock-out; *MMPs* matrix metalloproteinases; *MPO* myeloperoxidase; *N.D.* not determined tight junction molecules including occludin and ZO-1, [34–36], which are known substrates for MMP-9 [37, 38].

Neutrophil elastase: Activated neutrophils may damage endothelial cells by releasing inflammatory mediators and proteases including neutrophil elastase that degrades constituents of the endothelial basal lamina. The damaging effects of neutrophil elastase are best revealed by pharmacologic inhibition of this protease after spinal cord injury [39–41]. A variety of inhibitors of neutrophil elastase have been shown to improve neurologic recovery after spinal cord contusion and compression injuries (Table 16.3). This benefit is at least in part due to their ability to stabilize the endothelium, as evidenced by reduced trafficking of neutrophils [39–41] and vascular leakage [41], and greater preservation of endothelial cell integrity.

Myeloperoxidase: Upon their activation, neutrophils produce reactive oxygen species (ROS) through a process referred to as respiratory burst. The respiratory burst is primarily characterized by the production of the superoxide anion radical, the first ROS produced by neutrophils upon their interactions with a variety of stimuli, such as, cytokines, growth factors, and opsonins. Azurophilic granules contain MPO which reacts with hydrogen peroxide and converts it to hypochlorous acid, which is more reactive than superoxide. Early studies implicated ROS in endothelial injury [42] and barrier disruption after spinal cord injury [43]. As might be expected, spinal cord-injured MPO knock-out mice show reduced production of hypochlorous acid relative to wild-type controls (Table 16.3). Similar to studies of neutrophil elastase, MPO deficiency resulted in reduced trafficking of neutrophils [44]. Given the relationship between neutrophil transmigration and barrier disruption in pathologic states, it is likely that reduced trafficking of neutrophils into the injured cord is associated with stabilization of the barrier.

Neutrophil extracellular traps (NETs): Activated neutrophils release NETs, which consist of decondensed DNA associated with proteases [45] including elastase, cathepsin G, and proteinase-3 (found in azurophilic granules) and MMP-9 (found in gelatinase granules). While NETs have not been studied in spinal cord injury, there is evidence for their involvement in brain ischemia. A recent study by Allen and colleagues [45], using both in vivo (middle cerebral artery occlusion) and in vitro models, shows that IL-1-activated neutrophils transmigrate across the endothelium. In vitro studies using inhibitors to all of the proteases rescued neurotoxicity in the presence of transmigrated neutrophils. These studies suggest that proteolytically cleaved products generated by the proteases, and hence proteases themselves, are the main mediators of neurotoxicity.

There is in vitro evidence that endothelial cells, when activated, induce NETs and in fact are susceptible to NETosis-mediated cell death [46]. Whether NETs are generated in the injured cord has yet to be determined. However, given the potential importance of this emerging field to both vascular disruption and secondary damage, studies are warranted to advance this work in the context of spinal cord injury.

16.6 Strategies to Reduce Trafficking of Neutrophils into the Injured Spinal Cord

A number of genetic, pharmacologic, and immunologic-based studies have successfully reduced neutrophil trafficking into the injured spinal cord. Of the proteases listed in Table 16.3, deficiency/blockade of MMP-9 [32–36], MPO [44], or neutrophil elastase [39–41], all constituents of neutrophilic granules, results in improved neurological outcomes after contusion or compression injury to the spinal cord, and in the case of MMP-9 and neutrophil elastase, their deficiency stabilized the barrier.

The most studied transmembrane proteins are β2-integrins (CD18/CD11) and the immunoglobulin superfamily. Binding of Mac-1 (expressed on neutrophils) to ICAM-1 (expressed on endothelial cells) is involved in neutrophil rolling along the endothelium [47]. Notably, blockade of α4β1 integrins, which are expressed on leukocytes and regulate endothelial adhesion [48–50], reduces migration of these immune cells into the injured spinal cord and results in neuroprotection with enhanced neurologic recovery (Table 16.3) [51–53]. Similarly, administration of a monoclonal antibody against the CD11d subunit of the leukocyte CD11d/CD18 integrin in a murine model of spinal cord injury reduces neutrophil infiltration and is associated with reductions in MPO activity and ROS [54, 55] (Table 16.3).

ICAM-1 is integral to vascular dysfunction after spinal cord injury. Spinal cord-injured animals, deficient in ICAM-1 via the use of immunologic or genetic approaches, result in a marked decrease in trafficking of neutrophils into the injured cord [56, 57] (Table 16.3). In addition, there is a significant reduction in spinal cord edema, improved spinal cord blood flow, and enhanced neurological recovery [56] (Table 16.3).

16.7 Synergism Between Neutrophils and Monocytes in Vascular Destabilization

Neutrophils facilitate trafficking of monocytes into the injured cord and importantly, it is their collective presence that influences stabilization of the barrier after spinal cord injury.

Trafficking of leukocytes: Neutrophils, the first leukocytes to arrive at the site of injury, secrete cytokines and chemokines thus launching communication networks and issuing instructions to practically all other immune cells. Neutrophils recruit monocytes through several different mechanisms, including expression of chemoattractants and release of granular contents. The classical monocyte chemoattractants including MCP-1, MIP-1α, and MIP-3α [58, 59] are elevated in the injured rodent spinal cord [60, 61]. Moreover, proteins, stored in neutrophilic granules, such as proteinase 3, azurocidin, Ll-37, and cathepsin G, induce monocyte recruitment [19, 62–64]. Monocyte-chemotactic activity is not only predominantly in the defensin-containing fraction of the neutrophilic granules, but defensins themselves may play a role in the recruitment of monocytes by neutrophils [65]. Finally, neutrophils indirectly affect monocyte recruitment via upregulation of endothelial adhesion molecules, increase of transendothelial permeability, stimulation of expression of chemoattractants by other cells, and modulation of chemokine activity through the proteolytic processing by protease activity [19]. The importance of neutrophils in directing infiltration of monocytes is most evident in studies in which trafficking of monocytes was found to be reduced in neutropenic animals [64, 66]. This relationship has been confirmed in spinal cord injury where neutropenia, resulting from treatment with anti-Ly6G antibody, results in a reduction of infiltrating monocytes in the acutely injured cord [29].

There are several lines of evidence supporting synergism between neutrophils and monocytes in disrupting the barrier. The first key study involved a model of viral meningitis, which produces a massive recruitment of monocytes and neutrophils into the meninges as well as prominent abnormal vascular leakage [67]. Neutrophil and monocyte depletion studies were used to determine the dependency of vascular leakage on these leukocytes. Vascular stabilization was only achieved when both leukocyte populations were depleted. More recently, a similar strategy was used to determine the extent to which these leukocytes influenced barrier leakage after spinal cord injury [29]. Depletion of neither neutrophils nor monocytes stabilized the barrier. However, depletion of both neutrophils and monocytes resulted in partial preservation of barrier integrity. Such findings emphasize complementary pathogenic functions of neutrophils and monocytes in promoting vascular destabilization.

16.8 Neutrophils as Candidate Initiators of Angiogenesis

Background: Historically neutrophils have been thought to be key mediators of early secondary pathogenesis in the injured cord by virtue of their capacity to release damaging proteases [68]. Indeed, efforts to reduce infiltration of neutrophils or target these proteases using genetic, pharmacological, and immunological approaches have yielded beneficial neurological outcomes (Table 16.3). However, in the past several years, independent studies have examined neurological outcomes in neutropenic animals after spinal cord injury. Neutrophil depletion by anti-Gr-1 antibody treatment in the injured cord resulted in either no neurological benefit [29] or in fact was detrimental to neurological recovery [69]. These findings suggest that the adverse interactions of neutrophils in the injured cord may be countered by a more favorable contribution to other biologic events. One scenario may be neutrophil-directed angiogenesis.

The timing of infiltration of neutrophils and the products they secrete may be critical in initiating angiogenesis in the injured cord. Neutrophils accumulate within the acutely damaged cord at a time when angiogenesis is emerging [18, 31, 70]. This timing and proximity provides opportunity to influence the local environment. While neutropenia results in an early decrease in mRNA levels of growth factors including VEGF in the acutely injured cord [69], there have been no studies to date that have explored the impact of this on angiogenesis in the injured spinal cord.

Angiogenesis after spinal cord injury: After spinal cord injury there is loss of vessels at the lesioned epicenter by 1–3 days post-injury [18, 70, 71]. Thereafter, vascular density increases by 7 days to control, uninjured levels [31]. Increased vascular density is most likely due to angiogenesis [70], where microvessels exhibit irregular contours and are immature in that they are leaky and do not express the glucose-1 transporter [31].

The concept that neutrophils participate in angiogenesis is not novel. In fact, their proangiogenic role has been well established in tumor models [72, 73] as well as in a brain model of angiogenesis [74]. Neutrophils trigger angiogenesis by rapidly releasing their secretory granules filled with prestored cytokines and chemokines (i.e., CXCL1, CXCL8), proteases (i.e., MMP-9, neutrophil elastase) [75, 76], and growth factors (i.e., VEGF) [77, 78] (Fig. 16.2).

Benelli and colleagues showed that Gr-1-mediated neutrophil depletion reduced CXCL1- or CXCL8-driven angiogenesis in a murine Matrigel model [75]. These neutrophil-derived chemokines in addition to functioning as chemoattractants also induced endothelial cell proliferation and differentiation. Binding of CXCL8 stimulated fast degranulation of MMP-9 from neutrophils which in turn processed CXCL8 thus making it a more potent chemokine [79].

Neutrophils release proteases including neutrophil elastase and MMP-9 that degrade the extracellular matrix thus allowing endothelial cells to proliferate and migrate. Degradation of the extracellular matrix also serves to release a wide range of sequestered cytokines, chemokines, growth factors, and their cognate receptors resulting in their activation [80] (Fig. 16.2).

One of the most studied neutrophil-derived proteases in the context of angiogenesis is MMP-9. MMPs have broad substrate specificity and are associated with tissue inhibitors of metalloproteinases (TIMPs). Several key studies have confirmed neutrophil-derived MMP-9 as a major contributor to angiogenesis. In the first study, MMP-9, purified from neutrophils, was evaluated in an angiogenic model. A major finding was that neutrophil-derived MMP-9 had high angiogenic potency that is related to its unencumbered TIMP-free status [24]. Such a finding may have strong implications in the injured spinal cord as there is an elevation of MMP-9 in the acutely injured cord [32] which is most likely derived from neutrophils [29]. In further support of MMP-9 in neutrophil-directed angiogenesis are studies of focal angiogenesis in the brain [74]. Neutropenia resulted in reduced angiogenesis and MMP-9, with the latter finding consistent with neutrophils serving as the primary source of this protease in this brain model of angiogenesis. A final key finding comes from studies of a model of early stage pancreatic islet cell carcinogenesis. MMP-9 expressing neutrophils were involved in the initial angiogenic switch that occurs in previously nonangiogenic lesions [81]. This angiogenic switch is thought to occur through the activation/release of latent VEGF which is stored in the extracellular matrix [82].

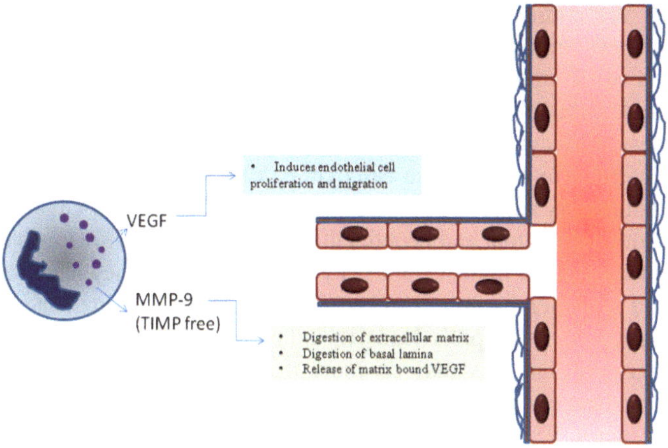

Fig. 16.2 Effects of neutrophil-derived products on angiogenesis. Neutrophil-derived MMP-9 is TIMP-free and hence bioactive. It digests the basal lamina, which is critical in the initial steps of angiogenesis that includes endothelial cell activation, migration, and proliferation. In addition, MMP-9 has the ability to digest the extracellular matrix and matrix-bound pro-angiogenic factors such as VEGF, making them bioavailable. VEGF is a potent angiogenic factor that promotes endothelial cell proliferation, migration, and tube formation and is presynthesized, stored, and released upon neutrophil activation

16.9 Summary

Our focus here has been on neutrophils as modifiers of vascular stability where we have addressed their involvement in both the acutely injured spinal cord as well as early angiogenesis. Neutrophils are the first innate responders to the injured spinal cord. Through a well-characterized series of molecular events, they home to the injured cord, where they come in contact with three major interfaces, namely, the glycocalyx, endothelium, and basal lamina. Of these structures, the least studied has been the glycocalyx, which completely covers the endothelial cells and bears an overall anionic surface charge, due to its carbohydrate moieties. In the injured spinal cord, there is a transient loss of the glycocalyx, which likely facilitates neutrophil interactions with the underlying endothelium. At and within each of these interfaces, neutrophils transiently disrupt the architecture during their transmigration across the vascular front which leads to vascular destabilization as evidenced by leakage to circulating proteins.

Neutrophils are a storehouse of proteases, including neutrophil elastase, myeloperoxidase, and MMP-9, that are released from granules during their transmigration across the endothelial front. Tight junction-related proteins and the basal lamina each contain substrates for these proteases, and as such transmigration is in part dependent on the degradation of these substrates.

While abnormal endothelial permeability is thought to arise from transmigrating neutrophils by virtue of disrupting endothelial tight junctional complexes, this is not the case for the injured spinal cord. In a series of experiments, employing techniques to deplete circulating neutrophils or monocytes, neither approach resulted in stabilization of the endothelium. Rather, partial stabilization was achieved when both leukocyte populations were depleted. These findings emphasize the synergism between these leukocytes and provide a much-needed awareness when considering strategies intended to stabilize the vasculature after spinal cord injury.

The inflammatory cascade has often been described as a two-edged sword, as it embodies both detrimental and beneficial functions in the CNS. However, rarely is this concept considered in the context of neutrophils where they have been primarily attributed to barrier disruption and secondary pathogenesis in the acutely injured cord. Depletion of neutrophils leads to unfavorable outcomes after spinal cord injury, suggesting that they may do more than initiate tissue damage. It is now becoming increasingly clear that these leukocytes play a key role in angiogenesis in a variety of non-CNS models as well as in the brain. The potential for a similar involvement in the injured cord has yet to be studied.

With the development of new genetic tools that influence functionality of activated neutrophils together with complimentary pharmacologic approaches, there is opportunity to study neutrophils in the complex environment of the injured cord where vascular destabilization and angiogenesis are superimposed upon a terrain of secondary tissue damage. With a long-term goal of developing therapies to improve recovery after spinal cord injury, a more detailed, mechanistic analysis of neutrophils will serve as a foundation for developing novel therapies that block the adverse effects of these leukocytes without interfering with putative beneficial roles in wound healing.

References

1. Beck KD et al (2010) Quantitative analysis of cellular inflammation after traumatic spinal cord injury: evidence for a multiphasic inflammatory response in the acute to chronic environment. Brain 133(Pt 2):433–447
2. Stirling DP, Yong VW (2008) Dynamics of the inflammatory response after murine spinal cord injury revealed by flow cytometry. J Neurosci Res 86(9):1944–1958
3. Hall JC et al (2012) Docosahexaenoic acid, but not eicosapentaenoic acid, reduces the early inflammatory response following compression spinal cord injury in the rat. J Neurochem 121(5): 738–750
4. Bartholdi D, Schwab ME (1997) Expression of pro-inflammatory cytokine and chemokine mRNA upon experimental spinal cord injury in mouse: an in situ hybridization study. Eur J Neurosci 9(7):1422–1438
5. Kigerl KA, McGaughy VM, Popovich PG (2006) Comparative analysis of lesion development and intraspinal inflammation in four strains of mice following spinal contusion injury. J Comp Neurol 494(4):578–594
6. Hawthorne AL, Popovich PG (2011) Emerging concepts in myeloid cell biology after spinal cord injury. Neurotherapeutics 8(2):252–261
7. Hawkins BT, Davis TP (2005) The blood-brain barrier/neurovascular unit in health and disease. Pharmacol Rev 57(2):173–185

8. Schmidt EP et al (2011) On, around, and through: neutrophil-endothelial interactions in innate immunity. Physiology (Bethesda) 26(5):334–347
9. Weinbaum S, Tarbell JM, Damiano ER (2007) The structure and function of the endothelial glycocalyx layer. Annu Rev Biomed Eng 9:121–167
10. Curry FR (2005) Microvascular solute and water transport. Microcirculation 12(1):17–31
11. Bazzoni G, Dejana E (2004) Endothelial cell-to-cell junctions: molecular organization and role in vascular homeostasis. Physiol Rev 84(3):869–901
12. Johnson-Leger C, Aurrand-Lions M, Imhof BA (2000) The parting of the endothelium: miracle, or simply a junctional affair? J Cell Sci 113(Pt 6):921–933
13. Noble LJ, Mautes AE, Hall JJ (1996) Characterization of the microvascular glycocalyx in normal and injured spinal cord in the rat. J Comp Neurol 376(4):542–556
14. Potter DR, Jiang J, Damiano ER (2009) The recovery time course of the endothelial cell glycocalyx in vivo and its implications in vitro. Circ Res 104(11):1318–1325
15. Gane J, Stockley R (2011) Mechanisms of neutrophil transmigration across the vascular endothelium in COPD. Thorax 67(6):553–561
16. Greenwood J et al (2011) Review: leucocyte-endothelial cell crosstalk at the blood-brain barrier: a prerequisite for successful immune cell entry to the brain. Neuropathol Appl Neurobiol 37(1):24–39
17. Anthony D et al (1998) CXC chemokines generate age-related increases in neutrophil-mediated brain inflammation and blood-brain barrier breakdown. Curr Biol 8(16):923–926
18. Benton RL et al (2008) Griffonia simplicifolia isolectin B4 identifies a specific subpopulation of angiogenic blood vessels following contusive spinal cord injury in the adult mouse. J Comp Neurol 507(1):1031–1052
19. Amulic B et al (2012) Neutrophil function: from mechanisms to disease. Annu Rev Immunol 30:459–489
20. Cepinskas G, Sandig M, Kvietys PR (1999) PAF-induced elastase-dependent neutrophil transendothelial migration is associated with the mobilization of elastase to the neutrophil surface and localization to the migrating front. J Cell Sci 112(Pt 12):1937–1945
21. Hermant B et al (2003) Identification of proteases involved in the proteolysis of vascular endothelium cadherin during neutrophil transmigration. J Biol Chem 278(16):14002–14012
22. Ionescu CV et al (2003) Neutrophils induce sequential focal changes in endothelial adherens junction components: role of elastase. Microcirculation 10(2):205–220
23. Borregaard N (2010) Neutrophils, from marrow to microbes. Immunity 33(5):657–670
24. Ardi VC et al (2007) Human neutrophils uniquely release TIMP-free MMP-9 to provide a potent catalytic stimulator of angiogenesis. Proc Natl Acad Sci U S A 104(51):20262–20267
25. Angelillo-Scherrer A (2012) Leukocyte-derived microparticles in vascular homeostasis. Circ Res 110(2):356–369
26. Reichel CA et al (2008) Gelatinases mediate neutrophil recruitment in vivo: evidence for stimulus specificity and a critical role in collagen IV remodeling. J Leukoc Biol 83(4):864–874
27. Rosenberg GA, Yang Y (2007) Vasogenic edema due to tight junction disruption by matrix metalloproteinases in cerebral ischemia. Neurosurg Focus 22(5):E4
28. de Castro RC Jr et al (2000) Metalloproteinase increases in the injured rat spinal cord. Neuroreport 11(16):3551–3554
29. Lee SM et al (2011) Prevention of both neutrophil and monocyte recruitment promotes recovery after spinal cord injury. J Neurotrauma 28(9):1893–1907
30. Fleming JC et al (2006) The cellular inflammatory response in human spinal cords after injury. Brain 129(Pt 12):3249–3269
31. Whetstone WD et al (2003) Blood-spinal cord barrier after spinal cord injury: relation to revascularization and wound healing. J Neurosci Res 74(2):227–239
32. Noble LJ et al (2002) Matrix metalloproteinases limit functional recovery after spinal cord injury by modulation of early vascular events. J Neurosci 22(17):7526–7535
33. Yu F et al (2008) Induction of mmp-9 expression and endothelial injury by oxidative stress after spinal cord injury. J Neurotrauma 25(3):184–195

34. Lee JY et al (2012) Fluoxetine inhibits matrix metalloprotease activation and prevents disruption of blood-spinal cord barrier after spinal cord injury. Brain 135(Pt 8):2375–2389
35. Lee JY et al (2012) Valproic acid attenuates blood-spinal cord barrier disruption by inhibiting matrix metalloprotease-9 activity and improves functional recovery after spinal cord injury. J Neurochem 121(5):818–829
36. Pannu R et al (2007) Post-trauma Lipitor treatment prevents endothelial dysfunction, facilitates neuroprotection, and promotes locomotor recovery following spinal cord injury. J Neurochem 101(1):182–200
37. Asahi M et al (2001) Effects of matrix metalloproteinase-9 gene knock-out on the proteolysis of blood-brain barrier and white matter components after cerebral ischemia. J Neurosci 21(19): 7724–7732
38. Caron A, Desrosiers RR, Beliveau R (2005) Ischemia injury alters endothelial cell properties of kidney cortex: stimulation of MMP-9. Exp Cell Res 310(1):105–116
39. Taoka Y et al (1998) Role of neutrophil elastase in compression-induced spinal cord injury in rats. Brain Res 799(2):264–269
40. Taoka Y et al (1997) Gabexate mesilate, a synthetic protease inhibitor, prevents compression-induced spinal cord injury by inhibiting activation of leukocytes in rats. Crit Care Med 25(5): 874–879
41. Tonai T et al (2001) A neutrophil elastase inhibitor (ONO-5046) reduces neurologic damage after spinal cord injury in rats. J Neurochem 78(5):1064–1072
42. Shappell SB et al (1990) Comparison of antioxidant and nonantioxidant lipoxygenase inhibitors on neutrophil function. Implications for pathogenesis of myocardial reperfusion injury. J Pharmacol Exp Ther 252(2):531–538
43. Nakauchi K et al (1996) Effects of lecithinized superoxide dismutase on rat spinal cord injury. J Neurotrauma 13(10):573–582
44. Kubota K et al (2012) Myeloperoxidase exacerbates secondary injury by generating highly reactive oxygen species and mediating neutrophil recruitment in experimental spinal cord injury. Spine (Phila Pa 1976) 37(16):1363–1369
45. Allen C et al (2012) Neutrophil cerebrovascular transmigration triggers rapid neurotoxicity through release of proteases associated with decondensed DNA. J Immunol 189(1):381–392
46. Gupta AK et al (2010) Activated endothelial cells induce neutrophil extracellular traps and are susceptible to NETosis-mediated cell death. FEBS Lett 584(14):3193–3197
47. Phillipson M et al (2006) Intraluminal crawling of neutrophils to emigration sites: a molecularly distinct process from adhesion in the recruitment cascade. J Exp Med 203(12):2569–2575
48. Davenpeck KL, Sterbinsky SA, Bochner BS (1998) Rat neutrophils express alpha4 and beta1 integrins and bind to vascular cell adhesion molecule-1 (VCAM-1) and mucosal addressin cell adhesion molecule-1 (MAdCAM-1). Blood 91(7):2341–2346
49. Shanley TP et al (1998) Requirements for alpha d in IgG immune complex-induced rat lung injury. J Immunol 160(2):1014–1020
50. Yednock TA et al (1995) Alpha 4 beta 1 integrin-dependent cell adhesion is regulated by a low affinity receptor pool that is conformationally responsive to ligand. J Biol Chem 270(48): 28740–28750
51. Bao F et al (2008) An integrin inhibiting molecule decreases oxidative damage and improves neurological function after spinal cord injury. Exp Neurol 214(2):160–167
52. Fleming JC et al (2008) Alpha4beta1 integrin blockade after spinal cord injury decreases damage and improves neurological function. Exp Neurol 214(2):147–159
53. Gris D et al (2004) Transient blockade of the CD11d/CD18 integrin reduces secondary damage after spinal cord injury, improving sensory, autonomic, and motor function. J Neurosci 24(16): 4043–4051
54. Bao F et al (2004) Early anti-inflammatory treatment reduces lipid peroxidation and protein nitration after spinal cord injury in rats. J Neurochem 88(6):1335–1344
55. Geremia NM et al (2012) CD11d Antibody Treatment Improves Recovery in Spinal Cord-Injured Mice. J Neurotrauma 29(3):539–550

56. Hamada Y et al (1996) Involvement of an intercellular adhesion molecule 1-dependent pathway in the pathogenesis of secondary changes after spinal cord injury in rats. J Neurochem 66(4):1525–1531
57. Isaksson J, Farooque M, Olsson Y (2000) Spinal cord injury in ICAM-1-deficient mice: assessment of functional and histopathological outcome. J Neurotrauma 17(4):333–344
58. Scapini P et al (2001) Neutrophils produce biologically active macrophage inflammatory protein-3alpha (MIP-3alpha)/CCL20 and MIP-3beta/CCL19. Eur J Immunol 31(7):1981–1988
59. Yoshimura T, Takahashi M (2007) IFN-gamma-mediated survival enables human neutrophils to produce MCP-1/CCL2 in response to activation by TLR ligands. J Immunol 179(3): 1942–1949
60. McTigue DM et al (1998) Selective chemokine mRNA accumulation in the rat spinal cord after contusion injury. J Neurosci Res 53(3):368–376
61. Stammers AT, Liu J, Kwon BK (2012) Expression of inflammatory cytokines following acute spinal cord injury in a rodent model. J Neurosci Res 90(4):782–790
62. Chertov O et al (1997) Identification of human neutrophil-derived cathepsin G and azurocidin/CAP37 as chemoattractants for mononuclear cells and neutrophils. J Exp Med 186(5):739–747
63. De Y et al (2000) LL-37, the neutrophil granule- and epithelial cell-derived cathelicidin, utilizes formyl peptide receptor-like 1 (FPRL1) as a receptor to chemoattract human peripheral blood neutrophils, monocytes, and T cells. J Exp Med 192(7):1069–1074
64. Soehnlein O et al (2008) Neutrophil secretion products pave the way for inflammatory monocytes. Blood 112(4):1461–1471
65. Territo MC et al (1989) Monocyte-chemotactic activity of defensins from human neutrophils. J Clin Invest 84(6):2017–2020
66. Janardhan KS, Sandhu SK, Singh B (2006) Neutrophil depletion inhibits early and late monocyte/macrophage increase in lung inflammation. Front Biosci 11:1569–1576
67. Kim JV et al (2009) Myelomonocytic cell recruitment causes fatal CNS vascular injury during acute viral meningitis. Nature 457(7226):191–195
68. Zhang H et al (2011) Role of matrix metalloproteinases and therapeutic benefits of their inhibition in spinal cord injury. Neurotherapeutics 8(2):206–220
69. Stirling DP et al (2009) Depletion of Ly6G/Gr-1 leukocytes after spinal cord injury in mice alters wound healing and worsens neurological outcome. J Neurosci 29(3):753–764
70. Casella GT et al (2002) New vascular tissue rapidly replaces neural parenchyma and vessels destroyed by a contusion injury to the rat spinal cord. Exp Neurol 173(1):63–76
71. Loy DN et al (2002) Temporal progression of angiogenesis and basal lamina deposition after contusive spinal cord injury in the adult rat. J Comp Neurol 445(4):308–324
72. Gregory AD, Houghton AM (2011) Tumor-associated neutrophils: new targets for cancer therapy. Cancer Res 71(7):2411–2416
73. Shojaei F et al (2008) Role of Bv8 in neutrophil-dependent angiogenesis in a transgenic model of cancer progression. Proc Natl Acad Sci U S A 105(7):2640–2645
74. Hao Q et al (2007) Neutrophil depletion decreases VEGF-induced focal angiogenesis in the mature mouse brain. J Cereb Blood Flow Metab 27(11):1853–1860
75. Benelli R et al (2002) Neutrophils as a key cellular target for angiostatin: implications for regulation of angiogenesis and inflammation. FASEB J 16(2):267–269
76. Coussens LM, Werb Z (2002) Inflammation and cancer. Nature 420(6917):860–867
77. McCourt M et al (1999) Proinflammatory mediators stimulate neutrophil-directed angiogenesis. Arch Surg 134(12):1325–1331, discussion 1331-2
78. Rennekampff HO et al (2000) Bioactive interleukin-8 is expressed in wounds and enhances wound healing. J Surg Res 93(1):41–54
79. Van den Steen PE et al (2000) Neutrophil gelatinase B potentiates interleukin-8 tenfold by aminoterminal processing, whereas it degrades CTAP-III, PF-4, and GRO-alpha and leaves RANTES and MCP-2 intact. Blood 96(8):2673–2681
80. Lee WL, Downey GP (2001) Leukocyte elastase: physiological functions and role in acute lung injury. Am J Respir Crit Care Med 164(5):896–904

81. Nozawa H, Chiu C, Hanahan D (2006) Infiltrating neutrophils mediate the initial angiogenic switch in a mouse model of multistage carcinogenesis. Proc Natl Acad Sci U S A 103(33): 12493–12498
82. Bergers G et al (2000) Matrix metalloproteinase-9 triggers the angiogenic switch during carcinogenesis. Nat Cell Biol 2(10):737–744
83. Popovich PG et al (1996) A quantitative spatial analysis of the blood-spinal cord barrier. I. Permeability changes after experimental spinal contusion injury. Exp Neurol 142(2): 258–275
84. Noble LJ, Wrathall JR (1988) Blood-spinal cord barrier disruption proximal to a spinal cord transection in the rat: time course and pathways associated with protein leakage. Exp Neurol 99(3):567–578
85. Noble LJ, Ellison JA (1989) Effect of transection on the blood-spinal cord barrier of the rat after isolation from descending sources. Brain Res 487(2):299–310
86. Noble LJ, Wrathall JR (1989) Distribution and time course of protein extravasation in the rat spinal cord after contusive injury. Brain Res 482(1):57–66
87. Cohen DM et al (2009) Blood-spinal cord barrier permeability in experimental spinal cord injury: dynamic contrast-enhanced MRI. NMR Biomed 22(3):332–341
88. Baldwin SA et al (1998) The presence of 4-hydroxynonenal/protein complex as an indicator of oxidative stress after experimental spinal cord contusion in a rat model. J Neurosurg 88(5): 874–883
89. Borregaard N, Sorensen OE, Theilgaard-Monch K (2007) Neutrophil granules: a library of innate immunity proteins. Trends Immunol 28(8):340–345
90. Ditor DS et al (2006) A therapeutic time window for anti-CD 11d monoclonal antibody treatment yielding reduced secondary tissue damage and enhanced behavioral recovery following severe spinal cord injury. J Neurosurg Spine 5(4):343–352
91. Saville LR et al (2004) A monoclonal antibody to CD11d reduces the inflammatory infiltrate into the injured spinal cord: a potential neuroprotective treatment. J Neuroimmunol 156(1–2): 42–57
92. Fleming JC et al (2009) Timing and duration of anti-alpha4beta1 integrin treatment after spinal cord injury: effect on therapeutic efficacy. J Neurosurg Spine 11(5):575–587

Chapter 17
Blood Biomarkers for Acute CNS Insults: Traumatic Brain Injury and Stroke

Olena Glushakova, Stefania Mondello, and Ronald L. Hayes

Abstract Proteins that are expressed in the central nervous system (CNS) are often detectable in peripheral blood and the search for biomarkers or biological signals for specific diseases and injury processes, including traumatic brain injury (TBI), has rapidly expanded. Currently no accurate biochemical assessment exists for objectively identifying the extent of damage following TBI. The ability to accurately determine the extent of initial severity of primary brain damage after TBI in the acute care setting is critical for the establishment of accurate neurologic prognosis and the guidance of appropriate therapeutic strategies (Nat Rev Neurol 9, 192–200, 2013; Neurotherapeutics 7, 100–114, 2010). This chapter will review the most current literature concerning the novel utility of blood biomarker assessment for the diagnosis and prognosis of patients with TBI.

17.1 Introduction

17.1.1 Traumatic Brain Injury and the Need for Biomarkers

According to the most recent data from the Center for Disease Control and Prevention, approximately 1.6–3.8 million people sustain a TBI in the United States alone, resulting in 52,000 deaths [1]. Moreover, TBI is the leading cause of combat deaths and it has been estimated that 15–25 % of all injuries sustained in warfare during the previous century involved TBI [2, 3]. In addition to severe TBI, more than 1.4 million people are treated and released from emergency departments [4] and the number of TBIs of mild to moderate severity far outnumbers those with severe injury [5–7]. Recent evidence suggests that participation in sports involving

R.L. Hayes (✉)
Banyan Biomarkers, Inc., 13400 Progress Blvd., Alachua, FL 32615, USA
e-mail: rhayes@banyanbio.com

E.H. Lo et al. (eds.), *Vascular Mechanisms in CNS Trauma*, Springer Series in Translational Stroke Research 5, DOI 10.1007/978-1-4614-8690-9_17,

contact and/or collisions may alter regional brain metabolic processes [8–10] and increase the risk of catastrophic neurodegenerative diseases [11–18], including chronic traumatic encephalopathy (CTE) which has been linked to repetitive concussion brain injuries [19–23]. Since diagnosis of CTE can only be made upon postmortem examination, the need for improved diagnosis and management of concussion in living athletes is also of great importance. Because the clinical symptomatology of concussion primarily reflects functional disturbances, the ability to detect and monitor TBI in living individuals is also essential to the design of therapeutic and rehabilitative strategies to improve posttraumatic outcome. It remains important to stress that TBI can no longer be regarded as a single clinical entity but rather as a spectrum of heterogeneous brain injuries that differ for each individual who requires individual assessment and appropriate personalized management to mitigate and/or prevent long-term neurological dysfunction.

During the past 2 decades, interest has increased exponentially in the elucidation and characterization of novel and selective biomarkers for TBI. Biomarkers that reliably reflect the extent of neuronal, axonal, and glial damage and/or microscopic pathologic events could conceptually become an important tool to both diagnose injury severity and predict clinical outcome in patients with TBI. Moreover, the impaired integrity of the blood–brain barrier (BBB) following moderate–severe TBI may facilitate the presence of detectable levels of biomarker proteins of brain damage in circulating blood. A PubMed search of "traumatic brain injury and biomarkers" on May 20, 2013, resulted in over 1,700 references. Since no single biomarker is likely to be diagnostic of TBI due to the complexity of the human nervous system and the heterogeneity of posttraumatic events following brain trauma, a combination of biomarkers that represent different brain cell populations with different subcellular origins will most likely provide the most optimal evidence of the underlying biochemical mechanisms involved with posttraumatic degenerative cascades and neurobehavioral dysfunction.

17.1.2 Blood Biomarkers of Acute Brain Injury

S-100B: Perhaps the most extensively studied TBI biomarker, S-100B, is a glial protein, one in a larger family of calcium-binding proteins (S100 proteins) that help regulate intracellular calcium concentrations [24]. S-100B is mainly expressed in astrocytes and has been postulated to be a marker of astrocyte injury or death [25]. Two homodimeric proteins, S100-A1 (possessing two alpha subunits) and S100-B (possessing two B subunits), have been identified, together with an aB heterodimer [26]. Increases in serum S-100B were found in amateur boxers who received direct blows to the head vs. those who had a history of blocking head blows [27], in female soccer players after a competitive game [29] and elevations reported in the acute phase after mild TBI [30–32]. In more severe TBI patients, elevated serum S-100B has also been shown in a number of studies to reliably correlate with GCS scores and neuroradiologic findings at time of hospital admission [33–35] as well as post-injury outcome [25, 35–48]. Two of these studies have suggested that assessment of serum S-100B can differentiate between patients with mild and severe TBIs [36, 37, 39, 43, 46, 49]. Recently, Defazio et al. [50] reported that serum S-100B obtained on admission and levels of S100B and plasma D-dimer obtained at 24 h were predictive for poor patient outcome status assessed 72 h post-admission. Moreover, in addition to S-100B, serum levels of S-100A1B and S100BB were also found to be elevated following pediatric TBI [51–54] in a manner related to outcome following severe TBI [55]. Although recent studies by Tavarez et al. [56] and Bouvier et al. [57] have suggested that S-100B may be a useful adjunct to the clinical evaluation and aid in minimizing the necessity of neuroimaging in pediatric patients with mild TBI, others [51] have suggested that S-100B may not be clinically useful as an independent screening test to select children with mild TBI who may need a cranial CT. Interestingly, a recent study by Lange et al. [58] suggested that the diagnostic utility of S-100B for detecting severity of TBI is highly accurate in sober brain-injured patients but compromised in the presence of acute alcohol intoxication. Despite the early interest in this protein as a sensitive biomarker of TBI-related astroglial damage, it has recently become clear that S-100B is also present in oligodendrocytes, microglia, and even extracerebral tissue including adipocytes and chondrocytes [59, 60]. Elevated levels of S-100B have also been observed in trauma patients not presenting with TBI [29, 62–64] and others presenting with hemorrhagic shock or circulatory arrest [65, 66]. It has also been reported that hospital stay in the NICU alone will induce significant increases in S100B when compared with healthy subjects [67]. Moreover, the measurement of S-100B has been shown not to be useful as a predictive biomarker of brain damage in pediatric TBI (children less than 2 years of age) based on its high normative values in this population [68–70]. Consequently, despite its early promise, the utility of this protein as a reliable TBI diagnostic biomarker remains uncertain.

Glial fibrillary acidic protein (*GFAP*): Glial fibrillary acidic protein, a type III intermediate filament, comprises the central cytoskeletal framework of astrocytes and is believed to be found only in central nervous system (CNS) glial cells and not outside the CNS [71, 72]. The reactive gliosis that is known to occur following CNS injury with concomitant upregulation of GFAP makes this protein an attractive candidate biomarker for TBI (see [73] for good review). Several studies have suggested that TBI results in elevated serum concentrations of GFAP [40, 47, 74–76] and that the measurement of this biomarker may predict both the severity of injury [25, 39, 40, 46, 77–79] and clinical outcome [25, 47, 75, 80]. Honda et al. [81] compared the sensitivity and specificity of GFAP, S-100B, and NSE in TBI patients (GCS ranging from 5 to 14) and reported that GFAP showed the most optimal specificity for injury severity (88.9 % when compared with 27.8 % for S-100B and 22.2 % for NSE). Although Metting et al. [30] found elevated levels of GFAP in patients with mild TBI and abnormal CT/MRI findings, serum levels of this biomarker did not predict patient outcome at 6 months post-injury. Circulating levels of GFAP have also been suggested to be reflective of injury severity following pediatric TBI [54, 82, 83].

Myelin basic protein (MBP): Myelin basic protein (MBP) is localized in the myelin sheath surrounding myelinated axons and was shown over 30 years ago to be released into the systemic circulation in individuals suffering from demyelinating diseases and in those sustaining TBI [84–86]. Serum MBP was recently shown to be elevated up to 2 weeks in an experimental model of TBI [87]. However, its sensitivity as a predictive biomarker of injury severity in pediatric TBI has been more recently challenged [52].

Neuron-specific enolase (NSE): Neuron-specific enolase, a protein localized in the cytoplasm of neurons and involved with regulating intraneuronal chloride levels during neural activity [88], has been proposed to be a specific biomarker for neuronal damage. Rapid elevations in serum NSE have been reported in TBI patients [36, 38, 47, 89]. Several studies have suggested that serum NSE may be useful as a predictor of short-term outcome in children with TBI [90–92]. Similarly, Ross et al. [93] suggested the utility of serum NSE as an aid to outcome prediction in adult TBI. Moreover, Berger et al. [68] evaluated serum NSE (as well as S100B and MBP) in children who sustained non-inflicted TBI (motor vehicle crashes and other accidents) and inflicted TBI (child abuse and shaken baby syndrome). These observations suggest that there may be an important ischemic component to TBI associated with child abuse vs. non-inflicted brain trauma [94]. A delayed increase in serum NSE, consistent with delayed neuronal death, was observed in those children with inflicted TBI but not after non-inflicted TBI. Increases in serum NSE (and S-100B) have been reported in amateur boxers [27], even after an extended post-bout period, suggestive of sustained release of this brain-specific protein [95, 96] while serum NSE elevations have been described in elite female soccer players after a competitive match [97]. Serum NSE has also been shown to add prognostic value to other neurobehavioral determinants of outcome assessed in the acute period following mild TBI [32]. However, methodological and technical issues associated with the pharmacokinetics of NSE (its slow elimination makes it difficult to distinguish between the extent of primary and secondary damage) have encumbered the accuracy of this protein as a diagnostic screening tool [93, 98]. Additionally, NSE has been shown to be released by blood during hemolysis, introducing a potentially significant artifact and source of error in trauma patients sustaining TBI [99, 100].

Ubiquitin C-terminal hydrolase L-1 (UCH-L1): Ubiquitin C-terminal hydrolase is a highly enriched and abundant neuronal protein that is resistant to degradation and able to diffuse into the systemic circulation after neuronal injury [101]. It is known to be involved in either the addition or removal of ubiquitin from metabolic proteins via the ATP-dependent pathway, thereby playing a critical role in ablation of damaged, misfolded, or overexpressed proteins in neurons both under normal conditions and in response to pathological insults [102, 103]. Serum UCH-L1 levels were found to be significantly increased in laboratory models of experimental TBI [104]. Elevated serum levels of UCH-L1 have been reported to be associated with abnormal BBB function after TBI [105] and several recent reports have also suggested that the elevated serum levels of UCH-L1 observed following human TBI are correlated with injury severity and clinical outcome, including GCS, evolving lesions on CT, and 6-week mortality [106]. Other studies have supported these observations in both human TBI [79, 107–111]. Mondello et al. [109] recently confirmed strong specificity and selectivity of serum UCH-L1 in the diagnosis of severe TBI (vs. controls) and the prognostic value of this biomarker in distinguishing severe TBI survivors from nonsurvivors. Early studies of mild to moderate TBI [111, 112] reported that UCH-L1 is detectable in serum within 1 h of injury and that its posttraumatic serum concentrations are predictive of injury severity and associated with GCS scores and the extent of lesions observed by brain imaging. Other than these recent reports, no persuasive studies have been conducted, to date, concerning UCH-L1 levels in peripheral blood following mild TBI.

Microtubule-associated protein-2 (*MAP-2*): Microtubule-associated protein-2 (MAP-2) is a dendritically enriched protein and marker of synaptic plasticity that has been observed to increase after TBI. Mondello et al. [113] have recently reported that severe TBI patients had significantly higher serum MAP-2 concentrations at 6 months post-injury when compared to normal controls. In these studies, increased plasma MAP-2 levels correlated with GOSE and LCFS scores at 6 months, while significantly lower serum levels of MAP-2 were observed in patients who showed worsened outcome (vegetative state). These data underscore the potential utility of circulating MAP-2 as a biomarker for emergence to higher levels of cognitive function and suggest that remodeling of synaptic junctions and neuroplasticity processes occur up to several months following severe TBI in humans.

Neurofilament (*NF-H*): Neurofilaments are comprised of neuron-specific intermediate filaments consisting of one light subunit (NF-L) plus either a medium chain subunit (NF-M) or heavy chain subunit (NF-H) [114]. Phosphorylated heavy chain neurofilament (pNF-H) is the extensively phosphorylated, axon-specific form of the NF-H subunit of neurofilament and represents one of the most abundantly distributed axonal proteins. Serum concentrations of NF-H have been reported to be unrelated to isolated changes in BBB integrity and the use of serum pNF-H as a biomarker of axonal loss released into the systemic circulation following TBI-associated white matter injury has been proposed [115, 116]. Serum levels of NF-H were found to be increased from 1 to 14 days in an experimental model of mild acceleration–deceleration TBI [87]. Serum NF-H concentrations have also been shown to remain elevated at 204 days following pediatric TBI in patients who had a poor prognosis while acute serum decreases were predictive of improved outcome [54]. These same authors have previously suggested that hyperphosphorylated NF-H may be a useful sensitive predictor of mortality following brain injury in children [117].

Tau- and amyloid-related proteins: Serum amyloid-A levels have been reported to be rapidly increased in brain trauma patients [118]. Tau protein is found abundantly in thin nonmyelinated axons of cortical interneurons [119]. Although increases in plasma tau protein have been reported to occur in comatose patients treated with hypothermia after cardiac arrest [120] and following hypoxic brain injury from cardiac arrest [121], few studies have evaluated serum tau levels after

TBI. Serum tau has been shown to increase from 1 to 14 days in a model of experimental acceleration–deceleration TBI [87]. A recent pilot study reported an increase in plasma tau concentrations of more than 3,005 over control levels in amateur boxers following a single bout [122]. Circulating tau may therefore represent a useful serum biomarker for brain injury associated with mild or repetitive TBI [19]. Despite the recent development of novel ultrasensitive digital ELISA techniques, capable of single-molecule arrays [121], little additional work has been performed to evaluate the potential utility of this biomarker in the diagnosis and treatment of TBI.

AlphaII-spectrin: AlphaII-spectrin is a major structural component of the cytoskeleton, particularly abundant in axons and in presynaptic terminals [123], and a major substrate for cysteine proteases calpains and caspase-3 involved in necrotic and apoptotic cell death. Proteolytic cleavage of αII-spectrin by calpain results in signature products of molecular weight 150 kDa (SBDP150) and 145 kDa (SBDP145), while the major cleavage product of αII-spectrin associated with caspase-3 proteolysis has been found to be a 120 kDa fragment (SBDP120) [123, 124]. Since calpain and caspase-3 represent two classes of the known major executioner proteins involved with necrosis and apoptosis following TBI [125–130], concurrent measurement of these SBDPs can provide important information concerning the underlying cell death mechanisms involved with TBI-associated axonal damage [25, 131].

Inflammatory cytokines and markers of metabolic/oxidative stress: Circulating cytokines and inflammatory proteins and acute phase reactant proteins have been observed to be elevated after CNS injury suggested to be biomarkers of CNS dysfunction and damage following TBI (see [132–134] for recent reviews). Although interleukin-1beta (IL-1β), a pro-inflammatory cytokine, has been shown to contribute to the development of posttraumatic astrogliosis [135], there is a paucity of studies reporting circulating IL-1 changes after TBI, perhaps because this protein has proved difficult to measure following human TBI [136]. Serum levels of IL-1β obtained within 6 h of injury were found to correlate with GCS after severe TBI [137]. Serum interleukin-6 (IL-6) has been shown to be elevated following TBI and useful for the differential diagnosis of elevated ICP in these patients [118, 138, 139]. Tumor necrosis factor alpha (TNF-alpha), another pro-inflammatory cytokine released by CNS microglia, astrocytes, neurons, and endothelial cells, is known to stimulate neutrophil and monocyte recruitment [28, 140–142]. Increased serum TNF-alpha levels have been reported in serum from TBI patients [143, 144]. Although Crespo et al. [145] found no correlation between acute increases in serum TNF-alpha and mortality in severely injured TBI patients, more recently, Stein et al. [146, 147] reported that increased serum TNF moderately correlated with subsequent increases in ICP or decreases in CPP following severe TBI.

Acute phase-reactive protein C-reactive protein has been shown to rapidly increase in serum of TBI patients [118]. Plasma levels of biomarkers of lipid peroxidation (thiobarbituric acid reactive species/TBARS) and protein oxidation (carbonyl) have been reported to increase significantly within the first 70 h after severe TBI but were not related to mortality or patient outcome [148]. MicroRNAs are small RNA molecules that are expressed endogenously in the CNS and play an important role in regulating gene expression. Recently, serum levels of the microRNA let-7i (and CSF levels as well) were found to be elevated in the acute post-injury period following experimental blast TBI [149]. Recently, serum levels of angiopoietins, important regulators of vascular structure and function, whose release into serum is believed to be hallmark indicators of vascular injury, have been suggested to be a promising biomarker for TBI [150]. These authors emphasize the study of the ratio of angiopoietin-1 (expressed in intracerebral and pial vessels) to angiopoietin-2 (minimal expression in brain microvessels) to assess posttraumatic BBB breakdown. Since injury to peripheral organs can cause increases in serum inflammatory markers, the measurement of these biomarkers may not provide high specificity for the extent of brain damage associated with TBI.

17.1.3 CSF Biomarkers of Acute Brain Injury

Since circulating cerebrospinal fluid (CSF) is in direct contact with the CNS extracellular space, biochemical sequelae of tissue injury should be accurately reflected by proteins released directly into the CSF. It is therefore generally assumed that the CSF represents an optimal source of biomarkers of TBI [131]. However, the invasiveness of lumbar puncture makes this sampling technique difficult (e.g., for concussed athletes) with the exception of those patients admitted for moderate–severe TBI where an accessible CSF drain may be inserted as part of routine clinical care.

S-100B: Unlike studies of serum S-100 proteins, there is a paucity of analyses of CSF levels of S-100 after TBI. Berger and colleagues [68, 151] reported elevations in CSF S-100B concentrations after severe TBI in infants and children. Using lumbar puncture after a bout, slightly elevated levels of s-100B have been recently reported in amateur boxers [152]. Acute elevations in S-11B concentrations in CSF (up to 3 days) have been reported to be predictive of clinical deterioration to brain death following severe TBI (but less so than NSE measurements) [153], and elevations in CSF S-100B were associated with episodes of intracranial hypertension and cerebral hypoperfusion during a week of ICP monitoring in patients sustaining severe TBI [146, 147]. CSF S-100B levels were found to be reflective also of the severity of posttraumatic secondary insults in severe TBI [154]. More recently, CSF levels of S100B measured over the first week following severe TBI in adults showed a superior predictive power regarding outcome over single point estimates [155]. In a novel biomarker study after experimental blast TBI, S-100B showed an acute increase at 6 h post-injury which returned to baseline values by 24 h [156].

GFAP: Neselius et al. [152] also evaluated GFAP in CSF obtained via lumbar puncture from amateur boxers and, similar to S-100B, observed the marginal elevation in CSF GFAP concentrations following a bout. Levels of GFAP in ventricular CSF have been shown to be markedly elevated in patients with severe TBI and analysis of this biomarker was recently used to improve the predictive power of the IMPACT outcome assessment [108]. Unlike S-100B, CSF levels of GFAP showed a biphasic pattern of increased concentrations, first increasing by 6 h, returning to baseline by 24 h and increasing again thereafter at 72 h post-injury, and remaining increased up to 2 weeks [156] following experimental blast TBI. Recently, novel antibodies detecting human GFAP and GFAP breakdown products have been developed and found sensitive to detect changes in CSF following penetrating experimental TBI and after severe TBI in human patients [157, 158].

Neuron-specific enolase: In both adult and pediatric patients with severe TBI and coma, CSF concentrations of NSE were found to correlate with mortality (higher levels in nonsurvivors than survivors) and/or with TBI severity scores, including GCS and GCS [93, 151, 159–163]. More recently, Bohmer et al. [153] reported that acute CSF levels of NSE (up to 3 days post-injury) were superior to S-100B and GFAP in predicting poor outcome following severe human TBI. Cerebrospinal NSE levels were also found to be associated with episodes of intracranial hypertension and cerebral hypoperfusion following severe TBI in human patients [147]. Following experimental blast TBI, CSF concentrations of NSE showed a biphasic pattern similar to GFAP, with levels significantly increased at 6 h, decreasing to baseline by 24 h, with a secondary increase at 72 h that remained significantly elevated up to 2 weeks post-injury [156]. However, as for serum measurement of NSE, CSF concentrations of NSE may be artificially elevated following hemolysis, thereby reducing the specificity of this marker for CNS damage following TBI.

UCH-L1: CSF concentrations of UCH-L1 have been shown to be elevated in laboratory models of experimental TBI [104] and in CSF obtained from patients with severe TBI [162]. Mondello and colleagues [39, 164, 165] reported that CSF concentrations of the deubiquitinase UCH-L1 could be used as a valuable prognostic indicator of patient outcome following severe TBI. Another recent study confirmed found that CSF levels of UCH-L1 were elevated following severe TBI and that higher peak UCH-L1 concentrations, measured by ELISA, were associated with GCS and evolving lesions on computer tomography. The magnitude of UCH-L1 elevation was also associated with mortality at 6 weeks post-injury [106].

Neurofilament-light (NF-L) and phosphorylated Tau (pTAU): Light chain neurofilament (NF-L) and tau are abundant in neuronal axons: NF-L is highly expressed in large-caliber myelinated axons projecting into deep brain layers [166], while tau protein is found abundantly in thin nonmyelinated axons of cortical interneurons [119]. CSF levels of these proteins have been proposed as possible sensitive biomarkers for axonal damage and degeneration [167–169]. Increases in CSF NF-L have been reported in amateur boxers that returned to normal only following a minimum of 3 months rest period [95]. Similar elevations of lesser magnitude were observed for CSF T-tau concentrations. These observations were confirmed in a more recent study on Olympic boxers [122]. The magnitude of NF-L elevation was also found to be correlated with number and severity of blows to the head and the reported intensity/severity of blows sustained [95]. The greater rise in CSF NF-L when compared with tau is suggestive of greater damage to long myelinated axons in white matter following repetitive TBI [48]. These observations suggest that assessment of CSF NF-L and tau proteins may have clinical utility for monitoring the extent of axonal damage in boxes and other athletes subjected to repetitive TBI [19]. Additionally, elevated levels of pNF-H have been reported in ventricular CSF in patients with severe TBI [162]. Elevated levels of T-tau in CSF following severe TBI have also been observed [170–172] and correlated with outcome at 1 year post-injury [171, 173].

Amyloid-related proteins: Intra-axonal accumulation of beta-amyloid (AB) and its precursor amyloid precursor protein (APP) has been reported after brain trauma in humans [174–187], even in long-term survivors [188]. Presumably, AB is released into the tissue surrounding damaged axons (where it may lead to posttraumatic plaque formation) and may leak into CSF via a compromised BBB. To this end, ventricular CSF levels of AB-40, AB-42, and APP have been shown to increase during the first week following severe TBI [189, 190], but not following milder TBI associated with boxers [95, 152]. Recently, Mondello et al. [191] reported that in severe TBI patients, increases in CSF concentrations of alpha-synuclein occur and suggested that these changes may reflect widespread neurodegeneration and secondary posttraumatic neuropathologic events.

AlphaII-spectrin breakdown products (SBDP): Elevated CSF levels of SBDPs have been reported following experimental TBI [125]. Similarly, elevated SBDPs were reported in CSF in humans [192, 193] following TBI that were associated with both severity of brain injury and outcome. These studies suggest that the biomarkers of proteolysis differentially associated with calpain and caspase-3 activity have distinct temporal profiles in CSF following TBI that are suggestive of a prominent role for calpain-induced proteolysis. Siman et al. [162] evaluated 11 neuron-enriched protein biomarkers in CSF from patients with severe TBI, including 14-3-3beta, 14-3-3zeta, UCH-L1, three distinct phosphoforms of NF-H, NSE, alpha spectrin, and three calpain- and caspase-derived fragments of alpha spectrin, and found that nine were significantly elevated between 24 and 96 h post-injury. Subsequent studies have shown that SBDPs in CSF are significantly elevated in TBI patients who died when compared with survivors [194, 195]. Increases in CSF levels of the biomarkers SBDP145 and SBDP120 were observed to be strongly predictive for mortality at 3 months post-injury.

Albumin: The CSF/serum ratio of albumin, a protein synthesized predominantly by the liver, is generally regarded as an accurate biomarker of BBB integrity [196]. Elevated concentrations of albumin in CSF, believed to represent a compromised or damaged BBB, have been reported following severe TBI [197].

Inflammatory cytokines and markers of metabolic, oxidative stress and cell death: TBI is known to initiate an acute and chronic inflammatory response in the injured brain that is reflected in increases in CSF concentrations of pro-inflammatory cytokines. Initial studies by McClain and colleagues [198, 199] originally described increases in CSF concentrations of interleukin-1beta (IL-1β) and interleukin-6 (IL-6) following severe TBI in human patients [200] reported increased CSF levels of IL-1 and IL-6 following severe TBI and observed that peak CSF IL-1 concentrations correlated with 3-month clinical outcome. In both adult and pediatric TBI, elevated CSF IL-1β concentrations were shown to be reflective of poor GOS [201] and other studies have reported that elevated CSF Il-1β concentrations are associated with poor outcome and increased ICP in severely brain-injured patients [202, 203]. Increases in CSF levels of IL-1β have been recently reported to be accurately reflective of injury severity in an experimental model of TBI [204]. Increases in CSF levels of IL-6 have been shown to occur between 3 and 6 days post-TBI [136] and have been related to functional outcome in pediatric TBI [205]. Similar elevations in CSF levels of TNF-alpha have been reported after severe TBI, but no correlation with outcome has been established [136, 144, 147, 202, 203]. Both acute and/or delayed elevations in levels of other CSF inflammatory proteins have subsequently been reported following TBI, including interleukin-8 (IL-8), interleukin-10 (IL-10), activin A, complement-derived soluble membrane attack complex (sC5b-9), and MCP-1 [61, 94, 136, 197, 200, 203, 206–217]. Anti-inflammatory cytokines (Il-10 and transforming growth factor-beta (TGF-β)) are known to counteract the deleterious effects of the pro-inflammatory cytokines and have also been shown to be elevated following TBI. Increases in CSF levels of IL-10 have been shown to occur within 24 h following severe TBI and are associated with a corresponding decrease in TNF-alpha levels [194, 218]. In children with severe TBI, IL-10 has been reported to correlate with outcome, including mortality [206, 207]. Levels of TGF-β in CSF have also been found to rapidly increase in the acute posttraumatic period and remain elevated for more than 3 weeks post-injury [219]. However, since plasma levels of inflammatory proteins are normally higher than CSF, passive leakage across a compromised BBB may result in elevated CSF levels in the absence of neuroinflammation [131], thereby potentially reducing the sensitivity of CSF pro- and anti-inflammatory biomarkers in the assessment of TBI severity. Other proteins related to metabolic and oxidative stress and/or cell death pathways (e.g., adenosine, Bcl-2, matrix metalloproteinases 3 and 9, adrenomedullin, and cytochrome C, a putative biomarker of apoptosis) have also been reported to increase in CSF following severe TBI in adults and children [209, 220–227]. Activated caspase-9 and cytochrome C were elevated in the CSF of patients with severe TBI with a weak correlation observed between CSF caspase-9 and neurologic outcome [228]. Wagner et al. [229, 230] more recently demonstrated the prognostic value of Bcl-2 and cytochrome C profiles, measured in the acute (first 6 days) posttraumatic period following severe TBI in adults. Similarly, CSF levels of cytochrome C and high-mobility group box 1 (HMGB1), a ubiquitous nuclear protein that is released from damaged/necrotic cells as well as from immune cells, have been recently reported to predict outcome after TBI in infants and children [231]. Evaluation of CSF levels of complement activation via analysis of the "membrane attack complex" C5b9 in CSF demonstrated that increases in C5b9 reflected serious secondary insults in patients with severe TBI [154]. Adamczak et al. [232] have also recently reported that biomarkers of the "inflammasome," a regulator of the immune response that activates caspase-1 and IL-1β and includes the inflammasome proteins ASC, caspase-1, and NALP-1, are elevated after severe TBI. Expression of each individual biomarker protein was correlated significantly with the GOS at 5 months post-injury and found to be significantly higher in patients with unfavorable outcome (death and severe disability).

17.1.4 Biomarkers in Acute Ischemic Brain Injury/Stroke

17.1.4.1 The Global Burden of Stroke

Stroke is the second leading cause of death worldwide and the third leading cause of death in the United States. In the United States, stroke is the leading cause of disability and the third leading cause of death, with 800,000 people experiencing strokes per year. Stroke remains an important global health burden [233]. As the third leading cause of death in the United States, stroke affects 700,000 Americans annually [234]. Every 45 s someone in America has a stroke and every 3 min someone dies of a stroke [234]. Stroke is likely to continue as the most prevalent cause of disability, despite the declining incidence of stroke, as the overall population ages [233].

17.1.4.2 Potential Uses of Biomarkers in Stroke Management

In coronary artery disease, the use of serum biomarkers, initially creatinine kinase and now troponin, has improved the diagnosis, triage, and disposition of patients with chest pain. The initial triage of the patient with chest pain is now done by emergency room personnel, often without cardiology consultation, with heavy reliance on ECG and biomarker activity. In a similar manner, a sensitive and specific cerebral biomarker could impact the delivery of stroke care in an equally important manner.

A clinically useful biomarker for the diagnosis of stroke does not currently exist. Unlike current diagnostic tests for stroke which include subjective neurological assessments and expensive neuroimaging scans, a low-cost, blood-based biomarker assay will potentially provide the first objective measure of stroke-associated brain injury. Rapid and objective discrimination of stroke type in acute care environments is likely to increase the number of ischemic patients receiving thrombolytic therapy, promoting their recovery and saving millions of healthcare dollars per year. Biomarkers could improve stroke care by allowing early diagnosis by nonexpert clinical providers, serial monitoring of patients, and rapid assessment of severity of brain injury.

A cerebral biomarker would complement present neuroimaging modalities for the diagnosis of stroke. The use of biomarkers would be of particular value in identifying patients with nonlocalizing or transient neurological symptoms or when neuroimaging cannot be obtained or is nondiagnostic. This would avoid delays in triaging stroke (or TIA) patients to appropriate care centers and allow expedited treatment of patients at high risk for early stroke recurrence.

Certain biomarker signatures could help the selection of appropriate treatment plans for patients with acute stroke. In "wake-up" stroke, where stroke onset is uncertain, low or, alternatively, high biomarker levels could identify recent or relatively remote stroke onset respectively. This information could be used in conjunction with acute neuroimaging patterns (e.g., diffusion perfusion mismatch) to determine if salvageable tissue is present and lead to a more appropriate therapy [235]. It is conceivable that an early high biomarker activity could identify patients at risk for secondary complications of stroke, particularly intracerebral hemorrhage [236]. In those patients caution may be warranted, while low levels of biomarkers may identify patients at lower risk of bleeding, who would benefit from more aggressive revascularization measures or antithrombotic regimens. Certain biomarker signatures may also identify patients at risk for developing malignant edema, for which presently no clinical or imaging predictors exist [236, 237].

Following biomarker activity during the first few days of hospitalization may lead to further insights into stroke progression. Often stroke symptoms worsen or fluctuate for unclear reason. Possible causes of worsening include infections, fever, metabolic derangements, edema, or continued ischemia. Serial monitoring of biomarker activity could potentially identify patients with ongoing ischemia who may benefit from more aggressive stroke management.

S100B: Several studies have demonstrated that serum concentrations of S100B are significantly increased following stroke with the serum concentration increasing up to 48 h after symptom onset and the peak occurring within the first 24 h after infarction. Significant correlations between S100B concentration in blood and the size of infarction area were also demonstrated in a variety of clinical or experimental studies [202]. While S100B might not be a useful single biomarker for differentiation between acute ischemic, hemorrhagic, and stroke mimics, it shows potential as clinical biomarker for stroke in a panel of biomarkers.

GFAP: Initial studies in stroke demonstrated increased serum concentrations of GFAP in ischemic stroke vs. controls [168]. In stroke a recent prospective study by Foerch et al. [27] included 93 patients with IS and 42 patients with ICH within 6 h after symptom onset. GFAP was detectable in serum of 81 % of ICH patients but only 5 % of IS patients. The mean GFAP concentration in serum was significantly higher in patients with ICH. A subsequent multicenter study [225] of S100B, neuron-specific enolase (NSE), GFAP, and activated protein C-protein C inhibitor complex (APC-PCI) demonstrated a significant ability of GFAP to distinguish ICH from ischemic stroke. GFAP has also been shown to be a sensitive serum biomarker of brain damage in patients with smaller lacunar lesions or minor stroke [19]. In a recent publication, GFAP showed a late upregulation at 24-h postischemia/reperfusion injury, indicating the presence of reactive gliosis in the middle cerebral artery territory [49].

αII-Spectrin BDPs: These signature cleavage products are apparent in brain tissue and CSF after brain ischemia [238, 239]. SBDP 145 and SBDP 150 provide a highly sensitive measure of calpain activation primarily associated with oncotic cell death. However, calpain can in some situations contribute to apoptotic cell death [132]. Thus, SBDP 150 is primarily associated with calpain-mediated oncosis but to a lesser extent also results from caspase-3-mediated apoptosis. SBDP 145 could result from some lesser amount of apoptotic cell death. SBDP120 is a sensitive measure of caspase-3 activation associated exclusively with apoptotic cell death.

The ability to monitor both calpain and caspase-3 activation during the acute period of CNS injury is a major advantage of αII-spectrin over other biomarkers. Indeed, preliminary data obtained from CSF of severely injured TBI patients indicate that accumulation of calpain- and caspase-3-mediated SBDPs ran show different temporal patterns of expression that vary in each patient [61, 115]. Similar patterns of changes in SBDP 150 and SBDP 120 were observed following sub-arachnoid hemorrhage [156]. This variability emphasizes the heterogeneous nature of CNS pathology after stroke and points to important implications for individualized treatment of brain-injured patients tailored to specific neurochemical cascades in the injured brain.

MAP-2: Dendritic damage after CNS injury, including stroke, has not been studied thoroughly. Mattson et.al. [240] reported that apoptotic changes can occur locally in synapses and dendrites, stressing the role of dendrites in propagating apoptotic signals to the soma and perhaps explaining the early dendritic degeneration seen in a variety of CNS disorders, including cerebral ischemia and Alzheimer's disease. MAP-2 loss has been documented following ischemic damage [19, 50] and excitotoxic lesioning [49]. Studies in our laboratory have documented MAP-2 loss, a dendritically enriched protein [241, 242].

MAP-2a/2b is largely located in neuronal cell bodies, dendrites, dendritic spines, and postsynaptic densities. MAP-2a/2b colocalizes with actin in dendritic spines and postsynaptic densities, which may modify microfilament stability. Although MAP-2 is not exclusively present in dendrites, it is important to note that it is highly enriched in the microtubule network in dendrites which are known to be vulnerable to ischemic damage.

UCH-L1: It has been suggested that UCH-L1 plays a critical role in the removal of excessive-oxidized or misfolded proteins both during normal and neuropathological conditions [67, 102, 143, 210]. Based on this important neuronal function and its high specificity and abundance in the CNS, we have selected UCH-L1 as a candidate biomarker for brain injury. Levels of UCH-L1 are elevated in CSF and serum following experimental middle cerebral artery occlusion [135].

βIII-tubulin: Class III tubulin is abundant in the brain and predominantly expressed during fetal and postnatal development [143]. During neurogenesis, the distribution of class III β-tubulin is neuron associated in progenitor cells in the cerebellum and in differentiating precursor cells in the subventricular zone. Although βIII tubulin has been identified as a biomarker in brain injury [210], little work has been performed to evaluate the utility of this biomarker in experimental models of stroke.

17.2 Inflammatory Cytokines, Markers of Metabolic Oxidative Stress and Cell Death

In the area of stroke research, Allard et al. have identified PARK7 (DJ-1) protein and nucleotide diphosphate kinase A protein as potential ischemic stroke markers, but their brain specificity and distribution have not been well characterized [243]. Endothelial monocyte-activating polypeptide II precursor (EMAP-II) is another potential microglia biomarker identified by differential neuroproteomics [244, 245], which is unregulated in the CSF and plasma after experimental stroke. These biomarkers could represent different pathways that can be at play at various time points after the initial injury. Finally, neuroinflammatory-linked cytokines [interleukin (IL)-6, IL-8, tumor necrosis factor-α (TNF-α), and MMP9] have been also studied in the area of stroke biomarkers. It is also appropriate to think of stroke biomarkers as a continuum of biomarkers that might be released at different time points following the initial brain injury event [246].

17.3 Biomarker-Based Stroke Management: Therapeutic Implications

Precise diagnosis of stroke patients is typically made by trained clinicians, supported by neuroimaging, usually a brain computed tomogram (CT), in some cases supplemented by diffusion- and perfusion-weighted magnetic resonance imaging (MRI). This allows ready differentiation of ischemic stroke from hemorrhagic ones. However, CT is often normal in the acute phase of stroke and negative in the presence of small ischemic lesions or in certain brain locations (posterior fossae). As a result, a diagnosis of stroke often requires clinical interpretation by highly trained personnel. Emergency room personnel, likely the first providers to see a patient with potential stroke, are less confident in making a diagnosis in the absence of objective laboratory confirmation. Stroke diagnosis is further complicated by the diversity of presenting symptoms. In the presence of focal deficits (i.e., weakness), stroke is relatively obvious, but with nonlocalizing symptoms such as delirium, seizures, dizziness/vertigo, or transient symptoms, the diagnosis can be more challenging. The presence of aphasia (left hemisphere) or profound neglect (right hemisphere) is often interpreted as confused delirium and is frequently misdiagnosed by nursing or primary care physicians even in hospitalized patients [247]. Stroke mimics include postictal states, hemiplegic migraines, brain tumors, epilepsy, encephalopathies, and at times metabolic derangements (e.g., hypoglycemia), all of which make the early diagnosis of stroke difficult. The diagnosis of transient symptoms, such as a transient ischemic attack (TIA), essentially viewed as a stroke equivalent, is difficult for trained clinicians, with substantial disagreement even between neurologists themselves [248, 249]. On the other hand, MRI can be helpful if ordered, but MRI is not readily available in all facilities and requires cooperation of frequently agitated patients and may be contraindicated in some patients. In addition, the need to identify patients with acute stroke or TIA is highlighted by the high incidence of early stroke recurrence. Approximately 11 % of TIA patients will experience a stroke at 90 days and half of those will occur within 48 h. Identifying such patients early would allow secondary stroke prevention treatment to be implemented more rapidly. The difficulties of stroke diagnosis are greatly increased in the acute settings. In the treatment of acute stroke, intravenous tissue plasminogen activator (i.v. tPA) has to be administered within 3 h of symptoms onset (although this window may be closer to 4.5 h). Pressed for time, a careful clinical diagnosis or complete neuroimaging evaluation may not be possible. Only 45 % of all stroke patients receive i.v. tPA, in part due to the reluctance of non-neurologically trained emergency room personnel to administer a potentially dangerous treatment to patients with unclear diagnosis [250]. In addition, the presence of intracranial hemorrhage, in which case these agents are contraindicated, must first be ruled out. Therefore, blood biomarker-based diagnostic test in the differential diagnosis of acute stroke will be highly valued in stroke clinical onset. In one recent stroke biomarker study which includes 100 stroke patients, it was reported that MMP-9 and D-dimer were found to be effective separately at differential diagnosis of ischemic–hemorrhagic stroke; there was no significance for S100β [251]. However, S100β and BNP have no place in the differentiation of hemorrhagic from ischemic stroke when used individually. However, when combined with BNP, D-dimer, MMP9, and S100β, it has more significance. Thus, it would be better to use a panel of biomarker tests rather than being used individually to differentiate hemorrhagic from ischemic stroke [252]. To this end, we have assessed a panel of the glial markers S100β and GFAP and UCH-L1 by ELISA in patient serum 1, 2, 3, 4, and 7 days after stroke onset. Of interest, all biomarkers increased poststroke indicative to be a good measure to evaluate brain damage [253].

17.4 Summary

In summary, ideal biomarkers for both TBI strokes that have been identified and validated with preclinical animal models need to be translated and validated in clinical studies as well. They should be tested in terms of their ability to detect injury magnitude as well as drug-based biomarker level reduction. A direct comparison of biomarker occurrence between preclinical models and biomarker data from human clinical studies would allow investigators to gain considerable insight into the validity (or challenges to the validity) of the employed preclinical animal models. Finally, the sensitive and specific cerebral biomarker could impact the delivery of TBI and stroke care in a critically important manner.

References

1. Langlois JA, Rutland-Brown W, Wald MM (2006) The epidemiology and impact of traumatic brain injury: a brief overview. J Head Trauma Rehabil 21:375–378
2. Okie S (2005) Traumatic brain injury in the war zone. N Engl J Med 352:2043–2047
3. Sapsford W (2003) Penetrating brain injury in military conflict: does it merit more research? J R Army Med Corps 149:5–14
4. Faul M, Wald MM, Sullivent EE, Sasser SM, Kapil V, Lerner EB, Hunt RC (2012) Large cost savings realized from the 2006 Field Triage Guideline: reduction in overtriage in U.S. trauma centers. Prehosp Emerg Care 16:222–229
5. Mccrory P, Meeuwisse W, Johnston K, Dvorak J, Aubry M, Molloy M, Cantu R (2009) Consensus statement on concussion in sport–the 3rd international conference on concussion in sport held in Zurich, November 2008. J Sci Med Sport 12:340–351
6. Vollmer DG, Dacey RG Jr (1991) The management of mild and moderate head injuries. Neurosurg Clin N Am 2:437–455
7. Yealy DM, Hogan DE (1991) Imaging after head trauma. Who needs what? Emerg Med Clin North Am 9:707–717
8. Giza CC, Hovda DA (2001) The neurometabolic cascade of concussion. J Athl Train 36:228–235
9. Vagnozzi R, Signoretti S, Cristofori L, Alessandrini F, Floris R, Isgro E, Ria A, Marziale S, Zoccatelli G, Tavazzi B, Del Bolgia F, Sorge R, Broglio SP, Mcintosh TK, Lazzarino G (2010) Assessment of metabolic brain damage and recovery following mild traumatic brain injury: a multicentre, proton magnetic resonance spectroscopic study in concussed patients. Brain 133:3232–3242
10. Vagnozzi R, Signoretti S, Tavazzi B, Floris R, Ludovici A, Marziali S, Tarascio G, Amorini AM, Di Pietro V, Delfini R, Lazzarino G (2008) Temporal window of metabolic brain vulnerability to concussion: a pilot 1H-magnetic resonance spectroscopic study in concussed athletes–part III. Neurosurgery 62:1286–1295; discussion 1295–1296
11. Guskiewicz KM, Marshall SW, Bailes J, Mccrea M, Cantu RC, Randolph C, Jordan BD (2005) Association between recurrent concussion and late-life cognitive impairment in retired professional football players. Neurosurgery 57:719–726; discussion 719–726
12. Guskiewicz KM, Marshall SW, Bailes J, Mccrea M, Harding HP Jr, Matthews A, Mihalik JR, Cantu RC (2007) Recurrent concussion and risk of depression in retired professional football players. Med Sci Sports Exerc 39:903–909
13. Lovell MR, Pardini JE, Welling J, Collins MW, Bakal J, Lazar N, Roush R, Eddy WF, Becker JT (2007) Functional brain abnormalities are related to clinical recovery and time to return-to-play in athletes. Neurosurgery 61:352–359; discussion 359–360
14. Mccrea M, Guskiewicz K, Randolph C, Barr WB, Hammeke TA, Marshall SW, Powell MR, Woo Ahn K, Wang Y, Kelly JP (2013) Incidence, clinical course, and predictors of prolonged recovery time following sport-related concussion in high school and college athletes. J Int Neuropsychol Soc 19:22–33
15. Mccrea M, Guskiewicz KM, Marshall SW, Barr W, Randolph C, Cantu RC, Onate JA, Yang J, Kelly JP (2003) Acute effects and recovery time following concussion in collegiate football players: the NCAA Concussion Study. JAMA 290:2556–2563
16. Mccrea M, Kelly JP, Randolph C, Cisler R, Berger L (2002) Immediate neurocognitive effects of concussion. Neurosurgery 50:1032–1040; discussion 1040–1042
17. Petraglia AL, Maroon JC, Bailes JE (2012) From the field of play to the field of combat: a review of the pharmacological management of concussion. Neurosurgery 70:1520–1533; discussion 1533
18. Jordan BD (2013) The clinical spectrum of sport-related traumatic brain injury. Nat Rev Neurol 9:222–230
19. Dekosky ST, Blennow K, Ikonomovic MD, Gandy S (2013) Acute and chronic traumatic encephalopathies: pathogenesis and biomarkers. Nat Rev Neurol 9:192–200

20. Goldstein LE, Fisher AM, Tagge CA, Zhang XL, Velisek L, Sullivan JA, Upreti C, Kracht JM, Ericsson M, Wojnarowicz MW, Goletiani CJ, Maglakelidze GM, Casey N, Moncaster JA, Minaeva O, Moir RD, Nowinski CJ, Stern RA, Cantu RC, Geiling J, Blusztajn JK, Wolozin BL, Ikezu T, Stein TD, Budson AE, Kowall NW, Chargin D, Sharon A, Saman S, Hall GF, Moss WC, Cleveland RO, Tanzi RE, Stanton PK, Mckee AC (2012) Chronic traumatic encephalopathy in blast-exposed military veterans and a blast neurotrauma mouse model. Sci Transl Med 4:134ra60
21. Mckee AC, Cantu RC, Nowinski CJ, Hedley-Whyte ET, Gavett BE, Budson AE, Santini VE, Lee HS, Kubilus CA, Stern RA (2009) Chronic traumatic encephalopathy in athletes: progressive tauopathy after repetitive head injury. J Neuropathol Exp Neurol 68:709–735
22. Mckee AC, Gavett BE, Stern RA, Nowinski CJ, Cantu RC, Kowall NW, Perl DP, Hedley-Whyte ET, Price B, Sullivan C, Morin P, Lee HS, Kubilus CA, Daneshvar DH, Wulff M, Budson AE (2010) TDP-43 proteinopathy and motor neuron disease in chronic traumatic encephalopathy. J Neuropathol Exp Neurol 69:918–929
23. Mckee AC, Stein TD, Nowinski CJ, Stern RA, Daneshvar DH, Alvarez VE, Lee HS, Hall G, Wojtowicz SM, Baugh CM, Riley DO, Kubilus CA, Cormier KA, Jacobs MA, Martin BR, Abraham CR, Ikezu T, Reichard RR, Wolozin BL, Budson AE, Goldstein LE, Kowall NW, Cantu RC (2013) The spectrum of disease in chronic traumatic encephalopathy. Brain 136:43–64
24. Xiong H, Liang WL, Wu XR (2000) Pathophysiological alterations in cultured astrocytes exposed to hypoxia/reoxygenation. Sheng Li Ke Xue Jin Zhan 31:217–221
25. Mondello S, Muller U, Jeromin A, Streeter J, Hayes RL, Wang KK (2011) Blood-based diagnostics of traumatic brain injuries. Expert Rev Mol Diagn 11:65–78
26. Isobe T, Ishioka N, Okuyama T (1981) Structural relation of two S-100 proteins in bovine brain; subunit composition of S-100a protein. Eur J Biochem 115:469–474
27. Graham MR, Myers T, Evans P, Davies B, Cooper SM, Bhattacharya K, Grace FM, Baker JS (2011) Direct hits to the head during amateur boxing is associated with a rise in serum biomarkers for brain injury. Int J Immunopathol Pharmacol 24:119–125
28. Sawada M, Kondo N, Suzumura A, Marunouchi T (1989) Production of tumor necrosis factor-alpha by microglia and astrocytes in culture. Brain Res 491:394–397
29. Stalnacke BM, Ohlsson A, Tegner Y, Sojka P (2006) Serum concentrations of two biochemical markers of brain tissue damage S-100B and neurone specific enolase are increased in elite female soccer players after a competitive game. Br J Sports Med 40:313–316
30. Metting Z, Wilczak N, Rodiger LA, Schaaf JM, van der Naalt J (2012) GFAP and S100B in the acute phase of mild traumatic brain injury. Neurology 78:1428–1433
31. Muller K, Townend W, Biasca N, Unden J, Waterloo K, Romner B, Ingebrigtsen T (2007) S100B serum level predicts computed tomography findings after minor head injury. J Trauma 62:1452–1456
32. Topolovec-Vranic J, Pollmann-Mudryj MA, Ouchterlony D, Klein D, Spence J, Romaschin A, Rhind S, Tien HC, Baker AJ (2011) The value of serum biomarkers in prediction models of outcome after mild traumatic brain injury. J Trauma 71:S478–S486
33. Raabe A, Menon DK, Gupta S, Czosnyka M, Pickard JD (1998) Jugular venous and arterial concentrations of serum S-100B protein in patients with severe head injury: a pilot study. J Neurol Neurosurg Psychiatry 65:930–932
34. Romner B, Ingebrigtsen T, Kongstad P, Borgesen SE (2000) Traumatic brain damage: serum S-100 protein measurements related to neuroradiological findings. J Neurotrauma 17: 641–647
35. Woertgen C, Rothoerl RD, Metz C, Brawanski A (1999) Comparison of clinical, radiologic, and serum marker as prognostic factors after severe head injury. J Trauma 47:1126–1130
36. Herrmann M, Curio N, Jost S, Wunderlich MT, Synowitz H, Wallesch CW (1999) Protein S-100B and neuron specific enolase as early neurobiochemical markers of the severity of traumatic brain injury. Restor Neurol Neurosci 14:109–114

37. Kovesdi E, Luckl J, Bukovics P, Farkas O, Pal J, Czeiter E, Szellar D, Doczi T, Komoly S, Buki A (2010) Update on protein biomarkers in traumatic brain injury with emphasis on clinical use in adults and pediatrics. Acta Neurochir (Wien) 152:1–17
38. Mckeating EG, Andrews PJ, Mascia L (1998) Relationship of neuron specific enolase and protein S-100 concentrations in systemic and jugular venous serum to injury severity and outcome after traumatic brain injury. Acta Neurochir Suppl 71:117–119
39. Mondello S, Papa L, Buki A, Bullock MR, Czeiter E, Tortella FC, Wang KK, Hayes RL (2011) Neuronal and glial markers are differently associated with computed tomography findings and outcome in patients with severe traumatic brain injury: a case control study. Crit Care 15:R156
40. Pelinka LE, Kroepfl A, Leixnering M, Buchinger W, Raabe A, Redl H (2004) GFAP versus S100B in serum after traumatic brain injury: relationship to brain damage and outcome. J Neurotrauma 21:1553–1561
41. Raabe A, Grolms C, Keller M, Dohnert J, Sorge O, Seifert V (1998) Correlation of computed tomography findings and serum brain damage markers following severe head injury. Acta Neurochir (Wien) 140:787–791; discussion 791–792
42. Raabe A, Grolms C, Sorge O, Zimmermann M, Seifert V (1999) Serum S-100B protein in severe head injury. Neurosurgery 45:477–483
43. Rothoerl RD, Woertgen C, Holzschuh M, Metz C, Brawanski A (1998) S-100 serum levels after minor and major head injury. J Trauma 45:765–767
44. Savola O, Pyhtinen J, Leino TK, Siitonen S, Niemela O, Hillbom M (2004) Effects of head and extracranial injuries on serum protein S100B levels in trauma patients. J Trauma 56: 1229–1234; discussion 1234
45. Unden J, Astrand R, Waterloo K, Ingebrigtsen T, Bellner J, Reinstrup P, Andsberg G, Romner B (2007) Clinical significance of serum S100B levels in neurointensive care. Neurocrit Care 6:94–99
46. Vos PE, Jacobs B, Andriessen TM, Lamers KJ, Borm GF, Beems T, Edwards M, Rosmalen CF, Vissers JL (2010) GFAP and S100B are biomarkers of traumatic brain injury: an observational cohort study. Neurology 75:1786–1793
47. Vos PE, Lamers KJ, Hendriks JC, van Haaren M, Beems T, Zimmerman C, van Geel W, de Reus H, Biert J, Verbeek MM (2004) Glial and neuronal proteins in serum predict outcome after severe traumatic brain injury. Neurology 62:1303–1310
48. Woertgen C, Rothoerl RD, Holzschuh M, Metz C, Brawanski A (1997) Comparison of serial S-100 and NSE serum measurements after severe head injury. Acta Neurochir (Wien) 139:1161–1164; discussion 1165
49. Egea-Guerrero JJ, Revuelto-Rey J, Murillo-Cabezas F, Munoz-Sanchez MA, Vilches-Arenas A, Sanchez-Linares P, Dominguez-Roldan JM, Leon-Carrion J (2012) Accuracy of the S100beta protein as a marker of brain damage in traumatic brain injury. Brain Inj 26:76–82
50. Defazio MV, Rammo RA, Robles JR, Bramlett HM, Dietrich WD, Bullock MR (2013) The potential utility of blood-derived biochemical markers as indicators of early clinical trends after severe traumatic brain injury. World Neurosurg
51. Babcock L, Byczkowski T, Mookerjee S, Bazarian JJ (2012) Ability of S100B to predict severity and cranial CT results in children with TBI. Brain Inj 26:1372–1380
52. Berger RP, Adelson PD, Pierce MC, Dulani T, Cassidy LD, Kochanek PM (2005) Serum neuron-specific enolase, S100B, and myelin basic protein concentrations after inflicted and noninflicted traumatic brain injury in children. J Neurosurg 103:61–68
53. Goyal A, Carter M, Niyonkuru C, Fabio A, Amin K, Berger RP, Wagner AMD (2013) S100b as a prognostic biomarker in outcome prediction for patients with severe TBI. J Neurotrauma 30:946–957
54. Zurek J, Fedora M (2012) The usefulness of S100B, NSE, GFAP, NF-H, secretagogin and Hsp70 as a predictive biomarker of outcome in children with traumatic brain injury. Acta Neurochir (Wien) 154:93–103; discussion 103

55. Nylen K, Ost M, Csajbok LZ, Nilsson I, Hall C, Blennow K, Nellgard B, Rosengren L (2008) Serum levels of S100B, S100A1B and S100BB are all related to outcome after severe traumatic brain injury. Acta Neurochir (Wien) 150:221–227; discussion 227
56. Tavarez MM, Atabaki SM, Teach SJ (2012) Acute evaluation of pediatric patients with minor traumatic brain injury. Curr Opin Pediatr 24:307–313
57. Bouvier D, Fournier M, Dauphin JB, Amat F, Ughetto S, Labbe A, Sapin V (2012) Serum S100B determination in the management of pediatric mild traumatic brain injury. Clin Chem 58:1116–1122
58. Lange RT, Iverson GL, Brubacher JR (2012) Clinical utility of the protein S100B to evaluate traumatic brain injury in the presence of acute alcohol intoxication. J Head Trauma Rehabil 27:123–134
59. Bloomfield SM, Mckinney J, Smith L, Brisman J (2007) Reliability of S100B in predicting severity of central nervous system injury. Neurocrit Care 6:121–138
60. Goncalves CA, Leite MC, Nardin P (2008) Biological and methodological features of the measurement of S100B, a putative marker of brain injury. Clin Biochem 41:755–763
61. Amick JE, Yandora KA, Bell MJ, Wisniewski SR, Adelson PD, Carcillo JA, Janesko KL, Dekosky ST, Carlos TM, Clark RS, Kochanek PM (2001) The Th1 versus Th2 cytokine profile in cerebrospinal fluid after severe traumatic brain injury in infants and children. Pediatr Crit Care Med 2:260–264
62. Mussack T, Kirchhoff C, Buhmann S, Biberthaler P, Ladurner R, Gippner-Steppert C, Mutschler W, Jochum M (2006) Significance of Elecsys S100 immunoassay for real-time assessment of traumatic brain damage in multiple trauma patients. Clin Chem Lab Med 44: 1140–1145
63. Romner B, Ingebrigtsen T (2001) High serum S100B levels for trauma patients without head injuries. Neurosurgery 49:1490; author reply 1492–1493
64. Rothoerl RD, Woertgen C (2001) High serum S100B levels for trauma patients without head injuries. Neurosurgery 49:1490–1491; author reply 1492–1493
65. Jonsson H, Johnsson P, Backstrom M, Alling C, Dautovic-Bergh C, Blomquist S (2004) Controversial significance of early S100B levels after cardiac surgery. BMC Neurol 4:24
66. Routsi C, Stamataki E, Nanas S, Psachoulia C, Stathopoulos A, Koroneos A, Zervou M, Jullien G, Roussos C (2006) Increased levels of serum S100B protein in critically ill patients without brain injury. Shock 26:20–24
67. Gonzclez-Mao MC, Reparaz-Andrade A, Del Campo-Perez V, Alvarez-Garcia E, Vara-Perez C, Andrade-Olivie MA (2011) Model predicting survival/exitus after traumatic brain injury: biomarker S100B 24h. Clin Lab 57:587–597
68. Berger RP, Adelson PD, Richichi R, Kochanek PM (2006) Serum biomarkers after traumatic and hypoxemic brain injuries: insight into the biochemical response of the pediatric brain to inflicted brain injury. Dev Neurosci 28:327–335
69. Berger RP, Dulani T, Adelson PD, Leventhal JM, Richichi R, Kochanek PM (2006) Identification of inflicted traumatic brain injury in well-appearing infants using serum and cerebrospinal markers: a possible screening tool. Pediatrics 117:325–332
70. Piazza O, Storti MP, Cotena S, Stoppa F, Perrotta D, Esposito G, Pirozzi N, Tufano R (2007) S100B is not a reliable prognostic index in paediatric TBI. Pediatr Neurosurg 43:258–264
71. Eng LF, Ghirnikar RS, Lee YL (2000) Glial fibrillary acidic protein: GFAP-thirty-one years (1969–2000). Neurochem Res 25:1439–1451
72. Eng LF, Vanderhaeghen JJ, Bignami A, Gerstl B (1971) An acidic protein isolated from fibrous astrocytes. Brain Res 28:351–354
73. Schiff L, Hadker N, Weiser S, Rausch C (2012) A literature review of the feasibility of glial fibrillary acidic protein as a biomarker for stroke and traumatic brain injury. Mol Diagn Ther 16:79–92
74. Missler U, Wiesmann M, Wittmann G, Magerkurth O, Hagenstrom H (1999) Measurement of glial fibrillary acidic protein in human blood: analytical method and preliminary clinical results. Clin Chem 45:138–141

75. Pelinka LE, Kroepfl A, Schmidhammer R, Krenn M, Buchinger W, Redl H, Raabe A (2004) Glial fibrillary acidic protein in serum after traumatic brain injury and multiple trauma. J Trauma 57:1006–1012
76. van Geel WJ, de Reus HP, Nijzing H, Verbeek MM, Vos PE, Lamers KJ (2002) Measurement of glial fibrillary acidic protein in blood: an analytical method. Clin Chim Acta 326:151–154
77. Mondello S, Jeromin A, Buki A, Bullock R, Czeiter E, Kovacs N, Barzo P, Schmid K, Tortella F, Wang KK, Hayes RL (2011) Glial neuronal ratio: a novel index for differentiating injury type in patients with severe traumatic brain injury. J Neurotrauma 29:1096–1104
78. O'Callaghan JP, Sriram K (2005) Glial fibrillary acidic protein and related glial proteins as biomarkers of neurotoxicity. Expert Opin Drug Saf 4:433–442
79. Papa L, Lewis LM, Falk JL, Zhang Z, Silvestri S, Giordano P, Brophy GM, Demery JA, Dixit NK, Ferguson I, Liu MC, Mo J, Akinyi L, Schmid K, Mondello S, Robertson CS, Tortella FC, Hayes RL, Wang KK (2011) Elevated levels of serum glial fibrillary acidic protein breakdown products in mild and moderate traumatic brain injury are associated with intracranial lesions and neurosurgical intervention. Ann Emerg Med 59:471–483
80. Nylen K, Ost M, Csajbok LZ, Nilsson I, Blennow K, Nellgard B, Rosengren L (2006) Increased serum-GFAP in patients with severe traumatic brain injury is related to outcome. J Neurol Sci 240:85–91
81. Honda M, Tsuruta R, Kaneko T, Kasaoka S, Yagi T, Todani M, Fujita M, Izumi T, Maekawa T (2010) Serum glial fibrillary acidic protein is a highly specific biomarker for traumatic brain injury in humans compared with S-100B and neuron-specific enolase. J Trauma 69:104–109
82. Fraser DD, Close TE, Rose KL, Ward R, Mehl M, Farrell C, Lacroix J, Creery D, Kesselman M, Stanimirovic D, Hutchison JS, Canadian Critical Care Translational Biology Group (2011) Severe traumatic brain injury in children elevates glial fibrillary acidic protein in cerebrospinal fluid and serum. Pediatr Crit Care Med 12:319–324
83. Hayes RL, Mondello S, Wang K (2011) Glial fibrillary acidic protein: a promising biomarker in pediatric brain injury. Pediatr Crit Care Med 12:603–604
84. Palfreyman JW, Thomas DG, Ratcliffe JG (1978) Radioimmunoassay of human myelin basic protein in tissue extract, cerebrospinal fluid and serum and its clinical application to patients with head injury. Clin Chim Acta 82:259–270
85. Thomas DG, Palfreyman JW, Ratcliffe JG (1978) Serum-myelin-basic-protein assay in diagnosis and prognosis of patients with head injury. Lancet 1:113–115
86. Thomas DG, Rabow L, Teasdale G (1979) Serum myelin basic protein, clinical responsiveness, and outcome of severe head injury. Acta Neurochir Suppl (Wien) 28:93–95
87. Rostami E, Davidsson J, Ng KC, Lu J, Gyorgy A, Walker J, Wingo D, Plantman S, Bellander BM, Agoston DV, Risling M (2012) A model for mild traumatic brain injury that induces limited transient memory impairment and increased levels of axon related serum biomarkers. Front Neurol 3:115
88. Marangos PJ, Schmechel DE (1987) Neuron specific enolase, a clinically useful marker for neurons and neuroendocrine cells. Annu Rev Neurosci 10:269–295
89. Skogseid IM, Nordby HK, Urdal P, Paus E, Lilleaas F (1992) Increased serum creatine kinase BB and neuron specific enolase following head injury indicates brain damage. Acta Neurochir (Wien) 115:106–111
90. Bandyopadhyay S, Hennes H, Gorelick MH, Wells RG, Walsh-Kelly CM (2005) Serum neuron-specific enolase as a predictor of short-term outcome in children with closed traumatic brain injury. Acad Emerg Med 12:732–738
91. Beers SR, Berger RP, Adelson PD (2007) Neurocognitive outcome and serum biomarkers in inflicted versus non-inflicted traumatic brain injury in young children. J Neurotrauma 24:97–105
92. Fridriksson T, Kini N, Walsh-Kelly C, Hennes H (2000) Serum neuron-specific enolase as a predictor of intracranial lesions in children with head trauma: a pilot study. Acad Emerg Med 7:816–820

93. Ross SA, Cunningham RT, Johnston CF, Rowlands BJ (1996) Neuron-specific enolase as an aid to outcome prediction in head injury. Br J Neurosurg 10:471–476
94. Kochanek PM, Berger RP, Bayir H, Wagner AK, Jenkins LW, Clark RS (2008) Biomarkers of primary and evolving damage in traumatic and ischemic brain injury: diagnosis, prognosis, probing mechanisms, and therapeutic decision making. Curr Opin Crit Care 14:135–141
95. Zetterberg H, Hietala MA, Jonsson M, Andreasen N, Styrud E, Karlsson I, Edman A, Popa C, Rasulzada A, Wahlund LO, Mehta PD, Rosengren L, Blennow K, Wallin A (2006) Neurochemical aftermath of amateur boxing. Arch Neurol 63:1277–1280
96. Zetterberg H, Tanriverdi F, Unluhizarci K, Selcuklu A, Kelestimur F, Blennow K (2009) Sustained release of neuron-specific enolase to serum in amateur boxers. Brain Inj 23: 723–726
97. Stalnacke BM, Sojka P (2008) [S100B for diagnosis and prognosis of sequelae following minor head injury. Contradictory results according to studies]. Lakartidningen 105: 1840–1845
98. Yamazaki Y, Yada K, Morii S, Kitahara T, Ohwada T (1995) Diagnostic significance of serum neuron-specific enolase and myelin basic protein assay in patients with acute head injury. Surg Neurol 43:267–270; discussion 270–271
99. Pelinka LE, Hertz H, Mauritz W, Harada N, Jafarmadar M, Albrecht M, Redl H, Bahrami S (2005) Nonspecific increase of systemic neuron-specific enolase after trauma: clinical and experimental findings. Shock 24:119–123
100. Pelinka LE, Jafarmadar M, Redl H, Bahrami S (2004) Neuron-specific-enolase is increased in plasma after hemorrhagic shock and after bilateral femur fracture without traumatic brain injury in the rat. Shock 22:88–91
101. Day IN, Thompson RJ (2010) UCHL1 (PGP 9.5): neuronal biomarker and ubiquitin system protein. Prog Neurobiol 90:327–362
102. Gong B, Leznik E (2007) The role of ubiquitin C-terminal hydrolase L1 in neurodegenerative disorders. Drug News Perspect 20:365–370
103. Tongaonkar P, Chen L, Lambertson D, Ko B, Madura K (2000) Evidence for an interaction between ubiquitin-conjugating enzymes and the 26S proteasome. Mol Cell Biol 20: 4691–4698
104. Liu MC, Akinyi L, Scharf D, Mo J, Larner SF, Muller U, Oli MW, Zheng W, Kobeissy F, Papa L, Lu XC, Dave JR, Tortella FC, Hayes RL, Wang KK (2010) Ubiquitin C-terminal hydrolase-L1 as a biomarker for ischemic and traumatic brain injury in rats. Eur J Neurosci 31:722–732
105. Blyth BJ, Farahvar A, He H, Nayak A, Yang C, Shaw G, Bazarian JJ (2011) Elevated serum ubiquitin carboxy-terminal hydrolase L1 is associated with abnormal blood–brain barrier function after traumatic brain injury. J Neurotrauma 28:2453–2462
106. Papa L, Akinyi L, Liu MC, Pineda JA, Tepas JJ III, Oli MW, Zheng W, Robinson G, Robicsek SA, Gabrielli A, Heaton SC, Hannay HJ, Demery JA, Brophy GM, Layon J, Robertson CS, Hayes RL, Wang KK (2010) Ubiquitin C-terminal hydrolase is a novel biomarker in humans for severe traumatic brain injury. Crit Care Med 38:138–144
107. Brophy GM, Mondello S, Papa L, Robicsek SA, Gabrielli A, Tepas J III, Buki A, Robertson C, Tortella FC, Hayes RL, Wang KK (2011) Biokinetic analysis of ubiquitin C-terminal hydrolase-L1 (UCH-L1) in severe traumatic brain injury patient biofluids. J Neurotrauma 28:861–870
108. Czeiter E, Mondello S, Kovacs N, Sandor J, Gabrielli A, Schmid K, Tortella F, Wang KK, Hayes RL, Barzo P, Ezer E, Doczi T, Buki A (2012) Brain injury biomarkers may improve the predictive power of the IMPACT outcome calculator. J Neurotrauma 29:1770–1778
109. Mondello S, Linnet A, Buki A, Robicsek S, Gabrielli A, Tepas J, Papa L, Brophy GM, Tortella F, Hayes RL, Wang KK (2012) Clinical utility of serum levels of ubiquitin C-terminal hydrolase as a biomarker for severe traumatic brain injury. Neurosurgery 70: 666–675

110. Papa L, Akinyi L, Liu MC, Pineda JA, Tepas JJIII, Oli MW, Zheng W, Robinson G, Robicsek SA, Gabrielli A, Heaton SC, Hannay HJ, Demery JA, Brophy GM, Layon J, Robertson CS, Hayes RL, Wang KK (2009) Ubiquitin C-terminal hydrolase is a novel biomarker in humans for severe traumatic brain injury. Crit Care Med 38:138–144
111. Papa L, Lewis LM, Silvestri S, Falk JL, Giordano P, Brophy GM, Demery JA, Liu MC, Mo J, Akinyi L, Mondello S, Schmid K, Robertson CS, Tortella FC, Hayes RL, Wang KK (2012) Serum levels of ubiquitin C-terminal hydrolase distinguish mild traumatic brain injury from trauma controls and are elevated in mild and moderate traumatic brain injury patients with intracranial lesions and neurosurgical intervention. J Trauma Acute Care Surg 72:1335–1344
112. Papa L, Ramia MM, Kelly JM, Burks SS, Pawlowicz A, Berger RP (2013) Systematic review of clinical research on biomarkers for pediatric traumatic brain injury. J Neurotrauma 30:324–338
113. Mondello S, Gabrielli A, Catani S, D'ippolito M, Jeromin A, Ciaramella A, Bossu P, Schmid K, Tortella F, Wang KK, Hayes RL, Formisano R (2012) Increased levels of serum MAP-2 at 6-months correlate with improved outcome in survivors of severe traumatic brain injury. Brain Inj 26:1629–1635
114. Liu Q, Xie F, Siedlak SL, Nunomura A, Honda K, Moreira PI, Zhua X, Smith MA, Perry G (2004) Neurofilament proteins in neurodegenerative diseases. Cell Mol Life Sci 61: 3057–3075
115. Anderson KJ, Scheff SW, Miller KM, Roberts KN, Gilmer LK, Yang C, Shaw G (2008) The phosphorylated axonal form of the neurofilament subunit NF-H (pNF-H) as a blood biomarker of traumatic brain injury. J Neurotrauma 25:1079–1085
116. Shaw G, Yang C, Ellis R, Anderson K, Parker Mickle J, Scheff S, Pike B, Anderson DK, Howland DR (2005) Hyperphosphorylated neurofilament NF-H is a serum biomarker of axonal injury. Biochem Biophys Res Commun 336:1268–1277
117. Zurek J, Bartlova L, Fedora M (2011) Hyperphosphorylated neurofilament NF-H as a predictor of mortality after brain injury in children. Brain Inj 25:221–226
118. Hergenroeder G, Redell JB, Moore AN, Dubinsky WP, Funk RT, Crommett J, Clifton GL, Levine R, Valadka A, Dash PK (2008) Identification of serum biomarkers in brain-injured adults: potential for predicting elevated intracranial pressure. J Neurotrauma 25:79–93
119. Trojanowski JQ, Schuck T, Schmidt ML, Lee VM (1989) Distribution of tau proteins in the normal human central and peripheral nervous system. J Histochem Cytochem 37:209–215
120. Mortberg E, Zetterberg H, Nordmark J, Blennow K, Catry C, Decraemer H, Vanmechelen E, Rubertsson S (2011) Plasma tau protein in comatose patients after cardiac arrest treated with therapeutic hypothermia. Acta Anaesthesiol Scand 55:1132–1138
121. Randall J, Mortberg E, Provuncher GK, Fournier DR, Duffy DC, Rubertsson S, Blennow K, Zetterberg H, Wilson DH (2013) Tau proteins in serum predict neurological outcome after hypoxic brain injury from cardiac arrest: results of a pilot study. Resuscitation 84:351–356
122. Neselius S, Zetterberg H, Blennow K, Randall J, Wilson D, Marcusson J, Brisby H (2013) Olympic boxing is associated with elevated levels of the neuronal protein tau in plasma. Brain Inj 27:425–433
123. Riederer BM, Zagon IS, Goodman SR (1986) Brain spectrin(240/235) and brain spectrin (240/235E): two distinct spectrin subtypes with different locations within mammalian neural cells. J Cell Biol 102:2088–2097
124. Zhang Z, Larner SF, Liu MC, Zheng W, Hayes RL, Wang KK (2009) Multiple alphaII-spectrin breakdown products distinguish calpain and caspase dominated necrotic and apoptotic cell death pathways. Apoptosis 14:1289–1298
125. Pike BR, Flint J, Dutta S, Johnson E, Wang KK, Hayes RL (2001) Accumulation of non-erythroid alpha II-spectrin and calpain-cleaved alpha II-spectrin breakdown products in cerebrospinal fluid after traumatic brain injury in rats. J Neurochem 78:1297–1306
126. Pike BR, Zhao X, Newcomb JK, Posmantur RM, Wang KK, Hayes RL (1998) Regional calpain and caspase-3 proteolysis of alpha-spectrin after traumatic brain injury. Neuroreport 9:2437–2442

127. Raghupathi R, Graham DI, Mcintosh TK (2000) Apoptosis after traumatic brain injury. J Neurotrauma 17:927–938
128. Saatman KE, Creed J, Raghupathi R (2010) Calpain as a therapeutic target in traumatic brain injury. Neurotherapeutics 7:31–42
129. Siman R, Mcintosh TK, Soltesz KM, Chen Z, Neumar RW, Roberts VL (2004) Proteins released from degenerating neurons are surrogate markers for acute brain damage. Neurobiol Dis 16:311–320
130. Wang KK (2000) Calpain and caspase: can you tell the difference? Trends Neurosci 23:20–26
131. Zetterberg H, Smith DH, Blennow K (2013) Biomarkers of mild traumatic brain injury in cerebrospinal fluid and blood. Nat Rev Neurol 9:201–210
132. Dash PK, Zhao J, Hergenroeder G, Moore AN (2010) Biomarkers for the diagnosis, prognosis and evaluation of treatment efficacy for traumatic brain injury. Neurotherapeutics 7:100–114
133. Schmid KE, Tortella FC (2012) The diagnosis of traumatic brain injury on the battlefield. Front Neurol 3:90
134. Woodcock T, Morganti-Kossmann MC (2013) The role of markers of inflammation in traumatic brain injury. Front Neurol 4:18
135. Giulian D, Lachman LB (1985) Interleukin-1 stimulation of astroglial proliferation after brain injury. Science 228:497–499
136. Kossmann T, Hans V, Imhof HG, Trentz O, Morganti-Kossmann MC (1996) Interleukin-6 released in human cerebrospinal fluid following traumatic brain injury may trigger nerve growth factor production in astrocytes. Brain Res 713:143–152
137. Tasci A, Okay O, Gezici AR, Ergun R, Ergungor F (2003) Prognostic value of interleukin-1 beta levels after acute brain injury. Neurol Res 25:871–874
138. Hergenroeder GW, Redell JB, Moore AN, Dash PK (2008) Biomarkers in the clinical diagnosis and management of traumatic brain injury. Mol Diagn Ther 12:345–358
139. Jeter CB, Hergenroeder GW, Hylin MJ, Redell JB, Moore AN, Dash PK (2013) Biomarkers for the diagnosis and prognosis of mild traumatic brain injury/concussion. J Neurotrauma 30: 657–670
140. Brenner T, Yamin A, Abramsky O, Gallily R (1993) Stimulation of tumor necrosis factor-alpha production by mycoplasmas and inhibition by dexamethasone in cultured astrocytes. Brain Res 608:273–279
141. Mier JW, Vachino G, van der Meer JW, Numerof RP, Adams S, Cannon JG, Bernheim HA, Atkins MB, Parkinson DR, Dinarello CA (1988) Induction of circulating tumor necrosis factor (TNF alpha) as the mechanism for the febrile response to interleukin-2 (IL-2) in cancer patients. J Clin Immunol 8:426–436
142. Shohami E, Gallily R, Mechoulam R, Bass R, Ben-Hur T (1997) Cytokine production in the brain following closed head injury: dexanabinol (HU-211) is a novel TNF-alpha inhibitor and an effective neuroprotectant. J Neuroimmunol 72:169–177
143. Goodman JC, Robertson CS, Grossman RG, Narayan RK (1990) Elevation of tumor necrosis factor in head injury. J Neuroimmunol 30:213–217
144. Ross SA, Halliday MI, Campbell GC, Byrnes DP, Rowlands BJ (1994) The presence of tumour necrosis factor in CSF and plasma after severe head injury. Br J Neurosurg 8:419–425
145. Crespo AR, Da Rocha AB, Jotz GP, Schneider RF, Grivicich I, Pinheiro K, Zanoni C, Regner A (2007) Increased serum sFas and TNFalpha following isolated severe head injury in males. Brain Inj 21:441–447
146. Stein DM, Kufera JA, Lindell A, Murdock KR, Menaker J, Bochicchio GV, Aarabi B, Scalea TM (2011) Association of CSF biomarkers and secondary insults following severe traumatic brain injury. Neurocrit Care 14:200–207
147. Stein DM, Lindell A, Murdock KR, Kufera JA, Menaker J, Keledjian K, Bochicchio GV, Aarabi B, Scalea TM (2011) Relationship of serum and cerebrospinal fluid biomarkers with intracranial hypertension and cerebral hypoperfusion after severe traumatic brain injury. J Trauma 70:1096–1103

148. Hohl A, Gullo Jda S, Silva CC, Bertotti MM, Felisberto F, Nunes JC, de Souza B, Petronilho F, Soares FM, Prediger RD, Dal-Pizzol F, Linhares MN, Walz R (2012) Plasma levels of oxidative stress biomarkers and hospital mortality in severe head injury: a multivariate analysis. J Crit Care 27(523):e11–e19
149. Balakathiresan N, Bhomia M, Chandran R, Chavko M, Mccarron RM, Maheshwari RK (2012) MicroRNA let-7i is a promising serum biomarker for blast-induced traumatic brain injury. J Neurotrauma 29:1379–1387
150. Chittiboina P, Ganta V, Monceaux CP, Scott LK, Nanda A, Alexander JS (2013) Angiopoietins as promising biomarkers and potential therapeutic targets in brain injury. Pathophysiology 20:15–21
151. Berger RP, Pierce MC, Wisniewski SR, Adelson PD, Clark RS, Ruppel RA, Kochanek PM (2002) Neuron-specific enolase and S100B in cerebrospinal fluid after severe traumatic brain injury in infants and children. Pediatrics 109:E31
152. Neselius S, Brisby H, Theodorsson A, Blennow K, Zetterberg H, Marcusson J (2012) CSF-biomarkers in Olympic boxing: diagnosis and effects of repetitive head trauma. PLoS One 7:e33606
153. Bohmer AE, Oses JP, Schmidt AP, Peron CS, Krebs CL, Oppitz PP, D'avila TT, Souza DO, Portela LV, Stefani MA (2011) Neuron-specific enolase, S100B, and glial fibrillary acidic protein levels as outcome predictors in patients with severe traumatic brain injury. Neurosurgery 68:1624–1630; discussion 1630–1631
154. Bellander BM, Olafsson IH, Ghatan PH, Bro Skejo HP, Hansson LO, Wanecek M, Svensson MA (2011) Secondary insults following traumatic brain injury enhance complement activation in the human brain and release of the tissue damage marker S100B. Acta Neurochir (Wien) 153:90–100
155. Niyonkuru C, Wagner AMD, Ozawa H, Amin K, Goyal A, Fabio A (2013) Group based trajectory analysis applications for prognostic biomarker model development in severe TBI: a practical example. J Neurotrauma 30:938–945
156. Ahmed F, Gyorgy A, Kamnaksh A, Ling G, Tong L, Parks S, Agoston D (2012) Time-dependent changes of protein biomarker levels in the cerebrospinal fluid after blast traumatic brain injury. Electrophoresis 33:3705–3711
157. Zoltewicz JS, Mondello S, Yang B, Newsom KJ, Kobeissy FH, Yao C, Lu XC, Dave JR, Shear DA, Schmid K, Rivera V, Cram T, Seaney J, Zhang Z, Wang KK, Hayes RL, Tortella FC (2013) Biomarkers track damage following graded injury severity in a rat model of penetrating brain injury. J Neurotrauma 30:1161–1169
158. Zoltewicz JS, Scharf D, Yang B, Chawla A, Newsom KJ, Fang L (2012) Characterization of antibodies that detect human GFAP after traumatic brain injury. Biomark Insights 7:71–79
159. Berger RP (2006) The use of serum biomarkers to predict outcome after traumatic brain injury in adults and children. J Head Trauma Rehabil 21:315–333
160. Chiaretti A, Barone G, Riccardi R, Antonelli A, Pezzotti P, Genovese O, Tortorolo L, Conti G (2009) NGF, DCX, and NSE upregulation correlates with severity and outcome of head trauma in children. Neurology 72:609–616
161. Scarna H, Delafosse B, Steinberg R, Debilly G, Mandrand B, Keller A, Pujol JF (1982) Neuron-specific enolase as a marker of neuronal lesions during various comas in man. Neurochem Int 4:405–411
162. Siman R, Toraskar N, Dang A, Mcneil E, Mcgarvey M, Plaum J, Maloney E, Grady MS (2009) A panel of neuron-enriched proteins as markers for traumatic brain injury in humans. J Neurotrauma 26:1867–1877
163. Varma S, Janesko KL, Wisniewski SR, Bayir H, Adelson PD, Thomas NJ, Kochanek PM (2003) F2-isoprostane and neuron-specific enolase in cerebrospinal fluid after severe traumatic brain injury in infants and children. J Neurotrauma 20:781–786
164. Mondello S, Jeromin A, Buki A, Bullock R, Czeiter E, Kovacs N, Barzo P, Schmid K, Tortella F, Wang KK, Hayes RL (2012) Glial neuronal ratio: a novel index for differentiating injury type in patients with severe traumatic brain injury. J Neurotrauma 29:1096–1104

165. Mondello S, Palmio J, Streeter J, Hayes RL, Peltola J, Jeromin A (2012) Ubiquitin carboxy-terminal hydrolase L1 (UCH-L1) is increased in cerebrospinal fluid and plasma of patients after epileptic seizure. BMC Neurol 12:85
166. Friede RL, Samorajski T (1970) Axon caliber related to neurofilaments and microtubules in sciatic nerve fibers of rats and mice. Anat Rec 167:379–387
167. Blennow K, Hardy J, Zetterberg H (2012) The neuropathology and neurobiology of traumatic brain injury. Neuron 76:886–899
168. Grady MS, Mclaughlin MR, Christman CW, Valadka AB, Fligner CL, Povlishock JT (1993) The use of antibodies targeted against the neurofilament subunits for the detection of diffuse axonal injury in humans. J Neuropathol Exp Neurol 52:143–152
169. Olsson B, Zetterberg H, Hampel H, Blennow K (2011) Biomarker-based dissection of neurodegenerative diseases. Prog Neurobiol 95:520–534
170. Blennow K, Nellgard B (2004) Amyloid beta 1–42 and tau in cerebrospinal fluid after severe traumatic brain injury. Neurology 62:159; author reply 159–160
171. Franz G, Beer R, Kampfl A, Engelhardt K, Schmutzhard E, Ulmer H, Deisenhammer F (2003) Amyloid beta 1–42 and tau in cerebrospinal fluid after severe traumatic brain injury. Neurology 60:1457–1461
172. Zemlan FP, Jauch EC, Mulchahey JJ, Gabbita SP, Rosenberg WS, Speciale SG, Zuccarello M (2002) C-tau biomarker of neuronal damage in severe brain injured patients: association with elevated intracranial pressure and clinical outcome. Brain Res 947:131–139
173. Ost M, Nylen K, Csajbok L, Ohrfelt AO, Tullberg M, Wikkelso C, Nellgard P, Rosengren L, Blennow K, Nellgard B (2006) Initial CSF total tau correlates with 1-year outcome in patients with traumatic brain injury. Neurology 67:1600–1604
174. Gentleman SM, Nash MJ, Sweeting CJ, Graham DI, Roberts GW (1993) Beta-amyloid precursor protein (beta APP) as a marker for axonal injury after head injury. Neurosci Lett 160:139–144
175. Gentleman SM, Roberts GW, Gennarelli TA, Maxwell WL, Adams JH, Kerr S, Graham DI (1995) Axonal injury: a universal consequence of fatal closed head injury? Acta Neuropathol 89:537–543
176. Graham DI, Adams JH, Nicoll JA, Maxwell WL, Gennarelli TA (1995) The nature, distribution and causes of traumatic brain injury. Brain Pathol 5:397–406
177. Graham DI, Gentleman SM, Lynch A, Roberts GW (1995) Distribution of beta-amyloid protein in the brain following severe head injury. Neuropathol Appl Neurobiol 21:27–34
178. Horsburgh K, Cole GM, Yang F, Savage MJ, Greenberg BD, Gentleman SM, Graham DI, Nicoll JA (2000) Beta-amyloid (Abeta)42(43), abeta42, abeta40 and apoE immunostaining of plaques in fatal head injury. Neuropathol Appl Neurobiol 26:124–132
179. Mckenzie JE, Gentleman SM, Roberts GW, Graham DI, Royston MC (1994) Increased numbers of beta APP-immunoreactive neurones in the entorhinal cortex after head injury. Neuroreport 6:161–164
180. Mckenzie KJ, Mclellan DR, Gentleman SM, Maxwell WL, Gennarelli TA, Graham DI (1996) Is beta-APP a marker of axonal damage in short-surviving head injury? Acta Neuropathol 92:608–613
181. Roberts GW, Gentleman SM, Lynch A, Graham DI (1991) Beta A4 amyloid protein deposition in brain after head trauma. Lancet 338:1422–1423
182. Sherriff FE, Bridges LR, Gentleman SM, Sivaloganathan S, Wilson S (1994) Markers of axonal injury in post mortem human brain. Acta Neuropathol 88:433–439
183. Sherriff FE, Bridges LR, Sivaloganathan S (1994) Early detection of axonal injury after human head trauma using immunocytochemistry for beta-amyloid precursor protein. Acta Neuropathol 87:55–62
184. Smith DH, Chen XH, Iwata A, Graham DI (2003) Amyloid beta accumulation in axons after traumatic brain injury in humans. J Neurosurg 98:1072–1077
185. Smith DH, Chen XH, Nonaka M, Trojanowski JQ, Lee VM, Saatman KE, Leoni MJ, Xu BN, Wolf JA, Meaney DF (1999) Accumulation of amyloid beta and tau and the formation of

neurofilament inclusions following diffuse brain injury in the pig. J Neuropathol Exp Neurol 58:982–992
186. Smith DH, Uryu K, Saatman KE, Trojanowski JQ, Mcintosh TK (2003) Protein accumulation in traumatic brain injury. Neuromolecular Med 4:59–72
187. Uryu K, Chen XH, Martinez D, Browne KD, Johnson VE, Graham DI, Lee VM, Trojanowski JQ, Smith DH (2007) Multiple proteins implicated in neurodegenerative diseases accumulate in axons after brain trauma in humans. Exp Neurol 208:185–192
188. Chen XH, Johnson VE, Uryu K, Trojanowski JQ, Smith DH (2009) A lack of amyloid beta plaques despite persistent accumulation of amyloid beta in axons of long-term survivors of traumatic brain injury. Brain Pathol 19:214–223
189. Olsson A, Csajbok L, Ost M, Hoglund K, Nylen K, Rosengren L, Nellgard B, Blennow K (2004) Marked increase of beta-amyloid(1–42) and amyloid precursor protein in ventricular cerebrospinal fluid after severe traumatic brain injury. J Neurol 251:870–876
190. Raby CA, Morganti-Kossmann MC, Kossmann T, Stahel PF, Watson MD, Evans LM, Mehta PD, Spiegel K, Kuo YM, Roher AE, Emmerling MR (1998) Traumatic brain injury increases beta-amyloid peptide 1–42 in cerebrospinal fluid. J Neurochem 71:2505–2509
191. Mondello S, Buki A, Italiano D, Jeromin A (2013) Alpha-synuclein in CSF of patients with severe traumatic brain injury. Neurology 80:1662–1668
192. Farkas O, Polgar B, Szekeres-Bartho J, Doczi T, Povlishock JT, Buki A (2005) Spectrin breakdown products in the cerebrospinal fluid in severe head injury–preliminary observations. Acta Neurochir (Wien) 147:855–861
193. Pineda JA, Lewis SB, Valadka AB, Papa L, Hannay HJ, Heaton SC, Demery JA, Liu MC, Aikman JM, Akle V, Brophy GM, Tepas JJ, Wang KK, Robertson CS, Hayes RL (2007) Clinical significance of alphaII-spectrin breakdown products in cerebrospinal fluid after severe traumatic brain injury. J Neurotrauma 24:354–366
194. Csuka E, Morganti-Kossmann MC, Lenzlinger PM, Joller H, Trentz O, Kossmann T (1999) IL-10 levels in cerebrospinal fluid and serum of patients with severe traumatic brain injury: relationship to IL-6, TNF-alpha, TGF-beta1 and blood–brain barrier function. J Neuroimmunol 101:211–221
195. Mondello S, Robicsek SA, Gabrielli A, Brophy GM, Papa L, Tepas J, Robertson C, Buki A, Scharf D, Jixiang M, Akinyi L, Muller U, Wang KK, Hayes RL (2010) AlphaII-spectrin breakdown products (SBDPs): diagnosis and outcome in severe traumatic brain injury patients. J Neurotrauma 27:1203–1213
196. Tibbling G, Link H, Ohman S (1977) Principles of albumin and IgG analyses in neurological disorders. I. Establishment of reference values. Scand J Clin Lab Invest 37:385–390
197. Kossmann T, Hans VH, Imhof HG, Stocker R, Grob P, Trentz O, Morganti-Kossmann C (1995) Intrathecal and serum interleukin-6 and the acute-phase response in patients with severe traumatic brain injuries. Shock 4:311–317
198. Mcclain C, Cohen D, Phillips R, Ott L, Young B (1991) Increased plasma and ventricular fluid interleukin-6 levels in patients with head injury. J Lab Clin Med 118:225–231
199. Mcclain CJ, Cohen D, Ott L, Dinarello CA, Young B (1987) Ventricular fluid interleukin-1 activity in patients with head injury. J Lab Clin Med 110:48–54
200. Singhal A, Baker AJ, Hare GM, Reinders FX, Schlichter LC, Moulton RJ (2002) Association between cerebrospinal fluid interleukin-6 concentrations and outcome after severe human traumatic brain injury. J Neurotrauma 19:929–937
201. Chiaretti A, Genovese O, Aloe L, Antonelli A, Piastra M, Polidori G, Di Rocco C (2005) Interleukin 1beta and interleukin 6 relationship with paediatric head trauma severity and outcome. Childs Nerv Syst 21:185–193; discussion 194
202. Hayakata T, Shiozaki T, Tasaki O, Ikegawa H, Inoue Y, Toshiyuki F, Hosotubo H, Kieko F, Yamashita T, Tanaka H, Shimazu T, Sugimoto H (2004) Changes in CSF S100B and cytokine concentrations in early-phase severe traumatic brain injury. Shock 22:102–107

203. Shiozaki T, Hayakata T, Tasaki O, Hosotubo H, Fuijita K, Mouri T, Tajima G, Kajino K, Nakae H, Tanaka H, Shimazu T, Sugimoto H (2005) Cerebrospinal fluid concentrations of anti-inflammatory mediators in early-phase severe traumatic brain injury. Shock 23:406–410
204. Rasouli J, Lekhraj R, White NM, Flamm ES, Pilla AA, Strauch B, Casper D (2012) Attenuation of interleukin-1beta by pulsed electromagnetic fields after traumatic brain injury. Neurosci Lett 519:4–8
205. Chiaretti A, Antonelli A, Mastrangelo A, Pezzotti P, Tortorolo L, Tosi F, Genovese O (2008) Interleukin-6 and nerve growth factor upregulation correlates with improved outcome in children with severe traumatic brain injury. J Neurotrauma 25:225–234
206. Bell MJ, Kochanek PM, Doughty LA, Carcillo JA, Adelson PD, Clark RS, Whalen MJ, Dekosky ST (1997) Comparison of the interleukin-6 and interleukin-10 response in children after severe traumatic brain injury or septic shock. Acta Neurochir Suppl 70:96–97
207. Bell MJ, Kochanek PM, Doughty LA, Carcillo JA, Adelson PD, Clark RS, Wisniewski SR, Whalen MJ, Dekosky ST (1997) Interleukin-6 and interleukin-10 in cerebrospinal fluid after severe traumatic brain injury in children. J Neurotrauma 14:451–457
208. Buttram SD, Wisniewski SR, Jackson EK, Adelson PD, Feldman K, Bayir H, Berger RP, Clark RS, Kochanek PM (2007) Multiplex assessment of cytokine and chemokine levels in cerebrospinal fluid following severe pediatric traumatic brain injury: effects of moderate hypothermia. J Neurotrauma 24:1707–1717
209. Clark RS, Carcillo JA, Kochanek PM, Obrist WD, Jackson EK, Mi Z, Wisneiwski SR, Bell MJ, Marion DW (1997) Cerebrospinal fluid adenosine concentration and uncoupling of cerebral blood flow and oxidative metabolism after severe head injury in humans. Neurosurgery 41: 1284–1292; discussion 1292–1293
210. Goodman JC, Van M, Gopinath SP, Robertson CS (2008) Pro-inflammatory and pro-apoptotic elements of the neuroinflammatory response are activated in traumatic brain injury. Acta Neurochir Suppl 102:437–439
211. Kirchhoff C, Buhmann S, Bogner V, Stegmaier J, Leidel BA, Braunstein V, Mutschler W, Biberthaler P (2008) Cerebrospinal IL-10 concentration is elevated in non-survivors as compared to survivors after severe traumatic brain injury. Eur J Med Res 13:464–468
212. Maier B, Lehnert M, Laurer HL, Mautes AE, Steudel WI, Marzi I (2006) Delayed elevation of soluble tumor necrosis factor receptors p75 and p55 in cerebrospinal fluid and plasma after traumatic brain injury. Shock 26:122–127
213. Phillips DJ, Nguyen P, Adamides AA, Bye N, Rosenfeld JV, Kossmann T, Vallance S, Murray L, Morganti-Kossmann MC (2006) Activin a release into cerebrospinal fluid in a subset of patients with severe traumatic brain injury. J Neurotrauma 23:1283–1294
214. Semple BD, Bye N, Rancan M, Ziebell JM, Morganti-Kossmann MC (2010) Role of CCL2 (MCP-1) in traumatic brain injury (TBI): evidence from severe TBI patients and CCL2−/− mice. J Cereb Blood Flow Metab 30:769–782
215. Stahel PF, Morganti-Kossmann MC, Perez D, Redaelli C, Gloor B, Trentz O, Kossmann T (2001) Intrathecal levels of complement-derived soluble membrane attack complex (sC5b-9) correlate with blood–brain barrier dysfunction in patients with traumatic brain injury. J Neurotrauma 18:773–781
216. Whalen MJ, Carlos TM, Kochanek PM, Wisniewski SR, Bell MJ, Clark RS, Dekosky ST, Marion DW, Adelson PD (2000) Interleukin-8 is increased in cerebrospinal fluid of children with severe head injury. Crit Care Med 28:929–934
217. Zedler S, Faist E (2006) The impact of endogenous triggers on trauma-associated inflammation. Curr Opin Crit Care 12:595–601
218. Morganti-Kossmann MC, Rancan M, Stahel PF, Kossmann T (2002) Inflammatory response in acute traumatic brain injury: a double-edged sword. Curr Opin Crit Care 8:101–105
219. Morganti-Kossmann MC, Hans VH, Lenzlinger PM, Dubs R, Ludwig E, Trentz O, Kossmann T (1999) TGF-beta is elevated in the CSF of patients with severe traumatic brain injuries and parallels blood–brain barrier function. J Neurotrauma 16:617–628

220. Bayir H, Kagan VE, Tyurina YY, Tyurin V, Ruppel RA, Adelson PD, Graham SH, Janesko K, Clark RS, Kochanek PM (2002) Assessment of antioxidant reserves and oxidative stress in cerebrospinal fluid after severe traumatic brain injury in infants and children. Pediatr Res 51:571–578
221. Bayir H, Kochanek PM, Clark RS (2003) Traumatic brain injury in infants and children: mechanisms of secondary damage and treatment in the intensive care unit. Crit Care Clin 19: 529–549
222. Bayir H, Kochanek PM, Liu SX, Arroyo A, Osipov A, Jiang J, Wisniewski S, Adelson PD, Graham SH, Kagan VE (2003) Increased S-nitrosothiols and S-nitrosoalbumin in cerebrospinal fluid after severe traumatic brain injury in infants and children: indirect association with intracranial pressure. J Cereb Blood Flow Metab 23:51–61
223. Clark RS, Kochanek PM, Adelson PD, Bell MJ, Carcillo JA, Chen M, Wisniewski SR, Janesko K, Whalen MJ, Graham SH (2000) Increases in bcl-2 protein in cerebrospinal fluid and evidence for programmed cell death in infants and children after severe traumatic brain injury. J Pediatr 137:197–204
224. Clark RS, Kochanek PM, Watkins SC, Chen M, Dixon CE, Seidberg NA, Melick J, Loeffert JE, Nathaniel PD, Jin KL, Graham SH (2000) Caspase-3 mediated neuronal death after traumatic brain injury in rats. J Neurochem 74:740–753
225. Grossetete M, Phelps J, Arko L, Yonas H, Rosenberg GA (2009) Elevation of matrix metalloproteinases 3 and 9 in cerebrospinal fluid and blood in patients with severe traumatic brain injury. Neurosurgery 65:702–708
226. Robertson CL, Minamino N, Ruppel RA, Kangawa K, Wisniewski SR, Tsuji T, Janesko KL, Ohta H, Adelson PD, Marion DW, Kochanek PM (2001) Increased adrenomedullin in cerebrospinal fluid after traumatic brain injury in infants and children. J Neurotrauma 18: 861–868
227. Satchell MA, Zhang X, Kochanek PM, Dixon CE, Jenkins LW, Melick J, Szabo C, Clark RS (2003) A dual role for poly-ADP-ribosylation in spatial memory acquisition after traumatic brain injury in mice involving NAD+ depletion and ribosylation of 14-3-3gamma. J Neurochem 85:697–708
228. Darwish RS, Amiridze NS (2010) Detectable levels of cytochrome C and activated caspase-9 in cerebrospinal fluid after human traumatic brain injury. Neurocrit Care 12:337–341
229. Wagner AK (2010) TBI translational rehabilitation research in the 21st Century: exploring a Rehabilomics research model. Eur J Phys Rehabil Med 46:549–556
230. Wagner AK, Mccullough EH, Niyonkuru C, Ozawa H, Loucks TL, Dobos JA, Brett CA, Santarsieri M, Dixon CE, Berga SL, Fabio A (2011) Acute serum hormone levels: characterization and prognosis after severe traumatic brain injury. J Neurotrauma 28:871–888
231. Au AK, Aneja RK, Bell MJ, Bayir H, Feldman K, Adelson PD, Fink EL, Kochanek PM, Clark RS (2012) Cerebrospinal fluid levels of high-mobility group box 1 and cytochrome C predict outcome after pediatric traumatic brain injury. J Neurotrauma 29:2013–2021
232. Adamczak S, Dale G, de Rivero Vaccari JP, Bullock MR, Dietrich WD, Keane RW (2012) Inflammasome proteins in cerebrospinal fluid of brain-injured patients as biomarkers of functional outcome: clinical article. J Neurosurg 117:1119–1125
233. Johnston SC, Mendis S, Mathers CD (2009) Global variation in stroke burden and mortality: estimates from monitoring, surveillance, and modelling. Lancet Neurol 8:345–354
234. Rosamond W, Flegal K, Furie K, Go A, Greenlund K, Haase N, Hailpern SM, Ho M, Howard V, Kissela B, Kittner S, Lloyd-Jones D, Mcdermott M, Meigs J, Moy C, Nichol G, O'donnell C, Roger V, Sorlie P, Steinberger J, Thom T, Wilson M, Hong Y (2008) Heart disease and stroke statistics–2008 update: a report from the American Heart Association Statistics Committee and Stroke Statistics Subcommittee. Circulation 117:e25–e146
235. Montaner J (2006) Stroke biomarkers: can they help us to guide stroke thrombolysis? Drug News Perspect 19:523–532
236. Castellanos M, Serena J (2007) Applicability of biomarkers in ischemic stroke. Cerebrovasc Dis 24(suppl 1):7–15

237. Serena J, Blanco M, Castellanos M, Silva Y, Vivancos J, Moro MA, Leira R, Lizasoain I, Castillo J, Davalos A (2005) The prediction of malignant cerebral infarction by molecular brain barrier disruption markers. Stroke 36:1921–1926
238. Pike BR, Flint J, Dave JR, Lu XC, Wang KK, Tortella FC, Hayes RL (2004) Accumulation of calpain and caspase-3 proteolytic fragments of brain-derived alphaII-spectrin in cerebral spinal fluid after middle cerebral artery occlusion in rats. J Cereb Blood Flow Metab 24: 98–106
239. Zhang C, Siman R, Xu YA, Mills AM, Frederick JR, Neumar RW (2002) Comparison of calpain and caspase activities in the adult rat brain after transient forebrain ischemia. Neurobiol Dis 10:289–295
240. Mattson MP, Keller JN, Begley JG (1998) Evidence for synaptic apoptosis. Exp Neurol 153: 35–48
241. Posmantur RM, Kampfl A, Taft WC, Bhattacharjee M, Dixon CE, Bao J, Hayes RL (1996) Diminished microtubule-associated protein 2 (MAP2) immunoreactivity following cortical impact brain injury. J Neurotrauma 13:125–137
242. Taft WC, Yang K, Dixon CE, Hayes RL (1992) Microtubule-associated protein 2 levels decrease in hippocampus following traumatic brain injury. J Neurotrauma 9:281–290
243. Allard L et al (2005) PARK7 and nucleoside diphosphate kinase A as plasma markers for the early diagnosis of stroke. Clin Chem 51(11):2043–2051
244. Yao C et al (2009) P43/pro-EMAPII: a potential biomarker for discriminating traumatic versus ischemic brain injury. J Neurotrauma 26(8):1295–1305
245. Yao X, Liu J, Mccabe JT (2008) Alterations of cerebral cortex and hippocampal proteasome subunit expression and function in a TBI rat model. J Neurochem 104(2):353–363
246. Lakhan SE, Kirchgessner A, Hofer M (2009) Inflammatory mechanisms in ischemic stroke: therapeutic approaches. J Transl Med 7:97
247. Hochholzer W, Morrow DA, Giugliano RP (2010) Novel biomarkers in cardiovascular disease. Am Heart J 160(4):538–594
248. Singh D et al (2010) Sputum neutrophils as a biomarker in COPD: findings from the ECLIPSE study. Respir Res 11:77
249. Berger RP et al (2007) Serum biomarker concentrations and outcome after pediatric traumatic brain injury. J Neurotrauma 24(12):1793–1801
250. Donnan GA et al (2011) How to make better use of thrombolytic therapy in acute ischemic stroke. Nat Rev Neurol 7(7):400–409
251. Whitleley W, Tseng MC, Sandercock P (2008) Blood biomarkers in the diagnosis of ischemic stroke: a systematic review. Stroke 39(10):2902–2909
252. Kavalci C et al (2011) Value of biomarker-based diagnostic test in differential diagnosis of hemorrhagic ischemic stroke. Bratisl Lek Listy 112(7):398–401
253. Herrmann M et al (2000) Release of glial tissue-specific proteins after acute stroke: a comparative analysis of serum concentrations of protein S-100B and glial fibrillary acidic protein. Stroke 31(11):2670–2677

Chapter 18
Biomaterials for CNS Injury

Teck Chuan Lim and Myron Spector

Abstract Given the complexity of the tissue environment after CNS injury, appropriate delivery and deployment of therapeutic agents (e.g. small molecules, nucleic acids, proteins and cells) are as critical as the identification of the therapeutic agents themselves. Biomaterials are non-viable materials devised to interact with biological systems. Taking a plethora of forms ranging from nanoparticles, microspheres, porous scaffolds and hydrogels, biomaterials can be designed to interact with the injured CNS on a molecular, cellular or even tissue level. They have naturally emerged as powerful tools that can navigate therapeutic agents through the spatial and temporal challenges of the ever-evolving milieu in the injured CNS. This chapter highlights the roles that biomaterials play in neuroprotection, repair and regeneration (by protecting molecules and targeting them toward the CNS, sustaining long-term release of drugs and providing structural support for endogenous/transplanted cells) and details the strategies they employ in each of these roles. Overall, the numerous applications of biomaterials in the injured CNS not only illustrate the state of the art but also reflect the trend of biomaterials becoming increasingly engaged in an intimate partnership with therapeutic agents to ultimately materialize effective treatment for CNS injury.

18.1 Introduction

A hallmark of the central nervous system (CNS) is a diverse array of cells playing intricate roles in a highly complex tissue organization. In the event of CNS injury, such complexity is further layered with dynamic processes including inflammation and injury-induced neurogenesis. Several chapters in this book have detailed the significant strides made in dissecting this complexity and unraveling the underlying

M. Spector (✉)
Tissue Engineering/Orthopedic Surgery, VA Boston Healthcare System/Brigham and Women's Hospital/Harvard Medical School, Boston, MA, USA
e-mail: mspector@rics.bwh.harvard.edu

E.H. Lo et al. (eds.), *Vascular Mechanisms in CNS Trauma*, Springer Series in Translational Stroke Research 5, DOI 10.1007/978-1-4614-8690-9_18,

mechanisms. The newfound knowledge provides the valuable basis for potential therapies, but translation to clinically relevant approaches has been challenged with its own complexities. Robust transport barriers, which exist to isolate and protect the CNS, have to be carefully penetrated or circumvented for the delivery of therapeutic agents administered systemically. The numerous processes that unfold after injury continue to evolve over time in an interconnected fashion and often cause therapeutic targets to shift in a dynamic and interwoven fashion. Distinctive CNS injury responses such as expression of growth inhibitors and stromal liquefaction further pose a challenging environment for transplanted or endogenous cells to operate in.

Just as extracellular matrices in the CNS are critical enablers of molecular and cellular functions, so too can biomaterials facilitate protective or reparative responses to CNS injury. Biomaterials—nonviable materials intended to interact with biological systems—can be prepared from a large variety of natural and synthetic polymers into diverse forms such as nanoparticles, microspheres, hydrogels, and spongelike scaffolds and can accommodate the wide array of therapeutic agents (e.g., small molecules, nucleic acids, proteins, and cells). With their various forms spanning across several orders of magnitude in their physical dimensions, biomaterials readily interact with elements of the CNS on a molecular, cellular, and tissue level. They therefore have the potential to steer therapeutic agents through the aforementioned complex aspects of the CNS injury milieu, ensuring that they preserve their integrity and function as well as meet spatial and temporal requisites. Therapeutic agents contribute to the various phases of CNS injury, namely, the acute phase of neuroprotection, the subacute to chronic phase of repair, and the long-term phase of remodeling and regeneration. In this chapter, we survey the various forms of biomaterials which can play helpful roles in each of these stages: (1) nanoparticulate materials that protect their drug cargo and direct their delivery into the CNS, (2) repository materials that provide sustained release of therapeutic drugs, and (3) scaffolding materials that provide structural support for endogenous/transplanted cells.

18.2 Nanoparticulate Materials for Efficient, Targeted Delivery

In the aftermath of CNS injuries, much attention is paid to the sizable area of tissues around the injury core that, albeit spared from the full brunt of the injury, steadily undergo degeneration under the influence of secondary damage mechanisms (e.g., excitotoxicity, ionic imbalance, oxidative stress). The potential aggravation of neurological impairment upon the loss of these compromised tissues, coupled with the limited prospects of regenerating tissues that can fully replace the highly specialized structure and functions, compel the need for neuroprotective agents to slow/halt the secondary damage mechanisms while time still allows. To administer neuroprotective agents, systemic delivery or intranasal routes are generally preferred due to concerns that other routes with an invasive nature may further perturb the injured CNS tissue and be counterproductive to any protective effect offered by the administered agents.

Delivery of any substance into the CNS via systemic delivery or intranasal routes is, however, challenging. Substances dispatched into the systemic circulation or nasal cavity may suffer stability issues and are subjected to efficient clearance by reticuloendothelial system (RES) or nasal mucociliary mechanisms, respectively. Blood–brain barrier (BBB) or nasal mucosa, both of which contain tight and adherent junction proteins, efflux transport proteins, and drug-metabolizing enzymes, further restricts the passage of therapeutic agents into the CNS such that only a minuscule proportion, typically less than 2 %, of the administered dose enters the CNS. Therapeutic regimens commonly resort to increasing doses to compensate for these impediments, but the required magnitude of increase often runs the risk of substantial side effects.

Nanoparticles have emerged as highly effective solutions to address these challenges (Fig. 18.1). With diameters that vary from 10 to 1,000 nm, nanoparticles are sufficiently sized to serve as protective vehicles, which can encapsulate drugs by adsorption or covalent linkage (Fig. 18.2a) and yet remain sufficiently small to traverse the BBB and nasal mucosa. In one study, nanoparticles prepared from a synthetic biodegradable polymer, poly(lactide-co-glycolide) (PLGA) were used to encapsulate and deliver superoxide dismutase (SOD), a free radical scavenger, in a focal cerebral ischemia-reperfusion injury model [1]. Freely soluble SOD, which is disadvantaged by its short half-life of 6 min in vivo and limited permeability across the BBB, conferred no benefit to the injury outcome and failed to prevent animal death. In contrast, SOD in nanoparticles displayed dramatically improved efficacy, leading to 65 % reduction in infarction volume and 75 % survival among the injured rats. In another instance, erythropoietin formulated into PLGA nanoparticles putatively underwent much enhanced stability and BBB-crossing to achieve therapeutic efficacy at a dosage that was 16 times lower than freely soluble erythropoietin and might potentially lessen potential side effects such as capillary sludging [2]. In addition to being prepared as nanospheres, nanoparticles can be formulated as nanocapsules, liposomes, dendrimers, and micelles (Fig. 18.1). Each form carries certain unique attributes and can suitably cater to the application of specific therapeutic agents. For example, when delivering hemoglobin as an oxygen carrier to the ischemic rat brain after stroke injury, liposomes not only prevented renal clearance, extravasation, and hypertensive response that were generally associated with free hemoglobin but also assumed a red blood cell-like structure that featured an aqueous core to house the protein and a thin lipid shell to facilitate efficient oxygen diffusion [3]. On the other hand, polymeric micelles, formed via the self-assembly of amphiphilic polymers into a hydrophobic core ensheathed by a hydrophilic corona, have been shown to aptly incorporate drugs with poor water solubility. When used to deliver methylprednisolone, a water-insoluble corticosteroid, to inhibit inflammation after spinal cord injuries in rabbits, polymeric micelles increased the half-life of the drug by sevenfold and helped avoid megadoses that could potentially lead to wound infections, pneumonia, and sepsis [4].

Nanoparticles can further host moieties via bulk entrapment or by adsorption or chemical conjugation to their surface to minimize their clearance and lengthen the

Fig. 18.1 Types of nanoparticles for drug delivery into the CNS. Liposomes are vesicles composed of an aqueous core enclosed in one or more phospholipid bilayers. Nanocapsules are similarly structured vesicles with a polymer-based shell and either an aqueous or hydrophobic core. On the other hand, nanospheres have a polymeric matrix throughout the entire volume. Dendrimers are macromolecules with a globular structure characterized by repeated branches. The highly ordered architecture on the surface of the dendrimers offers a molecular template that accommodates and precisely controls surface functionalization. Depending on the nature of the polymer used, nanospheres and dendrimers may possess with a hydrophilic or hydrophobic core. Micelles are another type of vesicles with a core-shell structure formed through the self-organization of amphiphilic molecules. In physiological environments that are usually aqueous in nature, hydrophobic regions of the molecules condense to form a core while hydrophilic regions are exposed in the exterior to form a shell

window of opportunity for the nanoparticles and their therapeutic cargo to enter the CNS (Fig. 18.2b). For nanoparticles entering the blood circulation en route to the CNS, the most common example is polyethylene glycol (PEG), an inert, hydrophilic polymer that has already been employed in several Food and Drug Administration (FDA)-approved protein conjugate drugs [5] and used widely with nanoparticles for CNS applications. The appeal of PEG largely lies in its ability to form a hydrophilic shield that (1) blocks the adsorption of opsonins, (2) reduces engulfment by macrophages, and (3) helps the nanoparticles to evade clearance by the RES [6]. The resulting benefit, as shown in an in vivo study investigating polycyanoacrylate nanoparticles in mice and rats [7], is evident—PEG increased the circulatory half-life of coated nanoparticles, affording them sufficient time to penetrate into the deep regions of the brain (e.g., striatum, hippocampus, hypothalamus, and thalamus) even without any modification to the BBB permeability. Also of note, related to penetration into the CNS, PEG has also been recently shown to render densely coated nanoparticles minimally adhesive, allowing them to spread rapidly within the brain extracellular space even when they are larger than 64 nm, the size limit for appreciable movement through the brain [8]. Nonetheless, PEG should be used with caution regarding (a) the possibility that depending on its coating density, PEG adopts a mushroomlike conformation, instead of a brushlike one, and becomes less effective in shielding the nanoparticle surface [9], and (b) the plausible elicitation of anti-PEG IgM, which actually triggers a counterproductive accelerated blood clearance phenomenon [10] and may necessitate a switch to alternative hydrophilic polymers such as poly(*N*-vinyl-2-pyrrolidone) [11]. For nanoparticles administered via the intranasal route, cationic polymers (e.g., chitosan [12]) and cationic molecules (e.g., stearylamine [13]) are typically used to cope with clearance by the nasal mucociliary mechanisms. Such cationic

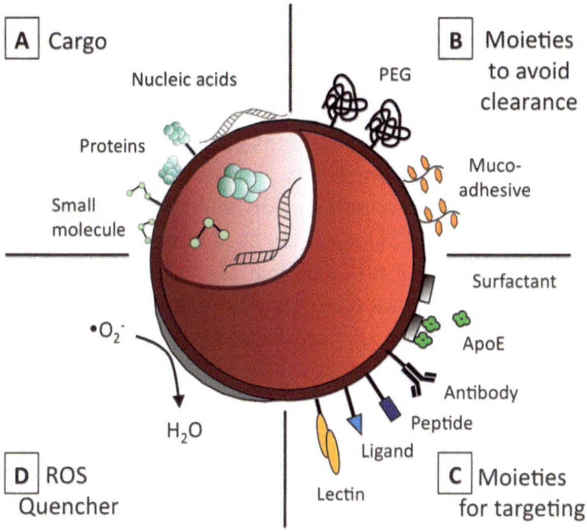

Fig. 18.2 Functional architecture of nanoparticles. (**a**) Therapeutic drugs, which can occur in the form of small hydrophilic or hydrophobic molecules, proteins (e.g., growth factors and antibodies), and nucleic acids (e.g., plasmid DNA and small-interfering RNA). (**b**) Nanoparticle surface may be functionalized with PEG to reduce opsonization and clearance by the RES or with mucoadhesives (such as chitosan or stearylamine) to bind with mucous proteins and prolong residence time within the nasal cavity. (**c**) Nanoparticle surface can also be functionalized with moieties to target the BBB or olfactory neurons. (**d**) Nanoparticles of select materials (e.g., platinum or carbon) can undertake functional roles such as quenching ROS to achieve neuroprotective effects

moieties serve as mucoadhesives that interact electrostatically with the negatively charged sialic acid residues in mucous proteins, thus prolonging the residence time of nanoparticles in the nasal cavity and enhancing their transport into the brain.

Nanoparticles can also steer their biodistribution toward the CNS and thereby minimize the side effects of their therapeutic cargo on non-CNS tissues. The general principle for this is to host targeting moieties that mediate the binding of the nanoparticles with receptors that are highly expressed on the BBB microvascular endothelial cells or olfactory nerves and to exploit the transcytosis of the receptors for nanoparticle transport (Fig. 18.2c). Targeting moieties for the BBB can take the following forms: (1) antibody (e.g., monoclonal antibody against transferrin receptor has been shown to enhance the transport of caspase inhibitor-loaded nanoparticles for neuroprotection after brain ischemia [14]); (2) ligand (e.g., insulin and thiamine for insulin receptors and thiamine transporters, respectively); (3) peptide, which can be subcategorized into viral capsid protein-derived cell-penetrating peptide (e.g., TAT peptide for lipid raft- and receptor-dependent endocytosis [15]), opioid mimic (e.g., synthetic opioid peptide with a single amino acid substitution to target δ/μ opioid receptors without the potential opioid effect [16]), and Kunitz domain-derived peptide (e.g., Angiopep to target low-density lipoprotein receptor-related protein 1 [17]); and (4) nonionic surfactants for promoting adsorption of endogenous targeting molecules (e.g., polysorbate-80 for the adsorption of apolipoprotein E which in turn targets low-density lipoprotein receptors [18]). On the other hand, targeting moieties for olfactory nerves include lectins (such as wheat germ agglutinin [19] and ulex europaeus agglutinin I [20]) that bind to the olfactory receptors. Not surprisingly, new targeting moieties are continuously being uncovered as part of the ongoing effort to reduce side effects. Among them, glia-specific TD2.2 peptide [21] as well as neuron-specific Tet1 peptide [22] represents a promising class of moieties that can enable specific targeting toward individual neural cell types. Antibodies for fibrin and the analogous peptide derivatives identified through phage display techniques are another class of interest given their potential in enabling the novel thrombolytic approaches where fibrinolytic nanoparticles home specifically toward thrombus and benefit ischemic strokes without the usual hemorrhagic risk [23].

Finally, while nanoparticles are commonly noted to be highly effective delivery vehicles, it is important to recognize that nanoparticles may also feature unique material attributes to undertake neuroprotective roles by themselves (Fig. 18.2d). For instance, nanoparticles made of platinum present the catalytic metal at a high surface/volume ratio, hold high electron density on their surface, and have acted as potent reactive oxygen species (ROS) scavenger to ameliorate brain damage and neurological impairment in ischemic and reperfusion injury [24]. Carbon nanoparticles also improve cerebrovascular dysfunction after controlled cortical impact-induced traumatic brain injury by providing graphitic structural domains that do not require supporting detoxifying molecules such as catalase or glutathione and completely annihilate ROS [25]. In another example, nanoparticles made of chitosan, a cationic biopolymer that has been shown to induce the fusion of small phospholipid bilayers [26], facilitate neuronal membrane sealing in addition to delivering hydralazine to trap toxic byproducts of lipid peroxidation after CNS trauma [27].

18.3 Repository Biomaterials for Local and Sustained Release of Therapeutic Drugs

As CNS injury progresses with time, the focus of intervention shifts from neuroprotection to remodeling, repair, and regeneration. Along with this shift come new challenges, the most prominent of which is the need to have a protracted presence of therapeutic agents to cater to the relatively prolonged nature of the underlying events. Rather than relying on repeated administration of nanoparticles, a preferable approach is to establish a local, sustained release mechanism at the site of interest. Such an approach bypasses the recurring difficulty of crossing the BBB or olfactory epithelium, offers a direct way of targeting a specific tissue region, and most importantly, delivers a steady stream of therapeutic drugs without large fluctuations between concentrations that are ineffectively low and prohibitively high.

Experimentally, this is frequently achieved via continuous infusion using osmotic pumps. For clinical applications, the use of osmotic pumps on a chronic basis is, however, fraught with difficulties. Much emphasis is understandably placed on biomaterials that can act as a depot for releasing therapeutic drugs with the desired kinetics.

Repository biomaterials commonly occur in the form of microparticles. In contrast to nanoparticles, they possess a significantly smaller surface area/volume ratio, which works in the favor of retarding the diffusive loss of their therapeutic cargo from their surface. In addition, at a length scale that is similar to or larger than most cells in the CNS, they tend to be more resistant against phagocytosis by individual cells than nanoparticles and are better suited for staying localized at their sites of application. Repository biomaterials may also be of an even larger length scale (~millimeters) by taking the form of hydrogels, which are physically or chemically cross-linked polymer networks with high water content. In addition to sharing the aforementioned advantages with microparticles, hydrogels have the further benefit of having the sheer size to remain stably anchored when applied outside the CNS parenchyma. As such, they can be deposited at intrathecal [28, 29] or epicortical [30, 31] locations (Fig. 18.3) to completely avoid any disturbance to the CNS parenchyma.

Through the calculated selection and use of different materials, microparticles and hydrogels can be tailored to adopt various strategies to store therapeutic drugs and release them locally in a sustained fashion in the CNS. One strategy is physical entrapment which usually involves the polymeric matrix within the microparticles or hydrogels impeding the diffusion of the loaded drug or to only allow drug efflux upon the degradation of the entrapping matrix (Fig. 18.3a). PLGA, a biodegradable polymer that undergoes hydrolysis into lactic and glycolic acids at a rate controlled by the lactide/glycolide ratio used during polymerization, is a classical example for this strategy. When formulated into microparticles for delivering nerve growth factor (NGF), it led to a continuous release of NGF for at least 6 weeks and could be introduced adequately close to axotomized neurons to promote their survival [32]. Recently, physical entrapment is also achieved by using lipid microtubules as hollow containers that limit drug efflux to only the miniscule openings on the ends of the tubular structures. The lipid microtubules were successfully employed after spinal cord hemisection injury to deliver chondroitinase ABC to remodel the inhibitory extracellular matrix and enhance axonal sprouting around the lesion [33].

Other than physical entrapment, affinity between the polymeric matrices and the loaded drugs can also extend drug efflux over time (Fig. 18.3b). Such affinity may be introduced by choosing a polymer that serves as a binding partner to the loaded drug. For instance, acidic gelatin, instead of its basic counterpart, was selected for its anionic nature to undergo electrostatic complexation with cationic growth factors [34]. When used in the form of microspheres to deliver insulin-like growth factor-1 or hepatocyte growth factor in the ischemic mouse brain, the sustained release effectively increased the number of new neurons in subventricular zone as well as their migration toward the injured striatum [35]. Alternatively, binding domains for the drug may be conjugated onto the polymeric matrices to institute

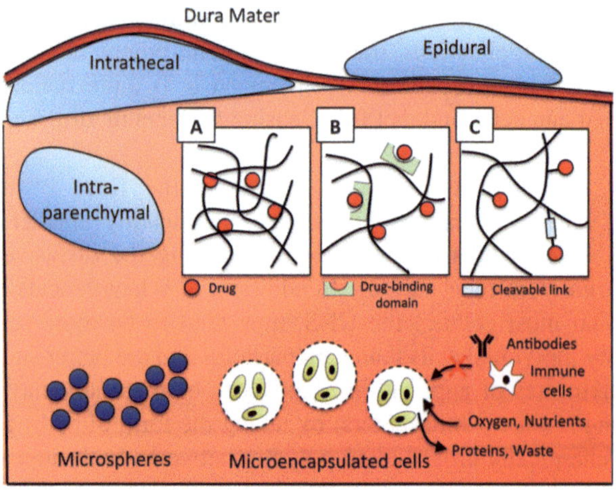

Fig. 18.3 Repository biomaterials for sustained delivery of drugs to a local tissue environment. Biomaterials, commonly in the form of microparticles and hydrogels, are implanted into targeted regions of the CNS parenchyma to store and release drugs over time. Hydrogels, in particular, may also be deposited at intrathecal and epidural locations. Strategies for sustained drug release include (**a**) physical entrapment, (**b**) drug-binding mechanisms, or (**c**) chemical links which tether drugs and release them only with material degradation or link cleavage. Chemical links may further incorporate domains to respond to specific stimuli. For highly extended protein release, biomaterials may take the role of an immunoisolating barrier to allow microencapsulation of protein-producing cells

affinity with the loaded drug. Using peptide engineering and a transglutaminase-based conjugation technique, heparin was covalently linked to fibrin matrices for the delivery of heparin-binding growth factors [36]. When employed in a spinal cord hemisection model, the heparin-based fibrin matrices controlled release of neurotrophin-3 and enhanced neural fiber sprouting by 54 % relative to unmodified fibrin matrices [37]. In the most extreme option for slowing or even halting drug diffusion, the polymeric matrices can be covalently conjugated with the drug such that drug release occurs solely by cleavage of the covalent links or degradation of the matrices (Fig. 18.3c). While this option runs the risk of weakening the bioactivity of the drug due to chemical modifications of the drug, it has been successfully demonstrated in hyaluronic acid hydrogels conjugated with Nogo-66 receptor antibody. In both brain ischemic and spinal cord hemisection injuries, the hydrogels enhanced axon regrowth, proving their biodegradability and their ability to release bioactive antibodies to block inhibitory signaling via the Nogo-66 receptor [38, 39].

Clearly, regardless of the choice of biomaterials, there is an inevitable limit to the amount of drugs that can be stored and to the duration over which their release can be sustained. In cases where the need for proteinaceous drugs extends into a timeframe measured in months and years, the logical inclination is to use repository biomaterials to encapsulate and immunoisolate protein-releasing cells instead (Fig. 18.3). In theory, cells may be genetically modified to overexpress any protein of interest and secrete them for indefinite periods of time as long as the cells remain uncompromised and have their metabolic demands well supported. Along with this change of approach for releasing proteins, different material characteristics are desired in repository biomaterials. For instance, the indispensable trait is selective permeability, which is extremely critical for the transport barrier around the protein-releasing cells to permit efficient exchange of secreted proteins and waste with nutrients and oxygen while preventing immune cells and antibodies from accessing the grafted cells. Polysulfone, a material whose selective permeability has been well tested in hemodialysis, is a natural candidate and has been used to encapsulate vascular endothelial growth factor (VEGF)-secreting hamster kidney cells in the ischemic rat brain to support post-injury angiogenesis [40]. Alginate/poly-L-ornithine also proved to be an adequate encapsulating matrix for genetically modified fibroblasts to survive at least 1 month after grafting into the injured spinal cord of adult rats and produce brain-derived neurotrophic factor (BDNF) for promoting axonal growth [41]. Additionally, while biodegradability is sought after in drug-storing biomaterials to ensure their disappearance after drug depletion, persistent materials are ideal here to ensure the integrity of the selectively permeable barrier over time. In an example where nondegradable poly(acrylonitrile-co-vinyl chloride) was used, NGF-secreting hamster kidney cells remained stably immunoisolated and active in NGF expression 13.5 months after grafting into the lateral ventricles of adult rats, thereby bringing about marked hypertrophy of cholinergic neurons [42].

The current state of repository biomaterials remains mostly with the capability to deliver a single type of therapeutic drug with predetermined release kinetics. Given that this approach falls far short of providing the multitude of signals typically involved in the contextually dependent progression of events after CNS injury, a long road lies ahead for repository biomaterials with much attention paid to devising programmable and stimuli-responsive delivery. Programmable delivery, which involves releasing multiple factors with a specific sequence of choice, is particularly important to handle the appearance of different intervention targets at varying times during injury progression. It has already been applied with promising outcomes in areas such as neovascularization and bone regeneration and has spurred the inception of several delivery systems that can potentially be applied to the injured CNS tissue. These include hydrogels (e.g., polymer blends of hyaluronan/methyl cellulose [43] and carbomer/agarose [44]) that engender fast and slow diffusion kinetics for small molecules and proteins, respectively, hydrogel-incorporating microspheres to each carry a protein and regulate its release kinetics differentially (e.g., hyaluronic acid hydrogels and PLGA microspheres to deliver BDNF and VEGF, respectively [45]), and hydrogels with distinct binding mechanisms for different proteins to independently vary their release kinetics (e.g., PEG hydrogels that bind to cationic proteins and hexahistidine tagged proteins via electrostatic interactions and metal ion chelation, respectively [46]). Stimuli-responsive delivery ensures the release of drugs only on demand, thereby minimizing nonspecific effects of the drugs and allowing the drugs to be conserved for use over a longer timeframe. A proof of concept has already been demonstrated with hydrogels that can release chemotherapeutic agents in response to the heightened

MMP activity [47] and can conceivably be extended to CNS injury once appropriate stimuli and their associated release mechanisms are identified.

18.4 Scaffolding Materials for Transplanted/Endogenous Cells

Often derived from polymers naturally present in extracellular matrices or amenable to physical and chemical modification to recapitulate features of ECM, biomaterials have widely been used as scaffolding materials in various tissues. Given the high tendency for stromal liquefaction and formation of structureless cavities after CNS injury (Fig. 18.4a), biomaterials become of paramount importance in terms of providing stromal support for endogenous or transplanted cells in CNS lesions. For example, in a rat model where focal cerebral ischemia led to a cystic cavity in the cerebral cortex, Matrigel, a hydrogel constituted from extracellular matrix proteins secreted by Engelbreth-Holm-Swarm mouse sarcoma, made the critical difference in allowing transplanted human neural precursor cells to survive 8 weeks after transplantation and mediate a ~50–60 % reduction in infarct cavity volume [48]. In another example, implantation of *N*-(2-hydroxypropyl)-methacrylamide hydrogels into a chronic spinal cord lesion (formed 5 weeks after a balloon-induced compression injury in adult rats) contributed to the survival of transplanted mesenchymal stem cells for as long as 5 months, permitting them to promote infiltration of myelinated axons and functional improvements [49].

A similar trend has been observed with endogenous cells in the injured CNS. The presence of a peptide nanofiber scaffold in the lesion caused by a 2-mm-deep transection of the optic tract in the hamster midbrain was deemed necessary as a bridge for axons to regenerate through the lesion, reconnect with target tissues, and restore functional vision [50]. By serving as structural support in surgically created cortical cavities, collagen-glycosaminoglycan scaffolds also made possible the infiltration of endogenous neuronal precursors which otherwise only migrated up to the lesion boundary [51]. Finally, the importance of scaffolding biomaterials was definitively illustrated in a study where polyglycolic acid scaffolds were implanted into cystic lesions resulting from a unilateral hypoxia-ischemic brain injury [52]. In this instance, only cystic lesions implanted with scaffolds enabled neurons derived from neural stem cells (NSCs) grafted into the lesion to innervate the contralateral hemisphere as well as host neurons from contralateral hemisphere to innervate the lesion.

With numerous possibilities for base materials (natural or synthetic polymers), multiple physical forms [preformed sponges, electrospun fibers, microspheres, and hydrogels (Fig. 18.4b–e)], and abundant ways to further chemically modify the scaffold surface, a huge number of permutations has been investigated for their applicability in supporting cells in CNS lesions. The collective knowledge generated from such studies has now made it possible for scaffolds to be engineered based not simply on a general notion of mimicking the brain ECM but on an array of

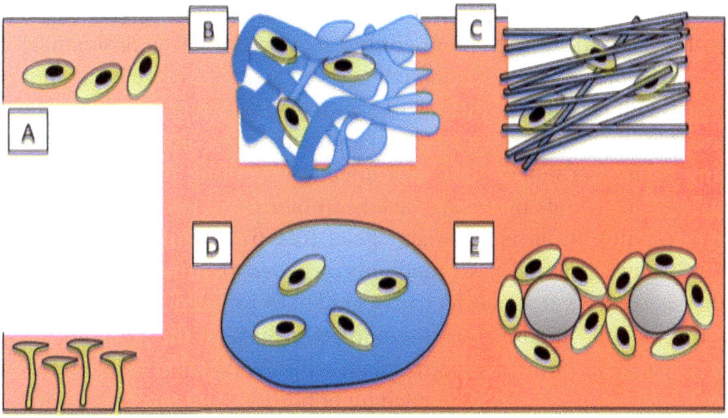

Fig. 18.4 Scaffolding materials in CNS lesions. Proteolysis of ECM in the injured tissue often results in stromal liquefaction and the formation of a cystic lesion. (**a**) Cells (e.g., transplanted or endogenous neural progenitors) and regenerating axons are strongly hindered in this structureless environment. Biomaterials can serve as scaffolding materials to introduce the crucial structural support in the cystic lesion. This may occur in the form of (**b**) preformed porous scaffolds, (**c**) electrospun fibers, (**d**) hydrogels, and (**e**) microspheres

design criteria. Of the essential properties that have emerged, the most familiar and routinely applied criterion is the cell adhesiveness of the scaffolds. This stems from the fact that cell attachment to the scaffolds is a fundamental requisite for both anti-anoikis signaling and cell migration and, thus, essential for scaffolds to succeed as a structural support in cystic CNS lesions. Clearly, the most straightforward strategy is to rely on the presence of cell-adhesive ECM proteins in the scaffolds. Along this line, collagen [53, 54], its denatured form gelatin [55], fibronectin [56], and laminin [57] have either been used as the base material or surface coating for scaffolds applied in the injured brain or spinal cord. The amino acid motifs Arg-Gly-Asp (RGD), which mediates binding of fibronectin with the integrin family of receptors, and Ile-Lys-Val-Ala-Val (IKVAV) and Tyr-Ile-Gly-Ser-Arg (YIGSR), which account for binding with laminin-binding proteins, have also been employed for the modification of scaffolds as alternatives to the full ECM proteins [58–60]. Their efficacy in accommodating neurite extension, axonal growth, and cell infiltration validates the ability of these molecular motifs to recapitulate the binding affinities of fibronectin and laminin with cell types found in the CNS and strongly signals their significance as a progressive way for decorating scaffolds and manipulating their cell adhesiveness with molecular precision. In fact, with the discovery of other domains that are involved or related to cell adhesion (e.g., Pro-His-Ser-Arg-Asn to synergize cell binding with RGD [61, 62]), it is reasonable to expect these molecular motifs to impart scaffolds with an increasingly thorough mimicry of ECM in their binding to transplanted or endogenous cells.

Another critical scaffold property is the directionality of scaffold features, i.e., pore orientation for porous preformed sponges and fiber alignment for electrospun fiber scaffold. Both pore orientation and fiber alignment have been areas of focus for scaffolds devised for the spinal cord in recognition of its pronounced level of directionality. Several types of uniaxially porous scaffolds have been prepared over the years through techniques such as freeze-drying with a uniaxial thermal gradient [54], oriented diffusion of ions into ionically cross-linkable hydrogels [63], templating with microfibers [64], and injection molding [65]. Similarly, aligned fiber scaffolds of various materials have been fabricated via collection of the fibers on a drum rotating at an appropriate speed [66, 67]. While axon regrowth after spinal cord injury is usually so randomly oriented that the axons fail to extend across the lesion [68], it was effectively guided to proceed in a linear fashion into these scaffolds and, in some cases, completely through the scaffold to reach the distal cord tissue. A further step was taken with the creation of a multicomponent porous scaffold featuring a NSC-seeded inner portion with randomly oriented pores to emulate the gray matter of spinal cord and an outer portion with uniaxially oriented pores to emulate the white matter and guide regenerating axons after a hemisection injury [69]. By better matching the region-specific pore orientation of the spinal cord, the scaffold created an appropriate combination of a NSC-mediated trophic support zone and a guidance zone for regenerating axons to produce persistent recovery in hindlimb function. Compared to the spinal cord, studies on the directionality of scaffold features are less strongly motivated and far fewer for the brain, since axons and neurites do not extend only in a single predominant direction throughout the brain. It has appeared that pore orientation and fiber alignment are still properties relevant for the brain given that they influence the accessibility of spaces into which endogenous cells can infiltrate. In a study examining scaffolds with different pore orientations in the healthy adult brain, scaffolds with pores oriented toward the apposing brain tissues increased infiltration of endogenous astrocytes [70]. On the other hand, for electrospun fiber scaffolds in the brain, randomly oriented fibers translated into porosity for accommodating neurite growth into the scaffold while aligned fibers prevented the neurites from penetrating into the scaffold [71].

In addition to the above, several noteworthy scaffold properties have emerged through extensive in vitro studies. One of them is surface topography, which involves micro-/nano-grooves, pillars, holes, and roughness. Through precise tuning of the feature attributes such as groove depth [72], groove width [73], and inter-pillar distance [74], surface topography can affect contact guidance of the cells and influence cell behaviors such as alignment of neuronal cell body and neurites. First shown to differentially influence the differentiation of mesenchymal stem cells into neurons, myoblasts, or osteoblasts across 0.1–40 kPa [75], substrate modulus is another scaffold property important to neural cells. Across the smaller range of 0.1–10 kPa, it has been demonstrated to affect differentiation of NSCs, with soft substrates favoring neuronal differentiation and hard substrates favoring glial differentiation [76, 77]. Given the electrical activity of neurons, electrical properties of scaffolds have also been found to be potent modulators of neural cells. In several studies, piezoelectric scaffolds composed of polyvinylidene fluoride-trifluoroethylene fibers induced human neural progenitors to undergo enhanced neuronal differentiation and neurite extension [78], while highly conductive carbon nanotubes interfaced tightly with neuronal membranes, favored back-propagating action potentials [79], and enabled neurons to modulate their synaptic strength [80]. Finally, the cross-linking process of scaffolds, especially injectable hydrogels that cross-link in situ in the presence of their cell cargo, has also been brought into attention as a novel scaffold property to influence transplanted cells positively in the injured CNS. As shown in injectable gelatin-hydroxyphenylpropionic acid hydrogels, the cross-linking process, which involves minute amount of oxidants being rapidly generated and consumed, preconditions encapsulated NSCs, markedly increases their oxidative stress resistance, and can potentially augment their survival in the injured CNS where free oxidative radicals become prevalent [81]. Overall, these scaffold properties are likely to open up new avenues of supporting and influencing transplanted and endogenous cells in the injured CNS as they shift from in vitro to in vivo applications.

It should be noted that the current state of the art for biomaterials has yet to resolve the mutual exclusivity between the injectability of the scaffolds and precise control over anisotropic and highly ordered features (i.e., oriented pores, aligned fibers, nano-/microgrooves or pillars with regular spacings). This has been a natural consequence of the fact that the fluidity often required for injection through a needle is not ideal for maintaining a highly ordered structure. The application of scaffolds in the injured CNS tissue should therefore involve deliberation over this constraint. For example, one may conclude that for a lesion deep within the injured brain, the use of injectable materials, such as hydrogels or microspheres, to minimize perturbation to healthy brain tissue surrounding the lesions exceeds the importance of the use of preformed scaffolds whose oriented pores or fibers may not be the most essential. New classes of biomaterials, such as the recently reported injectable cryogels that recover their macroporous architecture after compression and flow through a small-bore needle [82], may resolve this constraint in the not-too-distant future.

18.5 Combinatorial Approaches

Against the multifaceted nature of the injured CNS, biomaterials have demonstrated the ability to adopt a large assortment of functionalities and, thus, the versatility to fittingly handle each individual intervention target. However, many of these targets are intertwined so intimately that they often have to be addressed in a coordinated fashion to derive any measurable benefit. In face of this, one of the greatest strengths of biomaterials is their utility as a platform to integrate multiple functionalities and orchestrate a combinatorial approach. The importance of such biomaterial-mediated combinatorial approaches has already been underscored in several applications. In one of them, in which NSCs were transplanted into a spinal cord hemisection lesion as potential neuronal replacements, fibrin scaffolds conjugated with heparin for affinity-based controlled release of neurotrophin-3 and platelet-derived growth factor were used to concurrently address the absence of structural support in the lesion as well as the typical poor survival and uncontrolled differentiation of the transplanted stem cells [83]. Compared to groups with unmodified scaffolds and those without the scaffolds altogether, the combination of scaffolds with protein delivery systems was significantly more effective in enhancing the number of neural progenitor cell (NPC)-derived neurons and would be a useful approach to emulate in future attempts of stem cell transplantation. In another example, VEGF-releasing PLGA microspheres were employed to resolve the lack of structural support for transplanted NSCs and the absence of vasculature to sustain long-term viability of the transplanted cells in brain cavitary lesion resulting from ischemic strokes. This combinatorial approach successfully supported the transplanted NSCs to form a primitive de novo tissue and promoted its neovascularization [84]. For the spinal cord, therapeutic molecules have also been meaningfully integrated with scaffolds with directional features. In one instance, uniaxially porous collagen scaffolds releasing chondroitinase ABC for the remodeling of the growth inhibitory environment enabled the infiltration of growth-associated protein 43-positive axons into hemisection lesions [85]. In another instance, which is one of the most elaborate combinatorial approaches attempted to date, extramedullary chitosan channels, adult brain-derived NPCs, Nogo-66 receptor protein, and three growth factors (i.e., FGF-2, EGF, and PDGF) were pooled into a single biomaterial-facilitated treatment for a complete transection injury in spinal cord [86]. The various components present drove numerous events, namely, better NPC survival, enhanced differentiation toward oligodendrocyte, reduced differentiation into astrocytes, and increased number of myelinated fibers in the channel, which all culminated into increased area of bridging tissue between the spinal cord stumps.

18.6 Concluding Remarks

A therapy for sufficiently managing/improving the injured CNS remains elusive. Part of the reason lies in the many and myriad unknowns surrounding CNS injuries. The other part likely lies in the inability to cope with the exceedingly complicated CNS tissue environment and deploy therapeutic agents in a fully effective manner. Through rational material design and formulation, biomaterials provide the means to deliver therapeutic agents in a controllable fashion, provide a favorable environment for reparative cells, and thus optimize intervention for CNS injury as much as available knowledge allows. Emerging knowledge about the degenerative and regenerative processes following CNS injury is certain to inform synthesis and formulation of biomaterials, which in turn will provide better tools for further investigation as well as more effective treatments for the injured CNS.

The future of biomaterials for CNS injuries lies in three particular areas of development. First, biomaterials with established functionalities in the CNS, through the discovery of better alternative materials, can have their efficacy further optimized. Second, biomaterials can adopt increasingly advanced functionalities such as nanoparticle targeting with cellular or even subcellular specificity, programmable and stimuli-responsive delivery of multiple factors, and multifaceted scaffold effects that span across the chemical, geometrical, mechanical, and electrical domains. Third, combinatorial approaches can integrate a greater number of functionalities as well as present the functionalities with temporal precision and at carefully titrated levels. Through these areas, biomaterials are projected to become progressively more sophisticated to match the complexity of CNS injury and increasingly adept in managing the challenging tissue environment.

References

1. Reddy MK, Labhasetwar V (2009) Nanoparticle-mediated delivery of superoxide dismutase to the brain: an effective strategy to reduce ischemia-reperfusion injury. FASEB J 23(5): 1384–1395. doi:10.1096/fj.08-116947
2. Chen H, Spagnoli F, Burris M, Rolland WB, Fajilan A, Dou HY, Tang JP, Zhang JH (2012) Nanoerythropoietin is 10-times more effective than regular erythropoietin in neuroprotection in a neonatal rat model of hypoxia and ischemia. Stroke 43(3):884–887. doi:10.1161/Strokeaha.111.637090
3. Kawaguchi AT, Fukumoto D, Haida M, Ogata Y, Yamano M, Tsukada H (2007) Liposome-encapsulated hemoglobin reduces the size of cerebral infarction in the rat: evaluation with photochemically induced thrombosis of the middle cerebral artery. Stroke 38(5):1626–1632. doi:10.1161/STROKEAHA.106.467290
4. Chen CL, Chang SF, Lee D, Yang LY, Lee YH, Hsu CY, Lin SJ, Liaw J (2008) Bioavailability effect of methylprednisolone by polymeric micelles. Pharm Res 25(1):39–47. doi:10.1007/s11095-007-9484-0
5. Alconcel SNS, Baas AS, Maynard HD (2011) FDA-approved poly(ethylene glycol)-protein conjugate drugs. Polym Chem 2(7):1442–1448. doi:10.1039/C1py00034a
6. Owens DE, Peppas NA (2006) Opsonization, biodistribution, and pharmacokinetics of polymeric nanoparticles. Int J Pharm 307(1):93–102. doi:10.1016/J.Ijpharm.2005.10.010
7. Calvo P, Gouritin B, Chacun H, Desmaele D, D'Angelo J, Noel JP, Georgin D, Fattal E, Andreux JP, Couvreur P (2001) Long-circulating PEGylated polycyanoacrylate nanoparticles as new drug carrier for brain delivery. Pharm Res 18(8):1157–1166. doi:10.1023/A:1010931127745
8. Nance EA, Woodworth GF, Sailor KA, Shih TY, Xu QG, Swaminathan G, Xiang D, Eberhart C, Hanes J (2012) A dense poly(ethylene glycol) coating improves penetration of large polymeric nanoparticles within brain tissue. Sci Transl Med 4(149):149ra119
9. Gref R, Luck M, Quellec P, Marchand M, Dellacherie E, Harnisch S, Blunk T, Muller RH (2000) 'Stealth' corona-core nanoparticles surface modified by polyethylene glycol (PEG): influences of the corona (PEG chain length and surface density) and of the core composition on phagocytic uptake and plasma protein adsorption. Colloids Surf B Biointerfaces 18(3–4): 301–313
10. Ishida T, Ichihara M, Wang X, Yamamoto K, Kimura J, Majima E, Kiwada H (2006) Injection of PEGylated liposomes in rats elicits PEG-specific IgM, which is responsible for rapid elimination of a second dose of PEGylated liposomes. J Control Release 112(1):15–25. doi:10.1016/j.jconrel.2006.01.005
11. Ishihara T, Maeda T, Sakamoto H, Takasaki N, Shigyo M, Ishida T, Kiwada H, Mizushima Y, Mizushima T (2010) Evasion of the accelerated blood clearance phenomenon by coating of nanoparticles with various hydrophilic polymers. Biomacromolecules 11(10):2700–2706. doi:10.1021/Bm100754e

12. Wang X, Chi N, Tang X (2008) Preparation of estradiol chitosan nanoparticles for improving nasal absorption and brain targeting. Eur J Pharm Biopharm 70(3):735–740. doi:10.1016/j.ejpb.2008.07.005
13. Migliore MM, Vyas TK, Campbell RB, Amiji MM, Waszczak BL (2010) Brain delivery of proteins by the intranasal route of administration: a comparison of cationic liposomes versus aqueous solution formulations. J Pharm Sci 99(4):1745–1761. doi:10.1002/jps.21939
14. Karatas H, Aktas Y, Gursoy-Ozdemir Y, Bodur E, Yemisci M, Caban S, Vural A, Pinarbasli O, Capan Y, Fernandez-Megia E, Novoa-Carballal R, Riguera R, Andrieux K, Couvreur P, Dalkara T (2009) A nanomedicine transports a peptide caspase-3 inhibitor across the blood–brain barrier and provides neuroprotection. J Neurosci 29(44):13761–13769. doi:10.1523/JNEUROSCI.4246-09.2009
15. Liu L, Guo K, Lu J, Venkatraman SS, Luo D, Ng KC, Ling EA, Moochhala S, Yang YY (2008) Biologically active core/shell nanoparticles self-assembled from cholesterol-terminated PEG-TAT for drug delivery across the blood–brain barrier. Biomaterials 29(10):1509–1517. doi:10.1016/j.biomaterials.2007.11.014
16. Costantino L, Gandolfi F, Tosi G, Rivasi F, Vandelli MA, Forni F (2005) Peptide-derivatized biodegradable nanoparticles able to cross the blood–brain barrier. J Control Release 108(1):84–96. doi:10.1016/j.jconrel.2005.07.013
17. Ke W, Shao K, Huang R, Han L, Liu Y, Li J, Kuang Y, Ye L, Lou J, Jiang C (2009) Gene delivery targeted to the brain using an Angiopep-conjugated polyethyleneglycol-modified polyamidoamine dendrimer. Biomaterials 30(36):6976–6985. doi:10.1016/j.biomaterials.2009.08.049
18. Goppert TM, Muller RH (2005) Polysorbate-stabilized solid lipid nanoparticles as colloidal carriers for intravenous targeting of drugs to the brain: comparison of plasma protein adsorption patterns. J Drug Target 13(3):179–187. doi:10.1080/10611860500071292
19. Gao X, Tao W, Lu W, Zhang Q, Zhang Y, Jiang X, Fu S (2006) Lectin-conjugated PEG-PLA nanoparticles: preparation and brain delivery after intranasal administration. Biomaterials 27(18):3482–3490. doi:10.1016/j.biomaterials.2006.01.038
20. Gao X, Chen J, Tao W, Zhu J, Zhang Q, Chen H, Jiang X (2007) UEA I-bearing nanoparticles for brain delivery following intranasal administration. Int J Pharm 340(1–2):207–215. doi:10.1016/j.ijpharm.2007.03.039
21. Heffernan C, Sumer H, Guillemin GJ, Manuelpillai U, Verma PJ (2012) Design and screening of a glial cell-specific, cell penetrating peptide for therapeutic applications in multiple sclerosis. PLoS One 7(9):e45501. doi:10.1371/journal.pone.0045501
22. Liu JK, Teng Q, Garrity-Moses M, Federici T, Tanase D, Imperiale MJ, Boulis NM (2005) A novel peptide defined through phage display for therapeutic protein and vector neuronal targeting. Neurobiol Dis 19(3):407–418. doi:10.1016/j.nbd.2005.01.022
23. Lanza GM, Marsh JN, Hu G, Scott MJ, Schmieder AH, Caruthers SD, Pan D, Wickline SA (2010) Rationale for a nanomedicine approach to thrombolytic therapy. Stroke 41(10 suppl):S42–S44. doi:10.1161/STROKEAHA.110.598656
24. Takamiya M, Miyamoto Y, Yamashita T, Deguchi K, Ohta Y, Abe K (2012) Strong neuroprotection with a novel platinum nanoparticle against ischemic stroke- and tissue plasminogen activator-related brain damages in mice. Neuroscience 221:47–55. doi:10.1016/j.neuroscience.2012.06.060
25. Bitner BR, Marcano DC, Berlin JM, Fabian RH, Cherian L, Culver JC, Dickinson ME, Robertson CS, Pautler RG, Kent TA, Tour JM (2012) Antioxidant carbon particles improve cerebrovascular dysfunction following traumatic brain injury. ACS Nano 6(9):8007–8014. doi:10.1021/N0302615f
26. Pavinatto FJ, Pavinatto A, Caseli L, dos Santos DS, Nobre TM, Zaniquelli MED, Oliveira ON (2007) Interaction of chitosan with cell membrane models at the air-water interface. Biomacromolecules 8(5):1633–1640. doi:10.1021/Bm0701550

27. Cho Y, Shi R, Ben Borgens R (2010) Chitosan nanoparticle-based neuronal membrane sealing and neuroprotection following acrolein-induced cell injury. J Biol Eng 4(1):2. doi:10.1186/1754-1611-4-2
28. Gupta D, Tator CH, Shoichet MS (2006) Fast-gelling injectable blend of hyaluronan and methylcellulose for intrathecal, localized delivery to the injured spinal cord. Biomaterials 27 (11):2370–2379. doi:10.1016/J.Biomaterials.2005.11.015
29. Kang CE, Poon PC, Tator CH, Shoichet MS (2009) A new paradigm for local and sustained release of therapeutic molecules to the injured spinal cord for neuroprotection and tissue repair. Tissue Eng Part A 15(3):595–604. doi:10.1089/Ten.Tea.2007.0349
30. Cooke MJ, Wang YF, Morshead CM, Shoichet MS (2011) Controlled epi-cortical delivery of epidermal growth factor for the stimulation of endogenous neural stem cell proliferation in stroke-injured brain. Biomaterials 32(24):5688–5697. doi:10.1016/J.Biomaterials.2011.04.032
31. Wang YF, Cooke MJ, Morshead CM, Shoichet MS (2012) Hydrogel delivery of erythropoietin to the brain for endogenous stem cell stimulation after stroke injury. Biomaterials 33(9): 2681–2692. doi:10.1016/J.Biomaterials.2011.12.031
32. Pean JM, Menei P, Morel O, Montero-Menei CN, Benoit JP (2000) Intraseptal implantation of NGF-releasing microspheres promote the survival of axotomized cholinergic neurons. Biomaterials 21(20):2097–2101. doi:10.1016/S0142-9612(00)00141-1
33. Lee H, McKeon RJ, Bellamkonda RV (2010) Sustained delivery of thermostabilized chABC enhances axonal sprouting and functional recovery after spinal cord injury. Proc Natl Acad Sci U S A 107(8):3340–3345. doi:10.1073/Pnas.0905437106
34. Ikada Y, Tabata Y (1998) Protein release from gelatin matrices. Adv Drug Deliv Rev 31(3): 287–301
35. Nakaguchi K, Jinnou H, Kaneko N, Sawada M, Hikita T, Saitoh S, Tabata Y, Sawamoto K (2012) Growth factors released from gelatin hydrogel microspheres increase new neurons in the adult mouse brain. Stem Cells Int 2012:915160. doi:10.1155/2012/915160
36. Sakiyama-Elbert SE, Hubbell JA (2000) Development of fibrin derivatives for controlled release of heparin-binding growth factors. J Control Release 65(3):389–402
37. Taylor SJ, Rosenzweig ES, McDonald JW III, Sakiyama-Elbert SE (2006) Delivery of neurotrophin-3 from fibrin enhances neuronal fiber sprouting after spinal cord injury. J Control Release 113(3):226–235. doi:10.1016/j.jconrel.2006.05.005
38. Ma J, Tian WM, Hou SP, Xu QY, Spector M, Cui FZ (2007) An experimental test of stroke recovery by implanting a hyaluronic acid hydrogel carrying a Nogo receptor antibody in a rat model. Biomed Mater 2(4):233–240. doi:10.1088/1748-6041/2/4/005
39. Wei YT, He Y, Xu CL, Wang Y, Liu BF, Wang XM, Sun XD, Cui FZ, Xu QY (2010) Hyaluronic acid hydrogel modified with nogo-66 receptor antibody and poly-L-lysine to promote axon regrowth after spinal cord injury. J Biomed Mater Res B Appl Biomater 95(1):110–117. doi:10.1002/jbm.b.31689
40. Yano A, Shingo T, Takeuchi A, Yasuhara T, Kobayashi K, Takahashi K, Muraoka K, Matsui T, Miyoshi Y, Hamada H, Date I (2005) Encapsulated vascular endothelial growth factor-secreting cell grafts have neuroprotective and angiogenic effects on focal cerebral ischemia. J Neurosurg 103(1):104–114. doi:10.3171/jns.2005.103.1.0104
41. Tobias CA, Dhoot NO, Wheatley MA, Tessler A, Murray M, Fischer I (2001) Grafting of encapsulated BDNF-producing fibroblasts into the injured spinal cord without immune suppression in adult rats. J Neurotrauma 18(3):287–301. doi:10.1089/08977150151070937
42. Winn SR, Lindner MD, Lee A, Haggett G, Francis JM, Emerich DF (1996) Polymer-encapsulated genetically modified cells continue to secrete human nerve growth factor for over one year in rat ventricles: behavioral and anatomical consequences. Exp Neurol 140(2): 126–138. doi:10.1006/exnr.1996.0123
43. Baumann MD, Kang CE, Stanwick JC, Wang Y, Kim H, Lapitsky Y, Shoichet MS (2009) An injectable drug delivery platform for sustained combination therapy. J Control Release 138(3): 205–213. doi:10.1016/j.jconrel.2009.05.009

44. Perale G, Rossi F, Santoro M, Peviani M, Papa S, Llupi D, Torriani P, Micotti E, Previdi S, Cervo L, Sundstrom E, Boccaccini AR, Masi M, Forloni G, Veglianese P (2012) Multiple drug delivery hydrogel system for spinal cord injury repair strategies. J Control Release 159(2): 271–280. doi:10.1016/j.jconrel.2011.12.025
45. Wang Y, Wei YT, Zu ZH, Ju RK, Guo MY, Wang XM, Xu QY, Cui FZ (2011) Combination of hyaluronic acid hydrogel scaffold and PLGA microspheres for supporting survival of neural stem cells. Pharm Res 28(6):1406–1414. doi:10.1007/s11095-011-0452-3
46. Lin CC, Metters AT (2008) Bifunctional monolithic affinity hydrogels for dual-protein delivery. Biomacromolecules 9(3):789–795. doi:10.1021/bm700940w
47. Tauro JR, Gemeinhart RA (2005) Matrix metalloprotease triggered delivery of cancer chemotherapeutics from hydrogel matrixes. Bioconjug Chem 16(5):1133–1139. doi:10.1021/bc0501303
48. Jin K, Mao X, Xie L, Galvan V, Lai B, Wang Y, Gorostiza O, Wang X, Greenberg DA (2010) Transplantation of human neural precursor cells in Matrigel scaffolding improves outcome from focal cerebral ischemia after delayed postischemic treatment in rats. J Cereb Blood Flow Metab 30(3):534–544. doi:10.1038/jcbfm.2009.219
49. Hejcl A, Sedy J, Kapcalova M, Toro DA, Amemori T, Lesny P, Likavcanova-Masinova K, Krumbholcova E, Pradny M, Michalek J, Burian M, Hajek M, Jendelova P, Sykova E (2010) HPMA-RGD hydrogels seeded with mesenchymal stem cells improve functional outcome in chronic spinal cord injury. Stem Cells Dev 19(10):1535–1546. doi:10.1089/scd.2009.0378
50. Ellis-Behnke RG, Liang YX, You SW, Tay DK, Zhang S, So KF, Schneider GE (2006) Nano neuro knitting: peptide nanofiber scaffold for brain repair and axon regeneration with functional return of vision. Proc Natl Acad Sci U S A 103(13):5054–5059. doi:10.1073/pnas.0600559103
51. Huang KF, Hsu WC, Chiu WT, Wang JY (2012) Functional improvement and neurogenesis after collagen-GAG matrix implantation into surgical brain trauma. Biomaterials 33(7): 2067–2075. doi:10.1016/j.biomaterials.2011.11.040
52. Park KI, Teng YD, Snyder EY (2002) The injured brain interacts reciprocally with neural stem cells supported by scaffolds to reconstitute lost tissue. Nat Biotechnol 20(11):1111–1117. doi:10.1038/nbt751
53. Elias PZ, Spector M (2012) Implantation of a collagen scaffold seeded with adult rat hippocampal progenitors in a rat model of penetrating brain injury. J Neurosci Methods 209(1): 199–211. doi:10.1016/j.jneumeth.2012.06.003
54. Cholas RH, Hsu HP, Spector M (2012) The reparative response to cross-linked collagen-based scaffolds in a rat spinal cord gap model. Biomaterials 33(7):2050–2059. doi:10.1016/j.biomaterials.2011.11.028
55. Zeng X, Zeng YS, Ma YH, Lu LY, Du BL, Zhang W, Li Y, Chan WY (2011) Bone marrow mesenchymal stem cells in a three dimensional gelatin sponge scaffold attenuate inflammation. Promote angiogenesis and reduce cavity formation in experimental spinal cord injury. Cell Transplant 20(11–12):1881–1899. doi:10.3727/096368911X566181
56. Tate CC, Shear DA, Tate MC, Archer DR, Stein DG, LaPlaca MC (2009) Laminin and fibronectin scaffolds enhance neural stem cell transplantation into the injured brain. J Tissue Eng Regen Med 3(3):208–217. doi:10.1002/term.154
57. Hou S, Xu Q, Tian W, Cui F, Cai Q, Ma J, Lee IS (2005) The repair of brain lesion by implantation of hyaluronic acid hydrogels modified with laminin. J Neurosci Methods 148(1): 60–70. doi:10.1016/j.jneumeth.2005.04.016
58. Woerly S, Pinet E, de Robertis L, Van Diep D, Bousmina M (2001) Spinal cord repair with PHPMA hydrogel containing RGD peptides (NeuroGel). Biomaterials 22(10):1095–1111
59. Wei YT, Tian WM, Yu X, Cui FZ, Hou SP, Xu QY, Lee IS (2007) Hyaluronic acid hydrogels with IKVAV peptides for tissue repair and axonal regeneration in an injured rat brain. Biomed Mater 2(3):S142–S146

60. Fukushima K, Enomoto M, Tomizawa S, Takahashi M, Wakabayashi Y, Itoh S, Kuboki Y, Shinomiya K (2008) The axonal regeneration across a honeycomb collagen sponge applied to the transected spinal cord. J Med Dent Sci 55(1):71–79
61. Aota S, Nomizu M, Yamada KM (1994) The short amino acid sequence Pro-His-Ser-Arg-Asn in human fibronectin enhances cell-adhesive function. J Biol Chem 269(40):24756–24761
62. Potter W, Kalil RE, Kao WJ (2008) Biomimetic material systems for neural progenitor cell-based therapy. Front Biosci 13:806–821
63. Prang P, Muller R, Eljaouhari A, Heckmann K, Kunz W, Weber T, Faber C, Vroemen M, Bogdahn U, Weidner N (2006) The promotion of oriented axonal regrowth in the injured spinal cord by alginate-based anisotropic capillary hydrogels. Biomaterials 27(19):3560–3569. doi:10.1016/j.biomaterials.2006.01.053
64. Stokols S, Sakamoto J, Breckon C, Holt T, Weiss J, Tuszynski MH (2006) Templated agarose scaffolds support linear axonal regeneration. Tissue Eng 12(10):2777–2787. doi:10.1089/ten.2006.12.2777
65. Chen BK, Knight AM, de Ruiter GCW, Spinner RJ, Yaszemski MJ, Currier BL, Windebank AJ (2009) Axon regeneration through scaffold into distal spinal cord after transection. J Neurotrauma 26(10):1759–1771. doi:10.1089/Neu.2008.0610
66. Chow WN, Simpson DG, Bigbee JW, Colello RJ (2007) Evaluating neuronal and glial growth on electrospun polarized matrices: bridging the gap in percussive spinal cord injuries. Neuron Glia Biol 3:119–126. doi:10.1017/S1740925x07000580
67. Hurtado A, Cregg JM, Wang HB, Wendell DF, Oudega M, Gilbert RJ, McDonald JW (2011) Robust CNS regeneration after complete spinal cord transection using aligned poly-L-lactic acid microfibers. Biomaterials 32(26):6068–6079. doi:10.1016/J.Biomaterials.2011.05.006
68. Stokols S, Tuszynski MH (2004) The fabrication and characterization of linearly oriented nerve guidance scaffolds for spinal cord injury. Biomaterials 25(27):5839–5846. doi:10.1016/j.biomaterials.2004.01.041
69. Teng YD, Lavik EB, Qu X, Park KI, Ourednik J, Zurakowski D, Langer R, Snyder EY (2002) Functional recovery following traumatic spinal cord injury mediated by a unique polymer scaffold seeded with neural stem cells. Proc Natl Acad Sci U S A 99(5):3024–3029. doi:10.1073/pnas.052678899
70. Wong DY, Krebsbach PH, Hollister SJ (2008) Brain cortex regeneration affected by scaffold architectures. J Neurosurg 109(4):715–722. doi:10.3171/JNS/2008/109/10/0715
71. Nisbet DR, Rodda AE, Horne MK, Forsythe JS, Finkelstein DI (2009) Neurite infiltration and cellular response to electrospun polycaprolactone scaffolds implanted into the brain. Biomaterials 30(27):4573–4580
72. Miller C, Jeftinija S, Mallapragada S (2002) Synergistic effects of physical and chemical guidance cues on neurite alignment and outgrowth on biodegradable polymer substrates. Tissue Eng 8(3):367–378. doi:10.1089/107632702760184646
73. Mahoney MJ, Chen RR, Tan J, Saltzman WM (2005) The influence of microchannels on neurite growth and architecture. Biomaterials 26(7):771–778. doi:10.1016/j.biomaterials.2004.03.015
74. Dowell-Mesfin NM, Abdul-Karim MA, Turner AM, Schanz S, Craighead HG, Roysam B, Turner JN, Shain W (2004) Topographically modified surfaces affect orientation and growth of hippocampal neurons. J Neural Eng 1(2):78–90. doi:10.1088/1741-2560/1/2/003
75. Engler AJ, Sen S, Sweeney HL, Discher DE (2006) Matrix elasticity directs stem cell lineage specification. Cell 126(4):677–689. doi:10.1016/j.cell.2006.06.044
76. Saha K, Keung AJ, Irwin EF, Li Y, Little L, Schaffer DV, Healy KE (2008) Substrate modulus directs neural stem cell behavior. Biophys J 95(9):4426–4438. doi:10.1529/biophysj.108.132217
77. Leipzig ND, Shoichet MS (2009) The effect of substrate stiffness on adult neural stem cell behavior. Biomaterials 30(36):6867–6878. doi:10.1016/j.biomaterials.2009.09.002

78. Lee YS, Arinzeh TL (2012) The influence of piezoelectric scaffolds on neural differentiation of human neural stem/progenitor cells. Tissue Eng Part A 18(19–20):2063–2072. doi:10.1089/ten.TEA.2011.0540
79. Cellot G, Cilia E, Cipollone S, Rancic V, Sucapane A, Giordani S, Gambazzi L, Markram H, Grandolfo M, Scaini D, Gelain F, Casalis L, Prato M, Giugliano M, Ballerini L (2009) Carbon nanotubes might improve neuronal performance by favouring electrical shortcuts. Nat Nanotechnol 4(2):126–133. doi:10.1038/nnano.2008.374
80. Cellot G, Toma FM, Varley ZK, Laishram J, Villari A, Quintana M, Cipollone S, Prato M, Ballerini L (2011) Carbon nanotube scaffolds tune synaptic strength in cultured neural circuits: novel frontiers in nanomaterial-tissue interactions. J Neurosci 31(36):12945–12953. doi:10.1523/JNEUROSCI.1332-11.2011
81. Lim TC, Toh WS, Wang LS, Kurisawa M, Spector M (2012) The effect of injectable gelatin-hydroxyphenylpropionic acid hydrogel matrices on the proliferation, migration, differentiation and oxidative stress resistance of adult neural stem cells. Biomaterials 33(12):3446–3455. doi:10.1016/j.biomaterials.2012.01.037
82. Bencherif SA, Sands RW, Bhatta D, Arany P, Verbeke CS, Edwards DA, Mooney DJ (2012) Injectable preformed scaffolds with shape-memory properties. Proc Natl Acad Sci U S A 109(48): 19590–19595. doi:10.1073/pnas.1211516109
83. Johnson PJ, Tatara A, Shiu A, Sakiyama-Elbert SE (2010) Controlled release of neurotrophin-3 and platelet-derived growth factor from fibrin scaffolds containing neural progenitor cells enhances survival and differentiation into neurons in a subacute model of SCI. Cell Transplant 19(1):89–101. doi:10.3727/096368909X477273
84. Bible E, Qutachi O, Chau DY, Alexander MR, Shakesheff KM, Modo M (2012) Neo-vascularization of the stroke cavity by implantation of human neural stem cells on VEGF-releasing PLGA microparticles. Biomaterials 33(30):7435–7446. doi:10.1016/j.biomaterials.2012.06.085
85. Cholas R, Hsu HP, Spector M (2012) Collagen scaffolds incorporating select therapeutic agents to facilitate a reparative response in a standardized hemiresection defect in the rat spinal cord. Tissue Eng Part A 18(19–20):2158–2172. doi:10.1089/ten.TEA.2011.0577
86. Guo X, Zahir T, Mothe A, Shoichet MS, Morshead CM, Katayama Y, Tator CH (2012) The effect of growth factors and soluble nogo-66 receptor protein on transplanted neural stem/progenitor survival and axonal regeneration after complete transection of rat spinal cord. Cell Transplant 21(6):1177–1197. doi:10.3727/096368911X612503

Chapter 19
Isolated Blood Vessel Models for Studying Trauma

Eugene V. Golanov

Abstract Abnormalities of cerebrovascular circulation are one of the salient consequences of traumatic brain injury. Severity of cerebral blood flow disregulation is associated with the negative clinical outcome. Regulation of cerebral blood flow is complex and differs from regulation of blood flow in other vascular beds. Basic vascular tone formed by interaction of vascular smooth muscle cells and endothelium provides background for other regulatory mechanisms. Understanding of traumatic brain injury-induced abnormalities of basic vascular tone formation and its adjustment using isolated brain vessel model is important for unveiling the pathophysiological mechanisms of brain trauma and development of new therapeutic approaches.

19.1 Cerebral Blood Flow and Trauma

Cerebral blood flow (CBF) abnormalities are one of the salient consequences of traumatic brain injury (TBI) [8, 18, 19]. CBF disregulation regularly observed after TBI leads to metabolic crisis even without overt ischemia and to low levels of oxygenation especially after severe TBI [1, 3, 24, 59]. Abnormalities of CBF correlate with the negative outcome [10, 56].

Maintenance of normal brain activity critically depends on the continuous adequate blood supply. Total interruption of blood or oxygen supply to the brain results in loss of consciousness and cessation of EEG in less than 30 s [39, 47, 57] followed by neuronal anoxic depolarization within 2–3 min [2] and irreversible neuronal and brain damage [43, 44]. Complex system of cerebrovascular circulation control provides uninterrupted blood supply adequate to maintain normal brain activity and differs from circulation in other organs [20, 22, 34, 37].

E.V. Golanov (✉)
Feisntein Institute for Medical Research, 350 Community Drive, Manhasset, NY 11030, USA
e-mail: egolanov@nshs.edu

E.H. Lo et al. (eds.), *Vascular Mechanisms in CNS Trauma*, Springer Series in Translational Stroke Research 5, DOI 10.1007/978-1-4614-8690-9_19,

TBI results in damage of cerebral vasculature [46, 53, 60] and mechanisms of dynamic regulation of CBF. One of the fundamental mechanisms of CBF maintenance is a global adjustment of cerebrovascular resistance (CVR) in response to changes in systemic blood pressure, known as cerebral autoregulation. Autoregulation is determined by the basal tone of cerebral vessels [51]. Cerebral autoregulation arguably is the main mechanism regulating CVR to maintain CBF relatively stable within arterial pressure (AP) limits between ~50 and 160 mmHg [20]. Autoregulatory function is severely affected by TBI. Most pronounced feature of autoregulation disorder is its inability to maintain adequate level of CBF at lower levels of arterial pressure [9, 13, 14, 21, 48]. Another fundamental feature of CBF regulation adjustment of CBF in response to blood and brain tissue levels of carbon dioxide (CO_2) is also blunted [14, 30, 31, 52].

19.2 Nervous Tissue and Vascular Components

The size, vessel wall structure, and location allow to divide the brain vasculature in relation to the parenchyma into extrinsic and intrinsic components [42]. The extrinsic component includes larger conductance vessels, pial vessels, and penetrating arteries isolated from the parenchyma by Robin-Virchow space. In contrast to other organs, the brain surface and penetrating vessels constitute the main component, ~50–60 %, of total vascular resistance [23]. The extrinsic part of cerebral circulation receives heavy extrinsic autonomic innervation [20, 33, 34, 49]. The sympathetic innervation originates from the superior cervical ganglion. The major mediators released by sympathetic nerve endings are noradrenalin and neuropeptide Y and exert a vasoconstrictor effect. Parasympathetic innervation arises from the sphenopalatine, otic, and trigeminal ganglia. Parasympathetic nerve endings release acetylcholine, vasoactive intestinal polypeptide, nitric oxide, calcitonin gene-related peptide, substance P, neurokinin A, and pituitary adenylate cyclase-activating polypeptide [20, 33].

Intrinsic brain circulation does not receive direct innervation from extrinsic or intrinsic nerves fibers [12]. Basal lamina of intraparenchymal arteries comes into direct contact with brain parenchyma. Intracerebral arterioles and capillaries are practically completely ensheathed by astrocytic endfeet and pericytes [5, 36, 37]. The latter have been recognized to exert properties of contractile cells and participate in the regulation of capillary diameter [15, 61]. The coordinated complex of these closely apposed elements interacts functionally and determines capillary diameter and blood–brain barrier permeability, and has been termed "neurovascular unit" (NVU) [17, 36].

Astrocytes and pericytes ensheathing intraparenchymal vessels receive numerous contacts from intrinsic neurons. The intrinsic innervations alter the tone of brain microvessels by modulating the activity of NVU elements [33]. Distant intrinsic innervation arises from various structures including the nucleus basalis (cholinergic pathway), locus coeruleus (noradrenergic pathway), and raphe nuclei (serotoninergic pathway). There are data suggesting that the so-called subthalamic vasodilator area [26] may also directly innervate cortical neurons [50]. While the exact pathways remain to be established, the available data suggest the existence of direct or indirect intrinsic innervations of NVU resulting in changes in microvascular tone arising from fastigial cerebellar nucleus [4], nucleus solitary tract [26], and the medullary cerebrovasodilator area [25].

Thus, CBF is controlled by the myriad various mediators, which tremendously complicate the analysis of TBI-induced abnormalities. However, all these controlling mechanisms ultimately affect the principal basic system of vascular smooth muscle cell (VSMC) and endothelial cell forming a vascular wall. Endothelial-VSMC complex is a basic unit of circulation including cerebrovascular. Constant basal tone of the resistance arteries determined by pressure and flow provides a background for other vasoactive influences [35]. The distinctiveness of cerebrovascular circulation is that majority of the resistance is localized outside of the brain and rests with the extrinsic part of cerebral vasculature.

Basic vascular tone is formed by stretch of VSMC, mechanism of which is still poorly understood [41]. Coordinated response of VSMC is important for the maintenance of uniform tone of particular vessel segment. It is thought that synchronization of the vascular tone in the microcirculation is achieved through gap junctions, including endothelial and myoendothelial gap junctions [55]. Endothelium being responsive to sheer stress plays a critical role in controlling cerebrovascular tone interacting with VSMC through myoendothelial gap junctions [22, 58].

The activity of extrinsic and intrinsic components of the cerebral circulation is closely coordinated. Because extrinsic component of cerebral circulation is the major determinant of CBF and is responsible for about two thirds of overall cerebral vascular resistance [23], changes in the intrinsic circulation require adjustments of the resistance of external pial arteries. It is probably achieved through vascular-conducted response [7, 38, 40]. Gap junctions formed by one or more connexin proteins (Cx37, Cx40, Cx43, and Cx45) exist between the VSMC, endothelial cells, and astrocytes. These cells can communicate through hemichannels and convey the signal (such as Ca^{2+} or endothelial derived hyperpolarizing factor (EDHF), in the case of communication between VSMC and endothelial cells) triggering retrograde vasodilation/vasoconstriction in the respective surface resistance vessels [16].

19.3 Isolated Vascular Models and Mechanisms

Studying effects of TBI in isolated cerebral vessels allows to explore abnormalities of the basic VSMC-endothelial complex which provides background vascular tone of CBF and to unveil the mechanisms of TBI-induced CBF anomalies.

Myogenic tone provides basis for the phenomenon of CBF autoregulation, which is abnormal following TBI. Midline fluid percussion abolishes dilation of isolated middle cerebral arteries (MCA) in response to decrease of intravascular pressure as early as 5 min after the injury in rats [45]. Importantly, resting diameter of MCA harvested from injured rats is 40 % more which is in line with the recently reported observation of doubling of the diameter of pial vessels following TBI [54]. After severe controlled cortical impact (CCI), myogenic response as determined by the changes of the vessel diameter at the lower end of testing pressure (20–40 mmHg) is attenuated ipsilaterally at 2 and 24 h after the injury. At 24 h after the injury, MCA contralateral to the injury side is also affected. However, myogenic responses are not completely abolished, and damaged MCA is able to maintain constant diameter at increasing pressure. At 120 h no significant differences in responses are observed [27]. Mild CCI fails to significantly affect myogenic response. No differences are also observed between pressurized pial branches of MCA and penetrating vessels [29].

Cerebral blood flow response to changing levels of CO_2 is one of the basic cerebrovascular responses. Normal isolated cerebral vessels are also responsive to CO_2 changes. CO_2 reactivity is known to be attenuated after TBI in vivo. However, while mild CCI attenuates CO_2 responses in vivo by ~80 %, it fails to attenuate it in pressurized isolated MCA branches and penetrating vessels in rats [29]. Based on their observations, authors concluded that in vivo responses are mediated by the vascular "milieu."

Endothelial derived nitric oxide (NO) is one of the key factors regulating cerebral vessels tone [22]. Severe CCI does not affect the response of pressurized perfused MCA harvested 24 h after the injury. However, constricting effect of L-NAME, nitric oxide synthase (NOS) inhibitor, in contra- and ipsilateral vessels was diminished. Along with preservation of the dilatory response to NO donors, these observations suggest the decreased NO production following TBI. At the same time, vasodilator effect of P2Y1 receptor agonist increases [28]. Vasodilation in response to luminal application of ATP, which induces NO- and EDHF-dependent vasodilation, was not affected after 1 or 24 h after mild CCI. Effect of ATP is attenuated by combined application of L-NAME and indomethacin, but not in vessels harvested 24 h after CCI suggesting augmentation of EDHF mediated component of vasodilator response [32].

Similarly to CO_2 responses, responses to serotonin (5-HT) are also preserved when tested in MCA and posterior cerebral artery rings after fluid percussion [11]. Importantly, parenchymal arteries, which do not receive direct (extrinsic) innervation, are almost unresponsive to 5-HT application [12].

Unfortunately our understanding of the TBI effects on the functioning of the basic VSMC-endothelium unit is still very limited. Thus, recently it was established that TBI induces changes in expression of mRNA and protein of Cx43 and 45 in endothelial and VSMC [6], suggesting that normal interaction between VSMC and endothelial cells is modified in the aftermath of TBI.

19.4 Conclusions

Systematic exploration of the models of perfused brain vessels offer numerous advantages to unveil TBI effects on cerebral circulation. Analysis of the abnormalities of the basic system forming basal tone of cerebral vessels which provides background for the action of numerous other factors allows addressing the pathophysiological mechanisms of TBI-induced CBF abnormalities and discovering new ways to improve therapy.

References

1. Abate MG, Trivedi M, Fryer TD, Smielewski P, Chatfield DA, Williams GB, Aigbirhio F, Carpenter TA, Pickard JD, Menon DK, Coles JP (2008) Early derangements in oxygen and glucose metabolism following head injury: the ischemic penumbra and pathophysiological heterogeneity. Neurocrit Care 9:319–325
2. Aitken PG, Balestrino M, Somjen GG (1988) NMDA antagonists: lack of protective effect against hypoxic damage in CA1 region of hippocampal slices. Neurosci Lett 89:187–192
3. Alexander MJ, Martin NA, Khanna R, Caron M, Becker DP (1994) Regional cerebral blood flow trends in head injured patients with focal contusions and cerebral edema. Acta Neurochir Suppl (Wien) 60:479–481
4. Arneric SP, Iadecola C, Honig MA, Underwood MD, Reis DJ (1986) Local cholinergic mechanisms mediate the cortical vasodilation elicited by electrical stimulation of the fastigial nucleus. Acta Physiol Scand Suppl 552:70–73
5. Attwell D, Buchan AM, Charpak S, Lauritzen M, Macvicar BA, Newman EA (2010) Glial and neuronal control of brain blood flow. Nature 468:232–243
6. Avila MA, Sell SL, Hawkins BE, Hellmich HL, Boone DR, Crookshanks JM, Prough DS, Dewitt DS (2011) Cerebrovascular connexin expression: effects of traumatic brain injury. J Neurotrauma 28:1803–1811
7. Bagher P, Segal SS (2011) Regulation of blood flow in the microcirculation: role of conducted vasodilation. Acta Physiol (Oxf) 202:271–284
8. Barkhoudarian G, Hovda DA, Giza CC (2011) The molecular pathophysiology of concussive brain injury. Clin Sports Med 30:33
9. Bouma GJ, Muizelaar JP (1990) Relationship between cardiac output and cerebral blood flow in patients with intact and with impaired autoregulation. J Neurosurg 73:368–374
10. Bouma GJ, Muizelaar JP (1995) Cerebral blood flow in severe clinical head injury. New Horiz 3:384–394
11. Bukoski RD, Wang SN, Bian K, Dewitt DS (1997) Traumatic brain injury does not alter cerebral artery contractility. Am J Physiol 272:H1406–H1411
12. Cipolla MJ, Li R, Vitullo L (2004) Perivascular innervation of penetrating brain parenchymal arterioles. J Cardiovasc Pharmacol 44:1–8
13. Czosnyka M, Smielewski P, Kirkpatrick P, Menon DK, Pickard JD (1996) Monitoring of cerebral autoregulation in head-injured patients. Stroke 27:1829–1834
14. Czosnyka M, Brady K, Reinhard M, Smielewski P, Steiner LA (2009) Monitoring of cerebrovascular autoregulation: facts, myths, and missing links. Neurocrit Care 10:373–386
15. Dalkara T, Gursoy-Ozdemir Y, Yemisci M (2011) Brain microvascular pericytes in health and disease. Acta Neuropathol 122:1–9
16. De WC, Boettcher M, Schmidt VJ (2008) Signaling across myoendothelial gap junctions—fact or fiction? Cell Commun Adhes 15:231–245
17. Del Zoppo GJ (2008) Virchow's triad: the vascular basis of cerebral injury. Rev Neurol Dis 5 (Suppl 1):S12–S21
18. Dewitt DS, Prough DS (2003) Traumatic cerebral vascular injury: the effects of concussive brain injury on the cerebral vasculature. J Neurotrauma 20:795–825
19. Dewitt DS, Prough DS (2009) Blast-Induced Brain Injury and Posttraumatic Hypotension and Hypoxemia. J Neurotrauma 26:877–887
20. Edvinsson L, MacKenzie ET, McCulloch J (1993) Cerebral blood flow and metabolism. Raven, New York

21. Enevoldsen EM, Jensen FT (1978) Autoregulation and CO_2 responses of cerebral blood flow in patients with acute severe head injury. J Neurosurg 48:689–703
22. Faraci FM (2011) Protecting against vascular disease in brain. Am J Physiol Heart Circ Physiol 300:H1566–H1582
23. Faraci FM, Heistad DD (1990) Regulation of large cerebral arteries and cerebral microvascular pressure. Circ Res 66:8–17
24. Giri BK, Krishnappa IK, Bryan RM Jr, Robertson C, Watson J (2000) Regional cerebral blood flow after cortical impact injury complicated by a secondary insult in rats. Stroke 31:961–967
25. Golanov EV, Ruggiero DA, Reis DJ (2000) A brainstem area mediating cerebrovascular and EEG responses to hypoxic excitation of rostral ventrolateral medulla in rat. J Physiol 529: 413–429
26. Golanov EV, Christensen JRC, Reis DJ (2001) Neurons of a limited subthalamic area mediate elevations in cortical cerebral blood flow evoked by hypoxia and excitation of neurons of the rostral ventrolateral medulla. J Neurosci 21:4032–4041
27. Golding EM, Contant CF Jr, Robertson CS, Bryan RM Jr (1998) Temporal effect of severe controlled cortical impact injury in the rat on the myogenic response of the middle cerebral artery. J Neurotrauma 15:973–984
28. Golding EM, Steenberg ML, Cherian L, Marrelli SP, Robertson CS, Bryan RM Jr (1998) Endothelial-mediated dilations following severe controlled cortical impact injury in the rat middle cerebral artery. J Neurotrauma 15:635–644
29. Golding EM, Steenberg ML, Contant CF Jr, Krishnappa I, Robertson CS, Bryan RM Jr (1999) Cerebrovascular reactivity to CO(2) and hypotension after mild cortical impact injury. Am J Physiol 277:H1457–H1466
30. Golding EM, Steenberg ML, Contant CF, Krishnappa I, Robertson CS, Bryan RM (1999) Cerebrovascular reactivity to CO_2 and hypotension after mild cortical impact injury. Am J Physiol Heart Circ Physiol 277:H1457–H1466
31. Golding EM, Robertson CS, Bryan RM (2000) L-arginine partially restores the diminished CO_2 reactivity after mild controlled cortical impact injury in the adult rat. J Cereb Blood Flow Metab 20:820–828
32. Golding EM, You J, Robertson CS, Bryan RM Jr (2001) Potentiated endothelium-derived hyperpolarizing factor-mediated dilations in cerebral arteries following mild head injury. J Neurotrauma 18:691–697
33. Hamel E (2006) Perivascular nerves and the regulation of cerebrovascular tone. J Appl Physiol 100:1059–1064
34. Heistad DD, Kontos HA (1983) Cerebral circulation. In: Shepherd JT, Abboud FM (eds) Handbook of physiology. Circulation, vol III, Peripheral circulation and organ blood flow. American Physiological Society, Bethesda
35. Henrion D (2005) Pressure and flow-dependent tone in resistance arteries. Role of myogenic tone. Arch Mal Coeur Vaiss 98:913–921
36. Iadecola C (2004) Neurovascular regulation in the normal brain and in Alzheimer's disease. Nat Rev Neurosci 5:347–360
37. Iadecola C, Nedergaard M (2007) Glial regulation of the cerebral microvasculature. Nat Neurosci 2007(10/30):1369–1376
38. Iadecola C, Yang G, Ebner TJ, Chen G (1997) Local and propagated vascular responses evoked by focal synaptic activity in cerebellar cortex. J Neurophysiol 78:651–659
39. Iwama K (1950) The influence of oxygen lack on brain waves in man. Tohoku J Exp Med 52:63–68
40. Jensen LJ, Holstein-Rathlou NH (2013) The vascular conducted response in cerebral blood flow regulation. J Cereb Blood Flow Metab 33:649–656
41. Kauffenstein G, Laher I, Matrougui K, Guerineau NC, Henrion D (2012) Emerging role of G protein-coupled receptors in microvascular myogenic tone. Cardiovasc Res 95:223–232
42. Kulik T, Kusano Y, Aronhime S, Sandler AL, Winn HR (2008) Regulation of cerebral vasculature in normal and ischemic brain. Neuropharmacology 55:281–288

43. Lipton P (1999) Ischemic cell death in brain neurons [Review]. Physiol Rev 79:1431–1568
44. Martin RL, Lloyd HG, Cowan AI (1994) The early events of oxygen and glucose deprivation: setting the scene for neuronal death? Trends Neurosci 17:251–257
45. Mathew BP, Dewitt DS, Bryan RM Jr, Bukoski RD, Prough DS (1999) Traumatic brain injury reduces myogenic responses in pressurized rodent middle cerebral arteries. J Neurotrauma 16:1177–1186
46. Maxwell WL, Irvine A, Adams JH, Graham DI, Gennarelli TA (1988) Response of cerebral microvasculature to brain injury. J Pathol 155:327–335
47. Mayevsky A, Chance B (1975) Metabolic responses of the awake cerebral cortex to anoxia hypoxia spreading depression and epileptiform activity. Brain Res 98:149–165
48. Muizelaar JP, Ward JD, Marmarou A, Newlon PG, Wachi A (1989) Cerebral blood flow and metabolism in severely head-injured children. Part 2: Autoregulation. J Neurosurg 71:72–76
49. Owman C (1986) Neurogenic control of the vascular system: focus on cerebral circulation. In: Bloom FE (ed) Handbook of physiology, vol 4, The nervous system. American Physiological Society, Bethesda
50. Perkins EN, Parent AD, Golanov EV (2003) Morphological analysis of the SVA as a major relay of medullary vasodilator signals. Soc Neurosci 922:18
51. Peterson EC, Wang Z, Britz G (2011) Regulation of cerebral blood flow. Int J Vasc Med 2011:823525
52. Poon WS, Ng SC, Chan MT, Lam JM, Lam WW (2005) Cerebral blood flow (CBF)-directed management of ventilated head-injured patients. Acta Neurochir Suppl 95:9–11
53. Povlishock JT, Kontos HA, Wei EP, Rosenblum WI, Becker DP (1980) Changes in the cerebral vasculature after hypertension and trauma: a combined scanning and transmission electron microscopic analysis. Adv Exp Med Biol 131:227–241
54. Sangiorgi S, DE Benedictis A, Protasoni M, Manelli A, Reguzzoni M, Cividini A, Dell'orbo C, Tomei G, Balbi S (2013) Early-stage microvascular alterations of a new model of controlled cortical traumatic brain injury: 3D morphological analysis using scanning electron microscopy and corrosion casting. J Neurosurg 118:763–774
55. Schmidt VJ, Wolfle SE, Boettcher M, De Wit C (2008) Gap junctions synchronize vascular tone within the microcirculation. Pharmacol Rep 60:68–74
56. Soustiel JF, Glenn TC, Shik V, Boscardin J, Mahamid E, Zaaroor M (2005) Monitoring of cerebral blood flow and metabolism in traumatic brain injury. J Neurotrauma 22:955–965
57. Sugar O, Gerard RW (1938) Anoxia and brain potential. J Neurosci 1:558–572
58. Triggle CR, Samuel SM, Ravishankar S, Marei I, Arunachalam G, Ding H (2012) The endothelium: influencing vascular smooth muscle in many ways. Can J Physiol Pharmacol 90:713–738
59. Vespa P, Bergsneider M, Hattori N, Wu HM, Huang SC, Martin NA, Glenn TC, McArthur DL, Hovda DA (2005) Metabolic crisis without brain ischemia is common after traumatic brain injury: a combined microdialysis and positron emission tomography study. J Cereb Blood Flow Metab 25:763–774
60. Wei EP, Dietrich WD, Povlishock JT, Navari RM, Kontos HA (1980) Functional, morphological, and metabolic abnormalities of the cerebral microcirculation after concussive brain injury in cats. Circ Res 46:37–47
61. Winkler EA, Bell RD, Zlokovic BV (2011) Central nervous system pericytes in health and disease. Nat Neurosci 14(11):1398–1405

Part III
Clinical Challenges and Opportunities

Chapter 20
Managing Edema and Intracranial Pressure in the Intensive Care Unit

Brian M. Cummings, Phoebe H. Yager, Sarah A. Murphy, Brian Kalish, Chetan Bhupali, Rebecca Bell, Zenab Mansoor, Natan Noviski, and Michael J. Whalen

Abstract Despite several decades of intensive research efforts in the laboratory and the clinic, treatment for traumatic brain injury remains supportive and directed towards controlling intracranial pressure (ICP). First-line therapies for ICP control include positioning to promote jugular venous drainage, sedation and muscle relaxation, modest hyperventilation, and hyperosmolar therapy. Second-line therapies include barbiturate coma, systemic hypothermia, and decompressive craniectomy. Although all of these therapies can be used to treat intracranial hypertension, none have been shown to improve outcome in patients with TBI. This chapter reviews the current therapies used for ICP control with an emphasis on children with TBI.

20.1 Introduction

Case Presentation: A 17-year-old female involved in a high-speed motor vehicle accident was found by Emergency Medical Services personnel 30 ft from her car, unconscious. Her left pupil was dilated and nonreactive to light, and she was unresponsive to painful stimuli. Her physical examination and mechanism of injury suggested a high risk for brain edema and intracranial hypertension, which could result in brain herniation and death if left untreated. She was endotracheally intubated and hyperventilated, administered mannitol 1 g/kg intravenously, taken to a nearby hospital for further stabilizing treatment, and then transported by helicopter to a tertiary care facility for definitive treatment of severe traumatic brain injury. A head CT scan obtained on admission to the PICU disclosed significant cerebral edema (Fig. 20.1), and an intracranial pressure (ICP) monitor was placed revealing an opening pressure of 30 cm water. She was given

M.J. Whalen (✉)
Department of Pediatric Critical Care Medicine, Massachusetts General Hospital and Harvard Medical School, Boston, MA 02115, USA
e-mail: MWhalen@partners.org

E.H. Lo et al. (eds.), *Vascular Mechanisms in CNS Trauma*, Springer Series in Translational Stroke Research 5, DOI 10.1007/978-1-4614-8690-9_20,

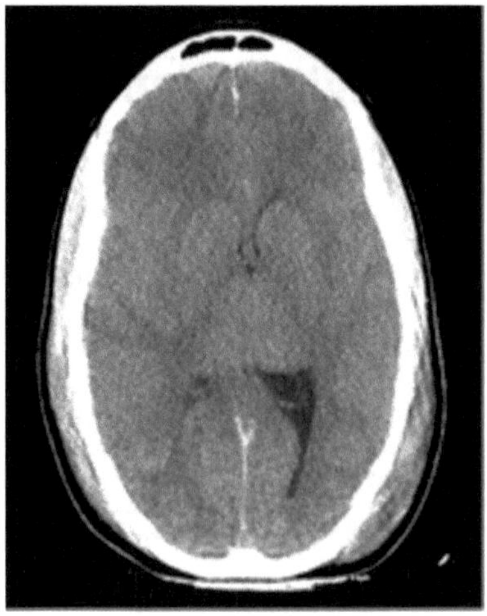

Fig. 20.1 Head computerized tomography scan obtained 6 h after injury. Note diffuse cerebral swelling and midline shift

intravenous fentanyl and propofol for sedation and vecuronium for muscle relaxation and hyperventilated to a $PaCO_2$ of 35. Intracranial hypertension persisted and she was administered hypertonic saline, followed by intravenous pentobarbital to induce coma. Despite these therapies intracranial hypertension (ICP >30) recurred and hypothermia was induced to a core temperature of 32 °C. ICP decreased rapidly and remained below 20 cm H_2O for 5 days. Upon rewarming, ICP increased to 30–40 cm H_2O and she underwent decompressive craniectomy. She eventually recovered with significant residual intellectual and emotional disabilities.

This case highlights a number of clinical therapies used in the intensive care unit (ICU) to control intracranial hypertension in patients with traumatic brain injury. The clinical sequelae of post-traumatic cerebral edema may range from increased ischemic brain injury to life-threatening cerebral herniation syndromes. Thus, most of the therapeutic measures in patients with TBI focus on management of cerebral edema and intracranial hypertension. Unfortunately, no controlled studies have shown that ICP monitoring improves outcome, and no studies have documented beneficial effects on outcome of any agent used to control intracranial hypertension in TBI patients.

20.2 Threshold for Treatment of Intracranial Hypertension

ICP is monitored using a number of invasive techniques, including intraparenchymal probes, subdural probes, and intraventricular catheters. Theoretically, treatment is initiated at a threshold ICP value at which cerebral perfusion is impaired or the risk of herniation increases.

Current guidelines recommend consideration of treatment of ICP at 20 mmHg [1]. In children, this recommendation is based on low-quality class III studies that demonstrate ICP levels greater than 20 mmHg are associated with poor outcome

[2–8]. The majority of pediatric studies have used 20 mmHg as a threshold for treatment of ICP and have assessed outcomes prospectively or retrospectively [2–4, 7, 8]. There have been some adult studies [9, 10] as well as a pediatric study [11] that suggested improved outcome when a lower treatment threshold of 15 mmHg was used. There have been no randomized controlled trials in children or adults examining outcome as a function of predefined ICP treatment values.

In children, optimal ICP treatment thresholds likely vary with age. Cerebral perfusion pressure (CPP) is defined by the mean arterial pressure (MAP) minus ICP. In young children, MAP is generally lower compared to adolescents and adults, and therefore, it is reasonable to expect that younger children may require lower ICP thresholds to maintain the same CPP. A study in healthy children has shown that the lower limit of pressure autoregulation is similar in young and older children [12]. This suggests that ICP thresholds may need to be lower in younger children compared to adults in order to preserve CPP above the autoregulatory limit.

In addition to variation with age, ICP treatment thresholds may vary with pathology. Herniation syndromes may occur at ICP levels below 20 mmHg, particularly in the case of bifrontal or temporal injuries [13]. Thus global ICP measurement may not accurately reflect the potential for cerebral herniation with focal lesions. Xenon-enhanced CT studies have demonstrated decreased cerebral blood flow (CBF) in and around contusions, suggesting enhanced susceptibility to ischemia [14]. Because we do not routinely measure CBF or cerebral ischemia in TBI patients, defining an optimal CPP and treatment threshold for ICP is problematic. CBF is often diminished early after severe TBI, whereas it may normalize or increase over several days. These dynamic changes in blood flow adaptation, along with age-based variation in cerebral vascular physiology, injury pathology, and the clinical exam, must all be considered when estimating the ICP threshold for treatment in a given patient.

20.3 Hyperventilation

For many years, hyperventilation was considered a mainstay of therapy for acute traumatic brain injury. This was largely based on a study in which 6 of 85 children with severe head injury demonstrated acute brain swelling associated with increased CBF despite a decrease in cerebral metabolic rate of oxygen consumption [15]. The authors hypothesized that this was due to a period of acute cerebral hyperemia and vasodilation placing patients at greater risk of intracranial hypertension. Hyperventilation lowers the partial pressure of carbon dioxide in the blood, which leads to cerebral vasoconstriction, a drop in cerebral blood volume, and reduced ICP. The ensuing increase in CPP was thought to promote blood flow to ischemic regions of the brain.

Subsequent to the aforementioned study, a number of pediatric studies examining the cerebrovascular response following severe brain injury found that cerebral metabolic autoregulation may remain intact in many patients. Sharples et al. [16] measured serial cerebrovascular resistance values in 17 children with severe head injuries (Glasgow coma score range 3–8) and reported that, in most cases, cerebrovascular resistance was either within or above normal range, refuting the hyperemia hypothesis [16]. Using xenon-enhanced CT, Stringer and colleagues showed that severe hyperventilation (end-tidal CO_2 as low as 8–19 mmHg) following acute brain injury induces secondary brain ischemia to injured as well as healthy regions of the brain [17]. Adelson and colleagues also employed xenon computed tomography to measure CBF in 30 children with severe traumatic brain injury over the first 9 days following injury [18]. They found that CBF less than or equal to 20 ml/100 g/min occurred frequently within the first 24 h and that this level of CBF was associated with poor outcome if it occurred at any time after injury [18]. These data suggest that hyperventilation early after severe TBI might promote increased cerebral ischemia in children with low CBF. Another study employing microdialysis and thermodiffusion blood flow probes found that even brief periods of hyperventilation induce a significant rise in lactate and glutamate (both mediators of secondary brain injury), particularly in the first 24–36 h after injury [19]. They also observed a decrease in CBF in the region adjacent to the primary injury. Interestingly, the authors noted a more significant decrease in local CBF during the later hyperventilation trials (3–4 days post injury) compared to the earlier trials (24–36 h post injury), suggesting that CBF alone may not be sufficient to detect local brain ischemia [19]. In adults, a classic study by Muizelaar et al. [20] suggested worse 3- and 6-month outcome in a subgroup of adults with TBI who were hyperventilated as early therapy for intracranial hypertension. Although outcome did not differ between hyperventilated and control groups at 12 month, this study widely influenced clinicians against routine hyperventilation in adults with TBI.

The cellular and molecular mechanisms responsible for the rapid cerebral vasoreactivity seen with abrupt changes in blood $PaCO_2$ are complex and not fully understood. Work by Kontos and Lassen supported the hypothesis that changes in CBF are mediated by changes in brain extracellular pH that occur with rapid diffusion of CO_2 across the blood–brain barrier [21, 22]. However, other studies suggest that alterations in blood $PaCO_2$ affect CBF via production of nitric oxide (NO), a powerful vasodilator. Fathi et al. [23] exposed endothelial cells and astrocytes to varying degrees of hypo- and hypercapnea while maintaining pH constant and found that NO production and cerebral blood vessel diameter increased with hypercapnic and decreased with hypocapnic conditions [23]. Interestingly, vasoreactivity observed with changes in $PaCO_2$ was short lived, in part due to rapid compensatory mechanisms that maintain pH within a tightly regulated range during persistent hypocarbia [23]. In rabbits, pial arteriolar vasoconstriction wanes significantly as arterial and cerebrospinal fluid pH normalizes over 24 h of induced hyperventilation ($PaCO_2$ 25 mmHg) [24]. From these observations, the authors concluded that preventive hyperventilation beyond 24 h may not sustain reduced ICP and may, in fact, lead to situations where unattainable pCO_2 is required to induce cerebral vasoconstriction during acute elevations in ICP.

Based on the limited available literature addressing the role of hyperventilation in management of acute pediatric brain trauma, the first edition of the *Guidelines for the Acute Medical Management of Severe Traumatic Brain Injury in Infants, Children, and Adolescents* was unable to offer a treatment standard or guideline regarding hyperventilation [25]. Options included (1) avoidance of mild or prophylactic hyperventilation, (2) consideration of mild hyperventilation for intracranial hypertension refractory to other therapies, and (3) aggressive hyperventilation as a second-tier option for refractory hypertension coupled with advanced monitoring to help identify cerebral ischemia [25]. Despite concerns raised in the 2003 guidelines regarding the use of hyperventilation, this therapy remains popular [26]. Unfortunately, in the period between 2003 and 2012, when the second edition of the Pediatric Brain Trauma guidelines was published, there have been no studies specifically comparing different degrees of hyperventilation on ICP or long-term outcomes nor have there been any studies examining the use of rescue hyperventilation for acute elevations in ICP. Therefore, the 2012 guidelines remain essentially unchanged. A 2008 retrospective cohort study before and after the 2003 Pediatric Brain Trauma Foundation guidelines did lend further support to the weak recommendations regarding avoidance of hypocarbia based on an association between incidences of severe hypocarbia (defined as $PaCO_2$ <30 mm) during the first 48 h after injury and risk of inpatient mortality [26].

In summary, hyperventilation is the first-line therapy for rescuing patients with impending cerebral herniation syndrome. In pediatric patients with cerebral hyperemia and raised ICP, hyperventilation may be better tolerated in terms of risk of cerebral ischemia compared to patients with significantly decreased CBF. In these patients, hyperventilation may induce cerebral vasoconstriction and decrease intracranial hypertension acutely, but may lead to secondary brain ischemia particularly around focal lesions. Protracted hyperventilation for periods greater than 24 h is not likely to confer a benefit as vasoreactivity wanes with the regulation of endogenous buffers to counter the hypocapnic environment. Additional studies are needed to determine the efficacy and safety of hyperventilation to treat acute ICP refractory to other therapies. If long-term hyperventilation is used, the Brain Trauma Committee recommends advanced neuromonitoring to assess for induced cerebral ischemia [1].

20.4 Hyperosmolar Therapy

Weed and colleagues first demonstrated the feasibility of intravenous hypertonic saline as hyperosmolar therapy for increased ICP in 1919 [27]. In 1961, Wise and Chater introduced mannitol into clinical practice, and mannitol replaced all other osmotic agents by the late 1970s [28]. Currently, both hypertonic saline and mannitol are used in the contemporary management of intracranial hypertension. Although both agents have been used extensively in children with TBI, few studies have rigorously investigated the role of intravenous hyperosmolar agents in reducing ICP in pediatric patients.

Mannitol is the most common agent used in pediatric ICUs for the treatment of intracranial hypertension. The use of mannitol for controlling ICP after severe TBI (at doses ranging from 0.25 to 1 g/kg of body weight) is largely based on adult studies (one class II and seven class III) [29]. Mannitol has a track record of clinical acceptance and safety without clear evidence of efficacy. A Cochrane review of mannitol administration to TBI patients found inadequate evidence for benefit compared to placebo [30].

Mannitol reduces ICP via several complementary physiological mechanisms. Mannitol induces a rapid but transient reduction in blood viscosity mediated by rheological effects on erythrocytes. In brain regions with intact flow autoregulation, reflex vasoconstriction occurs to maintain CBF in the setting of decreased cerebral blood volume [31–34]. Mannitol also induces a more gradual osmotic effect on tissues that results in movement of water from tissue parenchyma into the systemic circulation [35, 36]. An intact blood–brain barrier is necessary for maximum efficacy of mannitol to reduce ICP [35, 36]. In the injured brain, mannitol may accumulate in regions where the blood–brain barrier is broken, theoretically causing osmotic movement of water from the intravascular compartment to damaged brain parenchyma, leading to focal edema and possibly increasing ICP, especially if mannitol is used for extended time periods [37]. Despite limited evidence of this "reverse osmotic effect," some practitioners wean mannitol slowly after extended treatment periods to mitigate these possible detrimental effects. The current practice is to aim for a euvolemic hyperosmolar state in pediatric TBI patients. Limited evidence indicates a possible risk for acute tubular necrosis and renal failure with mannitol use when serum osmolarity levels are greater than 320 mOsm in adults [38, 39].

Hypertonic saline is used in conjunction with mannitol or as stand-alone therapy in the treatment of intracranial hypertension. Worthley et al. published two cases of small volume, 29 % saline used for refractory ICP elevation, and the use of hypertonic saline in TBI has since grown substantially [40]. Subsequent studies have employed a similar strategy of small volume hypertonic saline resuscitation in hemorrhagic shock and polytrauma [41–46]. The 2012 pediatric guidelines give a class II recommendation to the use of 3 % hypertonic saline for the acute treatment of intracranial hypertension and a class III recommendation to the use of continuous infusion of 3 % saline.

The mechanism of action of hypertonic saline in reducing ICP is multifactorial. The blood–brain barrier has limited permeability to sodium, and hypertonic saline establishes an osmolar gradient that promotes return to homeostatic cell volume [47, 48]. Hypertonic saline also reportedly stimulates the release of arterial natriuretic peptide and promotes cardiac output [49, 50]. Hypertonic saline may also confer anti-inflammatory effects through the inhibition of leukocyte adhesion [51–53].

In addition to the use of hypertonic saline to reduce ICP, hypertonic saline may also be used to treat post-traumatic hyponatremia, which can be secondary to cerebral salt wasting, the syndrome of inappropriate antidiuretic hormone secretion, or renal sodium losses. The diagnosis and treatment of hyponatremia is essential to prevent the potentially devastating sequelae of brain cell swelling and seizures [54]. Hyponatremia in cerebral salt wasting may be marked and is accompanied by elevated urinary sodium, hypovolemia, and polyuria [55]. There is no direct evidence for the optimal rate of sodium correction to avoid central pontine myelinolysis, but some have recommended an upper limit of correction of 12 mmol/L per day [25]. In addition to central pontine myelinolysis, the use of hypertonic saline has several other potential deleterious effects such as rebound intracranial hypertension, renal injury, natriuresis, and masking of diabetes insipidus [42]. Hypertonic saline may also increase the risk of subarachnoid hemorrhage and hyperchloremic acidosis [42].

A higher serum osmolarity may be acceptable in children treated with hypertonic saline as compared to mannitol [56, 57]. Despite reports of an increase in creatinine with serum sodium concentration greater than 160 mmol/L [58], the recommendation of an upper limit of serum osmolarity (360 mOsm/L) in the use of hypertonic saline in pediatric TBI is controversial [25, 59].

Currently, there is limited evidence to inform clinical decisions regarding the appropriate indications for use of hypertonic saline used to treat pediatric TBI, nor is there a clear evidence regarding ICP goals [1]. Studies published to date are limited by small sample size, inconsistent design, paucity of subject randomization, and/or unclear inclusion of children. In a double-blind crossover study, the use of 3 % saline versus 0.9 % saline in 18 children with severe TBI was associated with lower ICP and a reduction in need for other interventions [60]. A randomized controlled trial of 1.7 % saline versus lactated Ringer's solution (a hypotonic fluid) found that children with TBI who received 1.7 % saline required fewer interventions and had a shorter length of ICU stay [61]. Khanna et al. conducted a small prospective study in which ten children with TBI were given 3 % saline on a sliding scale to maintain ICP <20 mmHg [56]. The administration of hypertonic saline resulted in a decrease in ICP spikes and an increase in CPP. While both mannitol and hypertonic saline are widely used to control pediatric intracranial hypertension, there are many unanswered questions in terms of specific indications, comparative efficacy, and mechanism. For the most part, clinical experience rather than scientific evidence guides the use of hyperosmolar therapy in children with TBI.

20.5 Barbiturates

Barbiturates are used to lower ICP when first-tier therapies are inadequate. Despite evidence of their effectiveness in lowering ICP, there is limited evidence of improvement in clinical outcome [62]. Barbiturates have been used as prophylactic therapy after traumatic brain injury, but they are more commonly used as treatment of refractory intracranial hypertension. In a retrospective cohort study of children with refractory intracranial hypertension, Mellion et al. [63] found that the addition of high-dose barbiturates controlled ICP in nearly 30 % of patients, which was associated with increased likelihood of acceptable long-term outcome [63].

Pharmacokinetics and pharmacodynamics of barbiturates in head-injured patients are variable, and clearance may change with length of therapy. Although barbiturate levels can be measured, levels correlate poorly with brain electrical activity [64]. Therefore, EEG burst suppression is a better bedside method to measure therapeutic effect [64]. Maximal reduction in cerebral metabolism and resultant blood flow occurs at burst suppression [64]. However, cardiorespiratory side effects, including decreased cardiac output, hypotension, and intrapulmonary shunting, are common at therapeutic levels and may result in lower CPP and hypoxia.

The neuroprotective effect of barbiturates is may be partially attributable to the reduction in cerebral metabolic demand and alterations in vascular tone. Barbiturates also reduce lipid peroxidation, free radical formation, excitatory amino acids, and lactate [65, 66]. Modulation of CBF appears particularly important in barbiturate therapy, with evidence of both preserved autoregulation and improved coupling of metabolic demands and blood flow, with resultant higher brain oxygenation and reduced blood volume. Global cerebral metabolic rate of oxygen utilization ($CMRO_2$) is decreased by up to 50 % with barbiturate therapy, and as CBF is coupled to demand, there is a corresponding reduction in flow [67]. Decrease in $CMRO_2$ and CBF may have beneficial effects on threatened but non-damaged tissue. Decreased CBF may decrease cerebral blood volume, lower ICP, and improve global perfusion. Additionally, suppression of metabolic demands in well-perfused areas by barbiturates may divert blood flow through collateral channels to ischemic areas thereby stabilizing ischemic penumbra regions. These theoretical considerations have not been borne out in improved outcomes in patients receiving barbiturates, however.

There is clinical evidence for an improvement in oxygen delivery with barbiturate therapy. Chen and colleagues measured direct local brain tissue oxygenation ($PbtO_2$) before and after the administration of barbiturates to patients with intractable intracranial hypertension [68]. They found that pentobarbital administration was associated with a significant increase in $PbtO_2$ in 70 % of patients, independent of effects on ICP. However with later administration of barbiturates to sicker patients, a negative effect was observed, suggesting that barbiturate therapy applied in later stages may not be beneficial [68]. Although the majority of patients in this study died, many patients experienced improvement in $PbtO_2$ [68]. Thorat and colleagues explored the effects of barbiturates on tissue oxygen tension and cerebrovascular pressure reactivity (PRx) as an index of cerebral autoregulation in severe head injury patients [69]. Cerebrovascular PRx was measured as a moving correlation between ICP and arterial blood pressure. Barbiturate coma was instituted when ICP became refractory to other therapies (ICP >20 mmHg), with 75 % having reduction in ICP and improvement in $PbTO_2$. Favorable changes in ICP, PRx, and $PbTO_2$ with barbiturate coma were seen in those who survived, suggesting beneficial effects of barbiturates on cerebral autoregulation [69]. Despite these improvements in physiological parameters, no studies have shown a benefit of barbiturate therapy on clinical outcome after TBI.

20.6 CSF Drainage

CSF removal by catheter drainage is used in many centers to control intracranial hypertension. Removal of CSF, even in small amounts, can produce an immediate decrease in ICP and an increase in CPP. Intraventricular catheters are most commonly used for CSF drainage, but lumbar drains may also be used in cases of refractory elevated ICP. Jagannathan et al. retrospectively studied 96 children with severe TBI comparing management of ICP alone, ICP with surgical decompression via craniectomy, and external ventricular drainage [70]. Intracranial hypertension was defined as ICP >20 mmHg, and 20/23 patients with CSF drainage achieved ICP control [70]. In another study, Anderson et al. retrospectively reviewed 80 children with severe TBI, all of whom were treated with an ICP monitor, an external ventricular drain, or both [71]. EVD placement was associated with a threefold increased risk of hemorrhage when compared with placement of an intraparenchymal fiber optic monitor [71]. There was a fourfold increased risk of complications in children who received an EVD compared with those in whom a fiber optic monitor was placed. These complications included EVD malposition and hemorrhage. The authors reported a small number of infections in the intraventricular catheter group, but they could not draw conclusions about differences in infection risk between intraventricular and intraparenchymal monitors.

While there have been no randomized control trials studying the treatment of elevated ICP with and without CSF drainage, CSF removal has been shown to reduce ICP. This treatment modality is not without risks, including intracerebral hemorrhage and infection. The decision to place an intraparenchymal monitor, or an intraventricular catheter to monitor ICP and drain CSF, is often dependent on physician experience.

20.7 Decompressive Craniectomy

Decompressive craniectomy (DC) is increasingly being used as a means of treating or controlling increased ICP following severe TBI. The procedure may be performed solely for the purpose of preventing or treating cerebral herniation, or it may be performed in concert with evacuation of a space-occupying lesion to treat observed or anticipated brain swelling. The indications for the use of DC vary from center to center and clear guidelines for its use are lacking. In some centers, DC is used only as a last attempt at controlling ICP after other measures (such as hypothermia, paralysis, or pentobarbital therapy) have failed. In other centers, DC is used early in the management of patients who are thought to be at risk of developing high ICP. There are few clear guidelines to govern its use, but a number of publications have documented its use in recent years including case series, case reports, prospective and retrospective studies, and two large randomized controlled trials, one of which is still underway (RESCUE-ICP) [72–74]. The Brain Trauma Foundation guidelines and the recently updated Pediatric Brain Trauma Foundation guidelines both state that there are insufficient data to support a level I or level II recommendation. Nevertheless, DC remains a treatment option for managing medically refractory cerebral edema.

The described surgical techniques for decompressive craniectomy are also diverse. The technique may vary by location (frontal, temporal, parietal, occipital) or unilateral versus bilateral, by dural technique (scalp closure only or duraplasty with autologous or synthetic patch), by the size of the bony defect created, and by the timing of surgery. Studies have not demonstrated the superiority of unilateral or bilateral technique in controlling ICP. Although a clear relationship has been demonstrated between the size of the cranial defect and the ability to control ICP [75, 76], increased craniectomy size has also been associated with the risk of developing post-traumatic hydrocephalus [77, 78]. Additionally, the comparative benefit of early versus late decompression continues to be debated. There remains little consensus regarding the optimal technique, and for any particular patient, the benefit will likely be determined in part by the specifics of the underlying injury.

The aim of DC is to relieve intracranial hypertension and prevent or reverse cerebral herniation. Decompressive craniectomy has been demonstrated to decrease ICP and to improve the pressure–volume compensatory reserve after TBI [79, 80]. Few studies have addressed the effect of DC on cerebral hemodynamics, even though this may be an important element for its appropriate use. Multiple studies have shown an increase in CBF following DC. In one study using single-photon emission computed tomography (SPECT) to study regional CBF, hyperperfusion was seen ipsilateral to the DC in the area surrounding contusion immediately after decompression. The hyperperfused area increased in size at 1 week following surgery but returned to normal by 1 month [81]. The areas of increased CBF correlated with the cerebral edema seen on CT.

Several studies using transcranial Doppler (TCD) have also demonstrated increased cerebral blood flow velocity (CBFV) following DC. Bor-Seng-Shu et al. [79] and Daboussi et al. [82] found significant increases in CBFV and decreases in the pulsatility index, a marker of distal resistance to blood flow, in both hemispheres after surgical decompression [79, 82]. This effect was more pronounced ipsilateral to the site of DC, but the hemodynamic changes were not shown to correlate with outcomes. A more recent study used contrast-enhanced ultrasound to investigate changes in the microcirculation. Increased cerebral microvascular blood flow was seen over a 3-day period after DC and appeared to be unrelated to measured CPP [83]. The increased CBF seen following DC was greater when ICP was relatively low at the time of surgery, suggesting that late DC might not be as effective at restoring microcirculatory blood flow in the setting of prolonged increased ICP.

Vasomotor reactivity, an important component of cerebral autoregulation, has also been investigated in two studies with conflicting results. Timofeev et al. looked at both the pressure–volume compensatory reserve index (RAP), which is a measure of cerebral compliance, and the pressure reactivity index (PRx), a correlation between MAP and ICP and a measure of cerebral vasomotor reactivity [84]. They found that in spite of a sustained decrease in ICP and an improvement in cerebral compliance, there was impaired cerebrovascular pressure reactivity following DC. The findings of this study are in contrast to a study by Ho et al. in which a decreased PRx, reflecting restored cerebrovascular pressure reactivity, was seen following decompressive surgery [85]. The results of this study suggested a beneficial effect of DC on CBF–pressure autoregulation. Though derangements in PRx have been shown to correlate with outcome following TBI, it is debated if PRx measurements are accurate measures of cerebrovascular reactivity following craniectomy, and ICP measurements have not been well investigated in the setting of an open cranial vault [86].

We have an incomplete understanding of how DC affects cerebrovascular hemodynamics. However, studies of brain tissue oxygenation and brain tissue metabolism support the idea that CBF may be enhanced in some patients following DC. An increase in brain tissue oxygen tension ($PbtO_2$) from ischemic to normal levels has been observed following DC [87, 88]. In a subgroup of patients with a good outcome following DC, cerebral microdialysis markers have also shown improvement [85].

The recently published DECRA study found that in spite of lowering ICP and shortening key ICU variables such as length of stay and ventilator days, DC did not decrease mortality, nor did it improve outcome in adults with severe TBI [73]. The results of a second large multicenter trial, RESCUE-ICP, are eagerly anticipated to further our understanding of a therapy that is being used with increasing frequency. A more complete understanding of the alterations in cerebral hemodynamics that ensue following DC will be instrumental in better defining the role of DC in trauma, as well as other clinical scenarios.

References

1. Kochanek PM, Carney N, Adelson PD, Ashwal S, Bell MJ, Bratton S, Carson S, Chesnut RM, Ghajar J, Goldstein B, Grant GA, Kissoon N, Peterson K, Selden NR, Tasker RC, Tong KA, Vavilala MS, Wainwright MS, Warden CR, American Academy of Pediatrics-Section on Neurological Surgery, American Association of Neurological Surgeons/Congress of Neurological Surgeons, Child Neurology Society, European Society of Pediatric and Neonatal Intensive Care, Neurocritical Care Society, Pediatric Neurocritical Care Research Group, Society of Critical Care Medicine, Paediatric Intensive Care Society UK, Society for Neuroscience in Anesthesiology and Critical Care, World Federation of Pediatric Intensive and Critical Care Societies (2012) Guidelines for the acute medical management of severe traumatic brain injury in infants, children, and adolescents—second edition. Pediatr Crit Care Med 13(Suppl 1):S1–S82
2. Adelson PD, Ragheb J, Kanev P, Brockmeyer D, Beers SR, Brown SD, Cassidy LD, Chang Y, Levin H (2005) Phase II clinical trial of moderate hypothermia after severe traumatic brain injury in children. Neurosurgery 56:740–754, discussion 740–754
3. Alberico AM, Ward JD, Choi SC, Marmarou A, Young HF (1987) Outcome after severe head injury. Relationship to mass lesions, diffuse injury, and ICP course in pediatric and adult patients. J Neurosurg 67:648–656
4. Downard C, Hulka F, Mullins RJ, Piatt J, Chesnut R, Quint P, Mann NC (2000) Relationship of cerebral perfusion pressure and survival in pediatric brain-injured patients. J Trauma 49:654–658
5. Esparza J, M-Portillo J, Sarabia M, Yuste JA, Roger R, Lamas E (1985) Outcome in children with severe head injuries. Childs Nerv Syst 1:109–114
6. Pfenninger J, Kaiser G, Lütschg J, Sutter M (1983) Treatment and outcome of the severely head injured child. Intensive Care Med 9:13–16
7. Pfenninger J, Santi A (2002) Severe traumatic brain injury in children—are the results improving? Swiss Med Wkly 132:116–120
8. White JR, Farukhi Z, Bull C, Christensen J, Gordon T, Paidas C, Nichols DG (2001) Predictors of outcome in severely head-injured children. Crit Care Med 29:534–540

9. Marshall LF (1980) Treatment of brain swelling and brain edema in man. Adv Neurol 28: 459–469
10. Saul TG, Ducker TB (1982) Effect of intracranial pressure monitoring and aggressive treatment on mortality in severe head injury. J Neurosurg 56:498–503
11. Cruz J, Nakayama P, Imamura JH, Rosenfeld KGW, de Souza HS, Giorgetti GVF (2002) Cerebral extraction of oxygen and intracranial hypertension in severe, acute, pediatric brain trauma: preliminary novel management strategies. Neurosurgery 50:774–779, discussion 779–780
12. Vavilala MS, Lee LA, Lam AM (2003) The lower limit of cerebral autoregulation in children during sevoflurane anesthesia. J Neurosurg Anesthesiol 15:307–312
13. Andrews BT, Chiles BW, Olsen WL, Pitts LH (1988) The effect of intracerebral hematoma location on the risk of brain-stem compression and on clinical outcome. J Neurosurg 69: 518–522
14. McLaughlin MR, Marion DW (1996) Cerebral blood flow and vasoresponsivity within and around cerebral contusions. J Neurosurg 85:871–876
15. Bruce DA, Raphaely RC, Goldberg AI, Zimmerman RA, Bilaniuk LT, Schut L, Kuhl DE (1979) Pathophysiology, treatment and outcome following severe head injury in children. Childs Brain 5:174–191
16. Sharples PM, Matthews DS, Eyre JA (1995) Cerebral blood flow and metabolism in children with severe head injuries. Part 2: cerebrovascular resistance and its determinants. J Neurol Neurosurg Psychiatry 58:153–159
17. Stringer WA, Hasso AN, Thompson JR, Hinshaw DB, Jordan KG (1993) Hyperventilation-induced cerebral ischemia in patients with acute brain lesions: demonstration by xenon-enhanced CT. AJNR Am J Neuroradiol 14:475–484
18. Adelson PD, Clyde B, Kochanek PM, Wisniewski SR, Marion DW, Yonas H (1997) Cerebrovascular response in infants and young children following severe traumatic brain injury: a preliminary report. Pediatr Neurosurg 26:200–207
19. Marion DW, Puccio A, Wisniewski SR, Kochanek P, Dixon CE, Bullian L et al (2002) Effect of hyperventilation on extracellular concentrations of glutamate, lactate, pyruvate, and local cerebral blood flow in patients with severe traumatic brain injury. Crit Care Med 30: 2619–2625
20. Muizelaar JP, Marmarou A, Ward JD, Kontos HA, Choi SC, Becker DP, Gruemer H, Young HF (1991) Adverse effects of prolonged hyperventilation in patients with severe head injury: a randomized clinical trial. J Neurosurg 65:731–739
21. Kontos HA, Raper AJ, Patterson JL (1977) Analysis of vasoactivity of local pH, PCO2 and bicarbonate on pial vessels. Stroke 8:358–360
22. Lassen NA (1968) Brain extracellular pH: the main factor controlling cerebral blood flow. Scand J Clin Lab Invest 22:247–251. doi:10.3109/00365516809167060
23. Fathi AR, Yang C, Bakhtian KD, Qi M, Lonser RR, Pluta RM (2011) Carbon dioxide influence on nitric oxide production in endothelial cells and astrocytes: cellular mechanisms. Brain Res 1386:50–57. doi:10.1016/j.brainres.2011.02.066
24. Muizelaar JP, van der Poel HG, Li ZC, Kontos HA, Levasseur JE (1988) Pial arteriolar vessel diameter and CO2 reactivity during prolonged hyperventilation in the rabbit. J Neurosurg 69:923–927
25. Adelson PD, Bratton SL, Carney NA, Chesnut RM, du Coudray HE, Goldstein B, Kochanek PM, Miller HC, Partington MD, Selden NR, Warden CR, Wright DW, American Association for Surgery of Trauma, Child Neurology Society, International Society for Pediatric Neurosurgery, International Trauma Anesthesia and Critical Care Society, Society of Critical Care Medicine, World Federation of Pediatric Intensive and Critical Care Societies (2003) Guidelines for the acute medical management of severe traumatic brain injury in infants, children, and adolescents. Chapter 12. Use of hyperventilation in the acute management of severe pediatric traumatic brain injury. Pediatr Crit Care Med 4:S45–S48

26. Curry R, Hollingworth W, Ellenbogen RG, Vavilala MS (2008) Incidence of hypo- and hypercarbia in severe traumatic brain injury before and after 2003 pediatric guidelines. Pediatr Crit Care Med 9:141–146
27. Weed L, McKibben P (1919) Pressure changes in the cerebro-spinal fluid following intravenous injection of solutions of various concentrations. Am J Physiol 48:512–530
28. Wise BL, Chater N (1961) Use of hypertonic mannitol solutions to lower cerebrospinal fluid pressure and decrease brain bulk in man. Surg Forum 12:398–399
29. Brain Trauma Foundation, American Association of Neurological Surgeons, Congress of Neurological Surgeons, Joint Section on Neurotrauma and Critical Care, AANS/CNS, Bratton SL, Chestnut RM, Ghajar J, McConnell Hammond FF, Harris OA, Hartl R, Manley GT, Nemecek A, Newell DW, Rosenthal G, Schouten J, Shutter L, Timmons SD, Ullman JS, Videtta W, Wilberger JE, Wright DW (2007) Guidelines for the management of severe traumatic brain injury. II. Hyperosmolar therapy. J Neurotrauma 24(Suppl 1):S7–S13
30. Schierhout G, Roberts I (2000) Mannitol for acute traumatic brain injury. Cochrane Database Syst Rev (2):CD001049
31. Levin AB, Duff TA, Javid MJ (1979) Treatment of increased intracranial pressure: a comparison of different hyperosmotic agents and the use of thiopental. Neurosurgery 5:570–575
32. Muizelaar JP, Wei EP, Kontos HA, Becker DP (1983) Mannitol causes compensatory cerebral vasoconstriction and vasodilation in response to blood viscosity changes. J Neurosurg 59: 822–828
33. Muizelaar JP, Lutz HA, Becker DP (1984) Effect of mannitol on ICP and CBF and correlation with pressure autoregulation in severely head-injured patients. J Neurosurg 61:700–706
34. Muizelaar JP, Wei EP, Kontos HA, Becker DP (1986) Cerebral blood flow is regulated by changes in blood pressure and in blood viscosity alike. Stroke 17:44–48
35. Bouma GJ, Muizelaar JP (1992) Cerebral blood flow, cerebral blood volume, and cerebrovascular reactivity after severe head injury. J Neurotrauma 9(Suppl 1):S333–S348
36. James HE (1980) Methodology for the control of intracranial pressure with hypertonic mannitol. Acta Neurochir (Wien) 51:161–172
37. Kaufmann AM, Cardoso ER (1992) Aggravation of vasogenic cerebral edema by multiple-dose mannitol. J Neurosurg 77:584–589
38. Brain Trauma Foundation, The American Association of Neurological Surgeons, The Joint Section on Neurotrauma and Critical Care (2000) Use of mannitol. J Neurotrauma 17:521–525
39. Feig PU, McCurdy DK (1977) The hypertonic state. N Engl J Med 297:1444–1454
40. Worthley LI, Cooper DJ, Jones N (1988) Treatment of resistant intracranial hypertension with hypertonic saline. Report of two cases. J Neurosurg 68:478–481
41. Prough DS, Whitley JM, Taylor CL, Deal DD, DeWitt DS (1991) Regional cerebral blood flow following resuscitation from hemorrhagic shock with hypertonic saline. Influence of a subdural mass. Anesthesiology 75:319–327
42. Qureshi AI, Suarez JI (2000) Use of hypertonic saline solutions in treatment of cerebral edema and intracranial hypertension. Crit Care Med 28:3301–3313
43. Shackford SR, Bourguignon PR, Wald SL, Rogers FB, Osler TM, Clark DE (1998) Hypertonic saline resuscitation of patients with head injury: a prospective, randomized clinical trial. J Trauma 44:50–58
44. Vassar MJ, Fischer RP, O'Brien PE, Bachulis BL, Chambers JA, Hoyt DB, Holcroft JW (1993) A multicenter trial for resuscitation of injured patients with 7.5% sodium chloride. The effect of added dextran 70. The multicenter group for the study of hypertonic saline in trauma patients. Arch Surg 128:1003–1011, discussion 1011–1013
45. Walsh JC, Zhuang J, Shackford SR (1991) A comparison of hypertonic to isotonic fluid in the resuscitation of brain injury and hemorrhagic shock. J Surg Res 50:284–292
46. Zornow MH, Prough DS (1995) Fluid management in patients with traumatic brain injury. New Horiz 3:488–498
47. McManus ML, Soriano SG (1998) Rebound swelling of astroglial cells exposed to hypertonic mannitol. Anesthesiology 88:1586–1591

48. Nakayama S, Kramer GC, Carlsen RC, Holcroft JW (1985) Infusion of very hypertonic saline to bled rats: membrane potentials and fluid shifts. J Surg Res 38:180–186
49. Arjamaa O, Karlqvist K, Kanervo A, Vainionpää V, Vuolteenaho O, Leppäluoto J (1992) Plasma ANP during hypertonic NaCl infusion in man. Acta Physiol Scand 144:113–119
50. Moss GS, Gould SA (1988) Plasma expanders. An update. Am J Surg 155:425–434
51. Pascual JL, Khwaja KA, Chaudhury P, Christou NV (2003) Hypertonic saline and the microcirculation. J Trauma 54:S133–S140
52. Pascual JL, Khwaja KA, Ferri LE, Giannias B, Evans DC, Razek T, Michel RP, Christou NV (2003) Hypertonic saline resuscitation attenuates neutrophil lung sequestration and transmigration by diminishing leukocyte-endothelial interactions in a two-hit model of hemorrhagic shock and infection. J Trauma 54:121–130, discussion 130–132
53. Rizoli SB, Rhind SG, Shek PN, Inaba K, Filips D, Tien H, Brenneman F, Rotstein O (2006) The immunomodulatory effects of hypertonic saline resuscitation in patients sustaining traumatic hemorrhagic shock: a randomized, controlled, double-blinded trial. Ann Surg 243:47–57
54. Rivkees SA (2008) Differentiating appropriate antidiuretic hormone secretion, inappropriate antidiuretic hormone secretion and cerebral salt wasting: the common, uncommon, and misnamed. Curr Opin Pediatr 20:448–452
55. Carlotti AP, Bohn D, Rutka JT, Singh S, Berry WA, Sharman A, Cusimano M, Halperin ML (2001) A method to estimate urinary electrolyte excretion in patients at risk for developing cerebral salt wasting. J Neurosurg 95:420–424. doi:10.3171/jns.2001.95.3.0420
56. Khanna S, Davis D, Peterson B, Fisher B, Tung H, O'Quigley J et al (2000) Use of hypertonic saline in the treatment of severe refractory posttraumatic intracranial hypertension in pediatric traumatic brain injury. Crit Care Med 28:1144–1151
57. Peterson B, Khanna S, Fisher B, Marshall L (2000) Prolonged hypernatremia controls elevated intracranial pressure in head-injured pediatric patients. Crit Care Med 28:1136–1143
58. Dominguez TE, Priestley MA, Huh JW (2004) Caution should be exercised when maintaining a serum sodium level >160 meq/L. Crit Care Med 32:1438–1439
59. Dean NP, Boslaugh S, Adelson PD, Pineda JA, Leonard JR (2007) Physician agreement with evidence-based recommendations for the treatment of severe traumatic brain injury in children. J Neurosurg 107:387–391
60. Fisher B, Thomas D, Peterson B (1992) Hypertonic saline lowers raised intracranial pressure in children after head trauma. J Neurosurg Anesthesiol 4:4–10
61. Simma B, Burger R, Falk M, Sacher P, Fanconi S (1998) A prospective, randomized, and controlled study of fluid management in children with severe head injury: lactated Ringer's solution versus hypertonic saline. Crit Care Med 26:1265–1270
62. Roberts I, Sydenham E (2012) Barbiturates for acute traumatic brain injury. Cochrane Database Syst Rev (12):CD000033
63. Mellion SA, Bennett KS, Ellsworth GL, Moore K, Riva-Cambrin J, Metzger RR, Bratton SL (2013) High-dose barbiturates for refractory intracranial hypertension in children with severe traumatic brain injury. Pediatr Crit Care Med 14:239–247
64. Winer JW, Rosenwasser RH, Jimenez F (1991) Electroencephalographic activity and serum and cerebrospinal fluid pentobarbital levels in determining the therapeutic end point during barbiturate coma. Neurosurgery 29:739–741, discussion 741–742
65. Almaas R, Saugstad OD, Pleasure D, Rootwelt T (2000) Effect of barbiturates on hydroxyl radicals, lipid peroxidation, and hypoxic cell death in human NT2-N neurons. Anesthesiology 92:764–774
66. Goodman JC, Valadka AB, Gopinath SP, Cormio M, Robertson CS (1996) Lactate and excitatory amino acids measured by microdialysis are decreased by pentobarbital coma in head-injured patients. J Neurotrauma 13:549–556
67. Piatt JH, Schiff SJ (1984) High dose barbiturate therapy in neurosurgery and intensive care. Neurosurgery 15:427–444

68. Chen HI, Malhotra NR, Oddo M, Heuer GG, Levine JM, LeRoux PD (2008) Barbiturate infusion for intractable intracranial hypertension and its effect on brain oxygenation. Neurosurgery 63:880–886. doi:10.1227/01.NEU.0000327882.10629.06, discussion 886–887
69. Thorat JD, Wang EC, Lee KK, Seow WT, Ng I (2008) Barbiturate therapy for patients with refractory intracranial hypertension following severe traumatic brain injury: its effects on tissue oxygenation, brain temperature and autoregulation. J Clin Neurosci 15:143–148
70. Jagannathan J, Okonkwo DO, Yeoh HK, Dumont AS, Saulle D, Haizlip J, Barth JT, Jane JA (2008) Long-term outcomes and prognostic factors in pediatric patients with severe traumatic brain injury and elevated intracranial pressure. J Neurosurg Pediatr 2:240–249
71. Anderson RCE, Kan P, Klimo P, Brockmeyer DL, Walker ML, Kestle JRW (2004) Complications of intracranial pressure monitoring in children with head trauma. J Neurosurg 101:53–58
72. Bor-Seng-Shu E, Figueiredo EG, Amorim RLO, Teixeira MJ, Valbuza JS, de Oliveira MM, Panerai RB (2012) Decompressive craniectomy: a meta-analysis of influences on intracranial pressure and cerebral perfusion pressure in the treatment of traumatic brain injury. J Neurosurg 117:589–596
73. Cooper DJ, Rosenfeld JV, Murray L, Arabi YM, Davies AR, D'Urso P et al (2011) Decompressive craniectomy in diffuse traumatic brain injury. N Engl J Med 364(16): 1493–1502
74. Hutchinson PJ, Kirkpatrick PJ, RESCUEicp Central Study Team (2011) Craniectomy in diffuse traumatic brain injury. N Engl J Med 365:375–376
75. Qiu W, Guo C, Shen H, Chen K, Wen L, Huang H, Ding M, Sun L, Jiang Q, Wang W (2009) Effects of unilateral decompressive craniectomy on patients with unilateral acute post-traumatic brain swelling after severe traumatic brain injury. Crit Care 13:R185
76. Skoglund TS, Eriksson-Ritzén C, Jensen C, Rydenhag B (2006) Aspects on decompressive craniectomy in patients with traumatic head injuries. J Neurotrauma 23:1502–1509
77. Choi I, Park H-K, Chang J-C, Cho S-J, Choi S-K, Byun B-J (2008) Clinical factors for the development of posttraumatic hydrocephalus after decompressive craniectomy. J Korean Neurosurg Soc 43:227–231. doi:10.3340/jkns.2008.43.5.227
78. Li LM, Timofeev I, Czosnyka M, Hutchinson PJA (2010) Review article: the surgical approach to the management of increased intracranial pressure after traumatic brain injury. Anesth Analg 111:736–748
79. Bor-Seng-Shu E, Figueiredo EG, Fonoff ET, Fujimoto Y, Panerai RB, Teixeira MJ (2013) Decompressive craniectomy and head injury: brain morphometry, ICP, cerebral hemodynamics, cerebral microvascular reactivity, and neurochemistry. Neurosurg Rev 36(3):361–370
80. Olivecrona M, Rodling-Wahlström M, Naredi S, Koskinen L-OD (2007) Effective ICP reduction by decompressive craniectomy in patients with severe traumatic brain injury treated by an ICP-targeted therapy. J Neurotrauma 24:927–935
81. Yamakami I, Yamaura A (1993) Effects of decompressive craniectomy on regional cerebral blood flow in severe head trauma patients. Neurol Med Chir (Tokyo) 33:616–620
82. Daboussi A, Minville V, Leclerc-Foucras S, Geeraerts T, Esquerré JP, Payoux P, Fourcade O (2009) Cerebral hemodynamic changes in severe head injury patients undergoing decompressive craniectomy. J Neurosurg Anesthesiol 21:339–345
83. Heppner P, Elegala DB, Durieux M, Jane JA, Lindner JR (2006) Contrast ultrasonographic assessment of cerebral perfusion in patients undergoing decompressive craniectomy for traumatic brain injury. J Neurosurg 104:738–745
84. Timofeev I, Czosnyka M, Nortje J, Smielewski P, Kirkpatrick P, Gupta A, Hutchinson P (2008) Effect of decompressive craniectomy on intracranial pressure and cerebrospinal compensation following traumatic brain injury. J Neurosurg 108:66–73
85. Ho CL, Wang CM, Lee KK, Ng I, Ang BT (2008) Cerebral oxygenation, vascular reactivity, and neurochemistry following decompressive craniectomy for severe traumatic brain injury. J Neurosurg 108:943–949
86. Lazaridis C, Czosnyka M (2012) Cerebral blood flow, brain tissue oxygen, and metabolic effects of decompressive craniectomy. Neurocrit Care 16:478–484

87. Figaji AA, Fieggen AG, Sandler SJI, Argent AC, Le Roux PD, Peter JC (2007) Intracranial pressure and cerebral oxygenation changes after decompressive craniectomy in a child with traumatic brain swelling. Childs Nerv Syst 23:1331–1335
88. Jaeger M, Soehle M, Meixensberger J (2003) Effects of decompressive craniectomy on brain tissue oxygen in patients with intracranial hypertension. J Neurol Neurosurg Psychiatry 74:513–515

Chapter 21
Surgical Management of Traumatic Brain Edema

Takeshi Maeda, Tatsuro Kawamata, Atsuo Yoshino, and Yoichi Katayama

Abstract The early massive edema caused by severe cerebral contusion results in progressive intracranial pressure (ICP) elevation and clinical deterioration within 24–72 h post-trauma. Surgical excision of the necrotic brain tissue represents the only therapy, which can provide satisfactory control of the elevated ICP and clinical deterioration. In this chapter, we review the results of our clinical studies regarding the pathophysiology of contusion edema and evaluate the effects of surgical treatment.

21.1 Introduction

In patients with cerebral contusions, two types of edema can be clinically recognized. One is the early massive edema that occurs within the period of 24–72 h post-trauma. This type of edema creates strong mass effect resulting in progressive elevation of the intracranial pressure (ICP) and clinical deterioration known as talk and deteriorate [9]. The other is the delayed peri-contusion edema, which is typically seen in the white matter adjacent to the cerebral contusion, at several days post-trauma by T2-weighted magnetic resonance (MR) imaging. This type of edema rarely causes ICP elevation leading to fatal deterioration. Despite intensive medical therapy, the elevated ICP in patients with early massive edema is often uncontrollable and fatal. In such instances, surgical excision of the necrotic brain tissue is the only therapy, which can provide satisfactory control of the elevated ICP. The precise mechanism underlying such an early massive edema is not yet clearly understood. We review the results of our clinical studies [6–10, 12], which have provided several lines of evidence to suggest that a large amount of edema

T. Maeda (✉)
Departments of Neurological Surgery & Anesthesiology, Nihon University School of Medicine, Tokyo 173-8610, Japan
e-mail: maeda.takeshi@nihon-u.ac.jp

E.H. Lo et al. (eds.), *Vascular Mechanisms in CNS Trauma*, Springer Series in Translational Stroke Research 5, DOI 10.1007/978-1-4614-8690-9_21,

fluid is accumulated in the necrotic brain tissue within the central area of contusion, and this contributes to the early massive edema.

21.2 Histopathology of Cerebral Contusion

The classical histopathological study of Lindenberg and Freytag [14] demonstrated the presence of two components of cerebral contusion; one is the central (core) area in which cells undergo necrosis as the primary consequence of mechanical injury (contusion necrosis proper), and the other is the peripheral (rim) area in which cellular swelling occurs as a consequence of ischemia. A clear demarcation line separates these two components. The area of contusion necrosis is histopathologically evident as early as at 3 h post-trauma, as a primary brain damage [4]. The cellular elements in the central area, both neuronal and glial cells, uniformly undergo shrinkage and then disintegration, homogenization, and cyst formation eventually. In contrast, cell swelling is predominant in the peripheral area [14]. The ischemia in the peripheral area is largely attributable to microthrombosis, which is the main cause of secondary brain damages. Numerous clinical studies have demonstrated a decrease in cerebral blood flow in the central as well as the peripheral areas of contusion (e.g., [1]).

21.3 MR Diffusion Study

We have investigated the evolution of cerebral contusion by diffusion-weighted MR imaging and apparent diffusion coefficient (ADC) mapping in head trauma patients [10]. Following conventional T1- and T2-weighted MR imaging, diffusion-weighted MR images are obtained, and ADC mapping is computed from various b-factor diffusion images. The diffusion-weighted MR images demonstrate a low-intensity core in the central area, beginning at approximately 24 h post-trauma. The ADC value within this central area clearly increases during the period of 24–72 h post-trauma (Fig. 21.1; [10]). This elevated ADC value appears to represent the contusion necrosis proper, since cellular disintegration and homogenization in this area would result in an expansion of the extracellular space. The diffusion-weighted MR images demonstrate a high-intensity rim in the peripheral area of contusion beginning at approximately 24 h post-trauma, which corresponds with the timing of the elevated ICP and neurological deterioration. The combination of a low-intensity core and a high-intensity rim is a consistent finding in the cerebral contusion during this period, which can be termed a halo appearance (Fig. 21.2). The ADC value decreases in the peripheral area during the period of 24–72 h post-trauma [10]. This decreased ADC value appears to represent the cellular swelling, which would result in shrinkage of the extracellular space. The ADC values between the central and peripheral areas are maximally dissociated during the period of 24–72 h post-trauma. The ADC value in the peripheral area shifts from a decrease to an increase after 72 h post-trauma. At the same time, an increase in

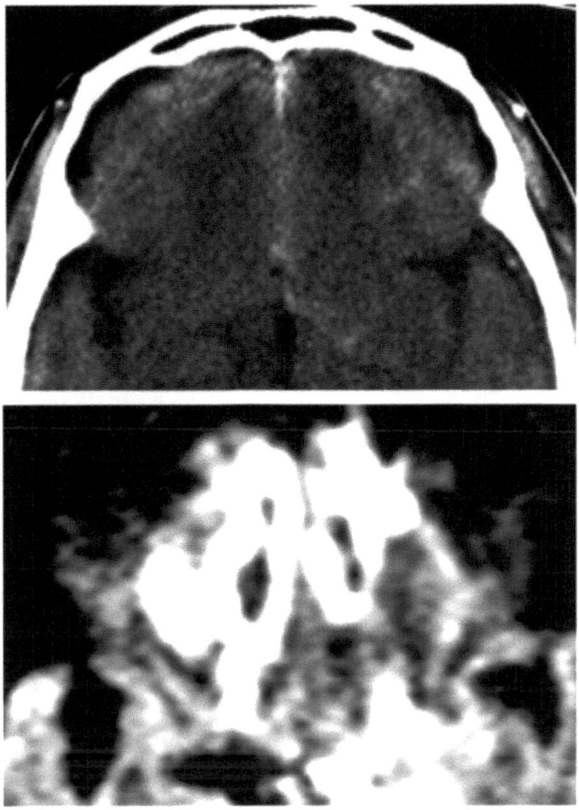

Fig. 21.1 A representative case of cerebral contusion in the acute phase, 33 h post-trauma. *Upper*: A CT scan revealed low-density area in bilateral frontal lobe. *Lower*: Diffusion image demonstrated low-intensity core with surrounding high-intensity rim, which was a typical pattern of diffusion image in the acute phase (<48 h post-trauma) of cerebral contusion without massive hemorrhage

ADC value becomes evident in the adjacent white matter. This increase in ADC value in the adjacent white matter appears to represent the delayed peri-contusion edema, which can commonly be seen on T2-weighted MR images as vasogenic edema.

21.4 Increased Cerebrovascular Permeability

Since gadolinium (Gd)-DTPA, like plasma protein, does not normally cross the blood–brain barrier, enhancement with Gd-DTPA, if observed in association with cerebral contusion, implies an increased cerebrovascular permeability. In a previous study [13], Gd-DTPA was administered intravenously by bolus injection, and MR imaging was undertaken soon after the administration. Such a procedure failed to detect any increase in cerebrovascular permeability associated with cerebral contusions during the initial few days post-trauma. Enhancement with Gd-DTPA has been reported to become detectable at 6–9 days post-trauma. Furthermore, an immunohistochemical study of the postmortem brain in such patients failed to reveal any plasma protein leakage around the contused brain areas during this

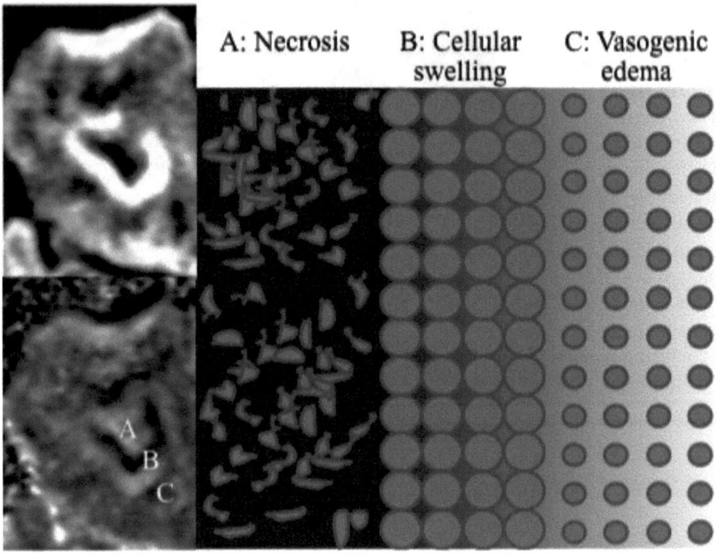

Fig. 21.2 Frontal contusion demonstrating a halo appearance with a combination of a low-intensity core and a high-intensity rim on diffusion-weighted image (*left upper*). On ADC mapping (*left lower*), the peripheral rim showed low ADC value representing the cellular swelling resulting from shrinkage of the extracellular space (*right*)

early period [17]. Evaluations of the vascular permeability by 99mTc pertechnetate single photon emission tomography failed to reveal any evidence of an increased vascular permeability within the area of contusion during the initial few days post-trauma [3]. Such findings appear to contradict the widely held view that an increased cerebrovascular permeability is responsible for the development of contusion edema [15]. We examined the changes in cerebrovascular permeability employing intravenous slow infusion of Gd-DTPA and delayed MR imaging [12]. In general, increases in cerebrovascular permeability can be detected more clearly by intravenous slow infusion of high-dose contrast medium and delayed neuroimaging [5]. This technique revealed that cerebral contusions can be enhanced at as early as 24–48 h post-trauma. In MR imaging undertaken at 2 h after Gd-DTPA administration, enhancement on T1-weighted images was observed in either the central area or the peripheral area, or both. It is evident that water supply from the blood vessels is not completely interrupted even in the central area of contusion.

The delayed peri-contusion edema has commonly been attributed to an increased cerebrovascular permeability. As mentioned above, enhancement with Gd-DTPA has been reported to become detectable at 6–9 days post-trauma. It is also possible, however, that resolution of the cellular swelling in the peripheral area of contusion might permit propagation of the edema fluid accumulated within the central area to the adjacent white matter. This would lead to edema appearance on CT and MR

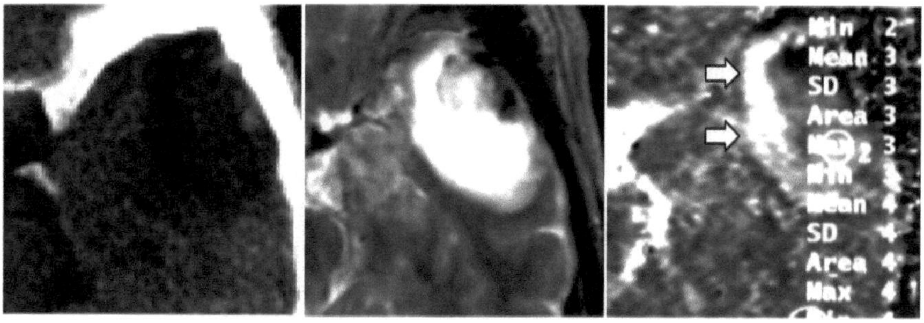

Fig. 21.3 Cerebral contusion in the temporal tip at 45 h post-trauma. *Left*: CT scan showed small LD area in the temporal tip. *Middle*: T2-weighted MRI revealed mixed intensity area in the core of contusion and surrounding edema formation. *Right*: ADC mapping demonstrated a crescent-shaped zone (*arrows*) of very high ADC value, as high as that of CSF, representing edema fluid accumulation within the central area of contusion

images which resembles to that induced by an increased cerebrovascular permeability, i.e., vasogenic edema.

21.5 Edema Fluid Accumulation in the Central Area

In approximately 50 % of patients with cerebral contusions, a crescent-shaped zone of very high ADC value develops at the border between the central and peripheral areas beginning at approximately 24 h post-trauma (Fig. 21.3; [10]). This crescent-shaped zone was always located within the central area of contusion. The very high ADC value appears to represent edema fluid accumulation within the central area of contusion. Edema fluid accumulation within the necrotic brain tissue is also suggested by the formation of a fluid–blood interface within cerebral contusions (Fig. 21.4). The fluid–blood interface can be formed without a fluid cavity. When we carry out surgery, we find no fluid cavity, but softened and water-rich necrotic brain tissue is present [8]. Within the contusion necrosis, hemorrhage undergoes enlargement within the initial 6 h post-trauma [9], and red blood cells diffusely permeate the softened necrotic brain tissue [14]. The fluid–blood interfaces observed within the central area of contusion may represent layering of red blood cells in the softened necrotic brain tissue, which has accumulated voluminous edema fluid.

21.6 Osmotic Potential of the Contusion Necrosis

We have reported that necrotic brain tissue sampled from the central area of contusion during surgery shows a very high osmolality, reaching 350–400 mOsm [7]. It is uncertain whether or not such a marked increase in osmolality is osmotically active and causes edema fluid accumulation. It appears that expansion of the

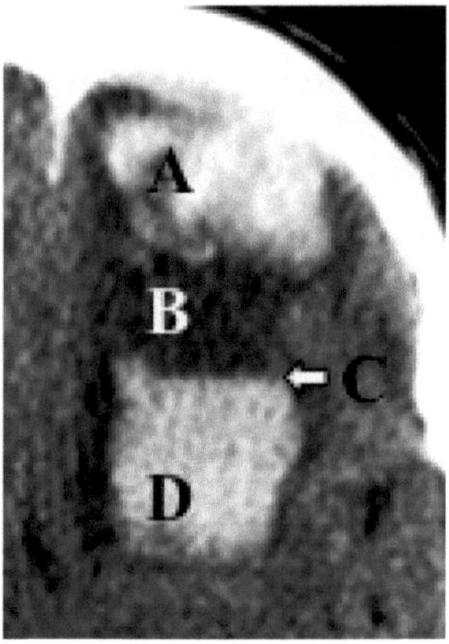

Fig. 21.4 CT scan of cerebral contusion in the frontal tip (30 h post injury), sowing fluid–blood interface in the area of contusion necrosis. (**a**) Hematoma; (**b**) edema fluid; (**c**) fluid–blood interface; (**d**) layering of red blood cells

extracellular space in the central area increases the capacitance for edema fluid accumulation. In contrast, shrinkage of the extracellular space in the peripheral area increases the resistance for edema fluid propagation or resolution. We hypothesize that the barrier formed by swollen cells in the peripheral area may prevent edema fluid propagation and also help to generate osmotic potentials across the central and peripheral areas. Since blood flow is greatly reduced but is not completely interrupted in the contused brain tissue, water is supplied from the blood vessels into the central area. We suggest that a combination of these events may facilitate edema fluid accumulation in the central area and contribute to the early massive edema of cerebral contusion.

21.7 Procedure for Contusion Necrotomy

A sufficiently large skin incision is made so that craniotomy which fully exposes the frontal or temporal tip in which the contusion is situated can be performed. According to the location of the contusion, a bifrontal incision, coronal incision, frontal–temporal incision, or combination of these is carried out. Craniotomy of sufficient size is performed, centering on the region of the contusion. However, it is important that the base of the depth of the craniotomy be such that the skull base of the frontal lobe or temporal lobe is exposed, since the contusion is often situated adjacent to the skull base. Especially for lesions in the frontal lobe, contusion of a cone, with the superior orbit side as the base, is formed in many cases (Fig. 21.5). First, the base of the cone which has appeared, i.e., a circle, is confirmed, because the cerebral contusion exists as a cone with the brain surface at the bottom. As mentioned craniotomy is performed so that the skull base can be exposed, since the

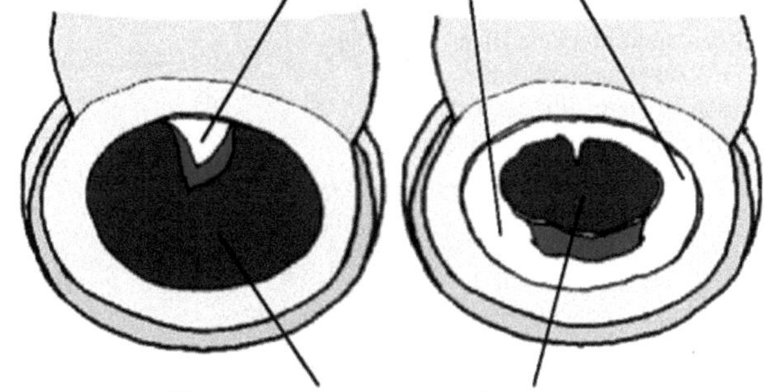

Fig. 21.5 Surgical removal of contusion necrosis is performed along the demarcation plane. In the first step, corticotomy is added on the surface of the necrosis to expose demarcation plane (*left*). Then the arachnoid membrane is cut in a circle along the demarcation plane, and the necrosis tissue is retracted and removed from the surrounding brain tissue (*right*)

bottom of the cone may be situated adjacent to the skull base. Corticotomy is added at the surface of the necrosis to expose the demarcation plane. The arachnoid membrane is cut in a circle along the demarcation plane, and the necrosis tissue is retracted and removed sharply from the surrounding brain tissue (Figs. 21.5 and 21.6). The operation is repeated over the whole circumference of the demarcation plane. The necrotic brain tissue tends to be very soggy, and water exudes from it continuously. Finally, the deeper white matter is exfoliated towards the vertex of the brain contusion, and the necrosis proper is separated as a lesion and removed.

21.8 Surgical Treatment

Kawamata and Katayama [11] investigated the effects of surgical excision of the necrotic brain tissue in patients with severe cerebral contusion. The data were collected from Japan Neurotrauma Data Bank (JNTDB) in which a total of 1,002 patients suffering severe traumatic brain injury who were registered during the period between 1998 and 2001 were analyzed [16]. Among these patients, 182 (18 %) demonstrated severe cerebral contusions as the major lesions contributing to their clinical status, 121 (66 %; group I) were treated conservatively, and the remaining 61 (34 %; group II) underwent surgery. There was, however, no significant difference in age between groups I and II (47.87 ± 23.8 vs. 54.47 ± 19.5 years). The surgical management involved internal decompression (complete excision of the necrotic brain tissue and evacuation of clots) with or without external decompression in most patients (90 %) of group II. The remaining patients underwent external decompression alone. Surgery was performed at 1.8–86.1 (19.57 ± 24.2) h, mostly (73 %) within 24 h post-trauma. Group I demonstrated a clearly poorer outcome on the Glasgow outcome scale (GOS) at 6 months post-trauma (Table 21.1). The mortality was clearly higher in group I, as compared with group II (48 % vs. 23 %; $p = 0.0001$; $n = 182$). A difference in mortality between the two groups was noted in patients who were scored at 8 or less on the GCS at the

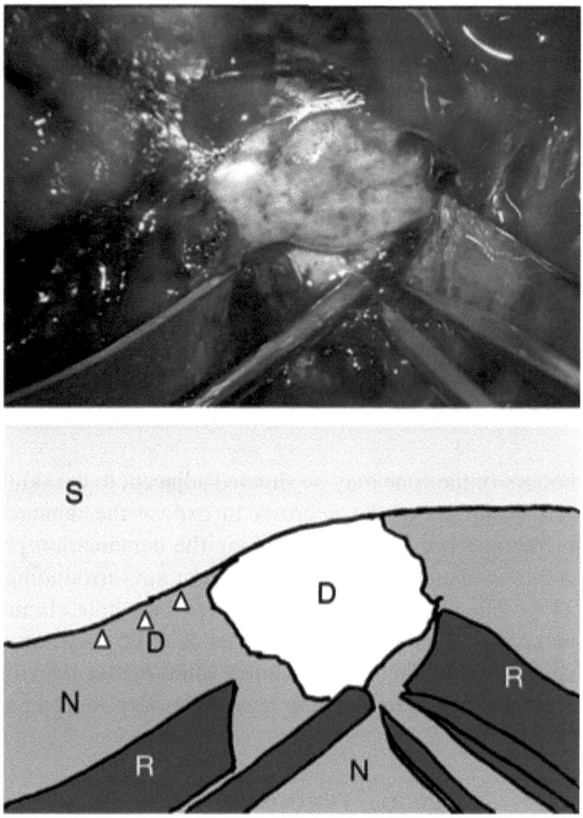

Fig. 21.6 The intraoperative photograph in contusion necrotomy. *D* demarcation plane (line △), *S* surrounding brain tissue, *N* contusion necrosis proper, *R* retractor

time of admission, but did not reach a statistically significant level (Table 21.1). The most striking difference was observed in patients who were scored at 9 or better on the GCS at the time of admission (Table 21.1). The mortality was clearly higher in group I, as compared with group II (56 % vs. 17 %; $p = 0.017$; $n = 45$). A clear difference in mortality between the two groups was observed even when the analysis was restricted to patients who definitely demonstrated "talk and deteriorate" (64 % vs. 22 %; $p = 0.026$; $n = 29$; Table 21.2).

The indications for surgical intervention in case of severe cerebral contusion remain controversial [2]. The huge variation in ratio of selecting surgical management among the present centers reflects diversity in management policy and an absence of consensus regarding the indications for surgery. The higher mortality in group I, as compared with group II, suggests that the surgery performed in group II patients helped to prevent their clinical deterioration and death. Such an effect of surgery was most striking in those patients who were scored at 9 or better at the time of admission. It is by no means certain whether patients in group II who were scored at 9 or better before surgery would have deteriorated or not if surgery had not been

Table 21.1 Outcome (6 months post-trauma) in patients treated with conservative therapies and surgical therapy

	GOS (%)					
	n	GR	MD	SD	VS	D
GCS on admission: 3–5						
Conservative	47	7	2	11	11	70
Surgical	11	9	9	27	0	55
GCS on admission: 6–8						
Conservative	58	29	21	10	10	29
Surgical	21	24	29	24	10	14
GCS on admission: 9–15						
Conservative	16	19	13	13	0	56*
Surgical	29	28	28	17	10	17
Total						
Conservative	121	19	12	12	9	48**
Surgical	61	23	25	21	8	23

GCS Glasgow coma scale, *GOS* Glasgow outcome scale
$^*p = 0.017$, $^{**}p = 0.0001$

Table 21.2 Outcome (6 months post-trauma) in patients demonstrating "talk and deteriorate"

	GOS (%)					
	n	GR	MD	SD	VS	D
Conservative	11	18	9	9	0	64*
Surgical	18	11	22	39	6	22

GCS Glasgow coma scale, *GOS* Glasgow outcome scale
$^*p = 0.026$

carried out. An effect of surgery on mortality is evident, however, since a difference in mortality was clearly observed even when the analysis was restricted to patients who definitely demonstrated "talk and deteriorate." This finding strongly suggests that the surgery itself was the major reason for the improved mortality in group II. In other words, death was probably prevented by the surgery in many patients of group II.

The present findings support our hypothesis that early massive edema is caused by cerebral contusion through the presence of necrotic brain tissue and indicate that surgical excision of the necrotic brain tissue is the only therapy which can provide satisfactory control of the progressive elevation of the ICP and clinical deterioration in many cases. Surgical intervention should be considered in patients with severe cerebral contusion who demonstrate "talk and deteriorate."

21.9 Conclusion

There is at present no established medical treatment which can effectively inhibit edema fluid accumulation within cerebral contusions. The most effective therapy for ameliorating the potentially fatal edema is surgical excision of the necrotic brain tissue. The effects of surgical excision of necrotic brain tissue have commonly been accounted for on the basis of an increased space compensation for mass lesions. It is possible, however, that excision of the necrotic brain tissue does mean elimination of the cause of edema fluid accumulation. If the ICP is elevated by early massive edema due to cerebral contusion and the elevated ICP is medically uncontrollable, surgical excision of the necrotic brain tissue would appear to represent the therapy of choice, regardless of the size of the associated hemorrhages. Surgery should be considered in patients who are scored at 9 or better at the time of admission, as soon as they show clinical deterioration.

References

1. Alexander MJ, Martin NA, Khanna M, Caron M, Becker DP (1994) Regional cerebral blood flow trends in head injured patients with focal contusions and cerebral edema. Acta Neurochir Suppl (Wien) 60:479–481
2. Bullock R, Golek J, Blake G (1989) Traumatic intracerebral hematoma which patients should undergo surgical evacuation? CT scan features and ICP monitoring as a basis for decision making. Surg Neurol 32:181–187
3. Bullock R, Statham J, Patterson D, Wyper D, Hadley D, Teasdale E (1990) The time course of vasogenic oedema after focal human head injury: evidence from SPECT mapping of blood brain barrier defects. Acta Neurochir Suppl (Wien) 51:286–288
4. Eriskat J, Schurer L, Kempski O, Baethmann A (1994) Growth kinetics of a primary brain tissue necrosis from focal lesion. Acta Neurochir Suppl (Wien) 60:425–427
5. Ito U, Reulen H-J, Tomita H, Ikeda J, Saito J, Maehara T (1988) Formation and propagation of brain oedema fluid around human brain metastasis: a CT study. Acta Neurochir 90:35–41
6. Katayama Y, Kawamata T (2003) Edema fluid accumulation within necrotic brain tissue as a cause of the mass effect of cerebral contusion in head trauma patient. Acta Neurochir Suppl 86: 323–327
7. Katayama Y, Mori T, Maeda T, Kawamata T (1998) Pathogenesis of the mass effect of cerebral contusions: a rapid increase in osmolality within the contusion necrosis. Acta Neurochir Suppl 71:289–292
8. Katayama Y, Tsubokawa T, Kinoshita K, Himi K (1992) Intra-parenchymal fluid-blood levels in traumatic intracerebral hematomas. Neuroradiology 34:381–383
9. Katayama Y, Tsubokawa T, Miyazaki S, Kawamata T, Yoshino A (1990) Oedema fluid formation within contused brain tissue as a cause of medically uncontrollable elevation of intracranial pressure in head trauma patients. Acta Neurochir Suppl (Wien) 51:308 310
10. Kawamata T, Katayama Y, Aoyama N, Mori T (2000) Heterogeneous mechanisms of early edema formation in cerebral contusion: diffusion MRI and ADC mapping study. Acta Neurochir Suppl 76:9–12
11. Kawamata T, Katayama Y (2007) Cerebral contusion: a role model for lesionprogression. Prog Brain Res 161:235–241
12. Kushi H, Katayama Y, Shibuya T, Tsubokawa T, Kuroha T (1994) Gd-DTPA enhanced magnetic resonance imaging of cerebral contusions. Acta Neurochir Suppl (Wien) 60: 472–474
13. Lang DA, Hadley DM, Teasdale GT, Macpherson P, Teasdale E (1991) Gadolinium DTPA enhanced magnetic resonance imaging in acute head injury. Acta Neurochir 109:5–11
14. Lindenberg R, Freytag E (1957) Morphology of cortical contusions. AMA Arch Pathol 63: 23–42

15. Marmarou A (2003) Pathophysiology of traumatic brain edema: current concepts. Acta Neurochir Suppl 86:7–10
16. Nakamura N, Yamaura A, Shigemori M, Ono J, Kawamata T, Sakamoto T; Japanese Data Bank Committee for Traumatic Brain Injury (2002) Epidemiology, prevention and countermeasures against severe traumatic brain injury in Japan and abroad. Neurol Res 24:45–53
17. Todd NV, Graham DI (1990) Blood–brain barrier damage in traumatic brain contusion. Acta Neurochir Suppl 51:296–299

Chapter 22
Optimizing Hemodynamics in the Clinical Setting

Jose Alberto Toranzo and Claudia S. Robertson

Abstract Hypotension is common following traumatic brain injury and can contribute to a poor outcome. Hypovolemia is the most common cause of hypotension in this setting, but myocardial dysfunction may also occur. Fluid resuscitation is the initial step in treating hypotension. Recent trials have failed to demonstrate an advantage of colloids or hypertonic saline over normal saline resuscitation, and albumin resuscitation was even associated with a higher mortality rate. Therefore normal saline remains the standard resuscitation fluid, even though colloids and hypertonic saline have some theoretical advantages following brain injury. The optimal hemoglobin concentration for the brain-injured patient remains controversial, and the decision to transfuse remains an assessment of the risk-benefit ratio of improving cerebral oxygen delivery with the risk of complications of transfusions. When blood pressure remains inadequate following fluid resuscitation, vasopressors may be needed. While definitive studies are lacking, available studies suggest that norepinephrine and phenylephrine have advantages over dopamine as the initial choice for vasopressor.

22.1 Introduction

Brain injury and hemodynamic instability have been areas of intense research in the past years. Brain injury accompanied by hypotension has been associated with poor outcome, and effort towards maintaining an adequate blood pressure is a key management principle. Topics of controversy include which resuscitation fluid is best, choice of vasopressor therapy which is often used concurrently with fluid resuscitation in patients with acute neurologic injury to maintain or augment perfusion to areas of injury, and when to transfuse blood. To be able to determine

C.S. Robertson (✉)
Department of Neurosurgery, Baylor College of Medicine, One Baylor Plaza, Houston, TX 77030, USA
e-mail: claudiar@bcm.edu

E.H. Lo et al. (eds.), *Vascular Mechanisms in CNS Trauma*, Springer Series in Translational Stroke Research 5, DOI 10.1007/978-1-4614-8690-9_22,

the optimal strategy, it is important also to determine the most appropriate measurements of responsiveness to these interventions.

22.2 Hemodynamic Instability in Patients with Brain Injury

Traumatic brain injury (TBI) is often accompanied by systemic hypotension, and both prehospital and in-hospital episodes of hypotension have been found to be important predictors of a poor neurological outcome. A single episode of hypotension (defined as systolic blood pressure <90) is among the five most powerful predictors of outcome after TBI [32]. Other important independent prognostic variables are age, Glasgow Coma Scale motor score, pupil response, and computerized tomography characteristics [32]. On analysis of the Traumatic Coma Data Bank database, occurrence of hypotension was associated with increased morbidity and a doubling of mortality as compared with a matched group of patients without hypotension [8, 29]. Analysis of the more recent IMPACT database showed similar results [7].

Hypotension in brain-injured patients is most commonly caused by substantial blood loss due to extracranial injuries. Intracranial injuries do not result in substantial blood loss unless surgical intervention is required. As intravascular volume decreases from active bleeding, the body tries to compensate by increasing the heart rate and constricting peripheral blood vessels. This compensatory mechanism is the reason why blood pressure will only start to decrease in adult patients after the loss of more than approximately 1.5 L of blood volume. But even before this occurs, cerebral perfusion and oxygenation may be compromised due to vasoconstriction and a decrease in oxygen-carrying capacity. Myocardial dysfunction can also be a cause of hypotension due to intense sympathetic hyperactivity caused by the brain injury [34].

Other studies have shown that there is a U-shaped relationship between both systolic and mean arterial blood pressure and outcome. Both lower and higher values of blood pressure have been associated with poorer outcome with a much better outcome occurring in patients with systolic blood pressures between approximately 120 and 150 [4] and between mean arterial blood pressures of 85–110 [7]. Hypertension could also cause additional brain injury, or this observed relationship may be because of other associated factors. Higher values of blood pressure (systolic blood pressure >150) are associated with lower motor scores, and as many as 40 % of patients with hypertension have an associated mass lesion. When hypertension occurs in the setting of a mass lesion, the higher blood pressure is critical in maintaining cerebral perfusion. Evacuating the mass lesion would therefore be the initial treatment.

Systolic and mean arterial blood pressures (MABP) are highly correlated, but MABP is more strongly influenced by diastolic blood pressure and measurements of MABP are more stable as they are less affected by measurement artifacts.

MABP can be calculated from systolic and diastolic blood pressures by the following formula:

$$\text{MABP} = (\text{systolic} + 2 \times \text{diastolic blood pressure})/3$$

In the early phase after trauma, systolic blood pressure may be of greater relevance than MABP due to the fact that stroke volume (the major determinant of systolic blood pressure) is decreased with blood loss, while systemic vascular resistance (major determinant of diastolic blood pressure and therefore of greater influence on MABP) may remain normal or even be increased during the initial compensatory phase of hemorrhagic shock. However, after initial hemodynamic stabilization, the emphasis should be on MABP because of its role in determining cerebral perfusion pressure (CPP).

The Guidelines for the Management of Severe TBI recommend that systolic blood pressure during the initial resuscitation should be kept at least 90 mmHg and that the goal for CPP should be at least 60 mmHg [3]. Selected patients may require a higher CPP to adequately perfuse the brain, but routinely maintaining a higher CPP is discouraged because of potential adverse effects of the additional fluid and vasopressors required to maintain the higher level of blood pressure [3, 9, 39].

22.3 Optimal Fluid Resuscitation in Brain-Injured Patients

The question of crystalloid versus colloid-based resuscitation strategies has been argued for many years. Trauma resuscitation protocols favor the use of crystalloid-based fluid strategies although the evidence supporting these strategies in cases of brain injury is limited. Current trauma protocols assume that prompt restoration of the volume of circulating blood and the prevention of hypotension may improve the outcome in patients with brain injury [51]. Colloid fluid-resuscitation strategies, including the use of albumin, have also been used based on physiological principles, aiming to maintain or augment plasma oncotic pressure and thereby minimize extravasation of intravascular fluid into the brain interstitium [17].

The Saline versus Albumin Fluid Evaluation Study (SAFE) has recently provided some information towards this question. The SAFE study compared the effect of fluid resuscitation with 4 % albumin or 0.9 % saline on mortality in a heterogeneous population of patients in intensive care units. Overall the study showed no significant difference in the risk of death among patients who received albumin as compared to those who received saline [15]. However, in a post hoc analysis of patients with TBI enrolled in the SAFE study, the rate of death was significantly higher among patients assigned to albumin than among those assigned to saline (41.8 % compared to 20.4 %) [33]. The reason for this difference in mortality may be related to intracranial pressure, which was higher during the first week post-injury in patients assigned to albumin [10].

Among the crystalloids available for resuscitation, hypertonic saline has potentially attractive characteristics for the brain-injured patient, including restoration of blood pressure while reducing brain edema and modulating the inflammatory response to injury [12, 19, 38]. Administration of hypertonic saline results in an increase serum osmotic pressure, which can draw fluid from the interstitial space into the intravascular space. A relatively small volume of fluid administered can result in significant volume expansion. It is estimated that administration of 250 mL of a 7.5 % saline solution results in an equivalent volume expansion to 3 L of normal saline [22].

The largest clinical trial of hypertonic saline for trauma resuscitation, conducted by the Resuscitation Outcomes Consortium, compared 250 cc 7.5 % saline, 7.5 % saline/6 % dextran 70, or 0.9 % saline as the initial resuscitation fluid administered in the prehospital setting following severe traumatic injury with evidence of either hypovolemic shock (SBP < 70 mmHg or 70–90 mmHg with a heart rate ≥ 108) or severe TBI (Glasgow coma score ≤ 8). There was no difference in the primary outcome among the treatment groups in either the shock cohort (28-day mortality) or the TBI cohort (6-month Glasgow outcome scale extended) [5, 6].

Despite the potential advantages of colloid or hypertonic saline as a resuscitation fluid in TBI patients, these large clinical trials have not demonstrated superiority and in the case of albumin have demonstrated an increase in mortality. Therefore, at the present time, normal saline remains the preferred resuscitation fluid for patients with TBI and hypotension.

22.4 The Role of Blood Transfusion in Hemodynamic Resuscitation

A need for blood transfusion is very common in the treatment of critically ill trauma patients. It is used in the acute setting as a means of correcting acute blood loss and later in the hospital setting for maintaining an adequate hematocrit. The objective of blood transfusion is to improve oxygen-carrying capacity. In healthy subjects, optimal cerebral oxygen delivery and cerebral blood flow are obtained at hemoglobin values near 42 % [16]. However, normovolemic anemia does not alter cerebral function until hemoglobin levels drop to 5–7 g/dl [52]. The optimal target hematocrit or hemoglobin concentration has not been clearly established in brain-injured patients. The decision lies in the relative risk-benefit ratio of potentially improving cerebral oxygenation and the complications associated with transfusion [11, 21, 26, 50].

The potential benefit of transfusing to a higher hematocrit for patients with brain injury includes improvement in cerebral oxygenation. Studies examining the effect of transfusion on cerebral oxygenation in the treatment of moderate anemia (hemoglobin concentration 7–10 g/dl) have demonstrated a small but consistent increase in cerebral oxygenation. In a small study of 35 patients, Smith et al. showed a small increase in brain tissue pO_2 ($PbtO_2$) with transfusion that was independent of changes in blood pressure or CPP [44]. Torella et al. had similar observations using near-infrared to examine changes in cerebral oxygenation [49]. In a larger study of 69 patients, Leal-Noval et al. showed a small but significant increase in $PbtO_2$ when the blood that was transfused was less than 19 days old [25]. When older blood was transfused, no change in $PbtO_2$ occurred.

In critically ill patients in a general intensive care unit setting, a large randomized clinical trial comparing liberal (for hemoglobin concentration <10 g/dl) and restrictive (for hemoglobin concentration <7 g/dl) transfusion practices found no difference in mortality rates [18]. In a subgroup analysis of the patients enrolled in this trial who had a closed head injury, the 30-day mortality rate was 17 % in the restricted transfusion group compared to 13 % in the liberal transfusion group ($p = 0.64$) [30]. However, the number of patients with TBI in this subgroup was small, and the long-term outcome that would have been of more interest than mortality was not available. The question of whether a restrictive transfusion strategy can be safely used in TBI patients has not yet been clarified and practice varies widely [43].

22.5 Vasopressor Therapy in Brain-Injured Patients

Fluid resuscitation is often successful in restoring blood pressure to adequate levels. However, if the patient remains hypotensive after volume resuscitation, vasopressors may be required. In addition to the treatment of hypotension, vasopressor therapy may also be used to elevate blood pressure in order to improve CPP. The response of cerebral perfusion may be determined by the choice of vasopressor. The ability of the brain to pressure autoregulate will also influence the cerebral hemodynamic response to vasopressor therapy [2, 20]. If the CPP changes are within a range where the brain is capable of autoregulating blood flow, cerebral perfusion does not change very much and intracranial pressure usually does not change or decreases. However, if the brain has impaired pressure autoregulation or if the CPP changes are below the limits of autoregulation, then cerebral blood flow increases with the improvement in CPP and intracranial pressure may also increase.

The best vasopressor for increasing MAP and CPP in TBI patients has not been clearly established. In experimental models of brain injury, a variety of vasopressors have been shown to improve cerebral perfusion and/or oxygenation, including norepinephrine, dopamine, phenylephrine, and vasopressin [13, 23, 24, 36, 47]. In a sheep blunt brain trauma model followed by infusion of endotoxin, Stubbe et al. compared resuscitation with fluids to infusion of norepinephrine. Both improved MABP and carotid blood flow, but only norepinephrine increased cerebral oxygenation [47]. In a fluid percussion injury/hemorrhagic shock model, Feinstein et al. showed that resuscitation with a combination of a pressor (either phenylephrine or vasopressin) with fluid restored blood pressure with lower ICP than fluid resuscitation alone [13]. In a rat cortical impact injury model, Kroppenstedt et al. found that elevating MABP to 120 mmHg with norepinephrine improved peri-contusional cerebral perfusion and tissue oxygenation [24]. The increase in cerebral perfusion was better with norepinephrine than with dopamine [23]. In contrast to these studies, in a rat weight-drop model complicated by hypoxia/hypotension, Ract et al. reported that neither norepinephrine nor dopamine infusion was effective in restoring MABP or cerebral perfusion and resulted instead in an increase in intracranial pressure [36].

In TBI patients, most of the available clinical studies have compared cerebral hemodynamic effects of norepinephrine and dopamine [35, 46] or represent retrospective analyses of clinical use of various pressors [45]. In a study by Ract et al., cerebral hemodynamics were compared while maintaining blood pressure alternately with dopamine and norepinephrine [35]. For the same MABP, ICP was significantly higher with dopamine than with norepinephrine. No other differences in hemodynamics were observed. Steiner et al. also compared the effects of dopamine and norepinephrine in stepwise increases in CPP from 65 to 85 mmHg [46]. Although there were no differences in the absolute values of cerebral blood flow velocity or intracranial pressure, the changes at each step increase in CPP with norepinephrine were more consistent. In a retrospective review of vasopressor usage at their institution, Sookplung et al. reported that patients receiving phenylephrine had a higher MABP than those on dopamine and a higher CPP than patients given norepinephrine [45]. All of these studies involve relatively small numbers of patients, the autoregulatory status is not reported making interpretation of the findings difficult, and none of the studies examine the effect of the various pressors on outcome. Although there are no definitive studies, the experimental and clinical studies available suggest some advantages of norepinephrine over dopamine as the initial choice of a vasopressor. Phenylephrine may also be an acceptable alternative.

22.6 Hemodynamic Parameters to Guide Fluid Resuscitation

In critically ill patients, it is important to be able to assess the intravascular volume, as uncorrected hypovolemia can lead to unnecessary infusions of vasopressor agents, which may cause an increase in organ hypoperfusion and ischemia. Likewise, excess intravascular volume is associated with complications, increased length of stay in the ICU, and increased mortality. Resuscitation with intravenous fluids should produce a demonstrable enhancement of perfusion. Individualized goal-directed therapy using functional hemodynamic parameters should optimize resuscitation efforts [37].

The tools available for assessing volume status and response to resuscitation have evolved in recent years. In the critically ill patient, physical examination is of limited use in predicting a patient's intravascular volume. There is poor interobserver agreement when assessing these variables using only physical exam assessments [41, 42]. Cardiac filling pressures including the central venous pressure (CVP) and pulmonary artery occlusion pressure (PAOP) are frequently used to assess intravascular volume. Cardiac filling pressures, however, have limitations in predicting fluid responsiveness, and volumetric measurements with echocardiography may be better in guiding resuscitation efforts [27, 40].

The goal of fluid resuscitation is to optimize preload. Preload is determined by left ventricular end-diastolic volume (LVEDV). According to the Frank-Starling principle, as the preload increases, left ventricular stroke volume increases until the optimal preload is achieved. At this point, further volume resuscitation has no little effect on cardiac output and acts only to increase tissue edema and to promote tissue hypoxia. In normal adults, both ventricles operate on the ascending portion of the Frank-Starling curve providing a functional reserve for the heart in situations of acute stress. As a result of altered left ventricular compliance and function, the position of an acutely ill patient on the Frank-Starling curve cannot be predicted from their preload alone. It is therefore important to determine not only the patient's preload but also their fluid responsiveness to whether the patient will increase his/her stroke volume or cardiac output with fluid loading.

CVP is widely used to guide fluid therapy assuming that it directly reflects intravascular volume. CVP is a good approximation of right atrial pressure, which is a major determinant of right ventricular filling. However, because of the changes in venous tone, intrathoracic pressures, left and right ventricular compliance, and geometry that occur in critically ill patients, there is poor relationship between CVP and right ventricular end-diastolic volume. Systematic review of available studies demonstrates that there is no association between CVP and circulating blood volume, that CVP is a poor indicator of left and right ventricular preload, and that CVP does not predict fluid responsiveness [27]. PAOP has been thought to be a more reliable indicator of left ventricular preload. However, recent studies have demonstrated that PAOP is also a poor predictor of preload and volume responsiveness. It is a measure of left ventricular end-diastolic pressure but does not directly assess LVEDV because of changes in left ventricular compliance in the critically ill patient.

Dynamic changes in systolic pressure, pulse pressure, and stroke volume in patients undergoing mechanical ventilation may be additional useful techniques to assess volume responsiveness. Suehiro et al. compare the ability of stroke volume variation (SVV) and CVP to predict the cardiac output response to fluid administration [48]. The area under the receiver operating characteristic curve was significantly better for SVV compared to CVP (0.793 for SVV and 0.442 for CVP) in this study. A recent review of the literature concluded that dynamic changes of arterial waveform-derived variables during mechanical ventilation are more accurate in predicting volume responsiveness in critically ill patients than traditional static indices of volume responsiveness [28]. One major limitation of this noninvasive method, however, is that it is limited to patients who are ventilated.

Bedside transthoracic echocardiography (TTE) is an alternative noninvasive method to assess preload by measuring ventricular volume changes, respiratory changes in inferior vena cava (IVC) diameter, or respiratory changes in aortic flow velocity [40]. Left ventricular function can also be readily assessed [31]. In studies where TTE has been applied in an ICU setting, new cardiac information was obtained and changed management in 37 % of patients examined [1]. Among the

TTE variables that can be assessed, IVC diameter has been validated as a good predictor of fluid responsiveness by determining FLAT (<2 cm) and FAT (≥ 2 cm) IVC diameter [14].

References

1. Benjamin E, Griffin K, Leibowitz AB et al (1998) Goal-directed transesophageal echocardiography performed by intensivists to assess left ventricular function: comparison with pulmonary artery catheterization. J Cardiothorac Vasc Anesth 12:10–15
2. Bouma GJ, Muizelaar JP, Bandoh K et al (1992) Blood pressure and intracranial pressure-volume dynamics in severe head injury: relationship with cerebral blood flow. J Neurosurg 77: 15–19
3. Bratton SL, Chesnut RM, Ghajar J et al (2007) Guidelines for the management of severe traumatic brain injury. J Neurotrauma 24(Suppl 1):S1–S106
4. Brenner M, Stein DM, Hu PF et al (2012) Traditional systolic blood pressure targets underestimate hypotension-induced secondary brain injury. J Trauma Acute Care Surg 72: 1135–1139
5. Bulger EM, May S, Brasel KJ et al (2010) Out-of-hospital hypertonic resuscitation following severe traumatic brain injury: a randomized controlled trial. JAMA 304:1455–1464
6. Bulger EM, May S, Kerby JD et al (2011) Out-of-hospital hypertonic resuscitation after traumatic hypovolemic shock: a randomized, placebo controlled trial. Ann Surg 253:431–441
7. Butcher I, Maas AI, Lu J et al (2007) Prognostic value of admission blood pressure in traumatic brain injury: results from the IMPACT study. J Neurotrauma 24:294–302
8. Chesnut RM, Marshall SB, Piek J et al (1993) Early and late systemic hypotension as a frequent and fundamental source of cerebral ischemia following severe brain injury in the Traumatic Coma Data Bank. Acta Neurochir Suppl 59:121–125
9. Contant CF, Valadka AB, Gopinath SP et al (2001) Adult respiratory distress syndrome: a complication of induced hypertension after severe head injury. J Neurosurg 95:560–568
10. Cooper DJ, Myburgh J, Heritier S et al (2013) Albumin resuscitation for traumatic brain injury: is intracranial hypertension the cause of increased mortality? J Neurotrauma 30:512–518
11. Desjardins P, Turgeon AF, Tremblay MH et al (2012) Hemoglobin levels and transfusions in neurocritically ill patients: a systematic review of comparative studies. Crit Care 16:R54
12. Doyle JA, Davis DP, Hoyt DB (2001) The use of hypertonic saline in the treatment of traumatic brain injury. J Trauma 50:367–383
13. Feinstein AJ, Patel MB, Sanui M et al (2005) Resuscitation with pressors after traumatic brain injury. J Am Coll Surg 201:536–545
14. Ferrada P, Anand RJ, Whelan J et al (2012) Qualitative assessment of the inferior vena cava: useful tool for the evaluation of fluid status in critically ill patients. Am Surg 78:468–470
15. Finfer S, Bellomo R, Boyce N et al (2004) A comparison of albumin and saline for fluid resuscitation in the intensive care unit. N Engl J Med 350:2247–2256
16. Gaehtgens P, Marx P (1987) Hemorheological aspects of the pathophysiology of cerebral ischemia. J Cereb Blood Flow Metab 7:259–265
17. Grande PO (2006) The "Lund Concept" for the treatment of severe head trauma–physiological principles and clinical application. Intensive Care Med 32:1475–1484
18. Hebert PC, Wells G, Blajchman MA et al (1999) A multicenter, randomized, controlled clinical trial of transfusion requirements in critical care. Transfusion Requirements in Critical Care Investigators, Canadian Critical Care Trials Group. N Engl J Med 340:409–417
19. Hinson HE, Stein D, Sheth KN (2013) Hypertonic saline and mannitol therapy in critical care neurology. J Intensive Care Med 28:3–11
20. Hlatky R, Valadka AB, Robertson CS (2005) Intracranial pressure response to induced hypertension: role of dynamic pressure autoregulation. Neurosurgery 57:917–923

21. Kramer AH, Le RP (2012) Red blood cell transfusion and transfusion alternatives in traumatic brain injury. Curr Treat Options Neurol (in press)
22. Kramer GC (2003) Hypertonic resuscitation: physiologic mechanisms and recommendations for trauma care. J Trauma 54:S89–S99
23. Kroppenstedt SN, Sakowitz OW, Thomale UW et al (2002) Norepinephrine is superior to dopamine in increasing cortical perfusion following controlled cortical impact injury in rats. Acta Neurochir Suppl 81:225–227
24. Kroppenstedt SN, Thomale UW, Griebenow M et al (2003) Effects of early and late intravenous norepinephrine infusion on cerebral perfusion, microcirculation, brain-tissue oxygenation, and edema formation in brain-injured rats. Crit Care Med 31:2211–2221
25. Leal-Noval SR, Munoz-Gomez M, Arellano-Orden V et al (2008) Impact of age of transfused blood on cerebral oxygenation in male patients with severe traumatic brain injury. Crit Care Med 36:1290–1296
26. Leal-Noval SR, Munoz-Gomez M, Murillo-Cabezas F (2008) Optimal hemoglobin concentration in patients with subarachnoid hemorrhage, acute ischemic stroke and traumatic brain injury. Curr Opin Crit Care 14:156–162
27. Marik PE, Baram M, Vahid B (2008) Does central venous pressure predict fluid responsiveness? A systematic review of the literature and the tale of seven mares. Chest 134:172–178
28. Marik PE, Cavallazzi R, Vasu T et al (2009) Dynamic changes in arterial waveform derived variables and fluid responsiveness in mechanically ventilated patients: a systematic review of the literature. Crit Care Med 37:2642–2647
29. Marmarou A, Anderson RL, Ward JD et al (1991) Impact of ICP instability and hypotension on outcome in patients with severe head injury. J Neurosurg 75:S59–S64
30. McIntyre LA, Fergusson DA, Hutchison JS et al (2006) Effect of a liberal versus restrictive transfusion strategy on mortality in patients with moderate to severe head injury. Neurocrit Care 5:4–9
31. Melamed R, Sprenkle MD, Ulstad VK et al (2009) Assessment of left ventricular function by intensivists using hand-held echocardiography. Chest 135:1416–1420
32. Murray GD, Butcher I, McHugh GS et al (2007) Multivariable prognostic analysis in traumatic brain injury: results from the IMPACT study. J Neurotrauma 24:329–337
33. Myburgh J, Cooper DJ, Finfer S et al (2007) Saline or albumin for fluid resuscitation in patients with traumatic brain injury. N Engl J Med 357:874–884
34. Nguyen H, Zaroff JG (2009) Neurogenic stunned myocardium. Curr Neurol Neurosci Rep 9: 486–491
35. Ract C, Vigue B (2001) Comparison of the cerebral effects of dopamine and norepinephrine in severely head-injured patients. Intensive Care Med 27:101–106
36. Ract C, Vigue B, Bodjarian N et al (2001) Comparison of dopamine and norepinephrine after traumatic brain injury and hypoxic-hypotensive insult. J Neurotrauma 18:1247–1254
37. Raghunathan K, McGee WT, Higgins T (2012) Importance of intravenous fluid dose and composition in surgical ICU patients. Curr Opin Crit Care 18:350–357
38. Rizoli SB, Rotstein OD, Parodo J et al (2000) Hypertonic inhibition of exocytosis in neutrophils: central role for osmotic actin skeleton remodeling. Am J Physiol Cell Physiol 279: C619–C633
39. Robertson CS, Valadka AB, Hannay HJ et al (1999) Prevention of secondary ischemic insults after severe head injury. Crit Care Med 27:2086–2095
40. Romero-Bermejo FJ, Ruiz-Bailen M, Guerrero-De-Mier M et al (2011) Echocardiographic hemodynamic monitoring in the critically ill patient. Curr Cardiol Rev 7:146–156
41. Saugel B, Kirsche SV, Hapfelmeier A et al (2013) Prediction of fluid responsiveness in patients admitted to the medical intensive care unit. J Crit Care 28:537.e1–537.e9
42. Saugel B, Ringmaier S, Holzapfel K et al (2011) Physical examination, central venous pressure, and chest radiography for the prediction of transpulmonary thermodilution-derived hemodynamic parameters in critically ill patients: a prospective trial. J Crit Care 26:402–410

43. Sena MJ, Rivers RM, Muizelaar JP et al (2009) Transfusion practices for acute traumatic brain injury: a survey of physicians at US trauma centers. Intensive Care Med 35:480–488
44. Smith MJ, Stiefel MF, Magge S et al (2005) Packed red blood cell transfusion increases local cerebral oxygenation. Crit Care Med 33:1104–1108
45. Sookplung P, Siriussawakul A, Malakouti A et al (2011) Vasopressor use and effect on blood pressure after severe adult traumatic brain injury. Neurocrit Care 15:46–54
46. Steiner LA, Johnston AJ, Czosnyka M et al (2004) Direct comparison of cerebrovascular effects of norepinephrine and dopamine in head-injured patients. Crit Care Med 32:1049–1054
47. Stubbe HD, Greiner C, Westphal M et al (2006) Cerebral response to norepinephrine compared with fluid resuscitation in ovine traumatic brain injury and systemic inflammation. Crit Care Med 34:2651–2657
48. Suehiro K, Rinka H, Ishikawa J et al (2012) Stroke volume variation as a predictor of fluid responsiveness in patients undergoing airway pressure release ventilation. Anaesth Intensive Care 40:767–772
49. Torella F, Haynes SL, McCollum CN (2003) Cerebral and peripheral oxygen saturation during red cell transfusion. J Surg Res 110:217–221
50. Utter GH, Shahlaie K, Zwienenberg-Lee M et al (2011) Anemia in the setting of traumatic brain injury: the arguments for and against liberal transfusion. J Neurotrauma 28:155–165
51. van den Elsen MJ, Leenen LP, Kesecioglu J (2010) Hemodynamic support of the trauma patient. Curr Opin Anaesthesiol 23:269–275
52. Weiskopf RB, Kramer JH, Viele M et al (2000) Acute severe isovolemic anemia impairs cognitive function and memory in humans. Anesthesiology 92:1646–1652

Chapter 23
Cerebrovascular Autoregulation and Monitoring of Cerebrovascular Reactivity

Philip M. Lewis, Marek Czosnyka, Piotr Smielewski, and John D. Pickard

Abstract Cerebrovascular autoregulation and reactivity are two important processes which maintain CBF at metabolically appropriate levels in response to fluctuations in cerebral perfusion pressure. Additionally, intact vascular reactivity protects the cerebral capillary bed against excessive hydrostatic pressures that may precipitate vasogenic oedema. The importance of vascular reactivity is amplified in the acute phase of injury by disruption to the integrity of the blood–brain barrier (BBB), known to occur even in mild cases of TBI. Monitoring of pressure reactivity in combination with measures of cerebrospinal compensatory reserve may assist in identifying those patients at risk for the development of vasodilatory, short-term increases in ICP. The state of pressure reactivity is also a robust, independent predictor of outcome after TBI. This chapter discusses CBF autoregulation, the careful regulation of vascular resistance, and the measurement of pressure reactivity in the cerebral vasculature.

23.1 Introduction

If one reviews the salient literature on the topic, it becomes clear that the term 'cerebral autoregulation' has many meanings. It may be used to describe the mechanisms responsible for maintenance of cerebral blood flow (CBF) at levels appropriate to the metabolic requirements of brain tissue [1]. It may also refer to the ability of the cerebral vasculature to protect the brain against hypotensive perfusion deficit or conversely hypertensive hyperaemia and cerebral swelling [2]. The more common definition is the maintenance of CBF despite fluctuations in cerebral perfusion pressure (CPP) [3], defined as the difference between the mean arterial (MAP) and intracranial pressures (ICP) [1, 4].

J.D. Pickard (✉)
Academic Neurosurgery Division, University of Cambridge, Cambridge, UK
e-mail: prof.jdp@medschl.cam.ac.uk

E.H. Lo et al. (eds.), *Vascular Mechanisms in CNS Trauma*, Springer Series in Translational Stroke Research 5, DOI 10.1007/978-1-4614-8690-9_23,

This definition of autoregulation has its origins in the seminal studies undertaken by early twentieth-century cerebrovascular researchers, who made extensive studies on the behaviour of the cerebral circulation in response to altered CPP. Techniques for global CBF measurement were unavailable at the time, with typical methods for CBF estimation including connecting extracorporeal circulatory systems that perfused the brain via the carotid [5] or vertebral [6] arteries, to which direct flow-measurement devices could be attached. A common technique was to study the behaviour of the cerebral circulation directly through a glass window implanted into the skull, a method first described in 1811 [7]. Improvements to this method, described by Forbes in 1928 [8], permitted more detailed investigations into the varied biochemical, neural and mechanical influences on CBF and particularly vessel diameter in vivo [9–15]. Relative changes in CBF were simply estimated by directly observing the movement of red blood cells under the microscope, during manipulations of systemic blood pressure or ICP [8, 13].

By the mid-1940s, there was a push to develop CBF measurement techniques that could be used in conscious humans [16]. Such a method was published by Seymour Kety and Carl Schmidt in 1945, based on a modified Fick principle and using nitrous oxide as the inert, diffusible gas [17]. This greatly expanded the opportunities for research into the physiology of CBF in health and disease, and by 1959 over 200 clinical articles on the topic had been published [1]. In his comprehensive review on CBF the same year, Lassen plotted the average CBF values of 11 human studies covering a wide range of blood pressures that were experimentally or pathologically induced [1]. This plot revealed that rather than being passively dependent on MAP, CBF was stable over a wide range of perfusion pressures [1]. Lassen's observations validated those previously made by Fog and Forbes who, using the cranial window method, observed changes in pial arteriolar calibres during hypotension that were consistent with a 'tonic regulation' [18] of CBF or constituted an 'important safety device' [13]. Lassen coined the term 'autoregulation' to describe the phenomenon, a term that stands today.

Modern imaging techniques have now established that CBF, averaged across the whole brain, is approximately 50 mL/100 g/min [19]. This is heterogeneously distributed throughout white and grey matter, with grey matter receiving approximately four times the blood flow of white matter [19]. Autoregulation is now defined as the maintenance of stable CBF across a wide range of perfusion pressures, ranging from 50 to 110 mmHg (Fig. 23.1). The upper limit, not visible on Lassen's plot and beyond which CBF increases rapidly in a pressure-passive manner, derives from the forced dilatation of cerebral vessels that is observed during stepwise increases in blood pressure to very high levels [11]. In chronic hypertension, adaptive vessel wall remodelling confers an ability to withstand higher transmural pressures before forced dilatation occurs [20]. This phenomenon explains the lack of an upper limit on Lassen's plot; all three data points in the upper pressure range were derived from studies of essential hypertension [1].

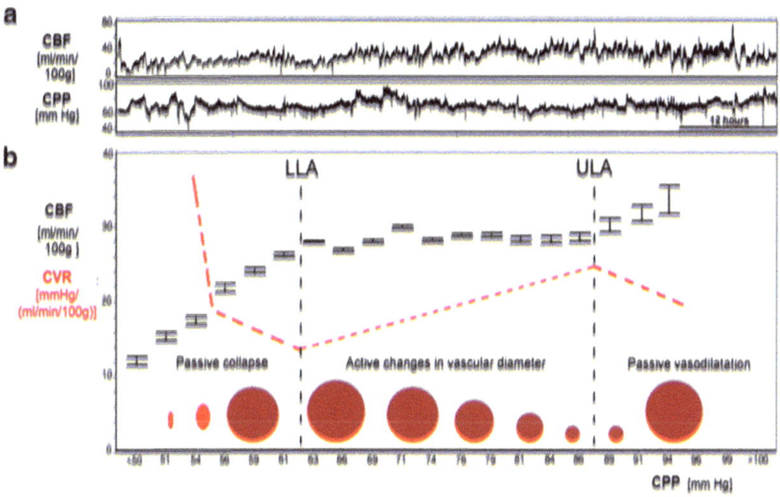

Fig. 23.1 (**a**) Cerebral perfusion pressure (CPP) and parenchymal cerebral blood flow (CBF), monitored over 3 days in a patient after head injury. (**b**): Error-bar plot of CBF averaged within approximately 2.5 mmHg CPP bins, illustrating Lassen's classic autoregulatory curve with clearly visible lower (LLA) and upper (ULA) breakpoints of autoregulation. Theoretical changes in the lumen of cerebral resistance arterioles are depicted as red circles, together with changes in cerebrovascular resistance (CVR, dashed line). Below the LLA, a sharp rise of CVR reflects the critical collapse of the arterial bed

23.2 Regulation of CBF

As shown by Cushing in 1902, when the brain is faced with critically low blood supply, systemic blood pressure is elevated in an attempt to restore perfusion [21]. While this mechanism is only brought into effect when perfusion pressure is critically low, the stability of CBF is maintained continuously across a wide range of perfusion pressures. This maintenance is achieved by the careful regulation of vascular resistance, itself largely a product of the radius of the cerebral vessels. Importantly, this relationship is commonly modelled using Poiseuille's law, for which the fundamental assumptions of laminar flow in straight tubes and an ideal Newtonian fluid are violated by blood flow through the cerebral vasculature. Nonetheless, the term for the radius of the vessel is raised to the fourth power; therefore, vascular resistance is exquisitely sensitive to changes in this parameter, and the brain has evolved numerous mechanisms for its regulation.

23.3 The Regulation of Cerebral Resistance Vessel Diameter

The regulation of vascular diameter is the product of a complex interplay between the myogenic, endothelial, neuronal [22, 23] and biochemical influences on myogenic tone. The myogenic component derives from the stretch-sensitive behaviour of cerebral vessels, a feature that is generally intrinsic to vascular smooth muscle throughout the body [24]. The cerebral endothelium exerts a profound influence over myogenic tone, with both pressure- and flow-dependent responses [25–27]. The cerebral vessels are also innervated by sympathetic and parasympathetic nerve fibres that project from upper cervical ganglia, terminating at the level of the pial arterioles and modulating myogenic tone [28]. Upon penetration into the brain substance, cerebral vessels receive input from intrinsic neurons and astrocytes that mediate the response of the cerebral circulation to increased metabolic demand [22, 28]. Finally the cerebral circulation is also profoundly influenced by alterations in the concentration of arterial carbon dioxide and oxygen [15, 29], which themselves may act via the influence of nitric oxide, a potent vasodilator [30]. All of the aforementioned mechanisms act via innumerable chemical messaging systems that it would be well beyond the scope of this chapter to discuss. The reader is directed to reviews by Attwell [22], Davis [24] and Toda [30] for further detail.

The myogenic or pressure-sensitive behaviour of cerebral vessels is not only central to autoregulation of CBF, but as pointed out earlier, it is intrinsic to vascular smooth muscle and therefore independent of other influences on vessel tone. The first evidence of this was given by William Bayliss at the turn of the twentieth century, when he filled an excised canine carotid artery with pressurised blood and observed a pronounced constriction [31]. The cranial window method employed by Forbes and Fog in the 1920s and 1930s enabled this observation to be extended to the cerebral circulation in vivo, after which in vitro investigations [27, 32–34] confirmed the independent pressure sensitivity of cerebral vessels across multiple species including humans. The term usually given to describe this pressure-responsiveness of cerebral vessels is 'cerebrovascular pressure reactivity' [35].

23.4 Autoregulation Versus Pressure Reactivity

The range of pressures over which cerebral vessels remain reactive extends beyond the classical limits of autoregulation of CBF. Numerous studies support this assertion, beginning with the early observations of Fog [36], who recorded the diameter of a 40 μm feline pial arteriole during a drop in MAP from 120 to 0 mmHg using the cranial window method. He observed that dilatation in this small pial arteriole commenced at a MAP of 60 mmHg and peaked at 20 mmHg, well outside the lower limit of CBF autoregulation for cats [37]. Later studies on cat and human arterioles [11, 33, 37] confirmed that during hypotension, vascular reactivity remains preserved beyond the pressure limits of stable CBF. Therefore, pressure reactivity and autoregulation, while related, are distinct phenomena.

23.5 Clinical Measurements of Autoregulation and Pressure Reactivity

Studies of cerebral vessel pressure reactivity demonstrate distinct patterns of dynamic and steady-state responses to transmural pressure change in vivo and in vitro [33, 34, 36]. Similarly, static and dynamic autoregulatory responses to blood pressure alterations can be seen during concurrent measurements in CBF. Measurements of global CBF can be performed using computed tomography (CT), positron emission tomography (PET) and magnetic resonance imaging (MRI); however, these techniques have poor temporal resolution rendering them of limited value in studying the dynamic components of autoregulation [19]. Techniques offering high temporal resolution include transcranial Doppler ultrasound (TCD) [3, 38, 39], near-infrared spectroscopy (NIRS) [40–42] and laser Doppler flowmetry [19, 43, 44] that afford more detailed investigation of both the time- and frequency-domain components of autoregulation.

One of the more popular tools for indirect measurement of CBF is TCD. Aaslid first introduced TCD in the early 1980s [45], with applications in subarachnoid haemorrhage, stroke and head injury management following shortly thereafter. TCD exploits the Doppler principle to permit recording of cerebral blood flow velocity (CBFV), with measurements most often being made from the middle cerebral artery (MCA). Two well-studied techniques for dynamic autoregulation assessment using TCD include the leg-cuff deflation test [3, 46] and transient hyperaemic response test (THRR) [47, 48]. These methods produce a sudden decrease in systemic blood pressure in the case of the leg-cuff or distal to a compressed carotid compression for the THRR. The strength of autoregulation can be gauged in numerous ways. Aaslid described the rate of regulation (RoR) by measuring the rate of change of cerebral vascular resistance (CVR) in response to a step change in blood pressure induced by leg-cuff release [3] (Fig. 23.2):

$$\mathrm{RoR} = (\Delta CVR/\Delta t) \div \Delta \mathrm{ABP}, \text{where } \mathrm{CVR} = \mathrm{ABP}/\mathrm{CBFV} \tag{23.1}$$

For the THRR (Fig. 23.3), the strength of autoregulation is computed as the ratio of the maximum post-release CBFV to the baseline value. This is referred to as the transient hyperaemic response ratio, or THRR [49]:

$$\mathrm{THRR} = \mathrm{CBFV}_{\mathrm{post}}/\mathrm{CBFV}_{\mathrm{pre}} \tag{23.2}$$

While autoregulation assessment using these methods is useful, they are interventional studies, requiring external manipulation of blood pressure. Moreover, in clinical settings where regular evaluation of autoregulatory capacity may be

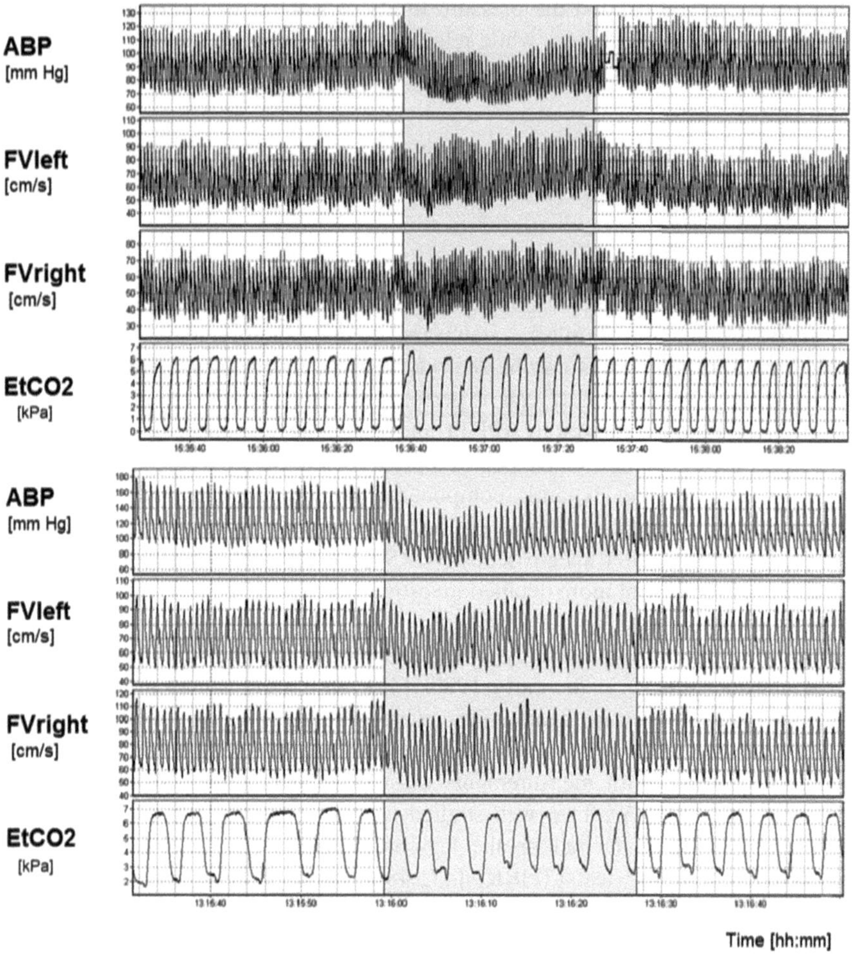

Fig. 23.2 Leg-cuff tests performed in healthy volunteers at two levels of end-tidal CO_2 ($EtCO_2$): 6 (*upper*) and 7 (*lower*) kiloPascals (kPa). The dependence of transcranial Doppler ultrasound-derived flow velocity in both the left (FVleft) and right (FVright) middle cerebral arteries on arterial blood pressure (ABP, obtained noninvasively with a Finapres monitor) is clearly visible at the higher $EtCO_2$ level (*lower plot*). Illustration reproduced from a student project conducted at the University of Cambridge in 1998. We thank Sarah, Xin and Andrew for their original contribution to this work

advantageous, repeated measurements using these techniques may be impractical. Alternatives to such manoeuvres chiefly rely on the presence of naturally occurring changes in CPP as autoregulatory stimuli. A method exploiting such intrinsic CPP fluctuations was described by Czosnyka et al. in 1996 [39], in which the averaged values of continuously recorded CBFV in the MCA were correlated with averaged values of CPP. This index of autoregulation, computed over 3-min periods and

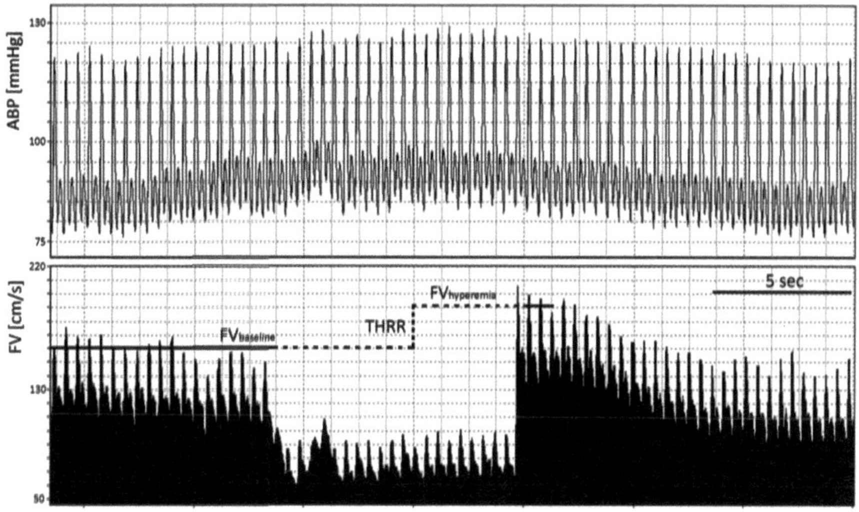

Fig. 23.3 Transient hyperaemic response test performed in a patient after subarachnoid haemorrhage. During a brief compression of the common carotid artery, flow velocity (FV) in the MCA decreases, with a brief elevation to hyperaemic levels after release. No change in arterial blood pressure (ABP) is seen that could confound the interpretation of the post-compression FV drop. The transient hyperaemic response ratio (THRR), being a ratio of post-compression hyperaemic FV to pre-compression baseline FV, is an index of cerebral autoregulation. We thank Mr. Karol Budohoski for granting permission to reproduce this graphic

termed 'mean index', abbreviated to Mx, has been validated against the RoR, correlating significantly (Fig. 23.4) [50].

23.6 Monitoring of Cerebrovascular Pressure Reactivity

Transcranial Doppler-based methods of autoregulation assessment, while reproducible and reliable, necessitate the availability of skilled operators and are generally impractical to perform continuously for periods longer than an hour or so.

As described previously, quantifying pressure-mediated changes in cerebral vessel diameter can give some insights into the state of autoregulation. Interpretation of the behaviour of the vascular bed must be interpreted carefully however, as autoregulation and vascular reactivity are not equivalent; autoregulation and therefore the stability of CBF may be impaired despite preserved vascular reactivity. Moreover, in vivo measurements of pial arteriolar diameter are difficult to obtain in humans. However, indirect evidence for pressure-mediated fluctuations in vascular diameter affecting cerebral blood volume is available in recordings of ICP [51].

Steady-state ICP is derived from a complex interaction between arterial and cerebral venous pressures, myogenic tone of the cerebral arterial bed, the dynamics of cerebrospinal fluid volume maintenance and circulation, and changes in the

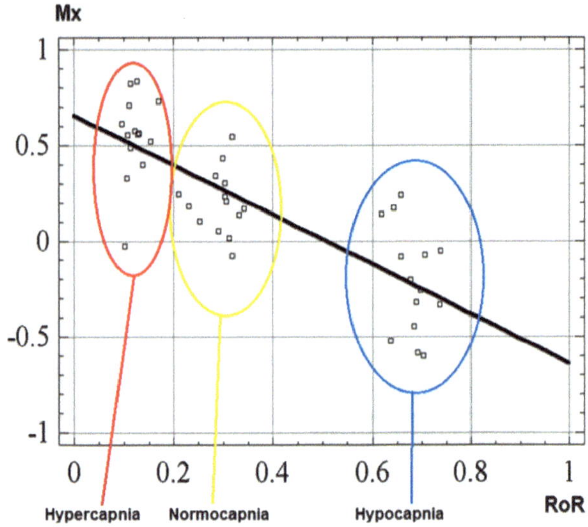

Fig. 23.4 A plot of mean index (Mx), an index of dynamic autoregulation, against the rate of autoregulation (RoR), calculated following the release of a leg-cuff at three different levels of end-tidal carbon dioxide ($EtCO_2$). The plot shows a clear trend towards deterioration of autoregulation with hypercapnia. Illustration reproduced from a student project conducted at the University of Cambridge in 1998. We thank Sarah, Xin and Andrew for their original contribution to this work

volume of brain parenchyma [52]. Over the time period consistent with blood pressure-mediated vasoregulation, the resultant changes in cerebral blood volume are reflected in the recordings of ICP (Fig. 23.5). This is evidenced by a variety of studies in which experimentally induced or naturally occurring alterations in pial vessel diameter correlate with changes in ICP [8, 14, 51, 53, 54]. Importantly, however, pial vessel diameter may not always reflect cerebral blood volume. As discussed previously, pial vessels are innervated by sympathetic and parasympathetic fibres that exert a modulating influence on vascular tone [23, 28]. However, this peripheral innervation is lost upon entry of the vessel into the brain substance, whereupon the vessels receive input directly from the brain itself [28]. Illustrating the importance of this concept, Gotoh et al. studied the influence of sympathetic stimulation on pial diameter and cerebral blood volume in the cat, demonstrating a dissociation between the two after approximately 3 min [55]. This was due to a period of secondary compensatory dilation or 'escape', likely reflecting a dominating metabolic influence on intraparenchymal vessels, an observation also made by others [9].

23.7 Techniques

Several methods for evaluating the reactivity of the vascular bed using ICP have been proposed previously. Daley et al. compared the ABP autocorrelation and ABP/ICP cross-correlation in 8 pigs before and after induction of severe hypercapnia, which produces failure of cerebrovascular reactivity by maximally dilating the cerebral vessels [56]. When reactivity was intact, the two cross-correlation

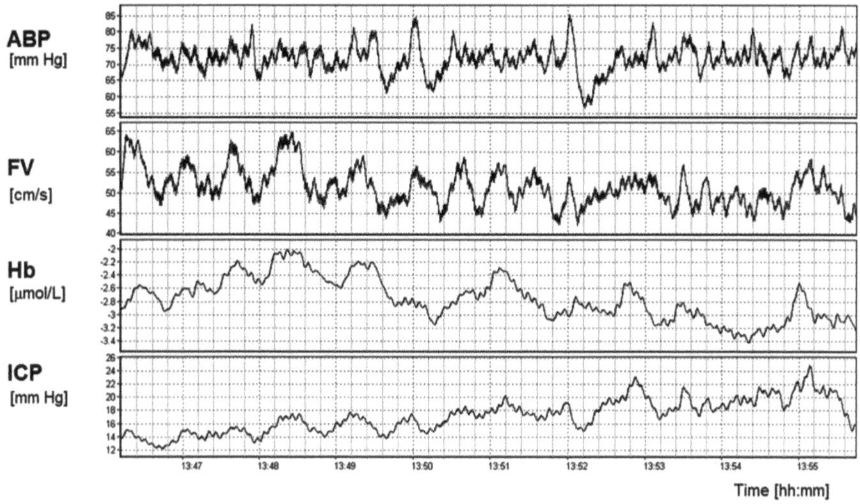

Fig. 23.5 Slow waves in intracranial pressure (ICP) and deoxygenated haemoglobin (Hb), measured using near-infrared spectroscopy (NIRS), are visibly coherent in this recording from a head-injured patient undergoing a lumbar infusion study. Faster oscillations appear in the arterial blood pressure (ABP) signal, with the transcranial Doppler flow velocity (FV) recording displaying components derived from both ABP and ICP/Hb. Thanks go to Dr. Zofia Czosnyka for the permission to reproduce this graphic

functions were dissimilar, with the ABP/ICP cross-correlation clearly showing a respiratory-derived oscillatory component. When reactivity was disturbed, this dissimilarity was lost, and the ICP and ABP waveforms became more coherent. Thus, pressure reactivity can be quantified by computing the least-squares distance between the ABP autocorrelation and ABP/ICP cross-correlation as a measure of their similarity. In a different approach, Steinmeier et al. recorded ABP and ICP in 16 adults with TBI, analysing the ABP/ICP cross-correlation after low-pass filtering with a cutoff frequency of 0.1 Hz to exclude respiratory and cardiac frequencies [57]. The authors observed a negative correlation and mean time delay of 6.89 s in 65/80 analyses in which stable ABP/ICP cross-correlations could be recorded (from a total of 122 recordings of ABP and ICP). This was taken to be indicative of active cerebrovascular reactivity. In the remaining 19/80 recordings, a short time delay (mean 0.62 s) and positive correlation was taken to indicate failed vascular reactivity. Another method described by Howells et al. [58] involves plotting hourly averages of ABP and ICP, with the authors quantifying pressure reactivity by computing the slope of the regression line between the two variables. Confounding the interpretation of the ABP/ICP regression slope as a measure of pressure reactivity is the fact that its value is not only influenced by the direction of ICP change in response to ABP fluctuations but also its magnitude; the second component is heavily influenced by intracranial compliance.

One method for continuously assessing cerebrovascular reactivity that has gained some traction in the research community was proposed in 1997 [59]. Similar to the method of Steinmeier et al., the authors described vascular reactivity as the correlation between slow waves (<0.1 Hz) in ABP and ICP, excluding both respiratory and cardiac components. The new index was called the 'pressure-reactivity index', which is abbreviated to PRx.

The precise method for computing PRx has evolved over time since its original publication. A typical method for computing the index is as follows:

- Prefilter the continuously recorded arterial and intracranial pressure signals to remove pulsatile and respiratory components. This can be achieved either using a simple moving average filter of approximately 10-s length or a higher-performance digital filter with a cutoff frequency of 0.05 Hz.
- Compute the Pearson's correlation coefficient between 30 × 10-s samples (5-min window) of filtered ABP and ICP data.
- Repeat this process every sample (i.e. every 10 s).

The computational requirements for calculating PRx are relatively minor; data collection and index calculations in the original PRx paper were performed using Amstrad 386 PCs [59]. Current-generation personal computer technology possesses substantially more computational power; therefore, this is no barrier to continuous vascular reactivity monitoring. Data capture can however be of concern. The neurointensive care environment is a complex milieu of medical technology, and methods of accessing raw data from monitoring systems can be highly variable. Capture software is often available from the vendors of clinical monitoring systems, and the data feed may be subsequently analysed using custom software developed using packages such as Matlab (Mathworks, Natick, Massachusetts, USA) or LabView (National Instruments Corporation, Austin, Texas, USA). Specific solutions combining multisystem data capture data processing with a focus on the monitoring of cerebral hemodynamic variables have been written [60], and such software has found application in the monitoring of cerebrovascular reactivity in particular [61].

23.8 Interpretation

The Pearson's correlation coefficient describes the goodness of fit of two variables on a linear regression line. The value varies from +1 to −1, with +1 indicating a positive, perfectly linear relationship between the two variables. A value of −1 therefore indicates a negative, perfectly linear relationship, with a correlation of 0 indicating statistically that the two parameters are independent of each other. Extending this logic to the correlation between ABP and ICP, a PRx of +1 signifies a complete dependence of ICP on ABP (in the absence of a Cushing response to cerebral hypoperfusion, in which case ABP is largely dependent on ICP) and complete absence of pressure reactivity. A PRx of zero or less reflects a gradually increasing phase shift between ABP and ICP, indicating intact pressure reactivity [62]. PRx has been carefully validated against PET-derived estimates of static [63] and dynamic [35] autoregulation in humans as well as laser Doppler flowmetry-based estimates of the lower limit of autoregulation in a piglet model [64].

23.9 Clinical Relevance of Autoregulation and Vascular Reactivity Impairment in Traumatic Brain Injury

23.9.1 Pathophysiology

Injury to the brain has been known to disrupt the regulation of CBF for over 50 years, confirmed using a variety of measurement techniques, across multiple species and injury severities [39, 65–69]. As alluded to previously, autoregulation not only serves to maintain CBF at metabolically appropriate levels in response to changing CPP, but intact vascular reactivity also protects the cerebral capillary bed against excessive hydrostatic pressures that may precipitate vasogenic oedema [70]. The importance of vascular reactivity is amplified in the acute phase of injury by disruption to the integrity of the blood–brain barrier (BBB), known to occur even in mild cases of TBI [71, 72]. Clearly illustrating the link between BBB disruption and brain swelling, Durward et al. injected hypertonic saline and Evans blue dye (EB) into the left carotid of rabbits, inducing both autoregulatory failure and BBB disruption [73]. The authors demonstrated extravasation of EB across a wide range of controlled CPP values (20–130 mmHg), with a moderate but significant correlation between the amount of extravasation and systemic blood pressure. These findings are supported by later studies demonstrating an 8 mmHg ICP rise after BBB disruption in the cat, despite no alteration in systemic blood pressure [74]. Plainly, a disrupted BBB in the setting of lost vascular reactivity can precipitate brain swelling, and this idea forms the basis of the Lund concept of ICP management in head injury [75]. However, the relative contribution of vasogenic versus cytotoxic oedema to the genesis of sustained intracranial hypertension is the subject of some debate. In experimental studies, traumatic BBB opening appears to be a transient phenomenon, although it may be prolonged if hypoxia or hypotension occurs immediately after injury [76, 77]. In a recent clinical study, however, MRI data suggested that vasogenic oedema may contribute less to ongoing brain swelling than cytotoxic oedema [78], although Marmarou et al. do not comment on the role of vascular reactivity impairment in this process.

Casting aside the role of a passive vascular bed in the genesis of cerebral oedema, the contribution of both active and passive blood volume fluctuations in the short-term regulation of ICP remains clinically important. A pressure-passive vascular bed responds to increasing blood pressure with vascular distension, an increase in cerebral blood volume and subsequently elevated ICP [79–82]. Importantly, the magnitude of ICP increase is dependent on intracranial compliance, itself partially dependent on the state of pressure reactivity [82, 83]. Conversely, loss of pressure reactivity renders the brain at risk for ischemia during fluctuations in CPP that may be within normal limits [84].

While vascular reactivity impairment is of obvious significance, it should be reinforced that preserved vascular reactivity is of clinical significance also, particularly for the management of ICP. A pressure-reactive vascular bed will respond to hypotension with vasodilatation, with serious consequences for ICP in the setting of a reduced cerebrospinal compensatory reserve. This concept is illustrated most effectively by the phenomenon of the ICP 'plateau wave', a term first coined by Lundberg in 1960 [85]. Plateau waves typically result from a brief reduction in CPP, sufficient to provoke a vascular response that itself reduces CPP further by increasing ICP (Fig. 23.6). The positive-feedback loop resulting in the sudden escalation of ICP to supranormal levels was described as a 'vasodilatory cascade' by Rosner [86] and is therefore largely indicative of intact pressure reactivity, as discussed by Castellani et al. [87]. While the occurrence of plateau waves has not been found to negatively impact outcome in TBI unless they become prolonged (>1/2 h duration) [87], their occurrence often provokes an emergent clinical response. Therefore, monitoring of pressure reactivity in combination with measures of cerebrospinal compensatory reserve [88] or even B-wave amplitude [51] may assist in identifying those patients at risk for the development of such vasodilatory, short-term increases in ICP.

23.9.2 Clinical Implications

The value of monitoring pressure reactivity, in particular the PRx, extends beyond its ability to provide insights into the pathophysiology of brain injury. The state of pressure reactivity is a robust, independent predictor of outcome after TBI, as shown in numerous studies of both adult [35, 58, 59, 89–96] and paediatric [97] head injury. In one study, the strength and statistical significance of the outcome correlation notably exceeded that of ICP [59]. Importantly, while PRx remains a strong predictor of outcome in most studies, this is not an unequivocal finding; it may be inferior to brain tissue oxygenation and direct monitoring of cerebral metabolites, although there is little data on this topic [98]. On the other hand, PRx is a global estimate of reactivity, while tissue-derived measures of oxygenation and metabolic state are highly focal. Thus, their clinical interpretation is directly tied to the proximity of the tissue probes to injured brain.

Outcome associations may depend on the length of time over which PRx is to be recorded; in one study comparing PRx and Mx, recordings were limited to approximately 1 h and were performed only once daily [35]. In this study, PRx was unable to discriminate between favourable and unfavourable outcomes unlike in previous reports [92]. In contrast to TCD-based indices of autoregulation, PRx can be measured continuously suggesting it offers the potential to guide clinical management. This view is reinforced by the observation that not only averaged but also temporal patterns of cerebrovascular reactivity impairment after TBI are of clinical

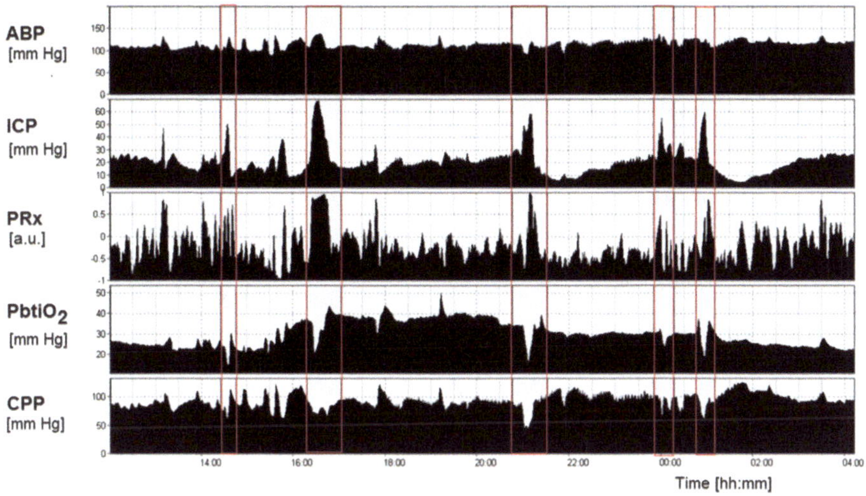

Fig. 23.6 Intracranial pressure (ICP) plateau waves (bounded by *red lines*) recorded in a patient with traumatic brain injury (TBI). Increases in ICP occur concurrently with decreases in cerebral perfusion pressure (CPP), decreases in brain tissue oxygenation ($PbtiO_2$) and increases in PRx. These changes are indicative of disturbances in pressure reactivity associated with maximal vasodilatation during plateau waves

significance; both early (<72 h) [35, 59, 94] and late [89] impairments of pressure reactivity have been shown to be predictive of either unfavourable outcome or death.

23.9.3 Dependence of PRx on CPP

Continuous monitoring of pressure reactivity reveals a dynamic dependence on perfusion pressure [59, 90, 97, 99]. Illustrating the relationship, a plot of PRx against CPP often demonstrates a U shape, implying that a region of CPP exists within which pressure reactivity is optimised (Fig. 23.7). Importantly, the location of this region is not fixed and may exist in a very narrow range of CPP values, illustrating both the variability and individuality of pressure reactivity disturbances after head injury [97, 99]. Moreover, in some patients, there may be no optimal CPP (CPP_{opt}) at all; in one study, only 60 % of head-injured patients exhibited the classical U-shaped dependence of PRx on CPP. PRx may also demonstrate a monotonically increasing or decreasing pattern of change with CPP or indeed may display a U shape that is inverted [99]. Importantly, a lack of CPP_{opt} may simply reflect the shifting of pressure reactivity to CPP values outside the clinically acceptable range.

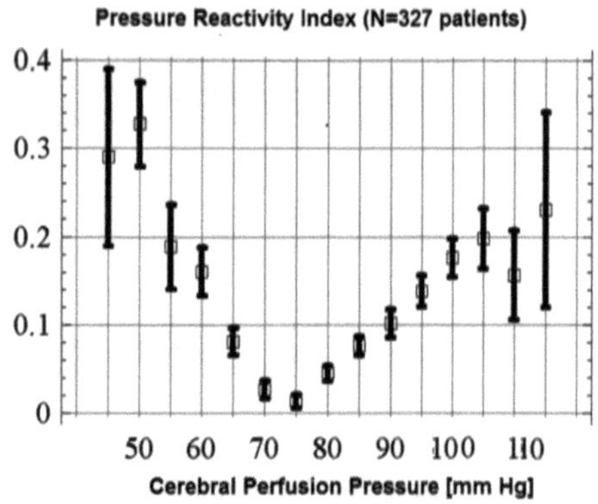

Fig. 23.7 The distribution of pressure-reactivity index (PRx) versus different levels of CPP in 327 patients after TBI shows a characteristic U-shaped curve. The minimum value of PRx, seen here at 75 mmHg, indicates the averaged 'optimal CPP' in this group of patients. Thanks to Dr. Marcel Aries for the permission to reproduce this figure

23.9.4 Individualising CPP-Targeted Therapy

The dependence of PRx on CPP implies that individualisation of perfusion pressure may both be possible using pressure reactivity as a guide and necessary to ensure that patients are maintained within their optimal range. Previous work has shown that TBI patients for whom CPP-directed management (CPP target ≥70 mmHg) resulted in a CPP within their 'optimal range' had improved outcome [99]. Importantly, the correlation between outcome (GOS) and the absolute difference between average CPP over the monitoring period and CPP_{opt} was moderate and highly significant, implying that both upper and lower limits of CPP could be incorporated into clinical management protocols. The period of data collection necessary to enable CPP_{opt} calculation is approximately 6–10 h.

23.9.5 Threshold Values in Outcome Correlations

The identification of critical thresholds in pressure reactivity monitoring is an ongoing process and is subject to change as new, larger datasets become available for analysis. A threshold PRx value for discriminating between favourable versus unfavourable outcomes was initially identified as PRx of 0.2 [59]; however, a recent study by the same group using a much larger dataset ($n = 459$) revised this threshold down to 0–0.05 [92]. Moreover, this study identified differences in PRx thresholds for discriminating between favourable/unfavourable and survivor/non-survivor groups, with the latter being set at 0.25. Based on this data, where possible, determination of optimal CPP should be based on the value of CPP at which PRx is <0.05. Well-conducted, prospective trials on CPP_{opt} should be undertaken to establish the feasibility of applying this concept in the critical care of TBI patients.

23.10 Conclusions and Cautions

Despite the growing interest in pressure reactivity monitoring and an evolving understanding of its implications for the management of patients with TBI, there remain questions to address and concerns to raise about its measurement [100]. ICP can be significantly altered by both carbon dioxide and brain metabolism on a timescale similar to that of the pressure-mediated changes that form the basis for reactivity monitoring. While this is important for monitoring in individual cases, it does not diminish the weight of evidence in support of the prognostic role for PRx in the general TBI population. Moreover, arterial carbon dioxide levels are usually well controlled by careful adjustment of ventilation, although periods of sudden end-tidal carbon dioxide fluctuation can be seen during suctioning or postural changes.

The signal-to-noise of PRx is typically quite poor. This can be improved by lengthy signal averaging (up to 1 h or more); however, this is at the expense of information on the genuine short-term variations in reactivity that occur throughout the monitoring period. Poor signal-to-noise can occur due to a low amplitude of either ABP or ICP oscillations. In the former, this is an unavoidable consequence of tight CPP control; however, incidental fluctuations in ABP are almost always present in the ICU patient, although these are often high frequency and will be filtered out by the PRx algorithm. On the other hand, a low amplitude of ICP oscillations is generally reflective of good intracranial compliance and may be an unavoidable circumstance in the generally well-managed patient or a patient having undergone decompressive craniectomy for intractable ICP. Confounding this picture even further, a recent study showed that low amplitude of ICP slow waves or a reduction in ICP signal 'complexity' [101] is predictive of poor outcome. Clearly, further research is required to clarify these disparate findings.

Despite the need for further work and an improved understanding of both the pathophysiology of and methods for monitoring pressure reactivity loss, the perceived value of pressure reactivity monitoring in the management of head injury by the general medical community is growing. This is indicated by the inclusion of the statement 'Patients with intact pressure autoregulation tolerate higher CPP values' in Chap. 9 of the most recent revision of the Brain Trauma Foundation guidelines for the management of head injury [102]. This suggests that centres specialising in the management of head injury, for whom ICP monitoring is standard practice, should institute the measurement of pressure reactivity also. Despite the recommendations, however, widespread adoption of novel neuromonitoring techniques is dependent on high levels of evidence obtained from well-conducted, multicentre prospective studies that are largely absent from this research domain. We hope and anticipate that continued effort on the part of neuroscientists and neurointensivists alike will address this deficit.

References

1. Lassen NA (1959) Cerebral blood flow and oxygen consumption in man. Physiol Rev 39 (2):183–238
2. Paulson OB, Strandgaard S, Edvinsson L (1990) Cerebral autoregulation. Cerebrovasc Brain Metab Rev 2(2):161–192
3. Aaslid R et al (1989) Cerebral autoregulation dynamics in humans. Stroke 20(1):45–52
4. Miller JD, Stanek A, Langfitt TW (1972) Concepts of cerebral perfusion pressure and vascular compression during intracranial hypertension. Prog Brain Res 35:411–432
5. Finesinger J, Putnam TJ (1933) Cerebral circulation XXIII—induced variations in volume flow through the brain perfused at constant pressure. Arch Neurol Psychiatr 30(4):775–794
6. Schmidt CF (1928) The influence of cerebral blood flow on respiration I: the respiratory responses to changes in cerebral blood flow. Am J Physiol 84(1):202–222
7. Feinsod M (2010) De Motu Cerebri: the history of the study of brain pulsations. Open Neurosurg J 3:10–16
8. Forbes HS (1928) The cerebral circulation I: observation and measurement of pial vessels. Arch Neurol Psychiatr 19(5):751–761
9. Auer LM, Ishiyama N (1986) Pial vascular behavior during bilateral and contralateral cervical sympathetic stimulation. J Cereb Blood Flow Metab 6(3):298–304
10. Wolff HG, Forbes HS (1928) The cerebral circulation V: observations of the pial circulation during changes in intracranial pressure. Arch Neurol Psychiatr 20(5):1035–1047
11. Kontos HA et al (1978) Responses of cerebral arteries and arterioles to acute hypotension and hypertension. Am J Physiol 234(4):H371–H383
12. Pucher R et al (1991) Cerebrovascular response to changes of cerebral venous pressure and cerebrospinal fluid pressure. Acta Neurochir (Wien) 109(1–2):52–56
13. Forbes HS, Nason GI, Wortman RC (1937) Cerebral circulation: XLIV. Vasodilation in the pia following stimulation of the vagus, aortic and carotid sinus nerves. Arch Neurol Psychiatr 37(2):334–350
14. Sjöstrand T (1948) Brain volume, diameter of the blood-vessels in the pia mater, and intracranial pressure in acute carbon monoxide poisoning. Acta Physiol Scand 15(4):351–361
15. Wei EP, Kontos HA, Patterson JL Jr (1980) Dependence of pial arteriolar response to hypercapnia on vessel size. Am J Physiol 238(5):697–703
16. Traystman RJ (2004) The paper that completely altered our thinking about cerebral blood flow measurement. J Appl Physiol 97(5):1601–1602
17. Kety SS, Schmidt CF (1945) The determination of cerebral blood flow in man by the use of nitrous oxide in low concentrations. Am J Physiol 143:53–66
18. Fog M (1939) Cerebral circulation II: reaction of pial arteries to increase in blood pressure. Arch Neurol Psychiatr 41(2):260–268
19. Dagal A, Lam AM (2011) Cerebral blood flow and the injured brain: how should we monitor and manipulate it? Curr Opin Anaesthesiol 24(2):131–137
20. Strandgaard S et al (1975) Upper limit of cerebral blood flow autoregulation in experimental renovascular hypertension in the baboon. Circ Res 37(2):164–167
21. Cushing HMD (1902) Some experimental and clinical observations concerning states of increased intracranial tension. Am J Med Sci 124(3):375–400
22. Attwell D et al (2010) Glial and neuronal control of brain blood flow. Nature 468 (7321):232–243
23. Zhang R et al (2002) Autonomic neural control of dynamic cerebral autoregulation in humans. Circulation 106(14):1814–1820
24. Davis MJ, Hill MA (1999) Signaling mechanisms underlying the vascular myogenic response. Physiol Rev 79(2):387–423
25. Davies PF, Spaan JA, Krams R (2005) Shear stress biology of the endothelium. Ann Biomed Eng 33(12):1714–1718

26. Faraci FM, Heistad DD (1998) Regulation of the cerebral circulation: role of endothelium and potassium channels. Physiol Rev 78(1):53–97
27. Toth P et al (2011) Isolated human and rat cerebral arteries constrict to increases in flow: role of 20-HETE and TP receptors. J Cereb Blood Flow Metab 31(10):2096–2105
28. Hamel E (2006) Perivascular nerves and the regulation of cerebrovascular tone. J Appl Physiol 100(3):1059–1064
29. Kontos HA, Wei EP (1985) Oxygen-dependent mechanisms in cerebral autoregulation. Ann Biomed Eng 13(3–4):329–334
30. Toda N, Ayajiki K, Okamura T (2009) Cerebral blood flow regulation by nitric oxide: recent advances. Pharmacol Rev 61(1):62–97
31. Bayliss WM (1902) On the local reactions of the arterial wall to changes of internal pressure. J Physiol 28(3):220–231
32. Harder DR (1985) A cellular mechanism for myogenic regulation of cat cerebral arteries. Ann Biomed Eng 13(3–4):335–339
33. Wallis SJ, Firth J, Dunn WR (1996) Pressure-induced myogenic responses in human isolated cerebral resistance arteries. Stroke 27(12):2287–2290; discussion 2291
34. Bevan JA, Hwa JJ (1985) Myogenic tone and cerebral vascular autoregulation: the role of a stretch-dependent mechanism. Ann Biomed Eng 13(3–4):281–286
35. Budohoski KP et al (2012) The relationship between cerebral blood flow autoregulation and cerebrovascular pressure reactivity after traumatic brain injury. Neurosurgery 71 (3):652–660; discussion 660–661
36. Fog M (1937) Cerebral circulation: the reaction of the pial vessels to a fall in blood pressure. Arch Neurol Psychiatr 37(2):351–364
37. MacKenzie ET et al (1979) Effects of hemorrhagic hypotension on the cerebral circulation. I. Cerebral blood flow and pial arteriolar caliber. Stroke 10(6):711–718
38. Giller CA (1990) The frequency-dependent behavior of cerebral autoregulation. Neurosurgery 27(3):362–368
39. Czosnyka M et al (1996) Monitoring of cerebral autoregulation in head-injured patients. Stroke 27(10):1829–1834
40. Elwell CE et al (1993) Measurement of adult cerebral haemodynamics using near infrared spectroscopy. Acta Neurochir Suppl (Wien) 59:74–80
41. Steiner LA et al (2009) Near-infrared spectroscopy can monitor dynamic cerebral autoregulation in adults. Neurocrit Care 10(1):122–128
42. Smielewski P et al (1997) Clinical evaluation of near-infrared spectroscopy for testing cerebrovascular reactivity in patients with carotid artery disease. Stroke 28(2):331–338
43. Tonnesen J et al (2005) Laser Doppler flowmetry is valid for measurement of cerebral blood flow autoregulation lower limit in rats. Exp Physiol 90(3):349–355
44. Czosnyka M et al (1994) Assessment of cerebral autoregulation with ultrasound and laser Doppler wave forms–an experimental study in anesthetized rabbits. Neurosurgery 35 (2):287–292; discussion 292–293
45. Aaslid R, Markwalder TM, Nornes H (1982) Noninvasive transcranial Doppler ultrasound recording of flow velocity in basal cerebral arteries. J Neurosurg 57(6):769–774
46. Hlatky R, Valadka AB, Robertson CS (2006) Analysis of dynamic autoregulation assessed by the cuff deflation method. Neurocrit Care 4(2):127–132
47. Smielewski P et al (1996) Assessment of cerebral autoregulation using carotid artery compression. Stroke 27(12):2197–2203
48. Giller CA (1991) A bedside test for cerebral autoregulation using transcranial Doppler ultrasound. Acta Neurochir (Wien) 108(1–2):7–14
49. Smielewski P et al (1995) Computerised transient hyperaemic response test–a method for the assessment of cerebral autoregulation. Ultrasound Med Biol 21(5):599–611
50. Piechnik SK et al (1999) The continuous assessment of cerebrovascular reactivity: a validation of the method in healthy volunteers. Anesth Analg 89(4):944–949

51. Auer LM, Sayama I (1983) Intracranial pressure oscillations (B-waves) caused by oscillations in cerebrovascular volume. Acta Neurochir (Wien) 68(1–2):93–100
52. Czosnyka M, Pickard JD (2004) Monitoring and interpretation of intracranial pressure. J Neurol Neurosurg Psychiatry 75(6):813–821
53. Wolff HG (1929) The cerebral circulation XIc. The action of Amyl Nitrite. Arch Neurol Psychiatr 22(4):695–699
54. Johansson BB, Auer LM, Trummer UG (1980) Pial vascular reaction to intravenous dihydralazine in the cat. Stroke 11(4):369–371
55. Gotoh F et al (1986) Comparison between pial and intraparenchymal vascular responses to cervical sympathetic stimulation in cats. Part 1. Under normal resting conditions. J Cereb Blood Flow Metab 6(3):342–347
56. Daley ML et al (1995) Detection of loss of cerebral vascular tone by correlation of arterial and intracranial pressure signals. IEEE Trans Biomed Eng 42(4):420–424
57. Steinmeier R et al (1996) Slow rhythmic oscillations of blood pressure, intracranial pressure, microcirculation, and cerebral oxygenation. Dynamic interrelation and time course in humans. Stroke 27(12):2236–2243
58. Howells T et al (2005) Pressure reactivity as a guide in the treatment of cerebral perfusion pressure in patients with brain trauma. J Neurosurg 102(2):311–317
59. Czosnyka M et al (1997) Continuous assessment of the cerebral vasomotor reactivity in head injury. Neurosurgery 41(1):11–17; discussion 17–19
60. Smielewski P et al (2005) ICM+: software for on-line analysis of bedside monitoring data after severe head trauma. Acta Neurochir Suppl 95:43–49
61. Guendling K et al (2006) Use of ICM+ software for on-line analysis of intracranial and arterial pressures in head-injured patients. Acta Neurochir Suppl 96:108–113
62. Czosnyka M (2000) Association between arterial and intracranial pressures. Br J Neurosurg 14(2):127–128
63. Steiner LA et al (2003) Assessment of cerebrovascular autoregulation in head-injured patients: a validation study. Stroke 34(10):2404–2409
64. Brady KM et al (2008) Continuous measurement of autoregulation by spontaneous fluctuations in cerebral perfusion pressure: comparison of 3 methods. Stroke 39(9):2531–2537
65. Enevoldsen EM, Jensen FT (1978) Autoregulation and CO2 responses of cerebral blood flow in patients with acute severe head injury. J Neurosurg 48(5):689–703
66. Bouma GJ, Muizelaar JP (1992) Cerebral blood flow, cerebral blood volume, and cerebrovascular reactivity after severe head injury. J Neurotrauma 9(suppl 1):S333–S348
67. Lewelt W, Jenkins LW, Miller JD (1980) Autoregulation of cerebral blood flow after experimental fluid percussion injury of the brain. J Neurosurg 53(4):500–511
68. Marshall WJ, Jackson JL, Langfitt TW (1969) Brain swelling caused by trauma and arterial hypertension. Hemodynamic aspects. Arch Neurol 21(5):545–553
69. Strebel S et al (1997) Impaired cerebral autoregulation after mild brain injury. Surg Neurol 47 (2):128–131
70. Faraci FM, Heistad DD (1990) Regulation of large cerebral arteries and cerebral microvascular pressure. Circ Res 66(1):8–17
71. Povlishock JT et al (1978) Vascular permeability alterations to horseradish peroxidase in experimental brain injury. Brain Res 153(2):223–239
72. Korn A et al (2005) Focal cortical dysfunction and blood-brain barrier disruption in patients with Postconcussion syndrome. J Clin Neurophysiol 22(1):1–9
73. Durward QJ et al (1983) The influence of systemic arterial pressure and intracranial pressure on the development of cerebral vasogenic edema. J Neurosurg 59(5):803–809
74. Kongstad L, Grande PO (2001) Arterial hypertension increases intracranial pressure in cat after opening of the blood-brain barrier. J Trauma 51(3):490–496
75. Grande PO, Asgeirsson B, Nordstrom CH (2002) Volume-targeted therapy of increased intracranial pressure: the Lund concept unifies surgical and non-surgical treatments. Acta Anaesthesiol Scand 46(8):929–941

76. Barzo P et al (1996) Magnetic resonance imaging-monitored acute blood-brain barrier changes in experimental traumatic brain injury. J Neurosurg 85(6):1113–1121
77. Tanno H et al (1992) Breakdown of the blood-brain barrier after fluid percussive brain injury in the rat. Part 1: distribution and time course of protein extravasation. J Neurotrauma 9 (1):21–32
78. Marmarou A et al (2006) Predominance of cellular edema in traumatic brain swelling in patients with severe head injuries. J Neurosurg 104(5):720–730
79. Bouma GJ et al (1992) Blood pressure and intracranial pressure-volume dynamics in severe head injury: relationship with cerebral blood flow. J Neurosurg 77(1):15–19
80. Figaji AA et al (2009) Pressure autoregulation, intracranial pressure, and brain tissue oxygenation in children with severe traumatic brain injury. J Neurosurg 4(5):420–428
81. Ter Minassian A et al (2002) Changes in intracranial pressure and cerebral autoregulation in patients with severe traumatic brain injury. Crit Care Med 30(7):1616–1622
82. Muizelaar JP et al (1989) Cerebral blood flow and metabolism in severely head-injured children. Part 2: autoregulation. J Neurosurg 71(1):72–76
83. Czosnyka M et al (1996) Significance of intracranial pressure waveform analysis after head injury. Acta Neurochir (Wien) 138(5):531–541; discussion 541–542
84. Lang EW, Chesnut RM (1995) Intracranial pressure and cerebral perfusion pressure in severe head injury. New Horiz 3(3):400–409
85. Lundberg N (1960) Continuous recording and control of ventricular fluid pressure in neurosurgical practice. Acta Psychiatr Scand Suppl 36(149):1–193
86. Rosner MJ (1986) The vasodilatory cascade and intracranial pressure. In: Miller JD et al (eds) Intracranial pressure IV. Springer, Berlin
87. Castellani G et al (2009) Plateau waves in head injured patients requiring neurocritical care. Neurocrit Care 11(2):143–150
88. Czosnyka M et al (1988) Analysis of intracranial pressure waveform during infusion test. Acta Neurochir (Wien) 93(3–4):140–145
89. Bowles AP et al (2012) Implications of neurophysiological parameters in persons with severe brain injury with respect to improved patient outcomes: a retrospective review. Brain Inj 26 (12):1415–1424
90. Zweifel C et al (2008) Continuous monitoring of cerebrovascular pressure reactivity in patients with head injury. Neurosurg Focus 25(4):E2
91. Thorat JD et al (2008) Barbiturate therapy for patients with refractory intracranial hypertension following severe traumatic brain injury: its effects on tissue oxygenation, brain temperature and autoregulation. J Clin Neurosci 15(2):143–148
92. Sorrentino E et al (2012) Critical thresholds for cerebrovascular reactivity after traumatic brain injury. Neurocrit Care 16(2):258–266
93. Lavinio A et al (2008) Cerebrovascular reactivity and autonomic drive following traumatic brain injury. Acta Neurochir Suppl 102:3–7
94. Ang BT et al (2007) Temporal changes in cerebral tissue oxygenation with cerebrovascular pressure reactivity in severe traumatic brain injury. J Neurol Neurosurg Psychiatry 78 (3):298–302
95. Kirkness CJ et al (2001) Cerebral autoregulation and outcome in acute brain injury. Biol Res Nurs 2(3):175–185
96. Eide PK et al (2007) Association between intracranial, arterial pulse pressure amplitudes and cerebral autoregulation in head injury patients. Neurol Res 29(6):578–582
97. Brady KM et al (2009) Continuous monitoring of cerebrovascular pressure reactivity after traumatic brain injury in children. Pediatrics 124(6):e1205–e1212
98. Low D et al (2009) Prediction of outcome utilizing both physiological and biochemical parameters in severe head injury. J Neurotrauma 26(8):1177–1182
99. Steiner LA et al (2002) Continuous monitoring of cerebrovascular pressure reactivity allows determination of optimal cerebral perfusion pressure in patients with traumatic brain injury. Crit Care Med 30(4):733–738

100. Diedler J et al (2012) Critical thresholds for cerebrovascular reactivity: facts, no fiction! Neurocrit Care 17:152–153
101. Lu CW et al (2012) Complexity of intracranial pressure correlates with outcome after traumatic brain injury. Brain 135(pt 8):2399–2408
102. Bratton SL et al (2007) Guidelines for the management of severe traumatic brain injury. IX. Cerebral perfusion thresholds. J Neurotrauma 24(suppl 1):S59–S64

Chapter 24
Cerebrovascular Responses After Pediatric Traumatic Brain Injury

Steven L. Shein, Nikki Miller Ferguson, and Michael J. Bell

Abstract In children, traumatic brain injury (TBI) is the leading cause of death between the ages of one and four. Injury to the brain as a result of TBI results from primary, secondary, and tertiary insults. Neurotrauma care focuses on minimizing secondary insults, with a major focus being the maintenance of adequate cerebral blood flow (CBF). TBI can disrupt the normal flow leading to increased and decreased amounts of CBF throughout the brain. In children, these changes are dependent on the age and stage of development. This chapter discusses the effect of developmental changes on CBF, cerebral perfusion pressure, and autoregulation of CBF in children of different age groups to further investigate an effective means of limiting the secondary effects of TBI.

Traumatic brain injury (TBI) is the leading cause of death and morbidity in children over the age of 1 year [1–4]. There are no proven therapies for improving these dire results, leading most of pediatric neurotrauma care focused on minimizing secondary insults—variously defined as intracranial hypertension, cerebral hypoxia/ischemia, metabolic disturbances (such as hypoglycemia), and secondary infections—and maintaining adequate cerebral perfusion [5]. For proper CNS functioning, maintenance of adequate cerebral blood flow (CBF) for metabolic demands is essential and even more fundamental than the secondary insults listed above.

M.J. Bell (✉)
Departments of Critical Care Medicine, University of Pittsburgh School of Medicine, Pittsburgh, PA, USA

Departments of Neurological Surgery, University of Pittsburgh School of Medicine, Pittsburgh, PA, USA

Departments of Pediatrics, University of Pittsburgh School of Medicine, Pittsburgh, PA, USA

Safar Center for Resuscitation Research, University of Pittsburgh School of Medicine, Pittsburgh, PA, USA
e-mail: bellmj4@upmc.edu

E.H. Lo et al. (eds.), *Vascular Mechanisms in CNS Trauma*, Springer Series in Translational Stroke Research 5, DOI 10.1007/978-1-4614-8690-9_24,

Insufficient CBF for metabolic demands quickly leads to hypoxia from inadequate oxygen and nutrient supply for the brain. And an oversupply of CBF can lead to the exacerbation of cerebral edema and intracranial hypertension. Therefore, understanding the normal cerebrovascular responses during development and the response of the vascular system to TBI is essential for caring for children with severe TBI. This chapter will focus on the cerebrovascular response after TBI, with particular attention to the effect of TBI on CBF. The CBF changes during development and after TBI will be reviewed, and since the cerebrovascular response plays an integral role in TBI-related therapeutic maneuvers, topics such as cerebral perfusion pressure (CPP), cerebral autoregulation and cerebral oxygenation will also be reviewed.

24.1 CBF and Metabolism During Child Development

CBF can be measured by a variety of means, including the Kety-Schmidt method of N_2O clearance, inhaled stable Xenon clearance (133Xe), and Xe-based CT scan. As stated above, CBF is closely linked to metabolic demands, and such demands are subject to developmental changes with all measurements demonstrating some variation based on the methodology used for its measurement. In one of the most comprehensive measures of normal CBF across age groups, Chiron and colleagues found that healthy neonates had similar CBF compared to adults, but nearly all other age groups were greater than their adult norm (see Table 24.1) [6]. Moreover, there were substantial variations between brain regions across most age groups, with areas containing predominantly grey matter exhibiting the largest values (particularly the thalamus). Takahashi and colleagues measured regional CBF using PET and observed that infants less than 1 year of age had decreased CBF relative to adults in all areas except the visual cortex [7]. Both of these studies found that CBF began to increase during infancy, ultimately peaking during childhood until slowly declining to adult values throughout adolescence. These findings confirmed a landmark study published in 1957 that used N_2O methodology for determining CBF, where children between the ages of 3 and 11 years had increased CBF compared to adults (106 mL/min/100 g vs. 60 mL/min/100 g) [8].

Evidence supports that these changes in CBF during childhood are largely mirrored by alterations in metabolism. Takahashi and colleagues reported that children less than 3 years old had decreased cerebral metabolic rate of oxygen consumption ($CMRO_2$) compared to adults, but that $CMRO_2$ increased during development in all areas of the brain. By 8 years of age, all areas of the brain had greater $CMRO_2$ than adults [7]. Kennedy and Sokoloff similarly found an increase in $CMRO_2$ in children ages 3–11 years (5.2 mL/min/100 g vs. 4.2 mL/min/100 g) [8]. Comparable developmental changes in cerebral metabolic rate of glucose (CMRglu) have been measured using PET. CMRglu levels were decreased at birth (13–25 μmol/min/100) compared to adults (19–33 μmol/min/100 g). However, values increased to adult levels by age 2 years and surpassed adult levels by

Table 24.1 Normative cerebral blood flow (CBF) values in healthy children and adults measured with 133Xe SPECT (adapted from Chiron et al. [6])

Age	Number of subjects studied	Mean CBF (in mL/100 g/min)
2–45 days	7	50 ± 3.4
2–7 months	7	55 ± 5.3
9–15 months	7	60 ± 6.7
16–22 months	5	62 ± 7.6
2–4 years	7	65 ± 7.8
6–19 years	10	62 ± 9.6
Adults	32	51 ± 7.7

age 4 years (49–65 μmol/min/100 g). CMRglu levels continued to increase until age 9, before finally decreasing to adult levels during late adolescence [9].

In summary, normal child development causes profound changes in CBF and metabolism which must be accounted for in considering therapies for children with TBI or other CNS disorders. The paucity of data regarding these values in children reflects the technical obstacles to obtaining this information in both healthy children and those with critical injuries.

24.2 CBF and Metabolism After Severe TBI in Children

Initial work suggested that childhood TBI was likely characterized as a state of relative excess blood flow for metabolic demands, so-called hyperemia [10, 11]. In this seminal work, Bruce and colleagues noted that the most common pattern of injury in children with severe TBI was diffuse cerebral swelling, and they argued that the predominant pathophysiological responses were due to excessive CBF for metabolic demands [10]. They based this interpretation on relatively few patients—a total of six adolescents between the ages of 16 and 21 with diffuse swelling—who demonstrated CBF of 75 mL/100 g/min (compared to values of 43 mL/100 g/min in similarly aged children without diffuse swelling). A more comprehensive study by Muizelaar and colleagues of 32 children demonstrated that (1) CBF tended to be decreased early in children with low GCS scores, (2) this association was reversed after 24 h (CBF was increased in more severely injured patients), and (3) 88 % of subjects demonstrated CBF values consistent with hyperemia at some point during the study [11]. However, this study did not find a relationship between intracranial pressure (ICP), pressure-volume index (PVI—a measure of cerebral compliance), and CBF—thereby, casting doubt on the established definition of hyperemia (defined as CBF 2 standard deviations greater than normal) and casting the hypothesis of hyperemia's importance after childhood TBI into doubt. Both of these studies were limited by the use of 133Xe techniques, which has limited ability for localization of CBF within specific brain regions.

More recent work on CBF after childhood TBI has challenged some of these findings while confirming others. Studying 18 children with over 150 CBF determinations over the first several days after TBI (and using the Kety-Schmidt technique of CBF determination with N_2O as the inert tracer), Sharples and colleagues demonstrated that hyperemia was relatively infrequent after childhood TBI—observed in only 7 % of the readings [12]. They reemphasized that the CBF response to TBI in children likely changes during the acute hospitalization and this should be accounted for in clinical decisions. Skippen and colleagues also demonstrated the temporal nature of the CBF response to childhood trauma, with the highest readings occurring on days 2 and 3 after injury in a study of 18 children who underwent Xe-CT determinations of CBF [13]. Most recently, Adelson and colleagues reported a series of over 140 measurements from 95 subjects over a 10-year period [14]. In this Xe-CT study, initial CBF readings—obtained directly from the Emergency Department in most cases—were lower than previously found (32.05 mL/100 g/min $\pm$ 21.45), while those from subsequent days were much higher (e.g., 55.36 mL/100 g/min $\pm$ 23.11 on post-trauma days 1–2). Importantly, this study demonstrated that CBF remains increased for up to 10 days after injury, although this obviously includes a selection bias for children who remain critically ill during this time period.

Associations between CBF and clinical variables and outcomes have been sought by most of the aforementioned studies, with a wide variety of associations found. Muizelaar and colleagues found an inverse correlation between initial GCS and CBF after 24 h, yet found no association between CBF and ICP [11]. Sharples and colleagues demonstrated a weak, negative correlation between ICP and CBF ($r = -0.24, p = 0.009$) but found no significant association between CBF and age or initial GCS in their relatively small sample size [12]. They did, however, demonstrate an association between early CBF and unfavorable outcome. In contrast, Skippen and colleagues failed to demonstrate an association between CBF with age, GCS, and time after injury [13]. In the most recent data, Adelson and colleagues found associations between CBF with time after injury as well as with overall outcomes [14]. Most intriguingly, they found that CBF values <20 mL/100 g/min in the initial 2 days after injury were associated with universally poor outcome ($n = 7$) and that children with favorable outcomes generally had increased global CBF readings during the initial days after TBI. Specifically, they found that mean CBF was increased in the favorable outcome group versus the unfavorable outcome group on post-injury day 0 (45.64 vs. 17.85 mL/100 g/min, $p = 0.001$) and post-injury days 1–2 (61.74 vs. 46.54 mL/100 g/min, $p = 0.01$) but not on later days, and this association was observed in children <2 and <5 years, when these groups were interrogated.

Less is known about how cerebral metabolism is altered in children after TBI. Muizelaar found that cerebral metabolism was always decreased after TBI in children and concluded that there was metabolic uncoupling after trauma based on the contemporary premise that hyperemia was prevalent [11]. This was later clarified when Sharples reported that $CMRO_2$ was normal in 17 of 21 children acutely after TBI, but decreased significantly from day 1 to day 3 after injury ($p = 0.001$). Mean $CMRO_2$ within 24 h of TBI was significantly increased in children with good/moderate outcome compared to those with poor outcome

($p = 0.017$). However, mean $CMRO_2$ values at later time points were not significantly associated with outcome [12].

It is reasonable to summarize that changes in CBF occur after childhood TBI and that these changes are both related to age and time after injury. While these findings from only several dozen children require confirmation with contemporary neurocritical care practices, it appears prudent to limit therapies that might reduce CBF early after TBI—particularly unnecessary or inadvertent hyperventilation in the field, in the Emergency Department, and during initial resuscitative efforts. The technical challenges related to measurements of CBF and metabolism remain formidable such that routine measurements are only possible in selected centers. As a result, surrogates of CBF discussed below are needed for clinical management at this time.

24.3 Cerebral Perfusion Pressure in Children with TBI

Calculation of cerebral perfusion, the mathematical difference between mean arterial blood pressure (MAP) and ICP, has been a staple of neurotrauma care for decades [15], although adequate CPP still may be insufficient to achieve what we believe to be normal CBF values [16]. Nonetheless, assessing CPP is likely the most common means to clinically assess the cerebrovascular response to pediatric TBI currently in practice. A number of studies have been performed to determine the optimal CPP for given ages, yet no prospective or randomized studies have been performed to address this critical issue. Instead, observational studies of CPPs that were achieved in various populations that have favorable and unfavorable outcomes currently serve as the best evidence for therapeutic targets.

Among these studies, a minimum CPP has been suggested by several studies. In a study that included 54 children with severe TBI, Barzilay and colleagues found that patients surviving to hospital discharge had a greater CPP than non-survivors (65 vs. 6 mmHg, $p < 0.01$), with the obvious limitation that the non-survivor CPP values were extremely low [17]. Kaiser and Pfenniger found that children with unfavorable outcome had a minimum CPP < 50 mmHg [18] and Figaji and colleagues found a minimum CPP of 44 mmHg in patients with favorable outcome versus 29 mmHg in those with unfavorable outcomes ($p < 0.05$) [17, 19]. Most recently, Kapapa and colleagues have demonstrated that a single instance of CPP below given age-related thresholds (0–1 month, CPP < 40 mmHg; 1 month–1 year, CPP < 45 mmHg; 1–7 years, CPP < 50 mmHg; >7 years, CPP < 55) was associated with worse outcome ($p = 0.013$) [20].

Another approach for analyzing CPP thresholds has included calculating mean CPP values over some period of time after injury and a series of studies have found this approach to be informative. Narotam and colleagues demonstrated that survivors had greater mean CPP values (81.5 mmHg ± 16.1 vs. 50.3 mmHg ± 31.7, $p < 0.33$). Chambers and colleagues found that more patients had poor outcomes when mean CPP was <40 mmHg [21]. And in two independent randomized controlled trials of therapeutic hypothermia, the effect of CPP was evident. Adelson and colleagues found that mean CPP was increased in patients with favorable outcome (69.19 mmHg ± 11.9 vs. 56.37 mmHg ± 20.8, $p = 0.0004$) [22], while Hutchison and colleagues found that the hypothermia group had decreased CPP during rewarming (60.8 mmHg ± 7.8 vs. 66.0 mmHg ± 10.8, $p < 0.001$) and the hypothermia group tended to have poorer long-term outcome [23, 24].

The sum of this literature led the Brain Trauma Foundation workgroup to provide two recommendations regarding CPP thresholds in the recently published guidelines—that a minimum CPP threshold of 40 mmHg may be considered and a CPP threshold of 40–50 mmHg may be considered with age-specific goals of younger children near the lower end of this range and those for older children at the higher end of this range [5]. This latter recommendation was based on a study where the mean CPP values in children 2–6 years of age were approximately 63 mmHg in those with favorable outcome and 48 mmHg in those with unfavorable outcomes. Mean CPP was increased in both outcome groups in children 7–10 years old (72 and 56 mmHg, respectively) [25]. Using the pressure–time integral, which accounts for duration and magnitude of CPP perturbation, the same author reported age-based CPP threshold values of 48 mmHg (2–6 years old), 54 mmHg (7–10 years old), and 58 mmHg (11–15 years old) [26]. The consensus of most practitioners is that maintenance of CPP at some level is important to maintaining adequate CBF after TBI, but there is still significant debate about what thresholds should be maintained. Moreover, it has not been proven that an intervention to achieve a given CPP has been either neuroprotective or affects mortality. Clearly, further study is needed in this area of cerebrovascular regulation after TBI.

24.4 Cerebral Oxygenation in Children with Severe TBI

In the past decade, technology has advanced to allow for the accurate measurement of partial pressure of brain oxygen in the interstitial space, termed PbO_2, which now provides the clinician with a bedside measure of the adequacy of the cerebrovascular response to TBI. PbO_2 is influenced by a number of factors, including the adequacy of the CBF in relation to metabolic demands, CPP, hemoglobin concentration, arterial partial pressure of oxygen (PaO_2), and temperature. In our experience, another valuable aspect of PbO_2 monitoring is the detection of inadvertent hyperventilation—as the subsequent vasoconstriction from hyperventilation results in a presumed decrease in CBF and an unanticipated decrease in PbO_2 that can be corrected by correcting the hyperventilation.

However, the main focus of research within centers using PbO_2 monitoring is to determine if a PbO_2-directed strategy can lead to improved neurological outcome or beneficial effects on other clinical parameters, since cerebral hypoxia is presumed to be a potential secondary insult that could be treated. For children with severe TBI, Stiefel and colleagues provided the first report on the utility of PbO_2 monitoring when they found that PbO_2 was higher in children during periods of ICP < 20 mmHg and CPP > 40 mmHg (29.29 mmHg $\pm$ 7.17 vs. 22.83 mmHg $\pm$ 13.85, $p < 0.01$ and 28.97 mmHg $\pm$ 7.85 vs. 2.53 mmHg $\pm$ 7.98, $p < 0.01$, respectively) [27]. Figaji and colleagues have provided a number of reports to demonstrate the utility of PbO_2 monitoring in a variety of ways [19, 28–31]. In one study, children with unfavorable outcome had a decreased mean PbO_2 in the first 24 h after injury compared to those with favorable outcome (21.9 vs. 30.4 mmHg). This study also showed that those with unfavorable outcome had a longer duration of time with low PbO_2 (<10 mmHg) and that PbO_2 is independently associated with mortality and unfavorable outcome. Another group demonstrated similar results, demonstrating that the mean PbO_2 at 2 h post-monitor placement was significantly increased in survivors compared to non-survivors (21.6 vs. 7.2 mmHg; $p = 0.009$) [32]. In yet another study by Figaji and colleagues, the relationship between PbO_2 and other pediatric neurocritical care targets (particularly ICP < 20 mmHg and CPP > 50 mmHg) was evaluated, and 80 % of the children had one or more episodes of cerebral hypoxia ($PbO_2 < 20$ mmHg) without concurrent intracranial hypertension or inadequate CPP [28]. They also found that the relationship between decreased PbO_2 and other physiologic/clinical factors that are routinely used to guide therapy in pediatric TBI was poorly predicted by these factors (ICP, CPP, PaO_2, and SaO_2). Based on these findings, the authors concluded that "brain hypoxia may still occur despite apparently adequate resuscitation and therapy of pediatric TBI based on current guidelines" [28]. Recently, we've also published data from our center which confirms that there are periods of intracranial hypertension and cerebral hypoperfusion that are surprisingly associated with increases in PbO_2, emphasizing that the entire neurocritical care protocol may need to be evaluated to determine the utility of this monitoring system [33]. In summary, we believe that it appears that the PbO_2 monitor may have some utility in managing the cerebrovascular response after TBI in children, especially since other modalities to assess the response (such as CBF measurements) are not readily available at the bedside. However, the use of the monitor will need to be put into some context, as other therapies to treat intracranial hypertension will have obvious effects on the cerebrovascular response measured by the PbO_2 monitoring.

24.5 Cerebral Autoregulation in Children with TBI

Cerebral autoregulation is defined in two contexts—the preservation of normal cerebral vasoconstriction/vasodilation in response to changes in acid/base balance associated with changes in $PaCO_2$ and the maintenance of normal CBF through a range of arterial blood pressures (normally defined as mean arterial blood pressure between 50 and 150 mmHg in adults). Both of these responses can be altered by TBI, and such dysfunction can be associated with adverse outcomes and may alter potential treatment plans.

Increases in $PaCO_2$ from hypoventilation from a number of causes (apnea directly after impact, lung disease as a result of multiple traumatic injuries, lung disease that evolves later after injury, or many others) normally cause acidosis within the CSF space and subsequent vasodilation of arterioles. Skippen and colleagues demonstrated that relatively severe hyperventilation was associated with marked increases in the brain regions that approached the ischemic threshold of tissues (CBF < 18 mL/100 g/min) [13]. The prevalence of ischemia was 28.9 % with $PaCO_2 > 35$ mmHg, 59.4 % with $PaCO_2$ 25–35 mmHg, and 73.1 % with $PaCO_2 < 25$ mmHg in this cohort of children. Adelson and colleagues also tested $PaCO_2$ reactivity in a subset of patients ($n = 38$) by measuring CO_2 vasoreactivity (CO_2VR; the increase in CBF per 1 mmHg increase in $PaCO_2$) [14]. In this study, 17 patients (44 %) had abnormal autoregulation (CO_2VR < 2 %). Absence of intact autoregulation during the first 48 h post-injury was significantly associated with an unfavorable outcome ($p = 0.029$). For all times after TBI, there was a trend towards increased CO_2VR (intact autoregulation) in patients with favorable outcome versus unfavorable (3.83 % vs. 0.72 %) that approached statistical significance ($p = 0.120$). In an earlier study of children $\leq$8 years old, CO_2VR < 2 % was 83 % sensitive and 100 % specific for unfavorable outcome [34].

Blood pressure autoregulation in children after TBI has also been studied, although a number of these studies have substituted the use of CBF velocity (as measured by transcranial Doppler) in place of CBF from other techniques outlined above [35–39]. Vavilala and colleagues have authored a number of reports, including one that reported 10 of 24 children with intact autoregulation had a favorable outcome as compared to only 1 of 12 children with TBI and disturbed autoregulation ($p = 0.04$). In another report, this group found that the association between impaired autoregulation and unfavorable outcome was statistically significant at 6-month follow-up. Similarly, Sharples used direct measurement of CBF to find that patients with a favorable outcome had a significant association between CPP and cerebrovascular resistance (CVR) ($p = 0.0002$), which implies intact autoregulation, but those with unfavorable outcome did not ($p = 0.22$). Doppler ultrasound has been used to calculate the autoregulatory index (ARI, the percent change in CVR divided by the percent change in CPP). Impaired BP autoregulation (ARI < 0.4) is more common in patients with unfavorable 6-month outcome (75 % vs. 23 %, $p = 0.03$) [39] and portends an increased risk of poor outcome (OR 23.1 [1.9–279.0]) [40]. However, lack of an association between BP autoregulation and outcome has been reported using direct [35] and indirect measures of CBF [38]. Factors associated with loss of BP autoregulation that have been investigated include injury severity (GCS score), ICP, CBF, age, and time after injury. Several studies found no association between GCS score and abnormal BP autoregulation [12, 35, 36], though an association between abnormal ARI and decreased GCS score at the time of measurement has been reported [39]. Similarly, patients with an abnormal ARI had increased ICP during measurement in one study (21 $\pm$ 15 vs. 14 $\pm$ 5, $p = 0.007$) [37]. In a prior publication, the same group showed that hyperemia was more common in patients with impaired ARI than in those with intact ARI (6/12 vs. 1/24, $p < 0.01$). Muizelaar also found associations between loss of BP autoregulation and both hyperemia and markedly decreased CBF (>2 SD from normal). ARI is more commonly impaired in children <4 years old (8/10 vs. 7/27, $p = 0.006$) [39] and normalizes approximately 1 week after TBI [36, 38].

24.6 Conclusion

As with most processes within the central nervous system, the cerebrovascular responses in children—both in normal development and after TBI—are remarkably complex. Most evidence suggests that cerebrovascular responses are age- and development-dependent and suggest that unique treatment strategies that would take the cerebrovascular responses into account might be different for children across the pediatric age spectrum. Our current tools to assess the response are relatively limited, but it is possible that new technologies will improve our ability to detect cerebrovascular disturbances in the future. Currently, it appears that clinicians caring for children with severe TBI should (1) guard against the potential of cerebral ischemia early after TBI, (2) consider age-adjusted goals for CPP, (3) consider use of PbO_2 monitoring to detect potential cerebral ischemia, and (4) judge whether children might benefit from testing of autoregulation abnormalities. Researchers will need to perform studies that might clarify the benefits of these techniques, using either trial designs that include randomization of overall TBI protocols that include these modalities or other methodologies that can determine the superiority of the various strategies being tested.

Acknowledgements Drs. Shein and Ferguson are supported by T32 HD040686. Dr. Bell is supported by U01 HD049981, R01 NS069247, and R01 NS072308. The authors report no conflicts of interest.

References

1. Faul M, Xu L, Wald MM, Coronado VG (2010) Traumatic brain injury in the United States: emergency department visits, hospitalizations and deaths 2002–2006. Centers for Disease Control and Prevention, National Center for Injury Prevention and Control, Atlanta
2. Naumann RB et al (2010) Incidence and total lifetime costs of motor vehicle-related fatal and nonfatal injury by road user type, United States, 2005. Traffic Inj Prev 11(4):353–360
3. Shi J et al (2009) Costs, mortality likelihood and outcomes of hospitalized US children with traumatic brain injuries. Brain Inj 23(7):602–611
4. Stanley RM et al (2012) US estimates of hospitalized children with severe traumatic brain injury: implications for clinical trials. Pediatrics 129(1):e24–e30
5. Kochanek PK et al (2012) Guidelines for the acute medical management of severe traumatic brain injury in infants, children and adolescents: second edition. Pediatr Crit Care Med 13 (1 suppl):S1–S82

6. Chiron C et al (1992) Changes in regional cerebral blood flow during brain maturation in children and adolescents. J Nucl Med 33(5):696–703
7. Takahashi T et al (1999) Developmental changes of cerebral blood flow and oxygen metabolism in children. AJNR Am J Neuroradiol 20(5):917–922
8. Kennedy C, Sokoloff L (1957) An adaptation of the nitrous oxide method to the study of the cerebral circulation in children; normal values for cerebral blood flow and cerebral metabolic rate in childhood. J Clin Invest 36(7):1130–1137
9. Chugani HT, Phelps ME, Mazziotta JC (1987) Positron emission tomography study of human brain functional development. Ann Neurol 22(4):487–497
10. Bruce DA et al (1981) Diffuse cerebral swelling following head injuries in children: the syndrome of "malignant brain edema". J Neurosurg 54(2):170–178
11. Muizelaar JP et al (1989) Cerebral blood flow and metabolism in severely head-injured children. Part 1: Relationship with GCS score, outcome, ICP, and PVI. J Neurosurg 71 (1):63–71
12. Sharples PM et al (1995) Cerebral blood flow and metabolism in children with severe head injury. Part 1: Relation to age, Glasgow coma score, outcome, intracranial pressure, and time after injury. J Neurol Neurosurg Psychiatry 58(2):145–152
13. Skippen P et al (1997) Effect of hyperventilation on regional cerebral blood flow in head-injured children. Crit Care Med 25(8):1402–1409
14. Adelson PD et al (2011) Cerebrovascular response in children following severe traumatic brain injury. Childs Nerv Syst 27(9):1465–1476
15. Dean NP et al (2007) Physician agreement with evidence-based recommendations for the treatment of severe traumatic brain injury in children. J Neurosurg 107(5):387–391
16. Philip S et al (2009) Cerebrovascular pathophysiology in pediatric traumatic brain injury. J Trauma 67(2 suppl):S128–S134
17. Barzilay Z et al (1988) Variables affecting outcome from severe brain injury in children. Intensive Care Med 14(4):417–421
18. Kaiser G, Pfenninger J (1984) Effect of neurointensive care upon outcome following severe head injuries in childhood–a preliminary report. Neuropediatrics 15(2):68–75
19. Figaji AA et al (2009) Brain tissue oxygen tension monitoring in pediatric severe traumatic brain injury. Part 1: Relationship with outcome. Childs Nerv Syst 25(10):1325–1333
20. Kapapa T et al (2010) Head trauma in children, part 2: course and discharge with outcome. J Child Neurol 25(3):274–283
21. Chambers IR, Treadwell L, Mendelow AD (2001) Determination of threshold levels of cerebral perfusion pressure and intracranial pressure in severe head injury by using receiver-operating characteristic curves: an observational study in 291 patients. J Neurosurg 94 (3):412–416
22. Adelson PD et al (2005) Phase II clinical trial of moderate hypothermia after severe traumatic brain injury in children. Neurosurgery 56(4):740–754; discussion 740–754
23. Hutchison JS et al (2010) Impact of hypotension and low cerebral perfusion pressure on outcomes in children treated with hypothermia therapy following severe traumatic brain injury: a post hoc analysis of the Hypothermia Pediatric Head Injury Trial. Dev Neurosci 32 (5–6):406–412
24. Hutchison JS et al (2008) Hypothermia therapy after traumatic brain injury in children. N Engl J Med 358(23):2447–2456
25. Chambers IR et al (2005) Age-related differences in intracranial pressure and cerebral perfusion pressure in the first 6 hours of monitoring after children's head injury: association with outcome. Childs Nerv Syst 21(3):195–199
26. Chambers IR et al (2006) Critical thresholds of intracranial pressure and cerebral perfusion pressure related to age in paediatric head injury. J Neurol Neurosurg Psychiatry 77(2):234–240
27. Stiefel MF et al (2006) Brain tissue oxygen monitoring in pediatric patients with severe traumatic brain injury. J Neurosurg 105(4 suppl):281–286

28. Figaji AA et al (2008) Does adherence to treatment targets in children with severe traumatic brain injury avoid brain hypoxia? A brain tissue oxygenation study. Neurosurgery 63 (1):83–91; discussion 91–92
29. Figaji AA et al (2007) Intracranial pressure and cerebral oxygenation changes after decompressive craniectomy in a child with traumatic brain swelling. Childs Nerv Syst 23 (11):1331–1335
30. Figaji AA et al (2009) Pressure autoregulation, intracranial pressure, and brain tissue oxygenation in children with severe traumatic brain injury. J Neurosurg Pediatr 4(5):420–428
31. Figaji AA et al (2010) The effect of increased inspired fraction of oxygen on brain tissue oxygen tension in children with severe traumatic brain injury. Neurocrit Care 12(3):430–437
32. Narotam PK et al (2006) Cerebral oxygenation in major pediatric trauma: its relevance to trauma severity and outcome. J Pediatr Surg 41(3):505–513
33. Stippler M et al (2012) Brain tissue oxygen monitoring after severe traumatic brain injury in children: relationship to outcome and association with other clinical parameters. J Neurosurg Pediatr 10(5):383–391
34. Adelson PD et al (1997) Cerebrovascular response in infants and young children following severe traumatic brain injury: a preliminary report. Pediatr Neurosurg 26(4):200–207
35. Muizelaar JP et al (1989) Cerebral blood flow and metabolism in severely head-injured children. Part 2: autoregulation. J Neurosurg 71(1):72–76
36. Vavilala MS et al (2004) Cerebral autoregulation in pediatric traumatic brain injury. Pediatr Crit Care Med 5(3):257–263
37. Vavilala MS et al (2006) Impaired cerebral autoregulation and 6-month outcome in children with severe traumatic brain injury: preliminary findings. Dev Neurosci 28(4–5):348–353
38. Tontisirin N et al (2007) Change in cerebral autoregulation as a function of time in children after severe traumatic brain injury: a case series. Childs Nerv Syst 23(10):1163–1169
39. Freeman SS et al (2008) Young age as a risk factor for impaired cerebral autoregulation after moderate to severe pediatric traumatic brain injury. Anesthesiology 108(4):588–595
40. Chaiwat O et al (2009) Cerebral hemodynamic predictors of poor 6-month Glasgow Outcome Score in severe pediatric traumatic brain injury. J Neurotrauma 26(5):657–663

Chapter 25
Subdural Hematoma in Non-accidental Head Injury

Jennifer C. Munoz Pareja, Josephine Lok, Natan Noviski, and Ann-Christine Duhaime

Abstract Subdural hematoma in infants and young children happens most frequently, but not exclusively, in the setting of non-accidental head injury. Irrespective of the etiology, this injury type can be associated with a variety of pathophysiologic changes, including those which appear to affect the relationship between substrate delivery and metabolic demand. The exact underpinnings and necessary conditions for these changes remain incompletely understood but appear to be specific for children during early development. This chapter will review the clinical presentation, spectrum of mechanisms, and neuroanatomic and cerebrovascular considerations for this common and often serious injury type.

25.1 Overview of Subdural Hematoma in Infants and Young Children

Non-accidental head injury (NAHI) in infants and young children encompasses a spectrum of clinical histories and physical findings, as well as a variety of imaging findings and clinical sequelae. Synonyms used for this spectrum also include non-accidental trauma (NAT), abusive head trauma (AHT), inflicted head injury, and other terms denoting specific mechanisms. While there is variability among patients, there are certain constellations of features which are typical and alert the clinician to the possibility of an inflicted etiology for the findings. The most common histories are descriptions of symptoms with no history of trauma, or a history of minor trauma such as a low-height fall. Presenting symptoms may include irritability, poor feeding, seizures, apnea, lethargy, or obtundation.

J.C. Munoz Pareja (✉)
Department of Pediatric Critical Care Medicine, Massachusetts General Hospital, Harvard Medical School, Boston, MA, USA
e-mail: jennifermunozpareja@partners.org

E.H. Lo et al. (eds.), *Vascular Mechanisms in CNS Trauma*, Springer Series in Translational Stroke Research 5, DOI 10.1007/978-1-4614-8690-9_25,

The most characteristic pathoanatomic injury type in NAT is subdural hematoma (SDH), which in this age group is highly associated with, though not diagnostic for, a non-accidental mechanism of injury. Extracranial trauma and retinal hemorrhages are common but also can occur in other accidental scenarios. Some medical conditions may mimic certain aspects of the findings associated with inflicted injuries, discussed in more detail below [1, 2].

In virtually all reported series of NAT, the most common and characteristic intracranial finding is SDH [3]. SDH can be described by location, size, associated mass effect, and chronicity. Acute SDH in the infant has been described as typically having a somewhat different pattern from that most often seen in the older child and adult. While SDH in older children and adults usually presents as a thick, unilateral space-occupying clot, in infants more often it has been described as a widespread, bilateral, thin film of blood [4]. In young children beyond the first year of life, either the bilateral or more unilateral form may occur. In both age groups, a propensity for SDH to be associated with blood in the interhemispheric fissure may be present as well.

Whether associated with inflicted injury or accidental trauma, acute SDH in infants and young children can be associated with profound cerebrovascular effects, and these will be explored in this chapter.

25.2 Epidemiology, Mechanisms, and Differential Diagnosis

The incidence of inflicted head injury is estimated at 15 per 1,000 children per year. Lethality and morbidity are significant with 12–30 % of infants dying and 60–70 % of survivors experiencing significant neurological deficits [3–6]. Approximately 1,000 deaths caused by inflicted injury are confirmed annually in the United States [7].

While both acute and chronic SDHs can occur in accidental injuries and in certain medical conditions, the most common association is with NAT. In 1946, Dr. John Caffey first recognized the telltale radiographic changes that came to characterize children potentially suffering from NAT [8]. In the following three decades, important work by Silverman [9], Ommaya et al. [10], Kempe et al. [11], and Guthkelch [12] contributed to the acknowledgement of child abuse as a distinct medical condition. In 1974 Caffey coined the term "whiplash shaken infant syndrome." The idea that shaking might be causative was first proposed by Guthkelch [3]; he suggested that shaking rather than striking the infant might be the cause, since not all infants with SDH had external injuries, and based on biomechanical studies on adult primates, he postulated the idea of angular deceleration as the causative mechanism of SDH. However, the magnitude of angular deceleration caused by shaking as opposed to other potential mechanisms was unknown at that time, as was the threshold for injury of specific types in young children.

The pathophysiology of the subdural bleeding caused by NAT often has been attributed to rapid acceleration-deceleration from severe blunt trauma or from forceful shaking of the infant, leading to a shearing of the bridging veins [13]. However, research using forces simulating an adult shaking an infant has shown that shaking alone generates substantially lower forces compared to those known to produce subdural bleeding in primates subjected to nonimpact acceleration [14, 15]. Duhaime et al. showed that the acceleration force generated by impact exceeded that caused by shaking by a factor of 50 [13]; however, the exact threshold of angular deceleration needed to tear bridging veins in infants and young children remains incompletely understood. Other researchers propose that the NAHI-induced SDH results from injury to intrinsic intradural vessels rather than to tearing of the bridging veins [33]. A review of the anatomy of the dura and the bridging veins is covered in more detail later in this chapter to highlight the issues under consideration.

When evaluating an infant with SDH, it is often challenging to determine whether the hemorrhage could be a result of NAHI. Accidental injury is another common cause, and much work has been done to attempt to differentiate these etiologies, though no method is definitive [6, 16, 17]. In young infants, birth subdural may be a consideration, as there is a high incidence of SDH from the birth process, documented by Looney in 2007 [18] and by Rooks in 2008 [19]. In screening 101 deliveries that were described as "normal," 46 % of the infants in Rooks' series were found to have SDH that formed in the perinatal period during labor. The location of the SDH included both supratentorial and infratentorial distributions in these studies. In Rooks' series, one of the neonates out of the 22 had developed a 1 cm extra-axial frontal collection at 26 days of life and underwent a complete evaluation which did not support NAHI as an etiology. In a separate review, Gebaeff [20] reported that a small percentage of neonates suffer birth-related complications and subsequently develop a chronic SDH. The duration of birth subdurals is generally considered to be short, resolving without complications by 5–8 weeks of age [19].

Another clinical condition that may be raised as a potential predisposition to the development of SDH, particularly of the more chronic-appearing or mixed forms, includes benign enlargement of the subarachnoid spaces ("benign external hydrocephalus" or "benign macrocranium of infancy") [21]. In this condition, children typically present with an enlarging head circumference, usually between the first few months of life up to the second year of age, and imaging demonstrates prominent extra-axial cerebrospinal fluid spaces. On rare occasion, follow-up imaging has been reported to demonstrate unilateral or bilateral small subdural hemorrhages, sometimes with a suggestion of rebleeding and subdural membranes [22]. Evidence of additional injuries concerning for physical abuse was identified in a quarter of the children with enlargement of the subarachnoid spaces and SDH, suggesting that the finding of enlarged subarachnoid spaces should not preclude an evaluation for other findings suggestive of NAT.

Other disease entities in which an association with SDH has been reported include coagulation defects, vitamin K deficiency [23], platelet dysfunction, hemophilia, von Willebrand disease, and acute leukemia [24, 25]. Other uncommon disorders include glutaric acidemia type I, Menkes disease [26], and Prader-Willi syndrome [22]. Von Willebrand disease and neuronal lipofuscinosis have also been associated with chronic subdural collections [22].

Spontaneous SDH in infants (SSDHI) is a rare occurrence which has been examined by Vinchon et al. [27]. In a series of 16 patients, 12 patients had idiopathic macrocrania, 7 of these being previously diagnosed with arachnoidomegaly on imaging. Five had risk for dehydration, including two with severe enteritis. The authors concluded that in these patients there were positive risk factors that may predispose to the development of SDH.

25.3 Anatomic Considerations and Pathophysiology of SDH in NAHI

While the exact mechanisms necessary to cause SDH in accidental and NAT in infants and young children remain incompletely understood, they may be influenced by the specific anatomy of the structures involved. In addition, once hemorrhage has occurred, events may unfold which influence the effect of the injury on the brain, including the development of seizures, apnea, and other pathophysiologic cascades. Some of these may contribute to the more extensive degrees of damage seen in specific cases of SDH in this age group.

25.4 The Dura

Our central nervous system is enclosed by three concentric membranes designated, from external to internal, as dura mater, arachnoid, and pia mater. At all times during development, the dura and arachnoid remain attached, and there is normally no anatomic subdural space. However, there is a distinct soft tissue layer at the dura-arachnoid junction that is easily disrupted, called the "dural border cell layer." This junctional soft tissue layer is characterized by flattened fibroblasts with sparse intercellular junctions, no extracellular collagen, and prominent extracellular spaces. Blood collecting in the dural border cell layer can easily dissect through the potential space in the subdural compartment [4], causing disruption of the loosely adherent cells of the dural border layer and accounting for the widespread distribution of SDH [4].

Between the periosteal and meningeal layers of the dura, there is a venous plexus which is very extensive in the fetus and neonate, occupying much larger areas than the sinuses of the adult dura [28, 29]. In the neonate the venous plexus is much more prominent than in the adult, forming sinuses in the tentorium, the posterior falx, and the dura of the floor of the posterior cranial fossa. Interestingly, infant SDH is predominantly found over the dural folds bearing these venous sinuses [30, 31], suggesting that these venous sinuses may affect subdural bleeds in the infant [29, 32]. Radiologically, the finding of a thin, linear, high signal between the hemispheres or over the tentorium has been shown to correlate with intradural bleeding/congestion rather than with subdural bleeding [33] has been used to lend support to the hypothesis that subdural hemorrhage of NAHI originates from the dural vasculature. However, subdural hemorrhage rarely, if ever, occurs as a consequence of isolated asphyxia, vomiting, or other nonhemorrhagic medical etiologies, so bleeding from the dura itself does not appear to be a significant cause of clinically apparent SDH in full-term infants and toddler-aged children [34].

25.5 The Bridging Veins

Separation of the venous drainage of the brain from the dural venous drainage gives rise to the specific anatomy of the bridging veins, which originate through venous cleavage in the first trimester. These superficial veins coalesce to form between 10 and 18 large bridging veins that will penetrate the dura-arachnoid interface layer and travel for a variable distance within the dura before entering the superior sagittal sinus [4, 35].

Rupture of the bridging veins as they cross from the subarachnoid space through the dural border cell layer continues to be the most accepted theory for the etiology of traumatic SDH. However, these veins are large caliber conduits and rupture only under considerable force [36, 37]. Additionally, rupture of such large vessels might seem unlikely to produce the thin film of hemorrhage characteristically seen in the young infant, unless inherent mechanisms for rapid and efficient hemostasis can effectively limit hemorrhage. It is also possible that incomplete rupture or oozing from congested bridging veins may cause small-volume subdural bleeding [4].

Recent studies of the bridging veins published by Vignes et al. [38] demonstrated the presence of a unique cuff of circular collagen fibers surrounding the intradural bridging veins at their entrance into the sagittal sinus. This muscle coat may act a sphincter regulating the blood flow from bridging veins and thus maintaining intravenous pressure when intracranial pressure rises [4, 39]. These fibers may be responsible for the terminal dilation of the intradural portion of the bridging veins as they enter the sagittal sinus, and it is unknown if the dilation in the intradural portion of the bridging veins makes the vessel prone to leaking after trauma.

25.6 Maturation-Dependent Response to SDH

Little is understood about the maturation-dependent pathophysiologic response to an SDH. Durham and Duhaime [40] attempted to address this issue by developing a piglet SDH model using subjects at 5 days, 1, and 4 months of age to approximate brain development of the human infant, toddler, and adolescent. A volume of blood equal to 10 % of the intracranial volume was injected into subdural space. The study revealed a maturation-dependent response of the immature brain to SDH. Surprisingly, the youngest animals experienced the least amount of ischemic damage from the presence of blood in the subdural space. In a related study, Duhaime et al. also found that the infant brain was relatively resistant to injury from mechanical deformation [41]. This is consistent with clinical experience in which young children recover from focal injury but may succumb or survive with significant sequelae to more diffuse injuries or to significant parenchymal damage that may be seen in many instances of inflicted injury [42].

In the situation of severe inflicted injury with bilateral parenchymal damage, survival occurs at a higher rate in the youngest infants, possibly due to the open cranial sutures which help to relieve intracranial pressure [43]. However, if they survive, infants with severe injuries with widespread parenchymal changes on CT scan or MRI have poor outcomes [42].

The worst outcomes are best predicted by the entity of the so-called big black brain, a pattern of tissue loss affecting the entire supratentorial hemisphere or both hemispheres in association with acute SDH [43–47]. The rapid appearance of extensive, diffuse supratentorial hypodensity with loss of gray-white differentiation in this entity is seen routinely only in infants and toddler-aged children. The mechanism and pathophysiology of this disease process are still elusive despite investigations by several groups [43–47]. Some workers [43] have hypothesized that the white matter is severely affected because of its increased vulnerability during early life, when the metabolic rate is relatively high in white matter but the vascular supply is immature and may respond inadequately to increased demand. This theory is consistent with the prevalence of white matter injury in infants with perinatal hypoxic-ischemic insults and prematurity [48]. It is possible that some synergistic effect of the hemorrhage, or the forces that caused it, and a secondary insult, such as apnea or subclinical seizures, result in some form of hypoxic-ischemic injury or perfusion-demand mismatch injury [49]. This process may involve a combination of insults that overwhelm the immature brain's ability to compensate, resulting in a mismatch between metabolic demand and substrate delivery. Such insults may include apnea or seizures. The observation that this pattern occurs uniquely in infants and toddlers may therefore be a reflection of their age-dependent brain response to the injury inflicted in NAHI or, in some instances, to accidental trauma resulting in SDH [40].

25.7 Hemostasis in Infancy

Hemostasis is a dynamic, evolving process that is age dependent. The physiology of hemostasis in pediatric patients differs from that in adults. In general, the hemostatic system is not fully mature until 3–6 months of age, and infants have lower physiological levels of a number of the coagulation factors. The levels of some fibrinolytic proteins, such as plasminogen and tissue plasminogen activator, and of coagulation inhibitors, such as antithrombin and proteins C and S, are also decreased [50].

The strongest age dependency is seen for proteins C and S, both significantly decreased in infancy and young childhood. Proteins C and S are vitamin K-dependent coagulation factors and remain 10–20 % reduced throughout childhood [51]. Antithrombin reaches adult levels by the age of 7–12 months and is also affected by inflammation. The levels of von Willebrand factor undergo several changes in infancy and childhood. Fibrinogen levels are generally lower in children under 6 months of age, although it may increase in response to inflammation.

These variations in the hemostatic system do not appear to manifest in a clinically significant risk of thrombosis or bleeding during normal infancy. However, these factors may have a role in the pathophysiology of SDH in the context of NAHI. However, most infants with traumatic brain injury of any etiology who demonstrate abnormal coagulation parameters do so because of the effects of injury, rather than having a preexisting coagulation deficiency [52, 53].

25.8 Effect on Cerebral Blood Flow and Metabolism

Previous studies have shown that cortical ischemic damage is due more to the chemical properties of the SDH or to its effect on focal cerebral blood flow rather than to its space-occupying capacity. Cortical ischemic damage is only minimal when an inert silicone gel mass with viscosity and volume similar to blood is injected into the subdural space. Kuroda et al. [54] investigated the effects of clotted blood in contact with the vessels of the cortical surface in rodents and demonstrated a significant enlargement of the zone of ischemic tissue under the hematoma. The ischemia and consequent cytotoxic edema in turn cause increased intracranial pressure. Removal of the hematoma restores cerebral blood flow in the contralateral hemisphere to near control values, but does not reverse focal progression of ischemia and edema.

Salvant et al. [55] compared regional cerebral blood flow in patients with severe closed head injury with and without SDH. They demonstrated significant reductions in cerebral blood flow as well as cerebral metabolic rate of oxygen consumption in brain areas adjacent to the SDH in the first 48 h after injury. Based on these observations, the authors suggest that SDH produces direct local effects on cerebral circulation and that these effects persist even after decompression.

Kuroda et al. have also postulated that clotted blood initiates a profound and sustained vasospasm [56] when it contacts vessels of the cortical surface. This vasospasm causes a zone of severe focal ischemia under the hematoma, resulting in neuronal infarction [57, 58]. The process, in this widely used rat model, is accompanied by a brief sevenfold increase in extracellular glutamate in the ischemic cortex, together with transient hypermetabolism for glucose [54–57], initiating an excitotoxic process which spreads outward from the ischemic core, causing progressive tissue damage [58]. The presence of clotted blood is associated with abnormalities in cerebral blood flow and cerebral metabolic rate of oxygen consumption [59], reduced metabolic demand [60], and post-decompressive hypoperfusion [61].

In vitro studies [62] show that the presence of purified hemoglobin in neuron-glia-blood clot cocultures produces an iron-dependent injury which is attenuated by iron chelators and antioxidants. Interestingly, this injury appears to be greater in the presence of neurons and glia. In a similar study in which blood clots were cocultured with primary neurons and glia, erythrocyte lysis was reduced by almost 80 % when neurons are absent from the cultures and by 90 % when both neurons and glia are absent. To isolate the effects of iron toxicity and neuronal death, addition of deferoxamine, Trolox, and the NMDA receptor antagonist MK-801 was studied. These agents prevented most neuronal death, but had no effect on hemolysis at neuroprotective concentrations. Additionally, the increase in culture fluid of malondialdehyde (by 27-fold) and in heme oxygenase-1 expression (by 5.8-fold) was also attenuated by deferoxamine and Trolox but not by MK-801. These results suggest that hemoglobin release from clotted blood is accelerated by adjacent neurons and glia and that the injury from acute SDH may result from both iron-dependent and excitotoxic injury pathways [63]. The effects of heme and iron toxicity include the formation of reactive oxygen species and increased oxidative damage to lipids, DNA, and proteins, leading to caspase activation and neuronal death. Additionally, damage to endothelial cells causes blood–brain barrier breakdown, resulting in vasogenic edema, increased ICP, as well as vasoconstriction and ischemia.

25.9 Complications of Acute SDH

A large SDH may constitute a significant space-occupying lesion and lead to increased intracranial pressure via a primary sustained volume effect, although this is more uncommon in infants in whom the cranial sutures are open. While bleeding likely causes a transient initial elevation, this can be compensated up to a point, but in many cases, a more delayed effect may occur. Thus, the presence of the SDH may lead to other cascades and complications, and acute SDH worsens outcome after traumatic brain injury, even if the hemorrhage is small. Meissner et al. [64] report that an acute SDH in a pig model of controlled cortical impact caused a drop in brain tissue oxygen ($ptiO_2$), a prolonged elevation in extracellular glutamate, as well as a reduction in somatosensory evoked potentials (SSEP).

In addition to the immediate deleterious physical and biochemical effects on the brain, SDH also stimulates an inflammatory reaction which may evolve over a significant period of time [65]. Golden et al. present evidence that the human dura contains a significant number of mast cells and that subdural hemorrhage is associated with an increase in mast cell density, which continues to increase over time [66]. In addition to contributing to neuro-inflammation, the authors propose that the upregulation of mast cells in SDH may result in stimulation of the trigeminal system and may alter vascular permeability, with the potential to cause sudden neurological deterioration [67].

Acute SDH may also occasionally result in the formation of a chronic hemorrhagic fluid collection. This has been described in patients with glutaric aciduria type I [68] and may be extrapolated to infants with accidental or NAHI [69]. Chronic SDH in the pediatric age group remains relatively uncommon, and most of the research on this entity has been on chronic SDHs in elderly adults. Healing of SDH occurs with the formation of a granulating membrane with dense vascularization, which may confer vulnerability to rebleeding from minor shearing forces. Additionally, impaired CSF resorption may contribute to the growth of the collection in the absence of further trauma [66]. These rebleeding episodes are often undetected as they do not usually present with significant acute neurological symptoms. However, in clinical experience, infants with SDH which do not result in severe immediate symptoms requiring acute medical care may be ascertained to have had, in retrospect, transient, nonspecific symptoms based on caretaker recall or records from prior medical evaluations [70]. Over the long term, chronic SDH in infants may lead to macrocephaly and a delay in psychomotor milestones, although the brain trauma itself is likely a more significant contributor to developmental delay. Patients that do become symptomatic may need to undergo neurosurgical drainage of the collection for management of progressive macrocephaly, irritability, or delayed development. Seizures also may be a presenting symptom.

In summary, SDH is a common feature of NAHI, with both deleterious acute and long-term consequences. Further research into the biochemical effects of SDH and of heme and iron toxicity on various CNS cell types, as well as the role of specific genetic or other host factors, will add to the understanding and management of subdural hemorrhage in the context of NAHI.

References

1. Harding B, Risdon RA, Krous HF (2004) Shaken baby syndrome. BMJ 328(7442):720–721
2. American Academy of Pediatrics, Committee on Child Abuse and Neglect (1993) Shaken baby syndrome: inflicted cerebral trauma. Pediatrics 92(6):872–875
3. Jayawant S, Rawlinson A, Gibbon F et al (1998) Subdural haemorrhages in infants: population based study. BMJ 317(7172):1558–1561
4. Mack J, Squier W, Eastman JT (2009) Anatomy and development of the meninges: implications for subdural collections and CSF circulation. Pediatr Radiol 39:200–210
5. Haviland J, Russell RI (1997) Outcome after severe non-accidental head injury. Arch Dis Child 77(6):504–507
6. Billmire ME, Myers PA (1985) Serious head injury in infants: accident or abuse? Pediatrics 75 (2):340–342
7. Duhaime AC, Christian C (2001) Child abuse. In: McLone DG (ed) Pediatric neurosurgery, 4th edn. WB Saunders, Philadelphia, pp 593–600
8. Caffey J (1946) Multiple fractures in the long bones of infants suffering from chronic subdural hematoma. AJR Am J Roentgenol 56:163–173

9. Silverman FN (1953) Roentgen manifestations of unrecognized skeletal trauma in infants. Am J Roentgenol Radium Ther Nucl Med 69:413–427
10. Ommaya AK, Faas F, Yarnell P (1968) Whiplash injury and brain damage: an experimental study. JAMA 204:285–289
11. Kempe CH, Silverman FN, Steele BF, Droegemueller W, Silver HK (1962) The battered-child syndrome. JAMA 181:17–24
12. Guthkelch AK (1971) Infantile subdural hematoma and its relationship to whiplash injuries. Br Med J 2:430–431
13. Duhaime AC, Gennarelli TA, Thibault LE, Bruce DA, Margulies SS, Wiser R (1987) The shaken baby syndrome: a clinical, pathological, and biomechanical study. J Neurosurg 66 (3):409–415
14. Cory CZ, Jones BM (2003) Can shaking alone cause fatal brain injury? A biomechanical assessment of the Duhaime shaken baby syndrome model. Med Sci Law 43:317–333
15. Prange MT, Coats B, Duhaime AC, Margulies SS (2003) Anthropomorphic simulations of falls, shakes, and inflicted impacts in infants. J Neurosurg 2003(99):143–150
16. Bechtel K, Stoessel K, Leventhal JM, Ogle E, Teague B, Lavietes S et al (2004) Characteristics that distinguish accidental from abusive injury in hospitalized young children with head trauma. Pediatrics 114(1):165–168
17. Duhaime AC, Alario AJ, Lewander WJ, Schut L, Sutton LN, Seidl TS et al (1992) Head injury in very young children: mechanisms, injury types, and ophthalmologic findings in 100 hospitalized patients younger than 2 years of age. Pediatrics 90(2 pt 1):179–185
18. Looney C, Smith J, Merck L et al (2007) Intracranial hemorrhage in asymptomatic neonates: prevalence on MRI and relationship to obstetric and neonatal risk factors. Radiology 242:535–541
19. Rooks VJ, Eaton JP, Ruess L, Petermann GW, Keck-Wherley J, Pedersen RC (2008) Prevalence and evolution of intracranial hemorrhage in asymptomatic term infants. AJNR Am J Neuroradiol (6):1082–1089
20. Gabaeff SC (2013) Investigating the possibility and probability of perinatal subdural hematoma progressing to chronic subdural hematoma, with and without complications, in neonates, and its potential relationship to the misdiagnosis of abusive head trauma. Leg Med (Tokyo) 15:177–192
21. McKeag H, Christian CW, Rubin D, Daymont C, Pollock AN, Wood J (2013) Subdural hemorrhage in pediatric patients with enlargement of the subarachnoid spaces. J Neurosurg Pediatr 11(4):438–444, Epub 2013 Feb 8
22. Carr RB, Khanna PC, Saneto RP (2012) Childhood subdural hemorrhage, macrocephaly, and coagulopathy associated with Prader-Willi syndrome: case report and review of the literature. Pediatr Neurol 47(1):59–61
23. Fujiwara T, Okuyama M, Miyasaka M (2008) Characteristics that distinguish abusive from nonabusive head trauma among young children who underwent head computed tomography in Japan. Pediatrics 122(4):e841–e847, Epub 2008 Sep 1
24. Stray-Pedersen A, Omland S, Nedregaard B, Klevberg S, Rognum T (2011) An infant with subdural hematoma and retinal hemorrhages: does von Willebrand disease explain the findings? Forensic Sci Med Pathol 7:37e41
25. Lin C, Hung G, Chang C, Chien J (2005) Subdural hemorrhage in a child with acute promyelocytic leukemia presenting as subtle headache. J Chin Med Assoc 68:437e40
26. Nassogne M, Sharrard M, Hertz-Pannier L et al (2002) Massive subdural haematomas in Menkes disease mimicking shaken baby syndrome. Childs Nerv Syst 18:729e31
27. Vinchon M, de Foort-Dhellemmes S, Desurmont M, Delestret I (2010) Confessed abuse versus witnessed accidents in infants: comparison of clinical, radiological, and ophthalmological data in corroborated cases. Childs Nerv Syst 26(5):637–645, Epub 2009 Nov 28
28. Nabeshima S, Reese TS, Landis DM et al (1975) Junctions in the meninges and marginal glia. J Comp Neurol 164:127–169

29. Frederickson RG (1991) The subdural space interpreted as a cellular layer of meninges. Anat Rec 230:38–51
30. Hellbusch LC (2007) Benign extracerebral fluid collections in infancy: clinical presentation and long-term follow-up. J Neurosurg 107:119–125
31. Haines DE, Harkey HL, Al-Mefty O (1993) The "subdural" space: a new look at an outdated concept. Neurosurgery 32(1):111–120
32. Rascol MM, Izard JY (1976) The subdural neurothelium of the cranial meninges in man. Anat Rec 186:429–436
33. Geddes J, Hackshaw A, Vowles G et al (2001) Neuropathology of inflicted head injury in children. I. Patterns of brain damage. Brain 124:1290–1298
34. Christian CW, Block R, Committee on Child Abuse and Neglect, American Academy of Pediatrics (2009) Abusive head trauma in infants and children. Pediatrics 123(5):1409–1411
35. Han H, Tao W, Zhang M (2007) The dural entrance of cerebral bridging veins into the superior sagittal sinus: an anatomical comparison between cadavers and digital subtraction angiography. Neuroradiology 49:169–175
36. Lowenheilm P (1974) Dynamic properties of the parasagittal bridging veins. Z Rechtsmed 74:55–62
37. Lee M-C, Haut RC (1989) Insensitivity of tensile failure properties of human bridging veins to strain rate: implications in biomechanics of subdural hematoma. J Biomech 22:537–542
38. Vignes JR, Dagain A, Guerin J et al (2007) A hypothesis of cerebral venous system regulation based on a study of the junction between the cortical bridging veins and the superior sagittal sinus. Laboratory investigation. J Neurosurg 107:1205–1210
39. Si Z, Luan L, Kong D et al (2008) MRI-based investigation on outflow segment of cerebral venous system under increased ICP condition. Eur J Med Res 13:121–126
40. Durham SR, Duhaime AC (2007) Basic science: maturation-dependent response of the immature brain to experimental subdural hematoma. J Neurotrauma 24(1):5–14
41. Duhaime AC, Hunter JV, Grate LL, Kim A, Golden J, Demidenko E, Harris C (2003) Magnetic resonance imaging studies of age-dependent responses to scaled focal brain injury in the piglet. J Neurosurg 99(3):542–548
42. Duhaime AC, Christian C, Moss E, Seidl T (1996) Long-term outcome in infants with the shaking-impact syndrome. Pediatr Neurosurg 24(6):292–298
43. Duhaime AC, Durham S (2007) Traumatic brain injury in infants: the phenomenon of subdural hemorrhage with hemispheric hypodensity ("Big Black Brain"). Prog Brain Res 161:293–302
44. Duhaime A, Bilabiuk I, Zimmerman R (1993) The "big black brain": radiographic changes after severe inflicted head injury in infancy. J Neurotrauma 10:s59
45. Ewing-Cobbs L, Kramer L, Prasad M, Canales DN, Louis PT, Fletcher JM et al (1998) Neuroimaging, physical, and developmental findings after inflicted and noninflicted traumatic brain injury in young children. Pediatrics 102(8):300–307
46. Giles E, Nelson M (1998) Cerebral complications of nonaccidental head injury in childhood. Pediatr Neurol 19:119–128
47. Dias M, Backstron J, Falk M et al (1990) Serial radiography in the infant shaken impact syndrome. Pediatr Neurosurg 29:77–85
48. Rorke LB (1992) Anatomical features of the developing brain implicated in pathogenesis of hypoxic-ischemic injury. Brain Pathol 2(3):211–221
49. Ichord RN, Naim M, Pollock AN, Nance ML, Margulies SS, Christian CW (2007) Hypoxic-ischemic injury complicates inflicted and accidental traumatic brain injury in young children: the role of diffusion-weighted imaging. J Neurotrauma 24(1):106–118
50. Monagle P, Ignjatovic V, Savoia H (2010) Hemostasis in neonates and children: pitfalls and dilemmas. Blood Rev 24(2):63–68, Epub 2010 Jan 13
51. Monagle P, Massicotte P (2011) Developmental haemostasis: secondary haemostasis. Semin Fetal Neonatal Med 16(6):294–300, Epub 2011 Aug 26

52. Hymel KP, Rumack CM, Hay TC, Strain JD, Jenny C (1997) Comparison of intracranial computed tomographic (CT) findings in pediatric abusive and accidental head trauma. Pediatr Radiol 27(9):743–747
53. Hymel KP, Jenny C, Block RW (2002) Intracranial hemorrhage and rebleeding in suspected victims of abusive head trauma: addressing the forensic controversies. Child Maltreat 7 (4):329–348
54. Kuroda Y, Inglis FM, Miller JD, McCulloch J, Graham DI, Bullock R (1992) Transient glucose hypermetabolism after acute subdural hematoma in the rat. J Neurosurg 76:471–477
55. Salvant JB Jr, Muizelaar JP (1993) Changes in cerebral blood flow and metabolism related to the presence of subdural hematoma. Neurosurgery 33:387–393
56. Suwanwela C, Suwanwela N (1972) Intracranial arterial narrowing and spasm in acute head injury. J Neurosurg 36:314–323
57. Bullock R, Butcher S, Chen MH, Kendall L, McCulloch J (1991) Extracellular glutamate concentrations correlates with the extent if blood flow reduction after subdural hematoma in the rat. J Neurosurg 74:714–802
58. Miller JD, Bullock R, Graham DI, Chen MH, Teasdale GM (1990) Ischemic damage in a model of acute subdural hematoma. Neurosurgery 27:433–439
59. Muizelaar JP, Ward JD, Marmarou A, Newlon PG, Wachi A (1989) Cerebral blood flow and metabolism in severely head injured children. Part 2: autoregulation. J Neurosurg 71:72–76
60. DeSalles AAF, Muizelaar JP, Young HF (1987) Hyperglycemia, cerebrospinal fluid lactic acidosis and cerebral blood flow in severely head injured patients. Neurosurgery 21:45–50
61. Clubb RJ, Maxwell RE, Chou SN (1980) Experimental brain injury in the dog. J Neurosurg 52:189–196
62. Regan RF, Guo YP (1988) Toxic effect of hemoglobin on spinal cord neurons in culture. J Neurotrauma 15:645–653
63. Jaremko KM, Cheng-Roetling J, Chen L, Regan RF (2010) Accelerated hemolysis and neurotoxicity in neuron-glia-blood clot co-cultures. J Neurochem 114:1063–1073
64. Timaru-Kast R, Meissner A, Heimann A, Hoelper B, Kempski O, Alessandri B (2008) Acute subdural hematoma in pigs: role of volume on multiparametric neuromonitoring and histology. J Neurotrauma 25(9):1107–1119
65. Golden N, Maliawan S (2005) Clinical analysis of non-accidental head injury in infants. J Clin Neurosci 12(3):235–239
66. Varatharaj A, Mack J, Davidson JR, Gutnikov A, Squier W (2012) Mast cells in the human dura: effects of age and dural bleeding. Childs Nerv Syst 28(4):541–545, Epub 2012 Jan 22
67. Superti-Furga A, Hoffmann GF (1997) Glutaric aciduria type 1 (glutaryl-CoA-dehydrogenase deficiency): advances and unanswered questions. Report from an international meeting. Eur J Pediatr 156:821e8
68. Hymel KP, Jenny C, Block RW (2002) Intracranial hemorrhage and rebleeding in suspected victims of abusive head trauma: addressing the forensic controversies. Child Maltreat 7:329e48
69. Wuerfel nee Tyslak E, Petersen D, Gottschalk S, Gerling I, Gliemroth J, Thyen U (2012) Progression of chronic subdural haematomas in an infant boy after abusive head trauma. Eur J Paediatr Neurol 16(6):736–739, Epub 2012 Mar 13
70. Jenny C (2009) Preventing head trauma from abuse in infants. CMAJ 180(7):703–704

Chapter 26
Blood Genomics After Brain Ischemia, Hemorrhage, and Trauma

Da Zhi Liu, Glen C. Jickling, Boryana Stamova, Xinhua Zhan, Bradley P. Ander, and Frank R. Sharp

Abstract Peripheral blood is routinely used for RNA expression studies. However, blood is a challenging tissue for studying gene expression due to the fact that blood has a variety of components, composed of plasma and multiple cell subsets (i.e., leukocytes, platelets, red blood cells). Most genome-wide expression studies of blood are based on analysis of leukocytes, because the leukocytes are able to recruit and migrate into the site of injury within the brain. Recently, circulating cell-free plasma RNAs have received more and more attentions for clinical applications, since increasing evidence supports that the release of RNA into plasma may be mediated by microvesicles and exosomes coming from cells undergoing necrosis and apoptosis, though the definite origin and release mechanisms of plasma RNA remain incompletely understood. Blood genomic studies will provide diagnostic, prognostic, and therapeutic markers and will advance our understanding of brain ischemia, hemorrhage, and trauma in humans. New techniques to measure all coding and noncoding RNAs along with alternatively spliced transcripts will markedly advance molecular studies of these acute brain injuries.

26.1 Introduction

The sequencing of the human genome greatly advances our knowledge about the causes and mechanisms involved in human diseases [1, 2]. There have been identified in human genome thousands of genes (mRNAs) and hundreds of non-coding RNAs, such as transfer RNA (tRNA), ribosomal (rRNA), siRNAs, snoRNAs, microRNAs (miRNAs), and piRNAs [3, 4]. Recent advances in micro-array technology and statistical algorithms facilitate large-scale genomic studies to compare mRNAs and miRNA (the most widely studied and characterized

D.Z. Liu (✉)
Department of Neurology and the MIND Institute, University of California at Davis, 2805 50th Street, Sacramento, CA 95817, USA
e-mail: dzliu@ucdavis.edu

E.H. Lo et al. (eds.), *Vascular Mechanisms in CNS Trauma*, Springer Series in Translational Stroke Research 5, DOI 10.1007/978-1-4614-8690-9_26,

noncoding RNA) expression from healthy individuals and patients. Genomic (i.e., mRNA, miRNA expression) studies in cancer tissue specimens have led the way and have indicated the potential of this methodology for diagnostic use in clinical applications [5–8]. Unlike cancer tissue specimens, sampling brain tissue from living human is not practical for most neurological diseases, thus investigators have used peripheral blood (i.e., blood cells, plasma/serum) as a widely accessible source of RNA to perform blood genomic studies [9]. This chapter introduces the rationale and potential clinical applications (i.e., subtype classification, severity evaluation, and outcome prediction) of blood genomics after acute brain injuries such as brain ischemia, hemorrhage, and trauma.

26.2 Peripheral Blood and Blood Components

Peripheral blood is routinely used for RNA expression studies because it can be easily collected. However, blood is a challenging tissue for studying gene expression by the fact that blood is a mixed tissue, composed of multiple cell subsets, so that differential expression profiles can reflect changes in cell subset proportions, changes in subset-specific gene expression, or both [10]. Moreover, RNA information primarily resides in the circulating blood leukocytes that account for small portion of the whole blood fraction [11]. Thus, it is important to understand the blood components that may affect the accuracy and consistency of gene expression measurement in blood.

Blood is made up of plasma (~55 %) and a mixture of multiple cell types at different stages of their life cycles [11]. The three primary blood cell types are red blood cells (RBC or erythrocytes) (~45 %), leukocytes or peripheral blood mononuclear cells (PBMCs) (~0.1 %), and platelets (thrombocytes) (~0.17 %) [11]. The leukocytes are commonly divided into myeloid and lymphoid cells: (1) myeloid cells include monocytes and their descendants, as well as granulocytes like neutrophils and basophils, and (2) lymphoid cells are primarily composed of B cells, T cells, and NK cells [10]. Proportions of these cells can vary widely between individuals, but T cells and B cells together usually make up ~75 % of PBMCs, while NK cells and monocytes make up around 10–15 % each. The remaining cell types, such as dendritic cells, are much more rare and account for <1 % of total PBMCs [12]. Neutrophils, which normally compose the majority of cells in a blood sample (40–80 %), are normally excluded by the methods used to isolate PBMCs, but may account for up to 20 % of a PBMC sample due to contamination [13].

26.3 Genomic Study in Blood Cells

Most genome-wide expression studies are based on analysis of leukocytes/PBMCs [10]. A conceptual issue in the field has been how leukocytes can report on or respond to brain infarction [14]. Though the answer is still unclear, leukocytes patrol the body and interact with cells from every tissue, endothelial cells of the vasculature, foreign organisms or cells, injured cells, and every element within blood [14]. In addition, leukocytes have a complement of expressed RNAs that reflect a combination of the genetics of these cells as well as the interactions of those cells with their environment [9].

26.3.1 Rationale of Blood Cell RNA for Genomic Study

Leukocytes of all types express various adhesion molecules on their surface and interact with normal or inflamed endothelium [15]. Thus, injured brain signals to endothelium to express different adhesion molecules on the luminal side of the vessel, which in turn signal to leukocytes [16, 17]. Leukocytes in blood are known to interact with both platelets and elements of the atherosclerotic plaque and may contribute to disease [14, 18, 19]. Thus, endothelial cell–platelet interactions signal to leukocytes and endothelial cell–atherosclerotic plaque interactions signal to leukocytes [20]. Moreover, blood clots that form in the heart or other sites are detected by circulating leukocytes. In addition, the platelets and clots that form on atherosclerotic plaques interact with leukocytes, which may contribute to instability of the potentially embolic elements [21]. Finally, leukocytes independent of the above influences respond to cytokines, chemokines, hormones, exosomal miRNAs, and other molecules in blood, which differ as a function of the genetics of each individual cell and their environment [14]. Indeed, each of the above leukocyte interactions results from a complex interplay of the individual environmental interactions of each cell and the genetic makeup of the cells [22]. Thus, though leukocytes themselves do not cause brain injury, they can sense and likely have specific intracellular signaling related to the main causes of brain injury [9].

26.3.2 Animal mRNA Profiling Studies in Blood Cells: Classification of Brain Ischemia, Hemorrhage, and Other Brain Injuries

Even assuming leukocytes detect all of the events that occur after brain injury as described above, we initially had no idea whether there would be enough reactive leukocytes in blood that could be detected by taking a single blood sample from an animal or human at a single point in time. After all, there are hundreds of thousands of leukocytes in peripheral blood, only a fraction of which might respond to factors related to stroke, potentially making it difficult or impossible to detect RNA changes in the few leukocytes collected in a single blood sample. To address this question, we performed the first study of its kind using several different experimental brain injury models in rats [23, 24].

Adult rats were subjected to experimental ischemic strokes, hemorrhagic strokes, kainic acid-induced status epilepticus, insulin-induced hypoglycemia, or hypoxia and compared with sham operated controls and to untouched, naive controls [23]. At 24 h, RNA from peripheral blood monocytes (PBMCs) was processed on Affymetrix microarrays. There were hundreds of upregulated and downregulated genes for each condition compared with sham or untouched controls [23]. This study demonstrated that (1) there were detectable changes of gene expression in blood 24 h after each injury, (2) no single gene was specific for a given injury, (3) there were groups or "profiles" of genes that distinguished each condition from the other, and (4) there is a peripheral blood genomic response to neuronal injury [23, 24]. Though there were genes shared by every injury, perhaps related to stress or similar mechanisms of injury, there were profiles specific for the injury [23]. These studies provided the first proof of principle that gene expression in blood following brain injury can be used as a marker of the neuronal damage, and specific gene expression profiles in blood were associated with different types of brain injury [23, 24].

In a follow-up study we demonstrated that there were specific gene profiles in brain for each of these injuries as well [25]. Just as in the blood, there were genes that were common to all of the injuries and could represent responses to stress, neuronal injury, or death and other factors common to each condition. Though many genes expressed in blood were also expressed in brain, the majorities were different [25]. Thus, one cannot necessarily use blood cell to infer changes of mRNA expression in brain [9].

26.3.3 Human mRNA Profiling Studies in Blood Cells: Assessing Diagnosis of Ischemic Stroke

The first human study to assess RNA expression in stroke was published by the Baird group [26]. Using PBMCs obtained 1–4 days following ischemic stroke, 190 genes were significantly regulated in 20 stroke compared with 20 control subjects [26]. A panel of 22 genes derived from the prediction analysis for microarrays algorithm in the index cohort ($n = 40$) classified stroke in the validation cohort ($n = 20$), with a sensitivity of 78 % and a specificity of 80 %. These findings were extremely important because they used the genes in one cohort to predict stroke in a second cohort—supporting the validity of the findings within this study [26], supporting the proof of principle in our previous rodent studies [23]. Notably, even though virtually identical methods were used in our prior rodent study [23] compared to this human study (both used PBMCs and Affymetrix arrays), very few genes were similarly regulated in the blood of rats [23] compared to the humans with ischemic strokes [26]. The explanation for this is unclear but is reviewed elsewhere [27]. It is possible that the experimental methods of producing ischemia (suture methods, anesthesia, young animals with no vascular disease) in animals simply result in different gene expression responses compared with stroke in humans or the immune responses in rodents may be different in humans following ischemic stroke. These and other possibilities need further study since they are relevant to understanding how well animal stroke studies "model" human stroke.

26.3.4 Human mRNA Profiling Studies in Whole Blood: Assessing Diagnosis of Ischemic Stroke

In our initial rodent study [23] and in the first human study [26], blood was drawn and then PBMCs were separated from the blood using a Ficoll gradient and centrifugation. This procedure in and of itself could affect gene expression in blood. If performed at different times after stroke, or if the methods were not identical, this could affect gene expression.

We therefore utilized PAXgene tubes to address this issue. These commercial vacutainer tubes lyse cells and stabilize RNA and have proven to be reliable for many clinical studies [28–31]. Blood from 15 patients was drawn into PAXgene tubes at <3, 5, and 24 h after ischemic stroke ($n = 45$ samples) and compared with 14 control samples [32]. RNA processed on Affymetrix U133 microarrays showed that over 1,000 genes were upregulated or downregulated in the blood of ischemic stroke compared with control subjects [32]. Most of the genes expressed at 2–3 h after stroke (before treatment) were also expressed at 5 and 24 h after the strokes [32]. Prediction analysis of microarrays derived the 25 probe sets for 18 genes that were most predictive of stroke [32]. The fold change of these genes varied from 1.6 to 6.8, and these genes correctly classified 10/15 patients at 2.4 h, 13/15 patients at 5 h, and 15/15 patients at 24 h after stroke [32]. When the results of this study were compared with the previous study from Moore et al., however, there were very few genes that were common to both. The explanations for these differences probably included bloods were drawn at different times after the stroke, differences in treatment, the use of PBMCs for the Moore et al. study and the use of whole blood/PAXgene tubes for the Tang et al. study, and the use of different RNA isolation and labeling methods and the use of different arrays for the two studies.

To address the issue of reproducibility, we have recently repeated our initial study in a larger cohort [33]. Thus, patients with ischemic stroke ($n = 70$, 199 samples) were compared with control subjects who were healthy ($n = 38$), controls with vascular risk factors ($n = 52$), and to subjects who had myocardial infarction ($n = 17$). Whole blood was drawn into PAXgene tubes at p3, 5, and 24 h after stroke onset and RNA processed on whole genome Affymetrix U133 microarrays.

The 25 probe sets previously reported in our study by Tang et al. predicted a new set of ischemic strokes with 93.5 % sensitivity and 89.5 % specificity. In order to derive profiles that would distinguish ischemic stroke from all control subjects, we derived 60 and 46 probe sets that differentiated control groups from 3 to 24 h ischemic stroke samples, respectively [33]. Thus, this study replicated our previously reported gene expression profile in a larger cohort and identified additional genes that discriminate ischemic stroke from relevant control groups [33].

Finally, another recent study from Barr supports the above findings. Whole blood was obtained in PAXgene tubes from 39 ischemic stroke patients and 25 healthy control subjects [34]. RNA was processed on Illumina HumanRef-8v2 bead chips. Among a large number of regulated genes, they identified a nine-gene profile that separated ischemic stroke patients compared with controls [34]. Moreover, five of these nine genes were identified in our previous study [32]. Thus, another group has confirmed at least a core set of genes, and we have replicated our own gene expression studies following ischemic stroke.

All of these studies, however, are confounded to some degree by various treatments, comparisons to healthy controls, variations in time after stroke, different risk factors between groups, and differences of age, race, and gender. Nonetheless, the first test of the technology has been achieved: replication of results and independent validation by at least two different groups. Such promising results provide strong support for further study.

26.3.5 Human mRNA Profiling Studies in Whole Blood: Subtype Classification of Ischemic Stroke

Early on it was apparent that developing a "diagnostic test for stroke" would be difficult and perhaps not of practical use unless it could be performed within the first few hours of stroke. In addition, a diagnostic test would not only have to diagnose ischemic stroke but rule out hemorrhagic stroke if it were used to guide acute stroke treatments such as tissue plasminogen activator (tPA).

Thus, we have approached different questions that could be rapidly translated to the care of stroke patients. We asked whether gene profiles in blood exist that are specific for the different causes of ischemic stroke. The main reason for developing such profiles would be to use them to diagnose the cause of ischemic stroke in those patients with "cryptogenic stroke" with no known cause who represent approximately one-third of all ischemic strokes.

In the first study, whole blood was collected in PAXgene tubes from acute ischemic stroke patients (<3, 5, and 24 h) and healthy controls. RNA was isolated and processed on Affymetrix Human U133 Plus 2.0 Arrays. Expression profiles in the blood of cardioembolic stroke patients differed from large-vessel atherosclerotic stroke patients [35]. Of the 77 genes that differed between the two groups (fold change >1.5, $P < 0.05$), a minimum number of 23 genes differentiated the two types of stroke with >90 % specificity and sensitivity [35]. Notably, some of the genes that distinguished cardioembolic from atherosclerotic stroke displayed little change over time [35]. These might be genes expressed differentially prior to stroke—and perhaps indicate risk of stroke. Other genes displayed significant change over time, suggesting that these time-dependent alterations in gene expression were associated with differential gene expression of immune cells due to the strokes caused by cardioembolism compared with atheroembolism [35].

We have confirmed these initial findings using identical methods to study 194 samples from 76 acute ischemic stroke patients [36]. A 40-gene profile differentiated cardioembolic stroke from large-vessel stroke with >90 % sensitivity and specificity [36]. A separate 37-gene profile differentiated cardioembolic stroke due to atrial fibrillation from nonatrial fibrillation causes with >90 % sensitivity and specificity [36]. When these profiles were applied to patients with cryptogenic stroke, 17 % were predicted to be large vessel and 41 % to be cardioembolic stroke. Of the cryptogenic strokes predicted to be cardioembolic, 27 % were predicted to have atrial fibrillation [36]. Moreover, our recent studies showed another gene expression profile to distinguish lacunar from nonlacunar stroke [37] and a separate gene expression profile (in conjunction with a measure of infarct location) to predict a probable cause in cryptogenic strokes [38].

26.3.6 Animal microRNA Profiling Studies in Whole Blood: Classification of Brain Ischemia, Hemorrhage, and Other Brain Injuries

The recent discovery of miRNAs, which are endogenous, noncoding, single-stranded RNA molecules of 19–25 nucleotides in length, has introduced a new level and mechanism of gene regulation. Moreover, the increasing reports have proved that the profiling miRNA expression in blood cells is diagnostically useful in clinic [39, 40]. We and others examined the miRNA expression in animal blood cells and brain in a variety of brain injuries, such as transient ischemic stroke [41], permanent ischemic stroke [42], hemorrhagic stroke [42], and kainic acid-induced status epilepticus [42].

Using an ischemic stroke model, Jeyaseelan et al. reported that (1) miR-19b, miR-290, and miR-292-5p were highly expressed, whereas miR-103 and miR-107 were poorly expressed 24 h after the reperfusion; (2) a new group of miRNAs (i.e., miR-150, miR-195, miR-352, miR-26b, miR-103, miR-107, miR-26a, let-7c) was observed in the 48-h-reperfusion blood samples; (3) among the 14 miRNAs that appeared at both time points, miR-150, miR-195, and miR-320 showed an opposite trend in expression at 24 and 48 h, whereas miR-103, miR-107, and miR-191 showed almost the same level of downregulation at both time points; (4) only a few miRNAs were changed in blood and brain after the reperfusion [41]. Using a different animal model, we assessed the whole blood and brain miRNA expression profiles 24 h after permanent ischemic stroke, hemorrhagic stroke, and kainic acid-induced status epilepticus using TaqMan rodent miRNA arrays [42]. Our miRNA profiling data showed that (1) the blood and brain miRNA response profiles were different for each condition; (2) many miRNAs changed more than 1.5-fold in blood and brain after each experimental manipulation, and several miRNAs were up- or downregulated in both brain and blood after a given injury; and (3) a few miRNAs (e.g., miR-298, miR-155, miR-362-3p) were up- or downregulated more than twofold in both brain and blood after several different injuries [42]. The results demonstrated that the possible use of blood miRNAs as biomarkers for brain injury [41, 42]. However, only a few miRNAs were changed in both blood and brain after each condition, since major changes of miRNA profiles were different in blood and brain [41, 42]. Thus, one cannot necessarily use blood cells to infer changes of miRNA expression in brain [42].

26.3.7 Human microRNA Profiling Studies in Whole Blood: Assessing Diagnosis of Ischemic Stroke

Following their animal studies, Jeyaseelan et al. first expanded miRNA profiling studies into human and described that miRNAs can be detected in total peripheral blood in human [43] as was demonstrated for rat's blood [41]. They selected the ischemic stroke patients aged between 18 and 49 years, characterized based on World Health Organization clinical criteria were further classified according to TOAST classification: (1) large-vessel atherosclerosis (LA), (2) small-vessel disease (SA), (3) cardioembolism (CEmb), and (4) undetermined cause (UND). The patients' functional status at the time of blood sampling (at the outpatient clinics) was evaluated with the modified Rankin Scale (mRS). Blood samples from normal individuals were used as controls. The data showed that miRNAs were differentially expressed between normal and stroke subjects [43]. Among the 157 miRNAs identified for total stroke samples, 138 miRNAs have been found to be highly expressed and 19 miRNAs have been found to have lower expression [43].

26.3.8 Human microRNA Profiling Studies in Whole Blood: Subtype Classification of Ischemic Stroke

In the stroke subtype classification study, Jeyaseelan group proved that within the 19 low expressed miRNAs, 8 miRNAs (hsa-let-7f, miR-126, miR-1259, miR-142-3p, miR-15b, miR-186, miR-519e, miR-768-5p) were common to the three subtypes of stroke (LA, SA, and CEmb) [43]. Similarly, among the highly expressed 138 miRNAs that were observed for all stroke patients, 17 miRNAs (hsa-let-7e, miR-1184, miR-1246, miR-1261, miR-1275, miR-1285, miR-1290, miR-181a, miR-25, miR-513a-5p, miR-550, miR-602, miR-665, miR-891a, miR-933, miR-939, miR-923) can also be identified as highly expressed in the subtypes [43].

26.3.9 Human microRNA Profiling Studies in Whole Blood: Outcome Prediction of Ischemic Stroke

Moreover, Jeyaseelan group demonstrated that miRNA expression profiling could predict clinical outcome (mRS) following stroke [43]. Principal component analysis (PCA) of the stroke samples based on the mRS showed that all samples of good outcome (mRS < 2; stroke, LA, and CEmb) have been found to cluster along the same panel with almost consistent distant between them [43]. The poor outcome (mRS > 2) stroke samples have been found to cluster away from the samples with good outcome (mRS < 2) [43].

26.4 Genomic Studies in Plasma/Serum

In recent years, increasing interest has been focused on the potential diagnostic and prognostic application of circulating cell-free plasma RNAs. It was reported that plasma mRNAs are usually present as short fragments of less than 1,000 nt [44], presumably because ribonuclease (RNase) activity is extremely high in blood plasma [45]. Aside from presence of entire mRNAs and their fragments in plasma of human blood [46, 47], there are various forms of cellular RNAs in the circulation, including rRNAs, tRNAs, snRNAs, snoRNAs, a tremendous variety of siRNAs, miRNAs, and noncoding regulatory RNAs [3, 4], among which miRNAs are the most widely studied and characterized. Compared to mRNAs, miRNAs in plasma are extremely stable, often found in association with Argonaute protein, microvesicles, or exosomes, and represent potentially informative biomarkers for a range of diseases mainly due to their high stability in plasma [7]. The plasma miRNAs, that have been wildly practiced in cancer diagnosis [48], can also be used as biomarkers of brain injuries, especially in combination with established clinical practices such as imaging, neurocognitive, and motor examinations, have the potential to improve brain injury patient classification and possibly management.

26.4.1 Rationale of Circulating Cell-Free RNA for Genomic Study

The changes of plasma RNA may infer the changes of RNA expression in injured brain cells, since increasing evidence supports that release of RNA into plasma may be mediated by the microvesicles (~100 nm to 1 μm) and exosomes (~30–100 nm) originating from cells undergoing necrosis and apoptosis [46, 47], though the definite origin and release mechanisms of plasma RNA have remained incompletely understood.

26.4.2 *Animal microRNA Profiling Studies in Plasma/Serum: Biomarkers of Ischemic Stroke and Blast Brain Trauma Injury*

Using the rat transient (60–90 min) and permanent middle cerebral artery occlusion (MCAO) stroke models, Laterza et al. proved that brain-specific miR-124 can be used to monitor ischemia-related brain injury starting at 8 h and peaking at 24 h after occlusion (an approximately 150-fold increase in plasma relative to the sham surgery control group) [49]. Balakathiresan et al. studied the effect of blast traumatic brain injury (TBI) on the miRNA signatures in the serum of rats [50] and described five miRNAs (such as miR-let-7i, miR-122, miR-340-5p, miR-200b, and miR-874) were significantly modulated in the serum samples of these animals at three time points post-blast TBI [50].

26.4.3 *Human microRNA Profiling Studies in Plasma: Severity Evaluation of Brain Trauma Injury*

Redell et al. examined the altered plasma miRNA levels in patients with TBI relative to matched healthy volunteers [51]. In severe TBI patients (Glasgow Coma Scale [GCS] score ≤8), miR-16, miR-92a, and miR-765 were increased [51]. In mild TBI patients (GCS score > 12), miR-765 levels were unchanged, while the plasma levels of miR-92a and miR-16 were significantly increased within the first 24 h of injury compared to healthy volunteers [51]. This study demonstrated that circulating miRNA levels in plasma were altered after TBI, providing a rich new source of potential molecular biomarkers.

26.5 Perspectives

The future blood cell genomic studies will require isolation of individual cell types including neutrophils, B and T lymphocytes, monocytes, and their many subtypes [9]. We and others have published gene expression profiles for these cell types [52–56]. The potential problem with these studies is that isolation may affect gene expression [54, 57–59]. This may be solved by microfluidic devices that use cell-specific antibodies that isolate specific cell types at the bed side [54].

Microarrays may eventually be replaced by next generation sequencing methods (RNA sequencing) to evaluate gene expression [60]. With this technology, the entire transcriptome (RNAs) of a given individual can be sequenced [61]. In theory, alternative splice variants of a given RNA can be quantified, as can the expression at the individual exon level [60, 61].

Profiling the expression of noncoding RNAs in blood cells or plasma/serum, including the recently discovered miRNAs, may be a growth field that will likely fuel disease-related research [60]. This may also apply to different types of brain injuries where types of noncoding RNA regulation will be specific for each type of brain injury [60].

References

1. Lander ES et al (2001) Initial sequencing and analysis of the human genome. Nature 409 (6822):860–921
2. International Human Genome Sequencing Consortium (2004) Finishing the euchromatic sequence of the human genome. Nature 431(7011):931–945
3. Pennisi E (2012) Genomics. ENCODE project writes eulogy for junk DNA. Science 337 (6099):1159, 1161
4. Eddy SR (2001) Non-coding RNA genes and the modern RNA world. Nat Rev Genet 2 (12):919–929
5. Yang B et al (2011) Using peripheral blood mRNA signature to distinguish between breast cancer and benign breast disease in non-conclusive mammography patients. Cancer Biol Ther 10(12):1235–1239
6. Olmos D et al (2012) Prognostic value of blood mRNA expression signatures in castration-resistant prostate cancer: a prospective, two-stage study. Lancet Oncol 13(11):1114–1124
7. Mitchell PS et al (2008) Circulating microRNAs as stable blood-based markers for cancer detection. Proc Natl Acad Sci U S A 105(30):10513–10518
8. Allegra A et al (2012) Circulating microRNAs: new biomarkers in diagnosis, prognosis and treatment of cancer (review). Int J Oncol 41(6):1897–1912
9. Sharp FR et al (2011) Molecular markers and mechanisms of stroke: RNA studies of blood in animals and humans. J Cereb Blood Flow Metab 31(7):1513–1531
10. Bolen CR, Uduman M, Kleinstein SH (2011) Cell subset prediction for blood genomic studies. BMC Bioinformatics 12:258
11. WTS Prohibited Blood Components. http://www.ajwrb.org/science/prohibit.html
12. Autissier P et al (2010) Evaluation of a 12-color flow cytometry panel to study lymphocyte, monocyte, and dendritic cell subsets in humans. Cytometry A 77(5):410–419
13. Schlenke P et al (1998) Evaluation of a novel mononuclear cell isolation procedure for serological HLA typing. Clin Diagn Lab Immunol 5(6):808–813
14. Franks ZG et al (2010) Platelet-leukocyte interactions link inflammatory and thromboembolic events in ischemic stroke. Ann N Y Acad Sci 1207:11–17
15. Pries AR, Kuebler WM (2006) Normal endothelium. Handb Exp Pharmacol 176(pt 1):1–40
16. Beck J et al (1997) Leukocyte-endothelium interactions in global cerebral ischemia. Acta Neurochir Suppl 70:53–55
17. Beck J et al (2007) Activation of leukocyte-endothelial interactions and reduction of selective neuronal death after global cerebral ischemia. Neurosci Lett 414(2):159–164

18. Caplan LR, Fisher M (2007) The endothelium, platelets, and brain ischemia. Rev Neurol Dis 4 (3):113–121
19. Htun P et al (2006) Course of platelet activation and platelet-leukocyte interaction in cerebrovascular ischemia. Stroke 37(9):2283–2287
20. Clark WM et al (1993) Circulating intercellular adhesion molecule-1 levels and neutrophil adhesion in stroke. J Neuroimmunol 44(1):123–125
21. Akopov S, Sercombe R, Seylaz J (1996) Cerebrovascular reactivity: role of endothelium/platelet/leukocyte interactions. Cerebrovasc Brain Metab Rev 8(1):11–94
22. Elneihoum AM et al (1996) Leukocyte activation detected by increased plasma levels of inflammatory mediators in patients with ischemic cerebrovascular diseases. Stroke 27 (10):1734–1738
23. Tang Y et al (2001) Blood genomic responses differ after stroke, seizures, hypoglycemia, and hypoxia: blood genomic fingerprints of disease. Ann Neurol 50(6):699–707
24. Tang Y et al (2003) Blood genomic expression profile for neuronal injury. J Cereb Blood Flow Metab 23(3):310–319
25. Tang Y et al (2002) Genomic responses of the brain to ischemic stroke, intracerebral haemorrhage, kainate seizures, hypoglycemia, and hypoxia. Eur J Neurosci 15(12):1937–1952
26. Moore DF et al (2005) Using peripheral blood mononuclear cells to determine a gene expression profile of acute ischemic stroke: a pilot investigation. Circulation 111(2):212–221
27. Turner RJ, Jickling GC, Sharp FR (2011) Are underlying assumptions of current animal models of human stroke correct: from STAIRs to high hurdles? Transl Stroke Res 2 (2):138–143
28. Chai V et al (2005) Optimization of the PAXgene blood RNA extraction system for gene expression analysis of clinical samples. J Clin Lab Anal 19(5):182–188
29. Thach DC et al (2003) Assessment of two methods for handling blood in collection tubes with RNA stabilizing agent for surveillance of gene expression profiles with high density microarrays. J Immunol Methods 283(1–2):269–279
30. Vartanian K et al (2009) Gene expression profiling of whole blood: comparison of target preparation methods for accurate and reproducible microarray analysis. BMC Genomics 10:2
31. Yamamoto T et al (2006) Examination of stability of bone marrow blood RNA in the PAXgene tube. Lab Hematol 12(3):143–147
32. Tang Y et al (2006) Gene expression in blood changes rapidly in neutrophils and monocytes after ischemic stroke in humans: a microarray study. J Cereb Blood Flow Metab 26 (8):1089–1102
33. Stamova B et al (2010) Gene expression profiling of blood for the prediction of ischemic stroke. Stroke 41(10):2171–2177
34. Barr TL et al (2010) Genomic biomarkers and cellular pathways of ischemic stroke by RNA gene expression profiling. Neurology 75(11):1009–1014
35. Xu H et al (2008) Gene expression in peripheral blood differs after cardioembolic compared with large-vessel atherosclerotic stroke: biomarkers for the etiology of ischemic stroke. J Cereb Blood Flow Metab 28(7):1320–1328
36. Jickling GC et al (2010) Signatures of cardioembolic and large-vessel ischemic stroke. Ann Neurol 68(5):681–692
37. Jickling GC et al (2011) Profiles of lacunar and nonlacunar stroke. Ann Neurol 70(3):477–485
38. Jickling GC et al (2012) Prediction of cardioembolic, arterial, and lacunar causes of cryptogenic stroke by gene expression and infarct location. Stroke 43(8):2036–2041
39. Lu J et al (2005) MicroRNA expression profiles classify human cancers. Nature 435 (7043):834–838
40. Guarnieri DJ, DiLeone RJ (2008) MicroRNAs: a new class of gene regulators. Ann Med 40 (3):197–208
41. Jeyaseelan K, Lim KY, Armugam A (2008) MicroRNA expression in the blood and brain of rats subjected to transient focal ischemia by middle cerebral artery occlusion. Stroke 39 (3):959–966

42. Liu DZ et al (2010) Brain and blood microRNA expression profiling of ischemic stroke, intracerebral hemorrhage, and kainate seizures. J Cereb Blood Flow Metab 30(1):92–101
43. Tan KS et al (2009) Expression profile of MicroRNAs in young stroke patients. PLoS One 4 (11):e7689
44. Cerkovnik P et al (2007) Optimization of an RNA isolation procedure from plasma samples. Int J Mol Med 20(3):293–300
45. Tsui NB, Ng EK, Lo YM (2002) Stability of endogenous and added RNA in blood specimens, serum, and plasma. Clin Chem 48(10):1647–1653
46. Tsang JC, Lo YM (2007) Circulating nucleic acids in plasma/serum. Pathology 39(2):197–207
47. Gahan PB, Swaminathan R (2008) Circulating nucleic acids in plasma and serum. Recent developments. Ann N Y Acad Sci 1137:1–6
48. Etheridge A et al (2011) Extracellular microRNA: a new source of biomarkers. Mutat Res 717 (1–2):85–90
49. Laterza OF et al (2009) Plasma MicroRNAs as sensitive and specific biomarkers of tissue injury. Clin Chem 55(11):1977–1983
50. Balakathiresan N et al (2012) MicroRNA let-7i is a promising serum biomarker for blast-induced traumatic brain injury. J Neurotrauma 29(7):1379–1387
51. Redell JB et al (2010) Human traumatic brain injury alters plasma microRNA levels. J Neurotrauma 27(12):2147–2156
52. Du X et al (2006) Genomic profiles for human peripheral blood T cells, B cells, natural killer cells, monocytes, and polymorphonuclear cells: comparisons to ischemic stroke, migraine, and Tourette syndrome. Genomics 87(6):693–703
53. Kobayashi SD, Sturdevant DE, DeLeo FR (2007) Genome-scale transcript analyses in human neutrophils. Methods Mol Biol 412:441–453
54. Kotz KT et al (2010) Clinical microfluidics for neutrophil genomics and proteomics. Nat Med 16(9):1042–1047
55. Robbins SH et al (2008) Novel insights into the relationships between dendritic cell subsets in human and mouse revealed by genome-wide expression profiling. Genome Biol 9(1):R17
56. Wong HR et al (2010) Leukocyte subset-derived genomewide expression profiles in pediatric septic shock. Pediatr Crit Care Med 11(3):349–355
57. Holmes D et al (2009) Leukocyte analysis and differentiation using high speed microfluidic single cell impedance cytometry. Lab Chip 9(20):2881–2889
58. Lee D, Chen PJ, Lee GB (2010) The evolution of real-time PCR machines to real-time PCR chips. Biosens Bioelectron 25(7):1820–1824
59. Shim JS, Browne AW, Ahn CH (2010) An on-chip whole blood/plasma separator with bead-packed microchannel on COC polymer. Biomed Microdevices 12(5):949–957
60. Hawkins RD, Hon GC, Ren B (2010) Next-generation genomics: an integrative approach. Nat Rev Genet 11(7):476–486
61. Werner T (2010) Next generation sequencing in functional genomics. Brief Bioinform 11 (5):499–511

Chapter 27
Molecular Biomarkers in Neurocritical Care: The Next Frontier

Sherry H.-Y. Chou, Eng H. Lo, and MingMing Ning

Abstract With modern advances in life support and resuscitation medicine, neurologic injury in critical illness has become the new and the last frontier in critical care medicine. In addition to preserving life, the "holy grail" of modern critical care is to preserve function and quality of life. Molecular biomarkers have revolutionalized modern medicine, leading to novel gold standard diagnostics such as troponin for myocardial infarction, new disease monitors such as tumor markers, and new "personalized medicine" tools for selecting patient likely to respond to certain therapy such as Imatinib (Gleevec) use in Philadelphina chromosome chronic myelogenous leukemia. The central nervous system (CNS) poses a special challenge for diagnostic and therapeutic treatments due to the skull being a barrier to brain monitoring and tissue sampling, the presence of the blood–brain barrier (BBB), the complex relationship between localization and function, and the frequently poor reflection of clinical disease in animal models. Novel molecular biomarkers may help reflect underlying pathophysiology, monitor disease progression, identify intermediate phenotypes for clinical trials, and improve prognostic accuracy and thereby revolutionize clinical practice in neurocritical care.

27.1 Introduction

With modern advances in life support and resuscitation medicine, neurologic injury in critical illness has become the new and the last frontier in critical care medicine. In addition to preserving life, the "holy grail" of modern critical care is to preserve function and quality of life. Molecular biomarkers have revolutionized modern medicine, leading to novel gold standard diagnostics such as troponin for myocardial infarction, new disease monitors such as tumor markers, and new "personalized

S.H.-Y. Chou (✉)
Brigham and Women's Hospital, Department of Neurology, Boston, MA, USA
e-mail: schou1@partners.org

E.H. Lo et al. (eds.), *Vascular Mechanisms in CNS Trauma*, Springer Series in Translational Stroke Research 5, DOI 10.1007/978-1-4614-8690-9_27,

medicine" tools for selecting patient likely to respond to certain therapy such as Imatinib (Gleevec) use in Philadelphina chromosome chronic myelogenous leukemia. The central nervous system (CNS) poses a special challenge for diagnostic and therapeutic due to the skull being a barrier to brain monitoring and tissue sampling, the presence of the blood–brain barrier (BBB), and the complex relationship between localization and function, and the frequently poor reflection of clinical disease in animal models. Novel molecular biomarkers may help reflect underlying pathophysiology, monitor disease progression, identify intermediate phenotypes for clinical trials, and improve prognostic accuracy and thereby revolutionize clinical practice in neurocritical care.

27.2 Unmet Need in Neurocritical Care

Neurocritical care, the specialized care of critically ill patients with neurologic dysfunction, began with intensive care of the neurosurgical patients in the 1930s and has since significantly expanded in scope. Modern neurocritical care units treat a wide variety of conditions, including ischemic stroke, intracranial hemorrhage (ICH), subarachnoid hemorrhage (SAH), and traumatic brain injury (TBI). Though advances over time have led to improved survival [1], there remains significant unmet need in neurocritical care. Specific challenges include:

- Need to salvage neurologic function in addition to prolonging survival.
- Lack of "gold standard" diagnostics that correlate with clinical outcome.
- Lack of brain-oriented therapies that correlate with clinical outcome.
- Lack of continuous "brain monitors" that reflect real-time pathophysiology.
- Lack of therapeutics that can either reverse injury or prevent further neurologic deterioration.
- Outcome from neurologic injuries is complex and highly variable.

27.3 Biomarkers in Critical Illness: Special Considerations

An ideal clinical biomarker needs to be easily measurable, accurate, reproducible, cost-effective, and detectable in early stage of pathology and has high sensitivity, specificity, and predictive value. An ideal molecular biomarker in brain injury needs to have the following additional characteristics:

- Non (or minimally)-invasive.
- Fast turnaround time to inform clinical decisions.
- Results can alter treatment.
- Correlates with clinical outcome.
- Reflects reversible process.
- Yields information on disease pathophysiology or mechanism.
- Follows disease progression.

Critical illness is characterized by high phenotypic heterogeneity and the simultaneous presence of multiple interacting and rapidly changing disease processes. Clinical samples in critical illness are often confounded by treatments such as transfusions, medications, and surgery or by parallel disease processes such as infections. Therefore, the choice of appropriate "control" subjects for biomarker studies in critical illness is of utmost importance. While disease-free control subjects may establish a normal baseline, the magnitude of changes that occur in critical illness often requires the comparison with an appropriate diseased control cohort in order to determine what levels of biomarker changes are clinically relevant. Furthermore, disease processes in critical CNS injuries are frequently multiphasic, therefore the biomarkers may be as well. Many molecules, such as endothelin-1 (ET-1) in TBI [2], have biphasic or multiphasic changes over the course of disease, and understanding changes in a biomarker over time is very important in conditions such as critical illness.

The source of a molecular biomarker also needs special consideration in neurocritical care. The unique presence of the BBB and its variable and dynamic injury in various types of critical CNS injury makes it difficult to model how molecular biomarkers in blood can reflect real-time processes in the CNS. However, blood is a much more practical and less invasive source of biomarkers in clinical care and can be a good source of biomarker for some CNS pathologies such as hypoxic-ischemic brain injury following cardiac arrest. Cerebrospinal fluid (CSF) is more proximal to the site of injury in CNS disease and has been the source of novel biomarkers in diseases such as Alzheimer's disease [3] and amyotrophic lateral sclerosis (ALS) [4]. Source of CSF sampling also requires special consideration in the analysis of CSF biomarkers because CSF protein contents can vary—lumbar CSF contains slightly higher protein content compared to suboccipital CSF [5], and brain-derived proteins such as neuron-specific enolase (NSE) and S100β are higher in ventricular CSF, whereas leptomeningeal proteins such as cystatin C have higher concentrations in lumbar CSF. These relative concentrations may change in the setting of blocked CSF flow [6]. Finally, there may be a role for combined blood and CSF biomarkers, as some disease processes may involve an interplay between pathologic processes on both sides of the BBB.

27.4 Important Molecules in Neurocritical Care

Here, we review select established clinical molecular biomarkers and discuss promising novel candidate biomarkers in critical CNS injuries.

27.4.1 Markers of Neurons, Astrocytes, and Glial Cell Death Turnover

S100β is a calcium-binding protein expressed mainly in astroglial cells, though it has been detected in extra-neural tissue [7]. Elevated S100β levels have been used as a marker of neuronal and astroglial cell injury as well as injury to the BBB [8]. Elevation of S100β in CSF and blood has been linked to the extent of brain injury in hypoxic-ischemic injury in children [9] and in adults post-cardiac arrest [10, 11] and is associated with poor outcome in TBI [12–15], SAH [16], stroke [17], post-cardiac surgery brain injury [18], CNS infection [19], and the development of malignant cerebral edema after ischemic stroke [20, 21]. While a normal S100β can reliably predict the absence of significant CNS injury, it has limited specificity. S100β can be elevated from injury to skeletal muscle and soft tissue [22] without brain injury and is elevated in patients with multisystem trauma without concurrent TBI [7, 23].

NSEs are glycolytic enzymes almost exclusively found in the cytoplasm of neurons and neuroendocrine cells [24]. NSE is the only biomarker specific to neurons and has higher specificity for CNS injury compared to S100β. Elevated blood NSE is associated with stroke, ICH, cardiac arrest, and TBI. Used together, blood NSE and S100β correlate with outcome after ischemic stroke [25], cardiac arrest [26], and TBI [27]. NSE may be associated with outcome of TBI [28–30], though results have been inconsistent [15, 31]. Clinically, NSE is most frequently used as a predictive biomarker for neurologic outcome after cardiac arrest [32], though its specificity and sensitivity are yet unknown in the new era of hypothermia use for neuroprotection after cardiac arrest. There is good evidence that hypothermia decreases NSE clearance, and hypothermia-treated patients with high NSE may still make good neurologic recovery [33]. An important consideration in NSE use is that erythrocytes contain large amounts of NSE and hemolysis can lead to significant NSE elevation in patients without brain injury [34].

Glial fibrillary acid protein (*GFAP*) is a brain-specific protein mostly expressed by astrocytes. GFAP does not have a significant extracranial source and may be a more specific marker of brain injury [17]. Clinical studies showed that elevated blood GFAP correlates with prognosis after stroke [17] and TBI [15, 28, 35] and that GFAP is not elevated in polytrauma without TBI, suggesting it may have better specificity than markers such as S100β.

Myelin basic protein (*MBP*) is one of the most abundant proteins in myelin in the CNS, and elevated serum MBP levels have been reported in TBI and ICH [36, 37].

Creatine kinase brain isoenzyme (*CKBB*) is an isoform of CK found in the CNS, mostly in astrocytes. Release of CKBB into CSF has been reported in conditions such as cardiac arrest and SAH [38], and elevated levels have been reported after TBI [37].

Ubiquitin C-terminal hydrolase L1 (*UCH-L1*) protein, also known as neuronal-specific protein gene product 9.5 (PGP9.5), is specifically expressed in high abundance in neurons. UCH-L1 is involved in the ubiquitination of abnormal or damaged proteins for proteasome degradation. UCH-L1 is present in almost all neurons, and its elevation in human CSF can be detected shortly after TBI [39]. Serum UCL-L1 also shows marked increase within 24 hours after severe TBI and correlates with poor outcome [40], making this a good future candidate clinical biomarker for TBI.

27.4.2 Markers of Inflammation and the BBB

Inflammation is an important mediator of secondary injury in numerous different forms of brain injury including TBI [41–44], SAH [45–48], ischemic stroke [49–54], ICH, and hypoxic-ischemic brain injury and may be a promising target in injury and repair of the neurovascular unit [55, 56]. Many emerging potential biomarkers are implicated in CNS inflammation.

Clinical observations and epidemiologic studies have repeatedly linked generalized leukocyte elevations and changes in leukocyte subpopulation with complication and poor outcome after critical brain injury such as SAH [57–65], ischemic stroke [66, 67], and TBI [68, 69]. One of the mechanisms through which peripheral leukocytes may facilitate CNS injury includes the activation and suppression of different cytokines [70, 71], and several cytokines are emerging as potential biomarkers of secondary complication or outcome following acute brain injury.

Elevation of pro-inflammatory cytokines Tumor necrosis factor alpha (*TNFα*) in CSF and in cerebral microdialysate is associated with delayed cerebral vasospasm and poor SAH outcome [72–74], and its elevation in blood is associated with poor long-term SAH outcome [75]. Interestingly, serum TNFα level is not associated with delayed cerebral ischemia following SAH [76], suggesting its association with SAH outcome may be through nonischemic mechanisms. Soluble TNFα receptor I (*sTNFR-1*) has also been detected in CSF of SAH patients, and its elevation is associated with initial SAH clinical severity and outcome [77].

Elevated CSF *interleukin-6* (*IL-6*) is associated with cerebral vasospasm after SAH [73], while its association with SAH outcome remains controversial [75]. An important limitation is that intracranial hypertension causes significant CSF IL-6 elevation [78], and this may confound most studies of CSF IL-6 levels and outcome after acute brain injuries. Other cytokines that have been associated with vasospasm and poor outcome in human SAH include IL-1 receptor antagonist (*IL-1Ra*) [74], *soluble endoglin* (*sEng*), and *transforming growth factor-β* (*TGFβ*) [79].

In TBI, elevation of blood TNFα and *IL-8* levels is associated with impending intracranial hypertension and cerebral hypoperfusion [80]. The association of cytokines in cerebral microdialysate and TBI outcome is less clear, partly due methodologic limitations [81]. Using inflammatory cytokines as potential biomarker of disease, clinical studies were able to demonstrate that prehospital hypertonic-saline resuscitation appears to reduce circulating levels of TNFα and *IL-10* and endothelial-derived soluble vascular cell adhesion molecule 1 (sVCAM-1) and soluble E-selectin (E-selectin) in patients with severe traumatic head injury

[82], which supports the hypothesis that hypertonic resuscitation may have beneficial anti-inflammatory effects compared with resuscitation with normal saline (0.9 % NaCl) in TBI.

27.5 Metalloproteinases and Other Markers of BBB Integrity

Inflammatory cells such as leukocytes are a major source of matrix *metalloproteinase* (*MMP*) release following brain injury [83]. MMPs are implicated in potential pathogenesis of acute brain injury through numerous mechanisms, including BBB disruption, progression of cerebral edema, worsening of cerebral ischemic injury [52], and disruption of neuron-extracellular matrix interaction leading to early brain injury in SAH [84, 85].

In human ischemic stroke, higher baseline levels of blood *MMP-9* are associated with increased risk of parenchymal hematoma after treatment with intravenous (IV) tPA [86–88] and with thrombolysis failure [89], and higher level of MMP9 mRNA expression is associated with poor outcome and mortality after ischemic stroke [90]. Elevated serum MMP-9 and *c-fibronectin* (*cFn*) have been associated with malignant cerebral edema [91]. Autopsy studies showed that *MMP-2* and MMP-9 are upregulated in human brain after ischemic stroke and ICH [92]. In ICH, increased plasma MMP-9 is significantly associated with ICH enlargement [93] and peri-hematoma edema [94, 95], while increased *MMP-3* is associated with higher mortality [95]. In SAH, elevated blood MMP-9 is associated with vasospasm [96] and with poor long-term SAH outcome [97]. Early elevation of MMP-9 in CSF of SAH patients is associated with poor 3- and 6-month outcome following SAH [97]. Across different types of acute brain injuries such as ICH and ischemic stroke, plasma MMP-9 elevation is associated with evidence of BBB disruption measured by the degree of hyperintense acute reperfusion injury on brain MRI [98].

Other potential biomarkers of the BBB include molecules involved in maintaining tight junctions. Elevated levels of tight junction proteins such as *occludin* (*OCLN*), *claudin 5* (*CLDN5*), and *zonula occludens 1* (*ZO1*) in plasma are associated with hemorrhagic transformation of acute ischemic stroke [99].

27.6 Endothelins

Endothelin-1 (*ET-1*) is implicated in acute brain injury through multiple different mechanisms [100]. It is the strongest vasoconstrictor in the CNS and has been implicated in the pathogenesis of cerebral vasospasm following SAH, and its overexpression is associated with increased cerebral edema [101]. ET-1 may be linked to inflammatory changes in acute brain injury through MMPs, which cleave *big endothelin* (*bET*) into vasoactive ET-1 [102, 103]. Human studies showed serum ET-1 levels of >5.5 fmol/mL and cFn >4.5 mg/L are associated with malignant cerebral edema in tPA-treated acute stroke patients [104]. Several clinical studies have shown contradictory results regarding whether ET-1 levels, in plasma or in CSF, are elevated after ischemic stroke [105, 106]. The largest study to date found no association between plasma ET-1 levels and ischemic stroke or its outcome [107].

ET-1 is considered one of the most important molecules in the pathogenesis of vasospasm following SAH [108]. Several human studies have shown that elevated CSF ET-1 levels are associated with vasospasm [109–111], though this remains controversial [112, 113]. Randomized clinical trials of selective ET-1_A antagonist use in SAH showed that, though ET-1_A antagonist reduced the incidence of angiographic vasospasm, it neither reduced delayed ischemic neurologic injury nor improved overall outcome in SAH [114], raising new questions about the role of ET-1 in SAH-related brain injury. Newer studies now begin to link CSF ET-1 and bET-1 [115] levels with SAH outcome but not with vasospasm [116].

27.6.1 Anti-inflammatory Markers

MMPs as endopeptidases also are known to cleave *plasma-type gelsolin* (*pGSN*) [117], an abundant protein in human plasma. While the function of pGSN remains poorly understood, it is known that pGSN scavenges extracellular actin [118], thereby mitigating downstream pro-inflammatory cascade injurious to the microvasculature [119]. In humans, decreased pGSN level is associated with increased mortality in sepsis and critical illness [120–122]. pGSN has been shown to have neuroprotective effects in ischemic stroke in animal models [123]. Recent data has shown that pGSN is decreased in SAH [124] and may be associated with mortality and with higher GCS score in TBI [125]. In fact, pGSN is present in the CSF, and novel pGSN fragments are found in SAH CSF but not in control subjects [124], suggesting these novel protein fragments may be a future target biomarker for SAH.

27.7 Future Directions

The "omic" revolution has shifted the paradigm of translational research by introducing connections between disease and genes, proteins, and other molecules that would not have come about with traditional approaches. We are now able to conduct multiplex, non-biased searches of potential novel biomarkers using new proteomics and metabolomics approaches. Formation of large collaborative consortiums and multidisciplinary research teams are vital steps towards the successful search and validation of novel biomarkers in the era of multiplex technology

[126]. To bring novel molecular biomarkers to the patients' bedside, we need multicenter consortium of patient samples with rigorous clinical phenotyping and outcome measures and standardized methods for sample collection, processing, and storage. The future of biomarker studies requires collaborative research teams with clinical and basic science experts, experts in translational science and molecular biology, and experts in bioinformatics and clinical trial design. All candidate molecular biomarkers found through these multiplex searches need to be validated in separate cohorts and with targeted confirmatory assays, and the underlying mechanism of association between a biomarker and disease needs to be elucidated and validated.

References

1. Suarez JI (2006) Outcome in neurocritical care: advances in monitoring and treatment and effect of a specialized neurocritical care team. Crit Care Med 34:S232–S238
2. Maier B, Lehnert M, Laurer HL et al (2007) Biphasic elevation in cerebrospinal fluid and plasma concentrations of endothelin 1 after traumatic brain injury in human patients. Shock 27:610–614
3. Mattsson N, Zetterberg H, Hansson O et al (2009) CSF biomarkers and incipient Alzheimer disease in patients with mild cognitive impairment. JAMA 302:385–393
4. Mitchell RM, Freeman WM, Randazzo WT et al (2009) A CSF biomarker panel for identification of patients with amyotrophic lateral sclerosis. Neurology 72:14–19
5. Maurer MH (2010) Proteomics of brain extracellular fluid (ECF) and cerebrospinal fluid (CSF). Mass Spectrom Rev 29:17–28
6. Reiber H (2001) Dynamics of brain-derived proteins in cerebrospinal fluid. Clin Chim Acta 310:173–186
7. Bloomfield SM, McKinney J, Smith L et al (2007) Reliability of S100B in predicting severity of central nervous system injury. Neurocrit Care 6:121–138
8. Michetti F, Corvino V, Geloso MC et al (2012) The S100B protein in biological fluids: more than a lifelong biomarker of brain distress. J Neurochem 120:644–659
9. Blennow M, Savman K, Ilves P et al (2001) Brain-specific proteins in the cerebrospinal fluid of severely asphyxiated newborn infants. Acta Paediatr 90:1171–1175
10. Bottiger BW, Mobes S, Glatzer R et al (2001) Astroglial protein S-100 is an early and sensitive marker of hypoxic brain damage and outcome after cardiac arrest in humans. Circulation 103:2694–2698
11. Rosen H, Sunnerhagen KS, Herlitz J et al (2001) Serum levels of the brain-derived proteins S-100 and NSE predict long-term outcome after cardiac arrest. Resuscitation 49:183–191
12. Pelinka LE, Toegel E, Mauritz W et al (2003) Serum S 100 B: a marker of brain damage in traumatic brain injury with and without multiple trauma. Shock 19:195–200
13. Berger RP, Adelson PD, Pierce MC et al (2005) Serum neuron-specific enolase, S100B, and myelin basic protein concentrations after inflicted and noninflicted traumatic brain injury in children. J Neurosurg 103:61–68
14. Woertgen C, Rothoerl RD, Metz C et al (1999) Comparison of clinical, radiologic, and serum marker as prognostic factors after severe head injury. J Trauma 47:1126–1130
15. Raabe A, Grolms C, Keller M et al (1998) Correlation of computed tomography findings and serum brain damage markers following severe head injury. Acta Neurochir (Wien) 140:787–791, discussion 791–792
16. Wiesmann M, Missler U, Hagenstrom H et al (1997) S-100 protein plasma levels after aneurysmal subarachnoid haemorrhage. Acta Neurochir (Wien) 139:1155–1160

17. Herrmann M, Vos P, Wunderlich MT et al (2000) Release of glial tissue-specific proteins after acute stroke: a comparative analysis of serum concentrations of protein S-100B and glial fibrillary acidic protein. Stroke 31:2670–2677
18. Raabe A (2001) High serum S100B levels for trauma patients without head injuries. Neurosurgery 49:1491–1492; author reply 1492–1493
19. Lins H, Wallesch CW, Wunderlich MT (2005) Sequential analyses of neurobiochemical markers of cerebral damage in cerebrospinal fluid and serum in CNS infections. Acta Neurol Scand 112:303–308
20. Foerch C, Otto B, Singer OC et al (2004) Serum S100B predicts a malignant course of infarction in patients with acute middle cerebral artery occlusion. Stroke 35:2160–2164
21. Castellanos M, Leira R, Serena J et al (2004) Plasma cellular-fibronectin concentration predicts hemorrhagic transformation after thrombolytic therapy in acute ischemic stroke. Stroke 35:1671–1676
22. Biberthaler P, Linsenmeier U, Pfeifer KJ et al (2006) Serum S-100B concentration provides additional information for the indication of computed tomography in patients after minor head injury: a prospective multicenter study. Shock 25:446–453
23. Romner B, Ingebrigtsen T (2001) High serum S100B levels for trauma patients without head injuries. Neurosurgery 49:1490; author reply 1492–1493
24. Pelinka LE, Hertz H, Mauritz W et al (2005) Nonspecific increase of systemic neuron-specific enolase after trauma: clinical and experimental findings. Shock 24:119–123
25. Missler U, Wiesmann M, Friedrich C et al (1997) S-100 protein and neuron-specific enolase concentrations in blood as indicators of infarction volume and prognosis in acute ischemic stroke. Stroke 28:1956–1960
26. Einav S, Kaufman N, Algur N et al (2012) Modeling serum biomarkers S100 beta and neuron-specific enolase as predictors of outcome after out-of-hospital cardiac arrest: an aid to clinical decision making. J Am Coll Cardiol 60:304–311
27. Chabok SY, Moghadam AD, Saneei Z et al (2012) Neuron-specific enolase and S100BB as outcome predictors in severe diffuse axonal injury. J Trauma Acute Care Surg 72:1654–1657
28. Vos PE, Lamers KJ, Hendriks JC et al (2004) Glial and neuronal proteins in serum predict outcome after severe traumatic brain injury. Neurology 62:1303–1310
29. Yamazaki Y, Yada K, Morii S et al (1995) Diagnostic significance of serum neuron-specific enolase and myelin basic protein assay in patients with acute head injury. Surg Neurol 43:267–270, discussion 270–271
30. Herrmann M, Curio N, Jost S et al (1999) Protein S-100B and neuron specific enolase as early neurobiochemical markers of the severity of traumatic brain injury. Restor Neurol Neurosci 14:109–114
31. Ross SA, Cunningham RT, Johnston CF et al (1996) Neuron-specific enolase as an aid to outcome prediction in head injury. Br J Neurosurg 10:471–476
32. Meynaar IA, Oudemans-van Straaten HM, van der Wetering J et al (2003) Serum neuron-specific enolase predicts outcome in post-anoxic coma: a prospective cohort study. Intensive Care Med 29:189–195
33. Rossetti AO, Oddo M, Logroscino G et al (2010) Prognostication after cardiac arrest and hypothermia: a prospective study. Ann Neurol 67:301–307
34. Schmitt B, Bauersfeld U, Schmid ER et al (1998) Serum and CSF levels of neuron-specific enolase (NSE) in cardiac surgery with cardiopulmonary bypass: a marker of brain injury? Brain Dev 20:536–539
35. Nylen K, Ost M, Csajbok LZ et al (2006) Increased serum-GFAP in patients with severe traumatic brain injury is related to outcome. J Neurol Sci 240:85–91
36. Thomas DG, Palfreyman JW, Ratcliffe JG (1978) Serum-myelin-basic-protein assay in diagnosis and prognosis of patients with head injury. Lancet 1:113–115
37. Ingebrigtsen T, Romner B (2002) Biochemical serum markers of traumatic brain injury. J Trauma 52:798–808

38. Coplin WM, Longstreth WT Jr, Lam AM et al (1999) Cerebrospinal fluid creatine kinase-BB isoenzyme activity and outcome after subarachnoid hemorrhage. Arch Neurol 56:1348–1352
39. Papa L, Akinyi L, Liu MC et al (2010) Ubiquitin C-terminal hydrolase is a novel biomarker in humans for severe traumatic brain injury. Crit Care Med 38:138–144
40. Mondello S, Papa L, Buki A et al (2011) Neuronal and glial markers are differently associated with computed tomography findings and outcome in patients with severe traumatic brain injury: a case control study. Crit Care 15:R156
41. Morganti-Kossmann MC, Rancan M, Otto VI et al (2001) Role of cerebral inflammation after traumatic brain injury: a revisited concept. Shock 16:165–177
42. Lenzlinger PM, Morganti-Kossmann MC, Laurer HL et al (2001) The duality of the inflammatory response to traumatic brain injury. Mol Neurobiol 24:169–181
43. Cederberg D, Siesjo P (2010) What has inflammation to do with traumatic brain injury? Childs Nerv Syst 26:221–226
44. Helmy A, De Simoni MG, Guilfoyle MR et al (2011) Cytokines and innate inflammation in the pathogenesis of human traumatic brain injury. Prog Neurobiol 95:352–372
45. Zhou Y, Martin RD, Zhang JH (2011) Advances in experimental subarachnoid hemorrhage. Acta Neurochir Suppl 110:15–21
46. Mashaly HA, Provencio JJ (2008) Inflammation as a link between brain injury and heart damage: the model of subarachnoid hemorrhage. Cleve Clin J Med 75(suppl 2):S26–S30
47. Dumont AS, Dumont RJ, Chow MM et al (2003) Cerebral vasospasm after subarachnoid hemorrhage: putative role of inflammation. Neurosurgery 53:123–133, discussion 133–135
48. Ostrowski RP, Colohan AR, Zhang JH (2006) Molecular mechanisms of early brain injury after subarachnoid hemorrhage. Neurol Res 28:399–414
49. Heiss WD (2012) The ischemic penumbra: how does tissue injury evolve? Ann N Y Acad Sci 1268:26–34
50. Kamel H, Iadecola C (2012) Brain-immune interactions and ischemic stroke: clinical implications. Arch Neurol 69:576–581
51. Xing C, Arai K, Lo EH et al (2012) Pathophysiologic cascades in ischemic stroke. Int J Stroke 7:378–385
52. Lo EH, Wang X, Cuzner ML (2002) Extracellular proteolysis in brain injury and inflammation: role for plasminogen activators and matrix metalloproteinases. J Neurosci Res 69:1–9
53. Petty MA, Lo EH (2002) Junctional complexes of the blood–brain barrier: permeability changes in neuroinflammation. Prog Neurobiol 68:311–323
54. Hayakawa K, Qiu J, Lo EH (2010) Biphasic actions of HMGB1 signaling in inflammation and recovery after stroke. Ann N Y Acad Sci 1207:50–57
55. Lo EH, Dalkara T, Moskowitz MA (2003) Mechanisms, challenges and opportunities in stroke. Nat Rev Neurosci 4:399–415
56. Xing C, Hayakawa K, Lok J et al (2012) Injury and repair in the neurovascular unit. Neurol Res 34:325–330
57. Chou SH-Y, Ning MM, Konigsberg RG, Loesch EC, Alpargu G, Chibnik L, De Jager PH, Feske SK, Lo EH (2010) Peripheral leukocyte count and matrix metalloproteinase-2 in cerebral vasospasm following subarachnoid hemorrhage. Neurology 74:A131
58. Sadamasa N, Yoshida K, Narumi O et al (2011) Prediction of mortality by hematological parameters on admission in patients with subarachnoid hemorrhage. Neurol Med Chir (Tokyo) 51:745–748
59. McGirt MJ, Mavropoulos JC, McGirt LY et al (2003) Leukocytosis as an independent risk factor for cerebral vasospasm following aneurysmal subarachnoid hemorrhage. J Neurosurg 98:1222–1226
60. Dhar R, Diringer MN (2008) The burden of the systemic inflammatory response predicts vasospasm and outcome after subarachnoid hemorrhage. Neurocrit Care 8:404–412
61. Yoshimoto Y, Tanaka Y, Hoya K (2001) Acute systemic inflammatory response syndrome in subarachnoid hemorrhage. Stroke 32:1989–1993

62. Weir B, Disney L, Grace M et al (1989) Daily trends in white blood cell count and temperature after subarachnoid hemorrhage from aneurysm. Neurosurgery 25:161–165
63. Maiuri F, Gallicchio B, Donati P et al (1987) The blood leukocyte count and its prognostic significance in subarachnoid hemorrhage. J Neurosurg Sci 31:45–48
64. Niikawa S, Hara S, Ohe N et al (1997) Correlation between blood parameters and symptomatic vasospasm in subarachnoid hemorrhage patients. Neurol Med Chir (Tokyo) 37:881–884, discussion 884–885
65. Spallone A, Acqui M, Pastore FS et al (1987) Relationship between leukocytosis and ischemic complications following aneurysmal subarachnoid hemorrhage. Surg Neurol 27:253–258
66. Kim J, Song TJ, Park JH et al (2012) Different prognostic value of white blood cell subtypes in patients with acute cerebral infarction. Atherosclerosis 222:464–467
67. Nardi K, Milia P, Eusebi P et al (2011) Admission leukocytosis in acute cerebral ischemia: influence on early outcome. J Stroke Cerebrovasc Dis 21:819–824
68. Gurkanlar D, Lakadamyali H, Ergun T et al (2009) Predictive value of leucocytosis in head trauma. Turk Neurosurg 19:211–215
69. Rovlias A, Kotsou S (2001) The blood leukocyte count and its prognostic significance in severe head injury. Surg Neurol 55:190–196
70. McKeating EG, Andrews PJ (1998) Cytokines and adhesion molecules in acute brain injury. Br J Anaesth 80:77–84
71. Provencio JJ, Vora N (2005) Subarachnoid hemorrhage and inflammation: bench to bedside and back. Semin Neurol 25:435–444
72. Hanafy KA, Grobelny B, Fernandez L et al (2010) Brain interstitial fluid TNF-alpha after subarachnoid hemorrhage. J Neurol Sci 291:69–73
73. Fassbender K, Hodapp B, Rossol S et al (2001) Inflammatory cytokines in subarachnoid haemorrhage: association with abnormal blood flow velocities in basal cerebral arteries. J Neurol Neurosurg Psychiatry 70:534–537
74. Mathiesen T, Edner G, Ulfarsson E et al (1997) Cerebrospinal fluid interleukin-1 receptor antagonist and tumor necrosis factor-alpha following subarachnoid hemorrhage. J Neurosurg 87:215–220
75. Chou SH, Feske SK, Atherton J, Konigsberg RG, De Jager PL, Du R, Ogilvy CS, Lo EH, Ning M (2012) Early elevation of serum tumor necrosis factor-α is associated with poor outcome in subarachnoid hemorrhage. J Investig Med 60(7):1054–8. PMID: 22918199
76. Beeftink MM, Ruigrok YM, Rinkel GJ et al (2011) Relation of serum TNF-alpha and TNF-alpha genotype with delayed cerebral ischemia and outcome in subarachnoid hemorrhage. Neurocrit Care 15:405–409
77. Gruber A, Rossler K, Graninger W et al (2000) Ventricular cerebrospinal fluid and serum concentrations of sTNFR-I, IL-1ra, and IL-6 after aneurysmal subarachnoid hemorrhage. J Neurosurg Anesthesiol 12:297–306
78. Graetz D, Nagel A, Schlenk F et al (2010) High ICP as trigger of proinflammatory IL-6 cytokine activation in aneurysmal subarachnoid hemorrhage. Neurol Res 32:728–735
79. Dietmann A, Lackner P, Fischer M et al (2012) Soluble endoglin and transforming growth factor-beta(1) and the development of vasospasm after spontaneous subarachnoid hemorrhage: a pilot study. Cerebrovasc Dis 33:16–22
80. Stein DM, Lindel AL, Murdock KR et al (2012) Use of serum biomarkers to predict secondary insults following severe traumatic brain injury. Shock 37:563–568
81. Perez-Barcena J, Ibanez J, Brell M et al (2011) Lack of correlation among intracerebral cytokines, intracranial pressure, and brain tissue oxygenation in patients with traumatic brain injury and diffuse lesions. Crit Care Med 39:533–540
82. Rhind SG, Crnko NT, Baker AJ et al (2010) Prehospital resuscitation with hypertonic saline-dextran modulates inflammatory, coagulation and endothelial activation marker profiles in severe traumatic brain injured patients. J Neuroinflammation 7:5

83. Cuzner ML, Opdenakker G (1999) Plasminogen activators and matrix metalloproteases, mediators of extracellular proteolysis in inflammatory demyelination of the central nervous system. J Neuroimmunol 94:1–14
84. Guo ZD, Sun XC, Zhang JH (2011) Mechanisms of early brain injury after SAH: matrix metalloproteinase 9. Acta Neurochir Suppl 110:63–65
85. Gu Z, Kaul M, Yan B et al (2002) S-nitrosylation of matrix metalloproteinases: signaling pathway to neuronal cell death. Science 297:1186–1190
86. Ning M, Furie KL, Koroshetz WJ et al (2006) Association between tPA therapy and raised early matrix metalloproteinase-9 in acute stroke. Neurology 66:1550–1555
87. Castellanos M, Sobrino T, Millan M et al (2007) Serum cellular fibronectin and matrix metalloproteinase-9 as screening biomarkers for the prediction of parenchymal hematoma after thrombolytic therapy in acute ischemic stroke: a multicenter confirmatory study. Stroke 38:1855–1859
88. Montaner J, Molina CA, Monasterio J et al (2003) Matrix metalloproteinase-9 pretreatment level predicts intracranial hemorrhagic complications after thrombolysis in human stroke. Circulation 107:598–603
89. Heo JH, Kim SH, Lee KY et al (2003) Increase in plasma matrix metalloproteinase-9 in acute stroke patients with thrombolysis failure. Stroke 34:e48–e50
90. Graham CA, Chan RW, Chan DY et al (2012) Matrix metalloproteinase 9 mRNA: an early prognostic marker for patients with acute stroke. Clin Biochem 45:352–355
91. Serena J, Blanco M, Castellanos M et al (2005) The prediction of malignant cerebral infarction by molecular brain barrier disruption markers. Stroke 36:1921–1926
92. Rosell A, Ortega-Aznar A, Alvarez-Sabin J et al (2006) Increased brain expression of matrix metalloproteinase-9 after ischemic and hemorrhagic human stroke. Stroke 37:1399–1406
93. Silva Y, Leira R, Tejada J et al (2005) Molecular signatures of vascular injury are associated with early growth of intracerebral hemorrhage. Stroke 36:86–91
94. Abilleira S, Montaner J, Molina CA et al (2003) Matrix metalloproteinase-9 concentration after spontaneous intracerebral hemorrhage. J Neurosurg 99:65–70
95. Alvarez-Sabin J, Delgado P, Abilleira S et al (2004) Temporal profile of matrix metalloproteinases and their inhibitors after spontaneous intracerebral hemorrhage: relationship to clinical and radiological outcome. Stroke 35:1316–1322
96. McGirt MJ, Lynch JR, Blessing R et al (2002) Serum von Willebrand factor, matrix metalloproteinase-9, and vascular endothelial growth factor levels predict the onset of cerebral vasospasm after aneurysmal subarachnoid hemorrhage. Neurosurgery 51:1128–1134, discussion 1134–1135
97. Chou SH-Y, Feske SK, Simmons SL, Konigsberg RG, Orzell SC, Marckmann A, Bourget G, Bauer DJ, De Jager PL, Du R, Arai K, Lo EH, Ning MM (2011) Elevated peripheral neutrophils and matrix metalloproteinase 9 as biomarkers of functional outcome following subarachnoid hemorrhage. Transl Stroke Res 2(4):600–607
98. Barr TL, Latour LL, Lee KY et al (2010) Blood–brain barrier disruption in humans is independently associated with increased matrix metalloproteinase-9. Stroke 41:e123–e128
99. Kazmierski R, Michalak S, Wencel-Warot A et al (2012) Serum tight-junction proteins predict hemorrhagic transformation in ischemic stroke patients. Neurology 79:1677–1685
100. Rubanyi GM, Polokoff MA (1994) Endothelins: molecular biology, biochemistry, pharmacology, physiology, and pathophysiology. Pharmacol Rev 46:325–415
101. Lo AC, Chen AY, Hung VK et al (2005) Endothelin-1 overexpression leads to further water accumulation and brain edema after middle cerebral artery occlusion via aquaporin 4 expression in astrocytic end-feet. J Cereb Blood Flow Metab 25:998–1011
102. Fernandez-Patron C, Radomski MW, Davidge ST (1999) Vascular matrix metalloproteinase-2 cleaves big endothelin-1 yielding a novel vasoconstrictor. Circ Res 85:906–911
103. Fernandez-Patron C, Zouki C, Whittal R et al (2001) Matrix metalloproteinases regulate neutrophil-endothelial cell adhesion through generation of endothelin-1[1–32]. FASEB J 15:2230–2240

104. Moldes O, Sobrino T, Millan M et al (2008) High serum levels of endothelin-1 predict severe cerebral edema in patients with acute ischemic stroke treated with t-PA. Stroke 39:2006–2010
105. Ziv I, Fleminger G, Djaldetti R et al (1992) Increased plasma endothelin-1 in acute ischemic stroke. Stroke 23:1014–1016
106. Lampl Y, Fleminger G, Gilad R et al (1997) Endothelin in cerebrospinal fluid and plasma of patients in the early stage of ischemic stroke. Stroke 28:1951–1955
107. Haapaniemi E, Tatlisumak T, Hamel K et al (2000) Plasma endothelin-1 levels neither increase nor correlate with neurological scores, stroke risk factors, or outcome in patients with ischemic stroke. Stroke 31:720–725
108. Kobayashi H, Hayashi M, Kobayashi S et al (1991) Cerebral vasospasm and vasoconstriction caused by endothelin. Neurosurgery 28:673–678, discussion 678–679
109. Seifert V, Loffler BM, Zimmermann M et al (1995) Endothelin concentrations in patients with aneurysmal subarachnoid hemorrhage. Correlation with cerebral vasospasm, delayed ischemic neurological deficits, and volume of hematoma. J Neurosurg 82:55–62
110. Masaoka H, Suzuki R, Hirata Y et al (1989) Raised plasma endothelin in aneurysmal subarachnoid haemorrhage. Lancet 2:1402
111. Suzuki R, Masaoka H, Hirata Y et al (1992) The role of endothelin-1 in the origin of cerebral vasospasm in patients with aneurysmal subarachnoid hemorrhage. J Neurosurg 77:96–100
112. Fujimori A, Yanagisawa M, Saito A et al (1990) Endothelin in plasma and cerebrospinal fluid of patients with subarachnoid haemorrhage. Lancet 336:633
113. Gaetani P, Rodriguez y Baena R, Grignani G et al (1994) Endothelin and aneurysmal subarachnoid haemorrhage: a study of subarachnoid cisternal cerebrospinal fluid. J Neurol Neurosurg Psychiatry 57:66–72
114. Macdonald RL, Higashida RT, Keller E et al (2012) Clazosentan, an endothelin receptor antagonist, in patients with aneurysmal subarachnoid haemorrhage undergoing surgical clipping: a randomised, double-blind, placebo-controlled phase 3 trial (CONSCIOUS-2). Lancet Neurol 10:618–625
115. Chou SH, Kuruppu S, Feske SK et al (2013) Increased big endothelin-1 in human cerebrospinal fluid is associated with vasospasm and poor 3-month outcome following subarachnoid hemorrhage. Stroke 44:AWMP114
116. Mascia L, Fedorko L, Stewart DJ et al (2001) Temporal relationship between endothelin-1 concentrations and cerebral vasospasm in patients with aneurysmal subarachnoid hemorrhage. Stroke 32:1185–1190
117. Park SM, Hwang IK, Kim SY et al (2006) Characterization of plasma gelsolin as a substrate for matrix metalloproteinases. Proteomics 6:1192–1199
118. Lind SE, Smith DB, Janmey PA et al (1986) Role of plasma gelsolin and the vitamin D-binding protein in clearing actin from the circulation. J Clin Invest 78:736–742
119. Haddad JG, Harper KD, Guoth M et al (1990) Angiopathic consequences of saturating the plasma scavenger system for actin. Proc Natl Acad Sci U S A 87:1381–1385
120. Lee PS, Sampath K, Karumanchi SA et al (2009) Plasma gelsolin and circulating actin correlate with hemodialysis mortality. J Am Soc Nephrol 20:1140–1148
121. Lee PS, Patel SR, Christiani DC et al (2008) Plasma gelsolin depletion and circulating actin in sepsis: a pilot study. PLoS One 3:e3712
122. Lee PS, Drager LR, Stossel TP et al (2006) Relationship of plasma gelsolin levels to outcomes in critically ill surgical patients. Ann Surg 243:399–403
123. Le HT, Hirko AC, Thinschmidt JS et al (2011) The protective effects of plasma gelsolin on stroke outcome in rats. Exp Transl Stroke Med 3:13
124. Chou SH, Lee PS, Konigsberg RG et al (2011) Plasma-type gelsolin is decreased in human blood and cerebrospinal fluid after subarachnoid hemorrhage. Stroke 42(12):3624–3627
125. Xu JF, Liu WG, Dong XQ et al (2011) Change in plasma gelsolin level after traumatic brain injury. J Trauma Acute Care Surg 72:491–496
126. Wijman CA, Smirnakis SM, Vespa P et al (2012) Research and technology in neurocritical care. Neurocrit Care 16:42–54

Chapter 28
Bedside Monitoring of Vascular Mechanisms in CNS Trauma: The Use of Near-Infrared Spectroscopy (NIRS) and Transcranial Doppler (TCD)

Sarah A. Murphy, Brian M. Cummings, David A. Boas, and Natan Noviski

Abstract Ischemia and adequacy of regional and global cerebral blood flow are important determinants of outcome in traumatic brain injury (TBI). Although brain ischemia may be a major common pathway of secondary brain damage following TBI, hyperemia and reperfusion injury may also occur and lead to elevated intracranial pressure and decreased cerebral perfusion pressure. Bedside monitors of cerebral ischemia include near-infrared spectroscopy (NIRS), transcranial Doppler ultrasound (TCD), continuous electroencephalography, and brain tissue microdialysis. This chapter will describe how NIRS and TCD enhance our understanding of vascular pathology following a brain injury and their potential applications in the acute management of TBI.

28.1 Introduction

Ischemia and adequacy of regional and global CBF are important determinants of outcome in traumatic brain injury (TBI). Although brain ischemia may be a major common pathway of secondary brain damage following TBI, hyperemia and reperfusion injury may also occur and lead to elevated intracranial pressure (ICP) and decreased cerebral perfusion pressure (CPP). Bedside monitors of cerebral ischemia include near-infrared spectroscopy (NIRS), transcranial Doppler ultrasound (TCD), continuous electroencephalography (cEEG), and brain tissue microdialysis. This chapter will describe how NIRS and TCD enhance our understanding of vascular pathology following a brain injury and their potential applications in the acute management of TBI.

D.A. Boas (✉)
Department of Radiology, Massachusetts General Hospital, Charlestown, MA 02129, USA
e-mail: dboas@nmr.mgh.harvard.edu

E.H. Lo et al. (eds.), *Vascular Mechanisms in CNS Trauma*, Springer Series in Translational Stroke Research 5, DOI 10.1007/978-1-4614-8690-9_28,

28.2 Near-Infrared Spectroscopy

While global measures of oxygen delivery and consumption are readily available, oxygen utilization at the tissue level has long been sought in clinical practice, particularly in regard to brain tissue. Near-infrared technology was introduced in the 1970s as a new modality to monitor oxygenation noninvasively when Jobsis demonstrated the ability to detect changes in the oxygenation state of tissues using NIRS in both feline brains and canine hearts [41]. Clinical use in humans, both adult and pediatric, soon followed [13, 28, 102]. Commercial cerebral oximeters have now been approved by the FDA, and multiple devices are available, with INVOS 5100 (Somanetics, Troy Michigan) the most commonly used in clinical settings [101]. The principle of tissue oxygen saturation using NIRS technology is briefly overviewed.

NIRS uses the modified Beer-Lambert law to measure the concentration of a substance via the scatter and absorption of light. The near-infrared spectrum includes wavelengths of light in the 700–1,300-nm range that pass through several centimeters of biological tissue. The primarily light-absorbing molecules in tissues are the metal complex chromophores, cytochrome oxidase, bilirubin, myoglobin, and hemoglobin. These molecules absorb light differently in oxidized and non-oxidized states. In particular, two molecules, cytochrome c and the oxygenated state of hemoglobin, are markers of oxidative metabolism and a reflection of the tissue energy state. These two molecules can thus be targeted for measurement. Because the concentration of cytochrome c is low and susceptible to algorithm errors [57], most commercial devices utilize the difference in absorption spectra between oxygenated and deoxygenated hemoglobin. Devices generally utilize the 700–900-nm wavelength range to discriminate the quantity of both molecules as their absorption is a minimum in this range and thus larger volumes of tissue can be interrogated. These absorption characteristics allow a calculation of tissue hemoglobin oxygen saturation, sometimes specified as a regional oxygen saturation (rSO_2), as oxygenated hemoglobin divided by the sum of oxygenated and deoxygenated hemoglobin. Additionally, the absorption allows quantification of the total amounts of hemoglobin in the tissue.

NIRS measurement reflects heterogeneous tissues and measures the mean tissue oxygen saturations. One concern for neurological monitoring is the depth of penetration and the contribution of overlying scalp, skull, and dura to the measurements [59]. Increasing the distance of the light receivers from the light transmitters allows deeper tissue penetration but is limited by the power of the transmitter and thermal injury [30]. An additional technique is utilizing two differently spaced receivers, with the closer receiver detecting superficial tissue and the farther receiver reflecting a combination of superficial and deeper tissues. A subtracting algorithm allows a measurement of tissue oxygenation at a penetration of 2 cm and in some models up to the level of the cerebral ventricles [69]. However, there is potential for artifact with extracerebral tissue that is thickened or edematous and in the presence of extracranial or subdural hematomas. An additional consideration is that the mean oxygen saturation is a sum of venous, capillary, and arterial blood. Based on correlations with PET scans, it has been estimated that the proportion of hemoglobin measured is approximately 70–80 % mixed capillary and venous and 15–25 % arterial, but biological variation exists [39, 99]. Thus, it is often recommended to use NIRS monitoring as a trend [92] and not an absolute reflection of venous values as jugular venous saturation correlation may be poor [103].

28.3 NIRS: General Literature Review

NIRS has found a variety of uses in multiple tissues and clinical scenarios. While validation of its use remains investigational, the ability to obtain tissue-specific perfusion has the attractive potential to direct therapy early, prior to clinically apparent deterioration. NIRS uses have been diverse, including resuscitation in trauma and massive transfusion [77], treatment of shock and sepsis [17, 24, 83], compartment syndrome [93], muscle and vascular disorders [46], and breast cancer detection [14]. Clinical reviews of neurological applications are available [64], and here we concentrate on some of the neurological applications of NIRS.

NIRS measures of cerebral hemoglobin oxygen saturation have been used to assess the delivery of oxygen to the brain during carotid endarterectomy and cardiac surgery with cardiopulmonary bypass. The use of intraoperative cerebral ischemia monitors is potentially useful to identify when compromise occurs. With carotid endarterectomy, cross clamping can induce ischemic damage and stroke in up to 5 % of patients. NIRS might have a useful role in determining which patients have insufficient collaterals prior to surgery. With carotid balloon occlusion, ipsilateral NIRS monitoring of the cortex is expected to decline as oxygen delivery is impeded. Patients with sufficient collateral circulation will resume normal values within a minute with an intact circle of Willis, whereas in those without sufficient collateral circulation, NIRS measurements remain low. During surgery, NIRS has been shown to be a valuable tool to detect cerebral ischemia, it is easy to use in comparison to other modalities, and it has the advantage of providing continuous monitoring [63, 70]. Additionally, it has been described as a guide for blood pressure management [32] or with shunt placement, although determining the threshold for shunt based on cerebral oximetry values remains unknown, with compromises in sensitivity and specificity based on thresholds of 12 or 20 % [61].

Similarly in cardiopulmonary bypass, the use of cerebral oximetry monitoring has been shown to improve the detection of intraoperative events and has been associated with improved patient outcome [98]. The use of cerebral oximetry was found to decrease the incidence of strokes after CPB [33]. A randomized, blinded study using NIRS in 200 coronary bypass cases showed that active monitoring and use of a treatment algorithm based on rSO_2 resulted in fewer instances of cerebral desaturation, less incidence of major organ dysfunction, and shorter ICU length of stay [65]. The use in deep hypothermic cardiac arrest is also potentially useful, since EEG can be attenuated at low temperatures and limit the usefulness of EEG as a monitoring tool. NIRS has been utilized successfully in pediatric cardiac surgeries [38], and low values appear to correlate with brain injury post-repair [22]. However, despite widespread adoption, whether monitoring improves neurodevelopment outcome remains an unanswered question. A recent study following neurocognitive outcomes 1 year after cardiac surgery failed to find a correlation between low intraoperative rSO_2 and poor outcome. However, few of the patients in the study were below previously described low threshold levels of <45 %, likely due to active monitoring and treatment response during surgery [47]. It may be that NIRS will contribute to patient benefit by playing a role in multimodal monitoring and early intervention [35].

The neonatal and premature patient population is another area of NIRS utilization, given the attractiveness of a non-radiating, noninvasive monitor with potential to detect neurological compromise and possibly predict neurodevelopmental outcomes. In the NICU, NIRS has been utilized in ECMO monitoring, PDA management, monitoring for intraventricular hemorrhage development, posthemorrhagic hydrocephalus drainage, blood pressure management, and therapeutic cooling [16]. NIRS may also have a role in predicting neurological outcome after perinatal asphyxia in the first 24 h during active cooling [5].

28.4 NIRS in TBI

While the use of NIRS in the bedside management of postoperative cardiac patients has become more commonplace, its use in patients with TBI remains investigational. Small studies have pointed to some potential applications in TBI, but a standard role in bedside clinical monitoring in this setting is yet to be established. In patients with TBI NIRS may serve as a noninvasive assessment of CPP or to monitor for increased ICP. Additionally, NIRS has recently been shown to be feasible in the prehospital and transport environment [100]. The interpretation of NIRS data in the setting of TBI should focus on trends rather than absolute values, as even deceased patients can have detectable NIRS readings as a reflection of death conditions, hemoglobin, and lack of metabolism [80].

In one small pilot study of four patients with severe TBI receiving CPP-directed care, NIRS values were shown to correlate significantly with the measured CPP. A noninvasive transcranial saturation ($StCO_2$) of greater than 75 correlated with adequate CPP, while values <55 were shown to correlate with lower CPP (<70 mmHg). However, in 13 % of the observations in which CPP was >70 mmHg, a low $StCO_2$ was found, raising the concern that tissue ischemia may persist in TBI even when "adequate" CPP is achieved [25]. This investigator also followed continuous NIRS recordings in 18 TBI patients for 6 days following injury. Both survival and good outcome correlated with NIRS measures of tissue oxygenation. Specifically, $StCO_2$ >70 was found to be independently associated with good survival and good neurological outcome [26].

Other studies have correlated noninvasive measurements of rSO_2 with direct measures of tissue cerebral oxygenation using microdialysis catheters [12]. It has been suggested that incorporating invasive measures of cerebral oxygenation into care protocols may lead to reduced mortality and improved clinical outcomes in severe traumatic injury [7, 66]. If noninvasive measures of cerebral oxygenation are able to faithfully replicate key information about brain tissue perfusion, they may also have a role in care protocols, though this has yet to be demonstrated.

Changes in NIRS-measured cerebral oximetry have been described in TBI patients with increased ICP. One group of investigators demonstrated both lower NIRS values in a group of patients with high ICP (>25 mmHg) as compared with the low ICP group (<25 mmHg) and a lack of improvement with hyper-oxygenation in the high ICP group only. NIRS may therefore be an important additional diagnostic tool in the evaluation of impaired cerebral microcirculation in patients with increased ICP [43]. An inverse correlation between NIRS measures of tissue oxygenation and elevated serologic neuron-specific markers of neuronal injury has also been shown [86].

NIRS has also found some limited use in patients with intracranial hemorrhage. Cerebral oximetry has reported good sensitivity in the detection of intracranial hematomas (subdural and epidural) in the ER or preoperative setting, although postoperative detection appears less reliable [42]. After TBI NIRS may be useful in detecting delayed hematomas. In patients who develop delayed traumatic hematoma, NIRS changes have been shown to precede changes in ICP [34].

28.5 NIRS: Future Applications

One area of current investigation in the use of NIRS is as a noninvasive bedside tool to assess cerebral autoregulation. Several different methods of testing autoregulation utilizing NIRS have been described. The cerebral oximetry index (COx) represents the relationship between cerebral oximetry and arterial blood pressure (ABP), based on the assumption that changes in tissue oxygen saturation are proportional to changes in CBF with stable cerebral metabolic rate. Brady et al. used an infant animal model utilizing NIRS and CPP by laser Doppler flowmetry to develop a cerebral oximetry index (COx). They found that COx is a sensitive tool to detect loss of autoregulation noninvasively and thus has potential to be a valuable adjunct in neurocritical care monitoring to dictate therapy [11]. The hemoglobin volume indices (HVx) use a relationship between relative total tissue hemoglobin (rTHb, measured by NIRS) and blood pressure, based on the assumption that autoregulatory-induced vasodilation and vasoconstriction produce changes in cerebral blood volume that are proportional to changes in rTHb [50]. This method of evaluating autoregulation has potential use as a noninvasive replacement of the pressure reactivity index, a measure derived from simultaneously evaluating ICP slow waves and blood pressure [23, 85].

Functional NIRS (fNIRS) and optical tomography are other areas of intense interest and active investigation in the field of neuromonitoring and neuroimaging, though their role in trauma has not yet been explored. Using NIRS monitoring it is possible to measure changes in oxygenated and deoxygenated hemoglobin in specific brain areas in response to various stimuli. Brain activity results in an increase in oxygenated hemoglobin and concomitant decrease in deoxygenated hemoglobin, changes which reflect a local increase in arteriolar vasodilatation, CBF, and cerebral blood volume resulting from neurovascular coupling [27]. Studies in humans have shown region-specific changes in oxygenation in the occipital lobes in response to photic and other visual stimulation and in the prefrontal cortex during the performance of calculation tasks [37]. Developmental neuroscientists, in particular, have employed fNIRS in order to better understand neural development from birth through early childhood, including the development and lateralization of language processing [6, 31].

When NIRS-measured cerebral hemodynamic changes are recorded simultaneously over multiple areas, a map of cortical hemodynamic response to different stimuli can be made. Multichannel systems with high temporal resolution are a more recent development that allow for these kinds of spatiotemporal reconstructions of brain activity, called diffuse optical tomography (DOT), and permit mapping of brain connectivity. Observed changes in hemodynamics are displayed as a map of cortical activation, and these physiologic images can be superimposed on structural brain imaging [18]. Using this technique, motor activity was shown to correlate significantly with increases in both oxygenated hemoglobin and total hemoglobin and decreases in deoxygenated hemoglobin in corresponding areas of motor cortex. Topograms of the cerebrovascular changes recorded by fNIRS localized the area of activation to the motor cortex using MRI. Regional changes in cerebral blood volume were found to overlap with the global change seen by fMRI around the motor cortex, demonstrating that NIR topography can be used effectively to observe human brain activity [53]. The potential clinical applications of noninvasive monitoring of brain activity are many, and studies investigating the potential uses of fNIRS and DOT are mounting. In the future, fNIRS may play a role in investigations into cognitive function and brain organization or reorganization following traumatic or nontraumatic injury. Such noninvasive, real-time techniques of looking at brain functionality may allow us to gain more information about neural function, recovery, connectivity, and reorganization in injured patients.

28.6 Transcranial Doppler Ultrasonography

There are an increasing number of applications of ultrasound technology in pediatric and adult acute and critical care settings. Advantages of ultrasonography include its portability—it can be performed at the bedside—and relative inexpense. In addition, it is noninvasive and non-radiating, it can usually be rapidly obtained, and it offers real-time information that can be very useful at the bedside and when performing invasive procedures. Neurological applications of ultrasound include simple ultrasonographic imaging of the brain—which is limited largely to the pediatric and neonatal populations in which a bone window is present through an open fontanelle—and transcranial Doppler (TCD) interrogation of cerebral vessels.

TCD provides a real-time, noninvasive measure of the cerebral blood flow velocity (CBFV) in the basal cerebral arteries. TCD utilizes a low-frequency ultrasonic beam (usually 2 MHz) produced from piezoelectric crystals that are electrically stimulated. The beam that is created is directed at the artery being insonated, and the sound waves reflect off insonated tissues. The reflected signal is received by the transducer. As blood moves away from the transducer, a phase shift occurs in the reflected signal, resulting in an increase or decrease in the frequency of the reflected signal as described by the Doppler principle. The change in frequencies is processed to calculate blood flow velocity and direction of flow and also allows for the calculation of other parameters that can be used in the evaluation of blood flow. The pulsatility index (PI), for example, can be calculated using the Gosling equation, PI = (peak systolic velocity − end-diastolic velocity)/mean velocity, and is a useful marker of distal resistance to flow.

The first successful measurement of intracranial artery velocity was described by Aaslid et al. in 1982 using a temporal window above the zygomatic arch [1]. The temporal window can be used to insonate the middle cerebral artery (MCA), the anterior cerebral artery (ACA), the posterior cerebral artery (PCA), and the distal internal carotid artery (DICA) at the bifurcation. Other approaches to intracerebral Doppler artery interrogation include a transorbital window which allows insonation of the ophthalmic artery as well as the internal carotid artery at the siphon level, the transforaminal (occipital) window which allows insonation of the distal vertebral arteries and basilar arteries, and the submandibular window which allows insonation of the more distal portions of the extracranial internal carotid artery [44]. Fast Fourier processing is used to transcribe the signal into a color-coded spectral analysis [58]. The normal depth, flow direction, and age-related flow velocities for each vessel have been established and can be used to correctly identify the insonated vessels. Alternatively, TCD can be performed with imaging (TCDI), allowing the vessels to be directly visualized while obtaining Doppler measurements. This is achieved using a phased-array 2–3-MHz sector transducer and special TCDI software. The potential benefits of this technique may include the ability to perform these studies with standard ultrasound machines and the ability to directly visualize and identify the vessel being interrogated. However, whether this method translates into providing a clinical advantage remains unclear [58].

There are two major limitations in the use of TCD. First, it is highly operator dependent. Second, temporal bone windows are inadequate for allowing imaging in 10–15 % of adult patients [91]. Under most conditions, the diameter of the large cerebral vessels remains relatively constant. If the angle of insonation is also constant, the measured flow velocity through the vessels will directly correlate with CBF. TCD is therefore used as a surrogate measure of CBF and is also used to diagnose complications that may occur in TBI including vasospasm, critical elevations of ICP, decreases in CPP, carotid dissection, and cerebral circulatory arrest (brain death).

28.7 TCD: General Literature Review

Recent practice standards published by a multispecialty panel of experts convened by the Clinical Practice Committee of the American Society of Neuroimaging set forth clinical indications and standards for the use of TCD in clinical practice [3]. Recognized routine indications include the monitoring and management of patients with suspected ischemic stroke, transient ischemic attack or carotid artery disease, sickle cell disease, and suspected brain death; the intraoperative and perioperative monitoring of selected vascular and endovascular procedures; and the acute management of subarachnoid hemorrhage (SAH).

In the setting of acute cerebral ischemia, TCD imaging may be used to augment CT or MRI studies to assess vessel patency, characterize the pathogenesis of acute ischemic stroke (e.g., artery-to-artery embolism), quantify distal cerebral arterial flow to confirm clinically significant steno-occlusive disease, or evaluate the sufficiency of collateral blood supply [3]. In addition, recent studies have suggested that the finding of undetectable residual flow at the site of an acute intracranial occlusion by TCD may predict a poor response to tPA therapy in patients with an acute ischemic stroke and therefore be useful in identifying patients in whom early endovascular therapy should be deployed [4, 78]. The vasomotor reactivity index, a measure of change in velocity in response to a vasodilatory stimulus (hypercapnia from breath-holding), has been used to assess the risk of stroke in patients with carotid artery disease [82, 95, 96]. Intraoperatively and perioperatively TCD is used to monitor for complications in patients undergoing carotid or cardiac surgery, including embolism, thrombosis, and hypo- or hyper-perfusion. The real-time flow changes detected by TCD precede the development of neurological deficits and changes detectable by electroencephalography.

One common critical care use of TCD is in monitoring for vasospasm in patients with aneurysmal SAH [45, 54, 56]. Angiographic evidence of vasospasm ("angiographic vasospasm") may occur in 50–70 % of patients following rupture of an aneurysm. It is estimated that half of these patients will have "clinical vasospasm," also known as delayed deterioration associated with vasospasm (DDAV), characterized by a neurological deficit that cannot be explained by other identifiable causes [45]. Ischemic brain damage and a delayed ischemic deficit are inculcated [91], though the exact pathophysiologic mechanisms culminating in this clinical end point are not well understood. Recent evidence suggests the mechanisms may include inflammation and microthrombi. Clinical signs of vasospasm typically develop 5–15 days following the initial subarachnoid bleed. Though vasospasm remains a clinical diagnosis, TCD is used for surveillance of this complication because it may enable one to diagnose clinically significant vasospasm and provide the opportunity to intervene before irreversible ischemia occurs. TCD has been shown to detect vasospasm in days 2–5, prior to the onset of clinical symptoms, and can be used to guide hemodynamic management in the intensive care unit [3].

The velocity of blood flow within an artery is directly proportional to the flow and inversely proportional to the cross-sectional area of that vessel. When a proximal vessel is narrowed, the mean flow velocity measured distal to that narrowing increases. In the MCA, a mean flow velocity of greater than 120 cm/s is associated with vasospasm [21]. Because both increased flow and decreased vessel diameter may contribute to increased mean flow velocity, the ratio of flow in the external and internal carotid arteries has been used to differentiate vasospasm from hyperemia. The Lindegaard ratio compares the mean flow velocity in the MCA as compared to the mean flow velocity in the extracranial carotid artery, a vessel that is not affected by vasospasm [51]. An elevated Lindegaard ratio (greater than 3) is consistent with proximal vessel vasospasm. The pulsatility index (PI), on the other hand, is a calculation of the difference between the peak systolic velocity and the end-diastolic velocity divided by the mean velocity. An elevated PI (>1.2) reflects increased distal cerebral vascular resistance and may be a sign of distal vasospasm.

28.8 TCD in TBI

The use of TCD in TBI remains investigational, as evidenced by its absence from the recent list of standard uses of TCD published by an expert committee on neuroimaging. However, publications related to the use of TCD in TBI are accumulating, and the potential applications of TCD ultrasonography in this setting are many. The remainder of this chapter will serve as a brief review of avenues of study and current trends related to TCD ultrasonography in TBI. These include its potential to detect increases in ICP, a potential role as a surrogate marker of CBF, an indicator of CPP, as a tool for detecting neurological deterioration, its use to guide early management of patients with TBI, its ability to assess autoregulation, and its use in diagnosing vasospasm following injury.

The pulsatility index (PI), a TCD marker of peripheral cerebral vascular resistance, has been explored as a potential noninvasive marker of increased ICP. In an initial publication from 2004, investigators found a strong correlation between calculated PI and measured ICP in a cohort of patients with mixed intracranial pathology in whom invasive ICP monitoring was being employed [9]. In a retrospective study of children with severe TBI, low admission MCA diastolic flow velocity (<25 cm/s) or elevated PI (>1.31) had a 94 % sensitivity in identifying increased ICP with a negative predictive value of 95 %, suggesting that TCD may be of value as a first-line screening examination to identify children who are in need urgent aggressive treatment [60]. However, other studies have not consistently replicated these results. In 2010 Behrens et al. published a series of ten patients in whom elevations in ICP were iatrogenically induced and maintained during a lumbar infusion test. Simultaneous intraparenchymal pressure monitoring and

TCD calculations of the PI were obtained. In this study, PI could not accurately predict elevations in ICP. Wide variability was found in the ICP-PI correlation, thought to be related to variability in vessel compliance, autoregulation, and ABP [8]. Similar findings were published by Figaji et al. in a series of children with severe TBI in which, again, only weak relationship between mean PI and mean ICP was found. In this study a threshold PI value of >1 was also investigated as a potential threshold screening value, but did not stand up as a reliable screening test. ICP was lower than 20 mmHg in 62.5 % of the patients with PI >1, and conversely, when ICP was 20 mmHg or higher, the PI was <1 in 75 % of the studies [29].

TCD may be a useful tool for identifying patients with TBI who are at risk for secondary neurological deterioration and death. A low admission mean MCA flow velocity has been shown to predict early mortality in patients with TBI [15]. More recently, low MCA diastolic flow velocity and elevated pulsatility index have also been shown to predict secondary neurological deterioration in patients admitted to the hospital with mild and moderate TBI (GCS 9–15 with no or mild lesions seen on CT scan) [10, 40, 90]. Similar suggestive findings have been found in patients interrogated in the prehospital setting [88]. Based on these findings, the incorporation of TCD indices into early goal-directed therapy management strategies in the triage and treatment of TBI has been advocated. In the published experience from one center, early TCD screening identified abnormal TCD variables in nearly half of severely brain-injured patients admitted to a surgical ICU. TCD was able to be obtained in less than 20 min from admission (18 $\pm$ 11 min), as compared with invasive intracranial monitoring which was obtained at 242 ($\pm$116) min. Patients with abnormal TCD indices were treated empirically with mannitol and/or norepinephrine. At the time that a follow-up TCD study was performed, when invasive intracranial monitoring was placed, TCD parameters had normalized and CPP and $SjvO_2$ were comparable between the patients with initially normal and initially abnormal TCD parameter [72]. This provides some preliminary evidence to support a potential role for early TCD to help guide initial acute management of patients with TBI.

TCD has also been employed to characterize altered cerebral hemodynamic profiles in TBI as well as to assess response to brain-specific therapies [9, 60, 71, 90, 94, 97]. Alterations in CBF following brain injury are heterogeneous and can range from low flow or ischemic states to high flow and hyperemia. Consequently, TCD studies following TBI have demonstrated normal, high, or low MCA flow velocity [81]. Both serial xenon CT and TCD studies in patients with severe head injury have shown a common pattern of blood flow changes following a severe TBI that is characterized by three distinct phases. An initial hypoperfusion phase may occur immediately following injury (day 0) and is defined by a low CBF. The cerebral metabolic rate of oxygen consumption (CMRO) is also depressed to approximately 50 % of normal during this phase. CMRO appears to remain depressed during the second and third phases as well. In the second phase, however, there may be hyperemia, or high CBF, and this can be followed by phase three in which vasospasm may occur [55]. Low MCA velocity has been associated with cerebral ischemia and poor outcome [94]. In contrast, high CBFV may be associated with diffuse axonal injury (DAI) [97]; result from hypoventilation, agitation, fever, or acidosis; and can lead to cerebral hyperemia and hemorrhage [71]. The severity of alterations in CBF following TBI has been shown to correlate with injury severity [76].

Following decompressive craniectomy for severe refractory intracranial hypertension, a significant and immediate normalization of CBF velocities has been demonstrated on TCD imaging [20]. In addition, patterns of TCD flow velocity changes that are consistent with a progressive reduction in CPP have been described. This includes an initial increase in systolic velocity and decrease in diastolic velocity, followed by an oscillatory velocity pattern, and finally total obliteration of the waveform consistent with cerebral circulatory arrest [73, 74].

Under normal physiologic conditions, CBF is autoregulated within a wide range of CPPs. Serial TCD studies demonstrate that autoregulation is frequently impaired or absent in moderate, severe, and even mild TBI. TCD interrogation can be used to characterize these changes in cerebral autoregulation by looking at either dynamic (fast response) or static (slow response) measures of autoregulation. Dynamic autoregulation (dAR) has been defined as the fast autoregulatory response that occurs within the first 30 s of a sudden change in mean arterial pressure (MAP). A well-described and commonly used technique for measuring dAR involves inflating and then rapidly deflating a thigh cuff to induce a transient drop in MAP. The change in MCA blood flow velocity and change in ABP following cuff release can be described using a calculated autoregulatory index (ARI) that reflects the change in cerebral vascular resistance per second in relation to the change in ABP [2, 52, 89]. If changes in CBFV passively follow the changes in ABP, the ARI is 0. Higher ARI values indicate better dynamic autoregulation. An ARI less than 4 is generally considered abnormal. Using this methodology, a high prevalence of impaired dAR following severe, moderate, and even mild TBI has been seen. These and other studies suggest a nadir in dynamic autoregulatory function at 2–4 days following injury [36, 87].

Static autoregulation is adjustments in CBF that occur within several minutes after a change in perfusion pressure. To calculate static autoregulatory index (sAR), norepinephrine infusion can be used to temporarily raise MAP 20 mmHg above baseline. CBFV is recorded at the elevated MAP level and then again 5 min after the MAP returns to baseline. The sAR is calculated as percentage change of calculated cerebrovascular resistance in relation to the percentage change in MAP. sAR is considered impaired at values below 50 %. Both dAR and sAR appear to be impaired after cerebral injury. The effect is more pronounced in injured, as compared with non-injured, hemispheres [62, 79]. Notably, impaired cerebral autoregulation has been associated with worse outcomes [19, 75].

At the bedside, the presence or absence of cerebral autoregulation may be used to guide clinical approach to management. Two competing concepts might govern management of CPP in the brain-injured patient. The Rosner concept argues that if cerebral autoregulation is intact, an increase in CPP will lead to vasoconstriction of the cerebral arteries and a decrease in cerebral blood volume and ICP. If cerebral autoregulation is intact, the Rosner concept would dictate that CPP should be kept high in order to optimally control ICP. On the other hand, if cerebral autoregulation is impaired, the Lund concept dictates that a lower CPP should be employed to avoid hyperemia and resultant increased ICP [79]. There is some hope that a bedside assessment of autoregulatory status using TCD parameters might be used to guide individualized management of CPP for and lead to improved patient outcomes [48, 84].

28.9 TCD and Future Directions

TCD might be used to monitor for and diagnose vasospasm in patients with TBI. In published series, vasospasm of the anterior or posterior circulation has been identified in 25–60 % of patients with moderate to severe head injury. In one large prospective study of adult patients with TBI, TCD screening for vasospasm was performed serially for 15 days. A 36 % incidence of MCA and 19 % incidence of basilar artery (BA) vasospasm were found [67, 68]. Vasospasm appears to develop most typically on post-injury days 2–4 and can persist for as long as 3 weeks after head injury. The degree of the spasm demonstrated in these studies can be as severe as that seen in patients with aneurysmal SAH. Some reports suggest that posttraumatic spasm is associated with poor clinical outcome. Notably, it may not be vasospasm per se but hemodynamically significant vasospasm in association with low CBF [55]. These findings together suggest that vasospasm might be an important posttraumatic secondary insult and lend support for the use of TCD to monitor patients following brain injury [49].

References

1. Aaslid R, Markwalder TM, Nornes H (1982) Noninvasive transcranial Doppler ultrasound recording of flow velocity in basal cerebral arteries. J Neurosurg 57(6):769–774
2. Aaslid R et al (1989) Cerebral autoregulation dynamics in humans. Stroke 20(1):45–52
3. Alexandrov AV et al (2012) Practice standards for transcranial Doppler (TCD) ultrasound. Part II. Clinical indications and expected outcomes. J Neuroimaging 22(3):215–224
4. Allendoerfer J et al (2006) Prognostic relevance of ultra-early Doppler sonography in acute ischaemic stroke: a prospective multicentre study. Lancet Neurol 5(10):835–840
5. Ancora G et al (2013) Early predictors of short term neurodevelopmental outcome in asphyxiated cooled infants. A combined brain amplitude integrated electroencephalography and near infrared spectroscopy study. Brain Dev 35(1):26–31
6. Aslin RN (2012) Questioning the questions that have been asked about the infant brain using near-infrared spectroscopy. Cogn Neuropsychol 29(1–2):7–33
7. Bardt TF et al (1998) Multimodal cerebral monitoring in comatose head-injured patients. Acta Neurochir 140(4):357–365
8. Behrens A et al (2010) Transcranial Doppler pulsatility index: not an accurate method to assess intracranial pressure. Neurosurgery 66(6):1050–1057
9. Bellner J et al (2004) Transcranial Doppler sonography pulsatility index (PI) reflects intracranial pressure (ICP). Surg Neurol 62(1):45–51; discussion 51

10. Bouzat P et al (2011) Transcranial Doppler to screen on admission patients with mild to moderate traumatic brain injury. Neurosurgery 68(6):1603–1610
11. Brady KM et al (2007) Continuous time-domain analysis of cerebrovascular autoregulation using near-infrared spectroscopy. Stroke 38(10):2818–2825
12. Brawanski A et al (2002) Comparison of near-infrared spectroscopy and tissue p(O2) time series in patients after severe head injury and aneurysmal subarachnoid hemorrhage. J Cereb Blood Flow Metab 22(5):605–611
13. Brazy JE et al (1985) Monitoring of cerebral oxygenation in the intensive care nursery. Adv Exp Med Biol 191:843–848
14. Cerussi A et al (2006) In vivo absorption, scattering, and physiologic properties of 58 malignant breast tumors determined by broadband diffuse optical spectroscopy. J Biomed Opt 11 (4):044005
15. Chan KH et al (1992) The effect of changes in cerebral perfusion pressure upon middle cerebral artery blood flow velocity and jugular bulb venous oxygen saturation after severe brain injury. J Neurosurg 77(1):55–61
16. Chang T, Plessis A (2012) Neurodiagnostic techniques in neonatal critical care. Curr Neurol Neurosci Rep 12(2):145–152
17. Creteur J et al (2007) The prognostic value of muscle StO2 in septic patients. Intensive Care Med 33(9):1549–1556
18. Custo A et al (2010) Anatomical atlas-guided diffuse optical tomography of brain activation. Neuroimage 49(1):561–567
19. Czosnyka M et al (2001) Cerebral autoregulation following head injury. J Neurosurg 95 (5):756–763
20. Daboussi A et al (2009) Cerebral hemodynamic changes in severe head injury patients undergoing decompressive craniectomy. J Neurosurg Anesthesiol 21(4):339–345
21. Dagal A, Lam AM (2011) Cerebral blood flow and the injured brain: how should we monitor and manipulate it? Curr Opin Anaesthesiol 24(2):131–137
22. Dent CL et al (2006) Brain magnetic resonance imaging abnormalities after the Norwood procedure using regional cerebral perfusion. J Thorac Cardiovasc Surg 131(1):190–197
23. Diedler J et al (2011) The limitations of near-infrared spectroscopy to assess cerebrovascular reactivity. Anesth Analg 113(4):849–857
24. Doerschug KC et al (2007) Impairments in microvascular reactivity are related to organ failure in human sepsis. Am J Physiol Heart Circ Physiol 293(2):H1065–H1071
25. Dunham CM et al (2002) Correlation of noninvasive cerebral oximetry with cerebral perfusion in the severe head injured patient: a pilot study. J Trauma 52(1):40–46
26. Dunham CM et al (2006) Severe brain injury ICU outcomes are associated with Cranial-Arterial Pressure Index and noninvasive Bispectral Index and transcranial oxygen saturation: a prospective, preliminary study. Crit Care 10(6):R159
27. Ferrari M, Quaresima V (2012) A brief review on the history of human functional near-infrared spectroscopy (fNIRS) development and fields of application. Neuroimage 63 (2):921–935
28. Ferrari M et al (1985) Continuous non invasive monitoring of human brain by near infrared spectroscopy. Adv Exp Med Biol 191:873–882
29. Figaji AA et al (2009) Transcranial Doppler pulsatility index is not a reliable indicator of intracranial pressure in children with severe traumatic brain injury. Surg Neurol 72 (4):389–394
30. Germon TJ et al (1999) Cerebral near infrared spectroscopy: emitter-detector separation must be increased. Br J Anaesth 82(6):831–837
31. Gervain J et al (2011) Near-infrared spectroscopy: a report from the McDonnell infant methodology consortium. Dev Cogn Neurosci 1(1):22–46
32. Giustiniano E et al (2010) Cerebral oximetry during carotid clamping: is blood pressure raising necessary? J Cardiovasc Med (Hagerstown) 11(7):522–528

33. Goldman S et al (2004) Optimizing intraoperative cerebral oxygen delivery using noninvasive cerebral oximetry decreases the incidence of stroke for cardiac surgical patients. Heart Surg Forum 7(5):E376–E381
34. Gopinath SP et al (1995) Early detection of delayed traumatic intracranial hematomas using near-infrared spectroscopy. J Neurosurg 83(3):438–444
35. Hirsch JC et al (2010) Near infrared spectroscopy (NIRS) should not be standard of care for postoperative management. Semin Thorac Cardiovasc Surg Pediatr Card Surg Annu 13 (1):51–54
36. Hlatky R et al (2002) Dynamic autoregulatory response after severe head injury. J Neurosurg 97(5):1054–1061
37. Hoshi Y, Tamura M (1993) Dynamic multichannel near-infrared optical imaging of human brain activity. J Appl Physiol 75(4):1842–1846
38. Ing RJ et al (2004) Detection of unintentional partial superior vena cava occlusion during a bidirectional cavopulmonary anastomosis. J Cardiothorac Vasc Anesth 18(4):472–474
39. Ito H et al (2005) Changes in the arterial fraction of human cerebral blood volume during hypercapnia and hypocapnia measured by positron emission tomography. J Cereb Blood Flow Metab 25(7):852–857
40. Jaffres P et al (2005) Transcranial Doppler to detect on admission patients at risk for neurological deterioration following mild and moderate brain trauma. Intensive Care Med 31(6):785–790
41. Jöbsis FF (1977) Non-invasive, infra-red monitoring of cerebral O_2 sufficiency, bloodvolume, HbO2-Hb shifts and bloodflow. Acta Neurol Scand Suppl 64:452–453
42. Kahraman S et al (2006) The accuracy of near-infrared spectroscopy in detection of subdural and epidural hematomas. J Trauma 61(6):1480–1483
43. Kampfl A et al (1997) Near infrared spectroscopy (NIRS) in patients with severe brain injury and elevated intracranial pressure. A pilot study. Acta Neurochir Suppl 70:112–114
44. Kassab MY et al (2007) Transcranial Doppler: an introduction for primary care physicians. J Am Board Fam Med 20(1):65–71
45. Keyrouz SG, Diringer MN (2007) Clinical review: prevention and therapy of vasospasm in subarachnoid hemorrhage. Crit Care 11(4):220
46. Komiyama T et al (2000) Near-infrared spectroscopy grades the severity of intermittent claudication in diabetics more accurately than ankle pressure measurement. Br J Surg 87 (4):459–466
47. Kussman BD et al (2009) Cerebral oximetry during infant cardiac surgery: evaluation and relationship to early postoperative outcome. Anesth Analg 108(4):1122–1131
48. Lang EW et al (2003) Noninvasive cerebrovascular autoregulation assessment in traumatic brain injury: validation and utility. J Neurotrauma 20(1):69–75
49. Lee JH et al (1997) Hemodynamically significant cerebral vasospasm and outcome after head injury: a prospective study. J Neurosurg 87(2):221–233
50. Lee JK et al (2009) Cerebrovascular reactivity measured by near-infrared spectroscopy. Stroke 40(5):1820–1826
51. Lindegaard KF et al (1987) Variations in middle cerebral artery blood flow investigated with noninvasive transcranial blood velocity measurements. Stroke 18(6):1025–1030
52. Mahony PJ et al (2000) Assessment of the thigh cuff technique for measurement of dynamic cerebral autoregulation. Stroke 31(2):476–480
53. Maki A et al (1995) Spatial and temporal analysis of human motor activity using noninvasive NIR topography. Med Phys 22(12):1997–2005
54. Marshall SA, Nyquist P, Ziai WC (2010) The role of transcranial Doppler ultrasonography in the diagnosis and management of vasospasm after aneurysmal subarachnoid hemorrhage. Neurosurg Clin N Am 21(2):291–303
55. Martin NA et al (1997) Characterization of cerebral hemodynamic phases following severe head trauma: hypoperfusion, hyperemia, and vasospasm. J Neurosurg 87(1):9–19

56. Mascia L et al (2003) The accuracy of transcranial Doppler to detect vasospasm in patients with aneurysmal subarachnoid hemorrhage. Intensive Care Med 29(7):1088–1094
57. Matcher SJ et al (1995) Performance comparison of several published tissue near-infrared spectroscopy algorithms. Anal Biochem 227(1):54–68
58. McCarville MB (2008) Comparison of duplex and nonduplex transcranial Doppler ultrasonography. Ultrasound Q 24(3):167–171
59. McCormick PW et al (1992) Intracerebral penetration of infrared light. Technical note. J Neurosurg 76(2):315–318
60. Melo JRT et al (2011) Transcranial Doppler can predict intracranial hypertension in children with severe traumatic brain injuries. Childs Nerv Syst 27(6):979–984
61. Mille T et al (2004) Near infrared spectroscopy monitoring during carotid endarterectomy: which threshold value is critical? Eur J Vasc Endovasc Surg 27(6):646–650
62. Moppett IK et al (2008) Effects of norepinephrine and glyceryl trinitrate on cerebral haemodynamics: transcranial Doppler study in healthy volunteers. Br J Anaesth 100 (2):240–244
63. Moritz S et al (2007) Accuracy of cerebral monitoring in detecting cerebral ischemia during carotid endarterectomy: a comparison of transcranial Doppler sonography, near-infrared spectroscopy, stump pressure, and somatosensory evoked potentials. Anesthesiology 107 (4):563–569
64. Murkin JM, Arango M (2009) Near-infrared spectroscopy as an index of brain and tissue oxygenation. Br J Anaesth 103(suppl 1):i3–i13
65. Murkin JM et al (2007) Monitoring brain oxygen saturation during coronary bypass surgery: a randomized, prospective study. Anesth Analg 104(1):51–58
66. Narotam PK, Morrison JF, Nathoo N (2009) Brain tissue oxygen monitoring in traumatic brain injury and major trauma: outcome analysis of a brain tissue oxygen-directed therapy. J Neurosurg 111(4):672–682
67. O'Brien NF et al (2010) Vasospasm in children with traumatic brain injury. Intensive Care Med 36(4):680–687
68. Oertel M et al (2005) Posttraumatic vasospasm: the epidemiology, severity, and time course of an underestimated phenomenon: a prospective study performed in 299 patients. J Neurosurg 103(5):812–824
69. Ohmae E et al (2006) Cerebral hemodynamics evaluation by near-infrared time-resolved spectroscopy: correlation with simultaneous positron emission tomography measurements. Neuroimage 29(3):697–705
70. Pennekamp CWA et al (2009) The value of near-infrared spectroscopy measured cerebral oximetry during carotid endarterectomy in perioperative stroke prevention. A review. Eur J Vasc Endovasc Surg 38(5):539–545
71. Philip S et al (2009) Cerebrovascular pathophysiology in pediatric traumatic brain injury. J Trauma 67(suppl):S128–S134
72. Ract C et al (2007) Transcranial Doppler ultrasound goal-directed therapy for the early management of severe traumatic brain injury. Intensive Care Med 33(4):645–651
73. Rao GSU, Durga P (2011) Changing trends in monitoring brain ischemia. Curr Opin Anaesthesiol 24(5):487–494
74. Rasulo FA, De Peri E, Lavinio A (2008) Transcranial Doppler ultrasonography in intensive care. Eur J Anaesthesiol Suppl 42:167–173
75. Reinhard M et al (2010) Secondary decline of cerebral autoregulation is associated with worse outcome after intracerebral hemorrhage. Intensive Care Med 36(2):264–271
76. Safin AM et al (2007) [Cerebral circulatory disorders in varying brain injury, as evidenced by transcranial Doppler study]. Zh Vopr Neirokhir Im N N Burdenko 2:16–20
77. Santora RJ, Moore FA (2009) Monitoring trauma and intensive care unit resuscitation with tissue hemoglobin oxygen saturation. Crit Care 13(suppl 5):S10
78. Saqqur M, Zygun D, Demchuk A (2007) Role of transcranial Doppler in neurocritical care. Crit Care Med 35(suppl):S216–S223

79. Schramm P et al (2011) Serial measurement of static and dynamic cerebrovascular autoregulation after brain injury. J Neurosurg Anesthesiol 23(1):41–44
80. Schwartz ES et al (2010) Magnetoencephalography. Pediatr Radiol 40(1):50–58
81. Sharples PM et al (1995) Cerebral blood flow and metabolism in children with severe head injury. Part 1: relation to age, Glasgow coma score, outcome, intracranial pressure, and time after injury. J Neurol Neurosurg Psychiatry 58(2):145–152
82. Silvestrini M et al (2000) Impaired cerebral vasoreactivity and risk of stroke in patients with asymptomatic carotid artery stenosis. JAMA 283(16):2122–2127
83. Skarda DE et al (2007) Dynamic near-infrared spectroscopy measurements in patients with severe sepsis. Shock 27(4):348–353
84. Steiner LA et al (2002) Continuous monitoring of cerebrovascular pressure reactivity allows determination of optimal cerebral perfusion pressure in patients with traumatic brain injury. Crit Care Med 30(4):733–738
85. Steiner LA et al (2009) Near-infrared spectroscopy can monitor dynamic cerebral autoregulation in adults. Neurocrit Care 10(1):122–128
86. Subbaswamy A et al (2009) Correlation of cerebral near-infrared spectroscopy (cNIRS) and neurological markers in critically ill children. Neurocrit Care 10(1):129–135
87. Sviri GE et al (2009) Time course for autoregulation recovery following severe traumatic brain injury. J Neurosurg 111(4):695–700
88. Tazarourte K (2010) Advocating for transcranial Doppler: a tool to detect early neurological deterioration. J Trauma 69(3):733–734
89. Tiecks FP et al (1995) Comparison of static and dynamic cerebral autoregulation measurements. Stroke 26(6):1014–1019
90. Trabold F et al (2004) The prognostic value of transcranial Doppler studies in children with moderate and severe head injury. Intensive Care Med 30(1):108–112
91. Tsivgoulis G, Alexandrov AV, Sloan MA (2009) Advances in transcranial Doppler ultrasonography. Curr Neurol Neurosci Rep 9(1):46–54
92. van Bel F, Lemmers P, Naulaers G (2008) Monitoring neonatal regional cerebral oxygen saturation in clinical practice: value and pitfalls. Neonatology 94(4):237–244
93. Varela JE et al (2001) Near-infrared spectroscopy reflects changes in mesenteric and systemic perfusion during abdominal compartment syndrome. Surgery 129(3):363–370
94. Vavilala MS et al (2004) Cerebral autoregulation in pediatric traumatic brain injury. Pediatr Crit Care Med 5(3):257–263
95. Vernieri F, Pasqualetti P, Diomedi M et al (2001) Cerebral hemodynamics in patients with carotid artery occlusion and contralateral moderate or severe internal carotid artery stenosis. J Neurosurg 94(4):559–564
96. Vernieri F, Pasqualetti P, Matteis M et al (2001) Effect of collateral blood flow and cerebral vasomotor reactivity on the outcome of carotid artery occlusion. Stroke 32(7):1552–1558
97. Visocchi M et al (2007) Haemodynamic patterns in children with posttraumatic diffuse brain swelling. A preliminary study in 6 cases with neuroradiological features consistent with diffuse axonal injury. Acta Neurochir 149(4):347–356
98. Vohra HA, Modi A, Ohri SK (2009) Does use of intra-operative cerebral regional oxygen saturation monitoring during cardiac surgery lead to improved clinical outcomes? Interact Cardiovasc Thorac Surg 9(2):318–322
99. Watzman HM et al (2000) Arterial and venous contributions to near-infrared cerebral oximetry. Anesthesiology 93(4):947–953
100. Weatherall A et al (2012) Feasibility of cerebral near-infrared spectroscopy monitoring in the pre-hospital environment. Acta Anaesthesiol Scand 56(2):172–177
101. Wolf M, Greisen G (2009) Advances in near-infrared spectroscopy to study the brain of the preterm and term neonate. Clin Perinatol 36(4):807–834
102. Wyatt JS et al (1986) Quantification of cerebral oxygenation and haemodynamics in sick newborn infants by near infrared spectrophotometry. Lancet 2(8515):1063–1066
103. Yoxall CW et al (1995) Measurement of cerebral venous oxyhemoglobin saturation in children by near-infrared spectroscopy and partial jugular venous occlusion. Pediatr Res 38 (3):319–323

Chapter 29
In Vivo MRI and MRS of Cerebrovascular Function Following Traumatic Brain Injury

Chandler Sours and Rao P. Gullapalli

Abstract Traumatic brain injury (TBI) is a leading cause of death and disability resulting in reduced quality of life for the patients and a significant economic burden to the society. Advances in magnetic resonance imaging techniques are enabling researchers to obtain critical insights into the pathophysiology of the TBI sequelae and also aiding clinicians in predicting long term outcome of TBI patients of all severities. Advanced MRI techniques are able to detect subtle changes in structural integrity of neural tissue as well as changes in functional neural networks. Multi-parametric approaches, including diffusion tensor imaging, functional MRI, magnetic resonance spectroscopy, and arterial spin labeling to non-invasively measure brain perfusion, show promise in detecting trauma-induced biophysical, biochemical and cerebrovascular alterations respectively. Together these techniques are bound to provide us with a better understanding of the sequelae of TBI non-invasively that may lead to better management of TBI patients.

29.1 Introduction

Traumatic brain injury (TBI) is a leading cause of death and lifelong disability among children and young adults throughout developed nations. The financial burden of TBI including both direct and indirect costs amounts to $60 billion each year [20]. The Centers for Disease Control and Prevention (CDC) estimates that each year 1.7 million Americans suffer a TBI, and 275,000 of these individuals are hospitalized [19]. These injuries involve impacts that cause sudden acceleration-deceleration forces on the brain tissue. This leads to linear, rotational or angular shearing injuries, especially at the boundaries of gray and white matter, resulting in diffuse axonal injury (DAI) caused by axonal stretching [61, 63].

R.P. Gullapalli (✉)
Department of Diagnostic Radiology & Nuclear Medicine, University of Maryland School of Medicine, UMMC N2W78, 22 South Greene St, Baltimore, MD 21201, USA
e-mail: rgullapalli@umm.edu

E.H. Lo et al. (eds.), *Vascular Mechanisms in CNS Trauma*, Springer Series in Translational Stroke Research 5, DOI 10.1007/978-1-4614-8690-9_29,

This damage can lead to myriad molecular events including edema, vascular dysregulation, ischemia, inflammation, disruption in plasma membrane and neurotransmitter release, mitochondrial dysfunction, production of reactive oxygen species, altered anaerobic metabolism and lactic acidosis. Together, all of the above conditions often lead to necrotic or apoptotic cell death contributing to long-term cognitive impairment [47, 60].

29.1.1 Classification of Head Injury

Head injury can be categorized through level of severity, level of consciousness, or mental status following head injury. TBI is often divided into mild, moderate and severe based on the Glasgow Coma Scale (GCS) [79]. While the GCS is generally accepted because it is easy to use and more or less correlates with the Glasgow Outcome Scale (GOS) and the Disability Rating Scale, the heterogeneous nature of the injury results in wide variability in patient outcomes [80]. Recently mild TBI has been subdivided into civilian, sports related, and explosive blast categories, all of which have unique features [49]. It is believed that a significant fraction (~40 %) of mild TBI patients will remain impaired for at least 3 months, and a substantial fraction of these patients will show deficits at 1 year leading to lost productivity and resultant socioeconomic consequences [2, 7, 13]. Furthermore, current research points to a cumulative effect of multiple concussive or sub-concussive events. Referred to as chronic traumatic encephalopathy (CTE), this form of neurodegeneration results from repetitive mild TBI and results in progressive memory loss and ultimately dementia similar to Alzheimer's disease [45]. Although attempts have been made to predict functional outcomes following TBI, due to the variable nature of this injury, it has been difficult to arrive at a consensus on the best way to manage TBI patients based on conventional imaging methods.

29.1.2 MRI in the Diagnosis of TBI

Computed tomography (CT) is the primary diagnostic tool used to classify and triage patients with TBI into three main classes including those with negative CT, those with focal injuries, and those who present with a diffuse pattern of parenchymal injury [16, 90]. It should be noted that the clinical presentation often does not match the presence of abnormalities seen on CT or conventional MR. While magnetic resonance imaging (MRI) is superior to CT scanning for detection of DAI, nevertheless conventional MR can also fail to detect DAI [1].

In recent years several advanced imaging techniques have been introduced that can probe the microstructural changes, and cerebrovascular changes that eventually lead to changes in cellular metabolism and changes in cortical and sub-cortical function. It is hoped that these advanced techniques applied during the acute stage will eventually have prognostic ability in determining long-term outcomes.

29.2 Conventional MR Imaging

A conventional MRI for trauma related injuries often consists of T1-weighted images to assess presence of focal injury, T2-weighted images to determine the extent of contusions and hemorrhages, a FLAIR image for parenchymal integrity, cortical surface lesion, brain stem and ventricular hemorrhage, a proton-density weighted image to assess white matter abnormalities and a susceptibility-weighted image (SWI) to assess the presence of micro-hemorrhages. In the past a T2*-weighted image was used but is being replaced by SWI, a 3D technique, that is more sensitive to micro-hemorrhages resulting from DAI [5, 32, 65]. For example, in a pediatric population of severe TBI patients, the extent of SWI lesions correlated with initial GCS, length of coma as well as neurologic impairments in memory and attention [81]. Furthermore, the clinical utility of SWI in a mild TBI population demonstrated a correlation between the aggregate SWI lesion volume and measures of clinical severity [9].

Following TBI, there is often an overall cerebral atrophy noted by loss of cerebral volume and enlargement of ventricles. Structural imaging using high resolution T1-weighted magnetization prepared-rapid acquisition gradient echo (MPRAGE) sequence is used to assess the time course of these volume changes. Volume loss is most prominent in the chronic stages of injury and includes both cortical and subcortical regions [85, 86]. Regional volume loss has been shown to correlate with tests of neurocognitive functioning, including a positive association of thalamic volume with processing speed, and hippocampal volume with memory and learning [85]. In addition, whole-brain and regional volumes are predictive of long-term disability as measured by the extended Glasgow Outcome Scale (GOS-E) [86] and are able to differentiate among levels of injury severity [46].

29.3 Diffusion Tensor Imaging

Diffusion tensor imaging (DTI) is used to assess changes in white matter structural integrity making it one of the leading ways to assess DAI. DTI entails measuring water diffusion in at least six directions to obtain an appropriate representation of the orientation of an axon by measuring the preferred direction of water diffusion. Measurements such as mean diffusivity or apparent diffusion coefficient (ADC) and fractional anisotropy (FA) can be measured from such an acquisition. Intact axons have high anisotropy while damaged axons have reduced anisotropy. Alterations in diffusion parameters following TBI have been found both at the whole brain level and region specific level and have shown to improve prognostic models of severe TBI [10].

Commonly damaged regions following mild TBI include the corpus callosum [8, 44, 53, 86], internal and external capsule [3, 8] and cingulum bundles [48]. Furthermore, time since injury greatly influences diffusion parameters, especially in the acute and sub-acute stages of injury. These changes have been attributed to axonal swelling or edema during the initial stages following injury, and inflammation and axonal damage in the later stages. For example, following blast TBI, various deficits in DTI parameters were observed from the sub-acute stage to chronic stage following injury. In the cingulum bundle an initial reduction in relative anisotropy (RA), and increases in radial (across axon) and mean diffusivities (RD & MD respectively) has been noted. However, in the follow up scans a partial normalization on RA values, reduced axial diffusivity (along the axon), and complete normalization of RD and MD were noted. The researchers interpret these results to indicate initial axonal injury and edema and inflammation, with ensuing resolution of the edema and inflammation at follow up visits [50].

Based on evidence from the literature it is largely believed that axonal injury leads to decreased FA and increased MD, indicative of loss of axonal integrity. However, many groups have found increased fractional anisotropy (FA) and decreased MD or RD values in the acute stages of injury in various regions including the corpus callosum [8, 53]. These changes have been attributed to axonal swelling as an early aspect of axonal injury. The exact time course of the primary and secondary injury associated with mild TBI and the effects of diffusion parameters remain an active area of research. Serial measurements of DTI parameters in the acute and sub-acute stages of injury, similar to those made by Wilde et al. [87], as well as studies in animal models of TBI, are needed to fully understand these changes. However, the heterogeneous nature of human TBI makes this line of research increasingly difficult.

Greater agreement regarding the time course of DTI changes exists for severe TBI patients. Severe TBI patients often demonstrate reduced FA especially in the corpus callosum and internal capsule. It is believed that these areas are especially vulnerable to the shearing forces from the injury, making them more susceptible to progressive degeneration as part of the secondary injury process which leads to a gradual loss of axonal integrity as represented by a decreased FA.

While DTI is now widely used to assess damage in white matter regions, it has had little utility in assessing alterations in gray matter structures following TBI. Recently, the ability of diffusion kurtosis imaging (DKI) to study the heterogeneity of the microenvironment of a tissue has gained significant attention. This technique overcomes a limitation of DTI which assumes water diffusion in tissue to have a Gaussian distribution. By introducing the non-Gaussian distribution of water diffusion into the estimation model, DKI is able to provide a sense for the heterogeneity of water diffusion, which indirectly represents the local microstructural conditions in the tissue. In addition to estimating the standard DTI parameters, DKI is able to assess alterations in gray matter structures [38]. In an animal model of TBI, DKI measures were found to be sensitive to reactive astrogliosis [89].

In addition, in chronic TBI, DKI parameters such as mean kurtosis (MK) in the thalamus and internal capsule positively correlate with cognitive measures of attention and processing speed [31]. While the DKI method is relatively new, it is quite promising as it allows for the in vivo analysis of microstructural changes in both gray and white matter. For example, high MK values noted around the location of the lesion in Fig. 29.1 may indicate the formation of glial scars. In addition, MK provides additional information to standard diffusion parameters such as MD and FA as indicated by different MK behavior in two regions which both showed low MD values. Other methods such as the q-ball method are also promising as it provides even finer details regarding changes in tissue microstructure. Unfortunately the acquisition time for the q-ball method is extremely long, making its use clinically impractical with the current hardware systems. However, future improvements in hardware and software may allow clinicians to use these methods to obtain tissue information at extremely high resolution in vivo.

29.4 Cerebral Blood Perfusion

CT perfusion has been used clinically for years; however given the high dose of ionizing radiation, use of this methodology is limited. A relatively new MRI technique such as arterial spin labeling (ASL) is able to measure cerebral blood perfusion (CBF) using endogenous blood contrast by MRI and has gained significant popularity. As one of the newer MRI techniques, ASL has only recently been applied to TBI in the chronic stages. Using ASL, severe TBI patients show a global hypoperfusion at rest [42] and mild TBI patients show reduced CBF in the thalamus [28]. As shown in Fig. 29.2, for example a severe TBI patient presenting with a left frontal lobe contusion demonstrates reduced resting CBF compared to the right frontal lobe at both 2 and 6 months following injury. Recently, investigators have used ASL to assess variations in task-induced perfusion. For example, an innovative study found that regions of task-related CBF changes in a severe TBI population during an N-back working memory task include similar regions that were hypoperfused at rest indicating cerebrovascular dysregulation at these cognitively relevant regions [43]. One study has also utilized ASL in a mouse model to study the effect of a combined injury model of hemorrhagic shock (HS) in conjunction with a controlled cortical impact (CCI) injury mimicking the scenario often faced by TBI patients. The CBF was reduced in many regions ipsilaterally and contralaterally for CCI alone and HS alone animals. However, animals that obtained CCI followed by HS, demonstrated a greater extent and duration of hypoperfusion, which had a subsequent influence on mediating the secondary injury process [21]. This study indicates that alterations in cerebrovascular dysfunction following injury may play an important role in the management of TBI patients and that non-invasive means of monitoring changes in CBF can be very valuable. It is likely that the application of ASL as an important prognostic tool in determining patient outcomes will be widely used in the future as it has implications for

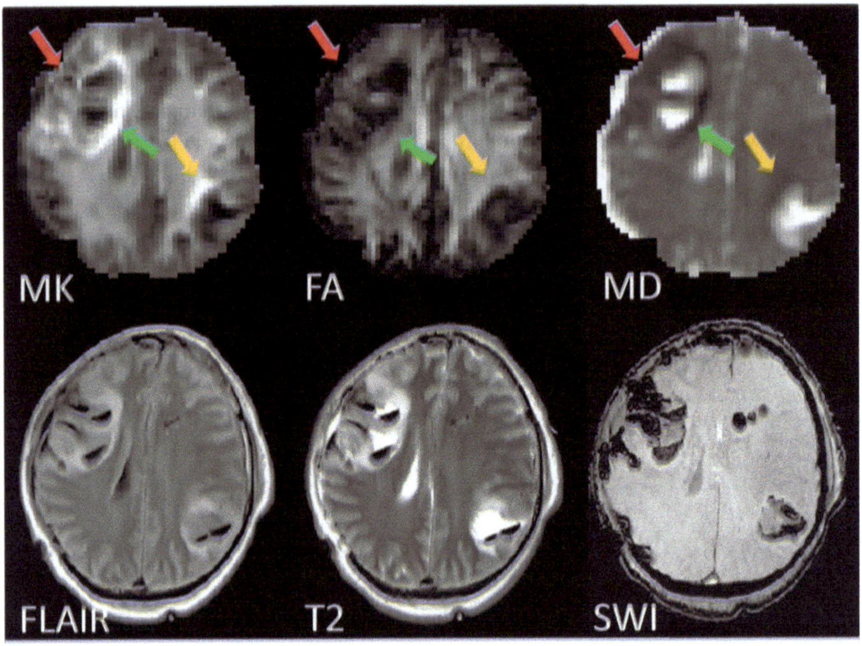

Fig. 29.1 DKI parameter maps (MK, FA, MD) and conventional MR maps of a severe TBI patient (40 years old male, GCS = 3T, scanned 2 days post injury) at the same axial location. High Ms may indicate formation of glial scars around the lesions (*yellow arrows*), which are otherwise not noticeable from conventional imaging. Notice the different MK behavior from *red* and *green arrows*, which both showed low MD values

metabolic integrity of the tissue which may eventually lead to changes in microstructure of the tissue.

29.5 Magnetic Resonance Spectroscopy

Magnetic resonance spectroscopy (MRS) offers a unique opportunity to non-invasively measure cellular biochemicals in vivo. In the brain, metabolites such as the neuronal and axonal marker, *N*-acetylaspartate (NAA), the myelin breakdown products choline (Cho) and phosphocholine, lactate (Lac) which is a marker of anaerobic metabolism, creatine (Cre), and myo-inositol (mI), can be measured to evaluate metabolic changes following brain injury. Both experimental and clinical studies have shown a significant decrease in NAA following TBI implying that neuronal injury through DAI is present in these regions [4, 17, 25, 26, 71]. What makes these findings more compelling is that the concentration of NAA was shown to correlate with patient outcome and severity of injury [66]. Elevated levels of Cho and mI have been found in the regions of normal-appearing

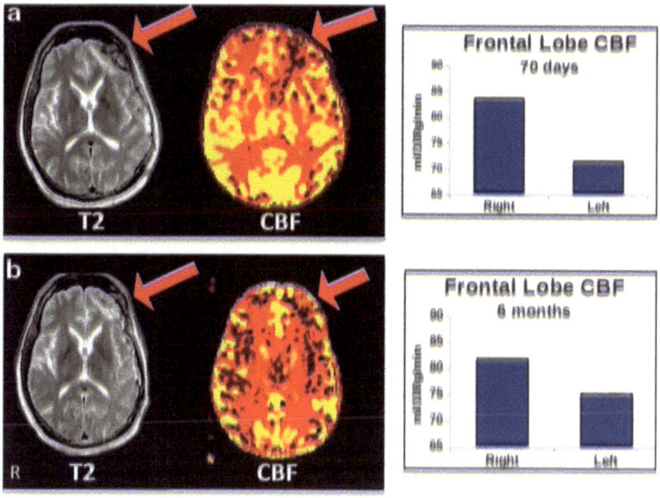

Fig. 29.2 Structural imaging and raw CBF maps from a pulsed arterial spin labeling (pASL) (TE = 11 ms, TR = 2,500 ms, FOV = 230 mm, resolution 64 × 64, 16 slices, sl.th. = 5 mm, 45 pairs of labeled and control volumes) scan from a severe TBI patient at (**a**) 70 days following injury and (**b**) 6 months following injury. Reduced CBF noted in left frontal lobe corresponding with the left frontal lobe contusion noted in the T2 weighted imaging as noted with the *red arrow*. Frontal lobe CBF was calculated using the WFU PickAtlas frontal lobe ROI

brain regions suggesting reactive astrocytosis that occurs in regions of DAI following trauma. The increase in Cho may also suggest a disruption of membranes in the normal appearing brain [25, 26]. Except in very severe TBI patients, few studies have found lactate in the brain following TBI. Ashwal et al. found that children with visible amount of lactate following TBI had poorer outcomes [4]. Several studies have reported reductions in NAA in white and gray matter and elevated Cho in various regions of the brain as shown. This reduction in NAA and increased choline can be visualized in Fig. 29.3 which shows an example of spectra from the splenium of the corpus callosum from a severe TBI patient compared to a healthy control. In addition, the TBI patient shows elevated lactate levels in this region compared to a healthy control.

Strong correlations have been observed between the metabolites and neuropsychological function suggesting that MRS could be a potential tool in determining the long-term outcome of TBI patients [6, 23, 24, 27, 62, 88]. Holhouser et al. [36] found a decrease in the NAA levels and an elevation in the Cho/Cr ratio, especially in the corpus callosum region. This predicted the long term outcome with 83 % accuracy. The fact that MRS is able to determine biochemical abnormalities among normal appearing brain tissues suggests that MRS may detect abnormalities long before morphological changes are seen using MRI, making it especially helpful in the assessment of mild TBI patients who have normal structural scans but persistent symptoms. Despite its strengths, MR spectroscopy in not routinely used in the acute setting of head injuries due the length of the MRS scans. However, recent advances

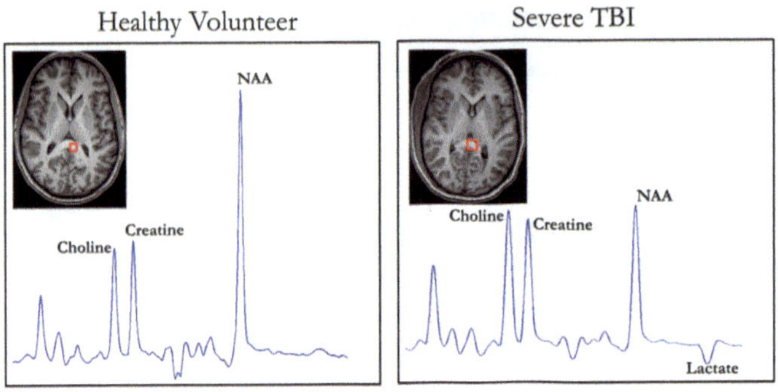

Fig. 29.3 MR spectroscopy imaging showing from a voxel in splenium of the corpus callosum in a severe TBI patient (21 year old male, GCS 4, scanned 1 day post injury) and healthy volunteer. Spectra demonstrate (a) reduced *N*-acetyl-aspartate (b) increased choline and (c) elevated lactate levels in severe TBI patient

have made spectroscopy scans faster. For example, 3D MRSI and echo-planar spectroscopic imaging (EPSI) in conjunction with parallel imaging are able to cover the entire brain within a clinically feasible time of 10–20 min. For example, spectra from a severe TBI patient compared a healthy control using the EPSI technique [29] clearly depicts the metabolic deficits in various regions of the brain including reduced NAA and an increase in the Cho levels. Establishing a link between local blood perfusion as measured by ASL and the changes in metabolic patterns may provide valuable insights into the cerebrovascular coupling with the metabolic changes.

29.6 Functional MRI

While most post-concussive symptoms resolve within the first few months following injury, impairment continues into the chronic stage in a significant portion of TBI patients. Two of the common findings following TBI are impaired information processing [34, 39] and memory deficits [55, 56]. Other common symptoms include fatigue, deficits in attention and executive function, headaches and chronic pain, sensory perception disorders, language, difficulty with socializing, depression, and anxiety [14, 37, 50, 58, 59].

Functional MRI (fMRI) is a valuable tool as it can identify the deficits in neural networks associated with various cognitive processes. It provides an indirect means to measure brain activation and is based on the signal differences between deoxygenated blood and oxygenated blood. When specific neurons for a given task are activated, there is an increase in freshly oxygenated blood to the local tissue to keep up with the increased neuronal activity. This change from deoxygenated blood to oxygenated blood in the activated region causes a change in the tissue signal as the local tissue changes from a predominantly paramagnetic state to diamagnetic state.

It is this change that is measured in fMRI and is called the blood oxygen level dependent (BOLD) signal.

29.7 Task Based fMRI

Task based fMRI studies in TBI have focused primarily on frontal lobe damage and executive functioning deficits noted in TBI. Multiple fMRI studies on patients with TBI have shown alterations in BOLD responses during tasks designed to probe spatial memory [72], working memory [15, 55, 67, 82], executive function [73], and sustained attention [52]. For example, McAllister et al. [55] found that during N-back task, chronic mTBI patients had increased activations during a moderate working memory load but had decreased activations during a high working memory load suggesting that mTBI populations have differing compensatory mechanisms based on differences in task difficulty.

While the use of task-based fMRI is difficult to perform during the acute stages of severe TBI, research in the chronic stage has led to many new insights in severe TBI. Kasahara et al. [41] found that chronic TBI patients had reduced performance and reduced activations during a working memory task indicating a failure of TBI patients to adequately activate the parietal regions of this task positive network (TPN). Furthermore, the default mode network (DMN) is a group of regions often deactivated during task related activities while remaining active during rest [30, 64]. Chronic TBI patients who failed to deactivate the DMN demonstrated impairments of attention [11]. This is also found in mild TBI populations as verified in Fig. 29.4 which depicts a failure of chronic mild TBI patients to deactivate the DMN during a 2-Back working memory paradigm compared to controls. Furthermore, this failure to deactivate the DMN has been associated with damage in white matter tracts connecting the right anterior insula to other nodes of the SN [12]. While subtle differences in study design have resulted in contradictory findings, future multi-center studies are likely to provide not only deeper insights into the injury process but also priceless information on the potential repair processes from rehabilitative and pharmacotherapeutic interventions.

29.8 Resting State fMRI

In addition to focusing on task based fMRI, there currently is an increased interest in looking into resting state brain networks to understand the interaction between global brain networks between disparate regions. Referred to as resting state functional connectivity (rs-FC), this method measures the efficiency and strength of interactions between these large-scale networks in the absence of a task [76, 83]. The DMN and the TPN are also functionally connected networks during resting conditions, and these two networks are anti-correlated at rest [22].

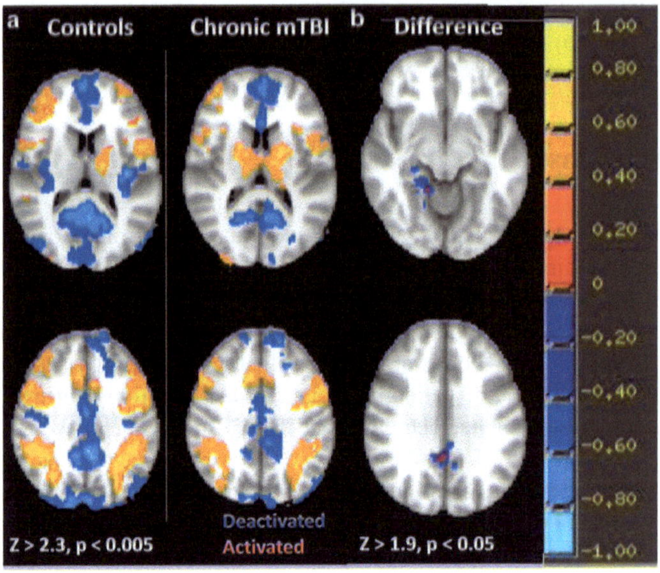

Fig. 29.4 Group results from a 2-back working memory fMRI paradigm in mTBI patients and controls. (**a**) Activations of the task positive network (TPN) in warm colors and deactivations of the Default Mode Network (DMN) in cool colors during 2-back paradigm. 18 mTBI patients in the chronic stage (6 months post injury) and 18 control. $Z > 2.3$, $p < 0.005$. (**b**) Control < mTBI cool colors. $Z > 1.9, p < 0.005$

The rs-FC within the DMN and the interplay between the two networks are associated with cognitive performance [18, 33, 84]. It also has been suggested that a third network, the Salience Network (SN), modulates the balance between the DMN and TPN [57, 68, 77]. Other networks include the motor and primary sensory networks. Altered rs-FC within these networks as a result of TBI is an active area of research as has the potential for providing valuable information on the cognitive condition of patients especially during the acute stage when they are unable to perform tasks or if they are in a vegetative state.

Both severe and mild TBI patients have demonstrated reduced inter-hemispheric rs-FC[51, 72], possibly due to the compromised integrity of the corpus callosum. However, TBI has also been shown to alter rs-FC in multiple networks that are not directly linked through the corpus callosum including the motor network [40, 70], thalamic network [78], TPN [35, 54, 70, 75], and the DMN [35, 54].

While it is often predicted that due to DAI, rs-FC will be reduced, there are some networks that demonstrate increased rs-FC following TBI. For example, groups found increased rs-FC in the left fronto-parietal network in mild TBI patents [70] and DMN in severe TBI patients [12, 69]. It was noted that TBI patients with higher rs-FC within the DMN had better cognitive performance suggesting the possibility of compensatory mechanisms within this chronic patient population [69]. Furthermore, this increase in rs-FC has been found between DMN and TPN in mild TBI patients [53]. In addition, another group recently reported increased rs-FC between

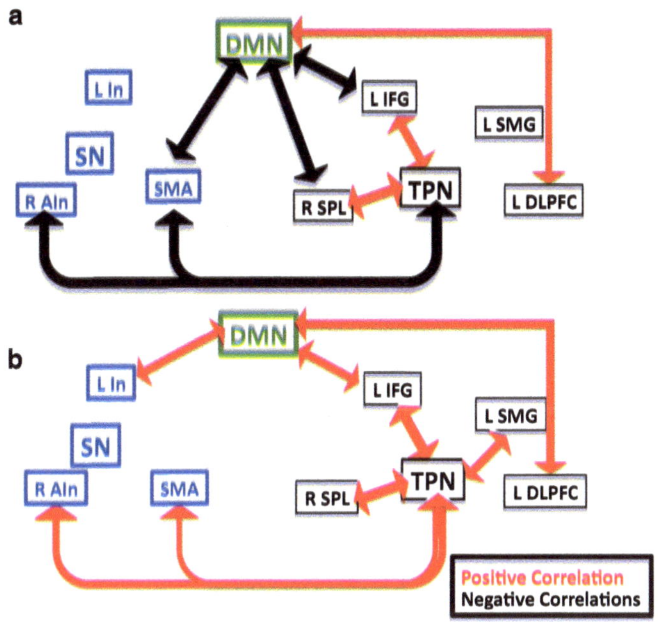

Fig. 29.5 Positive rs-FC (*red arrows*) and negative rs-FC (*black arrows*) connections between the TPN, DMN, and SN in (**a**). Mild TBI patients without memory complaints ($N = 10$). (**b**) Mild TBI patients with memory complaints ($N = 13$). *Arrows* represent connections with rs-FC significantly different than zero ($p = 0.05$) uncorrected. The *inset* shows the region in the right anterior insula which demonstrates increased rs-FC with the TPN in mild TBI patients with memory complaints compared to mild TBI patients without memory complaints. The SN includes: *L In* left insula, *R Ain* right anterior insula, *SMA* supplemental motor area. TPN includes: *L IFG* left inferior frontal gyrus, *L SMG* left supramarginal gyrus, *R SPL* right superior parietal lobule, *L DLPFC* left dorsolateral prefrontal cortex

the DMN, TPN and SN in mild TBI patients with memory complaints when compared to both control participants and mild TBI patients without memory complaints [74] as shown in Fig. 29.5. Furthermore, this increased rs-FC, specifically between the TPN and right anterior insula of the SN, was associated with reduced memory performance. Longitudinal studies examining resting state fMRI over the time course of injury are still needed for a more complete understanding of these alterations and the subsequent effects on recovery from injury.

29.9 Conclusions

The bulk of the pathologies experienced in TBI arise from secondary injuries and the sequelae includes DAI, inflammation, edema, apoptosis, excitotoxicity, mitochondrial dysfunction and neuro-metabolic alterations. Given the diffuse nature of these injuries and the various possible pathways that the injury could potentially take, no single imaging technique, however advanced, is likely to provide the sensitivity and specificity required to actively manage the negative consequences of such injury. While many advances in the understanding of the mechanisms of injury and recovery from TBI have been made using MRI, there is still much to be understood. Multi-parametric approaches using the most sensitive advanced imaging techniques that can measure cerebrovascular changes, biochemical and biophysical changes in conjunction with acute stage vital signs and neurocognitive assessments are needed. This will lead to a better understanding of the neurodegenerative processes, leading to the development of therapeutic interventions, while at the same time providing identification of imaging biomarkers, all of which are needed to actively manage consequences of TBI.

References

1. Adams JH, Graham DI, Gennarelli TA, Maxwell WL (1991) Pathology of non-missile head injury. J Neurol Neurosurg Psychiatry 54:481–483
2. Alves W, Macciocchi SN, Barth JT (1993) Postconcussive symptoms after uncomplicated mild head injury. J Head Trauma Rehabil 8(3):48–59
3. Arfanakis K, Haughton VM, Carew JD, Rogers BP, Dempsey RJ, Meyerand ME (2002) Diffusion tensor MR imaging in diffuse axonal injury. AJNR Am J Neuroradiol 23(5):794–802
4. Ashwal S, Holshouser BA, Shu SK et al (2000) Predictive value of proton magnetic resonance spectroscopy in pediatric closed head injury. Pediatr Neurol 23:114–125
5. Ashwal S, Holshouser BA, Tong KA (2006) Use of advanced neuroimaging techniques in the evaluation of pediatric traumatic brain injury. Dev Neurosci 28(4–5):309–326
6. Babikian T, Marion SD, Copeland S, Alger JR, O'Neill J, Cazalis F, Mink R, Giza CC, Vu JA, Hilleary SM, Kernan CL, Newman N, Asarnow RF (2010) Metabolic levels in the corpus callosum and their structural and behavioral correlates after moderate to severe pediatric TBI. J Neurotrauma 27(3):473–481
7. Bazarian JJ, Wong T, Harris M, Leahey N, Mookerjee S, Dombovy M (1999) Epidemiology and predictors of post-concussive syndrome after minor head injury in an emergency population. Brain Inj 13(3):173–189
8. Bazarian JJ, Zhong J, Blyth B, Zhu T, Kavcic V, Peterson D (2007) Diffusion tensor imaging detects clinically important axonal damage after mild traumatic brain injury: a pilot study. J Neurotrauma 24(9):1447–1459
9. Benson RR, Gattu R, Sewick B, Kou Z, Zakariah N, Cavanaugh JM, Haacke EM (2012) Detection of hemorrhagic and axonal pathology in mild traumatic brain injury using advanced MRI: implications for neurorehabilitation. NeuroRehabilitation 31(3):261–279
10. Betz J, Zhuo J, Roy A, Shanmuganathan K, Gullapalli RP (2012) Prognostic value of diffusion tensor imaging parameters in severe traumatic brain injury. J Neurotrauma 29(7):1292–1305
11. Bonnelle V, Leech R, Kinnunen KM, Ham TE, Beckmann CF, De Boissezon X, Greenwood RJ, Sharp DJ (2011) Default mode network connectivity predicts sustained attention deficits after traumatic brain injury. J Neurosci 31(38):13442–13451
12. Bonnelle V, Ham TE, Leech R, Kinnunen KM, Mehta M, Greenwood RJ, Sharp DJ (2012) Salience network integrity predicts default mode network function after traumatic brain injury. Proc Natl Acad Sci U S A 109(12):4690–4695
13. Centers for Disease Control and Prevention (CDC) (2003) National Center for Injury Prevention and Control. Report to Congress on mild traumatic brain injury in the United States: steps

to prevent a serious public health problem. Centers for Disease Control and Prevention, Atlanta, GA
14. Chaumet G, Quera-Salva MA, Macleod A, Hartley S, Taillard J, Sagaspe P, Mazaux JM, Azouvi P, Joseph PA, Guilleminault C, Bioulac B, Léger D, Philip P (2008) Is there a link between alertness and fatigue in patients with traumatic brain injury? Neurology 71 (20):1609–1613
15. Christodoulou C, DeLuca J, Ricker J, Madigan N, Bly B, Lange G, Kalnin A, Liu W, Steffener J, Diamond B, Ni A (2001) Functional magnetic resonance imaging of working memory impairment after traumatic brain injury. J Neurol Neurosurg Psychiatry 71 (2):161–168
16. Cihangiroglu M, Ramsey RG, Drohmann GJ (2002) Brain injury: analysis of imaging modalities. Neurol Res 24(1):7–18
17. Condon B, Oluoch Olunya D, Hadley D et al (1998) Early 1H magnetic resonance spectroscopy of acute head injury: four cases. J Neurotrauma 15:563–571
18. Esposito F, Aragri A, Vatorre V, Poplizio T, Scarabino T, Cirillo S, Marciano E, Tedeschi G, Di Salle F (2009) Does the default-mode functional connectivity of the brain correlate with working-memory performance? Arch Ital Biol 147:11–20
19. Faul M, Xu L, Wald MM, Coronado VG (2010) Traumatic brain injury in the United States: emergency department visits, hospitalizations, and deaths. Centers for Disease Control, Atlanta, GA
20. Finkelstein E, Corso P, Miller T et al (2006) The incidence and economic burden of injuries in the United States. Oxford University Press, New York, NY
21. Foley LM, Iqbal O'Meara AM, Wisniewski SR, Kevin Hitchens T, Melick JA, Ho C, Jenkins LW, Kochanek PM (2013) MRI assessment of cerebral blood flow after experimental traumatic brain injury combined with hemorrhagic shock in mice. J Cereb Blood Flow Metab 33 (1):129–136
22. Fox MD, Snyder AZ, Vincent JL, Corbetta M, Van Essen DC, Raichle ME (2005) The human brain is intrinsically organized into dynamic, anticorrelated functional networks. Proc Natl Acad Sci U S A 102(27):9673–9678
23. Friedman SD, Brooks WM, Jung RE, Hart BL, Yeo RA (1998) Proton MR spectroscopic findings correspond to neuropsychological function in traumatic brain injury. AJNR Am J Neuroradiol 19(10):1879–1885
24. Friedman SD, Brooks WM, Jung RE, Chiulli SJ, Sloan JH, Montoya BT, Hart BL, Yeo RA (1999) Quantitative proton MRS predicts outcome after traumatic brain injury. Neurology 52 (7):1384–1391
25. Garnett MR, Blamire AM, Rajagopalan B et al (2000) Evidence for cellular damage in normal appearing white matter correlates with injury severity in patients following traumatic brain injury: a magnetic resonance spectroscopy study. Brain 123:1403–1409
26. Garnett MR, Blamire AM, Corkill RG et al (2000) Early proton magnetic resonance spectroscopy in normal appearing brain correlates with outcome in patients following traumatic brain injury. Brain 123:2046–2054
27. Gasparovic C, Yeo R, Mannell M, Ling J, Elgie R, Phillips J, Doezema D, Mayer AR (2009) Neurometabolite concentrations in gray and white matter in mild traumatic brain injury: an 1H-magnetic resonance spectroscopy study. J Neurotrauma 26(10):1635–1643
28. Ge Y, Patel MB, Chen Q, Grossman EJ, Zhang K, Miles L, Babb JS, Reaume J, Grossman RI (2009) Assessment of thalamic perfusion in patients with mild traumatic brain injury by true FISP arterial spin labeling MR imaging at 3T. Brain Inj 23(7):666–674
29. Govind V, Gold S, Kaliannan K, Saigal G, Falcone S, Arheart KL, Harris L, Jagid J, Maudsley AA (2010) Whole-brain proton MR spectroscopic imaging of mild-to-moderate traumatic brain injury and correlation with neuropsychological deficits. J Neurotrauma 27(3):483–496
30. Greicius MD, Krasnow B, Reiss AL, Menon V (2003) Functional connectivity in the resting brain: a network analysis of the default mode hypothesis. Proc Natl Acad Sci U S A 100 (1):253–258

31. Grossman EJ, Ge Y, Jensen JH, Babb JS, Miles L, Reaume J, Silver JM, Grossman RI, Inglese M (2012) Thalamus and cognitive impairment in mild traumatic brain injury: a diffusion kurtosis imaging study. J Neurotrauma 29(13):2318–2327
32. Haacke EM, Xu Y, Cheng YC, Reichenbach JR (2004) Susceptibility weighted imaging (SWI). Magn Reson Med 52(3):612–618
33. Hampson M, Driesen N, Roth J, Gore J, Constable R (2010) Functional connectivity between task-positive and task-negative brain areas and its relation to working memory performance. Magnet Reson Imaging 28(8):1051–1057
34. Hillary FG, Genova HM, Medaglia JD, Fitzpatrick NM, Chiou KS, Wardecker BM, Franklin RG, Wang K, DeLuca J (2010) The nature of processing speed deficits in traumatic brain injury: is less brain more? Brain Imaging Behav 4(2):141–154
35. Hillary FG, Slocomb J, Hills EC, Fitzpatrick NM, Medaglia JD, Wang J, Good DC, Wylie GR (2011) Changes in resting connectivity during recovery from severe traumatic brain injury. Int J Psychophysiol 82(1):115–123
36. Holhouser BA, Tong KA, Ashwal S, Oyoyo U, Ghamsary M, Saunders D, Shutter L (2006) Prospective longitudinal proton magnetic resonance spectroscopic imaging in adult traumatic brain injury. J Magn Reson Imaging 24(1):33–40
37. Immonen RJ, Kharatishvili I, Gröhn H, Pitkänen A, Gröhn OH (2009) Quantitative MRI predicts long-term structural and functional outcome after experimental traumatic brain injury. Neuroimage 45(1):1–9
38. Jensen JH, Helpern JA (2010) MRI quantification of non-Gaussian water diffusion by kurtosis analysis. NMR Biomed 23(7):698–710
39. Johansson B, Berglund P, Rönnbäck L (2009) Mental fatigue and impaired information processing after mild and moderate traumatic brain injury. Brain Inj 23(13–14):1027–1040
40. Kasahara M, Menon D, Salmond C, Outtrim J, Taylor Tavares J, Carpenter T, Pickard J, Sahakian B, Stamatakis E (2010) Altered functional connectivity in the motor network after traumatic brain injury. Neurology 75(2):168–176
41. Kasahara M, Menon D, Salmond C, Outtrim J, Tavares J, Carpenter T, Pickard J, Sahakian B, Stamatakis E (2011) Traumatic brain injury alters the functional brain network mediating working memory. Brain Inj 25(12):1170–1187
42. Kim J, Whyte J, Patel S, Avants B, Europa E, Wang J, Slattery J, Gee JC, Coslett HB, Detre JA (2010) Resting cerebral blood flow alterations in chronic traumatic brain injury: an arterial spin labeling perfusion fMRI study. J Neurotrauma 27:1399–1411
43. Kim J, Whyte J, Patel S, Europa E, Slattery J, Coslett HB, Detre JA (2012) A perfusion fMRI study in the neural correlates of sustained-attention and working-memory deficits in chronic traumatic brain injury. Neurorehabil Neural Repair 26(7):870–880
44. Kumar R, Husain M, Gupta RK, Hasan KM, Haris M, Agarwal AK, Pandey CM, Narayana PA (2009) Serial changes in the white matter diffusion tensor imaging metrics in moderate traumatic brain injury and correlation with neuro-cognitive function. J Neurotrauma 26:481–495
45. Lakhan SE, Kirchgessner A (2012) Chronic traumatic encephalopathy: the dangers of getting "dinged". SpringerPlus 1:2
46. Levine B, Kovacevic N, Nica EI, Cheung G, Gao F, Schwartz ML, Black SE (2008) The Toronto traumatic brain injury study: injury severity and quantified MRI. Neurology 70:771–778
47. Lye TC, Shores EA (2000) Traumatic brain injury as a risk factor for Alzheimer's disease: a review. Neuropsychol Rev 10(2):115–129
48. Mac Donald CL, Johnson AM, Cooper D, Nelson EC, Werner NJ, Shimony JS, Snyder AZ, Raichle ME, Witherow JR, Fang R, Flaherty SF, Brody DL (2011) Detection of blast-related traumatic brain injury in U.S. Military personnel. N Engl J Med 364(22):2091–2100
49. Magnuson J, Leonessa F, Ling GSF (2012) Neuropathology of explosive blast traumatic brain injury. Curr Neurol Neurosci Rep 12:570–579
50. Makley MJ, English JB, Drubach DA, Kreuz AJ, Celnik PA, Tarwater PM (2008) Prevalence of sleep disturbance in closed head injury patients in a rehabilitation unit. Neurorehabil Neural Repair 22(4):341–347

51. Marquez de la Plata C, Garces J, Shokri Kojori E, Grinnan J, Krishnan K, Pidikiti R, Spence J, Devous M, Moore C, McColl R, Madden C, Diaz-Arrastia R (2011) Deficits in functional connectivity of hippocampal and frontal lobe circuits after traumatic axonal injury. Arch Neurol 68(1):74–84
52. Maruishi M, Miyatani M, Nakao T, Muranaka H (2007) Compensatory cortical activation during performance of an attention task by patients with diffuse axonal injury: a functional magnetic resonance imaging study. J Neurol Neurosurg Psychiatry 78:168–173
53. Mayer A, Ling J, Mannell M, Gasparovic C, Phillips JP, Doezema D, Reichard R, Yeo RA (2010) A prospective diffusion tensor imaging study in mild traumatic brain injury. Neurology 74(8):643–650
54. Mayer A, Mannell M, Ling J, Gasparovic C, Yeo R (2011) Functional connectivity in mild traumatic brain injury. Hum Brain Mapp 32(11):1825–1835
55. McAllister T, Sparling M, Flashman L, Guerin S, Mamourian A, Saykin A (2001) Differential working memory load effects after mild traumatic brain injury. Neuroimage 14:1004–1012
56. McDowell S, Whyte J, D'Esposito M (1997) Working memory impairments in traumatic brain injury: evidence from a dual-task paradigm. Neuropsychologia 35(10):1341–1353
57. Menon V, Uddin LQ (2010) Saliency, switching, attention, and control: a network model of insula function. Brain Struct Funct 214(5–6):655–667
58. Menzel JC (2008) Depression in the elderly after traumatic brain injury: a systematic review. Brain Inj 22(5):375–380
59. Nampiaparampil DE (2008) Prevalence of chronic pain after traumatic brain injury: a systematic review. JAMA 300(6):711–719
60. Nemetz PN, Leibson C, Naessens JM, Beard M, Kokmen E, Annegers JF, Kurland LT (1999) Traumatic brain injury and time to onset of Alzheimer's disease: a population-based study. Am J Epidemiol 149(1):32–40
61. Ommaya AK, Goldsmith W, Thibault L (2002) Biomechanics and neuropathology of adult and pediatric head injury. Br J Neurosurg 16:220–242
62. Parry L, Shores A, Rae C, Kemp A, Waugh MC, Chaseling R, Joy P (1994) An investigation of neuronal integrity in severe pediatric traumatic brain injury. Child Neuropsychol 10(4):248–261
63. Povlishock J, Becker DP, Cheng CL, Vaughan GW (1983) Axonal change in minor head injury. J Neuropathol Exp Neurol 42:225–242
64. Raichle ME, MacLeod M, Snyder Z, Powers WJ, Gusnard D, Shulman GL (2001) A default mode of brain function. Proc Natl Acad Sci U S A 98(2):676–682
65. Reichenbach JR, Venkatesan R, Schillinger DJ, Kido DK, Haacke EM (1997) Small vessels in the human brain: MR venography with deoxyhemoglobin as an intrinsic contrast agent. Radiology 204:272–277
66. Ross BD, Ernst T, Kries R et al (1998) 1H MRS in acute traumatic brain injury. J Magn Reson Imaging 8:829–840
67. Sanchez-Carrion R, Fernandez-Espejo D, Junque C, Falcon C, Bargallo N, Roig T, Bernabeu M, Tormos JM, Vendrell P (2008) A longitudinal fMRI study of working memory in severe TBI patients with diffuse axonal injury. Neuroimage 43:421–429
68. Seeley WW, Menon V, Schatzberg AF, Keller J, Glover GH, Kenna H, Reiss AL, Greicius MD (2007) Dissociable intrinsic connectivity networks for salience processing and executive control. J Neurosci 27(9):2349–2356
69. Sharp DJ, Beckmann CF, Greenwood R, Kinnunen KM, Bonnelle V, De Boissezon X, Powell JH, Counsell SJ, Patel MC, Leech R (2011) Default mode network functional and structural connectivity after traumatic brain injury. Brain 134(8):2233–2247
70. Shumskaya E, Andriessen TMJC, Norris DG, Vos PE (2012) Abnormal whole-brain functional networks in homogeneous acute mild traumatic brain injury. Neurology 79:175–182
71. Sinson G, Bagley LJ, Cecil KM et al (2001) Magnetization transfer imaging and proton MR spectroscopy in the evaluation of axonal injury: correlation with clinical outcome after traumatic brain injury. Am J Neuroradiol 22:143–151

72. Slobounov S, Zhang K, Pennell D, Ray W, Johnson B, Sebastianelli W (2010) Functional abnormalities in normally appearing athletes following mild traumatic brain injury: a functional MRI study. Exp Brain Res 202:341–354
73. Soeda A, Nakashima T, Okumura A, Kuwata K, Shinoda J, Iwama T (2005) Cognitive impairment after traumatic brain injury: a functional magnetic resonance imaging study using the Stroop task. Neuroradiology 47:501–506
74. Sours C, Zhuo J, Janowich J, Arabi B, Shanmuganathan K, Gullapalli RP (2013) Default mode network interference in mild traumatic brain injury – A pilot resting state study. Brain Res. in press
75. Sponheim SR, McGuire KA, Kang SS, Davenport ND, Aviyente S, Bernat EM, Lim KO (2011) Evidence of disrupted functional connectivity in the brain after combat-related blast injury. Neuroimage 54(suppl 1):S21–S29
76. Sporns O (2011) The non-random brain: efficiency, economy, and complex dynamics. Front Comput Neurosci 5:5
77. Sridharan D, Levitin DJ, Menon V (2008) A critical role for the right fronto-insular cortex in switching between central-executive and default-mode networks. Proc Natl Acad Sci USA 105 (34):12569–12574
78. Tang L, Yulin G, Sodickson DK, Miles L, Zhou Y, Reaume J, Grossman RI (2011) Thalamic resting-state functional networks: disruption in patients with mild traumatic brain injury. Radiology 260(3):831–840
79. Teasdale G, Jennett B (1974) Assessment of coma and impaired consciousness: a practical scale. Lancet 13:81–83
80. Thornhill S, Teasdale GM, Murray GD, McEwen J, Roy CW, Penny KI (2000) Disability in young people and adults one year after head injury: prospective cohort study. BMJ 320 (7250):1631–1635
81. Tong KA, Ashwal S, Holshouser BA, Nickerson JP, Wall CJ, Shutter LA, Osterdock RJ, Haacke EM, Kido D (2004) Diffuse axonal injury in children: clinical correlation with hemorrhagic lesions. Ann Neurol 56(1):36–50
82. Turner GR, Levine B (2008) Augmented neural activity during executive control processing following diffuse axonal injury. Neurology 71:812–818
83. van den Heuvel M, Hulshoff Pol H (2010) Exploring the brain network: a review on resting-state fMRI functional connectivity. Eur Neuropsychopharmacol 20(8):519–534
84. Wang L, LaViolette P, O'Keefe K, Putcha D, Bakkour A, Van Dijk KR, Pihlajamaki M, Dickerson BC, Sperling RA (2010) Intrinsic connectivity between the hippocampus and posteromedial cortex predicts memory performance in cognitively intact older individuals. Neuroimage 51(2):910–917
85. Warner MA, Marquez de la Plata C, Spence J, Wang JY, Harper C, Moore C, Devous M, Diaz-Arrastia R (2010) Assessing spatial relationships between axonal integrity, regional brain volumes, and neuropsychological outcomes after traumatic axonal injury. J Neurotrauma 27:1221–2130
86. Warner MA, Youn TS, Davis T, Chandra A, Marquez de la Plata C, Moore C, Harper C, Madden CJ, Spence J, McColl R, Devous M, King RD, Diaz-Arrastia R (2010) Regionally selective atrophy after traumatic axonal injury. Arch Neurol 67:1336–1344
87. Wilde EA, McCauley SR, Barnes A, Wu TC, Chu Z, Hunter JV, Bigler ED (2012) Serial measurement of memory and diffusion tensor imaging changes within the first week following uncomplicated mild traumatic injury. Brain Imaging Behav 6(2):319–328
88. Yeo RA, Phillips JP, Jung RE, Brown AJ, Campbell RC, Brooks WM (2006) Magnetic resonance spectroscopy detects brain injury and predicts cognitive functioning in children with brain injuries. J Neurotrauma 23(10):1427–1435
89. Zhuo J, Xu S, Proctor JL, Mullins RJ, Simon JZ, Fiskum G, Gullapalli RP (2012) Diffusion kurtosis as an in vivo imaging marker for reactive astrogliosis in traumatic brain injury. Neuroimage 59(1):467–477
90. Zimmerman RA (1999) Craniocerebral trauma. In: Lee SH, Rao KCVG, Zimmerman RA (eds) Cranial MRI and CT, 4th edn. McGraw-Hill, New York, pp 413–452

Index

E.H. Lo et al. (eds.), *Vascular Mechanisms in CNS Trauma*, Springer Series in Translational Stroke Research 5, DOI 10.1007/978-1-4614-8690-9,

O

P

T

MIX
Papier aus verantwortungsvollen Quellen
Paper from responsible sources
FSC® C105338
FSC
www.fsc.org

If you have any concerns about our products,
you can contact us on
ProductSafety@springernature.com

In case Publisher is established outside the EU,
the EU authorized representative is:
Springer Nature Customer Service Center GmbH
Europaplatz 3, 69115 Heidelberg, Germany

Printed by Libri Plureos GmbH
in Hamburg, Germany